dup

W9-CQP-744

2 bols
$100⁰⁰

The American petroleum industry

The age of illumination 1859–1899

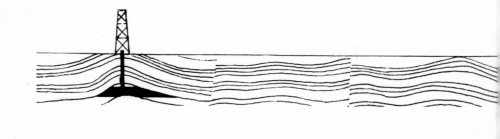

THE
AMERICAN
PETROLEUM
INDUSTRY

The age of illumination 1859-1899

by Harold F. Williamson
Arnold R. Daum

research associates Ralph L. Andreano
Gilbert C. Klose and Paul A. Weinstein

NORTHWESTERN UNIVERSITY PRESS
Evanston

Preface

This book traces the story of American petroleum, from its inception as the medicinal by-product of salt well operations through the peak of its subsequent development as a source of illumination for a large portion of the world's population. It terminates with an account of the industry at the close of the nineteenth century, when signs of petroleum's future role as a major source of energy rather than illumination were already visible.

The domestic petroleum industry was an integral part of the impressive expansion of the American economy during the post Civil War decades. Its members shared with other business enterprises the impact of periodic booms and depressions and the long-run secular increases in population, physical production, and capital formation that were characteristic of this period of American history. Oilmen contributed to the changes during these years when the United States shifted from an agrarian to an industrial economy and from predominantly rural to urban patterns of living.

Although petroleum had much in common with other American industries that emerged in the latter nineteenth century, it soon developed unusual, and in some instances unique, characteristics of its own. In many respects these distinguishing features reflected the nature of the basic raw material. Like most mineral deposits, petroleum has usually been discovered far from existing markets and distributing centers. New fields have almost invariably necessitated an extensive investment in transport and storage facilities for their successful exploitation. But unlike most minerals, petroleum is a liquid substance, usually found in conjunction with natural gas under pressure. Its recovery required quite different production methods than those employed in other types of mining operations. Because both crude and refined petroleum products (with few exceptions) retain their liquid form from well-heads to final consumers and are highly inflammable, it was also necessary to develop special methods of transport and storage for economical and effective handling.

Ever since the sinking of the Drake well in the remote section of western Pennsylvania in 1859, the discovery of new sources of petroleum has typically followed an erratic and generally unpredictable pattern, both in regard to their location and size. Once discoveries were made, however, the fugacious character of underground petroleum deposits coupled with a legal structure that gave owners full rights to mineral substances beneath the surface of their land, encouraged a rapid exploitation of new fields. As a result, the industry has at times been faced with supplies of petroleum above ground far in excess of the capacity of refiners to process or of distributors to market. On other occasions its members have been threatened with the possibility that a considerable portion of the industry's elaborate superstructure of transport, refining, and distributing facilities might become obsolete because of inadequate supplies of crude.

In contrast to most mineral based industrial operations during the latter nineteenth century, which supplied materials for other processers or fabricators, only a relatively small proportion of the output of petroleum refineries—chiefly lubricants, solvents, waxes, and fuel oil—was sold to industrial users. The bulk of their output in the form of illuminating oil was distributed through marketing channels that terminated with a final sale to household consumers. This characteristic of the petroleum industry had important implications for the competitive strategy followed by its members and the organization of the firms that attempted to expand their operations in order to protect or extend their marketing shares. Experience by 1900 clearly foreshadowed the future structure of the industry in the twentieth century when the major companies by definition were those which included marketing as well as production, transportation, and refining, among their operations.

Petroleum was also distinguished by the fact that, unlike other major American processors of raw materials, it sold the larger portion of its output to markets outside the United States. The importance of foreign sales reflected the ability of the domestic industry to supply a relatively inexpensive and efficient source of artificial illumination to a large proportion of the world's population. The early and continued attraction of the foreign demand for the American industry is revealed by the fact that from the mid-1860's to the turn of the century export markets absorbed well over half the domestic output of illuminating oil. Until the mid-1880's the United States remained virtually the world's only source of crude and refined petroleum. Thereafter the American industry competed with a rapidly expanding Russian production and beginning in the 1890's with outputs of crude and refined from the Dutch East Indies.

One further important characteristic of the nineteenth century American petroleum industry should be noted: its domination by a single firm, the Standard Oil Company. By the late 1870's, Standard, under the administration of John D. Rockefeller, had already achieved control over all but a small proportion of the domestic industry's refining capacity as well as the gathering lines, storage, and transport facilities utilized in moving crude oil from the fields to the refineries. The Standard management subsequently strengthened the organization's position in the industry by extending its operations forward into marketing and backward into production. Although these moves did not prevent the emergence of independent refiners and marketers in the domestic industry, they were sufficient to restrict severely the operations of independents. From Standard's point of view, the much more serious challenge to its position came from competition in foreign markets.

Petroleum was by no means the only American industry during the latter half of the nineteenth century to be dominated by one or more large firms. Yet Standard to an unusual degree became symbolic of the position and power of big business in the American economy during this period. The Standard Trust and Rockefeller were among the favorite targets of the "muckrakers" who did much to make oil and monopoly synonymous in the mind of the American public.

From its earliest beginnings, the more romantic and spectacular petroleum industry attracted the attention of journalists and popular writers. Stories of oil discoveries that almost overnight brought great wealth to

poor and sometimes deserving individuals had great appeal to most readers. So, too, did the descriptions of the boom towns that accompanied the
opening of new fields—each with a full complement of saloon keepers,
bawdy house proprietors, gamblers, and get-rich-quick artists. Accounts
of the personality of Rockefeller, the growth of Standard, and Standard's
effect on the fortunes of independents, dramatized the struggle between
a large, powerful adversary and its small, relatively weak opponents.

Not all writing covering the history of the industry prior to 1900 followed the romantic tradition. Historians with an antiquarian interest have
done much to recreate the circumstances surrounding the early growth of
the Pennsylvania oil fields. Most of the technical developments affecting
production, transportation, and refining up until the turn of the century
have been described and analyzed in various trade and scientific journals,
monographs, and chemical and engineering encyclopedias. Certain state
and federal agencies began early to gather statistical data covering certain phases of the industry's growth, which were incorporated in various
governmental reports. Standard's role in the industry has been the subject
of several notable and outstanding works covering the life of Rockefeller
and the early history of the organization.

This extensive body of literature was of great value to the authors in
the preparation of this volume. Yet because of its limited scope, much of
the existing literature had to be re-examined and reinterpreted in order
to provide the basis for an integrated analysis of the evolution of the industry. Only by a further intensive examination of basic sources was it
possible to treat certain phases of the industry's development that had
either been overvalued or neglected by previous writers. This was particularly true in respect to early advances in technology and the growth
and structure of markets both at home and abroad.

For purposes of description and analysis the history of the industry before 1900 was divided as follows: Part One traces the continued interest
in petroleum from antiquity to its early use as a medicinal in North America. Part Two indicates the background and the events which, by the
1850's, had brought revolutionary changes in the domestic supply and
uses of artificial illuminants—notably lard oil, camphene, manufactured
gas and coal oil. These changes in turn provided a rich technological and
marketing heritage for the infant petroleum industry. In addition to the
familiar account of Drake's discovery, Part Three describes the contributions of less-publicized pioneers who did much between 1859 and 1861 to
provide the foundations for a permanent industry. Part Four, covering

the years 1862–1872, carries the story of petroleum through its formative years which, against a background of intensive competition, saw impressive growth in crude supplies, transport and processing facilities, and the development of foreign and domestic markets. Following an examination in Part Five of the conditions that prompted Rockefeller and his associates to develop their plan to dominate the refining and transportation segments of the industry, the chapters in Part Six analyze the successful execution of this plan, which by mid-1880's had completely changed the competitive structure of the industry. Part Seven completes the history of American petroleum to 1900. It was during these years that production of crude was extended beyond the Appalachian fields into Ohio-Indiana, and American independents were able to grow despite Standard's general success in expanding and integrating its operations into marketing and production. It was also during this period that the American industry was faced with increasingly intensive competition in foreign markets from Russian and Dutch East Indian interests.

In the final preparation of the materials used in this volume, the authors' major purpose was to present an over-all, integrated, and objective account of the evolution of the American petroleum industry that general readers would find both interesting and understandable. At the same time, we have tried to make sure that our descriptions and analyses measure up to the professional standards of scholars and specialists in various phases of the industry's operations.

This study of the American petroleum industry was made possible by a grant to the Northwestern University Center for Social Research by the American Petroleum Institute. It was clearly understood that the authors were to have complete freedom in conducting their research and preparing the manuscript for publication. The terms of this agreement were fully met and the authors assume sole responsibility for the organization, contents, interpretations, and evaluations contained in this history.

We want at this point to acknowledge our special debt to our research associates, Ralph L. Andreano, Gilbert C. Klose, and Paul A. Weinstein, for their substantive contributions to this volume. Our obligations to the many other individuals whose co-operation made this study possible are specifically indicated in the Acknowledgments on page 733.

HAROLD F. WILLIAMSON
ARNOLD R. DAUM

Evanston, Illinois
June, 1959

Contents

Figures, Charts, Maps, and Tables

Appendix tables

Part ONE

Prologue

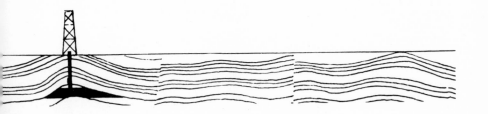

Chapter 1

Petroleum through the ages:
a 5,000-year preview

W ithin nine months after Colonel E. L. Drake drilled the first success-ful oil well near Titusville, Pennsylvania, in early fall, 1859, men began calling the new-born American petroleum industry the wonder of the nineteenth century.[1] Like all genuine wonders of the nineteenth century, the infant American industry was something new under the sun. Draw-ing on petroleum for their heat and power as well as light and lubrication, men in the twentieth century have with even greater reason regarded the petroleum industry as a unique phenomenon of the modern industrial age.

Though basically true, this modern conception of petroleum history re-quires qualification. The American petroleum industry was not the first, but the second petroleum industry ever to materialize. A span of five thou-

3

sand years separated the first petroleum industry from the second. Based on petroleum in the form of asphaltic bitumen, the first petroleum industry originated in Mesopotamia, cradle of western civilization. Before the rise of the American petroleum industry, this ancient predecessor represented the only certain historical record of an organized exploitation of petroleum resources into a sustained industry comprising a basic and integral part of the economy.

Petroleum in the ancient world

By contrast with the modern petroleum industry, now approaching one hundred years of age, the ancient industry thrived for nearly three thousand years. Beginning before 3000 B.C., the Sumerians, Assyrians, and Babylonians successively exploited seepages of asphaltic bitumen near Hit in Mesopotamia. Sometimes they supplemented these sources with the more difficult operation of rock asphalt mining. Faced with a scarcity of natural stone and timber in Mesopotamia, they employed asphalt extensively in their industrial arts. Mixed with sand and fibrous materials, asphalt was used as a mastic in architecture, hydraulics, and the waterproofing of ships. Still another major use of asphalt was in building roads.[2]

A host of minor uses for bitumen give further indication of the deep penetration of the petroleum industry into the economic life of these ancient peoples. Bitumen was used in their paints, and in waterproofing baskets, wicker, and mats. By Assyrian law, certain delinquents were forced to submit to having jars of molten bitumen poured over their heads. Medicinal and pharmaceutical uses for bitumen also abounded. Mixed with sulphur and other substances, solid bitumen was recommended for fumigation purposes and for the treatment of sores on hands and feet.[3] Fluxed with oil, it was prescribed for inflamed eyes; admixed with beer, it was offered as a cure for other ailments. In magic, it was used both to keep out evil spirits and to provide the favorite residence for the demon Asakku.

Bitumen figured prominently in the fine arts. It was a favorite cement for mosaic and inlaid work. Bitumen or mastic also entered warfare, in the construction of primitive bludgeons and dagger hafts similar to those still used by the Arabs in northern Syria.

Embracing a range of basic usages which would not be realized again until the last half of the nineteenth century, the early bitumen industry

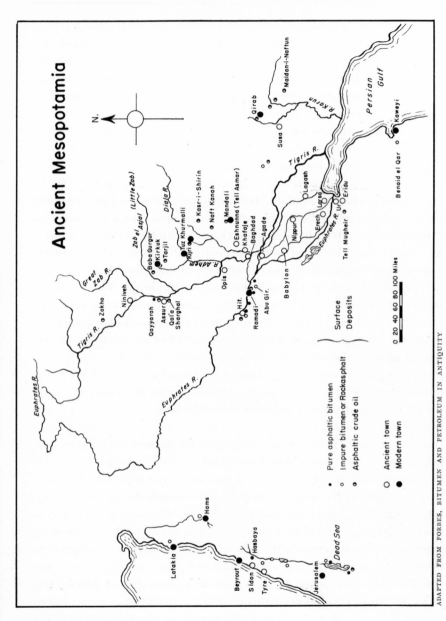

Ancient Mesopotamia

N.

Euphrates R.

Great Zab R.

Tigris R. ○ Zakho

Euphrates R.

Nineveh ●

Qayyarah ○
Assur ○
Qal'a ○
Sharghal ●

Zab el Asjel

Zab el (Little Zab)

Digla R.

Baba Gurgur ●
Kirkuk ○
Tarjil ●

Tuz Khurmalli ●
Kifri ●

Kasr-i-Shirin ○
Naft Kanah ○
Mandali ●
Eshnunna (Tell Asmar) ○
Khafaje ●
Baghdad ○
Agade ●

R. Adhem

Opis ○

Hit ●
Ramadi ○
Abu Gir.

Babylon ●

Nippur ○
Erech ○
Euphrates R. Ur ○
Tell Mugheir ●

Lagash ○
Lagra ○
Eridu ●

Qirab ●

Tigris R.

Susa ○

Benoid el Qar ●

Kaweyi ●

R. Karun

Persian Gulf

Maidani-Naftun ●

Surface
Deposits

0 20 40 60 80 100 Miles

● Pure asphaltic bitumen
○ Impure bitumen or Rockasphalt
◑ Asphaltic crude oil

○ Ancient town
● Modern town

Hams ●

Latakia ●
Beyrout ●
Sidon ●
Tyre ●

Hasbaya ○

Dead Sea

Jerusalem ●

ADAPTED FROM FORBES, BITUMEN AND PETROLEUM IN ANTIQUITY

MAP 1:1. *Ancient Mesopotamia.*

rapidly declined and lost its identity after the Persian conquest of the Neo-Babylonian kingdom, about 600 B.C. The Persians preferred their own methods of construction, and continued to use bituminous mastic only for caulking ships.[4] Herodotus and Ctesias, often erroneously cited as the earliest chroniclers of petroleum, were recording only the remnants of a vanished industry when they described bitumen in Persia about 450 B.C.[5]

Unfortunately, the ancient Greeks failed to transmit the technology of ancient bitumen through the Romans to western Europe. More interested in fine arts than technology, the Greeks did not assimilate the uses of bitumen into an important part of their lives. The abundance of marble and of the cedars of Lebanon, available both for construction and for obtaining cedar-oil pitch, removed any necessity to transplant bitumen techniques.

Many writers after Herodotus (such as, Aristotle, Strabo, Plutarch, Pliny, and others) described natural deposits of bitumen, including petroleum and gas. For the most part, these deposits were located in the Near East, especially in the region of Baku and the Euphrates. But combined vagaries of observation, translation, and interpretation often produced distorted notions which are still widespread. The sum and substance of the Roman heritage from the Greeks was a set of observations on the occurrence of natural gas, petroleum, and asphaltic materials and rocks as a curiosity of nature, and a vague tradition of random uses of bituminous products. Lost to western Europe was the knowledge of an industry in bitumen spanning almost three thousand years with a rich accumulation of techniques and products.

Blurred as their bituminous heritage was, the Romans might conceivably have revived the industry for road building and hydraulics, except for other circumstances. They possessed not only quarried stone in abundance, but a good binder for it in pozzolana, which was a loosely compacted rock from extinct volcanoes that the Romans mixed with lime to form excellent mortar and hydraulic cement. Known supplies of bitumen were relatively scarce within the Empire; main centers of production in Mesopotamia and Persia were either outside Roman jurisdiction or under it for limited periods only. In the range of hydrocarbon materials, tars and pitches from wood soon commanded most interest as a by-product of charcoal industries that sprang up in the numerous forested regions of the Empire. In the face of these influences, mineral bitumen was relegated largely to black magic and to the affiliated art of medicine, except for sporadic use in ship repairing and as paint.[6]

As the strands of a bitumen industry and technology were neglected in western Europe, a new technological continuity began taking form in the East that was of great importance to the ultimate rise of petroleum refining during the nineteenth century. Early in the Christian era, Arabs at Alexandria and elsewhere began to assimilate and build on the knowledge of chemistry and alchemy accumulated among the Alexandrian Greeks and probably to some extent among the Persians and Syrians.[7]

The interest of Arabs and Persians turned sharply away from the Babylonian civilization's preoccupation with asphaltic materials and concentrated on crude petroleum and the distillation of its light fractions into illuminants. The technology of distillation was carried forward to a point that inspired a search for new deposits of petroleum and asphalt in small seepages and shallow, man-made trenches. Due largely to the Arabic infiltration of Spain, a fairly mature industrial art of distillation became available in western Europe as early as the twelfth century.[8]

The various forms of distillation apparatus were developed in the Near East in the early centuries of the Christian era, first in Alexandria, then in Syria. About 100 A.D. a distillation method in which the vapor was delivered to a thin-necked vessel cooled with air or sponges was reported in the writings of the Coptic alchemists.

As distillation techniques slowly spread and improved, both crude oil and naphtha (terms used interchangeably in contemporary nomenclature) entered the preparation of "Greek Fire" for military use, probably no later than 750 A.D. The burning of Cairo, in 1077, furnished a measure of the advance in petroleum distillation. Twenty thousand jars and bottles of naphtha, equivalent to 1,200 barrels, added to the fury of the conflagration. In the vicinity of Damascus emerged impressionistic outlines of an industry in which distillation of naphtha was the main activity.[9]

Distilling equipment during this period approached full development. Utensils took on the necessary shapes, and their materials were of fired clay, stone, or lead. Although cooling still was done with wet rags or sponges, fractionating pipes, distilling heads, and other condensing apparatus were evolving. Some naphtha was even distilled from asphalt, usually a more viscous and stubborn raw material than crude oil to run in stills. With the commentaries of Marcus Graecus about 1250, the distillation of petroleum entered the literature of western Europe. Graecus reported that the formula for Greek Fire was employed with devastating effect by Arab and Mongol armies in grenades, flame throwers, and other war machines. The Crusades and reports of travelers like Marco Polo amplified this knowledge. About 1300, the retort was introduced from the East into western Europe, while almost at the same time Thaddeaus Alderotti of Florence reported using a tabular condenser in the distillation of wine, a pioneer innovation that paved the way for the modern method of cooling the vapor outside rather than inside the still-head.[10] The core of the necessary technology of distillation for ultimate realization of the value of petroleum and related bitumen had been sufficiently established in western Europe for development and dispersion to the New World.

Active exploitation of petroleum and bitumen had not started in western Europe when America was discovered, yet explorers in succeeding centuries maintained sharp vigilance for the occurrence and often the use of these materials. Behind this awareness was the infiltration of Arabic science into Spain and Italy, gradually arousing scientific curiosity and appreciation of new horizons. An equally powerful stimulant was the kindling of interest in the classics, as the Byzantine empire played an intermediary role in reawakening Europe to the glories of Greece and Rome.

From these various sources, literate men of Renaissance Europe could scarcely avoid having some familiarity with the occurrence of bitumen and some of its usable forms. Beginning with Marco Polo, who reported his visit to Baku in 1272 and his witnessing of streams of people going to seepages in the area to collect oil, many contemporary accounts of travels in Asia fortified awareness of bitumen. Finally, the high incidence of visible deposits in America combined with the lively interest in science and the industrial arts in Europe to encourage reports on bitumen and, eventually, to further the exploitation of American petroleum as a resource.[11]

Petroleum in the New World

Shortly after the discovery of Cuba, the Spanish found deposits of asphalt there, which were reported by Oviedo in 1526.[12] Havana soon acquired the nickname of "Carine" because all ships were careened, overhauled, and pitched there. Writing in 1566 about the most important products known to medicine imported from America, Monardes, a Spanish doctor, commented:

> There is in the Islande of Cuba, certaine Fountaines at the Sea side, that doth cast from them a kinde of blacke Pitch of a strong smell, whiche the Indians doe use, in their colde infirmities, oure people doe use it there to pitche their shippes withall, for it is well nere like unto Tarre, and they doe mingle

therewith Tallowe, for to make it Pitch the better. I do beleve that this is Napta, whiche the auncient writers doe speake of, Possidonio doth say, that there are two Fountaines in Babilon, one white and the other blacke.

This that they doe bryng from the Indias, we doe use it in griefes of the Mother . . .[13]

The first Englishman to discover the Pitch Lake of Trinidad, off the South American coast, recognized the chances of exploiting this asphaltic bitumen commercially for the waterproofing of ships. In his journal on March 22, 1595, Sir Walter Raleigh made this revealing entry:

> At this point called Tierra de Brea or Pitche there is that abundance of stone pitch that all the ships of the world may therewith loden from thence, and we made a trial of it in trimming our shippes to be most excellent good, and melteth not with the Sunne as the pitch of Norway, and therefore for shippes trading the South parts very profitable.[14]

Sahagun, Mexico's first historian, unfolded an engrossing account of asphalt (chapopotli), found in pools and deposits in the coastal district of Tampico. He reported that the native Totonacs not only used asphalt as medicine, but bartered it with the Aztec rulers of Mexico, who resided in the interior. Women gathered it on the beaches where it had been cast up from the waves. The asphalt, which they threw on the fire, was of two types: one, a perfume spreader; the other, a kind of pitch that the women called "chicle" (tzictli). Mixing the "chicle" with "axin," a resinous substance, to make it masticable, both women and men used it as a tooth paste and a chewing gum.[15]

Explorations in South America during the sixteenth century yielded a liberal quota of observations of petroleum and bitumen in Bolivia, Equador, and Peru.* In the interior of Peru, near Titicaca Lake, Monardes witnessed a local bitumen industry during the sixteenth century. His report is remarkable because it implies that, without any European influence, occurred the distillation ("destillatio per descensorium") of a bituminous rock to obtain a "licour whiche doeth come forthe of it, dooeth profite for many deseases, and especially when it dooeth depende of colde, or colde causes. . . . That whiche thei sente me is of a redde coullour, somewhat darke, and it hath a good smell." [17]

In North America during the seventeenth century, the English colonies

---

* Despite many exaggerated accounts to the contrary, verifiable reports clearly show that the Incas did not use bitumen as a material in road building or in paints; only use in medicine has validation.[16]

along the Atlantic seaboard were locked in a struggle for subsistence. England had been suffering decades of civil war, foreign wars, three over-throws of dynasties, and bitter trade rivalries. France seized the opportunity to stake out an empire in the New World, potentially almost a continent, and she sent out missionaries and hardy adventurers to explore the interior wilds of Canada, the Great Lakes region, and some areas southward. This wave of exploration swung the pendulum of reports on petroleum in North America to French sources.

Two French missionaries, Dollier and Gainee, put petroleum on the North American map. Drafting their map in 1670, they located an oil spring, "fountain de bitume," near what later became Cuba, New York.[18] Joseph Delaroch Daillon, another French missionary, probably was referring to the same source when, writing from the Lake Erie wilderness on July 18, 1627, he noted "a good kind of oil which the Indians called Antonontons." Penetrating the same region in 1642, several Jesuit missionaries "found a thick, oily, stagnant water, which would burn like brandy." [19] The first American to put petroleum on the map was the great pioneer geographer, Lewis Evans. In the vicinity of future Oil City, Pennsylvania, his map designated a location "Petroleum." [20]

During the eighteenth century, the mainstream of notices of petroleum shifted from the French to the British, their American colonials, and finally, after the conclusion of the Revolutionary War, to citizens of the new American nation. A different note began to creep into some of the notices during this period. Writers extended their observations beyond the mere occurrence of petroleum to methods of collection and use. An underlying circumstance of major importance to the eventual exploitation of petroleum as a resource was the new age of science then emerging in Europe. Alchemy was stripped of its Arabic prefix and its trappings of magic, and began to emerge as the empirical science of chemistry.

The new scientific enlightenment found its embodiment in America in the genius of Benjamin Franklin, who, typically, placed into the records a kind of oil entry that was different from his contemporaries. Beginning in 1757, he literally sought to pour oil on troubled waters for mariners in harbors or in heavy surfs where landings or delicate maneuverings were demanded. Intermittently over a decade, he conducted critical observations at sea, proving-ground tests on English ponds, and careful evaluation of laboratory reports, all of which culminated in a report read before the Royal Society in London on June 2, 1774, and later published in the Society's *Philosophical Transactions.*[21]

Franklin's test material was whale oil rather than petroleum, but he anticipated by more than a century intensive British efforts during the 1880's to employ petroleum for the same purpose, both in bombs and through pipelines installed in Aberdeen harbor.*

Numerous British and French military expeditions which pressed through the Allegheny region before and during the French and Indian War yielded no significant observations of petroleum, but a journey into the region during 1767 by a Moravian missionary, David Leisberger, resulted in a report of three different types of oil springs. Indians, he said, preferred the kind flowing into a creek without an outlet, enabling them to dip the brown fluid that had accumulated, stir the spring, then "fill their kettles with fresh oil." After boiling out the water, they used the oil medicinally as an ointment for toothache, headache, swellings, rheumatism and sprains; sometimes they took it internally. "It can also be used in lamps," he added.[23]

Oil Creek, Pennsylvania, a tributary of the Allegheny River and the site of the first commercial oil well, came in for direct mention in 1783. In a letter to Rev. Joseph Willard, president of the University of Cambridge, General Benjamin Lincoln reported that troops under his command halted at a spring in that vicinity. Collecting some of the "Barbadoes tar," which issued from it at a rate of several gallons daily, they "bathed their joints with it," which gave them "great relief, and freed them immediately from the rheumatic complaints . . ."[24]

General William Irvine, reporting from Carlisle in August, 1767, to His Excellency John Dickinson, Esq., made a similar commentary on petroleum traces in Oil Creek. Twenty-two years later George Henry Loskiel wrote at length in *Geschicte der Mission der Evangelischen Bruder Unter den Indianern in Nordamerika* of "two [springs] . . . discovered by the missionaries in the River Ohio." Finally, in 1795, the first *Gazetteer of the United States,* published by Joseph Scott of Philadelphia, took notice of Oil Creek, flowing "from a spring much celebrated for bitumen, resembling Barbadoes tar, and . . . known by the name of Seneca oil." A man in a single day could gather several gallons of this "sovereign remedy for various complaints."[25]

---

* Pipes were laid down in the port of Aberdeen in 1882 to supply oil to calm the waves. Experiments with Shield's apparatus for application, begun September 26, 1882, were reported successful December 4, 1882, at Folkestone harbor. Gordon's oil-shells, shot at Montrose, April 6, 1885, reportedly were effective calming agents. Adverse reports on oil's effectiveness were entered by Capt. Chetwind October, 1884.[22]

Oil springs and medicine

It was an interest in the applications of petroleum as a medicinal and pharmaceutic, in the guise of Seneca oil and a host of other localized names—fossil oil, Genesee bank oil, American oil, rock oil, to name just a few—which stimulated the only approximation of a commerce in petroleum prior to 1850. It was, moreover, a commerce in which crude oil occupied the unusual role of a retail commodity.

There was nothing unique in the reactions of early American settlers who imputed medicinal properties to petroleum. Men reacted similarly throughout history wherever and whenever they encountered petroleum. The processes of discovering these values remained remarkably similar. Men might observe healing effects on animals who bathed their wounds or burns in petroleum springs.° Or they might discover therapeutical benefits from petroleum by direct applications on themselves, made accidentally or intentionally.

The abundance of accounts by explorers and travelers during the seventeenth and eighteenth centuries, which mention medicinal administrations with petroleum among the Indians, tend to create the impression that white men slavishly imitated the Indian, as they did in their adoption of tobacco. Ledgers and account books of traders late in the eighteenth century fortified this impression, disclosing payments to Indians for "gallons" and "kegs" of Seneca oil. Seneca oil itself is believed to have acquired its name either from its use by Seneca Indians in western Pennsylvania, or from having been found at an early date, when petroleum was a distinct curiosity, in the vicinity of Seneca Lake, New York.[27]

The amount of transmission of Indian therapeutical practices can easily be exaggerated. Many white settlers observed and had their own direct experiences with petroleum used as a liniment and for a rheumatic remedy through the use of products such as "British oil," a remedy for muscular ailments that was widely sold in England and America. Settlers having any contact with densely populated parts of the Atlantic seaboard were apt to know several possible uses for petroleum.†

° Some such observations must have been operative in the vicinity of Hughes River in Wirt County, West Virginia, around 1825. About a mile below the junction of the north and south forks, there was considerable activity in digging sand-pits to depths of twelve feet for petroleum. "This Hughes River oil was used mainly as a liniment for burns, and was especially valued for horses."[26]

† British oil was patented in 1742 by Michael and Thomas Betton. "An Oyl extracted from a Flinty Rock, for the Cure of Rheumatick and Scorbutick and other Cases," it

Fortescue Cuming, reporting in 1808 on activities to collect Seneca oil along Oil Creek, intuitively used the well-known British oil as the standard of reference for his readers. "The virtues of Seneca oil," he wrote, "are similar to those of British oil and supposed to be equally valuable in the cures of rheumatic and other pains. Large quantities are being collected on Oil Creek . . . and sold at from one dollar and a half to two dollars per gallon." [29]

The prestige of British oil penetrated western Pennsylvania and impressed the citizens enough to inspire some druggists to appropriate the "British oil" trade-name for a product which they adulterated with Seneca oil. Meanwhile an extensive, if not as intensive, demand for Seneca oil had developed throughout most eastern states. Benjamin Silliman, one of the leading scientists of the time, reported widespread sales in 1833, and gave Oil Creek Springs as the source of the supply.[30] In 1830 the *New York Journal of Commerce* reported, "Considerable quantities are annually brought to this city and sold to apothecaries." [31] Apothecaries in New York and Philadelphia, with extensive experience in selling British oil, quite naturally gravitated to the sale of Seneca oil. By mid-century the American product had practically excluded British oil from the American market.[32]

The amount of pharmaceutical demand for Oil Creek petroleum defies accurate determination, but despite references to "large quantities" it must not have exceeded some dozens of barrels annually. Blanket-dip collections were made only in the dry season at Oil Creek, when, by one early estimate, a man might gather 20–30 gallons in two or three days.[33] As late as 1843, when the demand for Seneca oil had become solidly established, several landowners along Oil Creek were each collecting 2–12 barrels a season, "the quantity depending on the prevalence of dry weather and low water." The going price at Pittsburgh was 75¢ to $1.00 a gallon, in contrast to $16.00 some "forty or fifty years ago." [34]

Beyond the limitations of supply wrought by blanket-dip collection, there were market limitations. A few barrels sufficed to saturate the Pittsburgh market, where local demand was most intensive. A few barrels would have likewise sufficed pharmaceutical centers like Philadelphia and New York City. A single barrel would make up into a lot of half-pint bottles.

---

soon found its way into Lewis's authoritative *Materia Medica*, and gained widespread distribution both in England and America well before the close of the eighteenth century.[28]

Salt wells and medicine

Early in the nineteenth century the pharmaceutical demand for petroleum was being supplied from more abundant sources than oil springs or shallow digging operations. The new source was an unexpected and unwelcome by-product of drilling for salt wells. But salt-well drilling had an even more important significance for the petroleum industry. Out of its evolution came the experience and techniques that were already far advanced when the first attempts were made to obtain petroleum by drilling.

This branch of the industrial arts arose in the rich salt-producing areas of western Virginia where commercial salt production was already established before the end of the eighteenth century. Interest in salt production at this time is readily understandable. In an era without canning or universal refrigeration, supplies of salt became a condition of survival to man and beast alike. Townsfolk depended on salt no less than frontiersmen for the preservation of their meat—2 bushels to 1,000 pounds of pork, 500 pounds to a quarter of beef, and equally large portions for fish. For want of a better way to exclude air and provide a cushioning agent, salt found extensive use in packing eggs; it was the preservative of hides, vital for shoes and harness.[35]

Surmounting the mountain barrier of the Alleghenies in increasing numbers early in the nineteenth century, settlers of future oil country in Pennsylvania, western Virginia, and future Ohio, almost instinctively explored for salt. With these intensive explorations, greater attention was inevitably focused on petroleum and natural gas because of their frequent association with salt deposits, an association that more often than not interfered with the exploiting of those deposits.

Mere digging or extractions by boiling the waters of salt springs often gave yields that were unsatisfactory in quantity or purity. Drawing on native ingenuity and perhaps a dim knowledge of techniques employed on the seaboard, frontiersmen improvised both tools and methods of boring for salt, with the result that within one or two decades drilling for salt became fairly common. Foremost among these pioneer improvisations in the western country, and probably the earliest of them to yield petroleum as an unsought by-product, were those of two brothers, David and Joseph Ruffner.

The site of the Ruffner operations was a farm in Kanawha county in West Virginia, located at the junction of Cambell's Creek and the Great

Kanawha River. Purchased by their father Joseph Ruffner, Sr., in 1785, the property included "The Great Buffalo Lick," a salt lick some 12–14 rods in extent, and a spring which yielded a strong brine.[36]

Starting in 1806, the Ruffner brothers began a venture which would carry them through some eighteen gruelling months of discouragement and drain their limited technical resources. Determined to reach the bottom of the mire and quicksand through which the salt water flowed, they took a "straight, well-formed, hollow sycamore tree with four feet internal diameter, sawed off square at both ends. This was called the 'gum.'"[37] Setting the gum in an upright position, they propped it on four sides and provided a platform upon which two men could stand. They then erected a swape, or slanted pole, in a position with its fulcrum in a forked post set in solid ground. A rope at the other end from which a whiskey barrel was suspended as an improvised bucket completed the assembly.

Digging started, using six or seven men, one stationed in the hollow tree where he wielded pick, shovel, and crowbar. Two men mounted the platform, where they dumped and returned the bucket. Three or four more men manipulated the swape. After many delays and difficulties, they struck rock at 17 feet. New troubles arose. The square tree bottom striking the jagged surface of the jagged rock allowed water to infiltrate the tree "gum." Trimming of alternate sides of the trunk produced a smoother joint. Frequent insertion of wedges completed a makeshift approach to watertightness, so that bailing out the contents of the tree-hollow would establish whether the salt water came up through the rock. The water-yield was low, but salt content was higher than any yet encountered.

To go on seemed worthwhile, providing the stubborn rock could be coped with. Here the Ruffners hit upon an ingenious adaptation of rock-blasting technique when they made a long iron drill with a 2½-inch chisel and attached the upper end to a spring pole with a rope. By January of 1808, when they had drilled through 40 feet of the rock and reached a depth 58 feet down from the top of the gum, they stopped drilling, judging the flow of strong brine sufficient for their furnace.

Drilling stopped, but another formidable challenge menaced them. This was how to get the strong brine up without dilution by weaker brines at higher levels. The Ruffners found the most brilliant of all their solutions, for which they had no precedent, when they whittled out by hand the two halves of a wooden tube, 2½ inches in diameter, which were put together and wrapped with twine from end to end. Swathed in cloth at the lower end to buttress watertightness, the tube was lowered cautiously and painstakingly down into the top part of the rock boring. Brine began to flow copiously, and it now could be raised easily and in quantity by swape and bucket. This device worked even better when the brothers substituted a leather bag partially filled with flax seed for cloth around the bottom of the tubing. As the seeds became saturated with water, they swelled and tightly filled the space between the tubing and the rock.[38]

One last anticlimatic challenge presently confronted the Ruffners. Salt brine and profits continued to flow, but a contaminant to both began to appear along with them: brownish stuff which settlers in the region usually knew as Seneca oil. Disposal proved simple. They simply allowed the petroleum to flow over the top of the cistern into the river, a method which became increasingly common in succeeding decades wherever petroleum was encountered in salt borings.

Neighbors of the Ruffners, viewing their success with a "gold mine" of salt, began borings of their own. By 1817 some fifteen or twenty wells averaging 50–100 feet or more in depth were feeding thirty furnaces, enough brine to yield between 600,000 and 700,000 bushels of salt annually. According to one account, drilling had been extended in the area to some 500 feet by 1815; at this depth, however, the well was a failure as a salt producer, gas blew the tools out of the well, and it was abandoned.[39]

Chroniclers and oil historians usually ascribe the rapid migration of drilling technology to skilled drillers who developed in the Kanawha region and carried their craft elsewhere.[40] There is some reason to suppose, however, that many individuals without such experience were successful in utilizing the methods pioneered by the Ruffners.* But in any event, the Ruffner brothers, by freeing salt producers from dependence on flowing springs, laid the basis for a further evolution of techniques which could be applied with equal success to drilling for water, gas, and petroleum.

A number of improvements or refinements of their methods were not long in forthcoming. By 1815, for example, a tinsmith in Charleston, in western Virginia, began to fashion tin tubes of convenient lengths which replaced the crude wood and twine-wrapped tubing first used by the Ruffners. Within a decade this improvement was followed by copper tubing, permitting the first refinement of screw joints.[42] Whether copper

---

* In 1814, before Kanawha drillers had much opportunity to gain experience or travel, Robert McKee bored a salt well on Duck Creek near the Muskingum River, thirty miles north of Marietta, Ohio, to a depth of 475 feet. The account of McKee's well was reported by Dr. Hildreth, the Marietta, Ohio, physician whose observations of salt wells and petroleum phenomena during the first three decades of the century were usually informed and accurate.[41]

tubing was exclusively a Kanawha innovation cannot be determined, but it was clearly in general use elsewhere than in the Charleston area by 1828.[43] By the early 1830's, considerable labor savings had been realized. In contrast to the Ruffners' employment of six or seven men, two men standing regular hours of duty six hours each, night and day, now handled drilling operations. Their daily progress, depending on the density of the rock, varied from 1 inch to 5 feet or 6 feet. By this date, too, there were numerous instances of the use of a horse, working on an inclined treadmill, to supply the power for drilling operations, and an occasional reference to the use of steam power.[44]

Wells of 1,000 feet were bored during the 1820's, but to make drilling to depths of 1,500–2,000 feet cheap and simple, there was need to add another innovation to copper tubing and seed-bagging. This was the salt man's "slip," known later to oilmen as the "jar." In 1831 William Morris, a clever Kanawha driller, invented a successful slip to give the heavy sinker and bit a fast, clean-cutting fall, unencumbered by the slower motion of the auger poles above. Morris may not have been the only inventor of the slip nor the earliest, but his innovation is sufficiently authenticated to mark the time when the slip began coming into significant usage.[45]

Although petroleum had been an unwelcome by-product of many salt wells since the Ruffners' experiences early in the second decade of the century, the exudates of springs or trenching operations remained the source of most petroleum medicinal oils for almost two more decades. A noteworthy shift occurred with the organization of the American Medical Oil Company of Burkesville, Kentucky, late in the 1830's. Promoters there turned to a "ruined salt" well on a nearby farm for their medicinal petroleum, which in 1829 had reportedly gushed forth "thousands of barrels" of oil into the Cumberland River.* Nearly ten years later the well was still yielding enough petroleum for the promoters of the American Medical Oil Company to bottle and sell it for rheumatic and other ailments. Over the succeeding decade several hundred thousand bottles were sold in the United States and parts of Europe.[47]

Despite Pennsylvania's future destiny as the pioneer oil state, there were few reports of petroleum showings from salt borings in the Commonwealth during the first four decades of the century. But late in the 1830's, increased drilling activities at Tarentum, Pennsylvania, on the Al-

---

* When a small boy ignited the oily flotations two miles from their point of issue, he set off a conflagration that raged more than fifty miles downstream. It destroyed the vegetation along the banks leaving marks of its devastating path for many years.[46]

legheny River, paved the way for a sustained application of salt-boring exudates to commercial medicinals. Among the pioneers in drilling for salt in the area were Samuel Kier, operator of canal boats between Pittsburgh and Philadelphia, and his father Thomas, of Tarentum, who sometime about 1839 or 1840 summoned Josephy Doty, Sr., a veteran Kanawha borer, to drill for salt on lands leased from their neighbor, Lewis Peterson. Some months after striking salt water of good concentration at 465 feet, an unusual event in the area occurred: the appearance of small quantities of unwanted petroleum. For a time the Kiers and their neighbor Lewis

COURTESY DRAKE WELL MUSEUM
Samuel Kier

Peterson, whose salt well in 1844 also began producing small amounts of petroleum, simply drained off the oil and dumped it into the nearby Pennsylvania Canal.[48]

It was the illness of Samuel Kier's wife which served to link petroleum from these salt wells with commercial medicinal oils, which in turn initiated a chain of events of particular importance for the birth of the modern petroleum industry. Kier noted that the "American Oil" prescribed by his wife's doctor looked and smelled exactly like the crude waste from his own well. After linking this resemblance with some knowledge of the long-localized use of Seneca oil, he then arranged to lease the neighbor-

Five-gallon still used by Samuel Kier to separate petroleum from salt water.

PETROLEUM, OR ROCK OIL.

A NATURAL REMEDY!

PROCURED FROM A WELL IN ALLEGHENY COUNTY, PA.

Four hundred feet below the Earth's Surface!

PUT UP AND SOLD BY

SAMUEL M. KIER,

CANAL BASIN, SEVENTH STREET, PITTSBURGH, PA.

The healthful balm from Nature's secret spring,
The bloom of health, and life, to man will bring;
As from her depths the magic liquid flows,
To calm our sufferings, and assuage our woes.

CAUTION.—As many persons are now going about and vending an article of a spurious character, calling it Petroleum, or Rock Oil, we would caution the public against all preparations bearing that name not having the name of S. M. KIER written on the label of the bottle.

PETROLEUM.—It is necessary, upon the introduction of a new medicine to the notice of the public, that something should be said in relation to its powers in healing disease, and the manner in which it acts. Man's organization is a complicated one; and to understand the functions of each organ, requires the study of years. But to understand that certain remedies produce certain impressions upon these organs, may be learned by experience in a short time. It is by observation in watching the effects of various medicines, that we are enabled to increase the number of curative agents; and when we have discovered a new medicine and attested its merits, it is our duty to bring it before the public, so that the benefits to be derived from it may be more generally diffused, but have no right to hold back a remedy whose powers are calculated to remove pain and to alleviate human suffering and disease. THE PETROLEUM HAS BEEN FULLY TESTED! About one year ago, it was placed before the public as A REMEDY OF WONDERFUL EFFICACY. Every one not acquainted with its virtues, doubted its healing properties. The cry of humbug was raised against it. It had some friends,—those that were cured through its wonderful agency. These spoke out in its favor. The lame, through its instrumentality, were made to walk—the blind, to see. Those who had suffered for years under the torturing pains of RHEUMATISM, GOUT and NEURALGIA, were restored to health and usefulness. Several who were blind have been made to see, the evidence of which will be placed before you. If you still have doubts, go and ask those who have been cured! Some of them live in our midst, and can answer for themselves. In writing about a medicine, we are aware that we should make no statements that cannot be proved. We have the witnesses—crowds of them, who will testify in terms stronger than we can write them to the efficacy of this Remedy, who will testify that the PETROLEUM has done for them what no medicine ever could before—cases that were pronounced hopeless, and beyond the reach of remediate means—cases abandoned by Physicians of unquestioned celebrity, have been made to exclaim, "THIS IS THE MOST WONDERFUL REMEDY EVER DISCOVERED!" We will lay before you the certificates of some of the most remarkable cases; to give them all, would require more space than would be allowed by this circular. Since the introduction of the Petroleum, about one year ago, many Physicians have been convinced of its efficacy, and now recommend it in their practice; and we have no doubt that in another year it will stand at the head of the list of valuable Remedies. If the Physicians do not recommend it, the people will have it of themselves—for its transcendent power to heal, will and must become known and appreciated—when the voices of the cured speak out; when the cures themselves stand out in bold relief, and when he who for years has suffered with the tortures and pangs of an immedicable lesion, that has been shortening his days, and hastening him "to the narrow house appointed for all the living," when he speaks out in its praise, who will doubt it! THE PETROLEUM IS A NATURAL REMEDY—it is put up as it flows from the bosom of the earth, without anything being added to or taken from it.

Kier circular advertising medicinal

Rheumatism yields to the Power of Petroleum.

I feel myself under a debt of gratitude to the proprietor of the Petroleum—the use of which great medicine has entirely cured my wife of a violent attack of the Rheumatism. She had labored under an attack of the disease for about two months, suffering the most intense pain; the greater part of the time confined to her bed, and unable to do anything. The pain in the limbs was very great, attended with a great deal of swelling. The Petroleum had a happy effect; for the first two or three applications the swelling diminished, and the pain left her. I continued to apply the medicine daily for about two weeks, which entirely relieved her, and she is now as well as ever she was in her life. I feel confident that the Petroleum is one of the greatest medicines in the world for Rheumatic pains and swellings. (Signed) PETER PUHL, *Perry Street, Allegheny City.*

Read the following certificate. It is from Mr. Harvy Mann, the manufacturer of Mann's celebrated axes:

"There is nothing like it for Burns."

BELLEFONTE, February 21st, 1850.

I got three dozen of Mr. Kier's Petroleum of Mr. Howell, and I am disposed to think there is nothing equal to it for burns; having burnt my hand, and expecting nothing less than a long, tedious sore, but found no soreness at all from its only leaving a red scale, without any soreness when pressed upon, which quite surprised me. I therefore have great faith in its efficacy, so far at least, and will introduce it in this section. Yours, respectfully, &c.

HARVEY MANN.

———

MR. S. M. KIER:—Allow me to express to you my heart-felt thanks for the great benefit I have received from an article called Petroleum or Rock Oil, of which you are sole proprietor. I had occasion to use it about the 1st of January, in a violent attack of Rheumatism, which was very painful, flying about from place to place, accompanied with much swelling, so as to keep me in constant torture. I used the Petroleum externally, a few applications of which removed all pain, and every symptom of the disease. I am now entirely well, and would take this occasion to recommend the Petroleum to all who may be suffering under the agonizing pains of Rheumatism or kindred diseases.

Pittsburgh, February 12, 1850. ZIBEON WILBAR.

A Bad Case of Inflammatory Rheumatism.

The following extracts are from a letter received from a young gentleman of Middleburg, Summit County, Ohio. They are attested by Charles Belden and Dr. Elijah Curtis, of that place:

MR. KIER: *Dear Sir*,—Gratitude compels me to express my thanks for the discovery of your invaluable medicine I was afflicted with that awful and disheartening disease, the Inflammatory Rheumatism, on the 21st of September last, whilst I was in Cleveland; and was brought home from that place to Middleburg, where I lay on my back for over two months. For three weeks of this time, I was in the most intense suffering and agony. I lay with my limbs in one position, and could not stir a muscle, nor have my bed changed. My screams could be heard all over the neighborhood, so intense was my suffering. When every other remedy proved valueless, I had recourse to your invaluable Petroleum, of which I had made but three applications, until I commenced to get better. I used altogether but two bottles and a half, and am now well. I can get any number of persons to testify to the above facts, if necessary.

Attest. CHARLES BELDEN, (Signed) FREDERICK H. BLECKER.
DR. ELIJAH CURTIS.

CHRONIC COUGH CURED.

Public attention is most respectfully invited to the plain, unvarnished statement of John Watt, who was cured of an old cough by the use of the Petroleum:

"This may certify that I have been cured of an old Chronic Cough, by the use of four bottles of Petroleum. The cough attacked me a year ago last December, and I had lost all hopes of getting well, as I had taken the advice of several physicians without any benefit. I was benefitted almost instantly by the Petroleum. I coughed up, during the use of the Petroleum, a hard substance resembling bone. I make these statements without any solicitation from any one to do so, and solely for the purpose that others who are suffering may be benefited. You are at liberty to publish this certificate. I am an old citizen of Pittsburgh, having resided here 33 years. My residence, at this time, is in Second Street. JOHN WATT.

Pittsburgh, February 21st 1851."

value of rock oil with testimonials.

ing Peterson well and about 1849 opened an establishment in Pittsburgh for bottling and merchandising his oil.[49]

Whether his quick success or his advertising copy was the more sensational is hard to determine. Although he started out with an orthodox distribution, putting peddlers on the road in highly decorated wagons, his "Rock Oil" gained something approaching national notice.[50] Handbills simulating bank notes made some bold pronouncements. Three dosages a day internally might make the lame walk, the blind see; external applications would rout rheumatism, gout and neuralgia. Many former skeptics in the medical profession, it was claimed, had been converted, but the people would have used his oil whether or not the physicians recom-

COURTESY DRAKE WELL MUSEUM

Kier's petroleum bottle and the wrapper in which it was sold.

mended it.[51] There was also printed a stream of testimonials to Rock Oil's efficacy in curing cholera morbus, coughs and ague, toothaches, corns, neuralgia, rheumatism, piles, urinary disorders, indigestion, and liver complaint.[52]

For all of Kier's gifts as a huckster, his commercial success was apparently modest. Testifying in 1858, he stated that up to that time he had sold nearly 240,000 half-pint bottles of Rock Oil at a price of $1.00 per bottle. But because of heavy expenditures for advertising to establish his product in the public mind, his net profits had not been large.[53] In fact, his high costs of distribution appear to have prompted Kier about 1852 to alter his system of marketing. He withdrew his agents who had been touring the countryside in wagons decorated with pictures of the good Samaritan administering to a stricken victim in agonized contortions under a palm tree. In their stead he simply sold directly to drug stores.[54]

Although not an innovator in the narrow field of petroleum medicine, Kier was one of its most successful marketers, and through somewhat fortuitous circumstances, gained the distinction of becoming the unwitting instrument for the successful transfer of salt-drilling techniques to petroleum production. According to popular accounts, it took a remarkable contortion of destiny's finger to point in Kier's direction. Devising a simulated bank note in January, 1852, to be used as a broadside to promote sale of his Rock Oil medicine, Kier incorporated a picture of the Tarentum Oil well. Accompanying headlines boldly proclaimed: "1848 Discovered in

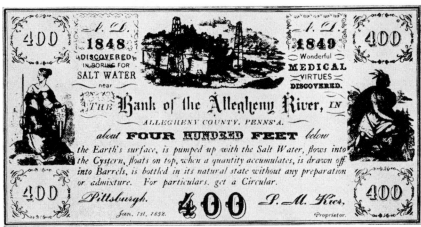

COURTESY OIL AND GAS JOURNAL

Kier advertisement simulating a state bank note.

Boring for Salt Water near the Bank of the Allegheny River about FOUR HUNDRED FEET below the Earth's Surface, [it] is pumped up with Salt Water . . ."[55]

As the story goes, it was one of these "bank notes" on display in a window of a New York drug store which caught the eye of George H. Bissell in the summer of 1856. From this chance observation Bissell presumably received the inspiration which led some three years later to the first successful drilling specifically for petroleum.[56]

Although widely accepted, this particular version of how the promoters of the Drake well conceived the notion of drilling for petroleum is probably apocryphal. It is true from Bissell's own testimony that he and his associates knew that Kier was obtaining small quantities of petroleum from one of his wells drilled for salt. And in 1858, when efforts to recover commercial quantities of petroleum by trenching at Titusville had failed, this fact, plus the knowledge that petroleum was commonly found in other salt wells, influenced their decision to proceed with drilling for petroleum.[57]

But to find the key to Bissell's interest in petroleum, it is necessary to look beyond pharmaceuticals and medicinals. He and his associates had no intention of competing with Kier and others in the marketing of petroleum medicinals. Living in New York City, these promotors were aware by 1854 that "Illuminating oil from coal was just beginning to be talked of, but very little was made then." Accordingly, as Bissell suggested in a later account, in deciding to lease the land upon which the Drake well was subsequently drilled, "We did not prospect the oil for medicinal purposes, but we believed it would be a good illuminator, and we sought it as an article of commerce."[58]

Back of this new appreciation of the properties of petroleum as a source of illumination, and to some extent of lubrication, lay slowly maturing developments tributary to the founding of modern chemistry. These developments played a significant role in making the 1850's a decade of revolution in all forms of illumination. It was this decade which, in turn, laid the groundwork and provided the momentum for the birth of the modern petroleum industry.

The Illumination Revolution of the 1850's

Chapter 2

Illumination in America

The intellectual atmosphere of the late eighteenth and early nineteenth centuries was marked by changing concepts of science as well as by an increased curiosity about these new concepts. Modern chemistry, born late in the eighteenth century, began to forge ahead rapidly when new methods of quantitative experimentation revealed the inadequacy of some of the old concepts and opened the door to a better understanding of many puzzling problems.

The beginnings of scientific chemistry

From the dawn of history, combustion with flame was of outstanding significance in the advance of civilization. Yet until late in the 1700's there was no adequate explanation of this common phenomenon. Early in the eighteenth century, the German physician, George Ernest Stall (1660–1734), advanced the theory that all combustible materials owed their inflammability to an invisible substance—the "principle of fire" which he called "phlogiston"—which was present in all combustible bodies in an amount proportional to their degree of combustibility. Coal, for example, was regarded as practically pure "phlogiston." [1]

Because "phlogiston" supplied a simple explanation of combustion and a number of other apparently unrelated chemical phenomena, most scholars of the period generally accepted this theory, even though it was inconsistent with the common knowledge that when metals were oxidized their weights increased.

It remained for Joseph Priestly (1733–1804) to provide the key to a better understanding of combustion and to lay the foundation of modern chemistry when, in 1774, he isolated oxygen by heating mercuric oxide. At first he described oxygen as "modified" or "phlogisticated nitrous air," or what is known as nitrous oxide today. He subsequently recognized that he had a unique gas, some five or six times more effective in supporting combustion than common air. But failing to see that he had turned a new page in science, he named his discovery "dephlogisticated" air. [2]

Building on the findings of Priestly in England and possibly Carl W. Scheele in Sweden, Antoine Lavoisier, "the father of modern chemistry," asserted that oxygen was an element and proved that the process of combustion, whether it takes place in the body, the furnace, or the engine, is nothing more than the uniting of other elements with oxygen. [3]

At the root of Lavoisier's contributions was his development of more delicate instruments for exact weighing. Coupled with meticulous application of quantitative methods, these enabled him to show that the quantity of matter remains unchanged in chemical processes regardless of changes in its form. [4] Lavoisier also undertook the task of establishing a new terminology which, published in the first textbook of the new chemistry, the *Traité Elementaire de Chimie* of 1789, still forms the basis of chemical language. [5]

The timing of this breaking through the age-old barriers surrounding chemical research to seek a better understanding of combustion and the nature of gases is a key to the revolutionary changes in illumination which marked the nineteenth century. It also helps explain much of the failure of the eighteenth century to make few significant advances in illuminating practices. There were few areas where this lack of progress toward better illumination was more obvious than in America.

American illumination to 1830

Until the early decades of the nineteenth century, illumination in America remained nearly as primitive as it had been in ancient Rome or Greece. Throughout the greater part of the colonial period most Americans depended on common tallow candles or plain open-dish lamps with little more than a rag stuck in them for a wick. Some use was made of iron or tin "Betty" lamps, with their pear-shaped bowls holding fish oils or greases, and with one or two spouts from which wicks were suspended (see Figure 2:1). During the latter part of the eighteenth century, the scarcity of whale oil in many localities led to a modification of the "Betty" lamp. This new development was the Cape Cod model, designed to burn lard oil and equipped with a charcoal brazier underneath to heat the lard and keep it fluid.

One improvement in the quality of illuminants resulted from the accidental discovery, in 1712, of the deep-water haunts of the sperm whale. Compared with the right whale, long a basis for the colonial whaling industry, the sperm whale was found to yield fats and oils richer in stearine and markedly superior in burning and lubricating qualities. Burning cleaner and brighter, it could be used in the old type of whale and lard oil lamps. The Carcel lamp, a leading French lamp suited for sperm oil, was introduced in 1800, but it remained rare in the early part of the century.* Following the discovery of the sperm whale, whaling quickly graduated to a deep sea industry. By 1775, Nantucket's fleet alone numbered 150 vessels and many other ports had entered the business.[7]

From the heads of the sperm whale also came spermaceti, a tasteless, odorless, sponge-like substance which, when frozen and pressed, proved an unrivalled material for candles. From the time of their first manufacture at Newport, Rhode Island, about 1750, sperm candles remained the standard for the best candlelight for over a century.[8] But sperm candles, like the sperm illuminating oil, remained a premium product, and their use was largely confined to the more wealthy consumers.

A basic step toward better lamplight was a burner usually, though not surely, ascribed to Benjamin Franklin late in the eighteenth century. With little information to guide him, Franklin decided that instead of one flame, two flames burning close together would increase the updraft and

---

* Like the better known American adaptations, the Diacon and Mechanical lamps, the Carcel lamp employed a complex clockwork mechanism to feed the oil, usually sperm oil, evenly to the wick.[6]

*Cross-section of American lamp evolution, Colonial period to 1860.
(1) Crusie, wrought iron, with half bail for suspension, and one or
more wick slots; among the earliest open grease lamps. (2) Betty Lamp,
wrought iron, seventeenth century to 1850; its half-tube wick support
and cover were improvements over the crusie. (3) Rushlight Holder:
inverted wrought iron pincers to hold reeds repeatedly dipped in fat,
making dripless rush candles. Cylindrical shades of pierced tin (not
shown) completed this early night light. (4) Spout Lamp, pewter, late
eighteenth century, with attached pick for cleaning crusted wicks.
(5) Spout Lamp, brass, a later model with double spout to catch and
return drippings to the font. (6) Time Lamp, pewter, early nineteenth
century, with Cardigan glass oil font around which pewter bands marked
off the hours. (7) Lard Oil Lamp, tin, about 1830, with characteristic
wide flat wick, loosely woven and long wick tube, to promote capillary
action. Other types used copper heat conductors, gravity, or pumping
devices to force lard or oil to the wick. (8) Whale Oil Lamp, pressed
glass, lyre pattern, one of the handsome patterns that quickly followed introduction of mechanical pressing of glass in 1826. This typical screw top burner, with two straight vertical tubes, closely spaced,*

heat sufficiently to give better combustion and intensify the light. He further increased the effectiveness of the burner by developing a loosely
braided cotton wick, which gave a more efficient capillary action in feeding the oil to the flame. Refusing to take out a patent, Franklin donated
his invention to the American people. The cleanest and most efficient
lamp then available, it quickly sprang into wide use.*

Meanwhile at Geneva, Switzerland, Aimé Argand, physician and practical chemist, also became interested in the possibility of improving lamp
performance by increasing the updraft. About 1783 he neared his objective by fitting a sheet-iron chimney over a metal tube extending through
the base of the lamp to admit air from below and encased with a circular
wick. This arrangement greatly improved the draft, but Argand's final
stroke of genius came when he substituted a glass chimney for the metal
one, thus extending the area of combustion without cutting off the light
from the flame.[10] The result, as described by a contemporary journalist,

---

* The full benefits of Franklin's suggestions were not realized until about 1800, when
the introduction of both single and multi-tubed drop burners equipped with a
collar that screwed tightly into the font of whale oil lamps, gave Americans their
first closed, oil-tight, genuinely agitable lamps.[9]

---

*made whale oil lamps the first closed oil-tight lamps (late eighteenth-
century model). (9) Camphene Lamp, pewter; actually a whale oil lamp
converted to the distinctive camphene burner. To increase air supply
and minimize hazards of explosion, the vertical tubes projected higher
above the burner collar than in whale oil burners, and angled away from
the volatile fuel supply. (10) Camphene Lamp, pressed glass, popular
loop design, 1840's, with twin extinguisher caps attached to the burner
to obviate the hazardous need to blow out the light. (11) Peg Lamp,
1850's, so named because its pressed cranberry glass font, typically turnip-shaped, was pegged into a brass candle stick. This model is a solar
lamp with an Argand burner distinctively modified by a concave plate
fitting with an opening that promoted a tall column of light. The stately
Argand chimney, a rarity earlier, aided this. The solar lamp, delivering
six candle power, was the greatest lamp consuming lard oil. This model
was one of the forerunners of early kerosene peg lamps. (12) Astral
Lamp, 1850's, with marble base, brass stem, top-shaped font of elegant
imported blue overlay glass, and an Argand burner distinctively modified to a ring shape to permit it to be set at oil level. The Astral shade
(not shown), though often pinch-topped, ranged widely in shapes and
types of glass, from plain ground to highly ornate.*

was a lamp with a "flame like the fire in a furnace" compared with common lamps which burned "like an open-air fire."[11]

Although all subsequent improvements in lamplight during the nineteenth century were elaborations of the Argand burner, its very simplicity resulted in many infringements and adaptations, and Argand received little financial reward from his patents, which he took out in France and England in 1784.[12] Nor did the Argand burner have any immediate impact on illumination in America, where until well into the first half of the nineteenth century, most lamps were open flame, poorly aerated, and fed

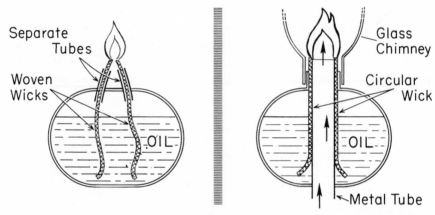

FIGURE 2:1. *Diagram of Franklin burner.*   FIGURE 2:2. *An illustration of the principle of the Argand burner.*

by inadequate wicks. Animal and vegetable fats remained the only sources of artificial light. Consequently, the use of artificial illuminants usually resulted in guttering, smoky, uneven flame; unpleasant odors; and the incessant chore of wick trimming.

One major innovation did brighten this murky pattern when pioneer experiments in England paved the way for gas manufacture to gain a permanent industrial footing in America. The fact that a combustible illuminating gas is produced during the decomposition of coal by heat was known in the seventeenth century. But the first commercial manufacture of gas did not come until 1802, when Thomas Murdock installed a coal-gas unit in the famous factory of Boulton & Watt, near Birmingham, England. Ten years later, following the organization of the London & Westminister Chartered Gas Light & Coke Company, the streets of London became the first in the world to be lighted by gas.[13]

The superior brilliance of the new illuminant quickly reflected across the Atlantic to Baltimore where the newly chartered Gas Light Company began producing coal gas on a modest scale in 1816, marking the first important use in America of a mineral as a source of illumination. By 1830, Boston and New York had also installed gasworks, but like the Baltimore company they confined the distribution of their output almost entirely to streetlights and public buildings.[14]

American illumination: 1830–1850

If illumination had developed slowly over the preceding two hundred years, it did not long remain unresponsive to the dynamic changes affecting American economic and social life during the second quarter of the nineteenth century. A near doubling of the population, a growing urban population, an expanding factory system, and the addition of some 9,000 miles of railroad, all quickened the need for more and better light.

Progress toward meeting this need took several different lines of development in America during this period: first, distillation of illuminating oil from turpentine; second, expansion and technological changes in the manufacture of lard oil; and third, growth of the domestic illuminating gas industry. A fourth line of development occurred in Europe and consisted of pioneering work in the production of coal oils. While this had relatively little immediate application in America it served to stimulate interest in the field of mineral lubricants. These developments combined by mid-century to sharply increase the variety and availability of illuminants and to lay the basis for even more impressive improvements to come.

Camphene: lamp oil from turpentine. The first major advance toward more lamplight for the common man occurred in 1830 [15] when Isaiah Jennings took out a patent for camphene, redistilled spirits of turpentine which, burned alone or mixed with alcohol, could be used as a lamp illuminant.[16]

Camphene, the only major innovation in illumination of native origin during the first half of the nineteenth century, was also the first synthetic oil lamp illuminant utilized in America. And though Jennings applied the art of distillation to a vegetable rather than a mineral material, he supplied a precedent for a transfer of techniques which would eventually be required in the effective processing of illuminants and lubricants from petroleum and solid mineral bitumens, such as coal.

Camphene, being much more volatile than animal oils, required a stronger draft to burn effectively. Within a decade lamp manufacturers found an adequate answer in simple modifications of the twin-tubed, screw-cap whale oil burner. Lamps 9 and 10 on page 30 show the greatly lengthened projection of the tube wicks above the burner to increase both air supply and safety. To add protection against the formation of explosive vapors in the oil font, the wick tubes were angled away from the fuel and tapered to improve wick security. Later lamp designers also sought to reduce explosive hazards by substituting smaller turnip-shaped fonts, a forerunner of kerosene lamp fonts, for the elongated fonts of whale oil lamps. Cheap conversion of whale oil lamps to camphene tube burners for 6–12 cents early in the 1840's sharply accelerated lamp availability and the consumption of camphene fluids. Consumers were attracted by the fact that camphene's light weight and low boiling point reduced the problem of effectively feeding heavy oils through the wick, particularly in cold weather. At the same time, camphene's low boiling and ignition points carried with them the ever-present danger of lamp explosions, the most serious objection to its use. But for most consumers this objection was largely offset by camphene's "remarkable intensity and high lighting power," "brilliant white flame," and above all, its "cheapness," which, in the opinion of contemporaries, presented "strong claims on the score of economy upon public notice." [17]

Aided by the growing demands of lubrication on animal oils, camphene became the dominant lamp illuminant during the 1840's. To the extent that lamp illumination had become democratized at mid-century, the average consumer owed his greatest debt to camphene fluids. His other debt was to technological innovations within the lard-oil industry.

Lard oil and stearin candles. The setting for the technological advances in the processing of lard oils was Cincinnati, the nation's pork packing center from about 1835 to the Civil War. In addition to slaughtering farm animals and packing meat products, Cincinnati also supported a specialized, by-product industry that made soap and candles.

Drawing on a process developed earlier in Europe, the Cincinnati soapmakers late in the 1830's found that by heating lard with soda alkali they obtained glycerin, stearic acid, and an improved lard oil with a melting point considerably below that of ordinary lard oil.[18]

Because of its hardness, stability, and resistance to heat, the waxy stearic acid proved an ideal material for candles. Cheaper than sperm candles but of equal quality, stearin or adamantine candles, made of one part

PURE LIGHT.

Where may be found a good assortment of

Hang, Side and Brackett Generator Lamps,
Hang and Side Camphene Lamps,
Hang & Side Solar do.
Girandoles,
Chandeliers,
Astral Generator Lamps,
Plated Glass Lamps,
Night Lamps,
Peg do.
Study Shades,
Patent Conductor Lanterns,

Together with a good assortment of

Fluid Lamp-tops, Brass and Britannia,
Hall and Hand Lanterns,
Brackett Lamps,
Study do.
Glass Chimnies,
Fluid Wicks,
Camphene Wicks,
Solar Wicks,
Tin Cans,
Brass Chain,
Glass Drops,
Patent New England Spring Bottom Lanterns,

J. F. DODGE,
AT 36 UNION STREET, BOSTON, MASS.,
Manufactures and offers at Wholesale and Retail,

PORTABLE BURNING FLUID,

AND

CHANDELIERS, LAMPS, GIRANDOLES, LANTERNS, &C. &C.

FOR USING THE SAME,

Adapted to the lighting of Churches, Halls, Factories, Stores, Railroad Stations. Cars, Dwellings, Workshops, Carriages, Streets, Ships, &c.

ALSO, CAMPHENE OF EXTRA QUALITY,

And a constant supply of

BLAKE'S PATENT

Arm or Conductor Lantern,

An article which should be in the possession of every person who wishes to carry his own light, and at the same time have both hands at liberty.

N. B.—Any article in the above line made to order at short notice.

Advertisement of Boston oil and lamp dealer, circa 1855.

tallow and two parts stearin, quickly came into wide use, particularly in the Midwest and the South.*

The new lard oil, with a melting point of 31° F, successfully challenged sperm oil, both as a lubricant and a lamp oil. With this new product as fuel, the solar lamp, introduced in 1841 by Robert Cornelius of the Philadelphia lamp house of Cornelius & Company, became the outstanding table lamp of antebellum America.[20] Consuming any viscous oil, but especially suited to lard oil, the solar lamp burned with a profuse, steady white light of six candle-power or more. Its performance compared favorably with that of the much more costly imported Carcel lamp, which consumed expensive sperm oil, as well as with those of the Diacon and Moderator lamps, also Argand burners with modified Carcel mechanisms for delivering oil to the wick top. The solar lamp's only distinctive modification of the Argand burner was a plate with a hole in it that fitted across the burner immediately above the wick. This saucer-shaped draft deflector shaded the flame into a tall column of light. Otherwise the solar lamp was indistinguishable from the astral lamp, the oldest modification of the Argand burner in America without a mechanical oil feed.

If at mid-century a "parlor lamp" was no longer the badge of social distinction it had been in Jacksonian days, it was mainly because of the improved lard oil and the Solar lamp. But while lard oil had been brought into competition with sperm oil, both types were responding to an expanding demand for lubricants. Neither could compete on a price basis with camphene in the growing demand for lamplight. The following tabulation, showing year-end prices, gives some impression of the relative costs per gallon during the 1850's of lard and sperm oil and the principal ingredients of camphene (turpentine and alcohol).

Year	Lard oil	Sperm oil	Turpentine	Alcohol
1852	$ .950	$1.340	$ .635	$ .546
1854	.855	1.970	.525	.909
1856	1.040	1.950	.460	.649
1858	.895	1.320	.505	.519
1860	.975	1.620	.355	.431

Source: Anne Bezanson, *Wholesale Prices in Philadelphia: 1852–1896* (Philadelphia: Univ. of Pa. Press, 1954), 3, 218, 223, 501.

But even the highly flammable and evil-smelling camphene, though it was relatively less expensive than lard or sperm oil, was retailed at prices

* By 1840 Cincinnati was producing about 2 million pounds of stearin annually from the slaughter of 500,000 hogs.[19]

as high as $2.00 or more per gallon.[21] Both cost and quality factors, there-fore, stimulated an interest in coal and other bitumens as a source of illu-mination.

European pioneers of coal-oil manufacture. In spite of the deficiencies in the illuminants in current use, signs of an impending turn to coal and other mineral bitumens for lamp illuminants were almost entirely absent from the American scene prior to 1850, even though domestic manufac-turers of gas had turned to the distillation of imported coal on an expand-ing scale during the late 1840's. Concerned primarily with obtaining maximum gas yields from their operations through intensive, high-tem-perature distillation, they displayed little interest in the range of oil prod-ucts resulting from low-temperature distillation.

In Europe, where an acute shortage of animal and vegetable fats stimu-lated an interest in alternative sources of oil, the situation was quite dif-ferent.[22] Out of extensive commercial experience with the distillation of coals, schists, and shales for illuminating gas, there gradually emerged an appreciation of the by-products made possible by distilling at lower tem-peratures.

Important work was done by many European chemists but the ac-complishments of A. F. Selligue of France and James Young of Scotland are of particular interest for their pioneer efforts in putting the production of coal oils on a commercial basis.

Selligue, working in the field of gas manufacture, took out a patent in 1834 for making water gas, a mixture of carbon monoxide and hydrogen produced by reacting steam with incandescent coke.[23] Because of the low heating and illuminating properties of water gas, he turned to the distilla-tion of Autun schists (shale) for an oil which he used to carburet and en-rich his gas. In 1838, Selligue patented the oil as a lamp illuminant and began producing it on a commercial basis.[24] By the mid-1840's, he was op-erating three refineries in France supplying oil both for gas manufacture and illuminants. By this time Selligue had also developed advanced treat-ing methods using sulphuric acid for deodorizing and alkalis for removing the acid and improving the color of his illuminating oil.[25]

James Young, trained in chemistry at Anderson College, Glasgow, be-gan his research about 1847 on the lubricating qualities of an oil which oozed at a daily rate of some 300 gallons from a "petroleum spring" in a coal mine at Derbyshire, England. Even before the oil supply began to wane, which was within two years, he turned to obtaining a similar oil by applying low-temperature distillation to coal. Scoring an unusual success from paraffin-rich Boghead coal from mines near Bathgate, Scotland, he obtained long-term contracts for its delivery and erected a refinery at Bathgate.

Meanwhile, Young obtained a patent in England in 1850, and in the United States in 1852, which he described as "improvements in the treatment of certain bitumenous mineral substances and in obtaining products therefrom." His patent explicitly covered the recovery of paraffin-type crude oils from coal by means of slow, low-temperature distillation. The patent also went on to develop the refinement of these oils through a series of redistilling and treating processes involving the use of oil of vitriol (sulphuric acid) to remove impurities, followed by caustic soda to remove the acid and remaining impurities. Young emphasized the importance in the initial distillation of bringing the temperature in the "common gas retorts" gradually up to "a low red heat," where it should be maintained until the volatile materials were removed. "Care must be taken," he stated, "to keep the temperature from rising above that of a low red heat so as to prevent . . . the desired products of the process being converted into permanent gases." [26]

Although Young specified the possible use of his paraffin oils as an illuminant, he did not turn to the production of coal oil as a lamp fuel until the mid-1850's. But in the interim he built up a highly successful business by marketing naphtha and lubricating oils, selling the latter at a handsome profit at about 5 s. a gallon in competition with animal and vegetable oils priced between 8 s. and 9 s. a gallon. [27]

Neither the work of Young nor Selligue had an immediate influence on the evolution of American illumination. But their accomplishments foreshadowed two lines of development which combined during the late 1850's to bring about the emergence of the coal-oil industry, the immediate predecessor and trail blazer for the petroleum industry. One line of development was reflected in a growing interest in producing a mineral-based lubricant by means of low-temperature distillation. The other development was an intensive drive to obtain materials for the enrichment of manufactured gas.

The growth of gas manufacture: 1830–1850. Probably the most spectacular development in lighting in America during the 1830's and 1840's came not in the field of lamp illuminants but in the growth of gas manufacture. By 1837 there had been a moderate expansion in the number of gas plants in the United States when, in addition to those plants already operating in Baltimore, New York and Boston in 1830, new companies were founded in Brooklyn, Bristol (Rhode Island), Louisville, New Orleans, Pittsburgh, and Philadelphia.[28] The panic of 1837 brought a halt to new construction, and it was not until the mid-1840's that the industry again began to accelerate. By 1850, approximately 56 plants were in operation.[29] In addition to large plants in urban centers, there was also an unrecorded number of smaller works which, by mid-century, were supplying factories in outlying districts with their own illuminating gas.[30]

Over the intervening years from the establishment in 1816 of the Baltimore Gas-Light Company, by no means all of the American gas companies followed the Maryland concern's example in using coal as a basis for their product. Various considerations, including relative costs and accessibility of supplies, led to experiments with a wide range of materials. But the main drive to find a substitute for coal arose from the impurities in coal gas-sulphur compounds, creosote, and similar substances that required elaborate and expensive equipment for their removal. The New York Gas Light Company, for example, began operations in 1823, using whale oil. When high prices of this material forced a change, the company turned to rosin oil rather than coal.[31]

By the late 1840's, however, technical advances pioneered chiefly by British producers tended to make cannel-type coals the most widely accepted material for gas manufacture. Chief among these improvements were better methods of purification, in which scrubbers replaced simple spray-washing equipment to remove ammonia and cyanogen (a poisonous gas), and iron oxide was added to lime to bring about a more effective elimination of sulphur compounds and carbon dioxide. Generally too expensive for installation in small plants, these improvements also provided large-scale operators with the basis for the recovery of by-products such as coke, coal tar, and ammonia liquor from manufacturing operations.[32]

These improvements, coupled with a reduction, in 1846, in the American tariff duties on imported coal from $1.75 a ton to 40¢ a ton, prompted most of the large gas companies on the Atlantic seaboard to operate on cannel coal, imported from England, until well down into the 1850's, when they began to use Albert bitumen from New Brunswick, Canada.[33] Meanwhile gas companies in the interior had begun to exploit the potentials of cannel coals in western Virginia, Kentucky, the Pittsburgh area of Pennsylvania, and even in Illinois.[34]

Other developments, such as improved burners and fixtures, added to the advantages of distributing this superior illuminant from large, central plants not only to public lighting facilities and industrial establishments, but to private residences as well. Of particular significance for the extension of service to households was the use of metering, invented in 1815 and put into general use during the 1840's.[35] In contrast to the previous system of nightly inspections to be sure that gas was being used only during hours contracted for by customers, metering service reduced labor costs, wastage, and customer complaints.

These developments all helped to make manufactured gas the largest and fastest-growing branch of the illuminating industry in America. Compared with any alternative form of artificial illumination, manufactured gas was widely preferred. In the households of the growing ranks of middle and upper-income classes in urban communities,[36] lamplight had already suffered a major eclipse, and it apparently would suffer a similar fate among the working classes once gas rates followed lowered manufacturing costs.[37]

But the majority of urban gas companies, partly because of the heavy capital investments involved and partly through inertia, were slow to extend their facilities and to pass cost savings on to their customers. Under these circumstances, there was a widespread and growing interest among consumers in cheaper methods or materials that might either prompt existing companies to reduce their rates or encourage competition.

One possibility which had a particular attraction both for urban and rural dwellers was a practical, portable, gas-producing unit. Such a development would enable individuals living in cities to escape dependence on central plants with their exclusive rights to pipe gas through the streets. Beyond the fifty or more cities burning public utility gas in 1850, and beyond those communities reached by private installations which brought gaslight to smaller towns and factories, stretched a twilight zone that required portable units and even individual lamps, if gas was to be made generally available.

Pioneer work by George Lowe in England, Selligue in France, and others, had made common knowledge of the potentialities of the vapors of benzole and other volatile fluids for enriching illuminating coal gas of poor quality, or carbureting water gas-hydrogen decomposed from water or atmospheric air.[38] But these products were generally too expensive to cut manufacturing costs materially. Thus when C. B. Mansfield in England developed and patented a process, in 1847, to obtain benzole cheaply from coal tar, an imposing barrier to expensive gas enrichment was apparently removed.[39] Moreover, Mansfield had also apparently broken the barrier to portable gas units with equipment to enrich coal gas or water gas, which he pointed out was "applicable on any scale, from the dimensions of town gasworks to the compass of a table lamp."[40]

Generally opposed by the large urban gas companies, which regarded the process as a threat to an extension of their own facilities, attempts to apply Mansfield's innovation to commercial operations proved highly disappointing. Benzole from coal tar turned out to be much too volatile and the safeguards against its evaporation far from adequate to utilize it as a gas enricher. During cold weather, in particular, gases would condense in the mains and pipes, suspending illumination and creating dangerous fire and explosive hazards.[41]

But if Mansfield's innovations failed to score a direct commercial success, their indirect effect was nonetheless significant. Modifying his suggestion to use benzole enrichment for table-sized lamps, several American manufacturers, for example, turned to the production of "spirit gas" lamps, which by 1850 "had become quite common."[42] These lamps burned gas that was vaporized from camphene rather than benzole, employed no wick except to bring the fluid to the heating surface through capillary action, and were ignited by burning a few drops of alcohol to start volatilization. For nearly two decades these "Lilliputian portable gasworks" occupied a secondary, though substantial, position in American lamp manufacturing.[43] Later in the century they evolved into the gasoline lamp, which gave a brilliant lamplight by employing gas vaporized from gasoline rather than camphene as a fuel.

An even more significant effect of the Mansfield system was to inspire an intensive, three-pronged drive to develop practical, portable gas units, to obtain more satisfactory materials for enriching coal gas, and to put the manufacture of water gas on a commercial basis. Although many difficulties had to be overcome, all three objectives were achieved during the next quarter-century.

These developments in gas manufacturing as they were extended into the 1850's were alone sufficient to mark that decade as perhaps the most revolutionary in the entire history of illumination. In distilling coal and other bituminous minerals on a large scale, gas manufacturing was exploiting a new source of illuminating materials at costs which at mid-century had already caused a partial eclipse of lamplight and candlelight derived from traditional animal and vegetable materials. But what was not equally plain to contemporaries until late in the decade was that coal-gas manufacturing was at the same time providing the incentive and the techniques for a counterrevolution in lamplight. Adaptation of high-temperature distillation in coal-gas manufacture to lower-temperature distillation of gas coals and then of petroleum brought improved and less costly lamp illuminants and a renaissance in lamplight that rapidly became world-wide during the latter half of the century.

Chapter 3

The American coal-oil industry

Coal-oil production, which began in America during the mid-1850's, had largely disappeared a half-dozen years later. Yet its influence on the history of American illumination was much more significant than its short life span would suggest. In adopting and modifying the techniques developed in connection with gas manufacture and lubricating oil production, coal-oil operators laid the essential basis for petroleum distillation and refining. No less important for the future petroleum industry were the commercial patterns developed during the coal-oil era.

Coal oil, derived from a mineral base, offered a significant new source for lamp illuminants, one that combined safety with the burning qualities of premium lard or sperm oils, yet could compete more effectively on a price basis with the highly dangerous camphene fluids. The marketing channels through which coal oil was distributed and lamps for its use developed and sold were a major part of petroleum's heritage.

The coal-oil illuminant itself owed much to the basic processes worked out by James Young, Selligue, and others. They developed the production of crude paraffin oils from a mineral base by relatively low-temperature destructive distillation in gas-type retorts. It was common knowledge that to produce a marketable naphtha or lubricating oil, crude coal oils had to be redistilled and chemically treated to rid them of impurities.

But no one outside France, prior to the mid-1850's, had followed Selligue's example of producing an illuminating oil on a commercial basis. Nor can any one person be credited with its subsequent "invention" in the United States. The inception of commercial illuminating oil must be related instead to the interplay of the work of several inventors and innovators and their financial backers during the late 1840's and early 1850's, none of whom initiated their initial research with the intention of obtaining a coal-oil illuminant.

There were two groups of particular significance in the development of the industry as it had emerged by the mid-1850's: one, Abraham Gesner and the backers of the New York Kerosene Company; the other, Samuel Downer, Joshua Merrill, and their associates at Boston.

Abraham Gesner

Abraham Gesner, a Canadian-born inventor who was trained in surgery in London, showed an early interest in the commercial value of the mineral deposits of Nova Scotia, where he began his medical practice late in the 1820's. Publication of his "Remarks on the Geology and Mineralogy of Nova Scotia" in 1836 led to his appointment two years later as Provincial Geologist of the Province of New Brunswick. In that capacity he conducted the first geological survey undertaken by any provincial government in Canada, and probably in any British colony.[1]

Stressing mineral deposits which might have potential commercial value, Gesner's *First Report on the Geological Survey of New Brunswick*, published in 1839, and four succeeding annual reports, inspired promoters to launch an ill-fated coal-mining and iron-smelting venture in Queens county, New Brunswick. The indignation of investors illogically fell on Gesner rather than the promoters, resulting in his dismissal as Provincial Geologist in 1843.[2]

Returning to Nova Scotia, Gesner resumed medical practice, but continued to experiment in the industrial arts. This interest led Gesner to experiment with the distillation of Trinidad pitch and Albert bitumen from mines in the Canadian province of New Brunswick. In August, 1846, he gave a demonstration, burning his oil in lamps at a public lecture on Prince Edward Island.[3]

Gesner's experiments soon attracted the attention of Thomas Cochrane, Tenth Earl of Dundonald and Admiral of the British North American and West Indian Station with headquarters at Halifax. Lord Dundonald had a particular interest in the commercial possibilities of Trinidad pitch, for which he expected to get a concession.[4] Gesner began working on Trinidad pitch for Dundonald about 1849 and by the end of the year had developed an improved process to distill illuminating gas directly from asphaltum. It was in connection with an American patent on this process, granted in January, 1850, that Gesner made his first visit to New York City early in that year.[5]

During his stay in New York, Gesner took the occasion to promote the use of Trinidad pitch, which he claimed could, by dry, low-temperature distillation, be made to yield large quantities of illuminating gas. Superior to cannel coals for large-scale gas manufacture, asphaltum in Gesner's process had the added virtue of being free from sulphur and other noxious materials. Its use, he pointed out, would eliminate the expense of purification, the high costs of maintenance induced by the corrosive action of coal impurities on metal, and the public nuisance of gasworks with their disagreeable odors released in coal-gas purification.[6]

Gesner also took occasion to call attention to another product of asphaltum. This was a liquid with qualities similar to paraffin oils, which he called "kerosene" (*keros*, Greek for wax; "ene" because it was a volatile oil resembling the popular "camphene" lamp illuminant).[7] When air was passed through or over it, the kerosene itself yielded a rich, kerosene gas.[8]

During his New York visit, Gesner met many important individuals in technological and commercial circles.[*]

Aware that Albert bitumen had properties closely similar to asphaltum pitch from Trinidad, Dundonald and Gesner in 1851 began laying plans to form a company to exploit the Albert mineral and the holdings Dundonald had acquired in Trinidad in that year. No formal action had been taken to organize the proposed concern when, in April, 1851, Dundonald left Halifax for England at the termination of his naval service. There he

---

[*] Toward the end of November, 1850, Gesner again visited New York City to apply for a United States patent "for one of the most valuable discoveries ever made in the manufacture of oil, rosin, or asphaltum gases."[9]

spent the remaining years of his life until his death in 1860 in futile attempts to interest British investors in the Trinidad pitch deposits. Evidence is lacking whether he and Gesner continued to advance plans for their proposed company by trans-Atlantic correspondence.[10]

In any event, a decision by the Canadian courts in July, 1852, put an end to any possibility of including Albert bitumen in their plans. The issue involved a definition of the Albert mineral. According to Canadian law, all coal-mining rights on the land involved were reserved to the Crown. In April, 1852, Gesner brought action in a Halifax court to show that Albert bitumen was asphaltum rather than coal, and therefore not subject to coal-mining rights.[11] Although the court initially ruled in favor of Gesner, William Cairns, who had already leased Albert county lands for coal mining, refused to recognize Gesner's claim, and brought action in the Albert county courts in July, 1852. Instructed by the judge that Cairns' license to mine coal included "other mines and minerals," the jury decided against Gesner.*

His cause seemingly lost in Canada, Gesner moved his family to New York City early in 1853.[13] There he was successful in getting financial backing to form a new company, the Asphalt Mining & Kerosene Gas Company (soon renamed the North American Gas Light Company), to work the combined patent rights he held jointly with Dundonald.[14] Typically, the prospectus for the new concern enumerated the broadest possible range of products which might be produced from working asphalt— mineral naphtha, railway grease, hydraulic concrete, mineral pitch, parafeni, coke, ashes for manure, even uncondensed gas "for lighting manufactory," and "above all" the production of kerosene for "burning fluids . . . which could be manufactured at a lower cost than the various burning fluids now most in use."[15]

Just when Gesner and his associates shifted their major attention from gaslight to lamplight is not clear. Apparently the shift was closely associated with Gesner's development of processes for refining crude oil distilled from asphalt or other bituminous materials.[16] From these processes, which Gesner patented in June, 1854, he was able to produce three distinguishable grades of oil which he described as Kerosene A, B, and C. Kerosene A, identifiable by weight, boiling point, and volatility as roughly equivalent to light fractions in gasoline today, was suitable for use as an

---

* Modern geology has upheld Gesner's contention, classifying the Albert coal as "albertite," a generic term for a whole class of asphaltic pyrobitumens.[12]

enricher for illuminating gas. Unfortunately for Gesner, few of the large coal-gas producers showed any interest in utilizing this product as a gas enricher. For maximum exploitation of the newly invented processes, markets had to be found for Kerosene B, the next heaviest fraction from distillation, and Kerosene C, the heaviest and least volatile of the three.

Kerosene C also presented special difficulties. Not being soluble in alcohol, it plainly had the characteristics of a solvent. But it was doubtful whether Kerosene C could compete in the markets for solvents when considered on a price basis with distilled turpentine. Gesner's patent hopefully suggested other possible uses in lubrication and illumination, either alone or mixed with the lighter fractions.[17]

Clues to a solution of the marketing problem for Kerosene B, however, were readily available. New York was in the midst of an intensive drive to get a reduction in gas rates. The proponents of cheaper gaslight all advanced one common, popular argument. Cheaper rates would place safe gaslight in practically every private home, and would provide "the means of preventing many of the casualties which are constantly occurring from the use of volatile, hydrocarbon fluids [camphene], and be a blessing to both rich and poor."[18]

The implications of this situation were not lost on Gesner and his associates. Camphene lamps of the Argand-button type were suitable for burning synthetic mineral oils, and in Kerosene B, Gesner had a product which was soluble in alcohol and at the same time less volatile and therefore less dangerous than camphene. Why not substitute Kerosene B for camphene as a superior burning fluid in camphene lamps, which outnumbered any others in city markets, to say nothing of a potential demand in small towns and rural areas?

Construction of a plant on Long Island, New York, delayed production for almost two years, but in March, 1856, the North American Kerosene Gas Light Company, subsequently renamed the New York Kerosene Company, announced that kerosene had been put on the market.[19] As a part of their sales promotion, Gesner and his associates hired a consulting chemist, Edward S. Kent, to make a comparative analysis of alternative sources of illumination. In his "testimonial letter," Kent stated that "Purified Kerosene" at $1.00 a gallon was cheaper, better, less liable to smoke, and less explosively hazardous than camphene. At current prices, light camphene cost 18 per cent more to produce than equal amounts of kerosene; whale oil cost three times as much; lard oil four times as much;

while sperm oil would be almost seven times as expensive. Even gas at $3.00 per thousand cubic feet would be almost twice as expensive to produce an equal intensity of light. Kent concluded:

> In view of the above facts, I am sanguine that your "Purified Kerosene" is destined to supersede all other oils or burning fluids, as a source of light for artificial illumination, and would recommend it as the most valuable material for the purpose with which I am acquainted.[20]

Despite these glowing claims, the New York Kerosene Company for a variety of reasons failed to achieve immediate commercial success. Gesner's talents as an engineer apparently did not match his prowess as an inventor. The equipment originally installed was relatively inefficient and yields from New Brunswick coal were disappointingly low. Moreover, the product had an offensive odor which proved difficult to eradicate, an experience which would be repeated in practically every new plant entering coal-oil manufacturing.[21]

Even though the commercial success of kerosene still lay in the future, the entry of the New York Kerosene Company into production of kerosene in 1856 established a valuable precedent. It was the first firm in America or Great Britain to turn to the manufacture of coal-oil illuminants. Not the least of the contributions of Gesner and his associates was to call attention to the properties of burning oils derived from a bituminous base and to arouse widespread interest in its potential use as a lamp illuminant.* While there is some question as to the fundamental character of Gesner's technical contribution, these achievements alone were sufficient to insure him a place in the gallery of pioneers in the illumination history of America.

Gesner suffered a fate not uncommon among innovators. Having assigned his patents in return for a salary, he profited little from his inventions, and after 1856 he played no significant role in the development of the coal-oil industry. As one of several steps to make the New York company thrive, Luther Atwood of the Downer organization became its chief chemist and operating head in 1857.[23] Relegated to an inactive role, Gesner made his living as a consulting chemist until 1863 when he retired to Halifax, where he died the following year at the age of sixty-seven.[24]

* It is also noteworthy that contrary to standard interpretation, kerosene did not result from a simple progression from sporadic experiments with asphaltum illuminants to a successful commercial enterprise within two years. Gesner's experiments extended on an intensive scale over a period of at least seven years, from 1849 to 1856. For the first four years of that period his primary interest lay in the commercial promotion of kerosene gas.[22]

Samuel Downer and Joshua Merrill

The unquestioned leader in the new coal-oil industry as it emerged during the late 1850's was the organization headed by Samuel Downer. This company had its origin in 1852, when three members of a prominent firm of Boston pharmaceutical manufacturers, Dr. Samuel R. Philbrick and Luther and William Atwood, formed the United States Chemical Manufacturing Company at nearby Waltham, Massachusetts.

The initial stimulus for the formation of the concern was to produce a

MC CLURE'S MAGAZINE, VOL. XXI, (1903)

Joshua Merrill

mineral lubricating oil from coal tar, a by-product of gas manufacture, under a process developed by Luther Atwood and patented in 1853. Naming his lubricant "Coup oil," Atwood indicated that it could either be used in its pure form or blended with any animal or vegetable-oil lubricants.[25] Handicapped from the outset by a disagreeable odor, coup oil had one characteristic that appealed to producers of lubricants for fine machinery: blended with fatty oils, it made them more fluid and less susceptible to gumming through the absorption of oxygen.[26]

In an attempt to improve their product and stimulate sales, the new company in 1853 hired Joshua Merrill, age 25, to work as a combination salesman and lubricating engineer. Merrill, the son of a Massachusetts clergyman, had no formal training beyond a grammar school education.[27] However, his native ability and keen interest in practical affairs combined over the next quarter-century to make him one of the outstanding technological innovators of his generation.

But even Merrill's competence did not prevent the firm from moving steadily toward bankruptcy. Blending skill could not suppress the bad odor of the oil. Sales lagged, debts rose, and working capital evaporated. It was at this point that Samuel Downer became interested. Since 1844, Downer had been sole owner and proprietor of Downer & Son, one of Boston's largest producers of sperm and whale oils and sperm candles. Anticipating that a further increase in the price of his whale oil and sperm products might well drive them from the market, he saw in coup-oil production an opportunity to diversify his investments.[28] Early in 1854 he purchased a controlling interest in the firm and took over active management.

In Downer, the organization acquired a chief executive whose talents as an entrepreneur matched the technological and innovative abilities of Merrill. This marked the beginning of an association between the two men that, over the years, kept their organization constantly in the vanguard, devising better methods of distilling and treating and improving the quality of both coal-oil and petroleum products.

It soon became apparent to Downer that, despite some improvement in its lubricating qualities, the persistent and pervasive odor of coup oil would prevent its ever becoming a commercial success. Early in 1855 he and his staff, the two Atwoods and Merrill, began experiments which, by the end of the year, covered "petroleums and bituments from nearly all the known sources, and many varieties of coals and shales," and had "succeeded in producing what they regarded at the time as a good lubricating oil from each of these sources." [29]

It was at this critical juncture that Downer received a request from Glasgow, Scotland, for permission to produce coup oil in Great Britain and for technical assistance in building a plant and installing the process. The applicant was George Miller & Company, large manufacturers of coal-tar naphthas, which was seeking an entry into the profitable lubricating-oil business, virtually monopolized by James Young with his tightly held patents on the production of paraffin oil.[30]

Luther Atwood arrived in Glasgow about mid-year, 1855, followed by Merrill some six months later. It was while they were completing their job of installing facilities for George Miller & Company that Luther Atwood completed an experiment which led both Downer in the United States and James Young in Britain to turn to the production of illuminating oils.

Atwood's experiment was with a few gallons of naphtha from Young's Bathgate plant. This material, used by George Miller & Company, was mixed with coal-tar naphthas to dissolve rubber for mackintoshes and other waterproof goods.[31] Atwood discovered that by redistilling the paraffin-based naphtha, he obtained a water-white oil which burned in a modified "moderator"-type lamp with a brilliant flame and without odor or other objectionable features.

It was while Young was visiting Miller's office that he noticed a lamp burning the new fuel. When Miller made the mistake of telling him the source, Young cut off the former's supplies of naphtha from Bathgate and shortly thereafter began marketing paraffin oils as a lamp illuminant. Over the next few years, Young built up his illuminating-oil business rapidly and by 1859 was manufacturing on a large scale, even selling a portion of his output for export.[32]

Though coup oil was no more successful in Britain than in America, George Miller & Company also began the manufacture of paraffin-oil lamp illuminants on a large scale. The business was highly successful until 1864, when Young brought suit for violation of his patents. The damages awarded Young were so heavy that George Miller & Company was forced permanently out of the business.[33]

Meanwhile Atwood and Merrill returned to Boston late in 1856 with a gallon of oil distilled from naphtha produced by Young and full of enthusiasm over the possibility of the new illuminant. An initial demonstration failed to impress Downer. "Our business is lubricating oil," he stated. "Illuminating oils don't amount to anything. You can never replace the lard or whale-oil lamp; they are the articles for illuminating purposes . . ."[34] But Downer could change his mind, and within a short time he decided to reorient his business to the production of illuminating oil. After further experimentation with materials, it was found that Albert coal from Canada gave high yields of crude oil, which could be refined into an illuminating oil of superior quality.

By the fall of 1857, the essentials of coal-oil refining which, with the exception of the first stage, would also be applied to petroleum refining, had been worked out at a new plant site in South Boston. Manufacturing began by breaking the coal into small pieces and feeding it into large, modified gas retorts. Here the coal was subjected to dry, destructive distillation at temperatures below 800° F, or about half those used in coal-gas manufacture, so as to promote maximum yields of volatile oils rather than gas. The oily vapors from the decomposition of the coal were condensed in water-cooled pipes into crude coal oil (the rough equivalent of crude petroleum) and transferred to receiving tanks. There the crude oil was heated to about 100° F to expedite the precipitation, settling, and removal of as much as 20 per cent water, in which there was ammonia, carbon particles, and various other impurities.[35]

To yield useful products, crude coal oil like crude petroleum required distillation to separate valuable oils from the crude, and distillation combined with chemical treatment in order to eliminate impurities impairing odor, color, burning, and other qualities of the products.

The first step was to charge the crude oil into a cast-iron still and to light the fires underneath it. An educt above the level of the oil charge in the still permitted the vapors from distillation to escape into a 100-foot water-cooled pipe, where they were condensed into liquid and passed to receiving tanks. The most volatile components of the crude with the lowest boiling points were vaporized and condensed first, then generally followed by the heavier components with higher boiling points. In 25-barrel stills the running times for this first rapid distillation approximated twelve hours.[36]

All oils issuing from the condenser up to the point where the temperature in the still had risen to about 600° F were piped to receiving tanks for further processing into burning or illuminating oils. The rest of the increasingly heavy distillates were routed to receiving tanks for paraffin-oil lubricants. The burning-oil portion varied, according to the type of coal and the skill of refiners in reducing the coal, from less than half to almost three quarters of the total distillate run.[37]

From the receiving tanks the burning-oil distillate was pumped to an iron-treating tank equipped with mechanical or compressed-air agitators. Five to 10 per cent of sulphuric acid was introduced and agitated for a couple of hours, after which the acid and the soluble impurities were allowed to settle for six or eight hours. The same procedure was repeated with caustic soda and water washes. A redistillation of the treated and once-distilled oil at slower rates of temperature elevation than on the first distillation completed the refining of burning oils.* The dense lubricating fractions of paraffin oil were refined in the same way, except that they sometimes required three distillations instead of two, and the alternate

---

* Any portions toward the end of the run which were too heavy for burning oils either were included in the next charge of burning-oil distillate being redistilled or transferred to the heavy lubricating oil tanks.[38]

chemical treatments between distillations also had to be much more intensive.

Starting in the fall of 1857 with six gas retorts, specially designed by Merrill, the Downer organization by the end of 1858 was operating fifty large retorts. By this time, the plant had an annual capacity of about 900,-000 gallons of crude oil and approximately 650,000 gallons of refined oils, including lubricants.[39] This output reflected yields from the Albert coal used by Downer and Merrill, which ran about 110 gallons of crude oil to the ton, and which in turn yielded approximately 65 gallons of illuminating oil and 5 gallons of lubricants. These rates of recovery were about as high as they or anyone else would achieve from Albert coal, although New York Kerosene Company and possibly a few others were subsequently able to match them from Boghead coal imported from Scotland or Breckenridge coal from Kentucky.[40]

The majority of coal-oil refiners, however, failed to achieve equal results because of the various types of coal utilized and the technological competence applied to them. Furthermore, most coal-oil refiners, because lubricants were more expensive to distill and treat and were in demand only as a cheap adulterant of animal and vegetable lubricants, concentrated on refining illuminating oil.

But their production of lubricants along with illuminating oil was not the only characteristic which made Downer and Merrill the leaders in the new industry. They had scarcely arrived at refining processes which became standard for the bulk of the industry, when they introduced two major innovations which remained unique.

The first of these involved treating. As early as 1857, Merrill developed and applied chemical treating as a final process without any further distillation, rather than as a process applied between the first and second distillation. Although final treating did not eliminate the need for two distillations, it did greatly reduce the quantities of chemicals needed, and improved the effectiveness of their application.[41]

The second innovation, developed by Merrill and Luther Atwood, was to subject the heavy paraffin-oil stocks to a revolutionary type of destructive distillation. Conducted slowly at temperatures above 600° F, this process changed the molecular structure of the heavier oils, breaking up or "cracking" some of the larger molecules into smaller ones with boiling points within the illuminating oil range. Of great significance for the petroleum industry, "cracking" immediately made it possible for the Downer organization (and the New York Kerosene Company) to step up their recovery of illuminating oils.[42]

Further evidence of the technological leadership of the Downer organization was an impressive list of by-products developed by the end of 1858. These included refined naphthas, several grades of paraffin-oil lubricants, and paraffin wax. Although a small minority of coal-oil refiners manufactured some of these specialties, the Downer organization remained predominant both in the development and commercial exploitation of by-products.[43]

Patent arrangements with James Young

In turning to Albert coal, Downer (whether he realized it or not) was violating the patents granted to James Young by the United States in 1852, covering the distillation of paraffin oils from coal. Young made little attempt before 1858 to enforce his American patents. However, sometime prior to that date, Downer acquired the right (for a royalty payment of 2¢ a gallon) to produce crude coal oil from coal by Young's processes—a privilege later extended by Young to any American refiner willing to pay the same royalty. But Downer also obtained about this time the exclusive American rights to Young's processes covering the refining of paraffin oils from crude, for which he agreed to pay an additional 2¢ royalty per gallon.[44]

Downer in turn was able, under his agreements with Young, to make an attractive arrangement with the New York Kerosene Company, which, in using Gesner's processes patented in 1854, was also infringing on Young's patents issued two years earlier. As a part of the agreement which sent Luther Atwood to take over the company's operations in 1857, Downer arranged for the New York Kerosene Company to utilize both Young's and his own patents in return for royalty payments and the exclusive right to share the New York concern's trademark, "kerosene," in the distribution and sale of his lamp oil.[45]

So far as the New York Kerosene Company was concerned, these agreements, coupled with the services of Luther Atwood, were of primary importance in its subsequent growth into a major factor in the coal-oil industry. Downer indicated the importance he attached to the kerosene trademark by renaming his company the Downer Kerosene Oil Company.

Downer's commercial success

Although his company became the recognized leader in the new coal-oil industry, Downer was not assured of commercial success until the end of the "burning-oil" season of 1858–59. During the summer of 1858, business was slow. Inventories, stored in iron tanks, amounted to some 200,-000 gallons by early September. For a time it appeared that the plant would have to be shut down, but orders began to mount and, by the following spring, sales at $1.35–$1.40 per gallon had brought in profits amounting to over $100,000.[46]

Much of this success was the result of Downer's marketing skill. In a circular issued to the trade in November, 1858, he emphasized the safety of his kerosene illuminant and its freedom from odor. But customers and competitors were warned, "Kerosene is the trademark of the New York and Boston companies, and all persons are cautioned against using the said trademark for other oils."

Above all, Downer paid particular attention to the quality of his product. By frequent inspection and sampling, he made sure that the oil shipped to customers measured up to the high standards he set for the illuminating qualities of his product. To make sure that his product would remain untainted and not subject to heavy loss through evaporation or leakage, Downer sold all but a small portion of his entire output during 1858–59 in tins holding three, five, or ten gallons each.[47] While tins were already in use for shipments of alcohol and turpentine, this represented their first use on any significant scale for illuminants. Subsequently Downer, along with the rest of the coal-oil industry, shipped the bulk of their coal oil in barrels. But his early use of tins set a precedent of great future significance when refined petroleum was shipped in tins over long distances and sold in tropical areas without incurring excessive losses from leakage and evaporation.

The growth of the coal-oil industry

While expansion did not become obvious until 1858, there were a number of projects outside New York and Boston which started earlier. In June, 1856, for example, the *Scientific American* reported the construction of facilities at Cloverport, Kentucky, to manufacture coal oil from Breckenridge coal.[48] Within two years this plant was distilling about 600,000 gallons of crude coal oil a week and shipping its finished products as far west as Chicago.[49] An Ohio plant, started about the same time, had by mid-1858 stepped up its production of various refined coal oils to 350 gallons a day, about half of which was sold for illumination.[50] In 1857 Downer, in co-operation with Dr. Samuel Philbrick, built a plant at Portland, Maine.[51]

Beginning in 1858 and extending over the next two years, expansion reached boom proportions. New York was the setting for an additional eight to ten plants.[52] Downer's position was challenged in the Boston area by a half-dozen or so new firms.[53] Cincinnati reported the construction of a large plant in the spring of 1859, followed later in the same year by others. This brought the total number of coal-oil refineries located in the area to ten.[54]

Perhaps the largest additions to coal-oil capacity came in the Pittsburgh area where, by the end of 1859, four large concerns, including the Lucesco, Aladdin, and North American refineries (all prominent later in petroleum refining) were in operation.[55]

A complex of factors operated to bring about this expansion. For some new entrants, coal-oil production offered an opportunity to exploit undeveloped cannel-coal deposits. Others must have been impressed with Downer's success. A further stimulus came in 1858 when James Young licensed some twenty-three plants to produce crude coal oil under his patents in return for a royalty payment of 2¢ per gallon.* Finally, there was a growing interest on the part of consumers for a safe illuminant. The encouragement given to new producers of camphene when the basic patents on its production ran out in 1856, coupled with rising prices of lard, whale, and sperm oils, contributed to the further dominance of this dangerous liquid in the illuminating markets.[57] A growing toll of deaths and accidents from lamp explosions and fires increased the public's interest in coal oil or any alternative illuminant that promised greater safety without serious sacrifice in economy and convenience in use. The press sought to stimulate this interest with editorial admonishments to "use coal oil, tallow candles, pine knots, anything rather than hazard life, limb, and property by the constant use of dangerous burning fluids." [58]

With so many new entrants, it is impossible to give any precise estimates of total production of coal oil during the fall and winter burning season of 1858–59. Incomplete returns suggest that daily outputs of finished products by the end of 1859 probably ranged between 22,000 and 23,000 gallons. Of this amount, the New England and New York City areas each contributed about 5,000 gallons; Cincinnati and the rest of Ohio, 2,500 gallons; Pittsburgh, 2,000 gallons; with the remainder scat-

* At the same time Young hired a New York law firm to take legal action against violators of his patents. Young's attempts to collect damages for violation of his patents created a tremendous resentment from American producers and, although his claims were upheld by the courts, his eventual collections amounted to about $27,-000, a fraction of what was really owed him and what he was usually reported as having "skimmed off the American coal oil industry." [56]

tered among plants in New York, Pennsylvania, western Virginia and Kentucky. This growth prompted the *Scientific American* early in 1859 to observe, "The manufacture of coal oil has become somewhat extensive in our country and the subject at this time is one of great importance." [59]

A year later, when daily production of coal oil was running about 30,-000 gallons, the same journal pointed out that the annual value of coal oil amounted to about $5 million; that the cost of chemicals to the industry was about $1,000 per day; that it directly employed 2,000 men; and that 700 miners were kept busy supplying the industry with cannel coal. In conclusion the editor stated:

> If we take into mind, that two years ago, there were only two or three oil-works in this country, the above statements form a strong illustration of the impetuous energy with which the American mind takes up any branch of industry that promises to pay well. As far as coal oil is concerned, the rapidity with which the manufacture of this beautiful illuminator has been propagated amounts (like the cultivation of the morus multicaulis [mulberry bush] some years ago) to a mania. [60]

Despite this sharp expansion the new coal-oil industry was dwarfed by the manufactured-gas industry, which by 1860 included approximately 400 plants in urban communities, representing a total investment in excess of $56 million, and with an output valued at about $17 million annually. [61] But if gas manufacture was by all odds the leading source of illumination in America at this time, earlier hopes that its expansion would bring gas rates down to the point where it could be afforded by every working man and mechanic fell far short of realization. [62]

For the great majority of Americans dependent on lamplight, coal oil offered the most attractive source of illumination because of its safety and the quality of its light. Furthermore, coal-oil refiners had shown unusual initiative in promoting the sale and distribution of their products. As early as 1857, they had stimulated the co-operation of the great Knapp lamp firm to design and produce a lamp successfully adapted to coal oil. [63] Not only did Dietz, Drake, and other firms promptly introduce successful models, too, but the response of lamp manufacturers was so great that they put their production on an interchangeable-part basis for the first time in turning out the new coal-oil lamps. Both prices and output promptly reflected this development. Coal-oil burners sold at retail as cheaply as $3.50 a dozen, with $7.00 to $8.00 being the typical price in dozen lots. By the end of 1859, an estimated 3–3.6 million burners had been placed on dealers shelves, of which 1.8 million burners had been sold to users.[64]

Pressure from refiners, fortuitously timed with developments in transportation, also had operated to give coal oil the first national distribution among illuminants. At first most coal oil moved through established channels—brokers, jobbers, and commission merchants in the regular oil trade, who handled animal and vegetable oils of all sorts, for lubricating and illuminating purposes. Of particular importance in achieving national distribution, however, was the swift transformation of wholesale and retail druggists and grocers from minor channels of distribution into major ones.[65] Underlying this new organization of distribution was the completion, in 1857, of through rail connections between all key eastern and western cities, welding the sections into a continuous economic empire.

The remarkable extent to which coal-oil distribution had penetrated western towns and cities was made evident by the publication, in 1860, of C. M. Wetherill's case study of illumination in Lafayette, Indiana, over the preceding two years.[66] By 1859, residents could procure locally at least eight leading makes of coal-oil lamps, in sizes ranging from 5–14 candle power. The Knapp, the Dietz, the Drake, the Excelsior, the Paragon, the Waterbury, the E. F. Jones, the Rufus Merrill, and the combined Jones-Merrill lamps could all be purchased locally. Leading brands of coal oil from New York, Pittsburgh, and various refining centers in Ohio and Kentucky likewise were retailed by lamp and oil dealers and grocers for $1.00 a gallon in 1859, except for kerosene from the New York Kerosene Oil Company, priced at $1.25. From Pittsburgh came the "Lucesco" and "Photogen," brands of the Lucesco and Thumm refineries respectively, in addition to Veeder & Adair's product. A Newark, Ohio, refinery supplied this local market with "Crysoline," and several Cincinnati refiners also were represented. Refiners at Maysville and Cloverport, Kentucky, also served this market regularly.

Coal oil was first introduced in Lafayette in 1858 by sample consignments from several Pittsburgh and Ohio refineries. All the samples were of a quality that encouraged a wave of lamp purchasing, but the new illuminant received a temporary setback when succeeding batches, without exception, were poor in odor and burning qualities. Before the end of 1859, lost ground had been more than made up with improved products from most of the original entries and numerous new ones besides.

Competition in quality and price from numerous sources had given consumers adequate assurance to establish coal oil as the dominant illuminant. Wetherill reported its performance greatly superior to all other lamp oils, including the best lard oil, and it was the most cheaply priced except for camphene fluids. But even this nominal differential disappeared early in 1860, when all coal oils dropped to 75¢ per gallon wholesale price. A unit of burning fluid then would deliver only 1.4 candle power in the best camphene lamp for a shorter time than a unit of coal oil at the same price would deliver 5 candle power. Coal oil also offered coal gas its first competition in small cities and towns, where rates commonly ran $4.00 or more per 1,000 cu ft for a gas much lower in candle power than that distributed in large cities. The fact of this competition was proved in the spring of 1860 when the local gas company lowered its rate to $3.50.

With allowances for temporary readjustments of supply and demand, prospects for a continued expansion of the coal-oil industry almost indefinitely looked promising. America's capacity to use an improved cheap lamplight was virtually untapped. For all practical purposes, assembly-line production had removed all limits to the supply of cheap lamps, while improved retorting of coal on a far larger scale than the dry distillation employed in gas manufacture had greatly relaxed early tightness in the supply of crude coal oil.

Better retorting and related treating methods also had widened the range of coals that might be profitably processed if priced to compensate for lower yields than those afforded by the expensive imported Albert and Boghead coals. New discoveries of domestic coals in the interior promised ample crude supplies for an orderly expansion accompanied by lowering costs. Specialized retorting at the mouth of mines had already given rise to a sufficient traffic in crude coal oil to permit the operation of small refiners, who confined themselves to simple distillation and treating of crude oils.[67] A lowering scale of costs was reflected in a wholesale price decline within a two-year period from $1.25 to 75¢ a gallon and, at times, even as low as 35¢.[68]

The possibility of using petroleum as a raw material, permitting the bypassing of the first dry-distillation stage of refining, had not escaped the coal-oil industry from the very beginning. Young's English and American patents and the earliest patents by Americans in the first part of the decade almost unfailingly included petroleum in the bituminous materials covered. Downer, Merrill, and others worked petroleum when available in commercial coal-oil stills. In the last two years of the decade, A. C.

Ferris of New York City successfully applied methods of distilling and treating coal oil to refining petroleum illuminants on a small scale in competition with coal oil.[69]

But in order for petroleum to become the raw material for the coal-oil industry, rather than a few units within that industry, it had to demonstrate a renewable sustained supply beyond the volume of animal oils still entering illuminating markets or even the much greater quantities of rectified turpentine and alcohol. Stated still more precisely, the issue hinged on whether petroleum could be produced in adequate supply more cheaply than coal could be mined and dry-distilled into crude oil. Proved markets with an enormous future potential, a greatly improved organization of distribution facilities, suitable lamps in unlimited supply, and advanced refining processes, plants, and equipment were almost equally ready for a continued expansion of the coal-oil industry or the conversion of that industry into a petroleum refining industry.

But as long as petroleum remained a by-product of salt wells and springs, there was little possibility of its making any significant contribution to the total supply of illuminants. It remained for the events following the discovery of the Drake well to make this possibility a reality.

Part THREE

The birth of
the modern industry:
1859–1861

Chapter 4

The first commercial oil well

Many extraordinary things had to happen in almost every field of human endeavor for a modern petroleum industry to be born. In many ways the most extraordinary thing of all was the demonstration of the existence of underground reservoirs of petroleum large enough to sustain an industry in raw materials.

The story of the first successful commercial oil well began in 1851 with a medical graduate of Dartmouth College, who abandoned his professional practice in Vermont to enter the lumber business in the isolated, tiny frontier village of Titusville. It was in these timbered foothills of northwestern Pennsylvania, on a Sunday afternoon eight years later, that E. L. Drake, a former railroad conductor, struck oil.

The intervening years were filled with many implausible events. The spark of life that arose in the venture and often kept only a flickering existence was dependent on arms of coincidence in a profusion that would discredit a third-rate piece of fiction. The fact that the former railroad man drilled where he did at one stage was contingent on his possession of a railroad pass that would save hard-up sponsors from paying in advance for transportation to the place of the proposed scene of action.

No less implausible was the fact that no one connected with the project had any experience in the commercial manufacture or sale of any kind of oil—animal, vegetable, or mineral. Viewed in this light, the story of the Drake well becomes the strongest kind of evidence that all the conditions were favorable for the birth of a petroleum industry. If E. L. Drake had not transferred techniques of boring for salt to drilling for petroleum, the probability is great that someone else soon would have done so. But fate, and Drake's enormous patience and perserverance, decreed that he would be the man.

Early promotion

Dr. Francis Beattie Brewer supplied the initial momentum to events. Soon after joining the lumber firm of Brewer, Watson & Company at Titusville in 1851, he spent considerable time examining an old oil spring on the company's property. It was from this spring that the lumber company was already drawing petroleum exudates, which it used as a lubricant and may even have employed as an illuminant in open places where smoke did not seriously matter.[1] Brewer was satisfying an old curiosity, for two years earlier, when still practicing medicine in northern Vermont, he had received from his father in Titusville five gallons of this creek oil.[2] Trial of the oil had verified his father's assurances that it had remarkable curative properties, and ". . . so long as I continued in my profession, I had it in constant use," Dr. Brewer reported.

Brewer's interest culminated on July 4, 1853, in what was probably the first lease looking toward petroleum development. The firm leased the spring to J. D. Angier of Titusville, who agreed to keep it in repair for five years, construct a new spring, and gather the oil, which was to be divided equally after deducting expenses from the proceeds. Following an orthodox trenching procedure, Angier dug ditches to carry the oil from the water with simple machinery that he erected at a cost of $200. Half of the yields of three, four, sometimes even six gallons, were consumed daily as lubricants, and in torches to illuminate open parts of the mill.[3]

Next fall, visiting friends and relatives in Hanover, New Hampshire,

Brewer took along a bottled sample which he showed to his uncle, Dr. Crosby, professor of surgery and obstetrics at Dartmouth. Professor O. P. Hubbard of the chemistry department also examined it. Evidently without any chemical analysis, the latter made a clairvoyant finding that it was valuable, but added that it could hardly become an article of commerce because of its limited quantities.[4] Brewer left the bottle with his uncle at the end of his visit.

A few weeks later the bottle came to the attention of George H. Bissell, a young New York lawyer, on a visit to the campus of his alma

COURTESY DRAKE WELL MUSEUM

Dr. Francis B. Brewer

mater. Currently engaged with his partner, J. G. Eveleth, in promoting the sale of stock in the American & Foreign Iron Pavement Company and the Safety Railway Switch Company, Bissell saw the possibilities of a new account. According to some reports, he was immediately struck by the similarity between coal oil and petroleum, a resemblance from which he deduced that petroleum might be exploited as an illuminant.[5] Bissell himself, in a statement more than a decade later, explained that he and his associates from the outset had prospected for petroleum as a medicinal, not as an illuminate.[6] It is possible that Bissell might have seen Dr. Abraham Gesner's prospectus for the North American Kerosene Company, issued in 1853.

In one way or another, Bissell had enough of a glimpse of a genie of wealth in the bottle to decide sometime during the summer of 1854 to pay the expenses of Dr. Crosby's son, Albert, for a visit to his cousin, Dr. Brewer, in Titusville, to look over the oil property. If young Crosby returned with a favorable report, Bissell and his business partner, Eveleth, agreed to form a company to acquire the land, develop the oil springs, and market petroleum.[7]

Young Crosby returned to New York early in September, full of enthusiasm over the prospects for oil and an agreement from Dr. Brewer to sell

HENRY, EARLY AND LATER HISTORY OF PETROLEUM

George H. Bissell

the 100 acres comprising the Hibbard farm for $5,000, in addition to oil rights on some 12,000 acres, if Messrs. Bissell and Eveleth, being represented by Crosby, would form a joint-stock company. This proposed company would be capitalized at $250,000, one-fifth to be assigned to Brewer, Watson and Company, another fifth as treasury stock, and three-fifths to be divided however the original purchasers might determine.

The two partners hardly had time to consider the preliminary steps involved in forming a new company, or to give serious thought to how it could be formed, when Dr. Brewer appeared late in September prepared to sign the contract. They understandably discounted the extravagant report by young Crosby on the abundance of petroleum on the Titusville properties and were reluctant to close the deal. Before Brewer left New York, however, they accepted his offer to keep the proposition open until one of the firm members personally looked over the Titusville properties. If the investigation confirmed Brewer's claims, the deal would be closed; if not, the Titusville lumber firm would foot expenses for the trip.[8]

The Pennsylvania Rock Oil Company of New York

At the suggestion of Anson Sheldon, a retired minister in New Haven, Connecticut, the partners journeyed to that city to seek out James M. Townsend, president of the local City Savings Bank, and other leading

COURTESY DRAKE WELL MUSEUM

James M. Townsend

citizens who conceivably might be prospective investors in the oil tract. Impressed by Bissell's account of the oil land's possibilities, the New Haven group asked as a condition to putting their money in the project that a firsthand inspection of the land be made to verify glowing preliminary reports and that some kind of scientific analysis of the oil should be provided to establish its economic value.

Eveleth filled the first condition by appointing himself a committee of one to visit Titusville.[9] Competent chemical consultants at mid-century still were not plentiful, but the partners filled the second condition with two happy choices. They arranged with Luther Atwood of Boston and Benjamin Silliman, Jr., of Yale University, to analyze and report on their oil samples. Atwood, already a renowned chemist and pharmaceutic manufacturer, had for the past two years been manufacturing his "coup" oil at New Waltham, Massachusetts. Silliman had just succeeded his father, (the initial occupant of the first chair of chemistry at Yale University, es-

Benjamin Silliman, Jr.

tablished in 1803) who, in 1833, had performed the first reported distillation of petroleum in America.[10]

Returning to New York from inspecting the Titusville oil tract, Eveleth was as jubilant over the prospects as young Crosby had been earlier, and rushed to complete details of the deal for organizing the first petroleum company. Dr. Brewer on November 10, 1854, deeded the Hibbard farm to Eveleth and Bissell for $5,000. No money passed, only joint and individual unsecured notes for the land.

On December 30, 1854, a certificate of incorporation was filed at Albany for the Pennsylvania Rock Oil Company of New York. Seventeen days later the organization was completed with the conveyance by Eveleth and Bissell of the Titusville property to the trustees. To make the high nominal capitalization of $250,000 look better, the land comprising the company's main asset was entered at $25,000 rather than at actual purchase price. Actually the land continued in their possession until the next fall due to failure to record the deed.[11]

A flurry of marketing efforts followed, but capital for the new concern was difficult to raise. Eveleth wrote complainingly to Dr. Brewer: "Money was never more in demand and businessmen are sorely pressed." [12] Ebenezer Brewer from his lofty post as local banker in Titusville warned his son: ". . . you are associated with a set of sharpers, and if they have not already ruined you, they will do so if you are foolish enough to let them do it." [13]

The Silliman report

Promotional efforts during early 1855 were further handicapped by the delay in the report from Silliman. Atwood's report, evidently confined to the results of a few routine distillations, had arrived promptly the preceeding November.[14] Although favorable, and pointing out some uses for the oil, it was little more than a supplement to the main report expected from Silliman.

Silliman, proceeding slowly and deliberately, completed his investigation in April, 1855, but not being one to compromise on money matters, he refused to release it until his bill of $526.08 had been paid. Perhaps the most epochal report in petroleum history was made available when Eveleth personally succeeded in raising the money to meet Silliman's bill.[15]

The report's optimistic findings, backed by Silliman's prestige, gave new life to the project. An incomplete but effective synthesis of readily available knowledge, the report did not include all the products and product uses for mineral oils contained, for example, in the prospectus issued in 1853 for the North American Kerosene Company. But the Silliman report did submit synthesized knowledge to laboratory analysis and confirmation.[16]

Silliman estimated that at least 50 per cent of the crude oil could be distilled into a satisfactory illuminant for camphene lamps. Stating that this could be done without any further preparation other than simple clarification by boiling with water, he greatly understated the problem of treating kerosene.[17] He correctly saw that illuminating gas might be made from crude oil, although he doubted the immediate ability of rock oil to compete with less expensive raw materials. The possibility of manufac-

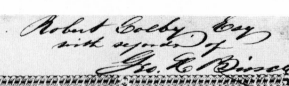

REPORT

ON THE

ROCK OIL, OR PETROLEUM,

FROM

VENANGO CO., PENNSYLVANIA,

WITH SPECIAL REFERENCE TO ITS USE FOR ILLUMINATION AND
OTHER PURPOSES.

BY B. SILLIMAN, JR.,

PROF. OF GENERAL AND APPLIED CHEMISTRY, YALE COLLEGE.

NEW HAVEN:

FROM J. H. BENHAM'S STEAM POWER PRESS.

1855.

Title page of Silliman Report.

turing illuminating gas from the light ends in the naphtha and gasoline range distilled from crude eluded him, but he did urge considerably more study of gasmaking applications of petroleum products.[18] He detected the possibility of extracting paraffin by high-temperature distillation, and recommended further study of the prospects for exploiting this material commercially in candle manufacture.[19]

In its evaluation of petroleum as a lubricant, Silliman's report was not especially perceptive. He had failed to receive any reports on samples that he had sent to Boston for testing in apparatus there, and his own findings that distilled products were noncorrosive, remained fluid at low temperatures, and had no characteristics of a drying oil, evidently applied to distillates in the kerosene range, without taking into account inflammability and other adverse characteristics of middle and light distillates.[20] There was as yet little about lubricating characteristics of mineral oils that he could appropriate from the fledgling coal-oil industry, and even as late as 1860 his endorsement of defecated whale oil as "unsurpassed" as a lubricant does not suggest a very profound knowledge of this phase of the subject.[21]

Whatever its shortcomings in the light of contemporary or subsequent knowledge, Silliman's report was a landmark in its timeliness and critical perceptions of petroleum's possibilities. Its assurances that no less than 50 per cent of the crude oil could be recovered as an illuminant of immediate commercial value, and 90 per cent in the form of distilled products holding commercial promise, sufficed to restore interest in developing petroleum as a commercial commodity.[22]

In the longer range view, Silliman's report established a permanent place in the literature of petroleum chemistry on one score alone. At a time when the status of chemical knowledge offered no substantial basis for such an assumption, Silliman, with a remarkable flash of intuitive insight, saw the probability that the various products of fractional distillation might be cracked. Reporting the boiling points of fluids obtained by his distillations, he observed:

> The uncertainty of the boiling points indicated that the products obtained at the temperatures named above were still mixtures of others, and the question forces itself upon us whether these several oils are to be regarded as educts (i.e., bodies previously existing, and simply separated in the process of distillation), or whether they are not rather produced by the heat and chemical change in the process of distillation. The continued application of an elevated temperature alone is sufficient to effect changes in the constitution

of many organic products, evolving new bodies not before existing in the original substance.[23]

Despite the impact of Silliman's report on New Haven people with surplus funds for investment, the days of the first petroleum company were numbered. Sad experiences with worthless issues of the New York & New Haven Railroad Company and the Western Empire Company, incorporated under New York laws (where the property of stockholders was liable for debts of a company), had made venture capital around New Haven extremely wary.[24] Soon the word went out from Sheldon that New Haven investors would be interested only if Bissell and Eveleth reorganized their company as a Connecticut firm, where stockholder responsibility was negligible.[25]

The Pennsylvania Rock Oil Company of Connecticut

Eveleth and Bissell acted promptly. Opening the books to a subscription to a new Connecticut corporation, they saw two-thirds of the stock taken by New Haven capitalists by June 25, 1855, a matter of a few weeks. Much now depended on the findings of a committee representing investors who inspected the Titusville springs early in July.[26] The committee's favorable report whetted subscribers' appetites for organization of the company and put Silliman's technical report in strong local demand.

Incorporation of the Pennsylvania Rock Oil Company of Connecticut occurred September 18, 1855. Capitalization was $300,000, portioned into 12,000 shares of $25.00 each. Bylaws provided that headquarters were to be at New Haven and a majority of the directors to be chosen from that community, although Eveleth and Bissell retained a controlling interest.[27] Arrangements were made to transfer the lease of the Titusville property to the new corporation, and on November 23, 1855, the stockholders of the Pennsylvania Rock Oil Company of New York met in New Haven, where in an atmosphere of optimism, they proceeded to wind up the affairs of the New York concern.[28]

Townsend and his associates now took more stock and raised enough cash to send Asahel Pierpont, a New Haven mechanic and stockholder, to Titusville to see if J. D. Angier's trenching operations could be improved. Dr. Brewer had written Bissell in October that Angier had recovered in one day 6 gallons from a new trench, but that an expenditure of $500 on trenching should increase the yield to 50 gallons or 100 gallons daily.[29]

Pierpont also recommended stepping up trenching operations. Instead, they were abandoned entirely and even Angier's services dispensed with. Dissension between New Haven and New York stockholders, which was to become perennial, made the treasury cupboard bare.

At the roots of the difficulty was the position into which Eveleth and Bissell had been maneuvered when they subscribed to the bylaws of the new company, for although they were majority stockholders, a majority of the board of directors was to be chosen from New Haven. Nor were relations improved when it was revealed that one of the New Haven stockholders, William A. Ives, had paid for his thousand shares with local securities that were totally worthless.[30] In the impasse that followed, there were no further payments from any source, in cash, notes, or in securities readily convertible into cash.

For Eveleth and Bissell this impasse threatened to bring to an end the whole notion of developing the Titusville property, and with it a total loss of the time and money they had put into promotion. Late in 1856, however, they turned up with a proposal that promised both to give new life to the project as well as to bring them personally an immediate financial return. This proposal arose out of a discussion of their problem with Renselaer H. Havens of the prominent New York real estate firm of Lyman and Havens. Havens, earlier identified with railroad construction in western Pennsylvania, became quite interested.[31] He offered the partners $500 if they would secure for him a lease of the Titusville property from the Pennsylvania Rock Oil Company of Connecticut.

With Haven's offer in hand, Eveleth and Bissell suggested to the New Haven stockholders that if they would purchase a block of stock from the partners, the latter would find someone to lease and exploit the property at Titusville. After some delay, negotiations were completed and the property was leased to Lyman and Havens for fifteen years on a royalty basis of 12¢ a gallon.[32]

It has been suggested that one reason for Haven's enthusiasm came directly from Bissell's chance look, during the summer of 1856, at a Kier handbill advertising rock oil "Discovered in Boring for Salt Water." From this account, Bissell was struck with the possibility of drilling for petroleum, an idea that he passed along to Havens.[33] As already noted, Bissell's own testimony casts a serious doubt upon the validity of this particular episode. Actually there is no clear evidence as to who originated this important idea, or when, or how. It is quite possible that, as with many innovations, a number of persons could have struck upon it independently. This may well have happened in Bissell's case in 1856. If so, it would place him chronologically first among the claimants.

Whatever Havens' ideas of exploiting the Titusville property, they were never put to the test. Hard hit by the panic of 1857, he and his associate took advantage of a legal technicality in the lease to break their contract. It seemed that the wives of the firm members of the Titusville lumber company had not signed the document that would convey the land.

This renewed interest in the property, however, made an impression on the New Haven banker, James M. Townsend, who had succeeded Benjamin Silliman, Jr. as president of the Pennsylvania Rock Oil Company of Connecticut. Townsend recognized several immediate needs if the property was to be developed. Namely, the title should be perfected and the land closely inspected again. If findings were favorable, a new company should be organized to take over the lease, start production, and exploit the oil.

An account of this stage of development, presumably prepared under the direction of Townsend, suggests both implicitly and explicity that he originated the idea of drilling for oil at this time. As reported, the reaction of his New Haven friends was one of incredulity, "Oh, Townsend, oil coming out of the ground, pumping oil out of the earth as you pump water? Nonsense! You're crazy."[34] Townsend purportedly even intended to hire a salt-well driller from Syracuse, New York, to proceed to Titusville and bore for oil, a plan that probably failed to materialize, if it ever existed, because of the shortage of funds.[35] A short time later Townsend's agent, proceeding to Titusville to perfect the title and look over the property, would stop en route at Syracuse, New York, to familiarize himself with salt borings there.[36] On such grounds, Townsend became the second claimant to the distinction of having conceived and executed the idea of boring for petroleum.

E. L. Drake

The events that carried the project to the next stage of development (and incidentally brought in another claimant to the idea of drilling) are somewhat clearer. Casting about for someone to perform his Titusville mission, Townsend chanced upon E. L. Drake, who had been residing in the Tontine Hotel at New Haven, with his child, since he had been widowed in 1854. A conductor on the New York & New Haven Railroad since

1849, Drake was recovering from an illness during the summer of 1857. Except for what he may have picked up at odd jobs during his first nineteen years on New York and Connecticut farms, Drake possessed no technical experience. His business experience had been confined to a few years in various kinds of clerking in a Michigan hotel and in dry goods stores in New Haven and in New York City. He had also done a stint as express agent at Springfield, Massachusetts, on the Boston & Albany Railroad.[37] Aged 38, Drake presented only two visible assets for the task: he

Edwin L. Drake

was immediately available, and his services would come cheaply, since he could obtain a railroad pass to get to Titusville.

Drake made the journey in December, 1857. After stopping at Syracuse, New York, possibly to see the salt wells there, he continued to Erie, where he transferred to a stage that took him the remaining forty miles to Titusville over rough dirt roads. Following Townsend's instructions literally, Drake's first act on arriving in Titusville brought him what was probably the cheapest, if not the most spurious, colonelcy ever acquired outside the state of Kentucky. Alert to the promotional value of a little showmanship to impress the local citizenry, Townsend had mailed the legal documents ahead to "Colonel" E. L. Drake in care of Brewer, Watson &

Company even before Drake had departed from New Haven. When Drake called for his mail, he found the townsfolk already interested in and receptive to the great man of affairs in their midst, and himself adorned with a new title that would remain with him for life.[38]

It took Drake less than three hours to get the signatures of Mrs. Brewer and Mrs. Rynd to the instrument of conveyance of the property, but there wouldn't be another return stage to Erie for three days.[39] He manifested a tourist's curiosity over the use of petroleum as a lubricant and crude illum-inant at the Brewer, Watson & Company lumber mill. When he expressed incredulity that the rock-oil medicine welled up from the ground under the Creek, Dr. Brewer escorted him to the site where this occurred. Ac-cording to his memoir, tinged with the doubtful accuracy of twelve elapsed years after the event, Drake had a blinding flash of inspiration to bore rather than dig for oil. He wrote that "within ten minutes after my arrival upon the ground with Dr. Brewer I had made up my mind that it [petroleum] could be obtained in large quantities by Boreing as for Salt Water. I also determined that I should be the one to do it." [40]

If this intuitive insight occurred, it was no normal mental operation for Drake. Most of his acts reflected mental processes that were methodical and deliberate. Drake's knowledge of petroleum or coal at this point was negligible, and ordinarily he was not given to sophisticated thought proc-esses or abstract reasoning. More characteristic was his reported determi-nation to proceed with drilling if anyone was going to do it. Given an idea, he could be tenacious to the point of stubbornness, and he did need work and income. Reliable contemporaneous evidence simply is lacking to establish precisely when, between 1856 and 1858, the notion of drilling for petroleum first crystallized, and in whose mind—Bissell's, Townsend's, Drake's, or perhaps even others.

The Seneca Oil Company

Whatever inspiration Drake may have received regarding the possibil-ity of drilling, like his predecessors who had examined the Titusville property, he too returned from his journey enthusiastic over the prospects of recovering petroleum in large quantities. Townsend and his associates moved quickly. Early in 1858, the Pennsylvania Rock Oil Company of Connecticut leased the property to Drake and E. B. Bowditch for a pe-riod of fifteen years in return for royalty payments of 12¢ per gallon.[41] Al-most immediately thereafter the New Haven group under the leadership of Townsend formed the Seneca Oil Company of Connecticut on March 23, 1858. A few days later this company took over the lease from Drake and Bowditch. Drake was appointed general agent of the new company at an annual salary of $1,000. It is possible to read into the report on Drake's instructions that he was "to raise and dispose of oil," an interpretation that the idea of drilling for petroleum had by this time occurred to someone in the organization.[42]

Early operations at Titusville

Drake took his family to Titusville in early May, 1858. During the next two months his activities gave few symptoms of sharp definition or purpose. He reactivated the machinery abandoned by Angier to collect oil by trenching, and hired a crew to open up new springs, recover oil, and prospect for coal, salt, and other minerals.

Heavy spring rains impeded operations, but with clearing skies in mid-June his daily collection of oil rose to 10 gallons. A second digging operation, begun at the main spring some 150 feet from Upper Mill, encountered a different complex of difficulties. The nearest adequate source of equipment including picks, shovels, chains and spikes was Erie, a 40-mile stagecoach journey away. Excavations halted after a few weeks when a vein of water was struck, forcing workmen to flee from the well.[43]

It was at this point that Drake determined to start drilling operations. He made a trip to Tarentum, some 100 miles towards Pittsburg and the scene of salt wells producing petroleum exudates for Kier and others, to line up a driller. Hearing that whiskey was a prominent part of the consumption pattern of most borers, Drake sought a rarity—a temperate or teetotaling driller. Finding a man who appeared to fill this requirement, Drake arranged to have him come to Titusville in mid-August.[44]

Returning to Titusville, Drake designed and completed the construction of an engine house for the six-horsepower "Long John" stationary engine, equipped with a tubular boiler of the type used on Ohio and Allegheny River steamers, which he had ordered to power the drilling. He also designed and built, with timber from Brewer, Watson & Company, a derrick in which to swing the drilling tools. Twelve feet square at the bottom, this structure of four timbers 30 feet long tapered to 3 square feet at the top. Two dozen volunteers from Titusville and Upper and Lower Mill donated their services to helping raise the assembled derrick from its prone position on the ground. With installation of the boiler and engine, he was all ready to start drilling—but there was no driller.[45] A return trip to Tarentum failed to locate the driller and when further search failed to find a replacement, Drake decided to suspend operations for the winter.[46]

The New Haven backers at this time were sympathetic with Drake's problem and at a stockholders' meeting in September, 1858, voted him an additional $1,000 to meet expenses. They also elected him president and a member of the board of directors.[47]

Early spring of 1859 brought Drake to the brink of despair. Seeking an early start, he made another futile trip to Tarentum in February to engage a driller; another broken contract resulted through failure to put in an appearance. In April, Drake received a welcome letter from Louis Peterson, Jr., Tarentum salt-well operator, who reported that W. A. Smith of Saline (a small community just outside Tarentum, in Butler county) probably could be hired.[48]

Drake rushed to Tarentum and succeeded in closing a deal with Smith for $2.50 a day, with his boy's services "thrown in." Looking back some years later, Drake wrote truly: "I could not have suited myself better if I could have had a man made to order." [49] Smith was more than an experienced borer. Part of his blacksmith's trade was making boring tools and salt pans for Kier, Peterson, and others around Tarentum, an invaluable skill in isolated Titusville where tools were unavailable. After finishing about four weeks of small jobs for which he was committed, Smith made drilling tools from the $30.80 worth of iron Drake supplied him with in his own shop, where the work could be done cheaper and easier than in Titusville. About mid-May, Drake sent a large spring wagon to bring the tools, Smith, his 15-year-old son Samuel, and daughter Margaret Jane to Titusville.[50]

Though, seemingly, the stage at last was set, appearances proved deceptive. Already Drake's workers had been digging a hole and cribbing it with timber, but water from nearby Oil Creek persistently seeped in knee-deep. Smith devised a pump so that digging could continue, but water rushed in faster than the pump could handle it. Frequent cave-ins completed the barriers to further progress. It became doubtful whether the well would ever get drilled following the orthodox salt-boring procedure of cribbing down to bedrock. Either Smith or Drake (by most accounts the latter) had a timely inspiration: why not stop trying to crib beyond the 16 feet already done, and drive an iron pipe the rest of the way through quicksands and clay to the rock?[51]

An initial experiment with cast-iron pipe, ½ inch thick and 3 inches in diameter failed when the pipe broke after being driven 10 feet down. Another try with soft iron 1½ inches thick was more successful.[52] With nothing but a white-oak battering ram controlled by a simple windlass that had been used in excavating surface dirt, this pipe drove successfully down to the bedrock at 32 feet. Smith shifted to steam power, proceeding to drill at a rate of about 3 feet a day.[53]

Long before this, the novelty of the project had worn off. Townsfolk no longer displayed even a sidewalk superintendent's interest. When they thought about him at all, they were apt to say that "Drake was fooling away his time and money." [54] New Haven stockholders felt even more strongly this way, after seeing $2,000 spent with only a mud hole to show for it. They voted Drake a final additional advance of $500 in April, 1859, which, with the $347 cash balance he still had on hand, gave him $847 with which to carry on his work that fateful summer.[55] Thereafter, James M. Townsend became the sole source of working funds, but late in August even these dried up. Reluctantly concluding that the Titusville operation would come to nothing, Townsend sent Drake a final remittance with which to pay off outstanding obligations and abandon the project.[56]

Discovery of oil

During this trying period, driller "Uncle Billy" Smith did not rest, Sundays or weekdays. Though his crew wasn't working, late in the afternoon of Sunday, August 28, he went out to the well for a routine inspection.[57] When they had knocked off work and pulled their tools out the previous night, the drill had reached 69 feet and a crevice which dropped it another 6 inches. Squinting into the pipe, Uncle Billy thought he caught a glimpse of some sort of dark film floating on the water, only a few feet below the level of the derrick floor. Grabbing a piece of tin rain spouting he plugged one end and rammed it down the narrow pipe. In the next tense moment, Smith and his son had the improvised ladle back to the surface and had found their answer. Gleefully they looked at the dirty greenish grease on their hands. Uncle Billy couldn't be fooled; it was oil, the same stuff he had seen a hundred times at the old salt wells he had worked at Tarentum.[58]

News of the strike spread quickly to the village by residents around the Upper and Lower Mills, so that there was a crowd of curious citizens and

The Drake Well in 1864, with "Uncle Billy" Smith in foreground.

well-wishers on hand when Drake arrived at the well on Monday morning. There he found Uncle Billy and his son already grappling with one of
many problems that was to characterize oil producing for years to come:
how to provide adequate storage for an unexpected strike in a remote locality. Uncle Billy and his boy already had mustered tubs, wash boilers,
and a few stray whiskey barrels to store the oil.

Seemingly pleased, but accepting the favorable turn of affairs matter-
of-factly, Drake lent a hand to the operations. He rigged an ordinary hand
pump to a 20-foot piece of pipe, attached the handle to the walking beam,
and the first oil well was on pump. The rate was probably an unprecedented 8–10 gallons a day, but in the excitement no one thought to gauge
the well.

In the mail on that day, or within the next few days, Drake received the
money order sent by the discouraged Townsend to close out accounts and
return to New Haven.[59] With a swifter postal delivery than the twice
weekly stage from Erie to Titusville provided, the birth of an industry at
69½ feet just outside of Titusville would not have occurred in the fall of

1859. To the very end the project to drill for oil was kept alive by a lengthy chain of extraordinary coincidences.

It is doubtful whether Drake gauged the significance of his accomplishment any better than the initial output of his well. There was little implicit or explicit in his attitude and actions suggestive of a man with an entrepreneurial vision. He was a man doggedly, almost stolidly, carrying out the specific project he had been employed to undertake. As a driller, Drake was a rank amateur. At a time when it usually took only six to eight months to bore 1,000 feet or more through solid rock in the great salt fields along the Kanawha River, it took him two years to drill 69½ feet. When Drake finally reached his peak rate of 3 feet a day in August, 1859, he was boring at only about one-half the Kanawha rate of a decade earlier.

But whatever Drake's shortcomings as driller, businessman, or entrepreneur, he was the first to demonstrate that petroleum could be drilled for and obtained in substantial quantities. By this demonstration he removed the major barrier to the rise of a new industry.*

---

* Events overwhelmed Drake after his well came in. While others scrambled to lease lands to drill on, Drake busied himself getting equipment for his well. On October 7, 1859, his storage and derrick were completely burned down. In the next few months he expended many futile efforts in trying to market his oil. He kept refusing offers to lease or purchase more oil land. In June, 1860, he was elected Justice of Peace and became an oil buyer for Schieffelin Brothers of New York City. Three years later he departed from the Oil Region with savings of $15,000 or $20,000 to become a partner of a Wall Street broker in oil stock. By 1866 he had dissipated his savings in speculation. He spent the rest of his life in ill-health, poverty, and obscurity. Awarded a pension of $1,500 annually by the State of Pennsylvania in 1873, he died in 1880.[60]

Chapter 5

Pioneer developments

Like the discovery of gold in California, silver in Nevada, and many other rich mineral deposits, Drake's well ignited a rush into western Pennsylvania that soon transformed petroleum production into a mining operation and quickly gave the area in and around Titusville all the characteristics associated with a mining boom. But it took more than the feverish activity of a mining community to establish the foundations of a full-fledged petroleum industry. There were many who questioned the permanence of the new source of raw materials for illuminants and lubricants. Their reactions were reflected in a failure to match a growing production of crude in the fields with adequate transport, marketing, and refining facilities that for the better part of two years kept the future of petroleum hanging in the balance.

Even before these doubts about the future of petroleum were resolved

in the fall of 1861, experience had already revealed many of the characteristics and patterns which were to distinguish the petroleum industry
in the years ahead. Pioneer drillers had not brought in many wells before
realizing at least subconsciously that they were founding a unique type of
mining industry, with distinctive features of its own.

No other mineral was liquid, highly inflammable, and fugacious in
character. This last quality, operators soon learned, built up a strong pressure for rapid exploitation. Failure to drill on one piece of property could
result in serious losses to wells on adjoining land. No later than the summer of 1860, petroleum producers also had ample confirmation of the unpredictable nature of petroleum supply. Repeatedly they saw production
expand overnight to levels well beyond their most liberal estimates of
needed storage and handling facilities.

Problems of transport

By no means least of the problems confronting inexperienced petroleum
producers was the very location of the Drake well, which struck off a ster-
eotype for many succeeding fields to the present day. More often than
not, new fields are discovered at inconvenient locations in relation to
transportation facilities and markets. Although an expanding railroad system had already made national distribution of lamp illuminants an accomplished fact several years earlier, a web of rails totaling 30,000 miles
in 1860 failed to provide the new oil territory with efficient outlets.

The closest railroad stations to Titusville, in Venango County, lay
twenty to twenty-five miles northward, at Garland, in Warren County,
slightly northeast, and at Corry and Union Mills in Erie County, the second county almost due north from Titusville (see Map 5:1). All of these
stations were situated on the Sunbury & Erie Railroad (soon known as the
Philadelphia & Erie), at this time an abbreviated line with sixty-six miles
of track which connected Warren with Erie, Pennsylvania's port on Lake
Erie, some forty miles almost due north of Titusville. Because of gauge
differences, any oil shipped east or west from Erie had to be reloaded
from the Sunbury & Erie.

Erie offered several keys to eastern shipments, if local transportation to
Erie could be solved. One possibility was to follow the route of the three
barrels of petroleum shipped from Titusville in 1853 to George Bissell:
drayage to Erie, steamer to Buffalo and the Erie Canal to New York City.
Alternatively, it was possible to ship via the Cleveland, Painesville, &

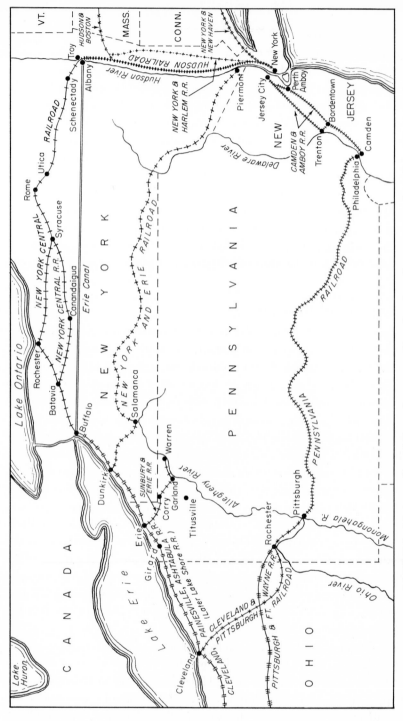

MAP 5:1. *Major railroad lines between Cleveland and eastern seaboard: 1860.*

Ashtabula Railroad's 4-foot 10-inch gauge track from Erie to Buffalo, a transfer of cargo to the New York Central's 4-foot 8½-inch track to Albany, and another transfer to the N.Y.C.'s connective with New York City over the Hudson River Road with its 4-foot 8¼-inch gauge track. A third and more popular alternative was transshipment at Dunkirk, New York, about midway between Erie and Buffalo, from the C.P.&A. to the 6-foot gauge track of the Erie Road, a trunk route to New York City. Westward from Erie, the C.P.&A. connected with Cleveland and, through various extensions, with Chicago.

Worse than the serious limitations of the rail facilities themselves was the problem of getting petroleum to them. Scarcely more than wagon trails, roads to the railroad stations twenty to forty miles away were barely negotiable under the most favorable circumstances. Exacting a heavy toll on horses and in damaged cargo, they became impassable in rain or snow. Making two round trips a week, wagons normally carried five or six barrels weighing about 300 pounds each.

If northerly facilities held little promise of moving substantial quantities of oil to markets, those to the south presented difficulties too. In this direction Pittsburgh offered the nearest rail outlet, 135 miles down the Allegheny River from its juncture with Oil Creek, located 16 miles below Titusville at the future site of Oil City.[1] The Pennsylvania Central's 4-foot 8½-inch gauge track afforded Pittsburgh with reasonably good transportation as far as Philadelphia, but cargoes destined for the New York area had to be broken at the Pennsylvania Railroad's terminus in Philadelphia for transfer to the 4-foot 10-inch gauge roads running north into New Jersey. Goods had to be drayed or moved by horse car through congested city streets over poorly maintained street rails either to Camden, across the Delaware River outside of Philadelphia and the southern terminus of the Camden & Amboy Railroad, or to Kinsington, just north of Philadelphia and the location of the Philadelphia & Trenton Railroad.[2] In addition, all through freight over these New Jersey railroads had to submit to an obnoxious tax of 6¢ per ton mile until 1868.[3] At their northern terminals at Amboy and Trenton again there were draying, canal, or ferrying charges into New York City. Many shippers understandably chose the alternative of transferring freight from the Pennsylvania Railroad to coastwise vessels at Philadelphia.

For western shipments Pittsburgh's facilities were unsurpassed. Since 1857 the Pittsburgh, Fort Wayne & Chicago Railroad's 4-foot 10-inch gauge track had provided a trunk line to Chicago. This facility was further improved in 1858 when the road bridged the Allegheny River into Pittsburgh, although through freight still usually had to be broken for transfer to the Pennsylvania's 4-foot 8½-inch gauge.[4] At Alliance, Ohio, and Rochester, Pennsylvania, the Cleveland & Pittsburgh road linked with this line. The Alleghany River provided Pittsburgh with a water highway to the west via the Ohio River.

Facilities for handling an extensive traffic between the oil fields and Pittsburgh by water, however, were almost as inadequate as draying service to northerly rail stations. Twice a week small passenger steamers carrying mail and small freight items from Pittsburgh passed the juncture of Oil Creek with the Allegheny River sixteen miles below Titusville.

Apart from a rail outlet from Corry to the seaboard in May, 1861, which had little immediate effect, no basic changes occurred in this transportation complex until the fall of 1861. Clearly, the mildest success in petroleum production would place a heavy burden on existing transport and handling facilities.

The rush for drilling sites

As would happen again repeatedly throughout the next hundred years, neither lack of transportation nor of markets visibly braked the extension of petroleum production. The process began the first day after the Drake well came in with a race between telegraphic communication and a man on a fast riding horse. In addition to giving two major producers of the decade their initial foothold and experience, the race unfolded the evolutionary beginnings of locational methods and leasing practices.

George Bissell in New York read a telegram telling the news of Drake's discovery, and immediately set in motion a program of leasing and land purchase. At the wellhead itself, Jonathan Watson, a member of the lumbering firm of Brewer, Watson Company, which had leased the mineral rights to the property on which Drake drilled, remained a curious bystander but briefly.[5] Mounting his fast riding horse, Watson plunged into narrow Oil Creek Valley, below Titusville, to obtain leases.

Drake's well was located on the flats near Oil Creek. Similar locations had marked nearly all salt wells with petroleum exudates in the Kanawha Valley, in the Tarentum vicinity, and in Kentucky and Ohio. With practical sense Watson chose this general guide heading into Oil Creek Valley, where a small colony of Germans and Scotch-Irish eeked out a subsistence existence on the forty-three farms that jigsawed haphazardly on the flats of the narrow valley, at some places no more than 40 rods wide. From Titusville the valley extended about sixteen miles south to where Oil Creek joined the Allegheny River. The narrow creek flats were flanked by precipitous hills rising 150 feet just below Titusville and 450–500 feet farther south near future Oil City.[6] (See Map 5:2) Dense stands of oak, maple, hickory, and cherry, plus a heavy matting of underbrush,

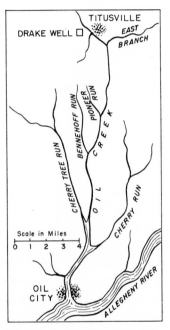

MAP 5:2. *Oil Creek Valley.*

made the steep hillsides almost impassable on foot and virtually forbade engineering operations of the simplest kind.

Watson that day completed one of the most profitable horseback rides in history, one that was to make him several times a millionaire. In the local lumbering business since 1845, Watson knew the geography of the valley and its residents intimately, and had no trouble winning landowners' confidence to obtain leases at bargain rates. By mid-September, 1859, he had leased or purchased much of the most promising land on the flats along Oil Creek. The Region's premier producer for more than a decade, Watson was credited with having drilled more than two thousand wells by 1871.[7]

Arriving four days after Watson's ride, Bissell was not far behind him in boldness of policy or results. With no particular regard to surface show-ings, he and his partner J. G. Evelyth boldly leased or bought up farms along Oil Creek and the Allegheny River, expending $200,000 that fall.[8] Investing $50,000 to become principal shareholder in the Pennsylvania Rock Oil Company, which in 1858 had leased the 105-acre tract to the Seneca Oil Company upon which the Drake well was drilled, Bissell also reached an agreement in 1860 by which the Seneca Oil Company took a small portion in fee of the land on the immediate site of the Drake well, while he acquired the lease on the rest, renamed the Bissell farm. In that year he brought in six wells on this property. In addition to shrewd purchases, leasing, and sub-leasing, he gained a marked advantage over many competing producers by erecting a large barrel factory at Frank-lin that permitted him to exploit markets when others were hamstrung by the area's perennial barrel shortage.*

Watson and Bissell were but the forerunners of hundreds of prospectors who flocked into the Oil Region. They came from a wide geographic area and from widely varied economic, social, educational, and occupational backgrounds. By far the greatest number of early oil producers came from Pennsylvania, New York, Ohio, and New England, in that order.

Transfers from the local lumbering industries represented the most im-portant single source of early producers. The most prominent among these producers were merchants or manufacturers of lumber, such as Jonathan Watson and Francis Brewer of Titusville. The highly mobile population of sawyers and ordinary laborers in lumbering also propelled numerous though less successful entries into oil producing. Representative of this class was William Kirkpatrick, head sawyer of Brewer, Watson Company's saw mill on Watson flats near the Drake well. Seeing his boss, Jonathan Watson, leave the throng of curious onlookers at the Drake well to embark on his "million dollar ride," Kirkpatrick in turn "dropped the saw levers forever" and by nightfall "was lucky enough to get two lots, numbered 66 and 67, on which to begin drilling." [10]

Salt borers, another floating population group, thronged to the Region to find more lucrative new employment, and often branched into drilling for themselves when they acquired capital or backers.[11] The attraction of

---

* Though he left the Regions with a large fortune in 1863, Bissell continued to be a force in oil production, in 1866 founding a bank at Petroleum Center that during the rest of the decade withstood the rigors of oil speculation that sent many other local institutions under.[9]

superior profit possibilities lured salt producers as well as borers to petroleum production to such an extent that the salt industries of Ohio and
Kentucky suffered a permanent decline.[12] Among the earliest Pittsburgh
capitalists to engage in the refining, producing, and eventual trading in
petroleum, were Charles Lockhart, and Samuel Kier, with his more than
a decade of experience in marketing petroleum medicinals.[13]

Entries into the new mining industry reflected the status of popular
education. College graduates, such as George Bissell, were rarities; the
grammar school marked the limits in education of most early producers.
Apart from those with experience in salt boring or lumbering, a surprisingly few early producers had significant backgrounds in the crafts and
skilled trades. Although liberal royalty payments made many local farmers wealthy, farmers who became operating producers were as rare as college graduates.

Mercantile pursuits formed the great common denominator of occupational experience among the early operators. Few lacked at least some
background as clerks in grocery and dry goods stores or in other similar
establishments where they gained the rudiments of both buying and selling, and of keeping accounts. A majority had also gained entrepreneurial
experience before entering oil production, either as partners in mercantile firms or as proprietors of their own establishments. Entrants were of
all ages, but most numerous of all were men in their thirties and forties
who saw in the new type of mining enterprise an opportunity to utilize
their business experience and moderate amounts of capital to achieve
financial success.

There certainly was no lack of opportunity to exercise entrepreneurial
experience and ingenuity in the pioneer phases of production. Apart from
salt-drilling techniques, there were few precedents to guide inexperienced producers in locating wells, in negotiating leasing terms, in appraising costs, and in handling, storing, and shipping an inflammable liquid in
bulk.

Location of drilling sites

Few phases of oil production were as lacking in reliable precedents as
the all-important matter of determining the most promising drilling sites.
At best, no less than $1,000 rode on each drilling decision, and very prob-
ably a good deal more. With so much at stake and few guideposts, pio-

neer operators were unavoidably receptive to any interpretation of petroleum showings that seemed to demonstrate any association with success.

Not only were locational methods few at the outset, but not many producers were in a position to use them. To locate according to the petroleum showings from salt wells was banned to all producers in the
Venango County region, where there was no salt production. Only a
limited number could acquire drilling sites adjacent to a few known petroleum springs or points on Oil Creek where oil flotations had been observed. The number who could locate in the immediate vicinity of
proved wells was also restricted. A majority had to generalize their locations along the flatlands of Oil Creek and the Allegheny River, where all
showings in the form of springs or wells had been observed. To select specific sites for boring on these generalized locations presented most producers with a need to refine and amplify their locational methods.

The services of professional geologists were at once almost wholly unavailable and useless. The most basic geological principles were in violent dispute, with Harvard University's renowned scientist, Swiss-born
Louis Agassiz, leading a strong group which denied the validity of evolutionary concepts and of needed revisions in the time-clock of the geological processes. Contradictory theories on the nature of petroleum, on its
origin (whether vegetable, marine, or mineral), on its method of formation (whether by gradual decomposition and fermentation or by distillation as the result of intensive subterranean heats), and perhaps most of
all its relation to coal—all such theories were in wide disagreement.[14]

To apply any of these theories to oil location was difficult and could
be misleading. For example, one widely held geological belief considered
petroleum nothing but the drippings of coal measures from adjacent
higher ground.[15] Thus geological theory as well as the difficulties of penetrating precipitous, densely wooded hills with scarce labor, pipe, and
steam power supported early operators in their preference for the low
flatlands. To cap this preference, the Creek flats from the very beginning
yielded more oil than markets and transportation facilities could assimilate.[16] Reasons other than naive location theories delayed drilling on high
grounds until the play in 1864 on Pioneer Run, a branch of Oil Creek.[17]

Efforts to improve locational methods beyond simple showings took
some odd turns. No later than the spring of 1860, location by divining
rod put in its appearance, followed by the use of oil smellers, spiritualists,
and others who claimed special powers in locating oils. One of the color-

ful eccentricities of the period, these individuals attracted the attention of journalists to a degree that probably exaggerated the amount of dependence upon mystical locational methods. Throughout the decade, however, divining remained an organized business, with fees ranging from $25.00 to $100, according to the diviner's record in locating wells, or his persuasiveness.[18]

Of all of these methods, the divining rod appeared the earliest, found the most devotees, and had the most rational basis. In France, during the 1850's, hydrology (employing divining rods to locate water) was accredited as a serious branch of the science of geology. For decades, frugal New England farmers had been expending $10.00 to $20.00 for twig-twirlers to locate drilling sites for water wells.

Using a forked twig of witch hazel or peach, the practioners of this art clutched the two ends or prongs, holding the rod vertically, and, as they passed over a vein of oil, the large end was supposed to dip downward, vibrating. Experts attributed their special powers with the device to their possession of an unusual positive charge of electricity, which acted upon the stick and gave it affinity with minerals below.[19]

If generally viewed with skepticism, claimants of spiritual powers in locating oil also attracted attention. Though used less extensively than wielders of the twig, these locational specialists found considerable employment throughout the decade. Some producers even retained "oil smellers" who claimed that by pressing their noses close enough to the ground like a hound picking up a rabbit's trail, they could smell oil veins several hundred feet underground.[20]

In all probability the most successful of these "consultants" were shrewd observers who anticipated the functions of oil scouts two decades later. To the extent their services were popular, they had abundant opportunities to observe and to learn information about latest locational and producing developments, a vital function in modern petroleum fields no less than in the early ones.

Within a few weeks after the Drake well, however, a few producers began to utilize more sophisticated and constructive locational methods than even a simple interpretation of surface showings. Both Bissell's and Watson's early land acquisitions, for example, extended beyond a narrow concept of locating wells immediately adjacent to oil springs. Within a few months numerous wells a half-mile distant from Oil Creek had further overturned a literal interpretation of surface indications as a sure clue to a promising location.[21]

The Barnsdall well, the second to be brought in on Watson's Flats, introduced the further notion of drilling deeper to improve yields. Yielding only five barrels a day in November, 1859, this well began pumping 10 barrels, and then 20 barrels a day after being drilled deeper to 112 feet, and eventually 75–80 barrels per day. Only 200 feet away from the Barnsdall well, the Williams well came in with a daily production of 12–14 barrels in June, 1860. Dissatisfied with the yield, the owners drilled deeper and within a few days obtained 80–90 barrels in fourteen hours, the most spectacular well to that date.[22] It soon became commonplace, either to improve a well's unsatisfactory yield or to restore a well that had stopped producing, for operators to drill moderate distances deeper in quest of a part of the producing vein that was not obstructed, or to find a new vein.

As early as 1860, dry holes as deep as 500 feet in Erie County and elsewhere testified to a second type of deep drilling.[23] The threshhold of an explicit recognition of distinct oil-bearing sands was reached by early 1861 when, failing to strike at the usual depths of 200–300 feet where the first two oil-bearing sands usually were encountered along Oil Creek, operators began probing deeper for an oil-bearing sand.[24]

Leasing arrangements

From the inception of petroleum production, leasing practices bore important relationships to well locations and production costs. The earliest leases were distinguished by a total absence of any tendency to standardize royalty payments at one-eighth, a trend that awaited the introduction during the mid-1870's of the practice of leasing large blocks of promising oil land for an annual rental.[25] Until accumulating experience gradually disclosed needs to incorporate special provisions, early leases typically set forth simple, generalized provisions, with payments to the lessor varying in accordance with the relative desirability of location, the land's proximity to the flats and Oil Creek, and to producing wells.

The length of leases varied from twenty to ninety-nine years to perpetuity. A term within which the lessee was obligated to begin drilling (six months to a year, or sometimes at "early convenience") usually was stipulated. In some instances, failure to perform called for cash payments; in others, a reversion of mineral rights to the lessor was a part of the agreement. Frequently 200 feet was established as the minimum drilling depth required, although depths of 500 feet already were called for in some leases during 1860. A few early leases stipulated royalty payments as low as one-twelfth or one-tenth in 1859, but these terms soon became rarities even for less desirable tracts. During the next two years the normal range of royalties ran one-quarter to one-third of the oil produced, to be delivered free in barrels to the landowner.[26]

Cash payments or bonuses flowered into prominence very early, and by 1861 two kinds had emerged as common features of leases. One was an outright payment made at the time the paper was executed, with amounts varying from $200 to $600. A second type of bonus became payable on some contingent happening. An example was the provision in G. P. Overton's lease in 1861, covering 1½ acres in Warren County. The lease stated that, in addition to a royalty payment of one-eighth of all oil barreled free, there would be a specified payment of $2,000 to the owner of the property if a well showed 6 barrels a day for six months.[27]

One of the dynamos for the astonishingly rapid pace of early petroleum development as well as its speculative excesses was the practice of selling fractional shares in leases, commonly called sub-leasing. Almost a standard feature by early 1860, sub-leases provided fortunes to speculative specialists in acquiring leases and subdividing them at huge profits to persons with limited resources seeking entry into production. For big producers and small ones it afforded a ready means of bolstering resources to continue or expand drilling activities. The sub-lease also encouraged the ruinous practice of spacing wells only a few feet apart, with as many as 150 wells clustered on a few acres and, coupled with random leasing to parties willing to pay large bonuses, also formed the insolvent base of many spurious stock-company promotions.[28]

The sale of shares in a lease provided another means of raising capital and spreading risk. For example, in 1862, J. W. Sherman of Cleveland arrived in the Oil Regions to seek his fortunes with few resources. Obtaining a lease on the Foster farm on Oil Creek, he used spring-pole power to get to the first sand, with a promising show, but was forced to shut down before reaching second sand for lack of horse or steampower with which to drill effectively. He disposed of one-sixteenth of his share for an old horse, which lasted a few weeks; he resumed drilling again by trading another sixteenth share for a second-hand small steam engine; lacking cash to buy coal, he traded still another sixteenth for $80.00 and a shotgun. On the verge of running out of sixteenth shares, he struck the fabulous Sherman well, whose oil yield was estimated at $1,700,000.[29]

Thus within a brief span of less than two years, oil leases had developed a remarkable flexibility. If this adaptability contained the weakness of encouraging many undesirable production practices, it also carried strengths that were critical for developing the new resource. Without leasing practices at once resilient and flexible, the changes in the earliest production, and still greater changes to come, would have been more difficult, perhaps impossible, to accommodate.

Drilling techniques

Drilling techniques used by Drake carried petroleum production through its first two years of extension, although deviations did appear. There were many who did not use rope as the cable from which the drilling tools were suspended, but employed instead the older "chain-gang" system of chain cables, despite drawbacks of excessive weight, poorer flexibility, and a horrendous clanging heard a half-mile away. To obtain better direction in drilling, a minority also simply used metal sucker rods as cables.[30] Drake's innovation of an iron driving pipe to reach bedrock was followed only where soil conditions and cave-ins dictated. With most shallow wells it was cheaper and easier to crib down to bedrock.

Few were as fortunate as Drake to have steam engines for drilling. Scarce and expensive engines had to be husbanded for the all-important operation of pumping. Some wells employed the old-fashioned spring-pole method, but generally one of two improved methods of manual power was adopted. "Jigging it down" probably was the more common. A small wooden platform was attached to the side of the derrick with a connection to the pole, or walking beam from which the cable and tools were suspended. To obtain the downward stroke of the drill, two men would step on the outer side of the platform, and then trod on the derrick side to bring the walking beam back into normal position. The jigging maneuvers of the men on the platform supplied this method with its name.[31] A third man guided the drill to make a round hole. The second method, "kicking it down," was faster. Near the fulcrum of a 10-foot or 15-foot pole, stirrups were arranged in which two men each placed a foot, and by kicking outward and downward made the pole or walking beam plunge downward and spring back. Experienced kickers could produce two strokes of the drill in a minute.

Horses were in primary demand for draying oil, but occasionally they were used to supply drilling power according to the plan used with

Sketch of early drilling operation.

threshing machines. The horses walked around the center and over a horizontal tumbling shaft that supplied the perpendicular motion to the drilling bit. Where water power was available naturally or by damming a small stream, it occasionally powered the drilling, but more often the more laborious job of pumping was employed.[32]

Seldom weighing more than 150 pounds, tool strings were the lightest in the history of petroleum drilling. At the bottom of the string was the steel center bit, about 3½ feet long, with a shaft 1½ inches in diameter that flared into a cutting edge about 3 inches wide. The threaded top end

Woodcut of driller guiding the stroke of the drill.

of the bit screwed into the auger stem, a device that permitted the driller to manipulate the cable so that each stroke of the bit would fall across the preceding cut. Usually becoming blunt after drilling 2 feet, the center bit would be interchanged with a larger and blunter reamer, a process requiring at least two laborious withdrawals of the entire tool string. If accumulated rock and water remained after reaming, two added tool string withdrawals were necessary to lower and raise a copper sand pump about 6 feet long and 1½ inches in diameter. Sometimes pumping had to be followed by another reaming operation before reinserting a sharpened center bit with which to continue drilling.[33]

Above the interchangeable bits and reamers ranged a series of vital components of the tool string. First was the auger stem, a shaft 20 feet long, strong and heavy enough to supply striking force to the percussion drilling. Attached to the auger stem were the "jars," two looped irons that gave a play of at least a foot to each drilling stroke. The jars reduced the vulnerability of the heavy tool string to breakage and supplied leverage for jarring loose tools stuck in the hole. The top of the tool string terminated with a rope socket with a metal clamp which attached the string to the cable.

Flooded holes, cave-ins, lost tools, broken strings, crooked holes, and many other perverse happenings made early drilling operations and costs highly uncertain and variable.* To drill to the prevalent common depths of 200 feet might cost $1,000 to $1,200 if reasonably favorable conditions prevailed, according to contemporary estimates, but unlucky operations or the need to go to greater depths raised many drilling outlays to $2,000 and more.[35] In soft slate, 5 feet or 6 feet could be bored in a day; in sand and rock conglomerate, only 2 inches or 3 inches. Three feet a day was judged a good average.[36]

But drilling costs generally comprised the least expensive and least variable part of early production costs. Indeed, considering the shallow depths of most wells, the pioneer period of pumping wells undoubtedly witnessed some of the highest production costs in relation to pay-off periods in the history of petroleum. Much drilling on unproved territory abnormally inflated exploration and development costs. The ratio of producing wells was no higher than one in twenty, and probably less, the lowest in the decade and beyond.[37] The low yields of pumping wells in a

---

* Using homemade tools and labor from within his own family, E. Evans, for example, brought in the first well at Franklin in 1860 at a depth of 72 feet, for a reported out-of-pocket expense of $200.[34]

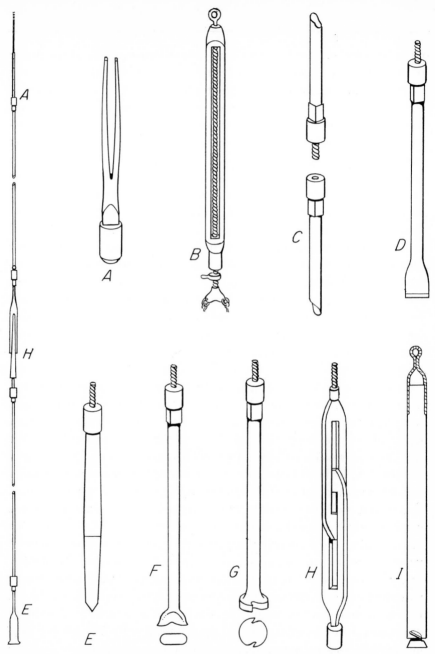

FIGURE 5:1. *Principal tools used in drilling.*

A. Rope socket   D. Club bit   G. Round reamer
B. Temper screw   E. Center bit   H. Jars
C. Drill stem   F. Reamer   I. Sand pump

period of declining prices and high transportation costs automatically prolonged pay-out periods excessively.

The inexperience of operators, the inadequacy of techniques, and soaring land values contributed to the same result.[38] As primitive and uncertain as the drilling techniques were, more time and expense usually was incurred in testing and proving a well than in drilling it. Even to determine at what depth to test a well was difficult. Early operators usually construed a show of oil in the mud brought up by the sand pump as a favorable sign, or a showing of gas that made the mud bubble. However, some also recognized that good wells had come in without these intermediate showings.

Once the position of the oil-bearing stratum was established, the first step in preparing the well for pumping was to tube it. Copper pipe, 2 inches or 3 inches in diameter, and in sections 12–14 feet long, was extended to the bottom of the well. To help keep the tubing in position, but primarily to seal off the well hole from water and sediment above, a seed bag was placed between the tubing and the sides of the well at a point below surface water streams or seepages (Fig. 7:4). As the final step, wooden sucker rods with a plunger or pumping apparatus attached to the lowest string were lowered inside the tubing to a position near the bottom of the well. With the stroke controlled from the derrick, the pump, operating on a vacuum principle, forced the oil through the tubing to the wellhead.

Seedbagging formed a critical part of the tubing operation. The leather bag was filled with a pint or a quart of wetted flax seed which within a few hours after placement in the hole would swell so as to shut out all waters from above. The bottom of the bag was firmly secured to the tubing, but the top was tied only by a loose knot that upon withdrawal of the tubing would turn the bag upside down, dumping its contents, so that the empty bag could be drawn to the surface without damage. The difficult crux of success was to tie the bag at the proper point on a section of tubing that would assure shutting out of all water from above.

Even in smooth, hard rock favorable to the seed bag's accomplishing its mission, a fracture in the rock or some other condition might let water in from above. Intensive pumping might then be resorted to for weeks, followed by the laborious task of raising the tubing and sucker rods section by section to readjust the seed bag and tubing. It was not uncommon to repeat fifty times the sequence of pumping and tubing readjustments, with no assurance of success in striking oil and putting the well on pump.

Apart from imperfections in seedbagging and tubing techniques, the shallower wells of the early years, before drilling to third-sand rock grew common, tended to yield proportionately greater quantities of water. Sometimes veins of water below one yielding oil were passed in drilling. Partly to save tubing costs, partly in the mistaken notion that surface waters were not harmful and even might be helpful, early operators did not tube near the surface. Successfully getting a well on pump by no means ended troubles; pumping wells often abruptly stopped producing. If the causes were cave-ins of walls or clogging of the producing vein, sometimes a remedy lay in withdrawing the tubing, lowering the sand-pump, and perhaps increasing the depth of the well. If particles of paraffin showed in the material brought to the surface, operators might try to steam out that material which was presumed to be clogging their pipes, or to pour distress naphtha down the plugged well in the hope it would exert a solvent action on the paraffin. Neither remedy brought frequent success.[39]

These considerations strongly suggest that contemporary estimates to evaluate various elements of production costs fell far short of realism. From an estimate of $5.00 a day operating costs, one account concluded that it cost only 1¢ a gallon to raise oil in a 12½-barrel well (500 gallons daily), and but one-third of a cent in a 37-barrel well (1500 gallons daily). Even in a 2½-barrel well, oil could be raised for 5¢ a gallon.[40]

It would be a mistake to regard all operators as entirely blind to production costs beyond the most obvious ones. But their warnings went largely unheeded in the mainstream of optimistic press accounts. As one observer noted, "no one should expend money in this enterprise without being able and willing to lose every dollar invested . . . [yet] so many rush into the thing without counting the cost, with inexperienced workmen and heavy outlays, with no calculations for accidents."[41]

Spread of production

Considering the available drilling techniques and uncertainties of bringing in profitable wells, exploration and drilling progressed with startling speed during 1860 and 1861. By January, 1860, the drill bit if not the actual strikes had demarcated all future districts of the Pennsylvania Oil Region until 1864. Operations had extended down Oil Creek to the Hamilton McClintock farm, within two miles of Oil City; to the areas of Tidioute, Irvine Warren, Tionesta, on the upper Allegheny River northeast of Oil

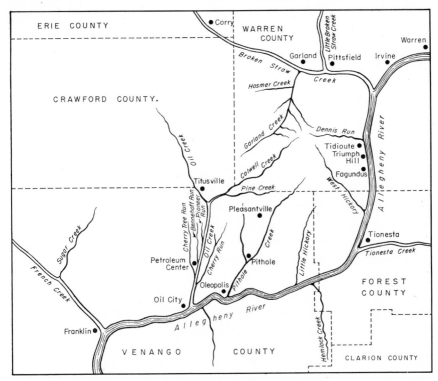

MAP 5:3. *Upper Oil Region.*

City; and to Franklin, west of Oil City at the juncture of French Creek with the Allegheny River. Before the year was up, operators pressed activities at various intermediate points where some production would appear in the future (notably around Reno, five or six miles up French Creek from Franklin; and at Henry's Bend, above Tidioute on the Allegheny).[42] Although the earliest production occurred on Watson's Flats in the southern tip of Crawford County around Titusville, most development along Oil Creek and the Franklin area lay in Venango County, essentially the primary Pennsylvania Oil Region. Tidioute and neighboring locations on the Allegheny River fell in Warren County.

Throughout 1860, drilling proceeded at an accelerating rate. By June, 1860, a journalist reported twenty-one wells along Oil Creek, five at Franklin and six at Tidioute. The combined daily output of these wells was 200 barrels.[43] In the summer, twelve drilling projects were under way at Franklin and many more at the new village of McClintockville, the first of many villages to spring up overnight as the result of drilling activities. Strikes by E. Evans and others inspired one hundred drilling projects within one mile of Franklin by August, and around Tidioute borings numbered fifty.[44] A rough estimate in November, 1860, placed the number of producing wells in the Venango oil region at seventy-four and their daily output at 1,165 barrels.[45] A far greater number of dry holes pockmarked the area. Two months earlier an observer had estimated the number of drilling projects at various stages of completion at two thousand.[46]

The earliest crude struck along Oil Creek was comparatively light and gave a high percentage of illuminating oil when refined. A heavier-type oil from a well at Franklin in the spring of 1860 brought the realization that crudes might differ widely in characteristics. This well and the output of wells in nearby French Creek and Sugar Creek were eventually to acquire special value as lubricants.[47] But for most operators largely preoccupied with illuminating oil production, they had little immediate attraction.

Interest in illuminating oil also delayed any extensive utilization of the heavy types of crude discovered in the two most important producing areas that emerged outside Pennsylvania during 1860 and 1861. This was particularly true of the new field around Mecca in Trumbell County, Ohio, where in the spring of 1860 it was reported that two wells yielding 5 barrels and 15 barrels daily were in production and fifty other drilling operations were in progress.[48] Almost black in color and less odorous than ordinary crudes, Mecca crude was so thick it would not flow in cold weather. Frenzied drilling brought in numerous small wells and many trials of the crude as a lubricant, but without any immediate commercial success.

While western Virginia, the second field to be brought into production during this period, was also a pioneer source of lubricating crudes, its output was initially important as a competitor of the Pennsylvania fields in burning oil crudes. Development of the western Virginia area began as early as November, 1859, on the Hughes River in Wirt County, where in the spring of 1860 J. D. Karns brought in a well pumping 37–50 barrels daily. Located on the Kanawha River eight miles from Elizabeth, Wirt County's seat, this district, known as the Burning Springs or the Kanawha District, attracted swarms of buyers and drillers when some months later a well was struck which produced 200–250 barrels daily of a light-type crude, attractive to illuminating-oil refiners.[49] By the spring of 1861, production had soared to an estimated 800 barrels daily.

Drilling activity in western Virginia, however, was more than matched in the Pennsylvania Regions where a report in May, 1861, estimated a total of 135 wells with combined daily production of 1,298 barrels: 61 barrels along Oil Creek, 59 barrels along the Allegheny, 9 barrels at Franklin, and 6 barrels on French Creek. An August press report counted 800 wells along Oil Creek, of which seven were flowing wells.[50] According to one eyewitness account, "At Tidioute there are some 200 wells in progress, and all the way from here [Oil City] thirty miles by raft, one is not out of sight of derricks and wells, hundreds and hundreds of them." [51]

The spread and intensity of drilling activity during 1860 and early 1861 resulted in an expanding output that put an ever-growing pressure on transport and marketing facilities and the prices paid for crude at the wellheads. It is impossible, however, to determine just how much was produced or to give any more than an impression of the trend of prices. The most generally accepted guesses of output for 1860, for example, range between 200,000 barrels and 500,000 barrels, most of which came in the latter part of the year.[52] Scattered price data suggest that prices of crude that were reported between $18.50 and $20.00 a barrel in January, had dropped to between $12.50 and $12.00 in March, to between $9.00 and $8.00 by July, and by fall had dropped as low as $4.00.[53]

Data available for 1861 are scarcely any more satisfactory. Recognizing that rates of producing wells varied widely, the Internal Revenue Commission, nevertheless, estimated daily production in Venango County at 750 barrels in January, 1861, followed by an increase to 1,200 barrels or 1,500 barrels by spring.[54] According to the same source, the advent of flowing wells had by the fall quarter of 1861 pushed up daily production to 8,000 barrels, although other estimates put the amount at 5,000–6,000 barrels daily.[55] Despite their uncertainty, these estimates suggest a severalfold increase of output for 1861 over 1860 that may well have carried production in 1861 to a commonly accepted figure in excess of 2 million barrels (2,113,609 barrels was the exact figure) plus a large wastage because of a lack of transport and storage facilities.[56]

Whatever the margin of error in this estimate, volume was sufficient to bring a further pressure on prices despite a temporary revival early in 1861 that brought some producers as much as $10.00 a barrel. But in March and April most prices ranged between $6.00 and $4.00, while during "the summer of 1861," $2.00 was apparently close to the ruling price.[57]

By the summer of 1861 it was questionable how much longer the new mining industry could survive on a pumping-oil basis in the face of declining prices. A combination of a high rate of dry holes, low average yields of producing wells, and $2.00 oil could well have marked the end of the boom sparked by Drake's success less than two years earlier. There was yet another uncertainty about the future of the petroleum industry that was not resolved until late in 1861. This arose from the glaring contrast between an impressive growth of production on the one hand and the frustrating difficulties of gaining markets and moving crude to them on the other.

Shipping difficulties and early petroleum refining

Coal-oil refineries, concentrated at large eastern metropolitan centers of burning-oil consumption, offered the most logical outlet for petroleum, indeed the only one already built and in operation. This industry had created a legacy of markets, distribution, and technology vitally important to the future of petroleum. National markets for burning oils, a distribution system, a lamp industry engaged in mass production of burners suited either to refined coal oil or refined petroleum, and a technology for refining hydrocarbon oils from mineral sources formed major constructive elements in that heritage. A minority of coal-oil refiners also had elaborated markets and a basic technology for by-products: lubricants, both for rough and finer duties; paraffin for candles and other uses; naphtha as a solvent in cleaning; lighter fractions like kersolene for gas-making and anaesthetics; and a projection of crude oil as a substitute for coal in marine engines.

This rich legacy contained some less-constructive elements as well. Among small coal-oil refiners were many who restricted their output to illuminating oils. They followed a variety of distilling and treating methods, chiefly distinguished by a random and imprecise character and low-quality illuminants. As prototypes for petroleum refining, they comprised a Pandora's box.

To gain access to coal-oil refining capacity and markets, and to lay claims on the coal-oil legacy, petroleum had to meet several requirements. For one, transportation and handling facilities had to be found to get petroleum continuously to markets in quantities needed and at delivered prices that would be competitive with coal at coal-oil refineries. Second, the refining techniques of the coal-oil industry had to be adapted to petroleum. Difficulties in meeting these requirements during 1860 and much of the following year prevented more than a small portion of the petroleum produced from reaching the large consuming markets. It encountered not only formidable barriers of costly and highly uncertain transportation facilities, but equally great uncertainties in men's minds at the centers of the oil trade whether petroleum production was more than a fleeting and geographically isolated phenomenon for all of its gains.

Data from the receiver's report on the New York Kerosene Company's bankruptcy in 1860 offered a rough gauge of prices that petroleum would have to meet to bring about conversion to petroleum on a significant scale. Cost figures based on extended runs of two hundred tons of Albert coal that yielded about 60 gallons of illuminating oil per ton indicated total processing costs per ton of approximately $11.35, or 18.6¢ per gallon. With coal costing between $10.00 and $15.00 a ton, the final cost range (raw materials plus processing) varied between 35.5¢ and 43.5¢ a gallon.*

The price at which petroleum would become competitive with coal as a source of illuminating oil depended largely upon yields. In actual practice at this time yields of illuminating oil of reasonably good quality from crude petroleum did not exceed 50 per cent, and for several years yields of 60 to 65 per cent were considered exceptional.[59] Assuming a yield of 50 per cent, 3 barrels of crude (40 gallons each) would be required to refine 60 gallons of petroleum kerosene. With crude petroleum priced at $7.60 a barrel (19¢ a gallon) the raw material costs of kerosene would have been 38¢ a gallon. Adding the usual processing costs of 5¢–6¢ a gallon, the total cost would have been about 43.5¢, comparable to the higher figure reported for the New York Kerosene Company.[60]

With yields of 75 per cent, which some proponents claimed early in the 1860's and by mid-decade became widely realized, only 2 barrels of crude (80 gallons) would be required to make 60 gallons of illuminating oil. At this rate of conversion, crude petroleum became competitive with coal (at $15.00 per ton) at a price of about 28.5¢ a gallon or $11.40 a barrel.

In reality, of course, these cost relationships were subject to a number of variables, such as differences in refining practices, the price and the quality of coal utilized, and variations in the cost of processing petroleum. During the years of gradual conversion, however, a figure of $12.00 a barrel (30¢ a gallon) for petroleum crude at the refinery came to be accepted by coal-oil refiners as the rough yardstick of whether to process petroleum or coal.[61]

---

* In this accounting, all processing and raw material costs were prorated exclusively to illuminating oil; none to lubricants, paraffin, naphthas, and other by-products. No charges were made for depreciation or interest on investment.[58]

There are abundant indications that during the first two years of petroleum production, transportation and handling charges alone usually
amounted to about two-thirds or more of the $12.00-per-barrel maxi-
mum delivered price that would make it an attractive substitute for crude
coal oil. More than half of these costs often accrued before crude oil could
leave the Region; they comprised expenses of collecting, barreling, storing, and teaming twenty to forty miles to railhead or wharf. The vagaries
of time in transit, whether a week or months, were almost infinite.

The irregular, episodic, and undependable character of transportation
restricted marketing no less than did excessive cost. All local transportation, whether overland or by water, was as seasonal and dependent on
weather as farm crops. During three winter months at the peak of the
burning oil season, the Allegheny River to Pittsburgh remained closed to
navigation by steamer, tug, barge, or raft. On Oil Creek water reached a
high enough stage to borrow the local lumbering industry's technique of
rafting and flatboating barreled oil less than six months a year.[62] An early
innovation of decking over the flimsy crafts reduced, but by no means
eliminated, appalling losses from swamped barrels or completely cap-
sized cargos. Organization of pond freshets, permitting some barging of
barrels during low-water periods, came only in the fall of 1861, two years
after the beginnings of oil production; likewise deferred was the organization of bulk shipments by barge on a significant scale.

In or out of navigation season on the creek and river, shippers had to
place their primary reliance on hauling by team to railheads or wharfs,
but teaming proved more weather-bound than boating. In winter, wagons
loaded with 5–7 barrels of oil weighing 360 pounds each slithered helplessly on the icy or snow-packed surfaces of the roads. During the extended fall and spring rains, the going was even worse. Wagons mired to
their hubs in muddy ooze slicked with oil leakage in which the straining
two-horse teams would slip and fall, frequently smothering to death.[63]

By mid-1860, the price to reach Union, the nearest of all railheads to
Titusville, had been advanced to $1.25 per barrel.[64] To Irvine, Corry,
and other more distant depots on the Philadelphia & Erie, teamsters
roughly scaled their charges higher in accordance with time, distance,
and very frequently, with what the traffic would bear. These charges during the early years usually ranged between $2.00 and $3.00, but in 1861
there were many instances of $5.00 fees to haul oil from the Tarr and
Blood farms on Oil Creek thirty-one miles northward to Corry, the most
used of all railheads.[65]

Early in the history of oil production, barrel costs pursued the same inflationary course as drayage charges, and for much the same reasons. Differences in quality of wood and staves supplied added variables. Only a few pumping wells in March, 1860, sufficed to hike the previous fall's prices of 50¢ to $1.60.[66] In September, when estimated daily production had risen to 1,250 barrels, an anonymous visitor to the Region reported: "Barrels, barrels, are the great want now, and much is lost daily by the scarcity of this article . . . The barrels are sold at $2.00 apiece, and there is already a demand for a thousand a day." [67] For several years thereafter, $2.00 remained a floor price for barrels of the poorest quality, and prices commonly ranged up to $3.00, and sometimes higher.[68]

Prices of barrels not only affected the cost of packaging oil for transit, but during the first two years of oil production they were also widely utilized to store oil at the wells. Their only supplement were a few square or oblong wooden vats of the type constructed by F. S. Tarbell at Rouseville in 1860. Ranging from four to several hundred barrels in capacity, box vats began slowly giving way in the following year to round wooden tanks, hooped with iron, that at first held only 6–12 barrels of oil.[69] Until litigation in 1862 began to bring a change in the practice, the insistence of landowners that their royalty payments include barrels aggravated the barrel crises for operators.[70]

Like most capitalists, railroads remained unimpressed by the potential in oil traffic and supplied only open flatcars upon which the barrels were shipped with no protective covering of any sort. This excessive exposure invited heavy losses in seepage from the unlined barrels and added to fire hazards, for which the railroads assumed no responsibility.[71] Because producers did not pump water out of the wells above the seed bag, few barrels arrived at their destinations without a thick layer of water in their bottom.[72]

It is impossible to give even approximate figures for the transport and handling costs of moving oil from the Region to the seaboard markets during 1860. The best guess would be that at times charges may have run as much as $11.00 a barrel or more. An estimate by the *Scientific American* early in 1862 of $8.00 a barrel gives a rough indication of costs of delivering oil to New York City that were apparently typical of average charges during 1861.[73] Shipments via the Allegheny River to Pittsburgh or to intermediate points on rail lines, such as Erie, Cleveland, or Philadelphia, would have been somewhat lower. But these differentials were moderate at best, since the charges for the use of water or rail facilities comprised only a small part of total costs.

The undependable character of transportation facilities inevitably created a wide gap between the impressive record in production and the volume that was marketed. There are abundant indications that only a fraction of the estimated production for 1860 of 200,000–500,000 barrels was ever shipped and those shipments suffered severe shrinkage en route. According to the most liberal estimates, crude receipts at Pittsburgh totaled only 17,161 barrels.[74] The Sunbury & Erie Railroad (reorganized as the Philadelphia & Erie in 1861), the sole rail outlet from the Regions, carried 22,119 barrels, most of it destined for New York City.[75] These shipments of just under 40,000 barrels of crude oil were the only important ones to major markets; production in western Virginia was still marginal, and the area worse transport-locked than Venango County.*

At most, this volume of shipments, coupled with high prices of crude in the Region, and transport and handling charges that frequently approached the delivered price of coal, had little impact on the coal-oil industry. Following a large expansion in 1859, coal-oil operators went on to achieve an all-time record of output during 1860. They began the year producing at an estimated rate of 8 million gallons (200,000 barrels) annually.[77] By the final quarter, output had reached an estimated rate of 10.5 million gallons (262,500 barrels) annually.[78] Even at the end of the third quarter of 1860, when the Cincinnati markets had yet to receive their first gallon of refined petroleum illuminants, two coal-oil refineries in the vicinity were operating at peak capacities, retailing burning oils at Cincinnati and points westward at prices ranging from 60¢–75¢ a gallon.[79] Not until September, 1860, did enough crude petroleum filter through sporadically to permit the first quotations of refined in New York markets. They began at 70¢–75¢ a gallon and, following the general range of coal-oil prices, fluctuated between 60¢ and 80¢ during the rest of the year.[80] In December, at the crest of the burning-oil season, inadequate transportation took its toll; there was not a drop of refined petroleum to be had in answer to inquiries.[81] On the informed testimony of Samuel Downer, major refiners in the New York area had not begun to "think petroleum as a competitor to Albert [coal]" in the fall of the year.[82]

At Pittsburgh, the metropolis most accessible to petroleum because of the Allegheny River waterway, the end of 1860 was much the same story. The Alladin, Lucesco, and North American plants, three of the largest in the coal-oil industry, were running at full capacity, as well as their numerous satellites, and demand was steadily increasing. They found profitable

---

* For these reasons, it would appear that estimates by the Philadelphia Board of Trade that between 50,000 barrels and 60,000 barrels were marketed during 1860 probably overstated the actual amounts shipped by a substantial margin.[76]

markets for crude coal oil at 26¢ a gallon wholesale, and for refined at 60¢ wholesale and 80¢ retail, in competition with refined petroleum at $1.04 retail.[83]

According to a survey made by Abraham Gesner, at the end of 1860, there were some fifty-six plants engaged in coal-oil production, with possibly a dozen or more small distillers of crude coal oil unreported.[84] But Gesner also noted the operation of fifteen or more small plants that had mushroomed during the year and were working on petroleum exclusively. Their locations were significantly concentrated at points where petroleum did not have far to travel: in the Oil Region, at rail junctions like Erie on the Region's perimeter; and at Pittsburgh, 135 miles down the Allegheny River from Oil Creek.[85]

Fifteen small distilleries formed an inadequate outlet for petroleum. Not only did their capacities range from a few barrels to 20 barrels daily, but refining and treating were in highly experimental stages. Treating methods rivalled the variety and ineptness of those of numerous inexperienced operators who had entered coal-oil refining on a small scale during 1859. Yet in this nucleus of small refiners, Gesner saw a promising future for petroleum illuminants, even though "Some time may elapse before their manufacture is brought to perfection, and the distilled hydrocarbon oils attain that commercial and economic value they are destined to reach." [86]

The portion of current output being moved from the Region before the advent of the flowing wells in the fall of 1861 put an added burden on transport and shipping facilities is impossible to determine. Year-end data, however, indicate that while total shipments for 1861 rose sharply compared to 1860, production continued by a substantial margin to outrun the volume going to markets. Aided by the organization of the pond freshets on Oil Creek in the fall, just over 94,000 barrels moved to Pittsburgh before the onset of winter closed the Allegheny River to navigation.[87] Another 135,000 barrels went by way of the Sunbury & Erie, most of it to New York but with some destined for Erie and Cleveland.[88] With the completion of an extension of its tracks from Salamanca, New York, to Corry, twenty-seven miles north of Titusville, the Atlantic & Great Western Railroad gradually organized its facilities sufficiently to carry 69,509 barrels by the year's end.[89] Before the opening of Civil War in April, 1861, Boston refiners and New York buyers were purchasing an indeterminate amount of crude oil from western Virginia.

These data covering the principal transportation routes give a rough indication of total shipments that could hardly have exceeded between

15–20 per cent of the estimated production of 2 million or more barrels of crude for 1861. It is obvious that there were formidable obstacles still to be overcome before transport and handling facilities could be brought into line with production.

At the same time, shipments ranging between 300,000 barrels and 400,000 barrels (even though a large portion came in the fall of 1861) and prices at the wellhead, often cheap enough to overcome transportation charges several times greater, set the stage for the emergence of a petroleum-refining industry in which the substance exceeded the shadow. These circumstances were sufficient to bring about a gradual conversion from coal to petroleum by large plants that required continuous operation to be run economically. They were also a primary factor in luring fresh entries into petroleum refining from outside the coal-oil industry.

Samuel Downer probably was the first major coal-oil refiner to lay plans for a possible conversion. Visiting the Oil Regions in the summer of 1860, he decided by October that he should prepare to convert his Boston plant at least partially if the gathering proportions of petroleum production continued.[90] Early in the next year he began purchasing crude in the new fields in West Virginia, which his superintendent Joshua Merrill soon began to utilize in the production of commercial mixtures of distilled petroleum and refined coal oil. With the exception of small coal-oil refiners specializing in lubricants, paraffin candles, and other paraffin by-products, who continued to work coal exclusively into the second half of the decade, nearly all refiners located near major markets in the northeast began to use appreciable amounts of petroleum during 1861. These refiners included the New York Kerosene Oil Company, which by the end of 1861 was working petroleum along with its Albert coal.[91]

By contrast, conversion to petroleum by large coal-oil refiners was slow at Pittsburgh, where the Lucesco, Alladin, and North American coal-oil works were still operating exclusively on coal at the end of 1861.[92] Large capital investments in coal-retorting equipment and long-range contracts for coal that were difficult to break or modify undoubtedly contributed to their tardiness in converting at a location where crude-petroleum supplies were the most reliable. There also remained many markets remote from the Oil Region where neither crude nor refined petroleum had yet made their entries. Limited quantities of the new commodity were introduced in Chicago, for example, only in 1863.[93]

This gradual and uneven conversion by coal-oil operators did much to widen the market for crude, which was moving from the Region in increasing volume during 1861. By the summer of that year, however, it would appear that a considerable portion of the oil shipped went to a growing number of new entries into petroleum refining. No doubt many of these were impressed by the propaganda that during 1860 emphasized the cheapness and ease of successful entry into refining. It was said that for only $200, you could profitably erect a still to refine 5 barrels a day of good burning oil that would cost only 5¢ or 6¢ a gallon to process.[94] Expenditure of $1,500 would give you a refinery "which will do a very efficient business," presumably 20 barrels or 30 barrels a day; for $4,000, you could obtain a unit producing more oil daily than the richest well, probably 75–100 barrels. A few hands could run these units, and fuel costs were nominal compared with coal-oil establishments costing $57,000 or $100,000, it was contended. Circulars elucidating this "rationale" were available for the price of a postage stamp.[95]

The broadcasting of these notions did much to encourage location of small units at points accessible to crude or at places where anyone thought a local market might exist. The impact was most readily observed near wells in the Region, at railheads like Erie, and at Pittsburgh where it was estimated that between thirty-five and fifty refineries were in operation by the end of 1861.[96] It was also felt at almost all metropolitan centers where supplies of crude if not of customers were considerably in doubt. In Cleveland, Boston, New York, and undoubtedly many other eastern locations, small stills were set up expressly to process petroleum.[97]

The year 1861 also brought some new elements to the character of new capacity erected expressly to process petroleum. During the previous year, limitations of time and resources had restricted most ventures to small stills of the alcohol type with capacities of 5 barrels or 10 barrels.[98] In January, William Abbott completed a $15,000 refinery at Titusville, comprising six stills and bleachers, the earliest multiple-still unit in the Region.[99] At Pittsburgh, William Frew and Charles Lockhart completed their plant begun in 1860, which remained the largest in the area until the three major coal-oil refiners shifted to petroleum during the following year.[100] At Erie, the partnership of Walker, Rust & Clarke built a plant with a capacity of 100 barrels a day,[101] and Charles Holmes erected a refinery of comparable size for the Erie Carbon Oil Company.

Flowing wells

Despite new petroleum-refining capacity and the gradual conversion by coal-oil refineries to the new material, there was no assurance until late in 1861 that the supplies of petroleum would ever be sufficient to

COURTESY DRAKE WELL MUSEUM

Empire well, Funkville, 1863.

challenge coal oil's position in the illuminating-oil markets. Most of the shallow wells pumping along Oil Creek and other parts of the Region were modest producers, while the high proportion of dry holes suggested that the limits of producing territory had been reached.

The first indication that drillers literally had only scratched the surface of the oil reserves of the Region came in April, 1861, with the advent of the first flowing well or gusher in the area. Located near the mouth of Oil Creek, the well had originally been brought in during 1860 at a depth of 150 feet. When pumping brought only moderate yields, Henry Rouse, one of the lessees of the property, undertook to improve output by drilling deeper.

Just at sunset on April 17, the drill simultaneously struck a huge oil vein and a large gas pocket at a depth of 300 feet. The intense gas pressure spouted the oil 60 feet into the air at an astounding rate of 3,000 barrels a day. While Rouse, his associates, and some spectators tried desperately to control the unprecedented deluge, gas evidently ignited when it came in contact with the boiler of a pumping well some 10 rods away. A cannon-like explosion set off a fire that fatally burned Rouse and eighteen others and blazed with terrifying magnificence for three days before it was finally smothered with manure and earth.[102]

Coming as it did less than a week after the firing on Fort Sumpter, the significance of the Rouse well was somewhat obscured by the panic which swept the country for some months following the declaration of war. Moreover, it took further demonstrations to convince most persons in the Region and elsewhere that flowing wells could be struck with any frequency or that their tremendous output could be sustained for any significant period of time.

Some of this skepticism disappeared when the Fountain well, located on the McElhenny farm some seven miles to the north on Oil Creek, was also brought in as a flowing well in May, 1861. Probably the first well drilled with the express purpose of probing the "third oil-bearing sand," the Fountain well struck oil at 460 feet.[103]

Following the Fountain well, the third sand of the "lower McElhenny farm" developed into one of the most prolific flowing-well localities on Oil Creek. In September, 1861, the mighty Empire well roared in from third sand with an initial flow of 2,500 barrels, which continued for eight months with only a gradual decline to 1,200 barrels before an abrupt termination in May, 1862. A succession of other flowing wells kept the daily output of this single farm at 5,000 barrels or 6,000 barrels from the fall of 1861 through the spring of 1863.[104]

Even the Empire well was eclipsed within a few weeks by the success of Phillips No. 2, located on the nearby Tarr farm, which started with a record-breaking flow of 3,000–4,000 barrels daily sometime in October and continued for a year at a rate of between 2,500 barrels and 3,000 barrels. The N. S. Woodward well and Elephant No. 2, both large-flowing producers brought in before the end of the year, established the Tarr farm as a second great flowing-well district along Oil Creek. In the next year several more farms entered the same ranks, notably the McClintock, Buchanan, and Blood farms.[105]

One immediate effect of the spectacular output from flowing wells was to drive previous pumping wells out of production.[106] Yet the elimination of these early pumpers, producing mostly from shallow, comparatively unproductive sands on a modest scale, did not mean the end of the pumping well as a permanent institution. Flowing wells continued as the more colorful aspect of production, and few outsiders thought of the petroleum industry except in terms of gushers and sudden wealth.

But flowing wells would seldom again account for such a large proportion of total output as they did during late 1861 and early 1862. For one thing, only a relatively small proportion of future wells were brought in as gushers.* Secondly, flowing wells typically followed a pattern, already apparent by June, 1862. However substantial the initial output, once the pressure that lifted crude naturally to the surface was dissipated, operations could only be continued on a pumping basis. The final factor which made pumping a permanent and significant institution in the industry was the fact that once brought into production, wells were continued in operation so long as output could be sold at prices covering out-of-pocket costs. In this respect the range of prices and demand conditions under which pumping wells could still be operated profitably was quite substantial.

Whatever their future role, the initial impact of gushers on the petroleum industry was unmistakable. Output from the flowing wells during the last quarter of 1861 contributed a substantial portion of the 2 million or more barrels of estimated production for the year. Existing transport and handling facilities were swamped. As a result, prices which had averaged about $2.00 a barrel at wellheads during the summer dropped to 10¢ a barrel, where they remained during the remainder of 1861.[108]

But even more important, this wide gap between prospective levels of crude production and existing transport and refining facilities brought a significant change in the attitude of businessmen and capitalists generally toward the future of the industry. It marked the beginning of a decade when many of the features emerged which were to characterize the industry throughout the nineteenth century and beyond. This was particularly true with respect to methods of exploration and drilling, the beginnings of bulk transport, the evolution of refining techniques, and a vigorous promotion of markets both at home and abroad. These developments combined to make petroleum an outstanding American industry by the early 1870's.

---

* It should be noted that not all flowing wells were large producers, in some instances yielding smaller outputs than pumping wells struck in the same fields. For instance, of the four famous Harmonial wells with which Abram James opened up the Pleasantville field in 1868, the first was a small gusher, flowing 100 barrels daily, but the other three as pumpers produced at identical rates.[107]

Part FOUR

The formative years:
1862–1873

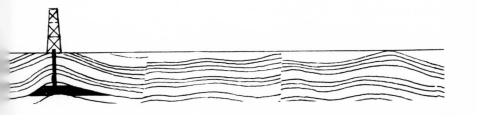

Chapter 6

Output of crude and the
spread of production

A general impression of the output of crude, which formed the base for the expansion of the industry over the period of 1862–73, may be gained from Chart 6:1. Although inadequate for some types of analysis, the charted data adequately reflect major trends.* With minor deviations, annual output of crude advanced gradually from about 3 million barrels in 1862 to over 5 million in 1870 and 1871. In 1873 it almost doubled to nearly 10 million barrels, a level nearly three times as great as the peak outputs in the years through 1868.

---

* Data on output and prices of crude oil throughout the 1860's have many limitations for purposes other than indicating general trends. There was no systematic gauging and reporting of the output of wells. Prices were those of only a fraction of the transactions at points along Oil Creek, and frequently deviated widely from what producers actually received at the wellhead. Barrels ranged from 40 to 50 gallons in size despite nominal standards of 40 gallons until 1866.[1]

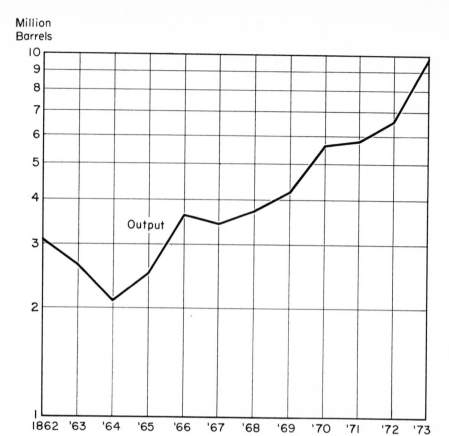

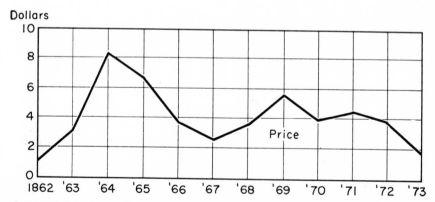

CHART 6:1. *Annual output of crude and average annual price per barrel in the Region: 1862–73.* (*Source:* Derrick's Hand-Book, *I, 711, 805.*)

Price behavior presented a sharp contrast to the gradual increase in output. Crude prices in the Region fluctuated widely, a fact suggesting that they were subject to many strong influences besides variations in production. Throughout the industry's formative period, however, two major trends stand out plainly: a gradual increase in production accompanied by wide variations in the prices of raw material.

Extension of producing territory likewise proceeded gradually during this period, and with a few minor exceptions was confined to western Pennsylvania.* Until 1864 the major producing territory not only fell in Venango County almost entirely, but in territory demarcated in the first two years after the Drake well.

The opening of a producing district in the spring of 1864 several miles up Cherry Run, an affluent of Oil Creek, marked the first important departure from pioneer territory in Venango County and from flatlands along streams. The Pithole district, (as shown in Map 6:1), which was entirely removed from the original producing territory, roared in with a production in 1865 that threatened Oil Creek's supremacy. Late in 1865 the opening of Pioneer and Bennehoff Runs, with important strikes on hilly terrain, definitely exploded notions that paying territory was confined to flats bordering streams.[3]

Triumph Hill at Tidioute in 1867 and the Shamburg and Pleasantville developments in 1868 strengthened the challenge to Oil Creek. In 1869 the tide of exploration and development began to move southward into Clarion, Armstrong, and Butler counties to create a new producing region known as the Lower Oil Region, that by 1873 had taken a decisive leadership in production over Venango County and the rest of the Upper Region[4] (see Map 6:2).

The extension of producing territory had another dimension. Operators gradually increased the depths to which they drilled. Until the fall of 1861, depths of 200 feet usually sufficed to probe the first and second oil-bearing sands along Oil Creek, though shafts of 500 feet were not uncommon.[5] Flowing wells then stimulated drilling to oil-bearing third sands, which along Oil Creek ranged between 400 feet and 500 feet.[6] At Pithole in 1865 the productive third sands required drilling to more than 600 feet,

* Early in the decade small amounts of production were reported at numerous locations outside western Pennsylvania. Those of greatest interest were in Ohio, around Mecca, in Kentucky, Tennessee, and California, in western Virginia (soon to become West Virginia), and in Enneskillen Township, Canada. Of these, only production in West Virginia, Ohio, and Canada assumed any commercial importance.[2]

and at Triumph Hill in 1867 they were struck at about 900 feet.[7] Wells at Pleasantville, Parker's Landing, and elsewhere in 1868 exceeded 800 feet and the prevalence of drilling on hills often added further to their depths.[8] In March, 1873, the Medoc district in Butler County was opened up with a third-sand flowing well at 1440 feet, and in the following November the first recognized fourth-sand flowing wells were brought in near Karns City at approximately the same depth.[9]

Boom and depression: 1862–66

With oil selling at 10¢ a barrel at the wells, and with the flowing wells showing no immediate signs of abating, prospects for profitable operations in the production of crude appeared remote during the early part of 1862. Total production by the end of the year, as already noted, reached an estimated figure of some 3 million barrels, a record not equaled or exceeded until 1866 (see Chart 6:1). But despite continued drilling that brought a number of new flowing well farms into production, there was an erratic but noticeable downward trend in the daily rates of output over the intervening months.[10] In part this trend reflected a drop in the output of some of the early gushers. But low prices also played a role in halting operations by many small pumpers and the stoppage of some flowing wells whose product could find no market.

In conjunction with growing domestic and foreign demand, the decline in output gave operators their most favorable outlook in more than a year as prices, after reaching a dollar a barrel in June, firmed to levels variously and somewhat contradictorily reported along Oil Creek at $2.50, $4.00, $5.50 and $6.00.[11] Although flowing wells still dominated output, the improvement in prices brought a number of the better pumping wells back into production.[12]

The following year brought further encouragement. Although some new flowing wells were brought into production, total output for 1863 fell to almost 2.6 million barrels as compared with 3 million barrels for 1862. To a limited extent this decline reflected a temporary drop in prices early in the year, which prompted some producers to reduce output.* The chief reason for the decline in output, however, was the exhaustion of earlier flowing wells that new gushers and a return of pumping wells

---

* During the early summer there was a short suspension of production when Robert E. Lee's army invaded Pennsylvania. But a week before the battle of Gettysburg on July 2, producers had returned to business as usual.[13]

did not fully offset.[14] An improvement in prices during the rest of the year was sufficient to bring the average for 1863 to about $3.50 a barrel, some three times the figure for 1862.[15]

The gradual improvement in outlook for producers during 1863 set the stage for a speculative boom which became a classic in petroleum history. Total production during 1864 dropped to a little more than 2 million barrels. With improved transport facilities, a growing foreign and domestic demand, and a wartime inflation that reached its peak during late 1864, the drop in output brought prices to levels not reached since 1860. Starting early in the year at $3.00 to $4.00 a barrel at the wellhead, quotations moved to a peak of $13.75 in July and closed for the year at levels ranging from $10.50 to $12.00.[16] At these prices many of the small pumping wells were brought back into production, drilling activity was stimulated in older territories, and "wildcatting" explorations were extended into new areas.

Much of the mounting enthusiasm for the development of production came in response to the examples of flowing-well owners who, during late 1863 and 1864, acquired great personal fortunes almost overnight. J. W. Sherman and his associates, for example, who brought in a well flowing 1,500 barrels a day in 1862, realized an estimated $1.7 million from their operations over the next two years.[17] By the mid-1860's Dr. M. C. Egbert, his brother A. C. Egbert, and their partner Charles Hyde, were reputed to have shared profits ranging between $8 million and $10 million from a 39-acre farm they had leased along Oil Creek in 1859.[18] Sale of a nearby farm in April, 1864, for $650,000 in greenback currency remained the largest amount of cash ever paid for oil land only until November, when a farm at Oil City sold for $750,000. A few months later one owner refused a cash offer of $800,000 for his farm.[19]

The record of a number of the early joint stock companies organized by local operators and Pittsburgh capitalists in order to obtain significant blocks of land and to distribute risks were equally impressive. The Columbia Oil Company, organized by purchasers of a 500-acre farm in May, 1861, with a capital of $200,000, paid dividends of $300,000 during the second half of 1863. In July, 1864, the Columbia Company paid out another $100,000 to its stockholders, while in August the Noble & Delameter Petroleum Company announced its fourth monthly dividend of 10 per cent. In the early part of the next year, the latter company sold its property for $300,000.[20]

The opening of Cherry Run by William Reed during the summer of

1864 triggered a wave of speculation that gripped the Oil Regions over the following two years. Drilling along the stream that flowed into Oil Creek about three miles north of Oil City, in an area generally considered most unpromising, Reed brought in a well on July 28 that flowed at a rate of 300 barrels a day. Within ninety days he and his two partners realized $785,000 from the sale of oil and leases.[21]

By late August, all land along Cherry Run and contiguous territory had been taken up for several miles; three hundred derricks lined the banks and hills; and output had mounted to 1,000 barrels a day.[22] The rush for drilling sites along the stream soon brought landowners bonuses of as high as $4,000 plus one-half of the oil produced in return for single-acre leases.

Cherry Run was merely the prelude to a speculative boom that reached classic proportions with the development of Pithole. By 1864, successes along the Allegheny stimulated some random leasing and purchasing a short distance up Pithole Creek, which ran roughly parallel with Oil Creek and emptied into the Allegheny River some eight miles east of Oil City. But not even tentative entry had been made into the adjacent densely wooded terrain, until I. G. Frazier, in the spring of 1864, following the advice of a professional "twig twirler," leased a portion of the Thomas Holmden farm, located seven miles up Pithole Creek. Early the following January, at a depth of 600 feet, he struck oil that came in with a daily flow of 650 barrels. Less than two weeks later, the twin wells of Kilgore and Keenan on the same farm roared in with 800 barrels daily.[23]

Like many events at Pithole, the reaction did not follow ordinary expectations. With oil priced $8.00 a barrel, news of the strike enormously inflated the price of stock in Frazier's company. Yet Frazier and others encountered frustration: an exceptionally severe winter, followed by disastrous spring floods, so curbed wagoning that Frazier could execute only two sales transactions. Early in May, developments remained virtually at a standstill.

At this point, Thomas Duncan and George Prather, who had purchased the Holmden farm for $25,000 subject to Frazier's lease, began to lay out future Pithole City on a hillside. On May 24, 1865, the town plan was completed and lots first offered for sale or lease. Within six days nearly two entire streets had been built up, and many buildings were under construction to accommodate the human avalanche that was to descend on Pithole City.[24] Included in those who were attracted to Pithole were hundreds of Civil War veterans who had not been assimilated into peacetime activities. They were joined by scores of nonveterans drawn from almost every walk of life.

Before the end of June, four gushers had boosted Pithole's daily production to more than 2,000 barrels daily.[25] A series of additional "spouters" throughout the summer brought output to a September zenith of 6,000 barrels daily, two-thirds of the Region's total output.[26] Pithole's production for the year, estimated at 900,000 barrels, contributed more than a third of the total output for 1865 (approximately 2.5 million barrels).[27]

The arrival of some 3,000 teamsters in June gave an added impetus to Pithole City's phenomenal growth. A month later a new post office was handling 5,000 letters a day, a volume exceeded in the state only by Pittsburgh and Philadelphia.[28] By September, Pithole reached its peak population of at least 15,000, and its community services included a daily newspaper, a waterworks system, and a fire company. Of churches, banks, telegraph offices, it had two each. Every other shop was a saloon; but there were scores of boarding houses, grocery stores, machine shops, and other business establishments, as well as dance halls and brothels. Of its hotels, the Morey, Chase, and Bonta offered the luxuries and conveniences of first-rank establishments.*

The appalling conditions of sanitation and the low moral standards inspired the *Nation's* correspondent to christen the brawling metropolis, with its daily quota of garroting and robberies, "the sewer city." He also termed it "a gigantic city of shred and patches" because of the nightmarish architecture of many of its improvised buildings.[30]

Pithole's speculative excesses defy adequate description. Hundreds of stock salesmen and promoters indiscriminately hawked stocks in legitimate and fly-by-night companies in streets 3 feet deep in mud and crowded with greasy, oil-smeared men who resembled "the lower class of scarecrows under the influence of a galvanic battery." [31] At first, leases averaged $2,000 an acre plus royalties of one-fourth to three-fourths of the oil which was soon reduced to one-half acre and increased in price to ranges of $4,000 to $10,000, with one-half the oil. Sixteenth and thirty-second interests in wells and leases sold for between $2,000 and $20,000, often without the purchaser inspecting the property. To give the "little man" a chance to risk his money, shares in wells were divided and subdi-

---

* When a new theatre seating 1,100 opened in September with a performance of Macbeth by a leading dramatic company, the leading lady experienced a unique curtain call in which the enthusiastic audience tossed $500 in coin and greenbacks at her feet.[29]

Street in Pithole City, 1865.

vided to as little as $\frac{1}{128}$ interest and sold at prices that frequently would have yielded no return even from a strike of 1,000 barrels daily.[32]

Pithole's army of sharpsters advanced the art of fraudulent practice in oil speculation to a stage that left few new tricks for the future. The doctoring of dry holes and abandoned wells with oil and sand to give them a productive appearance was commonplace. Crude oil piped in from concealed tanks sold dry holes—except when one careless promoter piped kerosene instead of crude oil. Operators sometimes sold seventeen or more shares of one-sixteenth.*

Struck by the speculative fever that had swept the Region, an agent of Jay Cooke & Company wrote from Titusville in March, 1865: "The people through this country can't talk connectedly about anything but oil. They tackle me with oil talk and I give them seven-thirty [government bonds], but they can't understand it. The bankers are all the same. No matter what you say it winds around to oil." [34]

---

* Women, too, sometimes participated in the "con man's art." One ingenious widow sold her farm near Pithole upon the basis of interesting surface indications artificially induced by dumping a barrel of petroleum in a spring. When the land unexpectedly proved productive, she sued the purchasers for taking advantage of a poor, guileless woman.[33]

Satire on speculative craze for oil stocks.

The mounting speculative fever quickly escaped the confines of the Region and created a nationwide "oil stock company epidemic" in 1864 and 1865.[35] Pennsylvania laws, permitting foreign corporations to operate in the state, were favorable to the swift spread of the contagion. By September, 1864, more than 100 stock companies had formed; by February, 1865, their numbers exceeded 500, with an aggregate capital of over $356 million.[36] With working men and all other classes catching the fever, stock subscriptions often were completed in less than a day.[37] To meet the demand for an organized exchange the Petroleum Board was formed in New York in the fall of 1864. In Washington, James A. Garfield, a future American president, and other members of Congress organized a company to speculate in oil land.[38] Jay Cooke, who judged nearly all oil stocks worthless, experienced the first severe test of his leadership in his banking house over the issue whether junior partners might lend their names to oil stock promotions.[39]

Extreme concentration of wells, leading to a rapid drop in gas pressures and a general disregard of flooding hazards, largely accounted for Pithole's fading almost as rapidly as it had flowered. The first signs of the field's impending decline came late in 1865, when six farms with 45 wells were still producing at a daily rate of 6,300 barrels, but some 341 others had been shut down in the face of a declining price for crude.[40] This was the beginning of a downward trend in Pithole's daily output, which slumped to 3,600 barrels in January, 1866, and to 2,000 barrels the following November. By the fall of 1867, production had practically stopped.[41]

The depopulation of Pithole City proceeded apace. Early in 1866 the exodus of transient speculators and operators began in earnest, soon followed by the departure of the several thousand teamsters whose services had been superseded by the introduction of pipelines as well as declining production.[42] Numerous major fires, causing a total property loss of $2.5 million, further promoted a drop in population to less than 3,000 by the end of the year. In June, 1868, the remaining citizenry moved almost en masse to the new Shamburg and Pleasantville fields.[43]

Because of the speculative excesses that marked the field's initial development, Pithole producers were particularly affected by the depression that hit the entire Oil Region in 1866. The first indication that the speculative boom in oil was about to collapse came early in 1865 when, for the first time since 1862, the export markets—largely because of the rise of the British shale oil industry—failed to increase.[44] When domestic demand during the closing months of the Civil War failed to take up the slack, prices of crude that had averaged between $10.00 and $12.00 a barrel during the last two months of 1864 began to taper off. They reached a low point of about $4.60 a barrel along Oil Creek in August, 1865, and closed the year at about $6.50.[45] To add to the difficulties of the producers, the federal government, beginning in May, 1865, imposed a tax of $1.00 a barrel on crude.

The gathering depression struck with full force in 1866. The signal for the total collapse of the speculative bubble occurred late in March, when Culver, Penn & Company of New York, operators of five banks in the Region and promoters of the Reno & Pithole Railroad, went bankrupt.[46] This brought about a wholesale collapse of the majority of the new oil companies. It was further reflected in a sharp decline in the total daily output of crude from the Region. Abandonment of an estimated 8,000 drilling projects plunged a large labor force into unemployment.[47] Even the repeal in May, 1866, of the federal tax of $1.00 a barrel on crude did not prevent crude prices—at times as low as $1.35 a barrel—from slumping to an average for the year of about $3.75 a barrel.[48]

The depression year of 1866 completed a cycle originating with the depression of 1862. Despite temporary stresses and strains, however, there were those who thought petroleum mining had achieved a degree of permanence, supported by improved transportation facilities and markets, that were only incipient some four years earlier. In announcing a change of its name to the *Mining and Manufacturing Journal* in June, 1866, the *Pittsburgh Oil News* expressed this feeling of a new permanence with the following comment:

> The oil business has settled down to an ordinary mining pursuit, and the interested public do not demand an organ exclusively devoted to petroleum. The speculation fever is past, the rapid sales and resales of oil lands, formation of oil companies, and stock speculations, have materially settled down to something substantial and ordinary, and there is neither so wide a field nor such an eager demand for minute and full details of everything relating to petroleum.[49]

Exploration and production in the upper region

Whether, as suggested by the *Mining and Manufacturing Journal*, the oil business had settled down to an "ordinary" mining pursuit, is open to question. The story of Pithole was to be repeated many times in the subsequent history of the industry. Events during late 1865 and 1866 indicated, however, a characteristic of the production side of the industry that was to become even more apparent in the years ahead. At a time when lowered prices curbed drilling activities and the abandonment of wells in established fields, exploration and development brought in two new producing districts in Venango and Warren counties. The combined output was more than sufficient to offset the rapid decline of Pithole, with the result that total output of crude for 1866 rose to almost 3.6 million barrels, breaking the previous all time record of 1862.

Farms in the hills west of Oil Creek along Bennehoff and Pioneer Runs, two small streams which flowed into Oil Creek just north of the village of Petroleum Center, provided the setting for the first development. It was here that the Ocean Oil Company of Philadelphia on September 1, 1865, brought in a gusher at a depth of 600 feet, with a daily flow of 300 barrels. Located on a hill side slightly north of Bennehoff Run at an elevation of 170 feet above the Oil Creek flats, the Ocean well marked the first successful attempt to find oil elsewhere than on flats adjacent to creeks.[50]

Experience in the area permanently removed all doubt about the exploration of high ground for oil when many of the heaviest-producing wells struck were at elevations 200–400 feet above Oil Creek. With some 150 drilling projects in operation, daily production along Bennehoff Run reached 2,100 barrels by early 1866.[51] Pioneer Run, which flowed into Oil Creek a short distance to the north, was a close competitor. There the striking of the Lady Brooks gusher early in April, 1866, launched a rush to the highland farms along the stream that soon converted Pioneer Run into a major producing district that was fused almost indistingishably with the Bennehoff Run district.[52]

Extension of the Oil Creek Railroad in July, 1866, to Boyd's farm near the twin producing districts spawned the small town Pioneer City and converted Petroleum Center over night into a large town of 3,000, with a bank, two churches, a theatre, a dozen grocery stores, and scores of offices of brokers, shippers, and producers. The timely arrival of proprietors of gambling and vice establishments who had fled waning Pithole also assured Petroleum Center a future as one of the roughest and most lawless communities in the Region.*

Occurring simultaneously with developments around Petroleum Center, hundreds of drills had been pounding industriously along the numerous tributaries of the Allegheny River in the Tidioute vicinity of Warren County. Bully Hollow, Grove Run, and Porcupine Creek entered the producing lists, but the valleys along West Hickory Creek and Dennis Run outranked them in importance.[54]

Early in January, 1866, only seven weeks after the first strike, there were seven producing wells and one hundred more being drilled along West Hickory Creek. This crude enjoyed an advantage over Pithole and Oil Creek crudes because of its good lubricating qualities and because low-cost wells hit oil at depths of only two or three hundred feet.[55] Commercial production on Dennis Run began in November, 1865, with strikes at depths of 325 feet; by late summer of 1866 its output had reached 2,700 barrels a day.[56] By August, 1867, however, Dennis Run wells were rapidly deteriorating, victims of being crowded three, four, and even five to the acre.[57]

The drive to extend producing territory became increasingly appar-

---

* Petroleum Center was the scene for the most famous robbery in the annals of the Regions. The victim, John Benninghoff, was a poor, illiterate German immigrant farmer who had realized a fortune of $500,000 or more from oil taken from the farm near Bennehoff Run which he had leased in 1866. Distrustful of banks, Benninghoff kept his wealth in a safe in his home. In January, 1868, his house was entered by several men, who, after forcing him and his wife to give them the combination to the safe, made off with $210,000 in greenbacks and government bonds.[53]

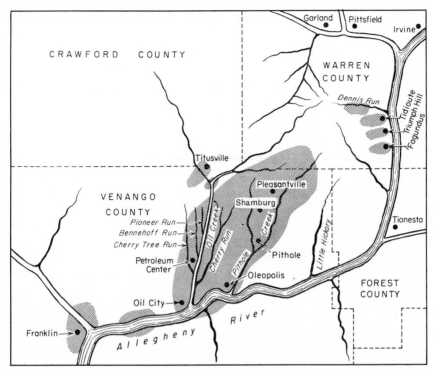

MAP 6:1. *Upper Region oil producing territories, 1860's.*

ent after 1866. Whether operating conditions were good or bad and whether the prospect was for high or low prices, nothing seemed to stop further exploration and development. Thus in 1867, total output was almost 3.4 million barrels—just below the record high of the preceding year, although crude prices averaging about $2.41 a barrel caused a wholesale abandonment of existing wells. As the production from the Bennehoff, Pioneer, and Dennis runs fell off, slack was largely taken up by new developments, chiefly at Triumph Hill near Tidioute. There, on farms located some 600 feet above the Allegheny River, strikes made in late 1866 brought Triumph Hill's output to an estimated 2,000 barrels a day by the summer of 1867. But like Dennis Run and Pithole, wells were clustered close together, and consequent flooding brought a sharp decline in production within a few months.*

---

* One significant exception, however, occurred in this repetitive pattern. By plugging up many of the wells, the Triumph Oil Company, which pioneered the field, succeeded in maintaining a large portion of the Hill as good producing property for five years, although it never again reached its output of the summer of 1867.[58]

The Shamburg and Pleasantville districts, only three miles apart at the headwaters of Cherry Run, crowned developments in 1868 and contributed significantly to a new total output record for the year of 1868 of over 3.6 million barrels. The Shamburg field was unique because it contained not a single dry hole. Timed with a moderate rise in prices that averaged about $3.63 a barrel for the year, operations were almost uniformly profitable at Shamburg. This fact probably encouraged the notion that petroleum mining had achieved a new plane of stability and scientific procedure. In ironic contrast to this notion, the Pleasantville field was opened up by a spiritualist, and unleashed a speculative tide exceeded only by Pithole.[59]

Before output at Shamburg began to decline in 1868, it lost the spotlight to Pleasantville, a sleepy farming hamlet in northeastern Venango County, six miles southeast of Titusville and five miles north of Pithole. The first well in the district was struck by Abram James, who claimed spiritual guidance in locating his Harmonial well Number 1, which started pumping at a rate of 130 barrels a day in February, 1868. James subsequently scored more than a dozen strikes with the aid of "spirit influences."

The ensuing rush to the new field patterned Pithole's experience. By summer of 1868, a pipeline built from Titusville in April was handling 3,000 barrels daily; by fall Pleasantville's population had expanded to 5,000.[60] As at Pithole earlier, land was sold or leased in progressively smaller parcels at soaring prices.

But the third-sand rock, found at a depth of around 900 feet in the Pleasantville field, was only 15–20 feet thick, and by 1870 production had declined to 1,500 barrels daily. Like Shamburg, the ratio of strikes to holes at Pleasantville was extremely high, and most operators made money, though in the district's later stages the life of overcrowded wells proved short.* The final curtain was drawn on Pleasantville as a producing district in December, 1871, when a fire destroyed three hotels and forty buildings in the once-peaceful little village.[62]

The year 1869 was not marked by the opening of any new fields comparable to Shamburg or Pleasantville. But further drilling in established areas, coupled with prices of crude that averaged about $5.60 a barrel for

---

* Visiting the scene in 1869, journalists Cone and Johns made a familiar observation: "So thickly have the wells been clustered together, that the whole supply of oil, or at least a greater portion, in the localities first developed, seems to have been exhausted by the pumps." [61]

the year, brought many pumping wells back into production. As a result, total output for the year rose to a new production record of just over 4.2 million barrels.

Prices averaged about $3.90 during the year as output again soared to a new record of over 5.2 million barrels during 1870. In part this expansion was the result of the last great renaissance of production along Oil Creek. Much of this activity occurred in the vicinity of Petroleum Center, where new flowing wells on one farm alone brought daily output to over 2,600 barrels by November, 1870. West Hickory, in the Tidioute district, also experienced an Indian summer of revived production that in December reached a high point of 3,657 barrels daily, followed by a rapid decline in the next year.[63]

The most active area in 1870, however, centered in the Fagundas field north of the Hickory district. There a syndicate purchased in April the Venture well, which opened up the district. Reflecting an increasing trend for large operators to purchase big blocks to permit more orderly development, the syndicate then bought five-sixths of John Fagundas's farm of 160 acres for $120,000 and a half interest in 90 acres owned by David Beatty for $90,000. They laid out Fagundas City, which within three or four months had a population of 2,500, centered in an area producing 3,000 barrels daily.[64] In 1871 production tailed off badly in both the Fagundas and neighboring Hickory districts, while a fire in May, wiping out 47 buildings in Fagundas City, signaled the end of another prominent producing area.[65]

Despite sharp declines in the West Hickory, Fagundas, and Petroleum Center districts, production in 1871 persisted above the level of 5 million barrels that brought producers an average price of $4.40 a barrel. Output from Cash-Up, in the Pithole vicinity, expanded sharply, but it was the developments in the Lower Region around Parker's Landing, Armstrong Run, and St. Petersburgh in Clarion County that brought a new chapter to production history.

Belt theories and the development of the lower region

Producers had first become interested in the Lower Region in 1868 when, in October of that year, a flowing well was struck near Parker's Landing, 52 miles south of Oil City on the west bank of the Allegheny (see Map 6:2). This strike was soon followed by activity at Brady's Bend, further south on the Allegheny, and at Foxburg, a few miles north, and at numerous intermediate points.[66] But it was not until Cyrus D. Angell began in 1871 to demonstrate the validity of his belt theory of locating oil wells that operators began to flock into the Lower Region.

Early in the decade, operators and journalists had referred frequently to the occurrence of petroleum in belts, but the origin of the concept remains obscure. Almost certainly the prevalence of closely spaced wells gave a belt-line appearance that suggested the idea to some. With many

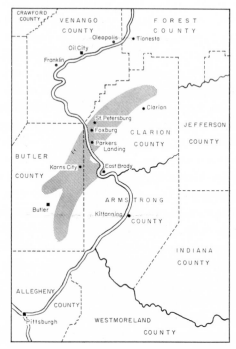

MAP 6:2. *Lower Region oil producing territories, 1869–73.*

the idea probably was a simple transference from coal terminology, in which reference to belts was frequent, especially in the coal fields of western Virginia, where visible outcrops in defined lines along valleys and mountains suggested belt lines to the naked eye.* Visiting the Oil Regions in the summer of 1868, J. R. G. Hazard reported a prevalent belief in an oil belt extending all the way from Titusville to Emleton in the still undeveloped Lower Region.[68]

---

* Visiting West Virginia's coal and oil region in 1865, a journalist, Edmund Morris, described both the coal and oil belts in the neighborhood of Burning Springs and extending to the vicinity of Marietta, Ohio, and he drew from them an analogy to a belt line that seemed apparent in western Pennsylvania between Franklin and McConnelsville.[67]

Although the theory of oil belts as developed and applied by Angell made its impact early in the 1870's, it was obviously much in the air in the preceding half decade and exerted a greater influence on locational methods than is generally appreciated. Leakage of news about Angell's own testing of his theories could not have accounted for all the earlier interest in the subject, which plainly derived from several sources; with or without Angell, the belt theory was evolving.

There is much to suggest that Angell was immediately influenced by his associate, Frederick Prentice. Forced by the Civil War to abandon extensive oil leases and land purchases in western Virginia, Prentice in 1861 returned to the Pennsylvania Oil Region with a clear concept that petroleum occurred in belts.[69] Assuming that the general direction of profitable development would be northeast and southwest, Prentice spent several thousand dollars over the next few years carefully surveying a line from Venango County, through West Virginia, Kentucky, and Tennessee. On the basis of this information he opened up the Foster field that eventually produced several million barrels. In operations below Franklin in 1867 and in many others early in the next decade, Prentice was associated with Cyrus Angell. Because the latter's experience in petroleum dated back only to his bankruptcy as a merchant in 1866, Prentice rather than Angell may well have been the real progenitor of the belt theory.[70]

Whatever his debt to Prentice for the belt concept, Angell did test, demonstrate, and make explicit its practical application. Starting with information from his own three producing wells at Belle Island in 1867, some eighteen miles below Franklin, on the Allegheny River, Angell began a tedious process of collecting fragmented information about other wells at different locations: depths, differences between upper surfaces of the different sand rocks, their thickness, quantity and quality of oil showings in second sand, and texture of third-sand rocks. At Foster Station, about nine miles north of Belle Island by river, he found striking similarities with his own wells. In 1868 and 1869 he hired a professional surveyor to aid him, and in the following year, in partnership with Frederick Prentice, he leased or purchased all land on the belt he formulated between Belle Island and Foster Station.[71]

About 30 yards wide, six miles long, and with third-sand depths of about 1,000 feet, Angell's belt created a sensation in 1871 as he hit well after well with such precision that 90 per cent of them produced abundantly. Operators began an extension of the belt beyond Foster and embraced the theory for application everywhere, often uncritically.[72]

Operators were also quick to follow Angell's lead in exploring the possibilities of a second belt that he proposed, with somewhat less accuracy than his first, during 1872. It extended from Scrubgrass southward and slightly westward, about thirty miles long and three to five miles wide, through Venango into Clarion, Butler, and Armstrong counties, where it penetrated or skirted most of the producing areas opened up or extended in 1872 and 1873 in the Lower Oil Region.[73]

The impact of these developments became apparent the following year. Although the Upper Region, with its numerous new small pools, was still in the ascendancy, Oil Creek's production of an estimated 1.7 million barrels during 1872 was for the first time equaled in Clarion and Butler counties in the Lower Region.[74] These outputs helped establish a new record of production—almost 6.3 million barrels—resulting in prices that sagged to a yearly average of $3.75 a barrel.

Much of the activity in the Lower Region was centered in the Parker's Landing district, where by June, 1872, daily production had reached 5,000 barrels, one-third of the entire Pennsylvania fields. Some seven miles to the south a strike during the spring had brought feverish drilling activity in the Petrolia field that by July was producing at a daily rate of 2,500 barrels. Petrolia City, one of the gayest of all oil towns, mushroomed in response.[75]

Exploration in the Lower Region received an added stimulus in 1873 with the extension of drilling to the fourth sands and the development of the renowned Armstrong-Butler Cross Belt territory. Drilling an abandoned well deeper on a farm near Karns City in Butler County, Charles Stewart brought in the first recognized fourth-sand flowing well in May, 1873; in September other operators on the adjoining Armstrong farm brought in the second, flowing 700 barrels daily.[76]

These strikes prompted Richard Jennings, onetime superintendent of the Brady's Bend Iron Works and principal owner of the Armstrong Run Oil Company, to develop a belt theory of his own. Using the Stewart and Armstrong wells as basing points, Jennings conducted surveys similar to those used by Angell. He concluded that oil could be found about 75 feet below the third sand in a territory that at its western end cut across Angell's belt at right angle. Taking its name from this characteristic, the Armstrong-Butler Cross Belt, a narrow strip of land some eleven miles long, extended from Brady's Bend in Armstrong County into Butler County. Jennings' first well, struck in the fourth sand in November, 1873, flowed some 30,000 barrels the first month. It was soon followed by others. Drilling to the fourth sands had become a permanent feature of exploration while Jennings' pioneering of the first cross belt found many counterparts.[77]

The year 1873 was a climactic one in oil production. Despite efforts by organized producers to curb production, output soared a third higher than the previous year's record high, hitting an all-time peak of nearly 10 million barrels, while prices plummeted from $3.75 to a demoralizing average of $1.80.

The seemingly irreversible momentum of extension of producing territory in the Lower Region contributed vitally to oversupply. The Lower Region contained more flowing wells than had been experienced in the entire history of production. Twenty-eight gushers alone yielded 8,883 barrels daily in the summer of 1873, an average of 315 barrels for each well. In Butler County, particularly, operators following the attractive Angell belt theory continued to press forward the frontiers of producing territory in the vicinities of Modoc, Karns, and Greece Cities and Millertown—all new communities springing out of the new developments.[78]

For the first time the center of production swung completely away from its initial moorings along Oil Creek and elsewhere in Venango County to the Lower Oil Region. Of a total production approaching 10 million barrels for the year, Oil Creek contributed only about 1.2 million barrels, while Butler and Armstrong counties alone in the Lower Region produced about 4.4 million barrels and Clarion County 2.5 million barrels.[79]

By 1873 one salient feature of petroleum production had been repeatedly and conclusively demonstrated: exploration and development were undertaken under the widest range of operating conditions and almost irrespective of the prevailing levels of demand, prices, output, or inventories. Nor was there any reason to believe that this tireless search for crude would change in the future; the business of finding and producing petroleum had become a built-in and self-contained part of the industry. But exploitation raised difficulties of its own. By the early 1870's, oilmen were aware that it could result in outputs which persisted in outrunning demand at prevailing price levels. How to prevent this from happening thus became a major concern. The rapid evolution of drilling and exploration techniques and the fugacious nature of crude petroleum itself intensified this concern.

Chapter 7

Changing techniques of production

$\mathbf{M}$uch of the permanence of the new petroleum mining industry that had emerged by the early 1870's was based on the rapid evolution of production techniques during the preceding decade. In reviewing the technological developments, J. T. Henry, writing in 1873, could say with a great deal of truth, "the business of operating in oil has been shorn of nearly all its drawbacks." [1]

Many of the changes relating to production came as practical oilmen responded to the particular problems encountered in the fields. Others had their origin in the mainstream of improvements affecting the economy as a whole. In either instance, innovation and adaptation left few immediate operating needs unsatisfied for long. New devices and mechanical applications were put into general practice with a speed that matched the reactions of operators to strikes in a new territory. There were few inhibitions about anything that seemed to promise a more effective exploitation of oil, whether it involved a modest improvement in a sucker rod or the use of nitroglycerin to blast out the bottom of a well.

Power applications

One major accomplishment of operators during the years between 1862 and 1873 was to evolve basic power applications and hookups for percussion drilling. Spurred by experience with flowing wells to employ steam power in drilling as well as pumping, operators retired not only the spring pole but its early substitute, the grasshopper walking beam, used by Drake to drill his well with steam power (see Figure 7:1).

In Drake's rigging, from a wheel and an axle at the base of the derrick (called the bull wheel), the cable and tool string was fed over the crown block at the top of the derrick, and down to the middle of the grasshopper walking beam, a wooden pole that ranged from 9–20 feet long, about 1 foot through at the middle, and tapered at its extremities.[2] Drake's bull wheel and later ones served two purposes: as a brake in operating the well, and to raise or lower the tool string or casing if the engine powering the walking beam stopped running.[3] The beam was hinged at the outer end to a support, termed "the Sampson post," a location that gave it a grasshopper motion when powered at the opposite end by a "pitman," or rod connected with a crank on the flywheel shaft of the steam engine. This arrangement so limited the location of the power assembly that the engine house had to be built against the derrick.[4] Apart from the inefficiency of this power hookup, which did not use the full leverage of the walking beam, the position of the engine and boiler increased the ever-present danger of fire.

Before the mid-1860's, operators moved the Sampson post to the center of the walking beam and suspended the tools from the end instead. The engine house was removed to any feasible belt-gearing distance by borrowing from threshing practice, in which the pitman at one end of the beam was connected with a crank on a bandwheel, which in turn was driven by a belt to the flywheel of the engine (see Figure 7:2).

Along with more effective power hookups came improvements in steam engines. These took several lines, including increase in horsepower, greater mobility, and more flexible control over the operation of the engines. Drake's engine was a primitive stationary-type engine currently used on steam boats and rated at 5 horsepower. Some of the very early

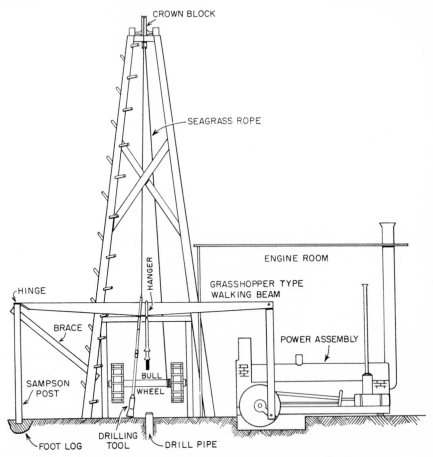

FIGURE 7:1. *Drake well rigging and power hookup.*

pumping wells were operated by Wallace engines, whose "rough as a rasp" pistons offered no marked improvement. The "Washington" engine manufactured at Newburgh, New York, and introduced about 1860, was the first engine with a portable boiler to come into general use for drilling.[5] It was soon exceeded in popularity, however, by the small portable engine introduced in 1861 by A. N. Wood & Company of Eaton, New York. Mounted on a compact tubular boiler that was in turn attached to a light wooden frame, the unit could be put into operation at a new location immediately upon unloading from a wagon. Later models, built with iron wheels attached, removed even the need for wagon transport.[6]

These early low-horsepower engines were reasonably adequate for

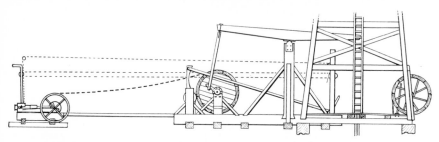

FIGURE 7:2. *New power hookup adopted by the mid-1860's.*

pumping or drilling to the shallow depths where the first flowing wells were struck. But when drilling began to go to 500 feet along Oil Creek and 700 feet or more at Cherry Run and Pithole, there was need for more power. This need was augmented when the close spacing of wells led to the growing practice of operating both drilling and pumping operations or several pumping wells from the same steam engine.

The demand was met when competitors of A. N. Wood & Company began supplying both portable and stationary engines that by 1865 were quite commonly rated at 12, 15, and 20 horsepower. Near Titusville one 40-horsepower engine was introduced to operate four pumping wells.[7] While a great variety of engines made their way into the Region by the mid-1860's, improvements in boiler construction generally resulted in actual horsepower delivered being greater than the indicated increases in horsepower.

The practice of gearing an engine to numerous pumping wells continued to grow in the postwar years, but by 1868 a strong countervailing trend had also set in. Hilly terrain often placed a limitation on multiple hookups, but another limitation proved more basic. Even the most closely spaced pumping wells often exhibited widely different behavior that required pumping at different rates and intervals. Adjusting the proper motion for one well disarranged the motion for another, and many observant operators concluded that it consumed more time and fuel attempting to maintain a proper syncronization from a single engine than the cost of a second engine. One or more large central boilers piped to the individual engines typified this plan of operation.*

While there was a notable increase in operating efficiency, engine prices

---

* Tallman farm, for example, had thirty-one engines for pumping twenty-nine wells producing from depths of 860 feet, while the well-managed Story farm in 1868 had twenty-nine engines on hand for its twenty-four pumping wells and drilling operations.[8]

reflected wartime inflation. In 1865, for example, the delivered price of 10-horsepower units was about $1,600, while 12- and 15-horsepower machines cost $2,200 and $2,500 respectively. Even at these prices boom conditions along Oil Creek and at Pithole brought about the delivery of some 3,000 engines to the Region in 1865 via the Atlantic & Great Western Railroad alone.[9]

Both the immediate available supply and the price of engines were affected by the collapse of Pithole in 1866–67, when many operators dumped their equipment on the market.[10] More important was the return of the iron manufacturers to peace-time operations. By the early 1870's, 10- and 12-horsepower engines were commonly sold at prices ranging between $600 and $1,000, less than half the costs some five years before.[11]

Further developments during the 1862–1873 period contributed to greater efficiency and flexibility in the operation of steam engines. The first of these involved a change in the fuels used to fire boilers. The majority of early engines ran on wood and could easily consume a cord, costing $7.00–$10.00, in twelve hours of pumping. Coal, priced at 60¢ a bushel in Oil City and as much as $1.25 along Oil Creek, was still more expensive.* In 1864 the superintendent at the Ladies well on Watson's Flats near Titusville turned to gas fuel with notable success, simply piping it through a barrel to the furnace and using it to light the engine house as well.† His example was soon followed by others, with gas being piped to as many as six engines from one flowing well.[14]

Extended use of gas received a temporary setback when it was discovered that recurring low pressures invited the flow from the boiler to back up into the pipe, exploding gas accumulated in the barrel reservoirs. By 1868, however, this difficulty had been largely overcome. The key to better control came with the installation of a check valve in the pipe leading to the furnace, allowing the gas to flow to the fire but automatically shutting off any reverse movement.[15]

Although the intensity of gas heat made it somewhat more injurious

---

* At times operators would substitute crude oil for wood or coal as fuel. But as was pointed out in 1869, "Oil can be used to advantage only when extremely low prices prevail. By the employment of the best apparatus for burning oil yet devised it requires three to four barrels of oil for fuel per *diem*."[12]

† The following contemporary account gives a more detailed description of the methods used about 1865 to recover gas from pumping wells: ". . . the mingled gas, oil, and water . . . is conducted by a pipe from the well tube into a tight barrel. The oil and water fall into the bottom of the barrel, and run off by a pipe into a huge tank or vat [for further separation] . . . The gas escapes by a small pipe at the top of the barrel, and is conducted into the furnace . . ."[13]

than coal to boiler bottoms, its cleanliness, convenience, and superior economy made its use distinctly advantageous.[16] In 1869, for example, when improved transportation had greatly reduced the price of coal delivered to the Region, its use to provide steam for drilling was estimated to cost about $15.00 a day per well.[17] Gas fuel, on the other hand, was practically free except for the nominal cost of putting in equipment necessary to use it. The superintendent of the Tallman farm, located on the headwaters of Cherry Run, also in 1869 reported that total pumping costs —using gas exclusively—of each of the twenty-eight producing wells on the farm averaged between $200 and $500 per month. A shift to coal, he estimated, would raise the average to between $500 and $700 per month.[18]

Of course, gas was not always available and supplies often fluctuated or ran out. But with the rapid advance of deeper drilling through casing, the availability of gas by the early 1870's was such that it was both the preferred and dominant fuel in old and new fields.

The final development of major significance during 1862–1873 affecting the efficiency and flexibility of steam engine operations in the oil fields came with the introduction and adaptation of variable or adjustable cutoff devices, developed initially for steamboats on the western rivers. All early low-pressure steam engines operated on a full stroke; that is, with the steam valve open throughout the entire stroke of the piston. At the end of the stroke, the steam was exhausted from the cylinder into the open air with a cannon-like report that, in the case of steamboats, announced their approach for several miles in advance.[19]

With the evolution of high-pressure engines there was growing recognition of fuel economies to be gained by using full steam pressure only during part of the stroke of the piston, letting expansion of the steam carry the stroke to completion.

By the 1840's, many western steamboats were equipped with two sets of cams to control steam intake valves. One cam was adjusted to provide maximum power when the steamboat was getting under way or running against a strong current. The other was set to cut off the intake of steam at points varying between three-eighths and three-quarters of the stroke, depending upon the amount of power needed for normal operating conditions.[20] The disadvantage of working with only two alternatives to the application of power led to developments by the 1850's of the adjustable or variable cutoff, essentially an eccentric-shaped cam that could be quickly shifted to a position giving the required power.

Contemporary accounts do not reveal when the adjustable or variable cutoff was first introduced into the Oil Region.* It is clear, however, that between the mid-1860's and the early 1870's the use of this device had become increasingly widespread. Its use was particularly advantageous where drilling and pumping operations were linked with a single engine, while full power was needed during drilling but only a fraction served to operate the pumps. It also helped to reduce fuel consumption where pumping operations were operated from a single power unit or

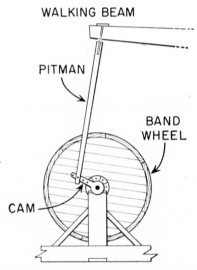

FIGURE 7:3. *Diagram illustrating principle of variable cutoff.*

where several engines drew upon a central boiler for their steam. In the former case wells could be added or withdrawn from pumping without loss in efficiency of operating the central power unit. In the latter instance adjustments could be made to meet the power requirements of each particular well.

Closer control over the operations of their engines was also an advantage to drillers, particularly as wells became deeper and tool strings heavier. But drillers needed an additional flexibility that the variable cutoff alone did not provide. To prevent damage to the tools when the drill struck hard materials, it was frequently necessary to shorten the stroke of the walking beam or reverse the movement of the tools.

The solution to this problem came with an adaptation of the variable

* One account credits A. N. Wood & Company, manufacturers of the early portable engines, with having introduced "The first eccentric link, reversible engines" into the oil country in the mid-1860's, but give no details on its construction.[21]

cutoff principle that seems to have originated in the Oil Regions. The essential features of this device are shown in Figure 7:3. A cam (with drilled holes) was attached to the axle of the band wheel. By changing the position of the lower end of the pitman, or rod, leading to the walking beam it was possible to shorten the stroke of the walking beam or to reverse its motion sufficiently (by half a stroke) to lift the tool string from the bottom of the well. The originator of this device is unknown; nor is it possible to determine when it was first applied to oil drilling operations. Contemporary evidence suggests, however, that by the early 1870's its use was common in the Regions.[22]

Improvements in tools

With one notable exception late in the period, there was little change in the design of the various components of the tool string between 1862 and 1873. But to meet the requirements of larger and deeper wells it was necessary to improve the wearing quality of certain pieces of equipment. This was especially the case in respect to the jars that provided flexibility to the tool string, reduced breakage, and gave leverage for jarring stuck tools loose. In 1866 George Smith of Rouseville introduced a major improvement in this piece of equipment by bringing out the first steel-lined jars. After a brief delay over patent litigation and initial difficulties in obtaining a firm welding, steel lined jars were generally adopted.[23]

As the center bit was increased from 2⅜ inches to 5½ inches in diameter, more generous use of steel in its construction added to efficiency. Tipped with more steel at their points, both round and flat reamers performed more effecitvely. Sand pumps were improved by two additions: one was a valve that automatically opened when the pump reached the bottom of the well; the other was a piston that kept its place at the bottom of the pump while being lowered, but upon being drawn up created a suction which filled the pump with loose debris and water. Pumping operations, too, gained in efficiency from design changes in the sucker rods and elimination of rivets which formerly had fastened the metal joints and sockets insecurely to the wooden frames.[24]

Early in the 1870's, George Koch, a practical driller, made, with his club bit, one of the most timely and important innovations in the design of the drill. Koch saw that the center bit, a thin, flat drill, was inefficient. In hard rock it was excessively vulnerable to breakage, and at best it became blunted within a few hours and had to be withdrawn for sharpening while the heavy, blunt reamer was substituted to pave the way for further drilling. This sequence required four laborious withdrawals and insertions of the tool strings. Koch's new bit combined the thin, flat center bit and blunt reamer into a single fluted drill that at once gave it greater strength and made it unnecessary to withdraw the tool string every few hours to exchange the bit for the reamer.*

There were few major advances in the devices generally credited to L. H. Luther, a blacksmith at Funkville on Oil Creek, for retrieving equipment stuck in the bottom of the wells. There were many factors that made the early fields populous graveyards of abandoned drilling equipment. The failure of the driller to manipulate the cable so that successive strokes of the bit fell slightly across preceding ones would result in holes not sufficiently round and perpendicular to give free play to the tool string. Delay or clumsiness in use of the reamer, or failure to temper the reamer to sufficient hardness in the nearby blacksmith shops also led to broken tool strings or a broken reamer in the hole.

Though he added many ingenious improvements to his repertoire of fishing tools, Luther's socket remained a standard implement for retrieving drilling bits from the bottom of the hole as late as the second decade of the present century. It comprised a hollow rim in which a pair of clamps was inserted. Designed to slip down over the tools at the bottom of the hole, when pulled up the clamp lips caught the shoulder of the bit and hauled it to the surface. His rope knife for cutting cable from stuck tools in the hole likewise performed yeoman service throughout nineteenth-century drilling.[26]

Casing

The early gushers, with their unpredictable eruptions of gas and oil, introduced unique problems and hazards, but few changes in the physical equipment. To handle the oil, gas, and water from flowing wells, an elbow, connected with a discharge pipe to receiving tanks, was frequently placed in the tubing about 8 feet above the derrick floor. To stop the flow after the initial violence of the gas pressure had subsided, operators sometimes resorted to the stopcock, a simple close-off valve long in use in the mechanical arts. The tubing was severed and the entire valve welded into

---

* Although the claim that it reduced the drilling time for typical wells from sixty to twenty days probably is exaggerated, the club bit did contribute importantly to reduced time and costs.[25]

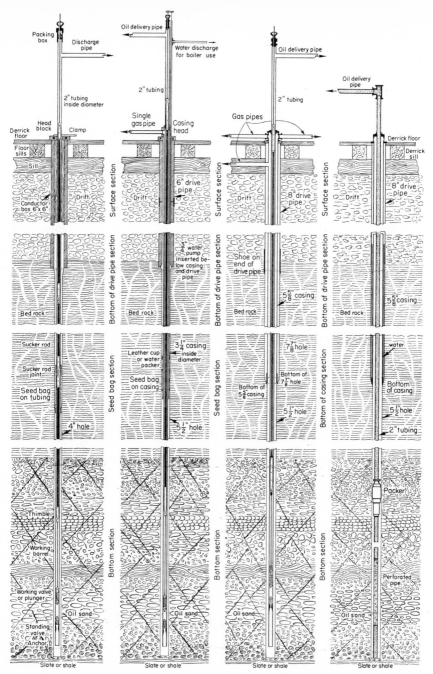

FIGURE 7:4. *Evolution of drilling techniques: 1861–1880. Left to right: uncased wet-drilled pumping well, 1861–1865; cased pumping well circa 1867–1870; pumping well dry-drilled through casing, 1870–1878; flowing well, 1880.*

it. The valve stem covered the entire interior diameter of the tubing, and had holes pierced in it in one direction, so that it could be opened or closed merely by a turn of the valve handle.[27]

Early in the war years, an accelerated evolution of tubing practices and materials occurred when it was found that copper tubing, such as was used by Drake, was too weak to withstand the pressures of third-sand drilling. Moreover, it was very expensive to tube deeper wells, either for testing or pumping.[28]

As a substitute, operators turned to 2-inch wrought-iron pipe and commonly tubed the entire well, which provided minimum but inadequate protection against flooding. Gas-tongs, a clamp-like instrument, reduced damage in handling and speeded screwing the 20-foot sections together. By 1865, handling had improved to a point where three men could tube a 600-foot well in a day, at a cost of 60¢ a foot for pipe, and an additional $149 for other necessary tools and equipment, including 600 feet of sucker rods, a pump barrel, and gas tongs.[28] A burst seed bag, faulty placement of the seed bag, leakage of tubing, and numerous other mishaps, still might occasion up to forty or fifty retubings in a period of six months, but improved tubing reduced the length and expense of testing and pumping.

Continued difficulty with flooding in third-sand wells reflected the current practice of drillers who made no effort to close out surface waters in the first two strata until drilling was completed and tubing with a seed bag attached to the lower end was inserted. Attempts to solve this problem led to the first use of casing, a major innovation in the evolution of drilling techniques. As developed by the mid-1860's, the procedure on old wells was to case or enclose the 4–5½-inch hole with heavy tubing to a point below the water-bearing rock strata. A seed bag attached to the outside of the lower end of the casing served to prevent much of the water that accumulated around the casing from flowing to the bottom of the well. The earliest casing was heavy artesian tubing in 20-foot sections, 3¼ inches in diameter, through which 2-inch tubing with a copper sand-pump chamber later could pass when testing and pumping began.[30]

This method of casing received its first extensive trial in March, 1865, on Tarr farm on Oil Creek, where every producing well had succumbed to flooding. While it did not completely stop flooding, the results of these early trials of casing were impressive. The moribund production of Tarr farm wells shot up to 1,000 barrels daily. The advantage of attaching the seed bag to the casing rather than the tubing alone would have encouraged the use of the new technique. There was no longer any need to remove the seed bag (which invited flooding) each time the tubing was withdrawn.[31]

Beginning to panic at rampant flooding and drastic declines in production, operators at Pithole early in 1866 extensively transplanted the new method from Tarr farm. In May, Pithole operators reported a large increase in the production of cased wells, in some instances to levels 70 per cent of their flush period.[32] Acute price depression, curbing drilling and pumping activities, slowed further progress for a time, but the new technique had made a lasting imprint that assured its further development with the return of better conditions.

Operations along Bennehoff Run in the summer of 1868 ushered in the next advance in the use of casing, known as "drilling through casing" or "dry drilling." The idea apparently suggested itself from the increased speed with which wells that had been "wet drilled" and then cased could be cleaned.[33] Its impact was so great that by the end of the year drilling through casing was widespread, and within another two years the practice became virtually universal.[34]

Previously wells had been drilled "wet" their entire depth. The new practice drastically restricted the depth of wet drilling and permitted dry drilling in the deeper parts of the hole below the casing. The first step was to drive a conductor or driving pipe of 6–6½ inches down to surface rock. Thereafter wet drilling proceeded at the same diameter to a point below the water courses. After pumping out the water, casing was then introduced to the bottom, with a seed bag or water packer, sometimes both, attached to its lower end to shut out surface waters from deeper drilling. From this shoulder, dry drilling then proceeded at slightly reduced diameters of 5½ inches to the oil sands. Two-inch tubing containing a sand pump was inserted through the casing to the bottom of the hole to test or pump the well.[35]

This arrangement permitted replacement of the head block at the floor of the derrick with a casing head that was screwed on the casing and provided a superior support for the tubing. Vents in the casing head simplified tapping of gas. By making the seed bag or water packer a permanent fixture, casing provided a stable protection against flooding from surface water at all times. It also removed the intermittent risk of flooding with each removal of the tubing, operations which frequently broke the seed bag and required its replacement. With casing, tubing could be introduced or drawn in a matter of hours rather than days thus facilitating repair of pump valves, steaming out of paraffin deposits, and countless other operations.[36]

Still other benefits accrued to casing and drydrilling below it. One was a drastic reduction in the time and inaccuracies of testing, which in uncased wells often consumed more time and expense than drilling. Instead of spending days or weeks to pump water from the well, testing now became an operation that could be completed in a few hours. Another benefit was reduction in tool losses in thick third-sand muds, which often oozed in with a consistency of cement to render uncased wells inoperative.[37]

While there was an almost unanimous opinion respecting the significant advantages of drilling through casing, there were also certain disadvantages. Apart from the price of casing, it cost about $150–$200 more to drill the larger hole required for the casing.[38] According to an estimate in 1870, the cost of 6-inch casing in wells 600–1100 feet deep ranged between $500 and $1,000.[39] A graver immediate disadvantage lay in an alarming increase in fire hazards arising from a cause probably not fully recognized at the time.

In wells drilled through casing there was no column of water to restrain gas pressure. As a result, there was an increasing number of violent eruptions of gas and oil when producing sands were penetrated. The difficulty of keeping uncontrolled flows of oil and gas from coming in contact with nearby fires inspired mounting protests from workers, who disliked the hazard of drilling through casing.[40]

To overcome these disadvantages required a third progressive step in drilling through casing. Essential components of this advance lay in wider holes, larger casing, extended areas of dry drilling, and an enlarged space for gas in the well. The widespread use of 6-inch casing and considerable drilling of 8-inch holes indicate that the advance already was well under way as early as 1870.[41] Without further refinements, the mere use of 6-inch casing (5⅝ inches inside diameter) in a hole drilled 8 inches in the cased area supplied a basic solution to the gas problem by greatly expanding the space for gas in a well.*

Larger holes and casing had several other distinct advantages beyond providing greater storage for gas space. Compared to a 5½-inch hole with casing 3¼ inches in diameter, the 8-inch hole with 6-inch casing was much

---

* A well cased 300 feet with the enlarged diameters of casing and holes in 1870 had ten times the gas space of wells using 3¼-inch casing in a 5½-inch hole to the same depth in 1868. Greater depths with enlarged casing and holes further increased the gain in gas space.

less liable to damage that might necessitate the withdrawal of the casing with the possibility of flooding the well. The 5⅝-inch inside diameter of the 8-inch casing made it possible to drill a hole this size through the encased portion of the well. The larger hole not only made it easier to manipulate fishing tools; it also expedited the subsequent deepening of wells by the use of a 5½-inch bit rather than one 3½ inches in diameter.

Further refinements of the third progressive step in permanent casing abolished both the seed bag and the leather packer. They were replaced by a steel shoe which could be ground securely into the sides of the 7⅝-inch hole on the beveled shoulder, where the casing stopped and drilling at a 5⅝-inch width through casing began. In addition, a similar shoe had previously been placed at the bottom of the 8-inch drive pipe above. This shoe for the first time provided a water shutout at bedrock, and it further improved drilling accuracy to the bottom of the cased area by providing a firmer lodgment at bedrock. These shoes also improved dry drilling by allowing only enough water to enter in order to assure efficient sand pumping.

Only when a flowing well was anticipated were further packing arrangements needed. In that event, a new type of rubber packer was introduced, not in the former position of the seed bag and leather water packer, at the bottom of the casing, but around the two-inch tubing immediately above the expected producing sand. The packer was constructed so that the tubing (which it encased) moved in a sliding joint, the upper section of which was securely fastened so that when it slid into the lower section it pressed the rubber with great weight against the sides of the hole. Prepared in this way, and properly connected with receiving tanks at the casing head, a flowing well required slight attention for months.[42]

Torpedoing

One of the problems that plagued producers was the clogging of tubing and well bottoms with paraffin from crude. Probably the earliest attempt to rid wells of clogging involved the use of steam. By inserting a small glass tube through the tubing and forcing down a jet of steam, it was possible under some circumstances to melt the paraffin. To be effective, however, it was necessary to remove almost all water from the well so that the steam jet could raise the temperature more rapidly to paraffin's melting point (112° F).[43]

Another very early practice was to dump naphtha, which Region refiners threw away, down plugged wells to dissolve accumulated paraffin. Although infrequently successful, this practice enjoyed a vogue for more than a decade.[44]

By the mid-1860's the battle against paraffin clogging widened to include cement-like accumulations of mud that flowed from third-sand veins into uncased wells. The improvement in packing, already noted in connection with the development of tubing and casing, gave some relief.

A second response resulted in the widespread use of blowers until dry drilling through casing reduced their utility. Frederick Crocker pioneered this field with his famed oil ejector, which in the summer of 1863 restored the defunct flowing wells, the Empire and the Buckeye, to a production of many thousands of barrels.[45] From a compressor at the top of the well, Crocker forced air through a small pipe lowered through the tubing in place of the lift pump. The lower end of the pipe turned up about 4 feet inside the tubing, terminating in a special valve through which oil or any other liquid in the well was discharged by direct action of the injected air. In addition to the direct action, the sudden injection of air guarded against paraffin formation in the tubing by elevating well temperatures. The ejector also made provisions for injecting steam when necessary.[46]

Toward the close of the first six years of oil production, solutions to problems of clogged wells and a cumulatively important chain of minor innovations in drilling reached their climax with a major innovation, the torpedoing of oil wells. Besides paraffin removal, torpedoing had other objectives. Prior to its introduction, tedious and expensive reaming for the frequent occurrence of crooked holes offered the only cure. Torpedoing further provided effects similar to those obtained in modern directional drilling. Early drilling experience was rife with straight holes in which the errant bit missed a strike by only a matter of feet. Properly tamped from above to concentrate its explosive force downward and laterally to the sides of the hole, a charge of powder or other explosives in an air-tight container lowered through the tubing to the bottom of either a crooked or straight hole might spell the difference between a "duster" and a profitable hit.

Torpedoing had its inception in hydrology, where the practice of exploding powder in holes drilled in the bottom of water wells, to increase the quantity of fluid or to tap new veins, could be traced back to early in the century. Borrowing on this technique, Henry Denis in 1860 exploded a 3-foot copper tube filled with rifle powder in the bottom of a water-filled well at Tidioute. John F. Harper failed to explode 5 pounds of powder in a well at Franklin in the same year because the water pressure collapsed the container, but following three successful explosions in wells on Watson flats, the Stackpole farm, and at Tidioute, he formed a partnership with William Skinner for "blasting oil wells." William F. Kingsburg also successfully exploded a charge at Tidioute.

Cheap prices of the early flowing well period, driving out pumping wells, caused a lapse in commercial efforts to revive wells by torpedoing until 1863, when William Reed, using electrical charges, exploded three charges in tin casings that withstood water pressure in the Criswell well on Cherry Run. In 1864 Frederick Crocker exploded a torpedo in the bottom of a well by means of a weight falling on a pistol cartridge inserted in the shell. By 1865, these individuals and others had fired at least thirty torpedoes in wells under water and early in that year several applications were filed for patents on particular torpedoing devices.[47]

It was Colonel E. A. L. Roberts, Civil War veteran, however, who emerged as the outstanding personality associated with oil well torpedoing.[*] On November 18, 1864, he applied for a patent for a "new and useful method of increasing the capacity of oil wells, and of restoring oil wells that have become clogged, to productiveness." His application was framed in the broadest possible terms and covered all types of containers, varieties of explosives, and methods of setting off the charges.[49]

In January, 1865, Roberts turned up in Titusville with a half-dozen torpedoes made for him by Elijah Brady of New York City. They were cast iron flasks, filled with gunpowder, and ignited by a weight that dropped along a suspension wire onto percussion caps in the flask.[50] On January 28, 1865, Roberts successfully discharged two of his 8-pound torpedoes in the Ladies well on Watson Flats, near Titusville. While the results were not spectacular, after the well was cleaned and retubed output that had dropped to a few barrels a day was brought up on a pumping basis to between 30 barrels and 40 barrels a day.[51]

In February, 1865, Roberts and his associates organized the Roberts Petroleum Torpedo Company under the laws of New York. In their pro-

---

[*] While there were many who challenged Roberts' claim that he had conceived the idea of shooting oil wells from his war experience, his story was not wholly implausible. Before the war he had displayed inventive proficiency. Engaged with his brother in manufacturing dental materials in New York City, Roberts had patented several inventions that won gold and silver awards from the American Institute.[48]

spectus they set a schedule of charges for torpedoing that ranged between $100 and $200, plus a royalty of one-fifteenth of the increased output following the torpedoing.[52] A shop for the assembly of torpedoes was set up near Titusville, and a force of agents was assembled who were to work on a 25 per cent commission of the gross income, including royalties, from each contract they sold.

In view of the long history of torpedoing water wells, there was a real question as to whether Roberts or anyone else could obtain a patent. After several hearings and appeals, however, Roberts' patent was granted, not as the invention of a new art, but because, in the view of the Patent Office, he had demonstrated the practical value of the method by having shot the Ladies well. While this ruling ignored the shooting experience of William Reed and others who had antedated Roberts by several years, an appeal to the United States Supreme Court in August, 1866, resulted in a confirmation of the Patent Office's decision.[53]

This decision for all practical purposes gave Roberts a patent on a process which, according to the reasoning of the Patent Office, could not be patented. But Roberts' initial application was so broad that it was virtually impossible for anyone to develop a type of torpedo that did not infringe on his patent.

Following the court's decision, the Roberts company set out to eliminate competitors who had swarmed into the Regions between 1864 and 1866. Some time before 1867 he added to his arsenal by obtaining an exclusive right from A. Nobel to use nitroglycerin, a much more effective explosive than gunpowder, in the oil fields.[54]

After 1864, the growing interest in torpedoing was in large part a result of the promotional activities of the Roberts organization. Intensive selling efforts, glowing press accounts of successful torpedoing, and pamphleteering took seed. In December, 1865, the last major remaining doubt was dispelled. The Woodin well, a dry hole on the Blood farm, came in pumping 80 barrels daily after two blasts from Roberts' powder-charged torpedoes. Torpedoing was more than a temporary rejuvenator of pumping wells; it was an instrument of directional drilling that could convert dry holes to producers.

Early in 1866, the Devene and Keystone wells on Tarr farm, pumping 3 barrels daily, were transformed by Roberts' torpedoes to flowing wells of 80–180 barrels daily.[55] Meanwhile, Pithole, desperate to remedy its persistent decline, experienced its first successful torpedoing at 610 feet in the Clara well on Morey farm in May, 1865.[56]

A torpedoed well.

Lifting the text of Roberts' brochures, Captain Jesse H. Lord, editor of the *Scientific American,* in July, 1866, termed Roberts' torpedo the most valuable device yet discovered for aiding petroleum production. Without naming Roberts as his source, he reported that competent judges had found that the torpedo had lifted daily production 1,400 barrels, valued at $1.4 million annually, in the past sixty days.[57]

Like Eli Whitney, who had attempted to enforce his patent claims on the cotton gin, Roberts found it difficult to control the manufacture and sale of a product that could be made entirely in a small machine shop or assembled from materials that were readily accessible. During 1866 and 1867 estimates of the total manufacturing cost of torpedoes ran about $20.00 for a six-foot torpedo and $15.00 for one five feet in length.

Roberts found it increasingly difficult to maintain his own price schedule of $125 to $200 for torpedoes (depending on length) against lower prices charged by competitors. When his share of the market declined further during 1867 and 1868, he cut prices and dropped the royalty feature of his contracts. In the latter year he brought action against two of his principal competitors, the Reed Torpedo Company and James Dickey, by seeking injunctions against them for violation of his patent.

Threatened by the possibility of the elimination of competition, the producers, through their recently organized Producers Union Association, raised between $40,000 and $60,000 to help defend the Reed company and Dickey in the courts.[58] Few cases in oil history had attracted as much interest or brought as much criticism until the U. S. District Court awarded Roberts a permanent injunction against his competitors in 1871.[59] By 1872 Roberts had expended more than $100,000 on some two hundred injunctions and other court actions. While he collected damages amounting to only about $60,000, he succeeded in establishing a legal monopoly in the business. In 1873 he was assured of a continuation of his position when he received a reissue of his basic patents.

The introductory years of torpedoing exacted a heavy toll in lives, both in factories and in the field. In 1868 and 1869 the proprietors of three factories in the Region met their deaths in explosions. Despite the beginnings of rudimentary precautions to store torpedoes in isolated excavations in the field, several agents and professional shooters of the Roberts and other companies suffered fatalities at field magazines. Though hazards were greatest in installing units in wells, fewer fatalities occurred there in these initial years.[60]

By late 1869 a growing use of nitroglycerin greatly magnified the perils

Wagon equipped to transport torpedoes.

of torpedoing. Though highly dangerous to transport, nitroglycerin was simple to make by reacting glycerin with nitric and sulphuric acids, chemicals widely employed in the Region's refineries. Its local availability quickly pressed nitroglycerin past the barrier of Roberts' exclusive license into general use, sometimes alone, but usually in shots of 4–6 quarts compounded with gunpowder or guncotton. Unlike powder, the new explosive produced lethal results if only a few drops were spilled.[61]

Meanwhile, Roberts' success in tying up the operations of leading competitors with restraining injunctions drove many professional operators from the field of legitimate operations.[62] In addition, he honeycombed the Region with agents to detect illicit torpedoing. By these actions, Roberts spawned a retaliatory army of "moonlighters," who shot wells illegally under the most hazardous of circumstances possible. With cans of nitroglycerin strapped to their backs, the iron-nerved "moonlighters" performed their delicate tasks of charging oil wells under the cloak of darkness, which made both their detection and safe operation more difficult. The ranks of this underground profession rapidly expanded and thrived for almost a decade.

While the "moonlighters" served to soften the effects of the monopoly, Roberts carried on a vigorous campaign against violators. When he died a millionaire in 1881, his expenditures in the courts to enforce his monopoly had swollen to about $250,000, and he had earned a reputation for having been responsible for more litigation than any other man in the United States.[63]

ROBERTS TORPEDO.

As litigation is much less than formerly, and hoping that it will entirely cease,

We Propose to Give the Producers the Benefit of the Following Schedule:

System of Orders for Roberts Torpedo

An order for $300 for $225............25 per cent. off
An order for $600 for $450............30 per cent. off.
An order for $1 200 for $800.......23½ per cent. off.
An o. der for $2,100 for $1.500......37½ per cent. off.
An order for $5,000 for $2,7 0........45 per cent. off.
An order for $10,000 for $5,000......50 per cent. off.

☞ Orders can be made up by two or more persons. All application for orders must be accompanied by the cash, and addressed to E. A. L. ROBERTS, Titusville, Pa., to receive prompt attention. No orders will be honored except signed by

E. A. L. Roberts, Titusville, Pa.

Advertisement, Titusville Morning Herald, *September 3, 1877.*

Drilling depths

By the early 1870's, torpedoing was an accepted addition to drilling technology, which made such striking gains over the preceding decade. There was much in this development to substantiate the unanimous opinion of informed contemporary observers that it was now possible to drill to double the depths of a few years earlier with no increase in costs. In the absence of reliable and comparable data, it is difficult to demonstrate statistically the extent of this advance; but the following examples, despite their deficiencies, give a reasonably reliable general impression of cost trends in drilling.

Prior to 1862, when wells seldom exceeded 200 feet in depth, drilling costs ranged between $1,200 and $1,500, although they could readily rise above $2,000 in the event of flooding, cave-ins, lost tools and the like—all suggested by the fact that the ratio of completions probably ran no higher than one well out of every twenty that were started.[64] The most expert driller could obtain but two strokes of the drill a minute, which in soft slate might advance him 5 feet or 6 feet a day, but only 2 inches or 3 inches in rock conglomerate. Three feet a day was judged an excellent average, but no measure of time progress. Testing consumed more time than drilling and usually had to be done repeatedly, even fifty times in unusual instances.[65] Wells pumping 20 barrels a day were excellent producers, with 50 barrels the upward limit. Most wells had a life of a few months.

By 1865–66 depths up to 600 feet became common. Deeper depths, heavier machinery, wartime inflation of materials and labor, compounded by speculative booms, all pressed drilling costs upward to an approximate range of $3,300 to $7,600, but there was more output over which to prorate such costs, for the ratio of completions had climbed to one well in five. Barring a somewhat more limited range of exceptional contingencies, ordinary drilling times had receded to a range of two to four months for a 500-foot well, which might cost $15.00 a foot, or $7,500. In that cost, $2,500 for a 15-horsepower engine and boiler and an equal amount for labor, fuel, and miscellaneous extras, were the largest items. By owning their own engines, prorating their costs to numerous drilling projects, and availing themselves of the economies of pumping several wells with one engine, operators might reduce these costs to $4,500 or $5,000.[66] Informed guesses indicated extensions in well life and higher output expectancies. Good wells lasted eighteen months and enjoyed nine months at peak producing rates, while wells which had come to be classed as "a second-class concern" pumped 25–100 barrels for twelve months.[67]

In 1868 the opening of the Pleasantville and Shamburg fields drove average drilling depths down to a range of 700–900 feet. The prevalence of inside drilling and exceptionally close spacing in these flush fields supplied a somewhat artificial stimulus to completion rates, which reached an estimated record high of five wells out of eight. While the general practice of drilling through casing raised costs, these were more than offset by the lowered incidence of flooding, the advantages of permanent casing, and the availability of steam engines at reduced prices that were reflected in drilling contracts that commonly ran about $2.30 a foot.*

---

* The superintendent of the Tallman farm reported presumably typical costs of wells averaging 860 feet at $4,800 broken down as follows: new engine, set up, $1,150; drilling rig, complete, $550; driving pipe, belting, related items, $200; drilling-contract price, $2,000; tubing, casing, sucker rods, and related items, $900. Average drilling time, forty days.[68]

A study of some 1,465 wells ranging in depths from 500–1,600 feet, completed in 1871, gives a somewhat more generalized picture of drilling costs in the early 1870's.[69] The average time to bring these wells to completion, ready for pumping, was 81 days, with individual wells ranging from 25–100 days. The effect of casing was reflected in the comparatively long life of these wells, which averaged some thirty-four months of production, despite the inclusion of dry holes.* The total average cost per well was estimated at about $6,100: labor and fuel, $2,100; engine, boiler, rig, and related items, $3,000; tubing, casing, sucker rods, and pumps, $1,000.[71]

Drilling and pumping costs

In the complex of such variables as the terms of leases, costs of drilling, operating expenses, the number of dry holes, the average life of producing wells and the price of crude, it is impossible to give more than the most general impression of the effects of improving technology on the costs and profitability of producing crude. In the excitement of the flowing wells, particularly toward the end of 1862, it is doubtful if cost calculations played much of a role in the minds of operators. Even leases that called for a large down payment and one-half or more of the oil as a royalty could pay off handsomely in a short time if the operator were fortunate enough to strike a sizeable gusher.

One estimate made in early 1865, when pumping wells were providing the larger portion of the output, gives some indication of the level of crude prices that would enable an "average" producer to break even on his operating costs.[72] Assuming a pumping well yielding 20 barrels a day for 313 working days a year, a royalty payment of one-half the oil, and a 10 per cent interest payment on the $4,055 cost of the well, total operating costs per "operating day" were estimated at $23.73, divided as follows: interest on cost of well at 10 per cent, $1.29; repairs, $2.00; labor, $6.00; ⅞ of a ton of coal at $16.50 per ton, $14.44. In the absence of the $1.00 tax per barrel on crude, the owner, after deducting the royalty of 10 barrels, could have met his expenses from the sale of the remaining 10 barrels at a price of $2.37 per barrel; with the tax the "break-even price" was $3.37 per barrel.

At prices ranging between a low of $4.00 and a high of $10.00 a barrel, these cost figures suggest a considerable margin of profit for the "average"

---

* The ratio of dry holes to successful producers during 1871 was running about five to eight (i.e., 540 dry holes and 925 successful wells).[70]

producer during the year. As an indication of over-all profitability of production, however, they must be qualified by taking into account the number of dry holes that at this time were running (about four out of every five wells drilled) and the average life expectancy of producing wells (eighteen months). Adjusting for these factors would raise the total "break-even" price per barrel of crude for the producer to about $6.35 without the tax, and to $7.35 with the tax.[*]

In 1871, the drilling costs of wells were averaging around $6,100, even though many of the effects of better power hookups, multiple pumping operations from a central power unit, and a reduction in repair costs, were reflected in estimates that the total out-of-pocket costs of pumping at that time "may be fairly averaged at $6.50 per day." [73]. Interest charges of 10 per cent on the cost of wells, plus an allowance for supervision, would add at most another $2.00 to daily operating costs. Thus an owner of a 20-barrel-a-day pumping well, still operating on a 50 per cent royalty basis in the early 1870's, could have broken even on his operating costs with crude selling at about 85¢ a barrel.[74]

These cost figures indicate why, at prices that ranged between $3.25 and $5.25 and averaged $4.40 for 1871, it paid to pump wells yielding sometimes less than 4 barrels per day. The fact that the average daily yield of some 3,275 producing wells in that year was between 5 barrels and 6 barrels suggests a large proportion of small pumpers in the total.[75]

The increase by 1871 in the average life expectancy of producing wells to more than three years and a lower percentage of dry holes—five out of eight wells drilled—reflected the more widespread use of casing and torpedoing. These techniques brought a sharp decline in the average total costs per barrel of producing oil. An allocation of the costs of five dry holes (at $6,000 each, less salvage of equipment of $500) to the three producing wells brought the total amount to be amortized to each producing well to about $15,000. Spreading this amount over three years meant a daily amortization charge of approximately $13.70. Pumping charges, overhead, and interest, as noted above, would add another $8.50 to bring total daily costs to $22.20. Thus on a 20-barrel-a-day pumping well, and with royalty still at 50 per cent, the owner could recoup all expenses, including his "share" of the incidence of dry holes, with crude selling at $2.20 per barrel, compared to $6.37 a barrel (without the tax) in 1865.

[*] Prorating the cost of five wells at $4,055 each, less an allowance of $2,275 for "salvage," over eighteen months (470 working days), would make the daily "amortization" cost, including an adjustment for interest, approximately $40.00.

In reality, of course, the effects of both operating and drilling costs were much more complex than any simplified "average" set of calculations would indicate. At any particular time, producers ran the gamut in respect to the profitability of their wells. At one end of the scale were those who had struck gushers or above-average pumping wells which enabled them to recoup their original investment long before their wells had settled down to a more moderate rate of output. For every barrel produced, the difference between operating costs and royalty deductions and selling price meant a clear profit on their ventures. At the other end of the scale were those who had no chance of regaining their original investment. Regardless of their position, however, all operators realized that once a well was brought into production there was little choice but to keep on producing as long as the price of oil was above out-of-pocket costs.

Producers were generally aware of the effect that advances in drilling and production technology had in improving the profitability of their operations. But they became increasingly concerned during the early 1870's, when declining crude prices began to wipe out these gains and threatened to leave them in a worse position than they found themselves at any time since late 1861 and early 1862. Average crude prices fell from $4.40 a barrel in 1871, to $3.75 in 1872, to $1.80 in 1873.[76]

Price declines obviously affected the profitability of all producing operations and closed down those pumping wells whose income from the sale of crude no longer covered out-of-pocket expenses. But abandonments were not sufficient to overcome various countervailing forces that, in the face of declining prices, brought about an expansion in the annual output of crude from about 5.7 million barrels during 1870 and 1871, to over 6.2 million barrels in 1872, and to approximately 9.9 million barrels in 1873.[77]

In part this expansion reflected one characteristic of petroleum production that was not uncommon to other mining pursuits. So long as there was a chance of making a strike in a new territory, operators would continue to move into new territory, as they did in the lower Region during 1870 and 1871.

The failure of prices to serve as a more effective governor in adjusting output to more favorable cost-price relationships also reflected certain characteristics of petroleum and its subsurface behavior that made the petroleum industry unique among mining operations. As early as 1860 many operators began to realize that oil was an unusual mineral, and that as a liquid, it had the unique characteristic of fugaciousness. Given a lower gas or water pressure point elsewhere, or some other attraction within the porous rock strata where it resided, oil readily migrated from one sub-surface location to another. This mobile quality gave petroleum a degree of elusiveness and concealment; uncertainties of location and recovery were heightened by the common-law rule of capture—the principle that the product belonged not to the owner of the property where it was orig-inally found, but to the owner of the land where it was intercepted or re-covered.*

In the era of small pumping wells, instances of encroachment of a neighboring well's output were numerous enough to reveal the fugacious character of oil. But the significance of this characteristic was lost in the contagion of a mining boom that in itself seemed to provide a sufficient explanation of the feverish rate of drilling despite persistently mounting overproduction.

A gradual decline in magnitude and frequency of flowing wells brought a clearer realization that there were more persistent and permanent forces leading to the rapid exploitation of oil properties. Operators be-came fully aware of the advantages of being the first to drill in a particu-lar locality when wells subsequently drilled on nearby properties began to affect their rates of output. Indeed, the location of drilling sites with studied intent to encroach on adjacent producing wells became common practice. When such attempts at encroachment failed to drain the output of the neighboring well, another tactic frequently brought results: the threat to withdraw tubing and to flood the producing well with water usually induced the blackmailed operator to buy the unproductive well at an exorbitant price or to grant the blackmailer a handsome share of the productive well's output.[78]

Countless wells also had their lives shortened without express or im-plied blackmail; they were flooded out when neighboring operators ir-responsibly withdrew the tubing from exhausted wells or dry holes.[79]

The rapid rate of exploitation and close crowding of wells no longer could be attributed solely to the excesses of speculation and a mining boom. An enlarged understanding of the properties of petroleum gener-ated a sustained race, almost without respect to prevailing prices, to drill leaseholds before rivals tapped the same oil. Similarly, self-interest en-couraged landowners to lease their property in the smallest possible units in order to obtain the largest number of wells on their land.† Conse-

---

* For a discussion of legal foundations and application of this principle, refer to Ap-pendix E.

† This probably was the largest factor accounting for the progressive reduction of typical leaseholds of from one to eight acres in the early years, to one-half acre, containing as many as three wells, at mid-decade.

quently, instances of wells lined in rows were far more common than wells spaced in accordance with geologic considerations.

Alarmed by the increase in wasteful practices and shortened well lives, responsible operators sought relief through changes in leasing terms. They specified the number of wells on leased lots, increased the size of lease-holds, and wrote in provisions restricting withdrawals of tubing and re-quiring abandoned wells to be plugged. Although these became common provisions in mid-decade leases, the results were disappointing, particu-larly as many absentee landowners, ignorant of the problems and inter-ested only in immediate returns, refused to participate in leasing re-forms.[80]

This was not true of all landowners, for by the late 1860's there were a number of well-run oil properties in the Region. One outstanding example was the Story farm, a 500-acre tract on Oil Creek that had been pur-chased by the Columbia Oil Company of Pittsburgh in 1861. Leasing a portion of the property to operators, the Columbia company realized large profits from flowing wells during 1863 and 1864.

As production settled down to a pumping-well stage, Columbia stopped leasing to outsiders and, as old leases were forfeited or cancelled, gradu-ally took over all operations on the farm including the drilling of new wells. According to the superintendent's report for 1868, all new wells after 1865, including eleven completed during 1868, had been drilled "on the hill, where great care is taken to protect the wells by locating them at least 300 feet apart." [81] It was further noted that "Five of the new wells were permanently cased and the water shut off before drilling into oil bearing rock." This cost $150 to $200 more per well, not including casing, but the advantages over ordinary drilling "are incalculable." [82] Daily out-put of the pumping wells, numbering twenty-three in December, 1868, averaged 374 barrels (about 17 barrels per well).[83] According to carefully kept accounts, this output was sufficient to bring the Columbia Oil Com-pany a gross profit for the year of just over $381,000.*

An agreement reached in 1865 by the leaseholders of the Tarr farm, be-low Titusville, also illustrated the possibility of bringing about reforms through voluntary co-operation. Their problem was acute; promiscuous and sometimes vindictive withdrawals of tubing had flooded some sixty producing wells on the farm and had brought production to a halt. After

---

* The management estimated the average "whole cost" of the approximate 137,000 barrels produced during 1868 at $1.71 per barrel, which they were able to market at an average price of $3.95 per barrel.[84]

several months of negotiation, an agreement was finally reached. It called for isolation and pumping of flooded areas; seedbagging and casing of all productive wells after they had been pumped dry of water; and synchronized operation of pumping wells. When the program quickly restored production to 1,000 barrels a day, it gained sustained support.[85]

Essentially the same reasons (absentee landlords and shortsighted operators) that limited the effectiveness of leasing reform also kept this early example of a successful attempt to unitize a field from being widely imitated. It was difficult enough to get agreement in the case of the Tarr farm, in which the crisis was obvious and most of the interested parties were local residents and available for negotiations.

Eventually, however, sentiment in the Regions began to press for legislative action to regulate well operations. Specifically it was proposed to compel operators to plug or case their wells before abandoning them and to finance a government police force by a tax either on each productive well or on the capital invested by each company, firm, or individual.[86] As a result of growing public demand, the state representative from Venango County introduced the first bill to regulate tubing and plugging of wells in the 1867 session of the legislature, only to have it tabled.[87] While this was the first move in subsequent agitation for legislation that resulted in the first plugging law passed in 1878, another half-century or more was to pass before effective measures for conservation came into operation.

Chapter 8

The transport of crude

The fourfold increase in production of crude that accompanied the discovery of flowing wells during 1861 and 1862 exerted a critical pressure to improve transport facilities and better methods of handling and storing oil. By the late 1860's response to these pressures brought about the adoption of two major innovations affecting the future transport and handling of petroleum: railroad tank cars and gathering pipelines. By this time also the Region was threaded with local railroad lines which, through their connections with the major trunk lines, offered ready access to major marketing areas. But to meet the immediate necessity of getting their oil to market during the intervening years, oilmen turned to the more promising all-water route down Oil Creek and on to Pittsburgh via the Allegheny River.

COURTESY STANDARD OIL COMPANY (NEW JERSEY)

Oil City, Pennsylvania, 1865.

Transport by water to Pittsburgh

Limited shipments of barreled oil during 1860 and 1861 had already demonstrated the feasibility of shipping crude from the Region to Pittsburgh via the Allegheny River. While the Allegheny was not fully navigable during the late summer and part of the winter, the principal bottleneck was Oil Creek, where much of the time the water level was too low to permit even shallow draft craft to negotiate its winding, narrow course from Titusville south to Oil City.

The key to the solution of this problem lay in the pond freshet. For years operators of sawmills on the main branches of Oil Creek, some of them as far as ten miles above Titusville, had made their "ponds fresh" by closing the floodgates on their dams during low-water periods. When enough water had accumulated, the floodgates were opened, raising the water level below sufficiently to float their lumber on down Oil Creek.[1]

The originator of the plan to adapt the pond freshets to petroleum transport remains anonymous, but in 1862 the producers and shippers appointed A. L. Dobbs superintendent of their organization to schedule the freshets, negotiate arrangements and fees with dam owners, and collect assessments from shippers.[2]

There was feverish activity along Oil Creek at the time of the freshets,

First-class passenger "packet" on Oil Creek.

which during peak periods were scheduled twice and occasionally three times weekly. Shipments averaging between 10,000 barrels and 20,000 barrels were handled by some two hundred to eight hundred boats and barges of all descriptions, which were laboriously towed upstream by mules and horses into position for loading, well in advance of each run. All craft had to be of extremely shallow draft, for the best freshet raised water levels only 22–30 inches above the highest rocks in the stream.[3] The boats, which varied widely in size, included: diminutive, scow-shaped "guipes;" small flatboats, 20–50 feet long, on which 25–50 barrels could be stowed; and "French Creekers" (barges formerly used to haul pig iron), 80–120 feet in length, 15–20 feet wide, with capacities of 1,000–1,200 barrels.[4]

Pond freshets brought hard-bitten rivermen into competition with teamsters to create wild, mining-camp lore. The rivermen literally murdered thousands of horses and mules with brutal treatment and overwork in dragging the empty barges up Oil Creek. At rates ranging between 15¢–75¢, and occasionally $1.00 a barrel, these skilled, hard-drinking pilots averaged between $100 and $200 for the trip down Oil

Loading barreled oil at Funkville, on Oil Creek, in the 1860's.

Creek. In some instances they made as much as $800 or $1,000 for the eleven-mile ride.[5]

The cry, "pond freshet," brought throngs to the creek banks and to the transfer point at Oil City to witness a colorful spectacle lasting three or four hours. Boatmen soon learned that it was best to cast loose just as the waters of the freshet began to recede, but anxiety to be among the first, and to avoid jams and crushes, upset many a pilot's sense of timing. Starting ahead of the rise, a laden boat would ground on the first shoal, and would usually swing helplessly broadside of the current, either to fill with water and sink, or to be smashed to kindling wood by the first collision with a larger boat. Several of the narrowest points on Oil Creek provided especially vulnerable places for jams and favorite gathering points for thrill-minded spectators.[6] The more frugal-minded improvised small dams at strategic locations along the creek to collect spilled oil; one man, when oil was $10 a barrel, was reputed to have earned $900 in a single afternoon with his improvised dam.

It was a rare freshet in which no oil was lost. Recoveries of barreled oil were usually substantial, but when a bulk cargo was wrecked there was little chance for the shipper to salvage oil dumped into the creek. Large-scale catastrophes affecting many shippers frequently occurred. In December, 1862, an ice gorge broke loose, crashing into 350 boats with 60,-000 barrels of crude oil, destroying half of the boats and their cargoes and causing property loss of $350,000.[7] The following spring an overturned lantern on the edge of a bulk barge caused a fire which ultimately consumed nearly a hundred boats, more than 8,000 barrels of oil, and destroyed the bridge over the Allegheny River at Franklin.[8] Despite these difficulties there were some 2,000 craft engaged in the oil traffic on Oil Creek and the Allegheny by 1865.[9] By the spring of 1862, some fifteen steamers and tow boats were already carrying crude between Oil City and Pittsburgh. During the eight or nine months the Allegheny was at a "good boating stage," steamers averaged three round trips a week, while flat-boats required three to four days to negotiate the one-way journey downstream.[10]

One of the significant features of water transportation down Oil Creek and the Allegheny was the early development of bulk handling. Initial attempts to ship in bulk proved impractical; the slightest motion of the water set the cargo to rocking, often capsizing the craft and spilling the oil. This hazard was greatly reduced late in 1861 when Richard Glyde of Pittsburgh designed and built an oil barge for use on the Allegheny with several compartments, each holding 80–100 barrels of oil, "the same being securely decked over."[11] Some of the smaller versions of Glyde's barges were used on Oil Creek and towed on down the Allegheny. More commonly, most of the oil, whether in barrels or bulk, was transferred at Oil City to craft more suited to river traffic.

The best of the bulk barges, however, were subject to considerable leakage plus the ever-present danger of loss through wreckage. For these reasons shippers generally preferred to use barrels. But when the price of oil was low, or barrels were scarce, or when demands coincided with good boating stages, shippers turned to bulk transport.

Among the pioneer shippers who did much to promote bulk handling of crude on the Allegheny was Captain Jacob Jan Vandergrift of Pittsburgh, destined to forge a career as refiner, producer, and most notably, pipeline executive with the Rockefeller organization. Vandergrift first entered the new oil business in November, 1861, when he towed 4,000 empty barrels on two coal-boat bottoms with his tug steamer, *Red Fox*, from Pitts-

COURTESY DRAKE WELL MUSEUM

Pond freshet disaster, Oil City, May 31, 1864.

burgh to Oil City.[12] There he bought 5,000 barrels of oil from the Mapleton Oil Company on the Blood farm, for delivery the following July.
Puzzled as to how he could best transport the oil to Pittsburgh, Vandergrift chanced to see one of the early bulk boats built by Richard Glyde.
He promptly had twelve duplicates built for himself, each 80 feet long,
14 feet wide, and 3 feet deep. With these and additions to his steamer
tugs, Vandergrift soon realized a small fortune.[13]

Though wasteful and subject to all of the hazards and uncertainties of
river transportation, boating and barging to Pittsburgh afforded the first
bulk shipping on a significant scale, so necessary to break the transportation impasse created by the surplus production of flowing wells. Flatboats manned by larger crews ordinarily charged from 40¢ to $1.00 a
barrel for the 133-mile haul to Pittsburgh; by steamer, the usual rate was
50¢–75¢ a barrel.[14]

One important effect was to make Pittsburgh the leading petroleum
warehouse during the first half of the decade. Deliveries of oil to Pittsburgh, principally crude (which amounted to nearly 172,000 barrels in
1862), grew in round numbers over the next few years as follows: 175,000
barrels in 1863; 268,700 barrels in 1864; 630,250 barrels in 1865; and
1,250,000 barrels in 1866, the high point for the decade.[15]

Railroad expansion in the Region

While the system of water transport linking Pittsburgh with Oil Creek continued to flourish through the mid-1860's, its relative importance declined after local railroads, closely linked with major trunk lines, began to thread the Region.

The significance of a more direct route to the Atlantic coast in affecting the plans of local railroad builders was illustrated by the Oil Creek Railroad. Chartered in 1860 and financed principally by producing interests in the Region, the Oil Creek Railroad was originally projected as a 4-foot 8½-inch gauge line to run between Oil City and Corry, where it would connect with the Philadelphia & Erie. But when the Atlantic & Great Western completed a wide, 6-foot gauge line from Salamanca, New York, to Corry in May, 1861 (offering a through route to the eastern seaboard via the Erie Railroad), the promoters of the Oil Creek Railroad shifted to the same gauge.[16] (See Map 8:1.)

Completed between Titusville and Corry in the fall of 1862, the Oil Creek Railroad was prosperous from the beginning. Between November, 1862, and January, 1864, its freight traffic included nearly 460,000 empty barrels hauled into the Region. And it shipped out over 430,000 barrels of oil, mostly to New York via the Atlantic & Great Western. Receipts from all sources for the period totaled almost $385,000, while operating expenses were low, being estimated at no more than 25 per cent of gross income.[17]

Plans to extend the Oil Creek Railroad south to Oil City and on to Franklin ran into difficulties in 1863, largely because of the problems of handling a large volume of traffic over a single track.[18] But in July, 1864, track was completed to the Shaffer farm eight miles below Titusville, and a year later extended to Petroleum Center three miles beyond, which became the southern terminus of the road.[19] Shaffer farm and the Miller farm, adjoining it on the north, became important loading stations for the Oil Creek Railroad, particularly when, as will be noted later, they became the terminals of several of the early gathering pipelines.

By adopting the same gauge as the Atlantic & Great Western, the Oil Creek Railroad became closely associated with a railroad destined to play an important role in the future trunk line competition for oil traffic. Shortly following the completion of the extension from Salamanca to Corry in 1861, the Atlantic & Great Western came under the financial management of a group of prominent British capitalists, including Sir

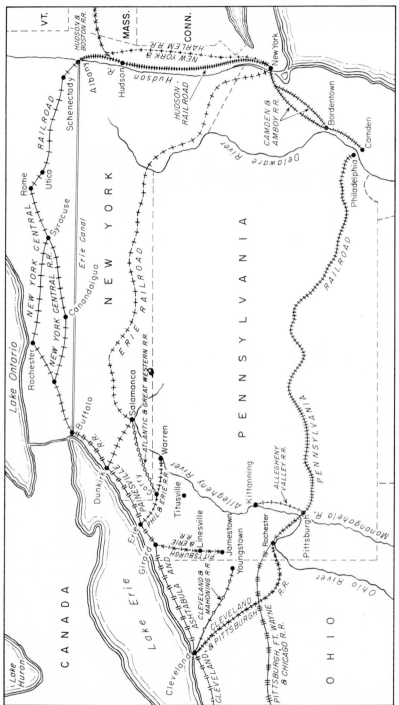

MAP 8:1. *Major railroad lines between Cleveland and eastern seaboard, 1861.*

Depot, Oil Creek Railroad, Titusville, 1865.

Morton Peto, world-famous railroad contractor and senior partner in the London investment banking house of Peto & Betts.[20]

With ample financial backing, the officials of the Atlantic & Great Western immediately launched a vigorous building program. By 1862 the road's main line was extended to Akron, Ohio, and on to Dayton a year later. By leasing the Cleveland & Mahoning in 1863, the Atlantic & Great Western was assured of a direct connection with Cleveland. Working arrangements also provided for an exchange of traffic with existing lines serving Cincinnati and St. Louis.[21]

Of more immediate interest to producers and shippers along Oil Creek below Titusville was the completion by the Atlantic & Great Western of a branch line from Corry to Meadville in the fall of 1862, and extended to Franklin the following July.[22] Late in February, 1865, the Franklin branch negotiated the final eight miles into Oil City.[23] (See Map 8:2.)

Both the Oil Creek Railroad and the Atlantic & Great Western's branch line, however, lay to the west of Pithole Creek, where developments during 1865, and the opening up of Bennehoff and Pioneer Runs off Oil Creek in 1866, added impetus to an extension of rail facilities. In 1865 construction was started on the Oil City & Pithole Railroad and a year later trains began running over the completed fifteen-mile track that connected Pithole with Oil City.[24] In 1866 another group of local capitalists began construction of the Farmers' Railroad which by the end of the year was operating along Oil Creek between Oil City and Petroleum Center some eight miles to the north.[25] Early in 1866, work was started on still a third local line, the Reno & Pithole Railroad, which was to extend from Reno, four miles west of Oil City, to Pithole, some eighteen miles to the north-

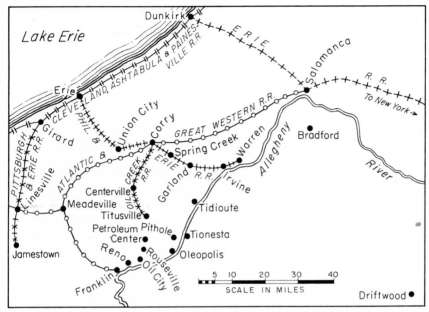

MAP 8:2. *Railroad facilities in the Region, early 1865.*

east. Despite extremely rugged terrain which slowed down construction, the line was completed by March, 1866.[26]

Railroad competition for oil traffic took on a new dimension in 1864 when the Pennsylvania Railroad, whose control over all the lines running between Pittsburgh and Philadelphia gave it a virtual monopoly of refined and crude oil shipped east from Pittsburgh, made its first move to acquire lines directly serving the Region. Indeed, Thomas A. Scott, the energetic and able vice president of the Pennsylvania, took the first step to divert traffic from the Atlantic & Great Western only a few weeks after the latter had completed its line to Corry in 1861.

Recognizing the great attraction to shippers of the Atlantic & Great Western's connection with the seaboard via the Erie, Scott set out to develop an alternative route over the Philadelphia & Erie. Under a long-term leasing arrangement, the Pennsylvania management agreed to finance the completion of the Philadelphia & Erie's tracks east from its terminus at Warren to Sunbury. Delays brought on in part by the outbreak of the Civil War held up construction for a time but the completion of the line by fall of 1864 provided through connections to the Atlantic seaboard over several alternative routes from the eastern terminal of the Philadelphia & Erie.[27] (See Map 8:3.)

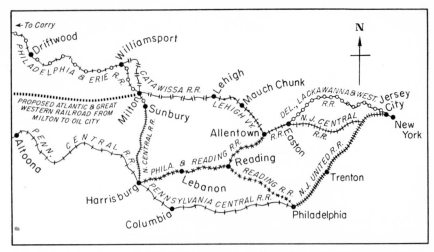

MAP 8:3. *Rail connections with Atlantic seaboard via Catawissa Railroad.*

The most direct route between Corry and Philadelphia was via the Philadelphia & Erie to Sunbury, the Northern Central to Harrisburg, and the main line of the Pennsylvania to Philadelphia. A second possible route branched off at Milton over the Catawissa to Mauch Chunk, the Lehigh Valley to Allentown, and the Reading to Philadelphia. Shipments to New York might follow either of these routes to Philadelphia and then go north by water or over the New Jersey roads already under contract to the Pennsylvania. Alternatively New York shipments could go from Milton via the Catawissa and Lehigh Valley to Easton, and on to New York over the New Jersey Central or the Delaware & Lackawanna to Jersey City.

As the Philadelphia & Erie neared completion in the summer of 1864, the Pennsylvania officers made their first move to secure a larger share of the traffic originating in the Region for the Philadelphia & Erie by acquiring a controlling interest in the Oil Creek Railroad. To facilitate direct transfer of rolling stock from the broad-gauged Oil Creek road to the narrower Philadelphia & Erie tracks, a third rail was added to the tracks of the former by the summer of 1865. Shortage of rolling stock and operational difficulties interfered with an immediate shift in traffic away from the Atlantic & Great Western, but that transition was only a question of time.[28]

In 1865 the Pennsylvania took a further step to build up traffic over the Philadelphia & Erie by organizing a "fast-freight" or freight-forwarding affiliate, the Empire Transportation Company. Under a contract drawn up in June, between the Pennsylvania and Empire, the latter was to provide a through freight service over the tracks of the Philadelphia & Erie between the lines connecting with the Philadelphia & Erie both east and west. Supplying its own rolling stock, the Empire was to pay the Philadelphia & Erie a regular toll for the use of its tracks, and in return received a mileage allowance plus a drawback of one-third of the toll charge.[29]

In making such an agreement with a separately incorporated transportation company, the Pennsylvania was following a precedent that antedated the Civil War. These companies, just beginning to develop in the late 1850's, were a reflection of the disintegrated character of railroading in an era when end-to-end lines were often separately owned and of widely varying gauge. They solicited their own freight, supplied part or all of their own rolling stock and, by equipping their cars with "adjustment wheels," were able to ship over differently gauged roads without breaking bulk. The saving in time and expense made their services so attractive on through shipments that they eventually became powerful enough to force some roads with widely deviant gauges to adjust their tracks.[30]

By mid-1860's several fast-freight companies were already operating over several railroads including the New York Central, the Erie, and the Atlantic & Great Western.* In the years that lay ahead they were to play an increasingly important part in the competition between the major trunk lines for traffic.

Early in 1866 the Pennsylvania management launched another flanking movement with the construction of the Warren & Franklin Railroad. Starting from Irvine, on the route of the Philadelphia & Erie where it crossed the Allegheny River, the Warren & Franklin followed the course of the Allegheny River south through Tidioute to Oleopolis, the main river shipping point for oil from the Pithole field. Completed in August, 1866, the Warren & Franklin almost immediately extended its trackage to Oil City by purchasing the Oil City & Pithole Railroad.[32] A few weeks later the Warren & Franklin Railroad executed a final coup by acquiring the Farmer's Railroad which was pushing its tracks north from Oil City

---

* Two years before Empire, the Pennsylvania had made a similar agreement with the Star Union Line to operate over its main tracks between Pittsburgh and Philadelphia.[31]

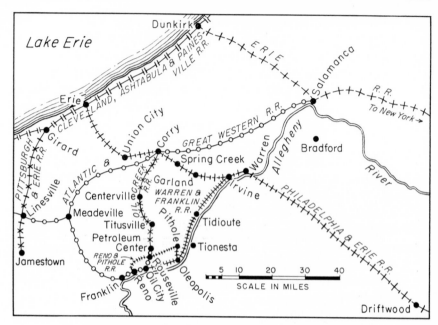

MAP 8:4. *Railroad facilities in the Region, 1866.*

to Petroleum Center. Completion of the Farmer's Railroad by late September, 1866, gave the Pennsylvania Railroad control of the Region's two great outlets at Corry and Irvine.[33] (See Map 8:4.)

The Atlantic & Great Western had too great a stake in the oil traffic to the eastern seaboard to allow these moves by the Pennsylvania to go unchallenged. But in attempting to strengthen the road's position *vis-a-vis* the Pennsylvania, the Atlantic & Great Western management operated under a severe handicap. This was particularly true in respect to its outlet to the eastern seaboard via the Erie. The Erie management was neither willing nor able to improve its single track line and to add to its rolling stock sufficiently to handle a growing volume of traffic. The Atlantic & Great Western might have overcome this handicap, in part at least, by cutting rates on oil shipped from Oil City via Franklin and Corry to Salamanca to a level equal to the rates from Corry east over the Philadelphia & Erie. But under an early contract with the Oil Creek Railroad, still in force, the Atlantic & Great Western had agreed not to carry oil from the Region to Corry at rates lower than those charged by the Oil Creek line.*

---

* At least as early as 1865, and possibly before the rate per barrel between the Region and Corry was established at 80¢ a barrel.[34]

Determined to improve its connections to the seaboard, the Atlantic & Great Western in 1865 tried without success to interest the Pennsylvania in a contract to share the Philadelphia & Erie tracks east from Corry. Turning to the Erie, the Atlantic & Great Western next made a proposal that the two lines—without losing their corporate identity—should function as a single trunk line between New York and Cincinnati and beyond. As a part of the arrangement, the Atlantic & Great Western offered to finance the double-tracking of the Erie between Salamanca and New York, and to double-track its own line west when it became necessary.[35] Apparently the prospect of occupying a subordinate position in the operation of the proposed trunk line did not appeal to the Erie management, which flatly rejected the offer.*

The Atlantic & Great Western then returned to the possibility of using the Philadelphia & Erie tracks in spite of objections by the Pennsylvania. In November, 1865, the Atlantic & Great Western leased the Catawissa Railroad which, as has already been noted (see Map 8:3), ran from Milton on the route of the Philadelphia & Erie to Mauch Chunk, where it connected with the Lehigh Valley Road. This was a bold move designed to outmaneuver the Pennsylvania, for in 1864 the Philadelphia & Erie and the Catawissa had entered into a long-term contract for an exchange of traffic. This contract was included in the agreement between the Catawissa and the Atlantic & Great Western. At the same time the Atlantic & Great Western began surveys for a direct line to the Oil Region to run between Franklin and Milton, which would have cut into the traffic of the Warren & Franklin extension that the Philadelphia & Erie was about to launch.[37]

The Pennsylvania's response was to resort to legal tactics. In January, 1866, the Pennsylvania's lawyers obtained a temporary injunction against the Atlantic & Great Western from a Justice of the Pennsylvania Supreme Court. They based their case on the grounds that the leasing arrangement between the Catawissa and the Atlantic & Great Western was invalid because of the latter's failure to register its charter with the Pennsylvania Secretary of State.[38] When the full court reviewed the case in June, 1866, the injunction was promptly thrown out. But this proved to be an empty victory for the Atlantic & Great Western. A panic in the London money market the preceding May had brought about the failure of the road's

---

* It is also possible that Cornelius Vanderbilt, who was represented on the Erie board, and had trunk-line ambitions of his own, did not want to encourage the British capitalists in their plan to develop a powerful competitor.[36]

chief financial backers, the firm of Peto and Betts.[39] All plans for an extensive building program by the Atlantic & Great Western, including a conversion to a 4-foot 8½-inch gauge to conform to the Philadelphia & Erie
and Catawissa tracks, were abandoned. Within two years, control of the
Atlantic & Great Western was taken over by the Erie.

But the defeat of the Atlantic & Great Western only brought to a close
the first round of a three-cornered rivalry for the oil trade that was soon to
emerge between Scott, Jay Gould, and Cornelius Vanderbilt and their
respective trunk lines. Development during the mid-1860's of the tank car
and gathering pipelines, were to have an important bearing on this conflict.

Bulk shipments of crude

From the beginning of the flowing-well period, pressure to improve the
handling of crude by rail had been enormous. Some improvement had
come after 1862 when interest in oil traffic prompted the railroads to begin supplying coal and box cars with some protective covering instead of
the open flatcars formerly allocated.[40] But this afforded little compensation
for the serious deficiencies of barrels as containers for crude oil. Shortages
and skyrocketing prices of barrels became commonplace in the Region.
Losses from leakage, evaporation, explosion, and fire en route were severe, and effective means to curb those losses were extremely difficult
to realize.*

In large part the difficulties stemmed from the nature of crude itself.
Practically all crudes contained salt water and other ingredients which
tended to dissolve the glues, shellacs, and other materials used to line barrels, thus magnifying losses through leakage and evaporation.[42] Crude had
another equally serious characteristic. Because its initial boiling point was
lower than that of refined illuminants, crude more readily developed gas
pressures in the barrel at dangerous levels, a condition which was heightened by the considerable amounts of gas in the output from most flowing
wells. As a result, both ordinary evaporation losses and those from explosions and fires were more severe with barreled crude than with refined
crude.[43]

These difficulties were compounded by the poor quality of available

---

* Some contemporary estimates placed losses at one-fifth to one-third of every barrel
of crude shipped out of the Region to domestic refineries. Others reported losses of
one-tenth in shipments to New York, an ordinary if not normal occurrence.[41]

Loading crude oil in barrels in the Region, 1865.

barrels. The initial demand for containers of any kind led shippers to ransack "cellars, barns, and trash heaps of western Pennsylvania for old vinegar and molasses barrels, beer and whiskey barrels, cider and turpentine barrels," which might be used to package crude for shipment.[44] By 1860, empty barrels of every description began making their way up the Allegheny to Oil City on barges and flatboats towed by river steamers or downriver on rafts from points along the upper Allegheny.

The supply was soon augmented from cooperage shops at Titusville and other nearby towns. But barrels made in the Region, principally during winter when farm labor was available, were of poor quality, even by contemporary standards. Cold, green wood was coopered without steaming, washing, or drying, and cold glue was applied but once, instead of being heated and reapplied several times, preferably under air pressure. The introduction of some coopering machinery without the complement of kiln drying speeded output of unseasoned staves, but did nothing to improve quality.*

---

* Pursuing advanced practices with an experienced imported force of twenty coopers, Samuel Downer's barreling establishment at his Corry refinery had no counterpart in the Region, yet it lacked a permanent brick kiln until 1864.[45]

After 1863, improved processes for heating, treating, and lining barrels with various compositions of glue, molasses, glycerin, and even soluble glass began to make moderate headway. But coopers in the Region did not begin to use these methods until 1866, when the transition to bulk shipments was well under way.[46]

Some barrels of better quality manufactured outside the Region were available, but because of the greater value of the product, refiners and commission merchants usually reserved the best barrels for the refined crude. There was an additional reason for not transferring premium barrels to the crude trade: once tainted with crude, the barrels became worthless for any other service. Thus for the most part only the greenest of new barrels or the poorest seconds found their way to the wells as replacements, returns, exchanges, or sales.[47] Leading exceptions were heavy-duty shipments overseas or to refineries over which merchants or refiners on the seaboard exercised close control.

The cost of barrels fluctuated considerably according to quality, the season, and variations in demand; but throughout the 1860's, prices in the Region seldom fell below $2.00 each and at times barrels sold at $4.00 each or more.* Even accepting a conservative estimate of an average price of $3.00 for the years between 1863 and 1866, the cost of barrels often exceeded the value of their contents. Cooperage and handling costs at shipping points and market terminals added approximately $1.00 to the cost of each barrel moving between wells and seaboard destinations.[49] Finally, there was an allowance for leakage charged against the shipper which amounted to 25¢ a barrel up until the mid-1860's, when a deduction of 2 gallons per barrel in favor of the buyer became general practice.[50]

With costs directly associated with their use amounting to $4.00 or more per barrel, it is not astonishing that shippers began to experiment with bulk handling as a means of eliminating barrels for the transport of crude. One of the earliest recorded attempts to substitute "bulk" for barrel shipments of crude by rail was made by the Pennsylvania Railroad, which early in 1865 introduced a "rotary oil car" into service between Pittsburgh and Philadelphia. This model consisted of two pairs of solid iron wheels, about 60 inches in diameter, attached to hollow axles, each

---

* In the summer of 1863, for example, quotations ranged between $3.00 and $3.25, having advanced from $2.75 to $3.00 the preceding December. A year later "prime new barrels" were bringing $2.50 to $2.75, while in fall of 1865 when the shift to bulk transport was well under way, empties sold at Titusville and Oil City at prices ranging between $3.25 and $3.60.[48]

COURTESY STANDARD OIL COMPANY (CALIFORNIA)

Densmore-type railroad tank car.

with a capacity of 500 gallons of oil. The axles in turn were mounted by a wooden platform on which other freight could be hauled.[51]

Credit for the first type of tank car to be widely adopted, however, goes to Amos Densmore, pioneer producer and member of the firm of Densmore, Watson & Company, crude-oil buyers in the Region. Late in the summer of 1865, Densmore mounted two wooden tanks, each containing 40–45 barrels of oil, on a flatcar belonging to the Atlantic & Great Western Railroad and sent them on their way to New York via the Oil Creek and the Atlantic & Great Western railroads. Wires from observers along the route at Salamanca and Elmira reported that no leakage was occurring and confirmed the success of the experiment.[52]

Although he attempted to keep his innovation a secret, other shippers quickly adopted Densmore's method of transporting crude. The railroads were also quick to appreciate the importance of supplying the services of tank cars to shippers as a means of protecting their share of the oil trade. By early 1866, the Pennsylvania's fast freight affiliate, the Empire Transportation Company, had begun adding tank cars to its equipment, while the Atlantic & Great Western had arranged with "oil tank companies to supply cars to carry oil in bulk." Early in 1867, the Atlantic & Great Western established its own subsidiary, the Broad Gauge Transit Company, primarily for the purpose of supplying cars upon which shippers could place their own tanks.[53]

Despite Densmore's early success, shippers found it extremely difficult to prevent wooden tanks from leaking. While this problem was reduced by replacing them with iron tanks, similar in design but securely riveted, it was soon discovered that this type of car had other and equally serious shortcomings. Being upright, the tanks were top-heavy when loaded, a condition which was further aggravated by lack of provision to prevent the oil from sloshing about in the tank. As a result, in collisions or in negotiating curves and grades, they were quite likely to tip over and catch fire. Moreover it was impossible to enlarge their capacity except by increasing their height. This not only raised the center of gravity, but ran into the further limitation of the clearance requirements of railroads for tunnels and bridges. Finally, neither the wooden nor the iron tub tanks provided any convenient way for the oil to expand. As a consequence there was an ever-present tendency to build up dangerous gas pressures, adding to the danger of explosions and leakage.[54]

A number of these difficulties were anticipated by J. F. Keeler of Pittsburgh who built a predecessor of the modern type tank car in the fall of 1865. Twenty-five feet long, 8 feet wide, shaped like a box car, with the bottom rounded like a U, the car was constructed of ⅜-inch wrought-iron plates that were riveted, well-braced, and stayed by angle irons. An unusual feature of its design was a false top or roof made of ½-inch board and suspended about 2 inches below the top plates of the car. Its purpose was to prevent the oil from surging when the car was filled and in motion and at the same time to permit the oil to expand, thus reducing the quantity of gas which otherwise would be formed.[55] Keeler's car apparently was never adopted on any extensive scale. Aside from its somewhat elaborate and expensive construction, the U-shaped design of its tank sacrificed the superior strength and space economies of cylindrical tanks.

These advantages were not fully realized until the first horizontal, boiler-type cars with the essential requisite of a low center of gravity and their dome to allow "the oil to expand without injury to the tank" were introduced into the Region in 1868.[56]

Although fiery wrecks continued at an alarming rate, the carriers moved slowly to replace the hundreds of tub-type wooden and iron tank cars they had in service with the more expensive boiler-type model. But after 1871 they had little choice. In that year an especially appalling wreck occurred on the Hudson River Railroad in which flaming petroleum took a heavy toll of lives. An aroused New York state legislature took immediate action banning all tank cars on the New York lines, except the horizontal-boiler type.[57]

For all their limitations, even the earliest tank cars brought significant savings to shippers through a reduction of losses from leakage and evaporation and, above all, the elimination of barrels costing several dollars each. Shippers also shared in some reduction of handling charges and of freight rates, turning principally on the difference between the 80-barrel capacity of a boxcar and the 80-barrel capacity, or more, of tank cars. By 1867 these rates were reflected in the New York market by quotations on crude in bulk (about 5¢ a gallon lower than those for barreled crude).[58]

However significant the contribution of tank cars to the elimination of barrels for rail shipments, a fully developed transport system whereby crude could be moved from wellheads to markets in bulk would have been impossible without the complementary development of gathering lines.

Gathering lines

Early promoters of petroleum pipelines were undoubtedly inspired by the fact that as early as 1850 the existence of more than eighty city water systems and over fifty gas-distributing systems had long since demonstrated the practicability of piping either liquids or gases over considerable distances.[59] Apparently the first person to consider the possibility of transporting petroleum by this method was S. F. Karns, pioneer operator in the western Virginia fields who, in 1860, proposed to lay a 6-inch line from his well at Burning Springs to the railhead at Parkersburg. But before Karns could start construction, the Civil War broke out and he left to join the Union Army.*

The experiences of J. L. Hutchings, a New Jersey inventor of a patented rotary pump, illustrate some of the technical difficulties involved in the early pipeline ventures in the Region. In the fall of 1862, Hutchings laid a string of tubing for Barrows & Company to connect their well on the Tarr farm to their refinery some 1,000 feet away.[61] This miniature pipeline, operating on a siphon principle, worked successfully. But the 2½-mile line from the Tarr farm to the Humbolt refinery at Plumer built by

---

* Hemus James, a neighbor of Karns in western Virginia who moved to Oil Creek after the outbreak of war, planned a 4-inch wooden line from Tarr farm to Oil City in November, 1862. Failing to secure a charter from the Pennsylvania legislature, he abandoned the project.[60]

Hutchings a year later was a different story. To force the oil over an intervening hill some 400 feet high, he used his own rotary pump. While the oil coursed through the pipes (refuting engineering predictions that friction would make pipelines impractical), the venture collapsed because of the poor quality of the cast iron pipes and a failure to make the joints tight against leakage.

In the winter of 1863–64, Hutchings corrected most of these difficulties in a second line from the big Sherman well to the Oil Creek Railroad at Miller farm by running a solid, cast-iron pipe connected with specially designed joints in a shallow ditch. But the jarring of the pumps and sudden variances in the velocity at which the oil was pumped kept the pipe in constant vibration, springing the joints and sometimes rupturing the pipe as well. Teamsters sealed the failure by tearing up the line.[62]

In 1864 the Western Transportation Company obtained a pipeline charter from the Pennsylvania legislature, but its first line between the Noble and Delemater well to the Shaffer farm also proved a failure. Laid on a regular grade like water pipes, its 5-inch cast-iron pipe leaked at its leaded joints "like a fifty-cent umbrella." [63]

Despite the great urge to move crude oil from wellheads to loading points more efficiently, the first technically successful pipeline was designed to move refined petroleum. In the spring or summer of 1864, the Warren refinery at Plumer laid and successfully operated a 2-inch wrought-iron line for forcing "distillate" from the refinery to the Allegheny River, a distance of about three miles.

It was the booming Pithole field with its enormous transportation congestion, however, that provided the setting for the first successful crude line. Samuel Van Syckle, an inventive Jerseyite who arrived in Titusville in the fall of 1864, soon became engrossed in Pithole's transport problem. Enlisting financial aid from New York, Cleveland, and Region capitalists, Van Syckle organized the Oil Transportation Association to carry out a $100,000 pipeline venture connecting Pithole with Miller's farm, on the line of the Oil Creek Railroad some five miles away.[64]

Van Syckle built his line with painstaking thoroughness. All joints of the 15-foot sections of 2-inch wrought-iron pipe were lap-welded and tested at a pressure of 900 pounds to the inch. Three pumps were used to force the oil through the pipe at a rate of 81 barrels an hour, and a telegraph line was installed to relay information about shipments. Despite sabotaging activities by teamsters, the line was completed, tested, and put into operation on October 10, 1865, a month after it was started.[65]

In contrast to prevailing charges by teamsters of $3.00 or more a barrel, the new line charged $1.00 for pipeage and realized a good profit. By early December, 1865, Van Syckle had completed a second line from Pithole to the Miller farm, giving him a combined daily capacity of 2,000 barrels.[66]

Van Syckle, however, did not remain long in the pipeline business. Although he stayed out of production, he did become involved in the buying and selling of oil on a large scale. When the oil boom collapsed late in 1865, he found himself deeply in debt to the Titusville National Bank, which in 1866 took over his interests in the pipeline company.[67] Thwarted in capitalizing on his pipeline innovation, Van Syckle obtained new financial backing and erected a refinery at Titusville, where he subsequently distinguished himself as a pioneer developer of continuous distillation.[68]

Meanwhile, Van Syckle's initial success ignited a small boom in the construction of local lines. The Warren brothers, whose "product" pipeline from Plumer to the Allegheny River had antedated Van Syckle's operations, completed a 2-inch crude line with a daily capacity of 2,000 barrels on October 24, 1865. Running from Pithole to Henry's Bend on the Allegheny River, the line netted its owners profits of $20,000 during the first six months of operation, even though output at Pithole was declining.[69]

In the summer of 1865, Thomas C. Bates, an entrepreneur from Rochester, New York, built a 6-inch line, seven miles long, from Pithole to the junction of Pithole Creek and the Allegheny River. A fall in grade of about 55 feet a mile made it possible to move some 7,000 barrels of oil daily without the use of pumps.* Completion of this line gave rise to the town of Oleopolis as a shipping and refining center.[71] Prior to the construction of the Warren & Franklin Railroad from Irvine to Oil City the following August, crude oil was shipped from the pipe not only in bulk boats downriver to Pittsburgh, but was also towed in barges upstream to Irvine where it was transferred to the Philadelphia & Erie Railroad.[72]

Alarmed at the inroads that new pipelines were making on the importance of Titusville as a shipping center, a group of local producers organized the Titusville Pipe Company. Starting in January, 1866, a 2-inch line was completed the following March between Titusville and Pithole at a cost of $150,000. The longest of the early lines, it required two pumps to force the oil through the nine-mile length of the pipe. Competition for Pit-

---

* Describing the advantages of the first gravity crude-oil line in the Region, one contemporary said: "It will eat no oats, burn no fuel, and be run by an immediate law of the Creator." [70]

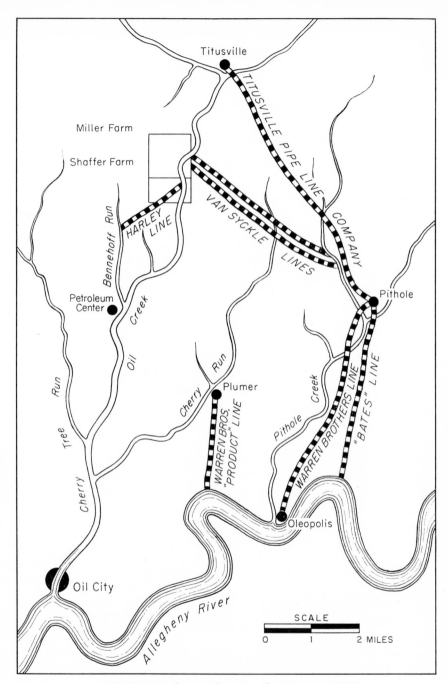

MAP 8:5. *Gathering lines in the Region, 1866.*

hole's declining production, coupled with the collapse of the oil boom in early 1866, cut earnings well below anticipated levels, and in June the line was sold to the Empire Transportation Company.[73]

The pioneer nucleus of gathering lines in the Region was completed in March, 1866, when Henry Harley built two short lines from Bennehoff Run to the Oil Creek Railroad's loading depot on Shaffer farm two miles to the east. See Map 8:5. The construction of these lines brought to a climax the bitter opposition of the teamsters who had from the start recognized the threat to their services posed by the gathering lines. Some four hundred teamsters, formerly engaged in hauling crude from Bennehoff Run to Shaffer farm, tried desperately to keep Harley's lines from operating. But a fire that destroyed a part of the loading equipment at Shaffer and the mob of seventy-five to one hundred men who a few nights later set fire to storage tanks delayed operations only temporarily. Further opposition disappeared when detectives employed by Harley arrested the ringleaders of the teamsters and had them jailed.[74]

Commenting on the current status of the gathering lines, the *Titusville Morning Herald* declared on April 21, 1866: "This process of moving oil has entirely superseded the old method of hauling oil in barrels from Bennehoff to Shaffer." The following September the *Scientific American* announced that because of the innovation, "Oil, 600 feet below the surface of the earth at Oil Creek, reaches Jersey City, a distance of over 400 miles, without having been touched by the hand of man." [75]

The innovation of gathering lines received one further elaboration in the summer of 1866 when Alfred W. Smiley, Van Syckle's buyer at Pithole, and George E. Coutant of New York built the first "accommodation pipeline" to connect the tanks at the wells with the dump tanks of the pipelines. Across the flats of Pithole they laid a network of four miles of 2-inch pipes; the rest of their outfit included a 20-horsepower boiler, a 12-inch by 8-inch pump, and two 250-barrel tanks. Priced only 25¢ a barrel, their service quickly dislodged teamsters, who had been charging from 50¢–$1.00 to move oil from the producers' tanks to pipeline dumps. Other established lines promptly adopted the service of accommodation lines.[76]

In connection with his accommodation line, Smiley also claimed to have devised the first means of accounting to the producer for the amount of oil shipped from his tanks. In the summer of 1866, he introduced the "run" ticket. Issued in duplicate, the ticket gave both the producer and the pipeline a record of the amount drawn off, as measured by gauging the producer's tanks before and after the pipeline run.[77]

By the end of 1866, gathering and accommodation lines had permanently established themselves as an integral part of oil transportation, freeing producers and shippers of the excessive rates and costly inefficiencies of teaming. At Bennehoff Run and at Pithole they had demonstrated themselves more flexible and efficient than either teaming or building rail spurs to serve new fields. This demonstration was especially timely in view of a growing appreciation by producers since the opening of Cherry Run in 1864 that drilling on the side hills and elevated lands held promise of finding oil.

No question remained of the technological and economic success of the new method of transport operating costs within a range certain to attract capital. Successful lines ranging from 2–6 inches in diameter had been built to lengths in excess of nine miles and spanning elevations up to 400 feet. Material and installation costs varied considerably but under favorable circumstances a 2-inch pipe could be built for a little over $3,000 per mile; approximately $2,000 for the pipe delivered along the route at 40¢ per foot and $1,000 for trenching and laying the pipe at the usual depth of about 30 inches. A 6-inch line, costing about $3.60 per foot for pipe, ran the costs per mile to over $18,000. Experience demonstrated that pumping stations costing approximately $2,800 ($800 for pumps and engines and $2,000 for a boiler) should not be located more than five miles apart. Pumping costs usually did not run more than 5¢ a barrel at each station when a constant supply of oil was run, while the number of stations did not often exceed two.[78]

Within less than a year, gathering lines had brought a sharp reduction in the costs of transporting oil from wellheads to shipping terminals. In the case of Pithole, for example, there was a drop in rates from $3.00 to $1.00 a barrel. The nine-mile line from Pithole to Oleopolis had proved profitable even after rates were reduced to 75¢, while Harley's line had made money charging 50¢ a barrel to move oil two miles from Bennehoff Run to Miller farm. In general the lines had pursued a policy of cutting rates enough to displace teaming; but beyond this no uniform pattern of rates had emerged. Thus while the year 1866 removed all doubt of striking new economies in local transport of oil for most shippers, it left the question open whether equitable rates for all shippers would result.

Events during 1866 foreshadowed one future characteristic of pipeline operations. In that year Henry Harley formed a partnership with William A. Abbott, pioneer refiner and producer in the Region. Purchasing Van Syckle's lines from the Titusville National Bank, they began building up storage capacity and loading facilities at Miller and Shaffer farms. Their subsequent expansion suggested a trend which would become increasingly important as proliferating independent lines were consolidated and merged under large operating companies.

Many relations between rail and pipelines remained undefined. Inspired by the great need for bulk transport both locally and to distant markets, the coincidence of the simultaneous development of tank cars and pipelines was both striking and fortunate. Although the economies of gathering lines alone would have been impressive without the supplement of tank cars, savings in barrels and handling costs would have been minimized. Linked to the tank car, the ease of constructing gathering lines to any railhead placed the rail lines in a better relative position to water transportation. Since a technological basis for trunk pipelines was totally lacking, there was no disposition on the part of railroads at this time to view pipeage as a potential competitive threat. If railroad interests gave serious thought to possible effects of gathering line systems on their traffic and rate structures, they did not evidence it by immediate direct action. By 1866 only the Empire Transportation Company had moved to acquire a gathering line.

Improvements and expansion of storage facilities

Despite their enormous significance for the oil industry, tank cars and gathering lines alone would not have brought the advantages of a fully developed system of bulk transportation without a complementary growth and improvement in storage and handling facilities. On the eve of the flowing-well period, bulk storage of crude had made scarcely more than an embryonic appearance in the Region. To receive oil from individual wells, owners typically constructed box-shaped, wooden vats, seldom exceeding 100 gallons in capacity. A few more venturesome operators, in order to supply tankage for several wells, expanded their storage capacity to several hundred gallons to make their oil more attractive to buyers supplying their own barrels.[79]

Essentially, however, the barrel served both as the primary storage retainer and as the package for handling oil in transit. Oil City, already the main transfer and shipping point for barreled oil destined for Pittsburgh, provided some auxiliary-vatted storage facilities at this time. But the railroad depots on the fringe of the Region were equipped only with loading platforms for receiving barrels. No auxiliary tanking facilities were available and the barreled oil waiting for shipment was stored in the open or placed under wooden sheds built by private parties, a situation which prevailed at these points to the mid-1860's.[80]

Storage and handling facilities at the key receiving centers at Pittsburgh and Philadelphia were scarcely any more adequate. Pittsburgh had yet to feel the pressure of shipments via the Allegheny which would soon make it the primary petroleum entrepot for the industry. Philadelphia was already operating under the limitations of abominable congestion at depots and in connecting streets, inadequate facilities for barrels or auxiliary tankage, and lack of a direct rail connection with the Oil Region, which were to handicap its growth as a distributing and export center until late 1864.[81]

Only at New York, long the coal-oil refining center of the United States, had the foundation for reasonably adequate storage and handling facilities been laid. These were provided chiefly at the Erie's depots and wharves at Jersey City and at the three yards of the Schieffelin Brothers in Brooklyn, which were well equipped to receive and store oil before re-barreling it for further shipment.[82] There is good reason to believe that the primary bulk storage facilities at this time were at the major coal-oil refineries in New York and Boston, which were in the process of converting to petroleum. In 1860, for example, the New York Kerosene Company was using a number of iron tanks with capacities of 15,000 barrels, and one which held 30,000 barrels. At his refinery in Boston, Samuel Downer also stored oil in tanks, each holding about 1,200 barrels.*

With the advent of flowing wells, the provision of bulk storage facilities assumed an urgency which was felt in the Region, at intermediate shipping points, and at the principal markets. The event which inspired the first real attempt to supply bulk storage at the wells came when the Elephant well on the Tarr farm began gushing oil at a rate of a thousand barrels a day in December, 1861. In a desperate effort to save a part of this wasting output, the Densmore brothers, operators of the well, hastily excavated a large earthen reservoir several feet deep and walled on the sides with 6-foot planks.[84] Operators of other spouting wells attempted to

---

* Downer insisted on using iron storage tanks of similar capacities for storage at the refinery he erected at Corry in 1862.[83]

salvage their oil by digging pits, 3–5 feet deep, sometimes covering an acre or more. Where gushers were located near ravines, it was possible to improvise dams at their mouths with planks and earth.*

Too hazardous and impractical for even the most reckless operators, open pits quickly gave way to wooden storage as soon as carpenters could be found. Cheaper, quicker and easier to construct than round tanks at this stage, box vats ranging in capacity from 100–1,200 barrels were commonly the initial replacements. Clustered thicker than derricks so as to cover several acres in some localities along Oil Creek, at Tidioute, and at French Creek near Franklin, these tanks soon numbered in the hundreds.[86]

Despite their inherent superiority over box vats, round wooden tanks hooped with iron, introduced in 1861, were somewhat slow to spread. But under the impetus of the speculative wave and the development of new fields in the succeeding two years, tanks of this type, ranging in size from 200–1,200 barrels, were installed by operators and shippers at every producing district and key shipping point along Oil Creek and upper Allegheny River valleys. Their cylindrical shape not only gave them greater strength and economy in space than box vats, but eliminated the vulnerable leakage points of jointed corners. But to be efficient, they required careful workmanship and well-seasoned pine staves and planks free of knots—all rare commodities in these early years.†

As oil began moving down Oil Creek in greater volume in late 1861, Oil City was quick to add to its storage and handling facilities. With the extension of the Atlantic & Great Western from Corry to Meadville in 1862 and on to Franklin the year following, Oil City also became an important outlet for rail shipments. In September, 1864, it was reported that one firm, Graff, Hasson & Company, had already built one large iron tank holding 8,000 barrels at Oil City and was planning to erect two more of the same size in the immediate future.[88] The completion of the Oil Creek Railroad between Corry and Titusville in 1862 and its subsequent exten-

---

* The first of these was constructed on lower McElhenny farm near Funkville in September, 1861, in an effort to salvage some of the famous Empire well's daily outpouring of 2,500 barrels. In 1862 another ground pool retrieved part of the Sherman well's daily flow of 2,000 barrels on Foster farm. Typically, this pool was replaced as soon as possible with the Sherman well's first permanent storage, wooden box vats, followed later by the stronger round wooden vats.[85]

† As these tanks were increased in size, flawed construction became a critical factor. Experience with large tanks of 1,500–2,000-barrel capacities which frequently burst under pressure, causing destructive fires at the wells, prompted most operators to accept 1,200 barrels as the maximum size within safety limits.[87]

sion to Petroleum Center in 1865 encouraged the growth of tankage on the strategically located Schaffer, Miller, and Tarr farms.

Pittsburgh's receipts of 100,000 barrels of oil via the Allegheny River in 1861 brought a quick response to the pressure for bulk storage needed for gauging and rebarreling. A part of this expansion was provided by commission merchants and refiners who bought property and erected barrel sheds and wooden tankage. But the bulk storage facilities which were to dwarf anything in the Region for some years to come materialized at dockside on Duquesne Way. There the addition of wooden tanks and the construction of the first of a number of iron tanks, with capacities of 4,500 barrels each, gave Pittsburgh the largest bulk storage facilities in the nation by spring of 1862.[89]

Although permanent iron storage was already installed at large refineries and a few terminals by 1862, wartime shortages of iron and difficulties of transporting heavy materials delayed its spread to the Region until after the mid-1860's, largely as a corollary of bulk transportation by tank cars and gathering lines. Indeed, a clear portent of the future relationship between bulk storage and bulk transportation arose in the very infancy of bulk transport developments when, in August, 1865, the Pennsylvania Tubing & Transport Company erected a 15,000-barrel iron tank at Oleopolis to receive oil from the newly installed pipeline from Pithole.

There were other influences which hastened the shift to iron tankage. These came with a mounting recognition of the serious limitations of the best types of round, wooden tanks, such as their impermanence, heavy maintenance costs, high losses through leakage and evaporation, and enormous fire hazards.[90]

First in a chronological sequence was the great flood of March 17, 1865, which moved down Oil Creek carrying with it all the wooden tankage at wellheads and shipping points along the entire valley, causing $5 million of property damage. The catastrophe created both a great need for new tankage and a desire for a more permanent and less vulnerable type of storage.[91] Operators were encouraged to turn to iron by more plentiful supplies of wrought iron made available by the return of the iron and steel industry to peacetime operations and the extension of rail lines into the Region. These factors were reflected in the reduction of the costs of constructing iron storage from about $2.50 a gallon in 1862 to $1.00 in 1865, only 35¢ more than wooden tankage.[92]

A further stimulus to convert to iron tanks came late in 1865 and early 1866 when a series of disastrous fires struck almost every major field in the

COURTESY DRAKE WELL MUSEUM

Storage tanks of Central Petroleum Company, Petroleum Center, 1866.

Region. Wooden storage huddled near wells was never absent as a primary or aggravating cause of such fires. But it took a great blaze on Bennehoff Run early in July, 1866, to solidify a growing conviction that oil should be piped to fire-resistant iron storage removed a safe distance from wells. The fire began when lightning struck a gas pipe in the Western Union Telegraph well, igniting an adjacent tank, which exploded, releasing a fiery mass of oil down the Run, destroying between twenty and thirty derricks and adjacent tankage in its path. The burning oil also enveloped the 3,000-barrel iron tank of the Ocean Oil Company at Petroleum Center, at the far end of the Run. But to the surprise of everyone, when the tank cooled off, its contents of 2,800 barrels of oil had not been harmed. This first tank in the entire Region to withstand the test of fire on its exterior made sensational news.[93]

The trend toward iron tankage received another boost from the depression which settled on the Region late in 1865. Rather than sell at what they considered ruinously low levels, operators began building up storage capacity to hold their oil for better prices.[94]

The experience of James McCray, a large well owner near Petroleum Center, illustrated the limitations of holding oil in wooden tanks over any extended period of time. Late in 1865 he erected new wooden tanks where he stored some 200,000 barrels of crude. Two years later, two-thirds of the oil had been lost through leakage, evaporation, and deterioration.[95]

For all these supplementary aids, it was bulk transport which supplied the major pressure for iron tankage. Relieved of the need to tie up a large portion of their capital in barrels, shippers found ample funds to undertake storage expansion.[96] The introduction of accommodation lines in 1866, connecting wells with pipeline dump tanks, speeded the conversion in districts still lacking rail connections.[97] The need for barrel storage disappeared as tankage was built up at the pipeline dump tanks on individual farms, at terminals of the gathering lines, and at receiving stations in major marketing areas.

There was a remarkable growth in the Region's iron storage capacity beginning in 1867. Estimates in February of that year placed the total tankage at 300,000 barrels; by August, 1868 it had climbed to over a million barrels; and by 1873 the Region's total had reached approximately 1.5 million barrels.[98]

The evolution of iron tankage brought significant improvements in the handling of oil: permanent storage and low maintenance cost, reduced danger from the hazards of fires and floods, and lower losses from leakage, evaporation, and deterioration. Moreover, adequate storage in the Region relieved producers of the pressure to turn their crude immediately over to buyers. For the first time, at least nominally, they could participate actively in the marketing of crude. Some, like James McCray at Petroleum Center, combined large storage with their operations as producers; but such instances were the exception. Most operators preferred to store oil with gathering-line companies, which by 1870 controlled most of the permanent tankage in the Region.

Railroad competition for the oil trade

One important effect of the reduction in transport costs made possible by the spread of bulk shipments after the mid-1860's was to make proximity to crude-oil supplies less significant in the determination of refinery locations. As a result, various centers far removed from the oil fields, such as Cleveland, Philadelphia, and New York, began to challenge the positions of Pittsburgh and the Region as the leading markets for crude. But lower rates on crude were but one of many factors affecting the intense competition between regions after 1866, among which none was more important than the rivalry among the oil-carrying trunk lines. This rivalry in turn had its roots in the broader trunk line struggle that had been developing since the late 1850's.

During the period of short, local lines that typified the seven thousand miles of railroads built in the United States before mid-century, competition between railroad companies was quite limited in scope. But the situation changed when future trunk lines, attracted by a growing volume of traffic between the Atlantic coast and the upper Mississippi Valley, began to reach common terminals or trading areas. The Erie, for example, completed a continuous line from New York to Dunkirk on Lake Erie in 1851. Two years later the New York Central was formed by consolidating eleven local lines between Albany and Buffalo, and by arrangements with independent connections at either end of its line began handling traffic moving between New York and Chicago. In 1857, the Pennsylvania opened its connections with the west by arranging to exchange traffic with the Pittsburgh, Ft. Wayne, & Chicago Railroad.

Whatever their other differences, these three roads all shared one thing in common: they had to compete with the Great Lakes-Erie Canal route between the Atlantic coast and the interior. Until well after the close of the Civil War the yearly opening of the canal sparked a reduction in railroad rates to meet the lower water rates. Even though railroad costs were still too high to challenge the canal for traffic in low-value grain and flour, these tariffs were pulled down along with those on passengers and high-value freight (chiefly westbound) from which the railroads had the most to gain. In competing with low water rates, the rail lines could scarcely avoid competing with one another. Railroading involved high fixed costs which had to be paid regardless of the volume of traffic. No one railroad could afford not to match or exceed rate reductions made by its competitors so long as the return covered operating expenses and a portion of overhead costs.

Experience with rate wars led very early to the first of a long series of "conventions" or meetings at which the heads of the Atlantic trunk lines (the Pennsylvania, the Erie, the New York Central, and the Baltimore & Ohio) attempted to agree on methods of controlling competition. But the pattern these conventions were to follow over the next twenty-five years became evident when agreements on rates and competitive practices made at the first convention held in 1858 and the one which followed it in 1860 were broken within weeks or months after they were signed.[99]

The instability of the agreements reflected rivalries among the trunk lines that went beyond the influence of water competition in affecting rates and rate structures. For example, the New York Central and the Erie, backed by the city's commercial interests, were under pressure to establish the rates between New York and the west as the minimum rates between the Atlantic seaboard and the upper Mississippi Valley. Since the distance from Chicago to Philadelphia was considerably less than to New York, the Pennsylvania, with equally strong commercial ties with the Quaker City, demanded a differential favoring Philadelphia. While the New York roads grudgingly accepted the differential in favor of Philadelphia, they did not hesitate to break it whenever they felt the Pennsylvania's revenues were increasing at their expense.

The mechanics of rate-making provided another source of friction. With rare exceptions, the main lines did not themselves reach areas in which east-bound traffic originated, but relied upon independent connections over which they seldom had close control. These connecting roads often made special through rates, and the main line, to keep its connections satisfied, usually acquiesced in validating rate cuts. As one witness was to explain later:

> The western roads have to meet competition between themselves, and on our behalf, against our competitors, and they must have the authority to act on the spot without being obliged to consult us; if they did not they would not get the business, nor we either.[100]

In this manner rate wars, begun among local lines, were frequently transmitted to the network as a whole.

But there were other ways of attracting shippers, which were ordinarily more difficult to detect than general rate cuts and much less likely to lead to open rate wars. These persuasions usually took the form of some sort of discrimination between shippers, by granting special and generally secret concessions to individuals or companies which offered a large volume of traffic or which were in a position to play one carrier off against another.* This method of diverting traffic from competitors was already widespread by the mid-1860's. Testifying in 1867, for example, Tom Scott admitted that the Pennsylvania had long allowed "drawbacks" (a refund of a por-

---

* Over the years discrimination between shippers took many forms other than drawbacks or rebates, already prevalent by the 1860's. It was not uncommon, for example, for railroads to underclassify the freight of a favored shipper and charge a rate which ordinarily applied to a low-value commodity to a high-value product. Similarly, by "underweighing" cargoes the carriers charged shippers for less weight or volume than was actually transported. Free cartage, free storage, special handling of equipment, or an excessive allowance for some service or facility supplied by the shipper were other ways of attracting business. Nor was the so-called "midnight tariff" unknown, under which rates were posted long enough for a favorite shipper to take advantage of them, but "corrected" before their application became general.

tion of the rates) to shippers and that "a great many of its competitors" had followed the same practice.[101]

As in the case of open-rate reductions, one railroad had little choice but to match or exceed special concessions offered to particular shippers by its competitors. These practices were often defended on the grounds that large shipments were more economical to transport than smaller ones. But there was abundant evidence that personal discrimination went beyond any economies involved and that its application served to reward those individuals or firms which were in a strategic position to bargain with the railroads and to penalize those which were not.

To offset losses incurred in maintaining a share of competitive traffic, railroads also engaged in "place discrimination" by charging higher rates to non-competitive, local shipments than on "through" traffic. Defending this practice, A. J. Cassatt, traffic manager of the Pennsylvania, explained,

> Through rates are governed by competition. If we want to remain in the market we have got to take what we can get. That forces us below paying rates sometime. If we put the local rates down to the same charge per ton mile, the road could not live. Another reason is, you can afford to carry cheaper for a long than short distance . . . because you have certain fixed terminal expenses which are less per mile on long than on short distances.[102]

Whatever rationalization was used, the general effect of place discrimination, like personal discrimination, was to encourage the economic growth of cities and towns served by competing railroads and to penalize those dependent on a single line.

Given the social and economic environment under which the trunk lines were developing, the unstable rate structures and competitive practices and abuses which emerged were almost inevitable. In the absence of federal regulation and with no general acceptance of the principle that railroads were common carriers, each railroad management attempted to build up business by any means at its disposal. Self-interest alone might have abated the worst of these abuses had the railroads been willing or able to live up to the terms of the compacts agreed upon at their conventions. But these conventions served primarily to postpone or bring an end to open rate wars. The compacts arrived at were essentially "gentlemen's agreements." Being in violation of common law principles prohibiting combinations or conspiracies in restraint of trade, they had no standing in the courts and violators could not be sued for breach of contract. Thus, whatever the nature of any existing agreements, there was the ever-present temptation to break the terms whenever there was the possibility of increasing the business of one road at the expense of others. Under these circumstances ambitious trunk-line managers did not depend on conventions to protect their competitive positions; instead they strove to enlarge their systems and to strengthen devices which would increase the volume of traffic moving over their lines.

Although transport demands during the Civil War deterred resort to intensive rate cutting, developments during and immediately following the war years did little to change the basic competitive relationships between the trunk lines. On the contrary, areas of potential conflict were widened as the various lines sought to enlarge their networks and to build up their traffic.

A part of this story as it affected the oil traffic has already been told in connection with the Atlantic & Great Western's first moves into the Region and the Pennsylvania's reaction, which resulted in the development of the Philadelphia & Erie as a through route to the Atlantic seaboard. The Pennsylvania's establishment of its fast-freight affiliates, the Star Union and Empire, was matched by the New York lines. In the case of the Erie–Atlantic & Great Western route, the American Express Company, which for some years had operated over the Erie, was supplemented in 1865 by a new firm to handle traffic originating on the western connections of the line. The Central Transit Company (better known as the White Line, because of the color of its cars) was set up in 1866 to operate over the New York Central and its connecting lines.[103]

By the mid-1860's, the fast-freight lines had thus become firmly imbedded in the system of trunk lines. Their attractions to the railroads were numerous. Supplying a part or all of their own rolling stock, they relieved the railroads of a direct financial burden of purchasing additional equipment. Charged with the principal responsibility for soliciting through freight traffic for their respective trunk-line associates, the fast-freight companies competed vigorously for business by offering shippers such services as through billing, minimum delays of goods in transit, and special equipment, such as tank cars for the transport of crude.

Despite these advantages, the fast-freight companies were open to criticism on the grounds that they served to strengthen the abuses that had been anticipated at the first trunk-line convention in 1858. While they customarily received a commission plus a share of the tolls fixed by the railroads over which they operated, the fast-freight company managers were in a position to make special concessions to favored shippers or shipping centers.

The fast-freight companies, however, were far too effective as a competitive arm of the railroads to be abandoned at this time. This fact was brought out at the third Atlantic Trunk Line Convention held in New York in spring of 1866. An agreement was soon reached on basic rate schedules affecting through traffic. But a strongly worded resolution calling for the abandonment of fast-freight affiliates, favored by the Baltimore & Ohio and the Erie, failed when Scott, representing the Pennsylvania, and Dean Richmond, president of the New York Central, voted in the negative.[104]

The area of potential conflict among the trunk lines was also widened by other events beyond the competitive friction that resulted from the fast-freight companies. Much of the intense competitive struggle after 1866 was attributable to the activities of two of the most famous men in railroad history, Jay Gould and Cornelius Vanderbilt. Commodore Vanderbilt, after a successful thirty-year career as a steamboat operator, had turned to railroading early in the 1860's. After purchasing the New York & Harlem in 1863 and the Hudson River in 1864, the two lines running between New York and Albany, he began laying plans to monopolize railroad transportation in New York by combining these roads with the New York Central and the Erie.

Vanderbilt achieved the first step toward this goal in 1867 when he was elected president of the New York Central. But his attempt to capture the Erie was blocked when Daniel Drew, treasurer of the Erie, proved more adept than the Commodore in the manipulation of stocks, bribing of New York legislators, and making deals with Tammany judges, which marked the classic struggle between the two men late in 1867 and early in 1868.[105]

Following this defeat, Vanderbilt turned to strengthening the New York Central's alliances with the lines running between Buffalo, Cleveland, and Chicago.* When these roads were combined in 1869 to form the Lake Shore & Michigan Southern, they became for all practical purposes a part of a single system, although formal incorporation of the Lake Shore into the New York Central system did not come until 1873.[106]

Jay Gould, who became a director of the Erie in 1867 and was elected president in 1868, had no less ambitious plans of his own. He immediately launched a badly needed rehabilitation program for improving the Erie's tracks and rolling stock. Old cars and worthless locomotives were scrapped

---

* These lines included the Buffalo & Erie; the Cleveland, Painesville & Ashtabula; the Cleveland & Toledo; and the Michigan Southern.

and replaced with new equipment; steel-topped and all-steel rails were substituted for iron rails on the line; and extensive yards and storage facilities were installed on three large tracts of land, ranging from sixty to eighty acres each, fronting on the New Jersey side of the Hudson River.[107]

Simultaneously Gould moved to consolidate and expand the Erie's western connections. A lease of the Atlantic & Great Western in 1868, with its lines into the Regions and a working agreement with the Cleveland & Mahoning, assured the Erie's trunk-line position between New York and Cleveland. But Gould's attempts to extend the Erie's connections to Chicago from Cleveland over the Michigan Southern were blocked when the latter became a part of the Lake Shore in 1869. Gould was likewise unsuccessful in a bold maneuver to obtain a Chicago connection by attempting to seize control of the Pittsburgh, Ft. Wayne & Chicago Railroad, which since 1857 had worked in close harmony with the Pennsylvania.[108]

Despite these failures, the Erie still occupied a strategic position (one of particular significance for the oil trade) in the Atlantic trunk-line system between the New York Central to the north and the Pennsylvania to the south.

The reaction of the Pennsylvania management to these moves by Vanderbilt and Gould was to step up its own program of expansion. By the early 1870's, the Pennsylvania owned or controlled lines reaching every important city in the broad area bounded by Cincinnati, Louisville, and St. Louis on the south and Chicago, Toledo, and Cleveland on the north.[109]

Nor did the Pennsylvania neglect its eastern connections. In 1867 an arrangement had been made for shipments moving over the main line of the Pennsylvania between Pittsburgh and Philadelphia to reach the New York harbor area via the Philadelphia & Trenton and the Camden & Amboy railroads. Two years later an agreement between the Philadelphia & Erie and the New Jersey Central provided for the use by the Empire Transportation Company of large oil handling and storage facilities which were built up at Communipaw, near Bayonne on the New Jersey side of the Hudson River. To make sure that its New York connections would not be raided by trunk-line rivals, the Pennsylvania in 1871 leased the United Companies of New Jersey, which in late 1867 had been formed by a consolidation of the four principal railroads operating in the state.*

---

* These four roads were: the Philadelphia & Trenton; the Camden & Amboy; the New Jersey Railroad & Transportation Company; and the Delaware, Raritan Canal Company.[110]

For the major trunk-line managements, the primary purpose for extending their respective systems was to share in expanding traffic in agricultural products between the upper Mississippi Valley and the seaboard. But the oil industry with its own rivalries between producers, shippers, and refining centers could hardly escape from being drawn into the tangle of unstable alliances, shifting nodes of power, and conflicting interests that characterized railroad competition and the oil trade after the mid-1860's.

Chapter 9

Refining technology

Petroleum refining may be said to have emerged as a defined branch of the industry by 1862. In that year the conversion of coal-oil refining centers at New York, Boston, and Pittsburgh, in that order of size of capacities, was completed or well advanced. Much new petroleum refining capacity had already made its appearance, particularly in Pittsburgh and the Region.

By this time it was clear that the coal-oil refining legacy could readily be probated to petroleum refining and treating. But it was also clear that many early claimants to that legacy were ill-prepared to take full advantage of their inheritance. Infected with the exuberant optimism of a mining boom, most early proponents of the new industry greatly underestimated the importance of skill and experience for successful refining operations. Some even accused established refiners of deliberately fostering false notions about the complexities of the art. For example, in 1860 Thomas Gale, one of the most enthusiastic of the early promoters of petroleum refining, in remarks directed to the major coal-oil refiners, wrote:

> Why envelop yourselves, gentlemen, in this ocean of fog, . . . that you may monopolize the profits which belong to those who produce the raw material? You think them too poor to go into the expense of refining, which you mystify and magnify that you, in alliance with a few wholesale merchants, may take the lion's share of the profits. Your keeping so dark proves that in clarifying you spend but little and pocket much. Others know just as much about the process as you do; and so far as circumstances seem to require, the companies who raise the raw material will do their own refining.[1]

Deluded by petroleum enthusiasts and still manufacturers as to the simplicity of refining, individuals inexperienced in any form of distillation flocked into the new business.

Even limited experience soon demonstrated that successful petroleum refining no less than coal-oil refining called for the utmost vigilance and precision in operating techniques. This was a necessity made all the greater by the many limitations of available equipment in providing close temperature controls since the nature of petroleum reduced the allowable margins of error at numerous operating points.

On these and related problems Samuel Downer and Joshua Merrill among others worked incessantly—experimenting, improvising, and drawing on every facet of their past experience. Even in the vanguard of petroleum refining the precise methods of distilling and treating remained in a state of flux for the better part of a decade. But by the early 1870's, the basis for the refining technology and practices which were to serve the American petroleum industry throughout the nineteenth century and beyond had been firmly laid.

Advances in chemistry

Considering the current status of chemical science, the advances in refining technology during the 1862–73 period were all the more impressive. Organic chemistry, the cornerstone of modern petroleum technology, had scarcely been born. Not until 1856, when the English chemist, W. H. Perkin, Sr., succeeded in synthesizing the first organic carbon compound (previously believed to form only in nature), was this now indispensable industry born. Distilling benzene from coal tar, he then synthesized from it, through a series of intermediate compounds, the dye, aniline purple, which he demonstrated to be identical with mauve, a scarce and expensive natural dye. Both a new industry and a new science sprang up almost overnight.[2]

Perkin nitrated the benzene with a mixture of nitric and sulphuric acids yielding nitro-benzene. Nitro-benzene was reduced with iron and acetic acid to produce the organic compound aniline. Aniline then was converted to aniline sulphate with sulphuric acid, then oxidized with bichromate of potassium to produce the purple aniline dye, shown to be identical with mauve, the scarce natural dye.[3]

Though widely regarded as an early date of unique importance in the rise of organic chemistry, Perkin's discovery, of course, cannot be taken too literally as the birthdate. Other compounds had been synthesized earlier. In 1826, Hennell, for example, succeeded in producing synthetic ethyl alcohol, and in 1828, Frederick Wöhler synthesized urea from ammonium cyanate. Among many others in succeeding years, the work of Justus von Leibig, Edmund Knowles Muspratt, and J. P. Hofman, who was W. H. Perkin's teacher, was important in opening up new fields of organic chemistry.[4]

In 1858, Frederick August Kekule of Darmstadt, Germany, discovered a base upon which a mighty superstructure in organic chemistry eventually was built. He demonstrated a signal structural difference between benzene and other hydrocarbons with the same number of molecular atoms. Benzene was an unusual type of hydrocarbon, in that its molecules were arranged in a ring rather than a chain. He named this type of hydrocarbon "aromatic compounds." All aromatic compounds contain at least one benzene or similar, unique ring. However, not until 1865 did he give adequate expression to his revolutionary closed-ring formula of molecules, after which there was still considerable lag in wide appreciation of the significance of his findings.[5]

A revealing consequence of the failure to realize the implications of Kekule's work occurred in the 1860's. Both in England and in the United States, petroleum benzine (a mixture predominantly of aliphatic compounds with a trace of aromatics all possessing a wide boiling point range) was widely believed to be identical with coal-tar benzene, an aromatic with a narrow boiling range. Repeated warnings in scientific and industrial art periodicals that the hydrocarbons were different failed to dispel the misconception that aniline dyes from coal-tar benzene were made from benzine, a petroleum by-product.[6]

Publication in 1867 of the brilliant researches of the French chemist,

M. P. E. Berthelot, on the action of heat on various hydrocarbons likewise came late in the day for petroleum refining. His findings proved that all hydrocarbons, heated in a sealed vessel long enough and to sufficiently high temperatures, decomposed or cracked into carbon and hydrogen, the basic atoms and atomic building blocks of all hydrocarbons. Under somewhat less severe conditions, this destructive distillation broke large molecules into smaller ones. Of equal importance, a reversible action occurred at high temperatures too. Small molecules built into bigger ones, or polymerized: for example, ethylene polymerized to tarry products with much higher molecular weights.[7]

In his renowned report on petroleum in 1855, Professor Benjamin Silliman, Jr., noted the highly irregular boiling points in petroleum under the most careful fractional distillation to separate its hydrocarbons.* Low-boiling products constantly occurred in the remaining portions boiling at high temperature. From these observations he made the remarkably brilliant intuitive deduction that petroleum was not composed merely of distillable components but also of very high molecular weight compounds that decomposed or cracked to smaller molecules before distilling.[8]

As will be described, refiners put into practice Silliman's deduction and were cracking petroleum long before Berthelot published his findings, and that fact generated important influences not only on refining location but on marketing as well. In doing this, however, they employed entirely empirical procedures, sometimes with amazing results. They aimed for a specific product, illuminating oil, then variously called kerosene, coal oil, carbon oil, and burning oil, and which was the rough equivalent of modern kerosene. They knew its major physical properties and characteristics, such as specific gravity, boiling range, and burning performance, but they proceeded in total blindness as to the chemistry and the chemical structures of the materials they were working with. Berthelot's experi-

---

* Unless otherwise noted, fractional distillation in this and the next two chapters refers to simple single stage distillation, and fractionation to only the rough separation obtained by simple distillation. It was only in this sense that Silliman and his contemporaries used and understood these terms. Fractional distillation as distinct from simple distillation did not enter American refining until the development of fractionating towers early in the twentieth century. It differed from simple or ordinary distillation in being multi-stage rather than single stage, in not leading the vapors directly to the condenser, but instead first passing them through some contacting device where they are intimately contacted with liquid (condensed vapor) flowing contracurrently. By means of this intimate contact between flows of liquid and vapors, modern fractional distillation achieves the equivalent of a number of simple distillations and a much closer separation of materials in narrow boiling ranges.

ments with pure compounds, which he converted by cracking to other pure compounds of known structure, provided the initial foundation upon which a framework slowly could be erected for interpreting what was happening in the fractional and destructive distillation of petroleum.

Nomenclature and standards

Petroleum refiners early adopted the basic nomenclature and measuring system employed in coal-oil refining, and much of it continues in use today. A basic need was some way of measuring the relative densities of crude oil, refined distillates, and still residues. These were and still are expressed in terms of specific gravity, the weight of a given volume of a substance in relation to the weight of the same volume of water at the same temperature, with water assigned a value of one. Since most oils are less dense than water, they are expressed in a descending order on the specific gravity scale in decimal fractions carried to three places: water equals 1.000; relatively dense paraffin or lubricating oil, 0.875; less dense kerosene, 0.795; still less dense gasoline (actually not yet a product in 1861), 0.650–0.635.

To abate the nuisance of carrying decimal fractions to four or five places in many calculations, however, refiners soon came to measure specific gravity in "degrees Beaumé," subsequently shortened to "Baume." * Invented by the French chemist Antoine Beaumé in the eighteenth century and used universally in the early sulphuric acid and caustic trade, the Baume scale was used in the oil trade until the 1920's, when the American Petroleum Institute slightly modified it into the A.P.I. gravity scale. The Baume scale assigned water a value of ten instead of one, while substances less dense than water scaled upward from ten in the order of their descending densities: water 10° B; paraffin or lubricating oil, 30° B; kerosene, 46° B; and gasoline, 85–90° B. Periodic sampling of the distillate with the hydrometer or gravity stick registering degrees Baume (usually manufactured by John Tagliabue, America's renowned Italian-born instrument-maker), was the refiner's main guide when to cut his products.[9]

In almost all cases the refiners supplemented the Baume test with observations of color. The lower-boiling distillates were lightest in color, ordinarily remaining almost white through the naphtha fractions, shading from yellow to progressively darker brown through the kerosene and par-

---

* B, rather than the standard Bé., will be used to abbreviate Baume in this volume.

affin-oil portions of the distillate.* Experienced early refiners equipped the pipe leading from the condenser with a sight glass or "look box" to expedite closer color observations.

Many inexperienced entrants into early refining all too often relied solely on color observation, a rough criterion at best. Too rapid or too high an elevation of temperatures, both common errors, brought about numerous disturbances, including the cracking of molecules before they vaporized, which falsified the normal color progressions.

By 1862 all but the totally unskilled refiners had a notion of at least part of the correlation between boiling points at which the components of crude oil vaporize in the still and the densities of the condensed vapors or distillates shown in Table 9:1. Products with the lowest densities and boiling points (gasoline, rhigolene and cymogene, in that order) had not yet been produced commercially, but refiners were aware that some components of crude vaporized at room temperatures or less. It was their purpose to get rid of these useless vapors as well as a good part of the comparatively worthless naphtha as fast as possible. Refiners also knew that naphtha, benzine, kerosene, and other products had no single boiling point but a range of boiling points.

There was, however, no uniform agreement on the average range of distillate densities which should enter the various products as might be implied from Table 9:1, which indicates what had come to be considered good practice by 1870. Both out of ignorance of the poor separation actually achieved and the desire to increase yields of burning oil, much kerosene distillate included portions of the highly inflammable fractions in the benzine and even naphtha range (60°–65° B and 65°–80° B, shown in Table 9:1), as well as some of the denser distillates with poor burning qualities below 38° B.[10]

To this generalized characterization of earliest refining knowledge and practice there were important exceptions. Even if petroleum had comprised only a few hydrocarbon compounds, instead of thousands then unsuspected, to separate them cleanly would have required a slow, even increase of still temperatures, for otherwise some of the components with different boiling ranges inevitably would vaporize simultaneously. The fact that the best coal-oil and early petroleum stills did not normally provide even temperatures throughout their heating surfaces dictated an

---

* With kerosene unimportant and gasoline the leading product, naphtha fractions have a much broader meaning in modern nomenclature. They include all fractions with boiling points up to about 450° F, or all fractions lighter than gas oil.

TABLE 9:1

Characteristics of products from fractional distillation and cracking of crude petroleum

Product	% Yield		Av. boiling point, F	Av. range density		Av. density	
	Fractional distillation	Cracking		Sp. gr.	Baume	Sp. gr.	Baume
Cymogene (butane)			32°	.572–.595	115–105	0.583	110
Rhigolene (pentane)			65°	.595–.622	105–95	0.608	100
Gasoline	1½		110°	.622–.667	95–80	.635–.650	90–85
Naptha, crude		20					
Naptha, refined	10		180°	.667–.717	80–65	.680–.695	76–71
Benzine	4		300°	.717–.738	65–60	.715–.730	65–62
Kerosene or refined petroleum	55		350°	.738–.832	60–38	.795	46
Burning oil		66					
Paraffin oil	19½		570°	.832–.903	38–25	.875	30
Coke, Gas and loss	10	14					

Source: C. F. Chandler, Report on Petroleum As An Illuminator (New York: New York Printing Company, 1871), 14.

extreme care in firing that ran counter to most operating schedules.

Another impediment to thorough separation lay in a fact scarcely understood at all. Not all types of hydrocarbons have the same molecular weights and structures in relation to their boiling points, as instanced by those in the aromatic and paraffin series. As a sole index of boiling range, density measurements were fallible. They became all the more so when distillates of high densities (around 38° B) were blended with much lower ones in the benzine range (60°–65° B) to make a burning oil with a good Baume reading of 46° B.

Finally, if distilling temperatures were permitted to go far above 550° F so as to induce destructive instead of fractional distillation, cracking the larger molecules into smaller ones before separating them, the density index became meaningless. As Berthelot reported in 1867, not only was there the major action of cracking large molecules into smaller ones, but often the counter one of polymerization, the building of smaller ones into bigger ones.[11]

In addition to the density index and color observation, pioneer refiners had the "dish test" to control the characteristic and uniformity of a product. Simply by waving a flame over a dish of kerosene it was possible to estimate at what temperature it would burn. With no uniformity in the size or shape of the vessel, the quantity of burning oil in it, distance of flame application, freedom from draft, and related conditions, the test had limited value. It failed entirely to indicate the lower ignition point or "flash point" (the temperature at which burning oil gave off sufficient vapors to ignite momentarily and to cause dangerous explosions in lamps).

To correct this, John Tagliabue introduced his famous closed-cup testing apparatus, in 1862.[12] (see Figure 9:1.) To determine the important but often-neglected flash point (the temperature at which an oil gives off enough vapors to form an explosive mixture with air and flash when ignited by a small flame), the test kerosene was placed in a standard closed cup, perforated to admit air but not drafts, and containing a mercury thermometer. The kerosene was heated by a surrounding water bath in a larger cup that received its heat from an alcohol lamp in the bottom of the apparatus. A wick or match was periodically inserted through an aperture at the top to test the flash point. To determine the higher ignition point (the temperature at which the oil would burn continuously) part of the lid of the closed cup was removed and the rest of the test conducted by open cup.

The apparatus provided adequate control under most important test-

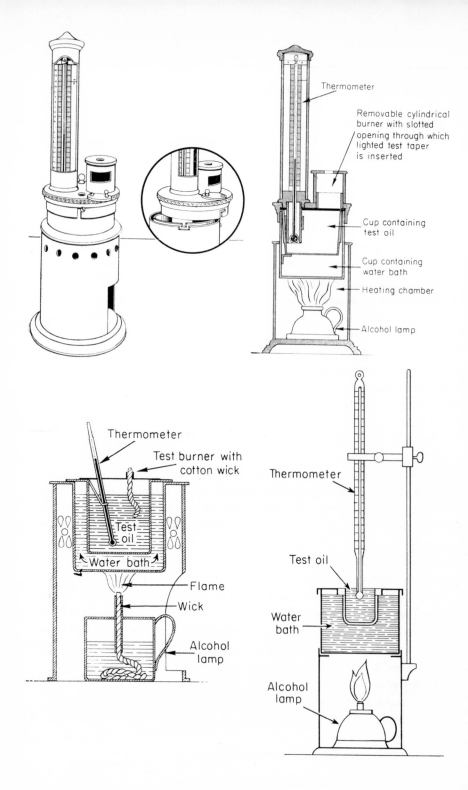

Thermometer

Removable cylindrical burner with slotted opening through which lighted test taper is inserted

Cup containing test oil

Cup containing water bath

Heating chamber

Alcohol lamp

Thermometer

Test burner with cotton wick

Test oil

Water bath

Flame

Wick

Alcohol lamp

Thermometer

Test oil

Water bath

Alcohol lamp

ing conditions, though it was criticized with some justification as placing the igniting point too low. This happened after the flash point had been obtained, when the massive heavy jacketing sometimes caused the water bath to heat the surface of the oil faster than the body of the oil where the thermometer was placed.[13] In order to meet this and other criticisms, Tagliabue in about 1868 replaced his closed-cup tester with an open-cup tester of both flash and igniting points. Liberalizing greatly in favor of oil, especially flash points, this instrument remained the leading testing instrument in America until 1879. Marketers and refiners selling in discriminating large foreign and domestic markets used the early closed-cup apparatus extensively and increasingly, although it did not displace the informal dish test immediately or completely until late in the decade.

Beginning in 1862 legislatures began passing fire-test laws, but they had little direct effect on refiners until the post-war period because of unrealistic provisions for enforcement and failure to specify testing procedure or flash point.[14] Apart from considerations of public safety, the Tagliabue apparatus gave the superior refiner another needed means to gain better and more uniform control of his product.

Direct-fire distillation

Largely through the coal-oil heritage many of the refining techniques basic to the petroleum industry throughout the nineteenth century and beyond had already been brought into commercial practice by the early 1860's. Indeed, by 1862 numerous broad alternatives had already opened to refiners. They included whether to choose a vertical or horizontal still; which of four types of heat applications to use in distillation; whether to engage in destructive distillation (cracking) which, though not understood at the time, changed the molecular structure of petroleum at high temperatures, and which of several treating methods to adopt.

In actual practice, choices were dictated by a variety of interrelated factors such as the sophistication of the refiners, refinery location, relative

---

FIGURE 9:1 (facing page). *Tagliabue's closed and open cut testers. Upper left, perspective view of closed cup tester, modified model, 1862. Upper right, vertical section of closed cup tester, modified model, 1862. Lower left, vertical section, original model, closed cup "coal oil" tester, U. S. Patent No. 36,488, Sept. 16, 1862. Lower right, vertical section of open cup tester, 1868–1879.*

costs, and the nature of the markets served. Nowhere was the influence of this complex of forces better illustrated than in respect to fractional distillation by direct firing which dominated petroleum refining throughout the 1860's and in all succeeding decades.

In its crudest, earliest form this method of refining involved little more than a 5-barrel, cast-iron still, in effect merely a closed kettle, in which the crude oil was cooked over an open fire. The rising vapors were piped from an opening near the top to a worm where they were condensed back into liquids. The typical worm condenser comprised about 100 feet of coiled-copper pipe submerged in water to obtain cooling. If the distiller embarked on refining, then understood to embrace the art of chemical treating, he added a tin- or zinc-lined agitating tank, and a wooden dasher, similar to that used in a churn, which sufficed to stir the mixture. If he did not undertake treating, he needed only a settling tank into which the condensed vapors could be collected. He might add a little hut, often of stone, to protect his apparatus from the elements, and perhaps fence it in with some rough boards.[15]

During the first two years or so of the decade such outfits mushroomed profusely in the Oil Region and nearby points like Erie, Dunkirk, and Union Mills. Shanty, still, worm, and fittings cost $200 to $300. Inclusion of an agitator added about $70 more to the investment.[16]

These petroleum distilleries were as portable and operated at about the same level as the numerous moonshiners' stills in western Pennsylvania and elsewhere. Indeed their operators took the same approach as alcohol distillers. Real separation of the various components of crude oil was no objective at all; their major purpose was simply to distill off the gases, gasoline and naphtha fractions as fast as heat and condensation could permit. All condensed liquid that conceivably could be fobbed off as burning oil (from densities of 65° B to 38° B and higher) was recovered and the tar residue was thrown away. Using such a loose gauge of burning oil that encroached on the lighter and explosive benzine and naphtha fractions on the one hand and the heavier paraffin oil fractions on the other, spuriously large yields of 50 per cent and higher of illuminating oil were reported.[17]

Only in the provincial isolation of the Oil Region and nearby locations did such outfits receive serious designations as petroleum refineries. Almost their only operating virtue was the fact that their picayunish batch of crude, some 5 barrels more or less, could be run within a ten-hour working day. They characterized the boom-inflated notions on the ease of

Monitor refinery on McClintock farm, near Oil Creek, 1861.

effective entry into refining for a trifling cost and with scarcely any more technical competence than was required to harness a team of horses. The prevalence of this type of operation gave the Region the poorest refining in any area of major refining concentration. But even here refining could scarcely have moved into any significant commercial importance without more adequate fractional distillation.

Until well into the 1860's this was provided predominantly by the up-right, oval-shaped still, common in coal-oil refining. Its charging capacity ranged from 20–40 barrels, although its working capacity was always about one-fourth less, to allow for expansion of the crude as temperatures increased. Direct-fired from the bottom, preferably with anthracite coal or wood, the oldest types were entirely of heavy cast iron, sometimes 6 inches thick at their concave or convex bottoms where the heat was most intense. Even before the introduction of petroleum a minority of coal-oil refiners had turned to wrought-iron or boiler-plate bottoms (about ⅜-inch thick) in the interest of conserving heat, providing a larger heating sur-face, expediting a more even heat distribution, and reducing cooling time after completing the distillation.

In contrast to the 5–10-barrel stills initially employed in the Region, the larger commercial stills were permanent installations, encased in brick masonry from bottom to top, both as an aid to conserving heat and as a protective measure against fire. They usually had a manhole located in the dome to permit stillmen to do repair work and clean out the tarry residuum after the stills had cooled at the end of a run.[18] The condensing worm comprised 100 feet of copper or iron pipe, tapering from 4 inches to 2¼ inches, immersed in cold water for cooling.

Another type of underfired-petroleum still, evolved from coal-oil refining, had almost as many proponents as the cylindrical-vertical still. This was the horizontal still, boiler-shaped, with its height, or diameter, usually about one-third its length. With working capacities of 40 barrels each, their most unusual feature was thin boiler plate construction rather than cast iron; ⅜-inch thick on their sides, and ½-inch fire-plates underneath.[19]

Between them, these two types of stills dominated refining by direct-fire distillation for decades. Each had relative advantages which remained inherent and more or less permanent. Variables in specific operating conditions often modified a refiner's choice. For complete distillation the vertical still was generally preferred.[20] But where the objective was to leave undistilled the heavier portions of the oil above the burning oil range, the horizontal still was more effective. Its longer underbelly simplified the task of supplying a more even, less concentrated heat by means of several furnaces. The evener, less-intense heat not only promoted better distillation but reduced losses in burnt-out fire-plates as well.[21]

On balance the choice was a narrow one and the controversy as to the superiority of the two types continued through the 1880's. By this time the horizontal cylindrical still was considered more economical on fuel and more easily repaired while the vertical or "cheese-box" still gave lighter gravity, less-colored distillates, and a larger yield of illuminating oil.[22]

With the earliest vertical and horizontal stills, perhaps the most central and influential characteristic in all nineteenth-century petroleum refining put in its emphatic yet largely unnoticed appearance. Petroleum could be distilled only in small batches and not continuously, and each batch tied up even the smallest stills for extended costly periods.

With minor variations, the typical operating schedule was as follows. Because petroleum required a slower distillation than coal oil, the stills had to be fired up more gradually over a period of four hours. To distill down as far as a gravity of 35° B required 40–48 hours. Fires then were drawn and a cooling period of 12 hours with cast iron (several hours less with wrought iron) was waited out anxiously, for it was during this period that the hazards of fractured still bottoms reached their maximum. Workmen then lowered themselves through the manhole to embark for eight hours or more on one of the hottest, dirtiest, most hazardous jobs in contemporary industry. This was to scrape out the tarry residuum, usually about 5 per cent of the original crude charge, chip out with a pick ax the coked encrustations, and make any necessary repairs.[23]

It is obvious from the performance of the smallest feasible commercial stills, with the shortest running times, that all direct-fire distillation was, with rare exceptions, limited to two batches a week. Few if any refineries operated around-the-clock in the 1860's, except during short emergency periods. A 300-day working year was also unusual if only because of the normal seasonal nature of demand. An even firmer limitation came from the nature of batch distillation; it was virtually impossible with the best scheduling of still and maintenance operations to run much beyond 75 per cent of rated still charging capacity.*

It was this background that spurred the minority of comparatively advanced refineries to explore every avenue of approach to greater efficiencies and economies in distillation more intensively during the 1860's than in succeeding decades when direct-fired distillation in larger batches became increasingly successful. It was this quest that brought all the main alternatives to direct-fired distillation into some degree of commercial application before the end of 1862.

Steam distillation

One of the earliest of these was fractional distillation with ordinary steam rather than direct fire, which had been widely practiced by large coal-oil refiners. Their usual procedure was to pass ordinary steam at temperatures of 212° F into the still to supply the heat for evaporating the liquid charge. In addition to supplying heat, steam co-distilling with naphtha, benzine, kerosene and other hydrocarbon components of the crude oil substantially lowered their boiling points and heat requirements.[24]

It is possible that the benefits of steam distillation at this early stage of petroleum refining and imperfect heat application outweighed its limita-

---

* For somewhat different reasons, this remains true to the present day. Despite continuous distillation, rarely do refining operations exceed 75 per cent of rated capacity because of the myriad products produced in varying volumes, much specialized equipment, inevitable amounts of standby capacity in various stages of repair and obsolescence, and many related factors of highly intricate patterns of scheduling.

tions more than at any subsequent time. One primary motivation for its use was the reduction in the fire hazards that were associated with direct firing. Steam distillation also brought substantial fuel economies and greatly extended the life of equipment, possibly fourfold, an especially persuasive consideration during the Civil War period when expensive still bottoms required replacement after 35 runs.*

Preventing overheating of the contents of the still, steam distillation gave a superior separation of the fractions, avoided the malodorous compounds formed by inadvertent cracking or destructive distillation of the heavier components, and altogether made a sweeter product with a smoother procedure than direct firing. Regulation of the distillation by steam instead of fire also simplified operating control problems, for with steam the stillman could put on a large amount of fuel at a time and direct most of his attention to the important task of checking the distillate coming off.[26]

Of the several limitations to steam distillation without the aid of direct firing, probably the most serious was that its utility was restricted to the lighter fractions, from naphtha through kerosene. It was not as effective in distilling the heavier paraffin-oil fractions above kerosene which required higher heat input. It also required additional equipment, including a boiler, a more efficient water-condenser to condense the water emerging or co-distilling with the distillate, and a water separator between the condenser and the final receiving tanks.[27]

Of itself, steam distillation was not conclusive evidence of superior refining technique. At least a minority of refiners made progress in direct-firing and other distillation methods during the decade. Linked with several other factors it does, however, serve as a rough gauge of the status of refining at various locations in this early period when detailed information is often fragmentary.†

* The impressive savings in fuel consumption from steam rather than direct-fire distillation are brought out by the following comparison. Assuming a 43-barrel still and yields of approximately 60 per cent or 27 barrels, distillation by direct firing used up about 1.4 tons of coal in contrast with steam which needed only 0.34 of a ton. With coal at $6.00 a ton, the delivered price in Pittsburgh throughout most of the decade, the fuel costs ran about $8.40 for direct firing and $2.04 for steam distillation, or prorated per gallon approximately ¾ of a cent and ¼ of a cent respectively.[25]

† It is of equal importance to keep in mind that the basic characteristics of steam distillation developed in this early period are useful in interpreting refining technology to the present day. Though never dominant, steam distillation has been practiced widely and in modified forms is still practiced.

Superheated steam distillation

Another basic approach to fractional distillation, designed to surmount the temperature limitations of ordinary steam, was also available by 1862. This process involved the substitution of superheated steam for ordinary steam which gave a better distillation of the heavier fractions of the crude. Advanced in a patent issued in 1853 to C. M. Brown and also suggested by Benjamin Silliman, Jr. in his report in 1855, superheated steam had been used to a limited extent in coal-oil refining.

Stanley Hope, an English banker associated with Clark & Sumner in erecting the Standard refinery at Pittsburgh, introduced superheated steam distillation into commercial petroleum refining practice before the close of 1862, subsequently extending the use of his process to ten smaller refiners.[28] His process was a modification of one developed by Dr. Herbert W. C. Tweddle, an English chemist whom Hope brought to the United States to superintend his refining enterprise. Tweddle met the problem of equalized distribution of superheated steam in the still with a blow-pipe distributing device whose radial arms and rim were perforated with holes.

"The object of my improvement," Dr. Tweddle stated in his patent application, "is to preclude the necessity of having fire in or near the building in which the operation of distilling is carried on, and to do that, which has been hitherto attended with considerable danger, perfectly safely."[29] Further claims included freedom from frequent and extended disruption of operations by cracking or burning out of the still and pipe leakages and arrangements permitting the stillman to watch the distillates closely without frequent interruptions for firing the still.

There were, however, several formidable obstacles to a widespread adoption of superheated steam to petroleum refining. The design of the superheaters was such that an overly elaborate equipment was needed to achieve distribution of the steam. The cost of this equipment narrowed the savings so as to make the investment marginally attractive at best in commercial practice. A more permanent disadvantage was the excessive corrosive effects of superheated steam on stills.[30]

In many instances, however, this disadvantage was more than compensated for by the fact that superheated steam resulted in a great reduction and softening of the heavy coke deposits which, in a run with direct-fired stills, sometimes accumulated to a depth of 1 foot.[31] This benefit proved so great that later in the decade many among the majority of refiners who did not distill with steam introduced it into the still three or four hours after the end of a run, the first successful method to save time in cooling the still without damage to the still bottoms.[32]

Vacuum distillation

Vacuum distillation was the fourth permanent approach to fractional distillation of petroleum to advance to commercial status by the end of 1862. Like superheated steam it was used primarily to secure a better distillation of the heavier components of crude oil which either decomposed or boiled at inconveniently high temperatures under atmospheric pressures. Vacuum distillation had its primary origin in the sugar refining industry in the United States. In 1860 George Wilson, a prominent manufacturer of stearic acid, patented an apparatus using superheated steam and a vacuum similar to that used in refining sugar for refining petroleum.[33]

It was Hope and Tweddle, however, who ushered vacuum distillation into commercial practice. After distilling off the lighter fractions ranging from naphtha through kerosene with superheated steam, they conducted the heavier fractions with higher boiling points to a second still equipped with an exhauster which utilized superheated steam to produce a vacuum.[34]

Vacuum distillation, like superheated steam distillation, required somewhat complicated apparatus and better condensers than ordinary steam. In the early decades of petroleum, when pressure reductions by partial vacuum were moderate, 200–500 mm. (¾–⅓ atmosphere absolute pressure),[35] this limitation was offset in many instances by the relative advantage of obtaining a cleaner fractional distillation of heavier components without decomposition.[36] As a result its use spread considerably in the 1860's.

Cracking

Modern petroleum technology owes much to the four major advances in distillation which had come into commercial operation by the close of 1862. It is equally indebted to the process of destructive distillation or "cracking" which was also transplanted from coal-oil practice into petroleum refining during the same period. Cracking affected the entire course of refining technology during the kerosene age; and again in the gasoline age which followed. The key difference is that early cracking at ordinary pressures was used to increase the yields of illuminating oil; modern thermal cracking under pressure (and later catalytic) was used to increase the yields of gasoline fuel.

As already noted, fractional distillation simply separates the various components of crude oil without affecting their molecular structure. Destructive distillation, through the application of high temperatures to the higher-boiling portions of the crude, breaks many of the larger molecules into smaller ones within the range of the desired product. With cracking, enough large molecules might be broken into smaller ones to increase yields of burning oil as much as 20 per cent or more, depending on the depth and extent of cracking and precision of the temperature control.

It will be recalled that Benjamin Silliman, Jr., in his report on petroleum in 1855, surmised that destructive distillation might boost yields of burning oils.[37] Within three years of Silliman's report, Joshua Merrill and Luther Atwood with Downer's Kerosene Oil Company began the commercial cracking of crude and heavy coal-oil distillates into burning oils. Assigned Luther Atwood's basic patent for cracking coal oil or petroleum taken out in 1860, Downer licensed the process for a royalty of 1¢ a gallon until 1862, when the huge volume of petroleum production influenced him to cut that royalty in half.[38]

Heavy uncontrolled cracking, yielding an abominable burning oil, began with the first refining in the Region in 1860 and continued on an extensive scale until at least 1865. It began as an unintentional process and was continued in ignorance of the factors of temperature control, and was compounded and perpetuated by a desire to obtain maximum yields of burning oil. A. N. Leet, superintendent of the Humboldt refinery at Plummer during 1862–66, was explicit about the prevailing practices of his neighboring refiners. As he noted, they did everything possible to hasten distillation; the refined oil was full of cracked products and paraffin oil, giving it a dark, yellowish-green color. When the gravity of the entire batch reached 45° B, they diverted the remaining heavy oil to a still and ran it off at 700° F.[39]

Samuel Downer, seeking to protect his patents, gave up as far as the myriad small distilleries mushrooming in the Region were concerned, but he left no doubt that cracking was their general practice. "Nobody," he wrote his Corry refinery superintendent, "ever ran petroleum down to coke or tar with a still outdoors that did not in winter more or less crack part of it." [40]

Cracking, contrary to general belief, did not emerge in the Region as the result of a technological migration of advanced methods from seaboard centers.[41] Heavy uncontrolled cracking originated within the Region from poorly reported information on coal-oil practice, imperfectly interpreted, and badly executed. It was the type of refining process implicit, for example, in the instructions on petroleum refining widely circulated by the Schieffelin Brothers, a New York pharmaceutical house, in the spring of 1860 to promote new entries into the industry.

They counseled a single distillation almost down to coke with the promise of burning oil yields of 85–90 per cent. The temperatures applied to the heavier portions of the oil did not, according to their specifications, necessarily have to exceed the boiling point of mercury, 357° C, or 674° F, but the elaborate precautions they advised to ventilate the worm condenser bespoke a heavy output of gases, a characteristic phenomenon of cracking.[42] The instructions also warned to keep the bottom of the condenser at 90° F or 100° F after two-thirds of the distillate had run through. Though unstated, the purpose plainly was to keep paraffin and wax above their melting point so as to avoid clogged condensers, a fertile source of hot spots within the stills sufficient to induce cracking and the release of highly explosive elements that made clogged condensers a frequent cause of fire.

Cracking at Pittsburgh was another matter, not accidental, but a calculated light cracking in violation of Downer's patent, probably picked up either from one of Downer's licensees or from independent experiments. Even before their conversion to petroleum in 1861, the major coal-oil companies were all cracking crude coal oil lightly, and by 1862 not only they, but numerous new petroleum refineries were, according to Downer, all cracking more or less. Moreover, he reported, they all know "they decompose the oil and use that term." [43] Unlike the thicket of small distillers in the Region, and somewhat to Downer's surprise, many Pittsburgh refiners were distilling slowly enough with sufficient temperature controls to obtain a merchantable product in a single distillation.

In January, 1862, Downer had wisely counseled W. H. L. Smith, superintendent of the refinery he was erecting at Corry, not to prosecute Pittsburgh coal-oil refiners and others for patent violation if it was at all possible to negotiate a licensing agreement, since, "No jury would ever convict in Pittsburgh." [44] Granted free access to all refineries during his visit there with Merrill in the following August, he pursued this conciliatory policy, but "we spoke to all openly about our patent and that it would be enforced." [45] He did hold some cards for bargaining, for Pittsburgh refiners did not yet have all the answers. Downer reported to Smith: "I am satisfied our own oil is no whiter, no sweeter, but burns very much better. They [Pittsburgh refiners] are almost right, but cannot be entirely so without adopting our stills and redistilling on alkali the heavier portions." [46]

In the New York area, the New York Kerosene Company had been a primary licensee of Downer's cracking process since 1860 and there were others with or without licenses who undoubtedly were attempting during 1861 and 1862 to crack petroleum with as yet indifferent success because of lack of experience. It is certain that this route direct from coal-oil cracking was one of the main ones leading to widespread successful cracking by refiners in the New York area by 1863.*

The development of cracking had important effects, both immediate and long range. The precedent was set to crack both kerosene and heavier fractions into gasoline in the twentieth century when the demand for motor fuel far outstripped that for burning oil. More immediately in the early 1860's when the weight losses in crude transported in barrels were excessive, cracking supplied refiners distant from oil fields and near large seaboard markets a margin which helped them to stay in business. On the less favorable side, cracking greatly aggravated the problems of treating refined oils, which delayed their solution and exerted an adverse influence on the quality of refined oils for at least two decades.

Treating

The overwhelming importance of treating petroleum stands out in the contemporary nomenclature itself. To distill either coal oil or petroleum was not to refine it, although distilling was an important preparatory step to refining, and one that admittedly could influence the quality of refined oil.† In actual practice, refining was synonymous with treating with chemicals even though in theory it included the use of additional agents, such as clay, minerals, or animal charcoal. Treating also represented a key cost in the refined oil, one rivaled only by the cost of fuels.

The general line of subsequent petroleum treating had been worked

---

* Cracking in the New York area may have also been stimulated by the accidental discovery of the process by a small Newark, New Jersey, refiner during the winter of 1861–1862. [47]

† One persistent line of development for more than two decades was to find a way to refine crude petroleum, omitting the preparatory step of fractional distillation.

out by Joshua Merrill and Samuel Downer at Boston as early as 1857 in connection with the refining of crude coal oil. As they subsequently discovered, crude coal oils because of their greater densities and much higher content of aromatic substances (such as phenolic substances, sulphur compounds, and the like) required two and sometimes three distillations and much heavier acid and alkali treatments than petroleum.[48] Pioneering the cracking of the heavier portions of both crude coal oil and petroleum, Merrill came to an early appreciation that cracked oils of either type required heavier acid and alkali treatments than those fractionally distilled.*

Departing from a somewhat common practice of coal-oil refiners of introducing acids and alkalis into the crude in the stills, Merrill further developed the treating of both coal oil and petroleum, whether fractionally distilled or cracked, as a final process after distillation. In this manner he improved both the economy and the effectiveness of the acid and alkali applications under most conditions.

By 1865, and in some instances sooner, various other experienced refiners outside the coal oil–kerosene group with which Merrill and Downer were associated gradually caught up with Merrill's methods. They arrived at the following sequence which became the normal treating practice for fractionally distilled Pennsylvania crude oil throughout the rest of the century. From the stills the condensed distillates were run into agitator tanks where they were mixed with about a 2 per cent by weight concentration of sulphuric acid (of about 60° B strength), usually followed by several water washes of fifteen minutes each, followed in turn by a solution of caustic soda of about the same concentration as the acid. Finally the treated distillate was drawn to the settling or bleaching tanks, shallow wooden tubs with metallic linings, where it stood sometimes for several days to improve color and allow any remaining volatile portions of the distillate to evaporate. After testing with a hydrometer to see if the treated and bleached distillate had a density of about 45° B, a fire test, usually with the Tagliabue cup instrument, was run to verify that the liquid would not ignite below 110° F or higher (125° F in the case of refined oil for export).

With organic chemistry in its infancy even the best of the coal-oil and

---

* In the late 1850's during the coal-oil era, Downer and Merrill also providently innovated a cheaper, easier, and more efficient alkali treatment with a caustic soda solution instead of soda ash, the more primitive method widely employed in homemade soap making and which survived for several years among the less skilled petroleum refiners.[49]

early petroleum refiners had only a limited understanding of the chemical reactions involved in treating. They did appreciate, however, that in the treating of fractional distillates, sulphuric acid served the primary function of deodorizing and a secondary one of improving color; that similarly, caustic soda primarily improved color, secondarily improved odor by removing certain types of smelly compounds which did not react with sulphuric acid; and finally, that a caustic soda wash was almost the only way to remove sulphuric acid and its cargo of petroleum sulphonic compounds from the treated distillate because it was the only thing in which they were readily soluble.*

Why cracked oils required a much heavier acid treatment, often running to 6–8 per cent by volume, than distillates from a careful fractional distillation, was not so well understood. The answer lay in several interrelated facts. Cracking increased certain sulphur compounds like hydrogen sulphide and mercaptans which remained impervious either to physical or chemical reactions with sulphuric acid, but upon which caustic soda exerted a solvent action. Of still greater importance, cracking created two other highly unstable and undesirable compounds seldom, if ever, present in fractionally distilled products to a significant degree. One type was olefins, unsaturated substances that polymerize and oxidize readily, giving smelly, colored, and frequently insoluble products. A second type was diolefins, still more unstable, with bad burning and gum-forming characteristics.[50] Since sulphuric acid of 66° B strength was the maximum commercially available, acid treatment probably reacted little or not at all with olefins, but it certainly reacted strongly with the much more offensive diolefins, effecting their removal.

In treating, the sulphonic acids resulting from the reaction of sulphuric acid with diolefins, polycyclic aromatics and mercaptans were removed partly as acid sludge and partly by caustic soda washes. Cracked oils required heavier caustic soda treatment not only to remove a much larger quantity of sulphonic acids but because of the secondary role of the alkalis in desulphurization.

Thorough caustic washes were of key importance, for any traces of sulphuric acid remaining in illuminating oil had a disastrous effect on its burning properties. Thorough water washes after caustic soda treatment

---

* A variety of acids other than sulphuric have been patented for use in treating petroleum but it has remained the basic acidic treating agent because of its unique action on aromatic hydrocarbons. Soda ash, lime and ammonia were among other commonly used basic treating agents, and while all function in nearly an equivalent fashion, economic considerations have dictated the use of caustic soda to near exclusion of other bases.

were no less significant. If the caustic soda was not removed the oil lost its color within a few days, while the caustic soda itself soon encrusted lamp wicks, impairing capillary action so that the flame wilted badly when half or less of the oil in the reservoir had been consumed. All these complaints, including color deterioration in storage and shipment, were common throughout the decade.

Whether processing cracked oils or ordinary distillates the more advanced refiners were aware of the importance of well-designed and constructed agitators for effective treating. Indeed no well-appointed refinery from the beginning of the decade was without an agitator which mixed the treating acids and alkalis with the distillates either mechanically or by compressed air.

A direct heritage of coal-oil refining, mechanical and air agitators changed little in basic construction throughout the century. The tanks holding the distillate to be treated were usually of light-boiler iron and of either horizontal or vertical design. A typical mechanical agitator comprised a shaft with fans of wood or iron, placed obliquely like a ship's propellor and turned by gearing powered by a small steam engine. Air agitators either introduced air blown from a fan through a 2-inch wrought-iron pump into the tank or the air was supplied from an air pump.

Opinion was divided on the relative merits of the two types. Some believed air agitation was more convenient and thorough.[51] Others correctly suspected that the air agitation often introduced moisture which weakened the strengths of the treating chemicals, boosting the cost and lowering the effectiveness of treating.[52] In some instances agitators were equipped with both mechanical and air devices, permitting the operator to select whichever mode seemed best under the particular operating conditions.

Of utmost importance with either type was a steam jacket at the bottom and lower sides of the agitating tank to keep the temperature of the distillate and chemicals around 80°–90° F. Temperatures higher than these promoted the formation of tarry by-products, heightening odors and poor color which treating was intended to improve, while lower temperatures impaired the effectiveness of the treating agents.* Failure

---

* A. N. Leet reported that the oil in treatment should never be colder than 60° F. At lower temperatures any water present in the oil will not precipitate at once so that an excessive amount of hydrogen remains suspended in the oil. He advised against heating it with steam because oxygenation often occurs, holding the watery vapor in suspension, diluting the acid, reducing its power, neutralizing its effects. Not always realizing the causes, refiners often reported using twice as much sulphuric acid in winter as in other seasons.[53]

to appreciate this comprised one of the main stumbling blocks for in-experienced refiners to competent treating for half a decade and more, although many also lacked the treating apparatus of well-appointed refineries.

But if the leaders among refiners had arrived at reasonably good treating practices by the mid-1860's, this was by no means true of the generality of refiners for whom such a goal was indeed elusive. Downer guarded his techniques carefully, licensing at most only selected parts of his treating processes. Unaware that petroleum generally contained much smaller percentages of aromatic substances than coal oil, a great many early petroleum refiners, following coal-oil-treating practices, invariably over-treated by adding up to 10 per cent dosages of acids and alkalis to the distillates. Often enough they compounded this wastage altogether needlessly by treating the crude oil as well.[54]

Those refiners who followed the instructions contained in the widely distributed Schieffelin Brothers' circular were led even further astray. They were advised, for example, to omit the sulphuric acid entirely from treating since its use would impair the burning qualities of the oil. Since the recommended single distillation process clearly involved cracking, this ommission virtually assured a poor-burning illuminating oil, filled with polycyclic aromatic substances and diolefins which only sulphuric acid could remove.[55] Sharpsters who did a lively business selling spurious treating formulae to credulous buyers only served to add to the confusion.[56]

With the best of contemporary information, it was not always possible to secure adequate treating. Neither personal observations of Ferris's treating nor written instructions, for example, enabled J. Cutler, a novitiate refiner in Boston, to turn out a suitable burning oil using A. C. Ferris' adequate acid-alkali treating method in 1862 without repeated redistillations.[57]

Cutler's difficulties no doubt reflected many of the variables which made operations even with an acceptable formula difficult. These ranged from the problem of obtaining proper temperature controls, working with acids of varying degrees of concentration, inadequate mixing, or inept use of equipment.[58]

Apart from deviations of this sort, there was no assurance that a particular treatment would be equally successful with different batches of crude. The speed and temperatures of successive distillations seldom remained constant. A heavy West Virginia crude, for example, or a Pennsylvania oil whose volatile components had evaporated during lengthy storage or exposure during a slow journey to distant refineries usually required extra heavy treatment.[59]

The difficulties and delays encountered by ordinary refiners in arriving at an adequate solution of treating problems are well illustrated by experiences within Downer's own organization. At the time Downer opened his refinery at Corry, Pennsylvania, in early 1863, W. H. L. Smith, the refinery superintendent, had great trouble applying the treating methods which had been so successfully worked out by Merrill at the Boston plant. Writing to Smith in April, 1863, Downer consoled him with the reminder that, "Agitation can only be learned by experience and practice and time . . . One must learn from all sources and then if they have judgment and industry and all that sort of thing, then they will succeed." [60]

Despite the lag of half a decade among the generality of Region refiners and many elsewhere, it still remained true that by 1863 the best refiners in Pittsburgh and on the Seaboard had caught up with Joshua Merrill. With many variations in efficiency and treating in one or several stages, they arrived at the basic broad approach that prevailed throughout the century, which was to treat with acid and soda as the final process after distillation.

It was about this time that Merrill, with one of his many brilliant insights, reversed his own previous practice which competitors were gradually catching up with. In 1863 he made a clean break-through to the best modern practice when he began giving all burning-oil distillates a final distillation after treating with acid and soda. Although yields were reduced 5–10 per cent by this method, he found that he had been gaining several other advantages. Redistillation obviated much of the expense of one to two days bleaching time, and it offered a margin of safety permitting reduction of treating agents to a minimum. It undoubtedly expedited the blending of straight-run distillates and cracked oils into a better balanced illuminating oil in uniformity of boiling range and burning qualities. Redistillation further assured that all traces of acid or alkali were removed, with the result that the last spoonful of oil in a lamp burned as well as the first. There was no deterioration of the refined oil in color, odor, or burning properties during lengthy intervals of storage from delayed reactions to invisible remnants of treating agents or traces of unremoved by-products. Redistillation, finally, made removal of the explosive light fractions in the benzine and naphtha range a virtual certainty. This was not only a vital step toward safe oils but better burning oils as well, for the dangerous light fractions, though burning with excessive white brilliance at the outset, also prematurely burned out the wicks.[61]

With Merrill's method of treating and final redistillation, Downer's kerosene established and maintained an unrivalled and deserved reputation for quality. It was part of the extra cost of a quality product that commanded a premium of 1.5¢–2¢ a gallon, although it also was used in Downer's two cheaper grades of kerosene. While there were other combinations of distilling and treating as a final process which adequately met contemporary premium standards, none was as effective as the processes used in the Downer plants.

Throughout the decade and beyond most refiners, however, were convinced that the cheapest way to refine an oil of reasonably good quality with a fire test of at least 100° F was to distill off the naphtha down to 58° B instead of 62° B or 65° B, and after treating to expose it to the sun in shallow bleaching tanks for one or two days to assure removal of any remaining naphtha.*

But these differences were only a reflection of a whole complex of forces which kept refining operation in a state of flux throughout the 1860's and early 1870's. As this brief review has indicated, the coal-oil heritage had an enormous influence in the development of processes by 1863 which were basic of the subsequent development of the American petroleum industry. One feature of the succeeding decade was the more widespread acceptance and utilization of distilling and treating methods which only the most advanced refiners had adopted by the early 1860's.

Scale of refining operations, yields, and refining costs at the mid-1860's

Though aggregate refining capacity had grown at Pittsburgh, the Region, New York, and Cleveland, the size of refineries remained modest to the mid-1860's. The receipt of crude in barrels alone narrowed opportunities for economies of scale. Apart from the costs of the barrels, manual handling characterized operations throughout, from storing the crude to moving it to stills and to treating batches of distillates. The requirements of batch distillation placed further limits both on the size of plants and on the desirability of working by-products. To maintain any given level of

---

* Charles Chandler in 1870 estimated that by following this final step to insure a good quality kerosene, refiners added between 3¢ to 4¢ per gallon to their costs.[62]

daily refined output, a refinery had to have at least three to four times as much still capacity that could be charged only twice a week, all distributed among small stills. Diversion of stills to redistillation of kerosene fractions or of by-product stocks sacrificed crude inputs and refined oil outputs.

The fact that the nation's largest three refineries, with weekly charging capacities of 2,000 barrels, and perhaps ten others approaching that capacity, were all built or in the process of construction before the end of 1862, suggests a scale of operations where diminishing returns rapidly set in. Pittsburgh had the largest concentration of refining, more than 50 plants with an aggregate capacity of 37,500 barrels weekly by 1866, yet only the Standard, Ardesco, and Petrolite charged 2,000 barrels weekly, the Brilliant 1,500 barrels, and three others 1,000 barrels. In the Region there were two atypical plants, the Humboldt at Plummer, charging 1,000 barrels weekly, and Downer's at Corry, charging 1,800 barrels, which was complemented by his original refinery of similar size at Boston. The New York Kerosene Company, a converted coal-oil refinery, about completed the plants in this class. In the next lower scale of refineries at Pittsburgh, only a few exceeded 200–600 barrels weekly.[63]

Still sizes likewise demonstrated little elasticity. The 5–10-barrel stills and the teakettle operations associated with them in the Region were no longer important. Stills running a charge of 25–50 barrels within a 48-hour period dominated, although stills running 100 barrels within the same time were becoming common. Most were of thick cast iron, though thinner, wrought-iron sides were used in most newer installations. The incidence of fractured still bottoms and sides continued high because of wartime deterioration in the quality of metal, reckless distillation down to coke, inadequate maintenance and shortened cooling periods, and failure to introduce superheated steam to shorten cooling times and expedite removal of cokings. One exception to the general pattern of still sizes came in 1865 when the Petrolite refinery at Pittsburgh installed a horizontal still, fired by twelve furnaces and capable of a 1,300-barrel charge of crude. There were no contestants to its claim of being the world's largest.[64]

Yields of kerosene continued to be almost universally reported as 75 per cent, and the reports continued almost meaningless. Fifty per cent was the maximum of a straight-run kerosene distillate cut at 60° B for its lowest boiling materials. Increasing the kerosene range further into both lower and higher boiling materials easily boosted the apparent yield

COURTESY DRAKE WELL MUSEUM

Humboldt refinery, Plumer, 1865.

of straight-run to 65 per cent and to 75 per cent when half of the remaining crude-naphtha distillate also was added to kerosene. Both practices were common.

The wide prevalence of cracking, fostered by pressure to compensate for high weight losses in shipping, further obscured the accuracy of reports of yields. To obtain cracked kerosene distillate of acceptable quality, destructive distillation had to proceed slowly at one-fifth or less the rate of ordinary distillation and at temperatures only moderately above those of the heavy, high-boiling fractions being cracked. Destructive distillation yielding 15 per cent of kerosene distillate thus took about as long as ordinary distillation yielding 50 per cent of straight-run kerosene.[65]

These requirements ran counter to the pressure of batch distillation for rapid running times, with the result that operators often succumbed to the temptation to speed destructive distillation at excessive temperatures. This resulted in a malodorous, badly colored, unstable produce, which required abnormal treating costs to make it marketable. The extent and quality of cracking obviously depended on a number of variables, including the cost of crude, the types of stills available, and the skill of refiners.

Van der Weyde among others well versed in the art of refining correctly reported in 1866 that 20 per cent was the maximum yield of acceptable products obtainable from cracking while 15 per cent was the more usual yield. In other words, a total yield of 65 per cent of straight run and cracked distillate was good; 70 per cent was exceptional.[66]

Like yields, manufacturing costs were not reported realistically. Almost invariably, 5¢ a gallon for refined was the only figure cited, regardless of locale or size of refinery. Left out of consideration were numerous variables, including costs of fuel and labor, types of stills—whether direct-fire or steam, costs of acid, the percentage acid used in treatment, whether refined was once or twice distilled, and whether the product met 100°-F ignitition test for domestic sales or the 115°-F test with a 6–8 per cent lower yield for the export market.[67]

While inadequate data and many variables rule out any close cost estimates, there is good reason to assume that the figure of 5¢ a gallon understated the costs of most Regional refiners by as much as one-half to two-thirds. It is also doubtful whether this amount adequately covered the processing costs of many refineries elsewhere.

This conclusion is borne out by cost estimates for 1863–65 from two of the most efficient plants in the industry, Samuel Downer's refineries at Boston and Corry, each of which had a weekly crude charging capacity of about 1,800 40-gallon barrels and employed a labor force of about two hundred men. Using simple distillation by steam, followed by a 2-percent-acid treatment, Downer's minimum refining cost at Boston was about 4¢ per gallon; at Corry about 7¢. With destructive distillation by direct fire followed by a 6-per-cent-acid treatment, his maximum costs at Boston ran about 6¢ a gallon; at Corry almost 11¢.

In computing the cost estimates shown in the following tabulation, it was assumed for Downer's plants: (1) that the yields were 60 per cent, or 43,200 gallons, of the weekly crude throughput of 72,000 gallons; (2) that 50 of the 200 workers were employed in the cooperage and machine shops not directly engaged in processing; and (3) that minimum costs involved the use of steam distillation and 2-per-cent-acid treatment, while maximum costs involved direct-fire distillation and 6-per-cent-acid treatment.

Costs per week	Boston		Corry	
	minimum	maximum	minimum	maximum
Labor [1]	$1,350.00	$1,350.00	$2,250.00	$2,250.00
Coal [2]	99.96	411.60	171.36	705.60
Sulphuric acid [3]	280.80	842.40	561.60	1,684.80
Total	$1,730.76	$2,604.00	$2,982.96	$4,640.40
Per gallon	4¢	6¢	7¢	11¢

[1] The daily wage rate in Boston was $1.50 per day; in Corry, $2.50.
[2] Coal is figured at $7.00 per ton in Boston; $12.00 per ton in Corry. Based on an average charge of crude of 43 barrels, 42 charges per week, direct-fire distillation required 1.4 tons of coal per charge; steam distillation .34 ton.
[3] Amount of acid used is figured on a weekly refined yield of 43,200 gallons, weighing 6.5 pounds per gallon. Acid is figured at 5¢ per pound in Boston; 10¢ at Corry.

In actual practice, gradations between these extreme costs were usual. For example, ordinary distillation was apt to be conducted by direct-fire and co-distillation with steam, while 6-per-cent-acid treatment was usually applied only to 15–20 per cent of the distillate representing the cracked stock.

In general, the most efficient refiners at New York and Pittsburgh probably came close to achieving costs similar to those of Downer's Boston plant, with Pittsburgh having a slight advantage in fuel and perhaps labor costs. Lacking local sources of chemicals, Cleveland shared some, but by no means all, of the Region's disadvantages in treating.

Differences in yields, refining costs, treating methods, and still types all reflected a complex of forces that kept refining operations in a state of flux throughout the 1860's and beyond. Diversity was the rule as refiners attempted to adjust the experience of the coal-oil period to the problems peculiar to the petroleum era.

During the latter half of the decade, refiners in general focused their attention on the problem of increasing the scale of their operations, principally by introducing larger-sized stills. A small but growing number were further interested in improving their methods of manufacturing an impressive list of petroleum by-products which had been brought into commercial production by the mid-1860's.

Chapter 10

Petroleum by-products

Most refiners were concerned only with the main kerosene fraction of their distillate and with enlarging this middle cut as much as possible. To accomplish this they usually extended ordinary distillation into destructive distillation of the heavier (higher-boiling) fractions above the kerosene range. However they proceeded, they inevitably obtained in ordinary distillation a large, lighter fraction before reaching the kerosene range. With the exception of those who cracked excessively down to coke, they also could not avoid producing a second rough fraction much heavier than kerosene. It was these two fractions that comprised by-product stock—that is, material from which useful by-products could be obtained by further distillation or treating.

The first light fraction of ordinary distillation was crude naphtha. Since it always contained some hydrocarbons boiling in the kerosene range and higher, this fraction required further distillation and treating. Though variable, its volume might equal 20 per cent or more of a low-gravity crude. The heavier fraction above the kerosene range, usually somewhat smaller, might serve as lubricating and paraffin wax stock. This stock could be further divided into lower-boiling constituents for light lubricants (spindle oils) and hard-crystalline paraffin wax and into higher-boiling oil for heavy lubricants (cylinder oils) and softer wax.

In the first part of the decade, opportunities and incentives to refine the naphtha, lubricating, and wax stocks remained so rare as to make the faintest of impacts on the organization of refining. With the expansion of throughput and markets in the postwar period, however, small refiners specializing in processing various by-product stocks became increasingly numerous, while a minority of kerosene refiners strategically located near large concentrated markets began to move into production of by-products. Intimately associated with this development, especially the refining of the heavier fractions that were the most difficult to process, was a spreading competence in co-distillation with steam, superheated steam, and in partial vacuum.

Various influences stimulated a rising interest in by-products. Perhaps the most important response came from skilled refiners located near industrial centers where specialized new types of demand were most likely to originate or concentrate. This was true of the Downer organization in Boston, with its unmatched experience in working every type of by-product stock and the technological services of Joshua Merrill. Some by-product uses involved innovations entirely external to refining, notably the utilization of by-product stocks in fueling marine boilers and in illuminating-gas manufacture.*

Though only relatively few refiners participated in it, the development of all petroleum by-products and their major uses throughout the century was a major achievement of innovation of the first decade of refining. Compared to kerosene, by-products did not loom high in total production, but this was no criterion of their qualitative contribution to American life, which often was inverse to their quantitative importance.

By-products from the light fractions

Naphthas. The first distillation of crude petroleum yielding 10–20 per cent by volume of crude naphtha exceeded yields from crude coal oil

---

* One important exception to the completion of by-product development in this decade was gas oil for coal-gas enrichment, but this fraction immediately above the kerosene range awaited commercial development of a water-gas process in the latter part of the 1880's. But coal-gas enrichment with crude naphtha late in the 1860's paved the way for even this development.

by several times, yet only a few petroleum refiners initially attempted to produce refined naphtha, which involved one or more redistillations and treatment with acid and caustic soda.

The principal markets for deodorized light naphthas at this time were for use as solvents in cleaning and to dissolve India rubber and gutta perchas for waterproofing materials. Most refiners, however, found no profit in diverting stills and expensive chemicals from refining burning oils. Accordingly, they routed what they could of the heaviest benzine portions of crude naphtha to burning oils and sometimes to boiler fuel, while throwing or giving away the rest, some of which was used in cleaning oil wells.*

The advent of the Civil War, which cut off supplies of crude turpentine from the South and drove up the price of alcohol (in part as the result of the imposition of a tax), sharply increased the demand for alternative solvents. Refined benzine was the chief beneficiary. It replaced spirits of turpentine in the manufacture of paints and varnishes in this country and in England and France. It substituted for alcohol in stiffening used by hatters, in furniture polish, in gums and glues applied to patterns and rollers in cotton mills and foundries, and in power-transmission belt dressings.[1]

After the Civil War the lighter naphthas entered important new uses as a solvent in the extraction of oils from vegetable seeds, notably flax, cotton, and the castor bean. Other extensive uses were to extract fats from the residuum of fat renderings, meat scraps, castor-oil seed cake, and cotton wastes. The recovery of wax from petroleum lubricating stock provided an additional use.[2] Although growing impressively, the demand for solvents failed to relieve the enormous surplus of crude and refined naphtha which accompanied the annual processing of 3–5 million barrels of crude oil in the latter part of the decade. The huge oversupply made adulteration of kerosene a tempting outlet for this otherwise worthless surplus product.

Gasoline. Gasoline, first of the new petroleum by-products, entered commercial usage no later than 1863. Who discovered it is difficult to establish, but Joshua Merrill's earlier innovation "keroselene," the most volatile product ever refined from coal oil, suggests he may also have originated gasoline. Merrill had obtained keroselene by steam distillation of

---

* Heaviest of the naphtha group was benzine, with an average boiling point of about 300° F and density of 60°–65° B. Lighter naphthas ranged in average boiling points to 180° F or lower and in densities from 65° B to 80° B. See Table 9:1, Chapter 9.

the lightest fractions of a thrice-distilled coal oil and applying salt and ice to the condenser. Finding keroselene (sp. gr. of .650) more volatile than ether and less toxic than chloroform, Dr. Henry Bigelow, a renowned Boston surgeon, introduced the extensive use of keroselene as a local anesthetic about 1861.

At the same time another market, somewhat less limited, opened up with the development of devices which evaporated volatile oils into air to give a "fixed gas" that could be piped and burned for gaslight throughout individual buildings removed from the distribution systems of gas utility companies. But this market was soon transferred to gasolines when Merrill applied procedures for refining keroselene to petroleum naphtha. Obtaining a 1.5 per cent yield of gasoline, Merrill in 1863 began the commercial production of gasoline at the Downer plant in Boston for the largest manufacturer of air-gas machines.[3]

The incentive for the production of gasoline apparently was to find a more satisfactory material for air-gas machines. Having a lower boiling point than benzole, naphtha, and most volatile oils, gasoline when vaporized was expected to have less tendency to condense, thus reducing the illuminating power of the gas. To help introduce the new product, Merrill and Downer sold gasoline for a time at 30¢ a gallon f.o.b. Boston, and 35¢ f.o.b. Corry, Pa., for sales in the New York market. Packaging the highly volatile product for shipment from Corry presented a problem which was solved by using only the stoutest beer barrels, either 25-gallon half-barrels or 40-gallon full barrels. Packaged in this form, the first regular shipments of gasoline began to move between Corry and New York City in 1863.

The demand for gasoline grew apace with the rapid improvement and spread of air-gas machines. Their sanction by insurance underwriters in 1867 was an important milestone in their progress. By 1872, 101 companies were engaged in the manufacture of gas machines, which, according to one account, were employed in all sections of the nation to illuminate mills and factories of all kinds as well as "public institutions, colleges, asylums, hotels, town halls, blocks of stores" in addition to "hundreds of elegant, private, country mansions."[4]

Although air-gas installations fell short of absorbing the entire output of light fractions from petroleum refining, they did provide numerous small refiners in Cleveland and elsewhere with a profitable specialty business. Certain of the large refiners in Pittsburgh were encouraged to extend their production beyond illuminating oil, lubricants and naphthas. In New York City, Charles Pratt & Company, in addition to leadership in the production of high-quality kerosene, had reputedly become the largest re-finer of gasoline in the world toward the close of the decade.[5]

Naphtha gas. The first large-scale demand for crude naphtha also arose out of gas illumination, when domestic manufacturers began to adopt the Gale-Rand naphtha gas process. Virtually unheard of in 1869, the Gale-Rand process within three years had revolutionized gas manufacture in large city systems, both as a direct competitor to coal gas and as an enricher of coal gas that often contributed more gas than the original coal-gas plant.

A. C. Rand was an extensive oil marketer and small refiner in New York who for several years had engaged in commercial ventures in petroleum gas.[6] In 1867 he introduced his "pneumatic gas machine," a portable unit for carbureting gasoline that embraced several patented improvements designed to improve the distribution of the gas by pipes.[7] In 1869, Rand and some financial backers joined with Dr. Leonard K. Gale to prove on a commercial scale at Saratoga, New York, a naphtha process for supplying large city systems. Evidently exerting both technological and financial leadership, Rand purchased Gale's two broad patents for "vaporizing of hydrocarbons below a red heat, and the decomposing of them above a red heat into fixed gases," and performed most of the engineering of the apparatus.* Though not completely successful, the promise of the Gale-Rand process after a year's operation was so encouraging that its patent and all related ones were purchased in 1870 by the Gas Light Company of America, a syndicate of financiers interested in gas investments.[9] This transaction evidently made available to Rand the funds and business connections to install for the Memphis Gas Works a plant serving naphtha gas only, and one to enrich coal gas for the Fort Wayne, Indiana, Gas-Light Company. Publications of comprehensive records of operating procedures and costs at once persuaded many in the gas business and related industrial arts that the Gale-Rand process had demonstrated its commercial feasibility, but it also touched off violent opposition from powerful coal-gas interests.

Stripped of controversial exaggerations, the feasibility of the process hinged on a few essentials. Over the previous two decades, naphtha gas processes had risen by the score. To make a rich gas from the volatile liquid was relatively easy, but all previous processes had failed to convert

---

* Fixed or permanent gases are gases which cannot be condensed by commercially available temperatures (usually 200° F or more below the 32° F, the freezing point of water) or pressures.[8]

that gas to a permanent gas capable of traveling through mains without excessive condensation and loss of illuminating power, or if used to enrich coal gas, that would not separate. It had long been recognized that if these limitations could be overcome, naphtha gas offered substantial economies for large plants which distributed their product through gas mains. To cite only a few examples, plant costs would be slashed by two-thirds because all need for elaborate purification equipment as well as for more than half the retorts in coal-gas manufacture would be eliminated. Labor requirements would lower in about the same ratio, along with maintenance. Space and tankage for storage would also be appreciably reduced. So long as a large new demand for naphtha did not drive its price far beyond 5¢ a gallon, the costs of raw materials would weigh heavily in favor of naphtha gas.

If the Gale-Rand process did not fully surmount previous shortcomings of naphtha-gas processes, its gains were sufficient to place it in strong commercial competition. The crude naphtha, usually of about 70° B gravity, initially was distilled with superheated steam at low temperature and pressure in an iron still. The vapors then were transferred to a clay retort of the type used in gas-making for conversion into permanent or fixed gas. At this point, an improvement in processing occurred. By passing the vapors through and over the red-hot retort at temperatures of coal-gas distillation (usually ranging above 2000° F), these temperatures apparently sufficed to decompose the vapors by cracking into fixed gases. The heated gases then were passed through condensers. The portions that condensed, not being fixed gases, were returned to the steam still for redistillation and then piped again to the red-hot retort for further conversion into permanent gases.[10]

Plants to supply naphtha gas for enrichment of coal gas in large city systems also produced the gas in much the same way, but they then diluted it to required standards. W. Henry White, engineer of the Citizens Gas-Light Company of Brooklyn, pioneer adapter of naphtha gas for large-scale enrichment, quickly discovered that mechanical admixture of atmospheric air as a diluent increased the excessive specific gravity of the gas. He remedied this difficulty by turning to George Olney's improved process, which introduced the diluent (whether air, water gas, or coal gas of low illuminating power) directly into the liquid in the still rather than at the later gas stage. The carbureted liquid then was carried to and through the heated retorts.[11]

Petroleum refiners hastened to assure interested persons of the sustained availability of crude naphtha at prevailing low prices. In 1870, Henry Edgerton was informed that the daily output of naphtha was 5,000 gallons and it sold for 4¢–7¢ a gallon, depending on the location of the buyer, while many refiners gladly gave it away.[12] N. P. Payne, of Clark, Payne & Company, Cleveland refiners, reported to Edgerton in 1871 a daily output of 300 barrels of gasoline and naphtha which he quoted at 90¢–$1.00 a barrel in bulk tank-car shipments. The output of all Cleveland refineries, Payne added, was 1,000 barrels daily, or enough to furnish the entire gas supply of New York, Philadelphia and Chicago. The annual output of benzine alone would suffice to produce 60 billion cubic feet of naphtha gas, Edgerton estimated, compared with a total annual production of coal gas of 12 billion cubic feet.[13] Charles Pratt in New York City offered long-term contracts to any and all local gas companies adopting the new process, at prices of 2½¢–4½¢ per gallon. Over a period of almost three years Pratt did not boost the price despite supplying four major companies.[14]

By 1873 naphtha gas in all the many variants of the Gale-Rand process had established firm commercial roots. The gas companies of fourteen middle-sized cities, including Chicago, Memphis, and Detroit, and smaller cities in Minnesota, Ohio, and Pennsylvania, had installed plants and were operating on naphtha gas exclusively, unmixed with coal gas or air. These companies supplied consumers with special burners which compensated for the extra richness of the gas by reducing the quantity delivered in standard burners. Eight companies in large cities, including four in New York, also had erected plants to make naphtha gas as an enricher of coal gas, in ratios ranging from 33–60 per cent, which they served to 28,500 consumers.

Naphtha gas gave the petroleum-refining industry its first large-volume market for by-product benzine and naphtha. Although not large enough to dry up the oversupply of cheap naphtha that made adulteration of kerosene a major problem, it made a valuable contribution toward a solution. Aided by succeeding improved processes, it furnished the largest outlet for naphtha, and one that continued to expand, for more than a decade.

Rhigolene and Cymogene: liquefied petroleum gases. Two new products lighter than gasoline, which appeared with dramatic suddenness in 1866, reflected the strong influence of proximity to centers of science and the industrial arts on innovations in refined by-products. Rhigolene, for example, was the direct outgrowth of Dr. Bigelow's success with keroselene as an anesthetic. Seeking a more volatile product with even greater freezing effects, Dr. Bigelow in 1866 asked Merrill to explore the possibilities of obtaining an improved anesthetic from petroleum.

Merrill responded by submitting extremely light gasoline of 85° B to a redistillation by steam. Applying ice and salt to the condenser, he obtained a 10 per cent yield of the lightest fluid then known, with a specific gravity of .625 and a boiling point only 65° F, compared with 96° F for ether. On April 9, 1866, Dr. Bigelow reported to the Boston Medical Society of a great advance in local anesthesia employing a new petroleum product which he named "rhigolene," after a Greek word meaning "extreme cold." [15] Had he employed modern nomenclature, Dr. Bigelow would have called the new product a mixture of pentane and butane.

For several years the unavailability of spraying equipment and a lack of adequate knowledge of administering techniques limited the use of rhigolene to the most advanced practitioners, but by 1871, several hundred gallons, used in minute dosages, had been purchased by doctors and dentists at $1.00 a gallon. For almost two more decades rhigolene remained one of the primary local anesthetics.[16]

Shortly after the introduction of rhigolene, Dr. P. H. Van der Weyde in the spring of 1866 announced the discovery of a still lighter product of gasoline distillation with a Baume gravity of 110° and a phenomenally low boiling point of about 40° F, which he named "chimogene," later "cymogene," after a Greek word root meaning "cold generator." This prominent Philadelphia physician, who had recently resigned his chair of chemistry at Girard College to become an industrial consultant to petroleum refiners, had discovered the liquefied petroleum gas known as "butane" in modern nomenclature.

Unlike Merrill, Van der Weyde was not specifically seeking a local anesthetic, although this possibility prompted him to write to the editors of *Dental Cosmos* in May, 1866, indicating the characteristics of cymogene which made it useful for this purpose. In subsequent reports and demonstrations before scientific groups Van der Weyde proposed the same use for cymogene to which Dr. Bigelow was putting rhigolene so successfully.[17] But because cymogene's extreme volatility made it too difficult to manage by prevailing techniques of handling and administering, it was never used extensively as an anesthetic.

Van der Weyde did, however, develop an outlet for his product in compression machines just coming into use in the manufacture of ice. Compared to ammonia, ether, and other refrigerants currently used in these machines, cymogene had the advantages of a much faster evaporating rate and a sharper cooling effect. To promote the new refrigerant, Van der Weyde designed and patented a compression ice-making machine, which he sold to Daniel L. Holden, who became a prominent figure in the new industry. By the end of the decade, Holden, obtaining his cymogene from Van der Weyde for $1.50 a gallon, had manufactured and installed several units in southern cities. In the next two decades of advance in artificial ice manufacture toward an important industrial status, compression machines supplied cymogene with its only market.[18]

By-products from heavy fractions

Fuel oil. The 1862–1873 period witnessed no more persistent efforts in by-product development than those to make petroleum a source of energy in competition with wood and coal. Coal-oil refiners had earlier projected crude coal oil as a commercially feasible and desirable substitute for coal in fueling marine boilers. Late in 1861 when the new phenomenon of gushers increased production and prices dropped to 40¢ a barrel, the proponents of liquid mineral fuels rapidly multiplied. In 1862 a war-minded Congress appropriated $2,000 to investigate the possibilities of petroleum as a fuel for naval vessels, and numerous inventors turned their attention to developing burning apparatus.

Perhaps the earliest significant use of petroleum as a fuel came in the Oil Regions at the Humboldt Refinery at Plummer near Oil Creek. Erected in 1862 when coal and wood were high in price and almost unobtainable, A. N. Leet, the plant's superintendent, turned to heavy crude and residuum as fuel for his boilers. For a year, Leet ran twenty stills and two boilers day and night by this plan, with no mishaps. But when the price of crude rose to about $4.00 a barrel late in 1863, he turned to coal.[19]

A device patented late in 1862 by two Philadelphians, Thomas Shaw and J. L. Linton, which they licensed to the French government, focused world-wide attention on the use of petroleum to generate steam for marine and locomotive boilers. Trial demonstrations at the Philadelphia Naval Yard in 1863 resulted in a favorable report from naval engineers recommending further experiments. Preliminary trials indicated that petroleum-fueled iron-clad warships might keep under steam at sea two or three times as long and with less labor and greater convenience than with coal.[20]

A patent taken out in 1863 for a process to use superheated steam to vaporize either crude or refined oil through burners brought a sharp reaction from critics of attempts to utilize petroleum as a fuel.[21] One report, for example, pointed out that while 1,252 pounds of crude petroleum gave the same heating power as 2,000 pounds of coal (with coal prices $10.00 a ton and crude petroleum at 40¢ a gallon in New York City), the cost of generating steam by petroleum was about ten times more expensive than with coal.[22]

Other experience, however, brought a renewed interest and in April, 1866, Congress appropriated $5,000 for a further testing of petroleum as a marine fuel, with B. F. Isherwood, Chief of the U. S. Navy's Bureau of Steam Engineering, in charge. In the spring of 1867 the Navy Department assigned the iron-clad steamer, *Palos*, a gunboat of the fourth class, to the Charleston Navy Yard in Boston for seagoing tests. Meanwhile, tests were also ordered to be run in three experimental boilers at the Navy Yard in Brooklyn.[23]

Glowing reports of wharf trials and cruises in June, 1867, received international coverage in the press. On her first cruise, the *Palos* steamed 25 nautical miles at a remarkable speed of 14 miles an hour, with only two firemen in the carpeted engine room instead of the usual complement of eight firing on a sooted grimy floor. Less than 4 barrels of petroleum did the work of 6–8 tons of coal with a 50 per cent gain in speed and no alterations of existing machinery. At the end of the trip, steam was being generated more rapidly than at the start, and fires were extinguished almost instantaneously.[24]

These accounts raised false hopes, for publication of Isherwood's report in the fall of 1867 put an end to any immediate prospects of the adoption of petroleum as a fuel by the U. S. Navy. It was admitted that the substitution of petroleum for coal would reduce both the weight and space required for storage by about one-third; that the number of firemen required could be reduced with petroleum by three-quarters; and that, with petroleum, boilers could be fired and extinguished almost instantaneously, whereas coal required firing and extinguishing periods of at least one hour each. But on the debit side it was pointed out that petroleum gases, highly explosive when admixed with air, were extremely hazardous in engine and boiler rooms where temperatures averaged 100° F.

Since even a medium-sized naval steamer would have to carry 250 tons of petroleum, it would be vulnerable to complete destruction by a single shot, no matter what protection was devised for the tanked petroleum.

Furthermore, the volatility of petroleum assured a big evaporation loss in stowage, offsetting part of its advantages in bulk, weight, and evaporating efficiency. In confined holds and poorly ventilated compartments of vessels, the smell of petroleum also would be intolerable. It was the question of relative costs, however, which made the case for marine use of petroleum hopeless. These costs, it was stated, were some eight times greater for petroleum than for coal in 1867. Any new or large demand for petroleum would only increase that disparity.[25]

The Isherwood report did much to broaden the skepticism already prevailing regarding the practicability of petroleum fuels. The withdrawal of the U. S. Navy from this field of interest eliminated the one agency qualified to develop more efficient apparatus for the use of petroleum fuels where costs were not the sole governing consideration. Thereafter leadership in developing petroleum fuels for naval vessels passed to the British, French, and Russians.*

Despite many attempts to utilize petroleum fuel in locomotives, on river steamboats, and for industrial purposes, it was generally recognized by the early 1870's that cost factors weighed heavily against its use, except under special circumstances. Crude petroleum neared a record low on the seaboard of $2.00 a barrel without stimulating a swing to petroleum fuels, while within the Oil Region where coal still was higher priced than elsewhere in Pennsylvania, ample experience demonstrated that refineries, factories, and other industrial establishments turned to oil fuel only when crude oil could be obtained for 50¢ a barrel or less.[26]

Though failing in their main objectives, the experiments and limited commercial uses in the 1860's produced a wealth of valuable experience in utilizing petroleum as an energy source. It remained for improved apparatus and more favorable economic conditions in the 1890's to allow its potential to be realized.

Petroleum lubricants. Destined to become, next to kerosene, the most valuable refined product, petroleum lubricants were relatively insignificant until the late 1860's. The heavier fractions of paraffin oil refined from petroleum had the same lubricating limitations as those from coal oil. They were generally deficient in body, viscosity, oiliness, odor and color, and they would ignite at unsatisfactorily low temperatures. These qualities automatically restricted markets to use as an adulterant in animal and vegetable oils or as a secondary ingredient in lubricating compounds de-

---

* Except for brief tests of petroleum in the late 1870's and again in the 1880's, the U. S. Navy did not again give serious attention to developing petroleum fuels until after 1900.

LUBRICENE

CYLINDER OIL.

CHARO'S CYLINDER OIL.

COURTESY STANDARD OIL COMPANY (NEW JERSEY)

Primitive distillation and treating of lubricating oils.

signed to exploit the best characteristics of several oils. The chief merits of paraffin oils for these purposes were cheapness, less tendency to oxidation and gum formation than natural oils, and, if they had been dewaxed, a lower pour point in cold temperatures. To enjoy relative success with these higher-boiling fractions required considerable refining skill.

Most of the small number of early refiners interested in lubricants turned instead to the production of "natural" lubricating oils directly from crude petroleum. A need for some sort of treatment similar to the clarification of natural oils comprised the only direct relationship to refining. The important thing was to select a heavy-gravity crude that seemed to be sweet, have good body, and other desirable characteristics and perhaps to steam it lightly to remove contaminants and volatile hydrocarbons. For the first half-decade, natural crude-oil lubricants did little to improve the popularity of petroleum lubricants. Their odor and color were worse than the refined product, and they frequently contained sand, grit and other contaminants. Failure to remove all volatile materials made them so hazardous that premium insurance rates usually resulted wherever they were put to heavy-duty use.

One route to improvement lay in filtration of the crude oil through animal charcoal and bone black, practiced for generations to purify and improve the color and odor of all kinds of oils. It remained for Robert Chesebrough, who entered coal-oil refining in 1858 and then converted to petroleum three years later, to exploit this process, widely assumed to have been in the public domain. In 1865–66, he was granted a series of patents covering all phases of filtration through animal charcoal or bone black of coal oil or petroleum, both crude and refined, either cold or passed through a steam-heated filter.[27] He also obtained a patent on "filtrene," obtained by filtering a sweet, heavy, West Virginia lubricating crude.

Though widely evaded for awhile, the validity of Chesebrough's patents was upheld in litigation. Beginning about 1870, he extensively licensed filtration for treating natural oils as well as lubricants refined from lubricating stock and residuum and petrolatum or petroleum jelly from unrefined residuum. These patents placed Chesebrough and eventually the Standard Oil Trust, which purchased a controlling interest in his company in 1881, in a strategic position in the development of both natural and refined lubricants.

A second line of improvements began about 1870 when a few manufacturers, prior to filtration, began more carefully to reduce their crudes, ridding them of volatile fractions and contaminants by light steam or superheated steam distillation, or by moderate ordinary distillation, or even by extensive exposure to sunlight.

The use of solid lubricating materials as additives completed the basic improvements in natural crude lubricants throughout the century. One of the best known was Galena oil, introduced about 1870 by Charles Miller, who operated the Galena Oil Works near Franklin, Pennsylvania. Miller earlier had discovered the value of Galena, or lead oxide, as a solid lubricant and patented a process for keeping it in suspension in reduced Franklin-district crude. One of the most durable lubricants throughout the century for heavy pressure duty, Galena oils were acquired by the Standard organization in 1878, and eventually were used in car journals by railroads representing 95 per cent of the total rail mileage.* A second crude lubricant with a high reputation in the same field employed plumbago or graphite as an additive. The merits of plumbago or graphite as a solid lubricant had been made known by early pioneers of studies in friction in the 1830's. Owner of a patent to keep the fine graphite in suspension in a reduced crude, the Johnson Graphite Oil Company of

---

* Miller prepared a lead soap by boiling oxide of lead with a saponifiable oil and then dissolved the soap in reduced Franklin crude.[28]

Rochester, New York, became the chief producer of plumbago lubricants.[29] Although the consumption of reduced crude lubricants never approached that of refined in the 1860's or later, they eventually carved a secure niche in markets for heavy cylinder oils.

Refined lubricants. Joshua Merrill at Downer's Boston refinery and William Atwood were the only important refiners of paraffin-oil lubricants until 1865, when a number of small seaboard refiners began to refine paraffin oil from residuum or still bottoms. A fine, tarry substance left in stills along with coke after the completion of distillation, residuum had remained largely a waste product except for occasional use as a boiler fuel. Its yield varied from a few per cent to 15–20 per cent of the still charge.

This new competitive lubricant was inferior to Atwood's and Merrill's lubricants, refined direct from paraffin-oil stock rather than residuum. Their procedure was to distill crude with as little cracking as possible to about a maximum gravity of 42° B.–36° B. For their heavier lubricating stock, distillation was continued to about 28° B. To obtain cylinder oils of approximately 30° B, the stock was in turn iced and pressed to remove the wax, followed by acid and alkali treatments until it was bright and clear. For lighter and higher quality spindle oil of about 32° B, the once-distilled lubricating fraction was treated and then subjected to destructive redistillation to run off the lighter kerosene fractions. This twice-distilled stock was once more chemically treated, and then chilled and expressed to produce the final product.[30] Such was the most advanced status of the art of producing petroleum lubricants as late as 1867. Even the best lacked numerous characteristics to make them a dominant lubricant rather than an adulterant or secondary ingredient in lubricating compounds.

Deodorized, neutral oils. The processes which made petroleum-paraffin oils competitive with animal and vegetable lubricants, like so many of the basic refining technologies of the decade, must be credited to Joshua Merrill. Throughout the mid-1860's he had experimented with methods of improving the quality of lubricants. The primary objective was to improve the odor, color, and viscosity without charring or cracking the heavier oils so they could not be bleached. It was against this background that a minor accident involving a clogged still condenser provided Merrill with the solution to the successful production of deodorized, neutral oils.[31]

The accident occurred when Merrill was working an intermediate stock, one too light for lubricants without further distillation, and too heavy for illuminating oils. Before distillation had gone beyond 600° F, a clogged condenser forced him to bank the fires and continue distillation at an abnormally slow rate. Though passing some light kerosene fractions overhead, several hours of slow distillation failed to reduce the still contents enough to avoid putting out the fires entirely. Returning two days later when the still had sufficiently cooled, Merrill was astonished to find the heavier fractions remaining in the still unlike any he had ever seen, "of a good yellow color, very odorless, and neutral in smell, and oily in body." He reasoned that the removal of the light oils "in the No. 2 oil was one of the causes of its neutrality and freedom from odor, and increased body, and that the yellow color was due to the fact that we had been obliged to use a very moderate fire in distilling." [32]

In other words, Merrill had discovered that the bad odor and inferior lubricating qualities of paraffin oils resulted from the presence of light oils, which were formed as a result of destructive distillation or cracking. By following a carefully controlled process of slow distillation, these deleterious components could be removed without further cracking, leaving a neutral, odorless oil with a fire-point of 500° F and excellent lubricating qualities. [33] To expedite closer work with the higher-boiling materials at reduced temperatures, he designed a special still which, after direct fire had raised the temperature to 300° F, continued further distillation with superheated steam at slowly rising temperatures. In 1868 Merrill put his deodorized, neutral oil on the market, and a year later, in May, 1869, he received a patent covering his oil and distilling apparatus. [34]

The market response to deodorized oils at 50¢ and 60¢ a gallon (appreciably higher than ordinary paraffin oil) was electric. Compared with a previous high of 195,000 gallons of ordinary paraffin oil in 1865, domestic sales of deodorized oil from the Downer plant totaled over 276,000 gallons in 1868 and over 465,000 gallons in 1869. Foreign sales, previously negligible, for 1868 and 1869 amounted to 74,400 gallons to Great Britain and 130,000 gallons to the continent. For the first time, American petroleum lubricants made a permanent entry into foreign trade and thereafter the value of this export was exceeded only by illuminating oils. [35]

Mineral "sperm oil." It was Joshua Merrill's familiarity with paraffin oils which also led him, in collaboration with his brother Rufus Merrill, an experienced refiner in his own right, to develop a new type of petroleum illuminant much safer than kerosene. There was a real need for such a product. Even the best kerosene was more dangerous to use than whale or lard oil, a danger which was multiplied by widespread adulteration of kerosene with naphtha and benzine. The need for a safer petroleum illuminant was dramatized during the mid-1860's by a series of train wrecks accompanied by a heavy loss of life and property from fires attributed to the use of kerosene illumination. Several state legislatures banned the use of kerosene by railraods while the federal government took similar action in respect to steamships.

It was while the two Merrills were experimenting during 1867 with an apparatus to burn paraffin wax that they happened to substitute a heavy-paraffin lubricating oil as a fuel. The unusual quality of the light suggested to Joshua Merrill the possibility of cracking this type of lubricating oil into light fractions suitable for use in lamps. The result was a product with a Baume gravity of 36°, a flash-point of 260° F, and a fire point of 300° F, proclaimed as "safe as whale oil" from which it took the name mineral sperm oil. Fluid enough to ascend lamp wicks, mineral sperm oil worked best in a specially designed dual burner Argand-type lamp, patented in 1867 along with the new oil.[36]

Selling at a price of 75¢ a gallon, compared to 50¢ for premium kerosenes and 40¢ for standard kerosenes, mineral sperm oil had only a limited market among householders. But its introduction did break the widespread ban on the use of petroleum illuminants on railroads and steamships. In 1871, for example, mineral sperm oil was cleared for illumination on federally regulated steamships while the Pullman Company began installing lamps to burn the fuel in their pullman cars. Within two years some 24 steamship lines and 27 major railroads, in addition to numerous mills and factories where safety was a major consideration, had installed the new illuminant.[37]

While the demand for mineral sperm oil never approached the volume of 160,000 gallons daily originally visualized by Merrill, for the better part of a decade it continued as a major by-product both in volume and value.* This demand for mineral sperm oil and for deodorized oils put such a drain on heavy-oil stocks that Merrill, for a considerable period, had to abandon cracking for kerosene fractions and purchase heavy-oil stocks from other refiners.[38]

Requiring unprecedented precision in cracking to a narrow boiling range, Merrill's innovation reflected and depended on great improvement in the most advanced cracking practice. The same thing is even more true of his eventual solution of his stock shortage. He acquired greater

---

* There are no statistics on total production of mineral sperm oil, but the best estimates suggest sales of possibly a million gallons annually during the 1870's.

flexibility so that he could crack not only the light but the heavy lube stocks for mineral sperm oil.[39] On the other hand, he could crack all of his heavy stock for kerosene and sperm mineral oil. Undoubtedly it was this need for maximum flexibility and selectivity in working more by-product stocks than any other manufacturer that in part dictated Merrill's and Downer's preference for small stills for cracking.

Paraffin wax. Paraffin wax, with the possible exception of naphtha, had the widest and most diversified use of any of the petroleum by-products. A joint product of distilling lubricating oil stocks, the processes for re-covering crude (scale) paraffin and refining it had been developed in coal-oil refining.

To obtain appreciable yields of scale paraffin from Pennsylvania crudes or crude coal oil required redistillation of the intermediate and heavier fractions (typically ranging in Baume gravity from 30° down and in boiling points from 500° F upward), followed by treatment with acid and caustic soda. The treated oil was barreled or tanked, and chilled for some ten days at temperatures under 40° F in the open during the winter, or in an ice house. It was then put into cotton bags and put under pressure to express or remove the paraffin-oil lubricating stocks, which in the Downer plants amounted to about 32 per cent of the chilled mass. This treatment of the intermediate stocks gave the hardest and most crystalline paraffin with the greatest value when refined. The heavier stock yielded softer, less-crystalline paraffin. Yields of crude paraffin probably averaged 4½ pounds to the barrel, though by the end of the decade, Downer and presumably other efficient refiners were getting 7–10 pounds.

Scale or crude paraffin had little commercial value until it had been refined. This involved repeated melting with live steam and acid treatment extending over a twelve-hour period, followed by subjecting the product again to heavy pressure and chilling. Yields of refined wax ran about 50 per cent or less by volume of the crude paraffin processed.[40]

Until the mid-1860's, rather little paraffin wax was supplied by the petroleum industry. Few early refiners worked their lube stocks, nor did they always find refining their by-product crude paraffin a profitable operation. Since yields from Pennsylvania crudes were less than from Albert and Boghead crude coal oils, the industry centered in small refineries located mainly in New England, which specialized in refining paraffin from coal. The principal exceptions were a group of small refineries at Marietta, Ohio, who were successful in refining paraffin from heavy Ohio and West Virginia crudes and Downer, who installed a refrigerating plant for processing paraffin oils at his Corry, Pennsylvania refinery in 1864.[41] At Boston, however, Downer produced his paraffin oil and wax stocks from coal until 1866.

It was a growing interest in paraffin-oil lubricants that made petroleum an increasingly important source of paraffin wax after 1865. By the early 1870's a number of refiners in New York and Philadelphia had joined Downer at Boston and the Portland Kerosene Company in the production of refined paraffin wax along with lubricating oils.

Illumination consumed by far the largest tonnage of paraffin. Already a leading material for candles in the coal-oil period, paraffin consolidated and extended its position throughout the decade. It early gained markets as an adulterant of adamantine candles and of expensive beeswax candles, widely used in religious services. In the last half of the decade, paraffin lamps became popular night lamps in bedrooms and nurseries, while, no later than 1862, match manufacturers started impregnating match sticks with paraffin to improve ignition and steadiness of burning. By the early 1870's one New York manufacturer alone was consuming 100,000 pounds annually for this purpose.[42]

The second largest market was distinctively American and elicited contemporary editoral comment that "the uses of paraffin are truly astonishing," for by 1865 paraffin chewing gum, "highly recommended for constant use in ladies' sewing circles," was absorbing the entire output of several small New England refineries. By 1870 one Maine chewing gum manufacturer was consuming 70,000 pounds annually.[43]

In an age lacking artificial refrigeration, demands for refined paraffin swiftly grew in food preservation. Use as a sealing agent for jellies and preserves was augmented in the second half of the decade by applications as a meat preservative, while bakers and confectioners employed paraffin to impart an attractive sheen to bon-bons and cake fillings, to prevent color deterioration, and to preserve both shape and edibility. Breweries and wineries used paraffin to seal barrels and casks.[44]

Pharmaceutics, medicine, surgery, dentistry, and electrical equipment all made early uses of this versatile material. In 1861 it was introduced into pharmaceutics as a substitute for wax, spermaceti, almond oil, and lard in cerates and salves, and as a coating for pills it scarcely had any competition. It was put to use as an acid-proof coating to insulate wires on electrical batteries employed in medicine, and in surgery it became invaluable as a coating for splints on fractures. Dentists worked with various refined paraffins in making false teeth and correcting faulty fits of plates. The expanding telegraph systems became important consumers of paraffin, both as an electrical insulator and to protect butts of poles from ground moisture.[45]

Uses as sizings and coatings sprang up in wide variety. By 1870 it had supplanted spermaceti as the main laundry sizing both in commercial establishments and households. Large textile manufacturers also adopted paraffin as a sizing for certain fabrics. The middle sixties witnessed its introduction as a coating for wrapping and writing papers. It entered postwar leather industries as a sizing and the lumber industries as a wood preservative. It partly displaced India rubber in waterproofing for military boots, coats, and tents.[46]

With the growing output of petroleum lubricants, this wide acceptance of paraffin by the early 1870's assured paraffin of a permanent place in the markets for petroleum by-products.

Petrolatum or petroleum jelly. A broadening experience with refining lubricants from heavy paraffin oils and residuum late in the decade led to the discovery of petroleum jelly or petrolatum, to the present day perhaps the most widely used petroleum by-product in pharmaceutics.

Robert Chesebrough pioneered this innovation in 1869 when he submitted the residuum from a low-temperature vacuum distillation of a selected heavy West Virginia crude to steam filtration through animal charcoal or bone black. Free of objectionable odors, Chesebrough's oily, pasty substance which he trademarked "Vaseline," became fluid and transparent at temperatures varying between 85° F and 110° F. It would not saponify and, unlike paraffin, would not crystallize.*

Invulnerability to oxidizing and becoming rancid like lard made petroleum jelly valuable in currying, stuffing, and oiling all kinds of leather, and as a lubricator of machines. Its unique characteristics made it a superior base for cerates, salves, unguents, pomades, lotions, and related pharmaceutical preparations.[47]

Undertaking to market his product, Chesebrough encountered resistance from wholesale and retail druggists. They were unfamiliar with the product, which they thought was nothing but a paraffin compounded with oil, and they were reluctant to substitute a semi-proprietary product not generally available for the lard base in their preparations. Chesebrough

---

* In obtaining his trade-mark, "Vaseline," Chesebrough claimed he had a unique substance. As evidence he pointed out that all destructively distilled paraffin oils contained paraffin which would crystallize, whereas his product from low-temperature vacuum distillation of residuum would not.

countered by sending out employees in peddler's carts to distribute direct to doctors and housewives free, one-ounce, sample bottles wrapped with instructions on the product's uses and value. By 1873 some 500,000 samples had been given away. By that time distribution had become sufficiently entrenched for Chesebrough to contract with Samuel Colgate, founder of Colgate & Company, assigning him all distribution rights in the United States. In 1874 he organized the Chesebrough Manufacturing Company to take over manufacturing and distributing abroad.[48]

No later than 1872 a solid core of medical acceptance of not only Vaseline petroleum jelly but a competing product, "Cosmoline," manufactured by E. F. Houghton and Co., Philadelphia, had formed. Numerous prominent doctors reported favorably on it in a leading professional medical journal, a circumstance that inspired A. W. Miller, a Philadelphia physician, to investigate the material and attempt "to contrive a formula" for it.[49] The way was paved for the rapid development of the first large market in pharmaceutics for a petroleum by-product.

Chapter 11

Postwar refining and refining organization

$\mathbf{B}$etween 1867 and 1873, pressure on refineries at large centers to increase the scale of their operations came from numerous sources. There was, for example, the impact of crude production which rose from about 3 million barrels in 1866 to nearly 10 million barrels in 1873. The introduction of tank cars about 1867 enabled refiners to obtain their crude in bulk. Finally, there was an approximate threefold increase in distribution of refined oils to rapidly expanding foreign and domestic markets.

A multiplication of stills, larger stills, and more extensive cracking were the principal methods available to refiners seeking to increase output. But each presented difficulties that had to be solved before it could be applied on a commercial basis. Expansion solely by multiplication of small stills, for example, soon brought diminishing returns, if for no other reason than the inconvenience of spreading operations over a wide plant area. High land prices in urban communities were also a barrier to this method of expansion. Expansion through use of larger stills presented difficulties too, apart from technological considerations.

Cracking could add 15 per cent, even 20 per cent, to yields of kerosene from ordinary distillation, while the spread of sulphuric-acid manufacture to all refining centers, accompanied by declining costs, provided an added incentive for extension of cracking. But destructive distillation, to be effective, could proceed no faster than one-fifth the rate of ordinary distillation in small stills. Performed in larger stills, cracking increased the problem of compressing running times to permit processing of two batches weekly. It also imposed a need for some innovation to obtain adequate separation when only half or less of the original charge was left in the still. To this was added the further limitation that, in large stills of any design, less of the total charge could be worked at the high temperatures required for destructive distillation without ruining still bottoms.

The new technology in stills

For these and many related reasons, all routes to expansion hinged on big increases in still sizes; this was probably the greatest technological challenge confronted by the refining industry during the entire century. By 1870, still sizes were successfully carried to capacities in excess of 3,000 barrels—the maximum that succeeding decades would develop. But quantitative figures do not reflect the true extent of the technological problems. The difficulties encountered in small stills, involving stresses on metals, heat control, and design all tended to increase geometrically in progressing to larger units.

Seamless wrought-iron or steel bottoms. Still bottoms, because of direct exposure to intensive heat, were particularly vulnerable to damage, even in small stills. A major improvement occurred in 1861 when Joshua Merrill substituted for the conventional riveted bottom of cast or wrought iron, a formed, seamless bottom cut from a single plate of flange iron or steel. The bottom was given a certain amount of curvature in order to keep contraction and expansion from straining the joints where it connected with the sides of the still.

Another important feature of Merrill's innovation was to design the bottom so it could be removed without risk of fracturing the sides of the stills. Many iron stills with sides and bottoms cast in one piece were in use, partly because those with separate, seamed bottoms (to expedite removal and replacement) had not proved successful. The main problem here was that the flanges to which the bottoms were attached were cast as part of the body; as a result, the flanges and the rest of the body were frequently fractured in the course of cutting off or forcing out the rivets to remove the bottoms. Merrill surmounted this difficulty by using separate angle-iron couplings, or by riveting bottoms directly to the still bodies.[1]

The Civil War delayed general adoption of Merrill's innovation and also led to deterioration in the quality of cast and wrought iron to a point where the life of the best bottoms on small stills declined from 100 runs to 35–50 runs. During the late 1860's, however, seamless-steel bottoms came into general use in stills with diameters of 25 feet and more. And by the early 1870's the use of seamless steel had increased the life span of still bottoms from an average of about 40 runs to an average of about 150 runs.

Air chambers, insulation, and cooling. In 1861, Merrill added to his basic advances in still design by inventing an air chamber to insulate the still during heating and cooling periods and to effect better temperature control. He accomplished this by enclosing the entire still in an iron casing to form an insulating air chamber. Openings with fitted coverings in the sides and top of the casing permitted air currents to be introduced quickly along the sides of the still to regulate temperature. The same iron casing arrangement also made it possible for the first time to remove an entire still and to replace it after repairs without tearing down the brick encasement.

In 1868, Charles Lockhart and John Gracie, pioneer Pittsburgh refiners, added further elaborations to Merrill's air chamber "for the purpose of protecting the still from the action of cold air during the distilling process, and also . . . of facilitating 'cooling off' the still after the distillation is completed." [2]

Aids to maintenance. An accumulation of minor improvements in operating practices, which reduced wear and tear on stills and shortened cooling and maintenance periods, counted heavily in the transition to large stills. Alkali treatment was not confined to the distillate in agitators. Often both crude and distillates were distilled over an addition of caustic soda or caustic potash in the bottom of the still, a practice contemporaries called subtreatment. In 1861 Merrill detected two bad results of this procedure: the layer of alkali, reacting on the fireplates, made them burn out faster; and the alkali also formed an insulating layer and so increased fuel requirements. Merrill corrected both difficulties by using a simple pan to hold solid alkali in contact with the oil, but free of the still bottom. Another minor innovation was Merrill's fire-and-oil-proof cement for making tight and lasting joints. Developed in 1857, this cement soon became a staple of still-repair work everywhere.[3]

In 1863, Lockhart and Gracie introduced mechanical scrapers which operated on a rotating shaft so that the residuum in still bottoms could be loosened and dropped into a deposit box before it had time to cool and harden. The stuffing box through which the shaft passed was water-cooled to resist the intense heat of the still and to permit the rotating operation to commence early in the cooling period. This and similar devices lightened cleaning labor, lessened the frequency of unsatisfactory distillations, and substantially reduced heat losses and the hazards of fire damage to plates.[4]

Steam cooling. A basic step in reducing cooling periods without damage to stills came into general practice in the second part of the decade and remains in modern refining today. This was the introduction of a jet of steam or superheated steam into the still about four hours after completion of a run. Exerting a gradual cooling effect, steam injection greatly reduced the waiting period before workmen could remove the manhole and enter the still. Steam also expedited the removal of tarry residuum and made it safer to crack deeper to coke. By creating an outward current through the condensing apparatus, it facilitated the removal of dangerous gases and it was also helpful for removing paraffin and other deposits— formerly a leading source of refinery fires. Coal-oil refiners had already put superheated steam to similar employments, but it required the increased availability of superheated steam and greater familiarity with its use during the immediate postwar period to make the practice general.[5]

Goosenecks and condensers. Re-design of condensers was a prerequisite of the move to larger stills. A small still of 50 barrels required 100 feet of 4-inch pipe, coiled in a water coolant; but a 600-barrel still required at least 500 feet of 6-inch tubing, an unwieldy length regardless of how it might be arranged. By the end of the decade, however, nearly all cylindrical stills had obtained reductions in large-pipe space consumption without loss in cooling effect. The large gooseneck exit pipe that conveyed vapors to the condenser in earlier stills was replaced by a condensing drum on top of the still, as shown in Figure 11:1. Big gains in cooling surfaces were realized by leading vapors from the drum into 40 or more small parallel pipes, 2–3 inches in diameter, cooled by water in the condensing tanks.[6]

The numerous separate coils might converge at the bottom of the con-

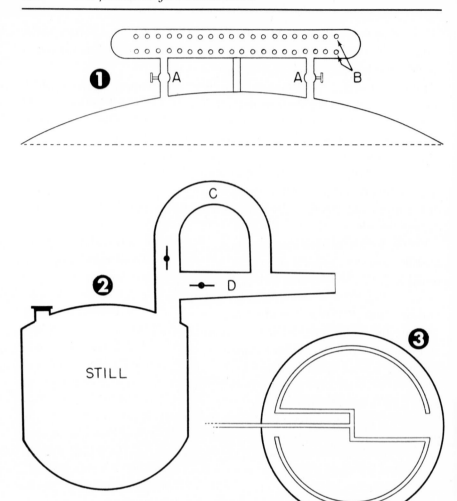

FIGURE 11:1. *Goosenecks and condensers. (1) section of condensing drum with two vapor outlets* (A) *controlled by stopcock and* (B) *multiple small pipes leading through water coolant; (2) still with conventional gooseneck outlet* (C) *and alternate outlet* (D) *introduced by John Gracie and Charles Lockhart in 1863; (3) section of curved perforated pipe for introducing steam through the still head.*

densing tank into one large exit pipe, or they might extend only to the top of the tank, there entering a large pipe about 12 inches in diameter and tapering to 3 or 4 inches at the bottom of the tank. Not all cylindrical stills used this type of drum. Many horizontal stills used only a drum and a single exit pipe to the condenser.

Throughout most of the decade, another phase of exit pipes for the vapors retarded progression to greater efficiency in large stills. Most contemporaries appreciated that large evaporating surfaces and proximity of exit pipes to those surfaces promoted vaporization. Moving toward larger stills, it was common to increase the size of exit pipes and often the number to two or three. As early as 1863, Lockhart and Gracie took the next step of providing an alternate outlet for vapors that would be close to the surface at advanced stages of distillation when the oil was low in the still, as shown in Figure 11:1(2). At the early stages of the run, when there was considerable difficulty with heavy oil passing over with the vapors into the condenser, the stillman used the ordinary gooseneck outlet, as shown in Figure 11:1(2C). Later, when the heavier vapors recondensed and fell back into the oil, he shut off the gooseneck outlet and opened a lower straight-pipe outlet, as shown in Figure 11:1(2D), that communicated at one end with the throat of the gooseneck and in its middle with the button-hook end of the gooseneck. In 1868 Lockhart and Gracie patented further improvements that sought to increase the selectivity of outlet heights and the number of condensers.[7]

Multiple outlets and condensers improved the potential rate of vaporization and separation in big stills, but not enough to approach separation performance in small stills. Vaporization at the middle range of distillation remained comparatively inefficient despite multiple outlets. At the last stage, entering destructive distillation, efficiency slumped and excessive cracking took place at the high heating rates.

Problems of entrainment became acute: vapors leaving the still absorbed droplets of boiling oil. To avoid this, reliance was placed almost entirely on the coalescence of the heavy vapors on the still walls. As late as 1871, a 475-barrel still installed at the newly erected Franklin Oil Works in Philadelphia was rimmed at its top by an inner ledge as a supposed aid to the return of liquid formed on the sides. Due to poor fractionation and excessive cracking, however, yields of lubricating stock and residuum were extremely low. Controlled cracking giving a high yield of kerosene distillate awaited installation of a new still equipped with what was described as greatly improved condensing arrangements.[8]

One improved condensing device was a dephlegmator: a simple coil, with water as a coolant, that was installed along the vapor line above the still. Its purpose was to produce a mild, cooling effect sufficient to condense the heavier, higher-boiling vapors so they would flow back into the still. The primary effect was to improve the separation of the different boiling fractions, but a commercially important secondary effect was to increase the cracking of the higher-boiling components.

Some reflux was, of course, inherent even in the early stills, and crude devices to increase it soon appeared. The common gooseneck, by providing some height for condensation, served the function of a dephlegmator, as did the design of tall, vertical stills with small diameter and evaporating surface in relation to their height. Inner ledges and other types of splash traps in the domes of stills probably were the immediate precursors of the cooling-coil dephlegmator, which had largely supplanted them by the early 1870's.

Coils for introducing steam into large stills. The view that dephlegmator devices represented the last major advance toward improved separation in large stills is perhaps an oversimplification.[9] But it carries a great deal of substance if it is extended to include another contemporary development of the late 1860's, which was equally simple but far more critical for obtaining adequate separation in large stills. The device in question, illustrated in Figure 11:1(3), consisted of a curved, perforated pipe that could be inserted through the still-head into almost any desired position within the still to introduce steam or superheated steam into the oil or above it. To minimize foaming, the steam was introduced into the vapors near the exit pipe rather than direct into the body of the oil. The effect of the steam was to reduce the boiling points of the hydrocarbons by reason of its mechanical sweeping action. In other words, using steam, vapors could be removed from the still below the normal boiling point of the oil. By thus decreasing the quantities of high-boiling, cracked components that passed over, steam or superheated steam greatly improved the color and odor of the condensate.

The same advantages applied in distilling and separating the low-boiling naphthas, but there were others besides. Lower temperatures reduced the entrainment of oils in the vapors at the early and middle stages of distillation, which at once bolstered heat efficiency and yields. They also added an important safety factor in working highly inflammable fractions. The effect of superheated steam on improving yields was most marked with the lower boiling naphtha fractions, but it extended through all the kerosene fractions in ordinary distillation with boiling points no higher than that of superheated steam.

The introduction of steam in destructive distillation reduced rather than increased the amount of cracking of high-boiling components; without it, however, controlled cracking in large stills could not have been accomplished. Dephlegmators aided both separation and yields of cracking, but it was the improved separation with steam, which made possible controlled cracking of the fractions boiling above the kerosene range into high yields of cracked kerosene distillate. Dephlegmators and steam represented the first and virtually the last improvements in separation to appear throughout the rest of the century, and in modified forms this application of steam remained basic in refining practice until recent decades, when it was supplanted by vacuum stills.[10]

Cracking performance in 500-barrel stills. Before the close of the decade, the influence of steam distillation had already become pronounced. Though not a proponent of large stills, Merrill reported that 500-barrel stills were in the best general practice among refiners interested in maximizing kerosene yields, and that running times were well within 48 hours. He described the typical performance as follows.

Some 15 per cent of naphtha was run off in a couple of hours with the aid of steam. Distillation continued rapidly from 60° B to 40° B, through the straight-run kerosene range. About 25 barrels of kerosene distillate came off hourly for a period of 10 hours, with a maximum yield of 50 per cent of the crude. Still fires then were dampened for the first slowed stage of destructive distillation of the denser fractions from 40° B to about 32° B, at temperatures above 600° F. About 5 barrels of cracked distillate came over hourly for about 10 hours, or a yield of 10 per cent. Guidelines to the rate of retardation were that the oils issuing from the condenser should be no heavier than 40° B, and as light in color as possible. The stream in the condenser then reduced to about 2.5 barrels hourly for another 10 hours and an additional 5 per cent of cracked distillate.

Diminishing returns often set in at about 15 per cent of cracked, but this yield could be and sometimes was increased to a maximum of 20 per cent by slowly cracking deeper for another ten hours or so, obtaining about 2.5 barrels hourly. The amounts of heavy gravity residuum (about 19° B) left in the still varied from 3–10 per cent. A majority of operators sold their residuum to refiners of lubricants and waxes, but some distilled it in a small still to separate the heavy paraffin oil, which they then added to the next batch of stock being cracked.[11]

Not all operators slowed up enough, but an increasing number of experienced ones did. As a result, they often escaped heavy treating expenses and obtained 100–200 runs before having to replace the seamless-steel still bottoms. These standards of performance tended to decline progressively in larger stills because of increased stresses, greater prob-

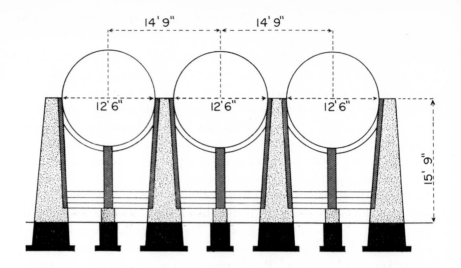

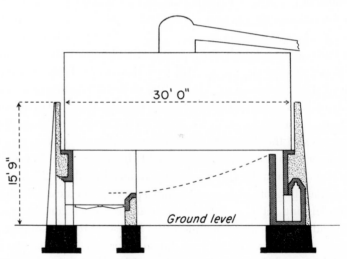

FIGURE 11:2. *Horizontal cylindrical stills, transverse (top) and horizontal (bottom) vertical sections.*

lems of entrainment and holding times, narrower margins at every point of operation, and greater temptation to compress running times. In large stills, damage to still bottoms usually is mentioned as the limitation on the extent of cracking; but limitations of holding long enough the fraction boiling immediately above kerosene in the still applied independently and with probably even greater force.

Cylinder stills. In the transition to large stills the longitudinal cylindrical stills had many proponents. Figure 11:2 shows stills with a 600-barrel capacity each, a typical size throughout the century, and the common arrangement of these stills in benches of three. Of particular interest is the extension of the fire-brick enclosure only to the center of the still, leaving the upper half of the still wholly uncovered or protected only by a thin, iron sheeting. Introduced in the second half of the 1860's when horizontal stills already had reached sizes of 830 barrels, exposure of the upper half of the still to the atmosphere promoted longer holding times of the oil in the still at advanced stages of ordinary distillation and in cracking up to temperatures of 1000° F. The mild cooling effect of the atmosphere aided a higher reflux of condensate from the heavier vapors in the still. Dephlegmators furthered this cause, while steam coils helped to control cracking and separation.[12]

Cheesebox stills. A cylindrical still, vertical rather than horizontal, the cheesebox still was built in intermediate as well as large sizes, but it predominated among direct-fired stills of 1,000–3,500 barrels, which began appearing in increasing numbers late in the 1860's. A typical 1,200-barrel model, 10 feet high, 30 feet in diameter, is illustrated in Figure 11:3.

The cheesebox, like other upright stills, had the natural advantage of uniform evaporating surfaces, but was designed much shorter and wider than orthodox uprights to provide two further advantages: a larger surface to promote a faster rate of evaporation; and a large dome that promoted condensation, refluxing, and longer holding times in compensation for the forfeited wall surfaces.

The cheesebox still incorporated all improvements in various types of cylindrical stills, including multiple-parallel condensing pipes, dephlegmators, and a steam coil for injecting superheated steam into the vapors. Though itself lacking unique basic features, it achieved a balanced alignment of these improvements that probably made it more responsible than any other still for bringing improved separation performance to large scale batch distillation. Its exceptional features included elaborate masonry arrangements to provide support for the heavy still, to expedite

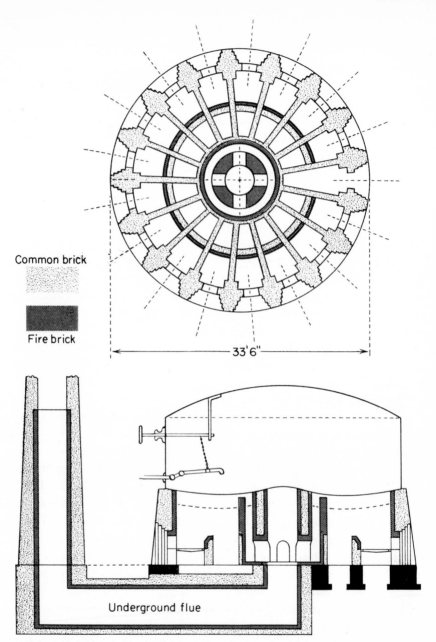

Common brick

Fire brick

33'6"

Underground flue

FIGURE 11:3. *Cheesebox still. Horizontal section (top) and vertical section of still setting (bottom).*

even heat distribution, and to route flue gases through the center tier underground to a chimney removed from the still. Proponents claimed that longer running times and higher maintenance expenses with the cheesebox still were offset by better quality of illuminating oils, lower-gravity products, and fuel savings.[13]

Continuous-operation and multiple-stage distillation

It is widely though incorrectly believed that American refining throughout the nineteenth century disregarded two important means for obtaining cheaper and larger outputs of improved product.[14] One was a system of continuous operation to produce a variety of products with a series of interconnected, simple stills, each producing an overhead distillate product and a bottom fraction. In this operation, the lead still receives fresh charging stock continuously, while the other stills operate on the bottom fractions from the adjacent stills. (See Figure 11:4.) In the 1880's, the Nobel Brothers in Russia introduced the first multiple-still system for continuous operation into extensive commercial practice for the production of kerosene, fuel oil, lubricants and other products. Later in the same decade, British shale-oil producers also employed a continuous system.[15]

The second process that American refiners reputedly neglected was the use of staging in the distillation operation to effect more precise fractionation. Although it had long been appreciated that redistillation could produce narrower fractions of products of improved quality, it is true that early petroleum refiners were slow to appreciate that the equivalent of several redistillations could be achieved in a single still through the introduction of a staging effect by utilizing tower packing, bubble trays, or other engineering expedients.*

The early development in America of these two processes can perhaps best be understood by reference to a present-day fractionator, incorporating both continuous operation and staging. A schematic drawing of a modern, multi-stage fractionator for splitting a wide boiling range of petroleum charge stock into a light product, a heart cut, and a heavy bottom fraction is shown in Figure 11:5. Such units operate continuously with regard to feed and the three products which are recovered as condensate, side-stream, and bottom fractions respectively.

---

* With the advent of staging, the term fractionation acquired a new connotation in the art of distillation. Whereas before, a rough split into components on the basis of volatility was frequently called fractionation, thereafter a more precise separation was implied, one that produced closer boiling fractions than simple distillation.

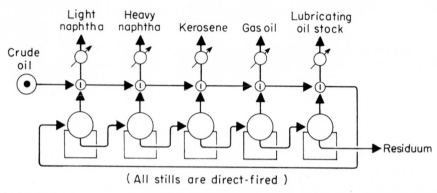

FIGURE 11:4. *Process of continuous distillation in a battery of shell stills.*

The unit is multi-stage, that is, it has a number of trays achieving in a single tower a separation efficiency equivalent to a number of distillations and redistillations. Any gas not condensed can be recovered in a refrigerated absorption system. Liquid and gas in mutual equilibrium are intimately contacted on each tray. Part of the condensate is returned as reflux to make sure liquid is present on each tray. A reboiler is provided at the bottom of the column to assure a copious upward flow of vapors. The column will effect separations with greater precision than simple single-stage distillation since its quantity of reflux and energy to the reboiler are increased up to the limiting point when the column floods. These expedients, however, decrease the throughput and the thermal efficiency, with the result that a practical limit is imposed before the flooding stage is reached. As a rule, product specifications establish the degree of fractionation required, and this in turn determines the number of stages, or trays, required.

Precedents for modern staging in the petroleum industry date back to at least 1830, when the Coffey "alcohol column" introduced the basic principles of batch distillation, with staging, into commercial practice in the alcohol and later in the coal-tar industries. Late in the nineteenth century, fractional distillation of alcohol in the modern sense was adapted to commercial petroleum refining in Europe, although the same practice was not utilized in the United States until early in the twentieth century. But while the indifference of American refining to continuous operation in a battery of stills and to the use of staged fractionating systems was characteristic of the late nineteenth century, perhaps the greatest claim of the 1860's to being a decade of innovation in refining lies in persistent

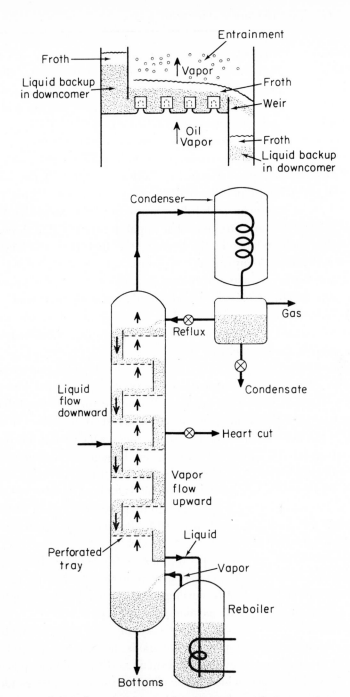

FIGURE 11:5. *Modern tray-type fractionator with inset of tray detail shown.*

work on these two interlocking problems. Practical motivations were many. No subsequent period was faced with as large a gap between the demands for throughput and the inadequacies of existing batch capacity. As to the advantages of some sort of fractionating system, refiners with any sense of informed self-interest could not be unmindful of them, for they were under fire almost constantly from the chemical fraternity, the public, and legislators for poor separation of light ends from kerosene.[16]

The first semi-continuous system in a battery of stills was patented in April, 1860 (even before petroleum distillation had established itself), by D. S. Stombs of Kentucky and Julius Brace from the Kanawha Valley coal-oil region of western Virginia.[17] The apparatus consisted of three horizontal stills, each independently heated: the first one, receiving the crude, was heated to temperatures appropriate to distilling off the light naphtha and kerosene fractions; the second one was for working lubricating oil; and the third one produced paraffin. Among numerous shortcomings of the system, a reliance solely on gravity to move the successively denser stocks to succeeding stills was apparently sufficient to keep it from operating effectively.

In 1866, Alexis Thirault of New York City developed a unit considerably more advanced in concept. It was a single still divided into three units: the first unit heated the crude and removed the light naphtha and benzine fractions; the second unit distilled kerosene; and the third unit redistilled fractions deep in the kerosene ranges which were too poor in color and odor to be suited for burning oil. Though not conquering the problem of continuous feed, Thirault recognized it more clearly than had Stombs and Brace. He also added an ingenious tar-cock to clear the still of tar and residuum. By scaling the sizes of individual stills and by designing the unit to produce kerosene only, Thirault's patented apparatus was much more practical than the earlier model.[18]

There were many early attempts to improve fractionation by two-stage distillation in a double still operated semi-continuously. During 1865 and 1866, for example, such a still, developed and patented by Adolph Millochau, attracted widespread acclaim and commercial interest. It was installed in at least one refinery in Jersey City, in a new refinery erected by Prather, Wadsword & Company at Oleopolis in 1866, and possibly others.[19] Millochau's design provided for a small, vertical still to be placed inside a much larger one which received the crude oil. The condensed vapors from the heated crude were returned to the outer still continuously until all of the easily volatilized materials were removed via a separate condenser, after which the heavier oil was returned to an inner still, which received its heat for further slow distillation from the heated crude and distillate in the outer still. As the volume of distillate in the outer still was lowered, arrangements permitted refilling with crude—in effect, a semi-continuous distillation.[20] For various reasons, chiefly associated with difficulties in control of temperatures and refluxes, Millochau's still, along with several others of similar design, failed in actual operation.*

These early failures in double stills were not entirely lost, however, for it was a desire to remedy their defects that led P. H. Van der Weyde, eminent Philadelphia physician and chemist, to design an improved double still, an important step in his progress toward a tower-tube still giving both selective condensation and continuous rather than semi-continuous distillation. In his improved double still, the arrangement of exit pipes near the surface of the oil to aid vaporization was a forerunner of designs in conventional large stills.†

At the next stage he invented a horizontal, tubular condenser consisting of a series of tubes, which permitted him to separate the fractions as they were distilled at various temperatures in a conventional still. In using

---

* The volume of the outer still had to be made too great for the inner still to be effective, at a great waste in space, heat, and equipment, while the inner still greatly cut down the total volume in the outer still. Running the cooled distillates of the outer still into the inner one sometimes so reduced the temperature of the outer still as to stop distillation. At the same time, the operator had no means of regulating the temperature of the inner still when it became excessively hot toward the end of the run. Because the vapors were run in a thin pipe to a highly elevated condenser, they frequently condensed en route in the pipe, falling back into the still. The same piping arrangement was extremely vulnerable to clogging with paraffin deposits, causing an explosion and splitting the still at its seams.

† Van der Weyde counteracted excessive waste of space, equipment, and heat in previous double stills by placing the second small still outside the first on the flue, and locating the first directly over the furnace. To obtain flexibility in drawing the contents of the second still, or running them back to the first still without redistilling them, he placed the second still sufficiently high so that its bottom was level with the surface of the oil in the first still when filled, and he connected the stills with a tube and stopcocks. He enabled the still operator to regulate different heats in the two stills by appropriate valves in the flue that even permitted all heat in the second still to be shut off when it was empty. He eliminated premature condensing of the vapors on their previously long routes to the condensers by rearranging the condensers as low as possible, only a few inches removed from the oil surfaces in the stills. He reduced the danger of explosion from clogging of condensers with paraffin at the end of the distillation by substituting for the long narrow tubing in the condensers a short wide gooseneck with an attached dome. Well protected from cooling influences by a felt covering, the dome was connected with the condensing worm by a slightly declining pipe.[21]

a minor vacuum to aid the removal of vapors to the condenser, he anticipated a feature of the continuous-distillation system developed by Samuel
Van Syckel in the 1870's. This invention probably marked Van der
Weyde's final step toward a break-through from semi-continuous to genuinely continuous distillation.[22]

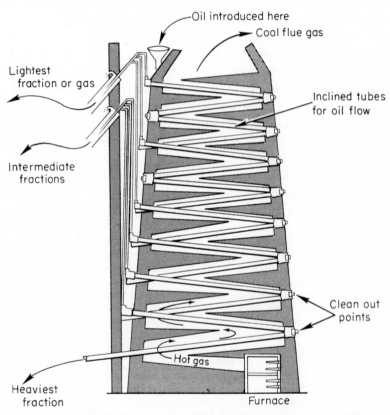

FIGURE 11:6. *Van der Weyde's tubular fractionating still.*

Merely by taking a leaf from the alcohol distiller's technique, he simplified and improved his engineering design in the construction of a continuous tubular fractionating still, having many of the basic advantages of
a modern continuous fractionating tower. (See Figure 11:6.) The zigzag
structure of the tube containing the oil and the flues surrounding it resembled a long, straight pipe that had been bent into a series of Z's and
enclosed with brick. A fireplace burning coal or wood at the bottom of
the tower supplied the heating gases which passed upward through the

zigzagging flue. Within the flue was the tubular distilling column. Each of the Z-joints on the right side of the still had clean-out points, and on the left side were exits for the vapors to pass into condensers. The crude oil to be distilled was fed through a funnel at the top of the column. As the liquid crude encountered increasing temperatures in its downward flow from the rising flue gases, fractions were removed through the vapor exits and condensed separately.

Van der Weyde's column had several things in common with the modern fractionator: it had a differentiated temperature from top to bottom and it supplied continuous operation; it guaranteed two phases, liquid and vapor, at all points; and it provided multiple withdrawal points for six or more different fractions and residuum.

Its differences from the modern fractionator all pivot around the vital point of controls. For control of the flow from one level or exit point to the next, the Van der Weyde column was solely dependent on the viscosity of the liquid. Modern bubble trays aid the control of flow from one tray to the next by the size of holes in the tray. The balancing of input of crude oil in the tubular column with the volume and temperature of flue gas was a difficult and tricky performance, which probably sufficed to prevent the still from coming into general use. Modern fractionators use reboilers at the bottom of the column, and a condenser at the top to cool whatever quantity of vapor may be needed as refluxing condensate. The column lacked a high reflux ratio, a shortcoming another innovator would soon correct, but at a cost in control of the vapors.[23]

In 1871, Henry Rogers, an experienced Oil-Region refiner currently in the employ of the Charles Pratt Manufacturing Company of New York and soon to become a Rockefeller man, carried the principles embodied in Van der Weyde's tubular still into a fractionating tower "similar to what is known as the column-still for distilling alcohol, but modified in all the details so as to make it available for distilling oils. In the spirit still it is only required to separate two principle liquids—alcohol and water. But in the oil still a long series of liquids is to be separated."[24]

The operation of Rogers' fractionating tower is shown in Figure 11:7. It had many features of a modern fractionator: for example, trays, partial condenser for reflux, the equivalent of a reboiler, and means for either pressure or vacuum operation. Rogers, however, mishandled his reflux liquid by returning it to the mid-point rather than to the top of the tray tower. Likewise, the liquid return to the still, which functioned as the reboiler, was a fraction of the condensate rather than liquid from the

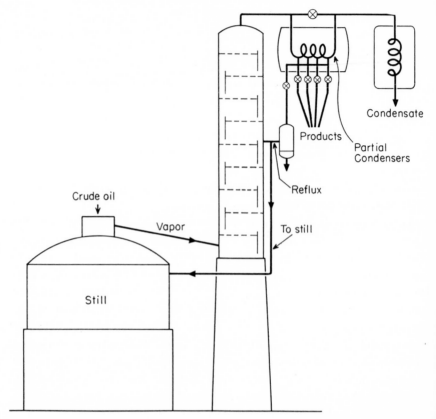

FIGURE 11:7. *Roger's tower still, 1871. (Adapted from U. S. Patent No. 120,539, issued Oct. 31, 1871).*

bottom tray. With simple mechanical changes, Rogers' apparatus would be the full equivalent of a modern continuous multi-stage fractionator such as shown in Figure 11:5. Like Van der Weyde, Rogers envisioned the simultaneous withdrawal of a number of products, continuous operation, and a temperature gradiant throughout his column. Rogers' products, however, would have been much narrower fractions because, unlike Van der Weyde, he employed a multi-stage tower, the lower portion of which received a controlled quantity of reflux. In addition, Rogers provided a method (not shown in Figure 11:7) for introducing an extraneous liquid to the several stages of his tower.

Within the Standard Oil Company, which Rogers joined in 1874, pressure for sharper, narrower fractions was insufficient to stimulate the mod-

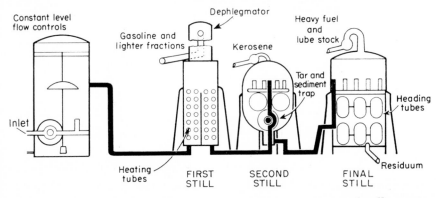

FIGURE 11:8. *Samuel Van Syckel's system of continuous distillation in battery of three shell stills. (Adapted from U. S. Patent No. 191,203, issued May 22, 1877).*

est mechanical modifications of his tower necessary to bring it into commercial use. Surpluses of both crude-oil and refining capacity, emphasis on throughput and kerosene fractions, and limited production of by-products in a general refinery were among the influences contributing to this neglect. Not until the introduction of J. W. Van Dyke's stone-packed tower in the first decade of the twentieth century by some of Standard's refineries did any American refiner turn to multi-stage distillation.

Brilliant development work in multi-stage distillation by Rogers early in the 1870's had its counterpart in the field of continuous distillation in a series of stills. At about the same time, the best features of various multiple stills with partially continuous operation, selective condensers, and other equipment designed over the previous decade to improve fractionation were being incorporated and brought into an effective new balance by Samuel Van Syckel in his continuous system. Van Syckel designed not only the first genuinely continuous system, but the only one to enter commercial operations on a limited scale in the United States in the nineteenth century.

Van Syckel's experiments with a continuous process began in 1862, when he owned and operated a successful refinery at Jersey City.[25] Selling out in 1864 to seek greater opportunities in the Oil Region, he gained renown in the fall of 1865 as innovator and operator of the first commercially successful gathering pipeline. Forced into insolvency and deprived of his pipeline during the general depression of 1866, he resumed his interest in improving refining processes. By 1871 Van Syckel had interested enough local capitalists in his plans to erect a small plant at Titusville.[26]

In the next year, the ill-luck which pursued him throughout most of his career struck twice with bad fires in his refinery.[27] This experience probably influenced him to incorporate flue heating with superheated steam or gases, rather than direct fire, in his continuous process. Persevering, by the end of 1873, Van Syckel had worked out the major features of his continuous-distillation system, upon which he was issued a patent in 1877.

Figure 11:8 illustrates the salient features of Van Syckel's process. The basic apparatus included three stills in a series preceded by a crude-oil feed tank. A distinct novelty was his application of float controls (common enough in modern industry) to the feed tank in order to maintain constant oil levels in the three stills. This governor allowed crude to flow into the stills at the same rate products were taken off overhead and the liquid contents were drawn off below.

His first still was not the primary one for kerosene, but was designed to remove gasoline and other products lighter than 80° B. Its most distinctive feature was a well-designed dephlegmator to minimize the loss of heavier fractions by entrainment. He recommended this still for adaptation to orthodox batch operations on the grounds that its utilization would improve yields by a reduction of entrainment.

His second still was his primary one for kerosene and deviated from conventional designs in two ways. First, it was equipped with a tar-and-sediment trap to expedite flow; and second, it incorporated provisions for internal heat exchange. These arrangements preheated feeds before mixing them with the contents of the still as a means of assuring smooth operations. In the third still, also equipped with heading tubes for internal heat exchange, the heavy fuel oil and lube stock normally was worked at temperatures of 400°–500° F. If cracking was desired, Van Syckel preferred to recycle stock of about 26° or 28° B to a fourth and final still whose contents were heated to 600° F, though temperatures could be greatly varied.

Heating arrangements were entirely without direct fire or any heat on still bottoms. Instead, heat was supplied by superheated steam or gases through flues surrounding and transversing the stills. The steam conductor to forward vapors to the spray condenser also served to create a low vacuum in the stills. Close temperature control, low fuel and labor requirements, and mininum fire hazards characterized this heating system.

Van Syckel clearly evolved a workable plan of continuous distillation, and claimed some degree of savings over batch operations. The savings which he claimed were undoubtedly realized. These advantages included savings of 25 per cent in fuel and labor costs, reductions of 75 per cent and more in investment in stills and equipment, improved yields of 5 and 9 per cent respectively for gasoline and kerosene, and greater uniformity of product and operating conditions throughout the distillation process. Of these claims, the probability is greatest that those for fuel and labor savings were realistic.

Beginning with the building of a test refinery in 1876 (a project that was never completed), Van Syckel spent eight years in promoting the commercial adoption of his process. Handicapped by a lack of capital, he charged with some evident justification that all commercial trials of his process were inadequate and even conducted in bad faith. Obsolete equipment in poor condition was used, permitting only improvised operations under adverse conditions. Official reports on yields and performance were at variance with the gauger's records and badly falsified, in Van Syckel's opinion. No reliable records survive upon which to base an informed judgment on the merits of the apparatus, although it is clear that published reports were wrong when they indicated that his float control in the feed tank was a failure.

The main technical shortcoming of Van Syckel's system was no doubt its reliance solely on gravity to feed the oil. This was a shortcoming shared by all early continuous systems, including the Nobels' operation in Russia, in large part because pumps had not been developed sufficiently to handle hot oils. The limited success and numerous difficulties experienced by Standard Oil Company in 1903, when it introduced Max Livingston's system of continuous operations in a series of shell stills depending on gravity feeding, suggests that Van Syckel's still twenty years earlier did not give sufficiently improved performance over batch operations to warrant its wide commercial adoption.

Refining organization on a new scale

By the early 1870's, refining had attained an entirely new scale in plant and still sizes. Characteristic sizes in stills had been defined for the rest of the century. Both refining practice and informed technological opinion had pretty well settled on stills of 500 barrels and 600 barrels as the most efficient large size. However, there were numerous deviations from this practice, including several outsize units of 3,000–3,500 barrels.[28]

The rate of increase in plant size necessarily lagged that in stills somewhat, but a new scale that would prevail for a decade had been reached.

Crude charging capacities of 500 barrels daily were typical in the sense of being the more numerous, but the typical size was undergoing rapid alteration because capacities of 1,000 barrels were the rule in new plants.[29]

Notions of the minimum size plant that could be operated economically also had altered considerably. In 1865, G. W. Gesner and others thought a capacity of 240 barrels of crude weekly might answer, distributed among four stills charging 25 barrels each. J. Lawrence Smith in the early 1870's placed the minimum size at 900 barrels weekly, divided among three stills with 160-barrel charging capacities and one naphtha still. Smith's model was a relatively smaller plant than Gesner's in 1865.[30]

The main change in refining organization over the decade, apart from those of scale, was a sharpening of the differentiation between two leading types of refiners and the emergence of a third. Pre-eminent as ever was, first of all, the class which concentrated exclusively on producing kerosene but obtained crude naphtha and residuum in the course of distillation. Secondly, and at the opposite pole from the kerosene refiners, was the general or complete refiner who manufactured an extensive line of by-products from the light and heavy fractions, in addition to kerosene from the middle range. Never numerous, but comprising refiners like Downer with large plants, these ranks of general refiners had thinned rather than increased; in fact, with possibly a few exceptions, complete refiners were confined to the Downer organization. The third type of refiner, of more recent origin, were numerous small refiners who subordinated kerosene production to specialization in one or more by-products, either from the light or heavy fractions.

By-products had made some imprint on all types of refiners, though on strictly kerosene producers it was faintest. With that class of refiners it was perhaps less a question of responding to strong, new incentives of demands for some by-products from naphtha and residuum than changes in technology that introduced a new flexibility in supplying such demands as appeared. Because of the low cost of better separation of light ends by steam distillation, kerosene refiners usually refined part of their crude naphtha, if only to obtain additional fractions within the kerosene range, so that little adjustment in normal operations was required to meet any market demands for refined naphthas and even gasoline.* To a lesser extent, adjustments to meet demands for lubricants were also reduced, for it had become general practice to distill some residuum to obtain paraffin

---

* In the Oil Region only a provident few salvaged a loss in kerosene fractions by returning heavy crude naphtha to the still charged with crude oil.[31]

oil for cracking stock and tar for admixture with coal as boiler fuel.[32] As a result, kerosene producers nearly all supplied refined naphthas when markets gave them the opportunity, and many more marketed some refined lubricants.

For the same reasons, no matter how extensive their conversion to large stills, nearly all kerosene refiners had some small naphtha and tar stills. Other small stills seldom were junked because of myriad uses to redistill various straight-run and cracked stocks, as stand-by capacity, or even as storage tanks. For these and similar reasons, slowness to retire obsolete equipment remained a permanent characteristic of petroleum refining.

Despite increased processing of crude naphthas and lubricants by kerosene refiners, greatly expanded throughputs placed large volumes of both light and heavy by-products stocks on the market. It was this that supplied the necessary minimum condition for the rapid rise of small refiners specializing in selected by-products. Relatively higher labor, packaging, and marketing costs prorated over smaller throughputs contributed further to this development, for they deterred kerosene refiners from diversifying extensively in by-product refining. Both the nature of demand and technological considerations limited refining of the heavy by-products to small stills of 50-barrel capacity, although these were considered giants when introduced in the manufacture of paraffin and petroleum jelly in the vicinity of Philadelphia early in the 1870's.[33] The main volume of paraffin, petroleum jelly, and lubricants was manufactured by small refiners concentrated at Philadelphia, New York, and a few other points where residuum stocks could be purchased from kerosene refiners. The low cost and effectiveness of steam distillation in working light ends facilitated entry at Cleveland and elsewhere, where ample crude naphtha stocks were procurable for a few cents a gallon.

The organization of a complete refinery

Naturally enough, the greatest impact of by-products was made on the few remaining complete refineries. The nature of that impact is shown in Figures 11:9 and 11:10, which present the contrasting flow charts of Downer's and Merrill's complete refinery in Boston and of a typical kerosene refinery in the early 1870's. Though cited for more than a decade to come as both a model and representative refinery, the Downer plant was not typical at any time.

Forced to pay higher prices for crude than any other refinery because

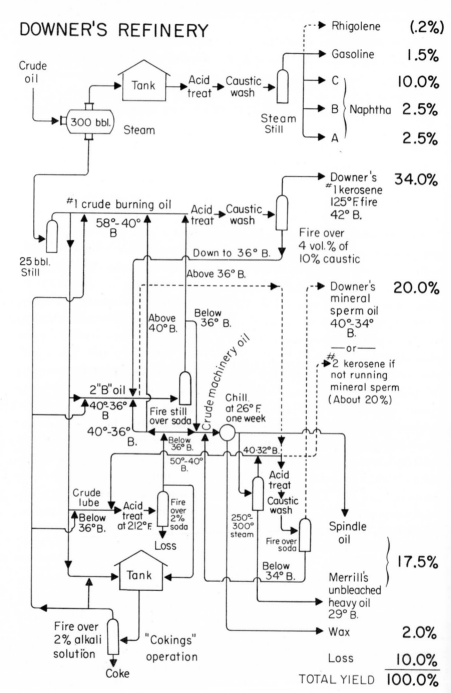

DOWNER'S REFINERY

Crude oil

300 bbl.   Steam

Tank → Acid treat → Caustic wash

Steam Still

Rhigolene   (.2%)
Gasoline   1.5%
C   10.0%
B } Naphtha   2.5%
A   2.5%

#1 crude burning oil
58°-40° B
Down to 36° B.
Above 36° B.

Acid treat → Caustic wash

25 bbl. Still

Downer's #1 kerosene 125°F. fire 42° B.   34.0%

Fire over 4 vol. % of 10% caustic

Above 40° B.
Below 36° B.

Downer's mineral sperm oil 40°-34° B.
—or—
#2 kerosene if not running mineral sperm (About 20%)   20.0%

2"B" oil
40°-36° B
40°-36° B.

Fire still over soda

Crude machinery oil

Chill. at 26° F. one week

Below 36° B.
50°-40° B.
40-32° B.

Acid treat
Caustic wash

Crude lube
Below 36° B.

Acid treat at 212° F.

Fire over 2% soda

Loss

250°-300° steam

Fire over soda

Below 34° B.

Spindle oil

Merrill's unbleached heavy oil 29° B.   } 17.5%

Tank

Fire over 2% alkali solution

"Cokings" operation

Coke

Wax   2.0%

Loss   10.0%

TOTAL YIELD   100.0%

FIGURE 11:9. *Flow chart of batch operation.*

of their location in New England, it was particularly important for Downer and Merrill to maximize yields and quality. Refining all by-products, they also were making more and narrower cuts than anyone else, a situation demanding flexibility in still capacity to realize economic utilization of equipment and to conserve by-product charging stocks. The last problem became particularly acute late in the 1860's when Downer and Merrill became exclusive refiners of deodorized lubricants and heavy illuminating oils, both products in high demand and drawing on intermediate oil stocks above the kerosene range.

After lengthy experiments with stills of every shape and size, they finally resolved these problems by almost complete reliance on small stills. Their only large still was 300 barrels, in which they steam-distilled crude to obtain good economic separation of crude naphtha. They conducted all further distillation of the topped crude in 25-barrel vertical stills, arranged in batteries of eight, all heated from a central furnace. Though they experienced slower rates of evaporation because of small diameters and evaporating surfaces, excessive running times were avoided because of the limited volume of the stills. At the same time, the upright construction, with an allowance of 6 feet above the surface of the oil for expansion at the outset of distillation, held entrainment problems to a minimum. At the later stages, when long holding times were needed, the large wall and dome spaces upon which the heavier vapors could condense and return to the still gave them a much greater reflux than larger stills of any design. As a result, aided by the introduction of super-heated steam, they were able to obtain better and closer separation of the higher boiling constituents both in ordinary and destructive distillation. Finally, the high reflux ratio and small size of the still enabled them to crack deeply to coke, obtaining a much greater yield of cracked stock than in larger stills.[34]

A potential problem of complete refineries with large markets for all by-products—competition from different products for identical stock—was reached in the Downer organization in the last two years of the decade. Refining of deodorized lubricants and of mineral sperm heavy illuminating oils, exclusive products in heavy demand, drained light and heavy lubricating stocks, forced Downer and Merrill to become heavy purchasers of residuum, and to abandon entirely cracking for illuminating oils. However, the flexibility of small stills and their own refining ingenuity allowed these refiners to resume cracking by 1872. Close control of destructive distillation enabled them to crack heavy illuminating oils from heavy rather than light lube stocks, relieving some of the pressure on the latter for deodorized oils and kerosene. Their facilities also permitted heavy cracking of selected batches of crude oil to obtain yields of 30 per cent and more of cracked kerosene stock.[35]

A generalized description of the operating plan in Downer's refinery follows. (See the flow chart in Figure 11:9.) A yield of 16–20 per cent of crude naphtha resulted from steam-distillation of the crude oil in large 300-barrel stills. The crude naphtha was redistilled in the large still, treated with acid and soda, and then redistilled over soda into 1–2 per cent gasoline and 12–15 per cent of three grades of naphtha. The remaining 85–88 per cent of the topped crude was transferred to direct-fire stills of 25-barrel capacity and distilled over a 2 per cent solution of caustic soda (14° B). About 50 per cent of No. 1 straight-run distillate (58–40° B) was treated with 2.8 per cent sulphuric acid (75 per cent strength) and caustic soda, followed by a redistillation in a fire still over 4 per cent of soda of 14° B. About 80 per cent of this (40 per cent of the topped crude charge and 34 per cent of the total crude charge to the naphtha still) remained as Downer's No. 1 standard kerosene with a fire test of 125° F and a gravity of 45° B.

The rest of the crude was cracked into No. 2 intermediate oils, of which about 20 per cent was No. 2 crude burning oil, a cracked distillate, and 25 per cent a light-lubricating stock. There was also a cracked No. 3 oil, a heavy-lubricating stock. Acid and alkali treatment and redistillation over soda of the No. 2 and No. 3 lubricating stocks, after they had been chilled and paraffin removed from them, brought further additions to the crude burning oils Nos. 1 and 2 and to the intermediate lubricating oils. To make his heavy-deodorized lubricating oil, Merrill subjected heavy distillate from which the paraffin had been removed to fire distillation to 300° F; he then introduced superheated steam through a perforated coil to avoid excessive cracking. To make his heavy-illuminating oil (mineral sperm) with an average gravity of 36° F, he lightly cracked in a fire still any distillate 40° B to 32° B.[36]

Yields varied, with perhaps a median of 16.5 per cent naphtha (of which 1.5 per cent was gasoline), 10 per cent light C naphtha, and 2.5 per cent each of the heavier B and A naphthas; 54 per cent was standard kerosene, unless 20 per cent of the stock was diverted to refining mineral sperm oil; 17.5 per cent was spindle and cylinder lubricating oils, 2 per cent paraffin, and about 10 per cent loss.[37]

The operations of Downer's refinery make clear why a trend toward complete refineries was at least two decades away, while the day of small refineries specializing in by-products was immediately at hand. It took an exceptional marketing and refining organization with fifteen years of experience, location in a concentrated industrial market, and a rare aptitude for strategic innovations, to assure sufficient demand for by-product to sustain this complete refinery. The depth and geographic distribution of demands for by-products were not yet great enough to justify many complete refineries under the most favorable circumstances. The diversity of facilities required obviously could be met only by the resources of a large refiner, yet the needs to organize capacity around small stills in order to assure adequate flexibility and quality entailed a sacrifice in throughput that was in direct conflict with the normal need of large refineries for maximum throughputs. More complex logistics of operation and greater demands on technological skills were additional deterrents to partial or complete refineries, except in the limited range of naphtha by-products, where steam distillation had removed barriers to easy entry.

On the other hand, minimum threats of strong competition from large refiners, perennial surpluses of cheaply priced crude naphthas and residuum, and low capital requirements for small stills and related facilities for processing individual by-products, were attractive incentives for small refiners to specialize in selected ranges of by-products. Among many examples of successful small enterprises in by-product manufacture were the Chesebrough Manufacturing Company in New York and E. F. Houghton and Company in Philadelphia, specialized refiners of petroleum jelly, waxes and lubricants, and the Danforth Company in Cleveland, refiners of gasoline.

Organization of a kerosene refinery

The sharply contrasting simplicity of a flow chart in a typical large kerosene refinery in the early 1870's is shown in Figure 11:10. Charles Pratt's greatly admired refinery in New York City offers an excellent model of the typical large kerosene refinery. The fact that he was also reputedly the largest manufacturer of gasoline in the world, and Merrill's and Downer's only strong competitor with a complete line of naphtha products, heightens rather than minimizes the representative character of his operations.

Pratt's refinery had a rated crude capacity of 1,500 barrels daily, which could yield a daily output of illuminating oils in excess of 1,100 barrels, or about 1,200 barrels of refined oils of all kinds, including naphthas. His main crude charging capacity was concentrated in a bank of four hori-

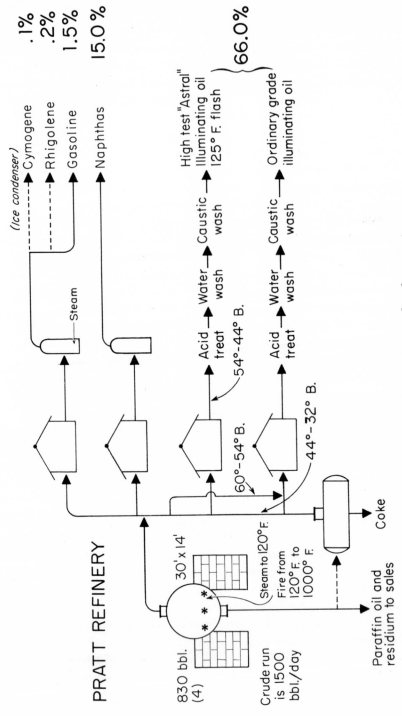

FIGURE 11:10. *Flow chart of batch operation.*

zontal cylindrical stills, direct-fired, each charging 830 barrels. This total combined charge of 3,320 barrels could be made and run in the large stills twice weekly. The brick masonry surrounding the stills stopped at the center line, leaving the upper part exposed to the atmosphere in order to promote condensation of the heavy vapors and refluxing during cracking. Six small, horizontal cylindrical stills also took charges of 50 barrels each. In addition, Pratt had some small steam stills which he used exclusively to refine gasoline and naphthas.

Unlike Merrill, Pratt conducted his entire distillation in his crude-oil stills. Instead of distilling crude naphtha in separate steam stills, he introduced steam into the crude still to aid separation. He began direct-fire distillation at 120° F and carried it to temperatures of 1000° F, cracking deeply to a few per cent of solid residuum. For his high-test Astral illuminating oil, with a flash-point of 125° F, he cut solely from the straight-run, between 54° B and 44° B. His ordinary-grade illuminating oil comprised the remainder of the straight run product mixed with the cracked distillate, parts of which might be redistilled before the mixture of straight-run and cracked was given acid and soda treatment. He did not refine much of his paraffin oil and residuum, yields of which were small due to extensive cracking, but sold them to refiners of lubricants and waxes. Most of his refining of by-product stocks was confined to crude naphtha in his steam gasoline stills.[38]

With respect to the excellent balance of its capacity, Pratt's refinery probably was not the most typical at the peak of the period of swift extravagant expansion in still and plant sizes early in the 1870's. Though concentrated in large stills, his capacity was distributed not only among four stills of identical size but of identical types and design. This distribution conferred obvious advantages in maintenance and repair and provided a hedge against disruption in throughput when a major still was out of commission. Newer large plants like the Octave refinery erected in the Region in 1869 characteristically divided up capacity irregularly in stills from 50 to 3,000 barrels, which lacked uniformity in type as well as size.[39] Others of more modest size but expecting to expand further, like the Franklin Oil Works at Philadelphia in 1871, had almost all capacity tied up in a single 475-barrel still, though expert opinion and, in time, actual practice preferred a division between three stills of 160 barrels each and the addition of a naphtha still.[40] Others like the Bennett, Warner & Company's refinery at Titusville attained a degree of balance by complementing one large still of 1,200 barrels with eight of 100 barrels each.[41] These diversified patterns of capacity were one of several sources of variation in the performances and costs of plants of comparable size.

Plant costs

Though long appreciated, the logistics of batch distillation exerted its most uncomfortable effects on plant costs during the period of precipitous expansion. Samuel Van Syckle undoubtedly characterized the situation correctly when he indicated that to turn out 500 barrels of refined daily by any arrangement of batch stills required 1,000 to 3,000 barrels of oils charging at various times in the stills. Among many examples this ratio was evidenced in the Pratt refinery and in a report of total refining capacity at Titusville in 1871: total crude capacity of stills was 12,155 barrels, and their daily capacity of refined oil was 2,856 barrels.[43] Lewis Emery stated that in the larger refineries, where there was the greatest concentration of large stills (with longer running times and extensive repair periods), charging capacity often exceeded daily refined output by four times or five times.[44]

There was no doubt that in new or greatly expanded plants the trend was toward daily refined outputs of 500 barrels or more a day and daily charging capacities of 800–1,000 barrels and more; but indicated costs varied considerably with locations, types of facilities, and methods of reporting costs. J. Lawrence Smith reported the costs of a small plant charging about 900 barrels a week in three stills of 110 barrels each, supplemented with a naphtha still, as about $30,000–$40,000. Samuel Van Syckle estimated the cost of a plant charging about 900 barrels daily, with a refined output of 500 barrels, at $100,000. Both estimates presumably included buildings but excluded costs of real estate, though neither expressly stated as much. Van Syckle gave no details of equipment except that it centered in large stills of 1,000 barrels and more, and he may have overstated costs somewhat, since he already was promoting his process of continuous distillation.[45]

Examination of numerous similar estimates, each having obvious shortcomings, suggest that the cost of $60,000–$80,000 (exclusive of land) may have been typical for plants charging 800–1,000 barrels daily; but there were many variations. The Octave Refinery, erected at Titusville in 1869, with a daily charging capacity of 800 barrels in stills with capacities from 50–3,000 barrels, was reported to have cost John D. Archbold and his associates $88,000. In Philadelphia, where land costs presumably were higher, William K. Harkness in 1870 built a plant capable of turning out 500 barrels of refined oil daily for a total investment of $130,000.[46] Bennett, Warner & Company's refinery at Titusville, which balanced one large 1,200-barrel still with eight charging 100 barrels each, involved an outlay of $70,000 in 1871, while a plant with about half that capacity (1,050 barrels), built in New York by Lombard, Ayers & Company during the same year, was reported to have cost about $40,000.[47] Among a few new plants that ran to much larger refining capacities was the Citizens Oil Refinery at Pittsburgh. Erected by August H. Tack in 1869 at a reported cost of $300,000, it had a daily capacity of 1,000 barrels of refined products. Tack's plant was a rarity among large units in turning out a wide range of by-products from gasoline to wax to lubricants, but it did not produce deodorized lubricants and heavy illuminating oils as Downer's plant did.[48] Its ratio of by-product output to kerosene was not comparable to Downer's and sufficed to give it only a nominal classification as a complete refinery.

Manufacturing costs

There is little question that the general trend of manufacturing costs was so steeply downward after 1865 as to make 5¢ a gallon, frequently cited since 1860, realistic and even liberal for the first time before the end of the decade. Happenings in nearly all phases of costs make it probable that in the early seventies the cost of a gallon of refined oil was closer to 3¢ than 5¢ in reasonably well-appointed and located refineries.

Expansion in the number of larger refineries promoted reductions in manufacturing costs. In part, these reductions came from flexibility in heating and treating made possible by a variety of stills of different sizes. To fire a still of 500 barrels or 1,500 barrels took no more man power than one of 100 barrels, usually one man and a helper. Because of the need for a somewhat larger force in cleaning, repairing and manning the machine shop, the common view of contemporaries that labor requirements with big stills were identical with those of small stills was not literally true.[49] But the fact that these costs were spread over a much larger volume of output indicated that petroleum refining had already established its permanent characteristic of relatively modest labor requirements in relation to capital investment and value of product.

The same kind of spread had occurred in fuel costs that probably overshadowed in importance the decline in absolute prices shown in Table

11:1. With that price decline had come an elimination of huge differentials in fuel prices at different locations, especially in the Oil Region, which was a major factor in the renaissance and expansion of refining there late in the sixties. The successful practice of mixing coal and residuum also reduced absolute costs for all refineries.[50]

But of far greater importance was the general extension of the four-fold economies of steam stills to units of intermediate size, and of co-distil-

TABLE 11:1

Year-end wholesale prices in Philadelphia of coal, sulphuric acid, and caustic soda: 1862–72

	Coal per ton	Sulphuric acid per lb.	Caustic soda per lb.
1860	$3.92	$.0213	$.0525
1861	3.50	.0212	.0563
1862	5.37	.0225	.0862
1863	7.50	.0250	.0825
1864	8.62	.0550	.1213
1865	8.75	.0500	.1050
1866	5.37	.0450	.0888
1867	4.63	.0300	.0850
1868	7.00	.0363	.0781
1869	6.13	.0256	.0738
1870	4.20	.0195	.0650
1871	5.08	.0281	.0650
1872	4.12	.0181	.0787
1873	4.60	.0300	.0813

Source: Anne Bezanson, *Wholesale Prices in Philadelphia, 1852–1896,* 2, 8, 51.

lation with superheated steam to the largest stills. Thinner metals, superior insulation with air chambers and other means, more effective heat distribution at still sides and center at the start of distillation followed by re-distribution to bottom heats at later stages, and many related developments applied to large batches must have at least halved fuel costs since 1865. Not all of the economies from these improvements were confined to large stills, as instanced by Merrill's and Downer's arrangements to heat eight small stills effectively from a central furnace.

The decline in absolute prices of sulphuric acid after 1867, shown in Table 11:1, are revealing of big reductions in treating costs. Extension of the acid industry to inland centers, including the Oil Region in 1871, reduced cost disadvantages that formerly stunted the growth of refining at those locations. A drop of more than 50 per cent in acid prices probably was a decisive factor in extending cracking to large-scale units.

Operating costs

Expansion in the number of larger-sized refineries was perhaps a bigger factor in lowering operating costs than in reducing strictly manufacturing costs. Receipt of crude in bulk rather than in barrels opened up opportunities for economies in scale. It not only wiped out the costs of barrels themselves while halving facilities and labor needed for cooperage, but linked with the growing postwar availability of iron tankage and steam power, it radiated influences throughout the refinery. By 1870, elimination of nearly all manual movements of oil distinguished not only large refineries like Charles Pratt's in New York City. The smallest decently appointed refinery with less than 1,000 barrels weekly capacity likewise had six steam pumps: to move the crude from tank car to storage tank and all other points; to pump water, distillate, and refined oil; and to power the air compressor for treating. Even in handling finished products where bulk distribution had not yet entered, there were savings, for small refineries as well as the large commonly employed apparatus for filling barrels automatically.[51]

On the periphery of manufacturing costs were such operating expenses as insurance. These too began to show signs of responding to improved organization of refining. Premiums varied from 25 per cent down to 5 per cent of valuation according to hazards of location and arrangements for fire protection that often were not accurately appraised. Among refiners there was a broader understanding of what comprised safeguards upon which to make a case for low rates. Location of tail-house and treating room at the fartherest possible distance from the stills, fire-bricking and adequate supports for furnaces, provisions for ventilation, steam distillation of inflammable naphtha fractions, steam pumps for blowing out clogged condensers (an acute problem at high temperatures because of paraffin formation and the great source of fires)—all these practices had become fairly routine. Gauges to register rises of pressure in stills, and provisions by Charles Pratt and other large experienced operators for steam-pump capacity readily diverted to fire-fighting, were commanding increasing attention among refiners.[52] Refiners needed the most advanced protective measures available and more, for few plants escaped decimating fires sooner or later. In 1871, for example, Samuel Van Syckle, an experienced refiner, watched his new refinery burn to the ground twice.[53] In the next year, Charles Pratt's refinery in New York, widely considered a model in safety as well as in other appointments, also was totally destroyed by fire. Like many other costs, the new scale of refining probably spread fire losses over much larger volumes of throughputs, but there seemed to be little sensible reduction in their frequency or their severity.

Chapter 12

Refining capacity, output, location, and
marketing of crude

Drake's discovery was the impetus for emergence of petroleum refining in widely scattered locations, including the area around Oil Creek, Pittsburgh, Cleveland, Erie, Buffalo, New York City, Portland, Maine, and Baltimore, in addition to smaller towns and cities. The nucleus of this early capacity was furnished by the major coal-oil plants, most of which shifted to petroleum after 1862.[1] But their ranks were swelled by new entrants, with the result that, according to one estimate, there were possibly 300 or more petroleum refineries in operation by the end of 1863.[2]

During what was described as the mushroom years of refining activity, when plants emerged like the "basidiomycetin fungi,"[3] much new capacity was represented by small installations, equipped with one or two 40–50-barrel stills, spasmodically operated, and producing for local markets. Yet there were new permanent entrants with plants of larger scale and modern design. The Brilliant Refinery at Pittsburgh, for example, was capable of charging 250 barrels daily. Others not as large but with average capacities between 100 and 150 barrels daily were numerous in all the refining centers.[4]

Capacity, output, and crude marketing to mid-1860

From the first foundling days until the mid-1860's the growth in refining capacity demonstrated the momentum of expansion which caught the industry. From literally a few hundred barrels during the aftermath of Drake's discovery, as shown in Table 12:1, total daily crude charging capacity at mid-decade was nearly 12,000 barrels. As a function of this enlarged capacity, refinery throughput (actual crude run through the stills), as shown in Chart 12:1, grew from slightly over a half-million barrels in 1862 to over 2 million barrels three years later, and 3.2 million in 1866. Even with product yields of 50–60 per cent, refined products manufactured kept pace with the growth of throughput and capacity by expanding from approximately 300,000 barrels in 1862 to over 2 million barrels four years later.

This approximate fourfold increase in shipments between 1862 and 1864–1865 put an increasing strain on crude marketing channels which remained relatively unchanged during these years. In the Region, many individual producers still engaged extensively in negotiating sales for their own and other producers' crude. Included in this group were such pioneer firms as Brewer, Watson & Company; William Abbott, A. P. Funk; William Barnsdall, and L. Haldeman.[5]

Another group of early marketers unique to the Region were the colorful "dump" men, perhaps the first oil speculators along the Creek. Operating from a "dump" (in contemporary nomenclature, any wooden tank holding from 10 to 600 barrels), they bought oil for cash from producers, who for various reasons were unwilling or unable to provide their own storage facilities. The dump men were persistent buyers, and their prices often furnished producers, ignorant of current market quotations, with a basis of comparison in negotiating contracts with other purchasing agents. In the opinion of Patrick Boyle, a contemporary oil journalist, it was the dump men who made the first prices in the oil fields.[6] Individual refiners or their agents, brokers, commission houses, and wholesale dealers in oils also combed the oil fields in search of sellers.

There was little organization to their purchases and nothing resembling an organized commodity market.* Most buyers in fact ". . . came into the field to seek the producer, and to buy the oil directly from him and pay him. Some days it would go up half a dollar a barrel when there was competition for it. . . . They would go right to Oil Creek to meet the producer and say: 'How much petroleum have you—a thousand barrels? I will give you $4.00 for it;' and make a contract right on the spot." [8]

But with the increase and geographic spread of production, the growth in refining capacity outside the Region, and improvement in the handling of transport of crude, the stage for the evolution of more efficient and centralized facilities for the buying and selling of crude was set.

Early refining location

Although refining capacity was still spread over a wide geographic area, there was already a trend toward concentration in certain areas by the mid 1860's. As shown in Table 12:1, in 1864–65 approximately 90 per cent of the industry's estimated total daily crude charging capacity of just under 12,000 barrels was located in Pittsburgh (4,500 barrels), the New York–New Jersey area (3,100 barrels), the Region (2,160 barrels), and Cleveland (750 barrels).

As this emerging pattern indicates, there was as yet no dominant combination of factors which dictated the choice of refining locations close to the sources of crude, at major distributing centers for refined products, or at intermediate points. But the concentration of nearly two-fifths of the industry's total capacity at Pittsburgh suggests that its location offered particular advantages to petroleum refiners.

Much of Pittsburgh's attraction stemmed from its importance as an industrial city close to the coal fields and with an available supply of skilled workers, abundant iron foundries, engine manufacturers, boiler makers, machine shops, and facilities for barrel and chemical manufacture. Petroleum refiners in the area also began with a rich technological legacy from the coal-oil era. But to the mid-1860's, Pittsburgh's strongest attraction was its position as the principal entrepot for crude oil moving down Oil Creek

---

* Mention should be made of the Pittsburgh Oil Exchange organized late in 1862. Its primary purpose, however, was not to serve as a commodity exchange but to set standards for gauging oil shipments and to supply marketing information to its members. According to one report, in February, 1863, "The Pittsburgh Oil Exchange is of great benefit to oil dealers and the trade. Its efficient secretary, George H. Thurston, and the board of directors give general satisfaction." [7]

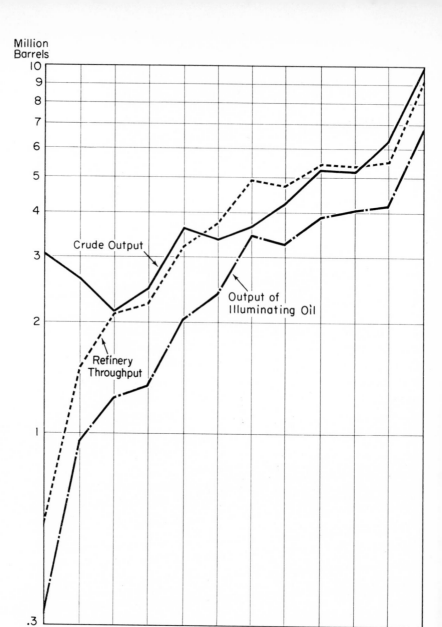

CHART 12:1 *Crude production and estimates of refinery throughputs of crude and output of illuminating oil, 1862–73 (in 40–42 gallon barrels). (Source: Appendix A.)*

and the Allegheny River and its rail connections with the seaboard and the west via the Pennsylvania Railroad.

Like Pittsburgh, the former coal-oil centers of Boston and New York also had readily accessible supplies of chemicals, labor, barrels, and the like, as well as excellent port facilities for serving the export demand. New York, in addition, was already established as a financial center and had well-developed agencies for distribution both at home and abroad.

TABLE 12:1

Location and daily crude charging capacity of major refining areas: 1864–73 (42-gallon barrels)

	1864–65	Per cent of total	1872–73	Per cent of total
Pittsburgh	4,500	39	10,000	21
Philadelphia	600	5	2,000	4
Boston	500	4	600	1
New York–New Jersey	3,100	26	10,000	21
Cleveland	800	6	12,500	26
Oil Region	2,160	19	9,200	20
Erie	. . . . .		1,200	3
Baltimore	20	1	1,200	3
Elsewhere	. . . . .		900	1
Total	11,680	100	47,600	100

Sources: G. H. Thurston, *Pittsburgh and Allegheny in the Centennial Year*, 205, 206, 208–11; Maybee, *Railroad Competition*, 235; Wright, *The Oil Regions of Pennsylvania*, 201–4; *Derrick's Hand-Book*, 1,780; J. T. Henry, *Early History*, 318; *Forty-First Annual Report of the Philadelphia Board of Trade*, 1873, 147.

But these advantages could only be partially exploited so long as crude, packaged in barrels, had either to be teamed from the Region to railheads some twenty or more miles distant, or shipped over the long route from Pittsburgh via the Pennsylvania Railroad to Philadelphia and north by rail or coastal vessel. Not only were transport costs high and losses through leakage excessive, but receipts of crude were too irregular and uncertain to sustain concentrated and continuous refining operations at the seaboard. Moreover, unless they had mastered the art of treating "cracked" kerosene fractions, seaboard refiners faced another handicap. With yields of illuminating oil from ordinary distillation of about 50 per cent, it was necessary to ship two barrels of crude for each barrel of refined output.

Refining in Cleveland began with even fewer locational advantages. It had only the most tenuous connective with coal-oil refining. A direct rail route with the Region awaited extension of the Atlantic & Great Western late in 1863. There was no local production of sulphuric acid until 1867, and until almost mid-decade, city authorities maintained an uncooperative attitude toward refiners. All of these factors inhibited the early growth of refining. By 1865, Cleveland's total daily crude capacity was only 800 barrels and most of the refineries were quite small and few, if any, operated full time.[9]

From the beginning of the industry, the Region's refiners were convinced that the ideal location for refining operations was close to supplies of crude. But as one observer pointed out in 1865, "The advantages and disadvantages of establishing refineries near the oil regions are many, and pretty nearly equally balanced. By being on the ground, one is enabled to take immediate advantage of every turn in the market in making purchases." Because of weight losses in refining, "There is also a saving in freight; though this is partly offset by the freight on chemicals which have to be sent on from the East . . . [while] real estate, fuel, labor, etc., are from fifty to one hundred per cent higher in Petrolia than elsewhere." [10]

These considerations were reflected in the general character of the refining capacity in the Regions, which was typically made up of semi-portable, relatively inefficient types of refinery installations. Without a ready market for by-products, many of the operators attempted to maximize their yields of illuminating oil by "cracking." The relatively high cost of treating materials plus the slow spread of the techniques of adequate treating resulted in a generally inferior grade of illuminating oil.

For these reasons the proportion of total capacity of approximately one-fifth in 1864–65 undoubtedly over-emphasizes the relative importance of the Region as a refining center. The principal exceptions to the general run of small, inefficient, and almost mercurial refineries in the area were Downer's refinery and the Warren Brothers' plant, both at Corry, and the Humboldt Refinery at Plummer. Unique in the Region, the Downer and Humboldt plants were among the few complete refineries anywhere, engaging extensively in the manufacture of by-products. But most of the Region refiners found it difficult to maintain continuous operations. Many worked only to fill specific orders, while others waited for favorable shifts in crude prices to secure a limited supply of raw material for processing.*

* Under the circumstances, a number of Region refiners insured their survival by turning to a more limited operation of steaming or skimming crude to remove the volatile gases and the naphtha and benzine fractions before shipping it to points outside the Region for distillation and treating. In addition to reducing the weight of the oil to be shipped by some 15–20 per cent, elimination of the light ends re-

Refining capacity and output to the early 1870's

Refining was no less responsive than other branches of the industry to the annual outputs of crude which rose from about 2.3 million barrels in 1864–65 to nearly 9.9 million barrels in 1872–73.[12] This approximate four-fold increase in production, as shown in Chart 12:1, was matched by an equivalent expansion in daily refining capacity, from just under 12,000 barrels to nearly 48,000 barrels.

A rough measure of the increase in the average scale of refining operations is indicated by the decline in the number of plants from the 300 or more reported for 1863 to approximately 100 in 1872–73.[13] There was still a considerable variation in the average size of refining operations in the major refining centers. Cleveland's approximate 12,500-barrel daily capacity in 1872–73, for example, was spread among six establishments. By contrast, some 27 plants shared the Region's capacity of just under 10,-000 barrels. New York–New Jersey with 15 plants and Pittsburgh with 22 plants, ranked next to Cleveland in the average size of their individual refining operations.*

With expanding refining capacity there was an equally impressive increase in the volume of crude processed and refinery outputs. As indicated in Chart 12:1, estimated refinery throughputs of crude, which averaged between 2.1 and 2.4 million barrels in 1864–65, rose to just under 5 million barrels in 1868, and to approximately 9.5 million barrels in 1873. It may be further noted that during 1867–71 it was necessary to supplement current production of crude by drawing on crude inventories, in order to meet the demand of the refineries for raw material.

As a function of the volumes of crude processed, outputs of illuminating oil followed essentially the same pattern of expansion, modified by improvements in refining operations that increased yields from about 65 per cent in 1864–65 to 75 per cent in 1872–73.† The result was an approximate fivefold growth in the production of illuminating oil between 1864 and 1873; from about 1.27 million barrels to 6.75 million barrels.

---

moved a major cause of fires and explosions while the oil was in transit. It also took out most of the offensive sulphur compounds, which were a source of much public complaint and, because of their corrosive action on barrels and tanks, added to the problems of leakage.[11]

* For certain of the other refining centers listed in Table 12:1, the number of plants (and total daily refining capacity) in 1873 were: Philadelphia, 12 plants (2,000 barrels); Erie, 6 plants (1,200 barrels); Baltimore, 8 plants (1,200 barrels).[14]

† See Appendix A.

Changes in the marketing of crude

Both producers and refiners benefited after the mid-1860's from improvements in the methods of handling and marketing of crude. With the proliferation of gathering lines, producers began to deliver their crude from wells directly to the storage tanks of the pipeline companies, usually located at railroad sidings. Buyers of crude no longer found it necessary to seek out sellers at wellheads. Instead they could travel by train over any of the several railroads serving the Region, stopping off at the various "oil farms" along the route where they purchased oil credited to the accounts of producers by the pipeline companies. These trains, in fact, became a sort of traveling oil exchange, the headquarters for both buyers and sellers.

According to one account, for example, when the Farmers' Railroad was completed to Oil City in 1867, ". . . it brought so many operators to town that a car was assigned them, in which they bought and sold 'spot,' 'regular,' and 'future' oil. There were no certificates, no written obligations, no margins to bind a bargain but everything was done on honor and no man's word was broken." [15]

As the volume of transactions increased, there was a growing need for more adequate facilities than were offered by the informal traveling exchanges. In December, 1869, the first formally organized oil exchange was established at Oil City. By the early 1870's, similar organizations were operating in the principal towns in the Region and in Pittsburgh and New York. [16]

While refiners' agents, brokers, and other middlemen, including dump men and producer-marketers, continued to operate, the bulk of transactions took place in the exchanges. As one historian of crude marketing pointed out: "For the greater part the members of the exchange were speculators or brokers their clients being in all parts of the country and included capitalists, the larger oil companies, refiners, bankers and brokers." [17]

Coupled with the gradual assumption of the storage function by pipeline companies and their practice of issuing pipeline certificates to producers for each 1,000 barrels held in storage, the organized exchanges greatly simplified the buying and selling of crude. For the producers the ability to hold inventories in certificate form, although they had to pay storage charges, presented an opportunity to act in accordance with their personal expectations respecting the future supply and price of oil. Whenever they decided to sell they were assured of finding a buyer in the exchanges. For refiners, who needed a continuous flow of raw materials for the efficient operation of their plants, the opportunity of buying certificates in the exchanges was also a welcome innovation.

Refinery locations to the early 1870's

All of the locations listed in Table 12:1 for 1864–65 shared to a greater or less extent in the subsequent expansion of total refining capacity. And as had been true earlier, there was no dominant combination of forces dictating refinery location at any particular area; some nine-tenths of total capacity in 1872–73 was still concentrated in Pittsburgh, New York–New Jersey, the Region, and Cleveland. The big change during the intervening years was the emergence of Cleveland with an estimated daily capacity of 12,500 barrels as the number one refining center and the relative decline of Pittsburgh with its capacity of approximately 10,000 barrels, which was about equal the capacity in New York–New Jersey and only slightly larger than the amount located in the Region.

Reflected in this shift in the relative positions of the major refining centers were a complex of developments affecting refinery location during the postwar years. Refiners in the Region, for example, benefited most from the extension of rail facilities to the area, which were in turn connected with the major trunk lines. It was no longer necessary to ship out refined oil or team in supplies of fuel, treating materials, and equipment, over dirt roads linking refineries with railheads twenty or more miles away.

For refiners in the New York–New Jersey area the extension of rail facilities into the Region assured them of regular shipments of crude, adequate to sustain concentrated and continuous operations. They also benefited from improvements in refining efficiency and bulk transport which favored location at major distributing centers. As yields of illuminating oil were increased by cracking, or alternately by-products became more important, the weight loss in refining operations was correspondingly reduced. In contrast to the mid-1860's when it had been necesssary to ship almost two barrels of crude to obtain one barrel of refined products, by the early 1870's the ratio was under 1½ barrels to 1 barrel. Bulk transport of crude by rail worked in the same direction. Not only did it cut handling and special packaging costs, but because of the weight of the barrels alone it was more economical to ship equivalent amounts of oil by tank car.

Although transport developments during the late 1860's were of great benefit to the New York–New Jersey area (and Cleveland), the reverse

was generally true of Pittsburgh, which was slow to promote an extension
of rail connections to the Region. Until mid-1860's, Pittsburgh refiners
generally felt little need for such facilities. With the greater portion of
their refining capacity located on the lower stretches of the Allegheny,
they received their crude from river boats (much of it by bulk) at costs
which made them almost immune to competition. But with the expansion
of the northern and eastern roads into the Region, drawing oil directly to
New York, Philadelphia and Cleveland, Pittsburgh refiners began to lose
their complacency. Year-round rail shipments had obvious advantages

TABLE 12:2

*Crude receipts at major refining areas: 1865–71
(millions of 42-gallon barrels)*

Year	Pittsburgh	Cleveland	New York	Others
1865	.63	.23	..	..
1866	1.25	.61	..	..
1867	.73	.69	..	..
1868	1.06	.96	..	..
1869	1.03	1.12	1.15	.46
1870	1.04	2.00	1.32	.51
1871	1.15	1.85	1.54	.59

Source: Maybee, *Railroad Competition*, 189, 224.

over river traffic, which was shut off during the winter months and a part
of the summer. The balance swung even further to rail shipments with the
introduction of gathering lines and bulk shipments by tank car.

The extension late in 1867 of the Allegheny Valley Railroad (connecting Pittsburgh with Kittaning) on to Venango City across the Allegheny
River from Oil City offered a partial solution to the problem of supplying
Pittsburgh refiners with an all-weather route. See Map 12:1. By early
1868, the road had about 100 tank cars in operation, each holding 83 barrels of crude and loaded directly from storage tanks in Oil City by several
3-inch pipelines laid across the bed of the river. Of the approximate one
million barrels of crude received at Pittsburgh in 1868, for example, over
500,000 barrels were shipped over the Allegheny Valley line.[18] But while
Pittsburgh, as shown in Table 12:2, maintained crude receipts at about
one million barrels annually between 1868 and 1871, Cleveland's receipts,
already close to one million barrels in 1868, averaged nearly twice that
amount in 1870 and 1871.

In part the growth of shipments to Cleveland was in response to the same factors which had improved the position of the New York–New Jersey area as a refining center: the extension of rail facilities to the Region, the introduction of tank-car shipments, and higher yields of refined products from inputs of crude. But these factors alone fail to account for the fact that Cleveland refiners, who by the late 1860's were selling a substantial portion of their illuminating oil for export, were able to compete so successfully with refiners much more favorably located with respect to the seaboard.[19]

The basic answer to this question lay in the tangle of conflicting interests, between the four principal refining centers, and their relationships with the major trunk lines—the Pennsylvania, New York Central, and the Erie—which competed for the oil trade in the postwar years.

Railroads and the oil trade to the early 1870's

The main trunk line connections linking the Region with the principal refining centers as they developed after 1867 are shown on Map 12:1. The Pennsylvania, already piloted by Tom Scott (although he did not become president until 1874), tapped the oil country from two directions. To the south, shipments starting from Venango City (across the river from Oil City) moved over the Allegheny Valley line to Freeport and on to Pittsburgh. Alternately, they might bypass Pittsburgh and reach the Pennsylvania's main line by way of the West Pennsylvania cut-off from Freeport to Blairsville. From Philadelphia, the Pennsylvania had rail access to the New York area through its leases of the New Jersey lines. To the north, the Pennsylvania-leased Philadelphia & Erie line connecting Erie City with Sunbury and Harrisburg, drew oil from the Region, either over the Oil Creek line, extending from Corry to Oil City, or the Warren & Franklin, which angled northeast from Oil City to Warren.

Jay Gould's Erie railway depended for its oil traffic, as it did for much of the other freight outside New York, on its lease in 1868 of the Atlantic & Great Western. The latter drew oil from the Region over its branch from Oil City to Franklin and Meadville, where shipments might either be routed west to Cleveland or east to Salamanca, and on to New York via the Erie.

Vanderbilt's New York Central entered the oil trade through close working relationships with the Lake Shore, which provided Cleveland with a through route to New York. A lease of the Jamestown & Franklin road, completed in 1867, gave the New York Central–Lake Shore system a connection with Franklin over the Erie & Pittsburgh to the main line of the Lake Shore at Girard. It was not until 1870, when the Jamestown & Franklin was extended to Oil City, that the New York Central–Lake Shore management was in a position to compete vigorously for traffic in crude oil.[20]

This pattern of railway connections linking the Region with the major markets for refined products provided the background for an intensive struggle between the trunk lines for the oil traffic. How the various refining centers would fare in such a struggle depended largely on their opportunity and ability to bargain with the carriers for special concessions.

As the situation developed during the late 1860's and early 1870's, Pittsburgh found itself in the weakest bargaining position of any of the four major refining centers. The principal reason was that while Pittsburgh refiners were entirely dependent on the Pennsylvania for their connections with the seaboard, the Pennsylvania, particularly with its lease of the Philadelphia & Erie, was not equally dependent on Pittsburgh for its oil traffic. Nor was the Pennsylvania under any particular pressure to build up Philadelphia as a refining center, since its main line virtually monopolized all traffic between Pittsburgh and Philadelphia. Under these circumstances, the Pennsylvania classified all petroleum shipments over its main line as "local traffic," subject to higher rates than "through traffic" originating west of Pittsburgh.[21]

Compared to Pittsburgh and Philadelphia, refiners in the Region were in a somewhat better position to obtain concessions from the carriers by playing the Pennsylvania's Philadelphia & Erie line against the Atlantic & Great Western. In fact, once refined oil reached Corry or Warren on the Philadelphia & Erie or Corry by way of the Atlantic & Great Western, it moved on to market as "through traffic" at lower rates than those applied to "local traffic." But this advantage was largely offset by the high "local" rates on shipments from the Region to Corry or Warren over the Oil Creek and Warren & Franklin roads. The alternative of shipping via the Atlantic & Great Western's line from Oil City to Salamanca offered no relief, since the Atlantic & Great Western was bound by an agreement with the Oil Creek line not to deliver oil to Corry via Meadville below the rates set by the Oil Creek road.[22]

Located at the eastern terminals of the New York Central, the Erie, and the lines connecting the Philadelphia & Erie with the seaboard, refiners in the New York–New Jersey area benefited both from the "competitive"

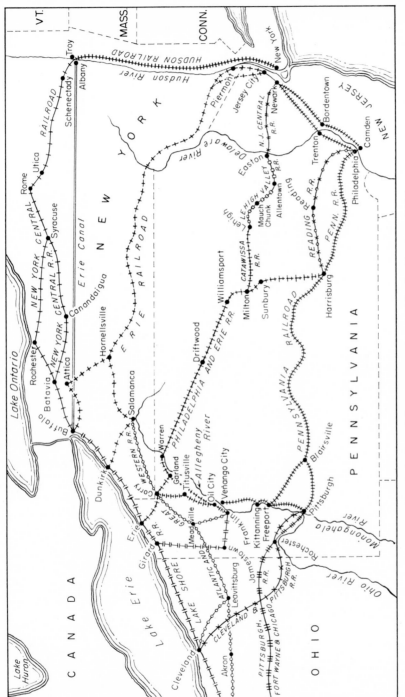

MAP 12:1. *Trunk line connections with the Region, 1864.*

through rates on oil and the growing use of the tank cars that stimulated bulk shipments of crude.

Yet it was Cleveland, farthest removed of all the refining centers from the eastern markets, that had the greatest incentive to obtain rate concessions and was also in the best bargaining position to obtain them. Served by both the Erie–Atlantic & Great Western and the New York–Lake Shore systems, Cleveland refiners had the added alternative of reaching New York by way of the Great Lakes and Erie canal. Although this route was not used extensively, the fact that it provided a cheaper alternative brought pressure on the Atlantic & Great Western and the Lake Shore to grant rates to Cleveland refiners that not only matched but were frequently lower than rates paid by shippers closer to the seaboard.*

In 1867, for example, one Pittsburgh oilman testified that he was paying ". . . the Pennsylvania Railroad $1.75 per barrel on oil from Pittsburgh to Philadelphia. The Atlantic & Great Western was carrying oil at the same time from Cleveland to New York for $1.53 a barrel" despite the fact that "the distance from Pittsburgh to Philadelphia is 355 miles; Cleveland to New York is 629 miles." [24] Four years later a Cleveland refiner, paying open rail rates, was able to haul the crude equivalent of a barrel of kerosene 148 miles from the Region and ship the resultant refined product 629 miles to New York for $1.80. To move an equivalent amount of crude from the Region over the 500-mile route of the Atlantic & Great Western and Erie, a New York firm paid, on a basis of open rates, between $1.75 and $2.10 a barrel. [25]

Other refining centers protested continually against a rate structure that eliminated their geographic advantage opposite Cleveland. The New York Central and the Erie, with generally strong alliances with New York, were under particular pressure from refiners in the area to cease favoring Cleveland as a refining location. Yet each line was understandably reluctant to give up its share of the oil traffic from the Ohio city, for fear that it would strengthen the competitive position of its rival. Vanderbilt and Gould were more than ordinarily aware of this point, for the Lake Shore and the Atlantic & Great Western, their respective connections to Cleveland, were crucial links in their through routes to Chicago. Any significant weakening along these routes might lead to a deterioration of their position with respect to the important grain and merchandise trade to and from the west.

---

* In 1871, only 230,000 barrels of Cleveland's total shipments of 1.75 million barrels went by canal. [23]

The general role played by the railroads in the interregional competition of the oil industry was reflected in variations in the structure of "open rates" during this period. But few paid the "open rates" and not all paid the same net rates from any one point. Beneath the rivalry of one region against another there was also competition of firm against firm and one of the easiest ways to undersell competitors or increase profits was to obtain lower freight rates than rivals.

In seeking personal concessions, refiners were no different than shippers of other types of merchandise. Nor was the practice among refiners unique to any one refining center. Yet the same intensive carrier rivalry that served to strengthen Cleveland's position as a refining center offered similar opportunities for its leading refiners to obtain special concessions from the rail lines that gave them an advantage over their competitors. Of all the Cleveland group none showed greater skill in utilizing this device than John D. Rockefeller, the future head of the Standard Oil Company.

John D. Rockefeller and the Standard Oil Company of Ohio

John D. Rockefeller, born in the small town of Richford in central New York in 1839, moved with his family to Strongsville just outside Cleveland in 1853. Upon graduation from high school two years later at the age of 16, he took a job as assistant bookkeeper with the Cleveland firm of Hewitt & Tuttle, commission merchants and produce shippers.[26] Three years later Rockefeller decided to go into business for himself and formed a partnership with a fellow Clevelander, Maurice B. Clark.

The firm of Clark & Rockefeller, produce merchants, began operations in March, 1859, and was financially successful from the beginning. With the outbreak of the Civil War the partnership shared in the expanded shipments that went through Cleveland and profited from rapidly rising prices. By the end of the war both members were moderately wealthy.

The move that brought Rockefeller into the oil business came late in 1862 when he and Clark were asked by Samuel Andrews, an employee of a local refinery, to join in financing a refinery. A new partnership, Clark, Andrews & Company, was formed early in 1863 and shortly thereafter began operating a new plant, known as the Excelsior Works, with Andrews in charge and Clark's two younger brothers handling purchases and sales.

During the next two years Rockefeller became increasingly attracted by the refining business. He encouraged Andrews in his efforts to improve refining methods and to cut costs. By early 1865, Clark, Andrews & Company had one of the best equipped refineries in the Cleveland area.

By this time Rockefeller decided to get out of the produce business and devote full time to refining. Unimpressed by Clark's brothers, he arranged with his partner to auction the oil properties between themselves; the buyer to take the refinery, leaving the commission business to the other. In February, 1865, Rockefeller outbid Clark to acquire the refining properties for a payment of $72,500. A few days later an announcement in the Cleveland papers stated that Clark, Andrews & Company had been dissolved and its former property, the Excelsior Oil Works, would continue in business under the name of the new owners, Rockefeller & Andrews.[27]

With a characteristic confidence in his own judgment, Rockefeller pushed the new partnership ahead rapidly. In 1866, Andrews' brother was hired and sent to the Region to purchase crude and William Rockefeller, younger brother of John, was brought into a new firm, Rockefeller & Company, which built an additional refinery in Cleveland. Late that same year it was decided to invade the east coast and engage in foreign marketing. William moved to New York City, where he set up an office to handle export sales.[28]

Early in 1867, Rockefeller persuaded Henry M. Flagler and Stephen Harkness to join his organization. Flagler, with a background of business experience as a grain and produce merchant, immediately took an active role in the management in a new firm, Rockefeller, Andrews & Flagler. Harkness, Flagler's father-in-law, who had made a fortune in the California gold rush days, was less active, but as a silent partner made a substantial investment in the organization.

As the business continued to expand, it was decided, late in 1869, to consolidate the various partnerships and, in January, 1870, the Standard Oil Company of Ohio was incorporated, capitalized at $1 million. Its officers were John D. Rockefeller, president; William Rockefeller, vice-president; Flagler, secretary; and Andrews, superintendent. Ten thousand shares of common stock were divided; 2,667 to John D. Rockefeller, 1,334 to Stephen Harkness, 1,333 shares each to Flagler, Andrews, and William Rockefeller, 1,000 to the firm of Rockefeller, Andrews, & Flagler, and 1,000 to O. B. Jennings. Jennings, a resident of New York and the only newcomer in the group, was a brother-in-law of William Rockefeller. The holdings of the new corporation were impressive. They included

. . . sixty acres in Cleveland, two great refineries, a huge barrel making plant, lake facilities, a fleet of tank cars, sidings and warehouses in the Oil Regions, timberlands for staves, warehouses in the New York area, and lighters in New York harbor.[29]

Given the abilities of John D. Rockefeller and his associates, the growth of the Standard organization in a situation that favored the rise of Cleveland as a refining center is understandable. But Rockefeller's chief competitors in the Cleveland area, according to Ida Tarbell,

. . . began to suspect something. John Rockefeller might get his oil cheaper now and then, they said, but he could not do it so often. He might make close contracts for which they had neither the patience or the stomach. He might have a unusual mechanical genius in his partner. But these things could not explain all. They believed they bought, on the whole, almost as cheaply as he, and they knew they made as good oil and with as great, or nearly as great, economy. He could sell at no better price than they. Where was his advantage?[30]

There was a growing conviction that the answer lay in special concessions obtained by Rockefeller on rail shipments of his oil.

Considering Rockefeller's attention to all phases of the oil business it was highly unlikely that he would overlook such an opportunity. Although there is no evidence that he and his partners received any special rates before 1867, in that year Flagler approached General James H. Devereux, newly appointed vice-president of the Lake Shore Railroad, which had just completed its extension from the Erie & Pittsburgh Railroad at Jamestown to Franklin. Devereux, eager to increase the Lake Shore's traffic in crude over its new extension to Cleveland, which up to this time had received its oil from the Region largely by way of the Atlantic & Great Western, granted Flagler a rebate, estimated at 15¢ per barrel. There is every reason to suppose that the Atlantic & Great Western granted a similar rebate. Moreover, as Flagler discovered later, he and Rockefeller were not the only Cleveland refiners to receive the rebate.[31]

In 1868, however, Rockefeller and Flagler joined two other leading Cleveland refiners, Clark, Payne & Company and Westlake, Hutchins & Company, in a move designed to insure them of special consideration from the railroads. They began by acquiring a one-fourth interest in the Allegheny Transportation Company, a gathering line in the Oil Regions which had an exclusive contract to deliver crude to the Atlantic & Great Western. Under this contract the transport company received a drawback "of $5.00 per car for each and every carload of bulk oil loaded for Cleveland at points between Oil City and Titusville inclusive, passing over the Atlantic & Great Western Railway." The transport company also had a contract which provided for a drawback of $12.00 per car for all bulk oil shipped via the Atlantic & Great Western and the Erie Railway to the east coast.[32]

On their own crude shipped over the Atlantic & Great Western the three refiners, under their contract with the Allegheny company, were to receive three-quarters of the drawback of $5.00 per car paid by the railroad to the transport company. In addition, their one-quarter interest gave them a proportionate share of the drawbacks on all other shipments, including the $12.00 per car going to the eastern markets.

A subsequent contract, signed on June 5, 1868, between the Atlantic & Great Western and the Allegheny Transportation Company, brought even better terms. The open rate on crude shipped from the Region to Cleveland was not to be less than 60¢ per barrel and the railroad further promised to "make no special rate of Freight and . . . will pay no draw back to any person or persons doing business in the City of Cleveland Ohio or their Agents or to anyone . . ."[33] The advantage of this arrangement to the refiners with the agreement and part ownership in the Allegheny company is obvious. Oil shipped on their account to Cleveland over the Atlantic & Great Western in cars holding not over 85 barrels was to be billed as cars having only 80 barrels.[34] At 60¢ a barrel this meant an additional "saving" to the three refiners of $3.00 per car.

On their part, the Allegheny company, Rockefeller, Andrews & Flagler, Clark, Payne & Company and Westlake, Hutchins & Company agreed not to ". . . build a Pipe Line from Oil City to Franklin or directly or indirectly be engaged therein and in case such Pipe Line shall be built by anyone else they agree not to connect with the source."[35] The obvious purpose of the Atlantic & Great Western's inclusion of this clause in the agreement was to limit the movement of crude to Franklin, the terminus of the Lake Shore's extension from Jamestown.

The contract, dated June 5, 1868, was for one year, but on August 27, 1868, it was extended to four years by an agreement signed by The Erie Railway Company, which in the interim had leased the Atlantic & Great Western. At the same time the drawback to the Allegheny company on crude shipped over the Atlantic & Great Western and the Erie to the east coast was raised from $12.00 to $15.00 per car. The Erie also undertook to build an oil-receiving depot at Penhorn, New Jersey ". . . where they shall receive all bulk oil transported over their road and from thence to be distributed to other points as may be required." [36]

The final item in this series of contracts was an agreement made between the Allegheny Transportation Company and the Standard Oil Company of Ohio, on March 19, 1870. Standard promised not to ship crude oil to Cleveland over any other line but the Atlantic & Great Western in return for a rebate of "five (5) cents per barrel of forty-five gallons moved for their use in the business of refining in Cleveland over the Atlantic & Great Western Railway for the period of one year." [37]

These agreements suggest that Rockefeller's competitors among the Cleveland refiners had good reason to suspect that his organization, along with Clark, Payne & Company and Westlake, Hutchins & Company, was getting preferential treatment in respect to freight rates on crude. As the terms of the contracts indicate, the special concessions obtained through the Allegheny Company were not shared by anyone else in the Cleveland area. This was especially important, for from 1865 to 1870, despite the attempt by the Lake Shore to compete, the Atlantic & Great Western, with the only through line to Oil City handled ". . . between 95 and 97 per cent of the crude oil to Cleveland." [38]

How much these arrangements contributed to the growth of Standard's refining capacity (estimated at about 1,500 barrels of crude daily in 1870) is impossible to measure. [39] But whatever their influence, the effects tended to be cumulative. Every increase in refining capacity further strengthened the firm's bargaining position in maintaining or increasing its special concessions from transportation companies.

Standard's bargaining strength by 1870 was well illustrated by negotiations between Flagler and Devereux of the Lake Shore on freight rates to the seaboard, particularly on refined oil. It is quite likely that none of the Cleveland refiners was paying the full, published rate of $2.00 a barrel on oil shipped to New York. [40]

But Flagler proposed that in return for guaranteed large, steady shipments, Standard be given a rate of $1.30 per barrel on refined products shipped to New York, and 35¢ on crude from the Regions to Cleveland. [41] Having just completed the Lake Shore's extension from Franklin to Oil City, Devereux was quite willing to enter such an agreement. As he explained later,

> . . . Mr. Flagler's proposition offered to the railroad company a larger measure of profit than would or could ensue from any business to be carried under the old arrangements, and such proved to be pre-eminently the case; that the

proposition of Mr. Flagler was therefore accepted, and in affiant's [Devereux's] judgment this was the turning-point which secured to Cleveland a considerable portion of the export traffic.[42]

In defending his action so far as the railroad was concerned, Devereux pointed out

> . . . that the then average time for a round trip from Cleveland to New York for a freight car was thirty days; to carry sixty cars per day would require 1,800 cars at an average cost of $500 each, making an investment of $900,000 necessary to do this business, as the ordinary freight business had to be done; but affiant found that if sixty carloads could be assured with absolute regularity each and every day, the time for a round trip from Cleveland to New York and return could be reduced to ten days, by moving these cars in solid trains instead of mixing oil cars in other trains, as would be necessary when transported in small quantities and by moving the oil trains steadily without regard to other cars; that by thus reducing the time to ten days for a round trip, only six hundred cars would be necessary to do this business with an investment therefore of only $300,000. That the regularity of the traffic would insure promptness in the unloading and return of the cars; . . .[43]

When the other Cleveland refiners heard of this arrangement they demanded similar concessions of the Lake Shore. They were told:

> That this arrangement was at all times open to any and all parties who would secure or guarantee a like amount of traffic or an amount sufficient to be treated and handled in the same speedy and economical way, the charges for transportation being always necessarily based upon the actual cost of the service to the railroad, and whenever any shipper or shippers will unite to reduce the cost of transportation to the railroad, to refuse to give them the benefit of such reduction would be to the detriment of the public, the consumers, who in the end pay the transportation charges.[44]

Instability in the early 1870's

"It would seem," Ida Tarbell observed, "as if the one man in the Cleveland oil trade in 1870 who ought to have been satisfied was Mr. Rockefeller."[45] Yet 1871 found the head of Standard increasingly concerned over the conditions of the industry and the possible effects on his organization.

The basis of Rockefeller's concern lay in the immediate outlook for profitable refining operations, although at the beginning of 1871 the trade press viewed the industry's prospects as a whole with considerable optimism. In January the *New York Bulletin* pointed out that while the

Franco-Prussian war had cut seriously into export demand, an expected rise in domestic consumption would no doubt take up the slack. Moreover, an anticipated expansion of at least 4,000 barrels per day in refining capacity during 1871 would require larger amounts of oil than "have been previously used." [46]

Yet the "very discouraging" tendency to increase refining capacity "ad infinitum," noted by the *Pittsburgh Evening Chronicle* in 1870, continued into the following year. [47] One estimate put the industry's daily refining capacity at the end of 1871 at approximately two-and-one-half times the rate of crude production. [48] With the average price of illuminating oil running about 2¢ per gallon lower in 1871 than in 1870, the margin squeeze brought cries of anguish from virtually all refining centers. In July, 1871, *The Pittsburgh Gazette* described the situation as "not very promising" in that city nor did advice from Cleveland and the seaboard offer much more encouragement. [49] Looking back over 1871, the *Titusville Herald* stated, "But little profit was realized by the refineries during the year." [50]

Refiners tended to blame the producers for their problems. Said *The Pittsburgh Gazette*, " a general feeling here [is] that there must be a decline in crude, though the producers continue obstinate and refuse to make [price] concessions sufficient to justify refiners in buying." [51] In February, 1872, the *Cleveland Leader* charged, "For eighteen months past the producers have carried a high head, and Cleveland and Pittsburgh . . . were forced to dance as the men of the Creek might choose to fiddle." [52]

In reality, the picture was much more complex than the refiners indicated. Their difficulties during the early 1870's were chiefly the result of impersonal, often ruthless, market forces taking their toll. Refinery suspensions, threatened early in 1871 "unless the price of crude declines," became prevalent later in the year. In September the owners of the Alexander, Scofield works in the Region "declined to run at a pecuniary loss" and shut down. When three Titusville refineries, with a combined daily capacity of 2,000 barrels, suspended operations in November, the *Oil City Derrick* intimated that this was only one instance of a nation-wide movement. [53]

But excess refining capacity was not the only threat to the stability of the petroleum industry early in the 1870's. It was at best an uneasy truce that followed the furious rate war that had broken out between the major trunk lines in 1869. According to James Devereux, the Pennsylvania Railroad reacted to the Lake Shore's extension of its lines from Franklin to Oil City in 1870 "with such rates and arrangements . . . that it was publicly proclaimed in the public print in Oil City, Titusville, and other places, that Cleveland was to be wiped out as a refining center as with a sponge." [54] In May, 1871, 40¢ was cut from the Pennsylvania's rate of $1.90 per barrel on crude to New York—a move which the New York press rightly interpreted as a renewed bid for the city's oil freight; in July, the rate was reduced another 25¢.

Although, as the *New York Bulletin* observed, these rates were no doubt still well above costs, they were clearly symbolic of Scott's interest. The *Cleveland Leader*, however, noted, "Fortunately . . . the Erie managers have a keen eye on Mr. Scott and we may rely upon them to so adapt their oil rates to the market, that Cleveland's refining interest will not seriously suffer." [55]

But there was no certainty that Cleveland's relative position would be unaffected, should an all-out rate war again break out between the carriers. Late in 1872, such a possibility appeared imminent.

Chapter 13

Foreign and domestic markets and marketing channels

The advent of flowing wells during the spring and summer of 1861 removed most doubts about supplies of crude sufficient to support an expanding petroleum-refining industry and to justify the development of more adequate transport facilities. But if potential crude supplies offered no long run obstacle to expansion, the outbreak of the Civil War in April, 1861, did make it uncertain whether marketing outlets would be available for any significant increase in the distribution of petroleum products.

The secession of eleven Confederate states with a population of some nine million cut off almost one-third of the domestic market for illuminating oil. Another threat to the new industry came when Congress early in 1862 announced its intention of levying heavy taxes on both crude and refined petroleum. Congress waived taxation of crude, but under legislation effective September 1, 1862, a tax of 10¢ a gallon was imposed on petroleum illuminants. In the same period, abandonment of the gold standard added another uncertainty by setting the stage for a subsequent domestic currency inflation that raised the wholesale price level to a peak in 1864, some 80 per cent above the level in 1860.[1]

Fortunately for the petroleum industry, the burdens of tax and economic events fell in such a way as to eliminate any serious competition in lamp illumination. With shipments of crude turpentine from North Carolina suspended by wartime blockade, camphene was already short in supply. A tax of 40¢ a gallon on alcohol, the main ingredient in camphene and similar types of burning fluids, made the elimination of the prime competitor of petroleum illuminants conclusive. Shortages of turpentine and the tax on alcohol also yielded the petroleum industry an additional dividend by encouraging the substitution of petroleum naphthas for spirits of turpentine in paint and varnish manufacture and for alcohol for use as solvents.

The fact that the remaining types of illuminants, including animal oils, manufactured illuminating gas, and candles, were taxed equitably relative to petroleum did much to hasten the acceptance of refined petroleum as a lamp illuminant. Shortages and expanded uses as lubricants had already priced lard, sperm, and whale oils out of all but the most marginal illuminating markets. Illuminating gas, only a peripheral competitor, was taxed at least as heavily as petroleum illuminants and priced higher. Candles provided an important outlet for paraffin, enabling a few small coal-oil refineries to survive, and also offered a by-product market that petroleum refiners would later assimilate. In deference to higher manufacturing costs, the tax on coal oil was 8¢ a gallon instead of the 10¢ levied on petroleum illuminants. But this differential did not begin to make coal-oil production competitive with refined petroleum under the scale of crude oil prices that prevailed after 1861.

Abundant supplies of crude at low prices after 1861 also raised the possibility of developing an important market for petroleum outside the United States, an opportunity that, because of its high costs, had been closed to the coal-oil industry. By fall of 1861 several experienced refiners, including Samuel Downer, had already anticipated a substantial export business in both crude petroleum and refined illuminants. Because of the greater hazards and weight losses in shipping and processing crude, they further anticipated that the bulk of the export trade would be in refined products.[2]

Wartime abnormalities during 1861 and 1862, however, made it difficult for the members of the American industry to push petroleum exports. Shipping space was particularly in short supply, even for the coastal trade, as the American government pressed many ships into service, including 150 vessels from the whaling fleet. Added to this difficulty was the menace of Confederate privateers, which led to war-risk insurance rates so high that American vessels became virtually unavailable for foreign trade. Owners of foreign ships were also reluctant to expose their craft to the hazards of fire and explosion by hauling petroleum.

The course of diplomatic relations was a further threat to trade relations in a new commodity. The Trent affair, originating in December, 1861, with the seizure of two Confederate envoys from a British vessel by an American sea captain, brought England and the United States to the brink of war. The incident intensified a period of turbulent diplomatic relations with both England and France, particularly over the possibility of their recognition of the Confederacy.

By fall of 1862, these tensions had been resolved sufficiently to remove any lingering doubts about the possibility of developing an important export market for the American petroleum industry. Exports of crude and refined that had amounted to about 37,000 barrels in 1861 rose to over 277,000 barrels in 1862.[3] Yet the domestic market continued to absorb the larger share of the refinery output. It was not until 1866 that petroleum refiners achieved a position unique among American manufacturers by selling more of their product abroad than at home.

Domestic markets and marketing to the mid-1860's

Among the initial factors that contributed to an expanding domestic demand for petroleum during the early 1860's was the ready, almost automatic acceptance of the petroleum illuminant by consumers. Many of those who bought kerosene, aladdin oil, and other prominent brands of illuminating oil after 1862 were scarcely aware that these products were currently being distilled from petroleum rather than coal. Indeed, the terms "coal oil," "burning oil," and "kerosene" soon became identified in the minds of the public as pertaining to petroleum illuminants.

The availability of coal-oil lamps capable of burning petroleum illuminants also played an important role in stimulating domestic sales. American lamp manufacturers, by adopting mass production methods, had sold some 1.8 million coal-oil burners during 1859 at prices averaging between

57¢ and 67¢ each. By the end of 1862, there were perhaps between four and five million lamps of this type in American homes.[4]

Domestic lamp manufacturers continued to make inexpensive lamps available to users of petroleum illuminants. But in supplying this market they found it necessary, in order to obtain a better performance, to modify the construction of the coal-oil type lamps because of petroleum's greater inflammability and faster rate of combustion compared to coal oil. Used with petroleum kerosene, for example, the all-metal burners (fuel reser-

FIGURE 13:1. *Sketch of Meucci's chimneyless kerosene handle lamp, 1862.*

FIGURE 13:2. *Sketch of Brown's metal-capped kerosene lamp chimney, 1862.*

voir and wick holder), typical of coal-oil lamps, quickly heated to the point where explosive mixtures of air and vapor were formed in lamp reservoirs when as little as half the fuel had been consumed. Thus, unless the lamp was allowed to cool several hours, refueling, trimming wicks, or the removal of chimneys for cleaning were hazardous. To extinguish a lamp which had been burning for several hours by blowing down the chimney was reasonably safe with animal, vegetable, or even coal oil; with petroleum kerosene it was an invitation to sudden death.[5] The more intense heat generated by petroleum also affected the glass chimneys, leading to excessive breakage and a greater frequency of accidents.

Lamp designers and manufacturers attempted with only partial success to remedy these difficulties. One approach was to reduce the transference of heat to the lamp reservoir and chimney by using porcelain, glass, and other non-conductors between them and the metal burner. A second was to improve the design of clasps and fasteners, which would permit the removal of a hot chimney without touching it, thus reducing the hazards of breakage and accident each time the chimney was removed to extinguish the flame. A third basic approach was to strengthen the chimney itself by the use of mica glass, metal-caps, or bracket protectors. None of these devices, however, gave adequate protection, and all tended to cast shadows, which reduced illumination. Designers were no more successful in their attempts to introduce specialized types of purportedly stronger glass, or to cast chimneys into new shapes minimizing stress.

Although it was the cheaply priced, all-metallic burners, with all their shortcomings, which served as the primary vehicle for the rapid popularization of petroleum illuminants, by mid-decade lamps with glass reservoirs had become the more numerous. Their aesthetic appeal, which glass manufacturers exploited to the utmost, was one reason for this growing popularity. Another was the success of glass manufacturers in lowering costs sufficiently to produce models at prices competitive with metal burners. Finally, even though many authorities questioned their safety, consumers were strongly attracted to glass reservoirs, which made it easy to determine when half of the oil in the reservoir had been consumed and refueling was required to avoid critical dangers from explosion.[7]

The most widely accepted glass lamps were made of pressed glass, generally referred to as "peg-bowl lamps" because of a glass peg attached to the bottom of the oil reservoir for anchorage and for insertion into a mounting made of milk glass. A brass collar united the mounting with a bowl-shaped glass reservoir. A brass, flat-wick burner was attached to the reservoir in such a way as to be easily unscrewed for refueling and cleaning. Attached to the wick were holders for a clear glass chimney which completed the assembly.

Glass bowls from the same molds as popular tableware—Bellflower, Honeycomb, Prism, Ivy, Argus, Diamond Point, Sawtooth, Horn of Plenty, Bull's Eye with Fleur de Lis—added to the popularity of both small table models, 6–8 inches high, and the large lamps, 12–16 inches high. By mid-decade the pressed-glass industry had achieved in many localities its widely advertised objective of putting "a lamp in every room." Sandwich of New England, Cornelius of Philadelphia, Holmes, Booth & Hayden of Boston, were among the prominent American manufacturers who catered to this market. By offering a product that had a combined aesthetic and utilitarian appeal they contributed significantly to the demand for kerosene, particularly among the more well-to-do consumers for whom the all metal burners had little appeal.[8]

Lamps in the kerosene age. (1) Early kerosene peg lamp about 1865 of pink bristol glass with marble base, brass stem, and font. (2) Peg lamp about 1870 featuring recently popularized milk glass in base and stem, a pressed glass font in grape pattern, and brass collar. (3) Early handle lamp about 1870 of bottle-green pressed glass, in attractive sweetheart design (chimney not shown). (4) Tin handle lamps with chimney made their entry in early 1870's. This is a rare original of the famous Mei-foo tin lamp manufactured and distributed by Standard Oil of New York to introduce kerosene into the Orient later in the century. It was painted red, a color signifying health and prosperity to the Chinese. Mei-foo—meaning literally an institution so large the sun never sets on it—was the word used by the Chinese to designate the oil company. (5) China stem lamp of the late 1870's, with black iron base, china stem in soft-hued, moss rose pattern, and pressed glass font in frosted leaf design. (6) The Student Lamp, introduced in 1868; Victorian adaptation of the original Argand lamp, and first great lamp of the kerosene age. Salient features with which it reconciled problems of safety, shadow and burning efficiency were (a) genuine Argand burner with circular wick encasing an air tube; (b) removal of brass oil font away from the burner by means of a tubed arm in order to obtain safe

Shortcomings in the construction and design of kerosene burning lamps made knowledge of their correct maintenance and use of particular importance at a time when a sharp expansion in the volume of illuminating oils distributed tended to impersonalize the relations of refiners, marketers, and lamp dealers with their customers. Consumer education in the care and use of new types of lamps and illuminant inevitably suffered with the result that carelessness and poor lamps probably contributed more to the mounting toll of accidents from the use of kerosene than at any subsequent period in the history of the industry. No informed person, however, shifted the basic cause of accidents away from the bad quality of kerosene. The axiom that "no lamp is safe with dangerous oil, and every lamp is safe with safe oil," applied with the same force then as when it was made famous later in the decade by C. F. Chandler, the crusading chemist for safer illumination.[9]

The fundamental reasons for the growing acceptance of the petroleum illuminants were well brought out in a widely publicized report in 1862 by two Philadelphia chemists, Professors James C. Booth and Thomas H. Garrett. On the basis of careful experiments, they demonstrated that none of the alternative sources of illuminants, including candles and

---

distance from burner heat, and gravity flow and automatic control of quantity of fuel fed to the burner; (c) brass oil font containing a capped, oil-tight steel cylinder, which could be simply removed for safe and convenient refilling. In double-burner models (not shown), the font usually was centered at the lamp stem and the two burners separated from it by tubular arms. (7) Brass parlor-size Student Lamp. (8) Early minature lamp, 1870–75, with one-piece base, stem, and font in bull's-eye pattern. (9) Early nutmeg patent miniature (1871), with brass handle, banded to a blue pressed glass font, in nutmeg pattern. (10) More elaborate miniature of the 1880's with matching globe, bowl and font of milk glass, Cosmos pattern. (11) Large table lamp of the 1880's, frosted glass base and stem in hand pattern, frosted globe etched with flowers, and clear glass font. (12) Removable Font or "Gone With the Wind" parlor and banquet lamp introduced in the late 1880's, though anachronistically placed in the Civil War era by the popular novel and movie after whose title the lamp has recently been renamed. Actually these flower-decorated, ball-shaped lamps marked the twilight of the kerosene age, not its beginning.

manufactured gas, could compete with refined petroleum on the basis of cost. They were of the opinion that, where it was available, "savings of labor, convenience, and greater safety, will cause gas still to dominate over all other sources of illumination." [10] To those who objected to throwing away their camphene burners, however, Garrett and Booth suggested that if they ". . . shifted in a single day to petroleum and coal-oil burners but half the size of the older types, they would get the same amount of light as they now do, without any more smoke, and at one-half the present cost."

They further pointed out that while the petroleum product was much less inflammable than burning fluids with an alcohol base, the failure of refiners to separate benzine and other volatile naphthas made petroleum illuminants dangerous to use. This danger, they felt, would undoubtedly remain for some time to come, ". . . in consequence of the want of skill . . . [or knowledge] to remove the volatile fluids, or by reason of manufacturing cupidity, which would prefer allowing them to remain in the oil, in order to increase the quantity sold. . . . Until the character of large dealers shall have been established for the sale of oil free from danger," they concluded, "we will content ourselves with this general advice to those who use such oils: never to fill a lamp at night, and not to store the oil where it can become heated." [11]

Unsafe oils, overloaded with highly inflammable naphthas, continued to characterize much of the refined petroleum distributed in the domestic markets throughout the 1860's and beyond. At no time did their sale seriously restrict demand, but a mounting number of lamp explosions and fires caused many individuals to conclude that aside from "a cheaper strong light" the petroleum illuminant offered no particular advantage over the dangerous camphene-burning fluids. [12]

This sentiment was strong enough during 1862 and 1863 to inspire a growing demand for regulatory legislation of the sale of refined petroleum in municipalities and state legislatures everywhere. Insurance companies, enlisting strong support in business and sometimes residential communities, spearheaded the move for zoning regulations covering storage and sales. Legislators, responsive to these and other interests, pressed for laws to provide for inspection and testing of products.

Although forced by this agitation to give serious consideration to the problem of adulteration, most petroleum refiners and marketers actively opposed any attempt to impose legislative regulations on the trade. As a result laws affecting the sale and distribution of refined products were enacted in only a handful of state legislatures where oil interests were not strong. In 1862, for example, Kentucky banned the sale of illuminating oil with a fire-test of less than 150° F. But with no provision for effective enforcement, the law had little effect except to draw a response from New York dealers. The latter, engaged in combatting a proposed ordinance restricting storage to refined products with a fire-point of 120° F or higher, stated that the application of the Kentucky standard would wipe them out of business entirely.[13] Ohio, where the oil interests were stronger, enacted a moderate statute in 1862 establishing a minimum fire test of 100° F for illuminating oil sold in the state. A year later, Indiana adopted the same minimum standard, adding stiff penalties for violations, including fines of $100 to $500 or six months in jail and machinery to set up an inspection system, financed by fees of 5¢ a barrel. Apathy on the part of state officials, however, kept the Indiana regulations from ever going into effect.[14]

An enactment by the Pennsylvania General Assembly in 1865 made it illegal to store more than one barrel of crude or 25 barrels of refined anywhere in the city of Philadelphia, except in zoned areas where refining and storage were permitted under supervision of the fire marshal. In the same year the New York General Assembly enacted a similar statute designed to strengthen local zoning ordinances in New York City.[15]

Effective attempts to regulate the petroleum trade during the first half of the decade were thus confined to a few municipal zoning ordinances affecting its storage. Despite considerable agitation for inspecting and testing the quality of refined products by public authorities, this method of protecting the public from unsafe oils made no advance beyond setting a preliminary pattern for more effective future legislation.

An expansion in the output of refineries and the growing impersonalization of the market also brought with them a necessity for establishing accurate nomenclature, standards, and grades in order to facilitate trade. Quotations for petroleum illuminating oil began in September, 1860, and although the 40-gallon barrel and gauging practices of the coal-oil industry were adopted automatically, quantities were initially so limited that all quotations were in gallons. There was no differentiation by grade, and prices throughout the rest of the year simply followed those of standard-grade coal oil. The term "refined petroleum," however, was adopted and, unless specifically qualified, was used for a decade or more to designate illuminating oil.[16]

The advent of flowing wells added new reasons in the fall of 1861 for continuing to quote prices in gallons, since the shortage of barrels was so acute that all sizes and types were pressed into service. Because of this variation all quotations were also made subject to the addition of the costs of packages. It was not until November, 1862, that the barrel shortage had eased sufficiently for the trade to adopt a resolution that thereafter all quotations were to include the cost of packages.[17]

Growing difficulties with establishing satisfactory standards to measure the quality and uniformity of petroleum illuminants in the important New York market were reflected in several ways during 1862. First, because of variations in quality all sales were by warrant or sample. Second, only two grades were quoted, which led to confusing overlaps in designations. The first grade, for example, was variously known as "white," "prime white," or "water white"; the second grade was known as "colored," "dark," "yellow," or "straw." Other deviations included "city refined," and "Pittsburgh refined," a carry-over from the coal-oil practice of designating a premium product locally refined, and "western refined," a term applied to the output of small refineries in the Oil Region and other inland centers which was usually priced below any of the second-grade oils.[18]

The imposition of the 10¢ tax on refined illuminants September 1, 1862, introduced two market classifications of considerable importance until the repeal of the tax in July, 1868. "Refined free" designated illuminating oil for sale in domestic markets upon which the refiner had paid the domestic tax. "Refined in bond" referred to illuminants for tax-free export sale, which was held in storage under bond until shipped abroad.

The leading channels through which petroleum was marketed for some two decades were already well established by the early 1860's. In the populous Middle Atlantic and New England states, and to a considerable extent elsewhere, the largest volume of illuminants was distributed by large wholesale grocers to their retail customers. In New England and New York the trade was commonly divided between country and city grocers, each allied to a different group of wholesalers.[19]

Next in importance were the wholesale druggists, whose prominence in the oil trade was exemplified by the pioneer entry of the Schieffelin Brothers of New York into petroleum marketing and refining in 1860. Along the seaboard in the southern states and in certain localities in the Midwest, druggists were more in evidence than grocers as wholesalers and retailers of illuminants.[20] Almost as distinct and important a group in many localities were wholesalers and retailers of oils, paints, and varnishes. Smaller volumes were also sold at wholesale, jobbing, and retail levels by chandlers, lamp houses, and various types of specialized and general oil dealers.

In some instances large wholesale grocers and drug distributors bought oil directly from refineries or refinery agents.* More commonly, they and other wholesale distributors obtained their oil from firms that by the mid-1860's were specializing in the petroleum trade. While much of their business was done on a commission basis, it was unusual for these houses to confine their operations to brokerage services; they frequently took title to the product and in many instances sold on consignment to customers. With some exceptions, the larger houses also dealt in both crude and refined oil. In large part this was to maintain close relations with refiners who preferred to sell refinery output to firms that could in turn supply them with crude during periods of scarcity.† From the outset these large middlemen specialists probably transacted a greater volume of business in major eastern markets than was sold directly or through agents by refiners. With the rapid growth in refined output, the need to specialize both in refining and marketing tended to strengthen their position as petroleum distributors by mid-decade.

The year 1863 brought considerable clarification of grades and quality standards in the New York market which, because of its relative importance as a distributing center, set the pattern followed in the other marketing centers. Early in January, all sales by sample stopped permanently when a large volume of foreign orders stimulated so much contracting for future delivery and resale that enough samples could not be made available to meet the volume of transactions.[23]

To meet the emergency, the New York Petroleum Association, an organization of refiners and dealers, met in September, 1863, to establish more definite regulations covering the contract terms for the sale of refined petroleum. Five grades were initially specified: "strictly white," a top premium oil, followed by "white," "light straw," "straw" and "dark straw." ‡ In practice, however, three grades instead of five emerged from the regulations. The standard grade was "light straw to white," with a fire

---

* Following a practice instituted in the coal-oil era, the Aladdin Oil Works of Pittsburgh was represented during the 1860's by Robert Chesebrough in New York and Gapen, Sherman & Company in Boston.[21]

† A temporary incentive to deal both in crude and refined came from foreign buyers during the early years who looked to dealers in crude for their supplies of refined.[22]

‡ As noted by Samuel Downer, the emphasis on color as a measure of quality was somewhat misplaced. As he pointed out, it put too little stress on the burning qualities of the oils and opened the doors to adulteration with cheap naphtha.[24]

test (ignition point) of 100° F. The premium grade was "strictly white," with a fire test of 115° F or higher, which usually sold at 1½–2 cents more a gallon than standard. "Straw to dark straw" was used to designate the third grade. In January, 1867, "standard white" with a fire test of about 110° F became the standard for all transactions, unless otherwise specified in the contract, throughout the rest of the decade and beyond, while "water white," with a fire test of 115° F, became the premium grade.[25]

The development of these regulations and standards was a major step in the evolution of an effective organization for the distribution of petroleum products. It provided the basis for marketing specialists to serve as the link between refiners and customers far removed from refining centers. This evolution was of primary importance both for supplying the export trade and the widening domestic markets.

Early distribution of petroleum illuminants in the American market was largely concentrated in New England and the Middle Atlantic states, with an extension into the Middle West as far as the frontier states of Iowa, Nebraska, and Kansas. Throughout most of the war the Ohio River marked the southern boundaries of distribution except for the border states Kentucky, Tennessee, and Missouri. Dealers in California received shipments from Philadelphia and New York by boat "around the Horn."

Nearly everywhere the distribution of petroleum kerosene proceeded fastest in large cities where the demand for artificial light was the strongest, costs were lowest, and established outlets most plentiful. In New England and other seaboard states with dense rural populations and short distances between settlements, however, there was no appreciable lag in rural distribution. It was from observations in this area, for example, that Dr. John Draper, distinguished New York City chemist, noted as early as 1864,

> Kerosene has, in one sense, increased the length of life among the agricultural population. Those who, on account of the dearness or inefficiency of whale oil, were accustomed to go to bed soon after sunset and spend almost half their time in sleep, now occupy a portion of the night in reading and other amusements; and this is more particularly true of the winter seasons.[26]

For various reasons, distribution of petroleum illuminants was slow to develop west of the seaboard states. One retarding influence was the wartime disruption of traffic on the Ohio and Mississippi rivers. A second was the higher costs of distribution in large, sparsely settled areas. A third factor was the quality of the refined oil distributed in western markets, which for some time were supplied with low-grade "western refined" from refineries in or near the Oil Region and Cleveland.

In 1863, the recorded shipments westward from Pittsburgh totaled only 32,365 barrels, of which 16,968 barrels were refined. All were barged to Cincinnati, a small refining center and a secondary distributing center for Chicago and other markets. As late as 1865 shipments to Cincinnati totaled only 100,870 barrels.[27] Although not recorded, there also undoubtedly was some distribution in these years from Pittsburgh to Chicago, St. Louis, and other cities with direct-rail connections, but it was not extensive. Apparently the first kerosene was introduced into the Chicago market in the fall of 1862.

Since the Oil Region and Cleveland lacked a strong group of middlemen dealers in refined, the initiative for developing western markets necessarily fell upon small refiners with limited resources. Refiners or their employees personally took to the road to canvass small wholesale and retail dealers along the byways. One employee of an early Cleveland refiner described this as a difficult and expensive procedure. Not only did the cost of travel frequently exceed the value of scattered, small orders obtained, but sudden changes in prices of crude and refined unknown to the man on the road not infrequently resulted in sales at a loss.

With the completion of the Atlantic & Great Western rail connection with the Region in November, 1863, Cleveland became the major center serving the Midwestern markets. By 1865, it sold 129,195 barrels of refined domestically while exporting 24,805 barrels. Though having direct-rail connections with Detroit and Chicago, it still placed a heavy reliance on water shipments via the Great Lakes. Of a total of 68,000 barrels shipped to lake ports in 1865, 58,753 barrels went to western ports: 38,324 barrels to Detroit; 10,294 to Chicago; 2,097 barrels to Milwaukee; and 207 barrels to Marquette.[28]

It is doubtful whether the combined distribution from Cleveland and Cincinnati to the areas west of Ohio in 1865 was much in excess of 100,000 barrels, or about one-seventh of the total domestic sales in that year. Yet even in this relatively retarded market, the growth in kerosene consumption was impressive and gave evidence of an important potential demand in rural areas as well as in cities.

It was against this background that domestic sales of illuminating oil climbed from an estimated 240,000 barrels in 1862 to 633,000 barrels in 1866, the first complete postwar year. (See Table 13:1.) A lack of significant alternative supplies of illuminants was unquestionably the most important factor contributing to this expansion. Yet consumption rose in spite of the imposition of a tax of 10¢ a gallon on domestic consumption of kerosene in 1862 (raised to 20¢ in April, 1864), which contributed to an

TABLE 13:1

Distribution of refined output: 1862–66
(thousands of barrels, calendar-year basis)

Year	Total refined output	Domestic consumption	Exported	Per cent exported
1862	335	240	95	28
1863	855	350	505	59
1864	1,266	722	544	43
1865	1,336	727	609	46
1866	2,049	633	1,416	69

Source: Appendix A.

increase in domestic wholesale prices (in currency) from about 36¢ a gallon in 1862 to nearly 75¢ in 1865.* See Table 13:3. With this record of sales during wartime, members of the industry had good reason to assume that domestic distribution of their products would continue to expand in the post-war years. Prospects for export sales looked even better.

Foreign markets and marketing to the mid-1860's

There were good reasons for the early interest of consumers outside the United States in petroleum. This was particularly true of Europe where, with the exception of the British Isles, there had been an acute shortage of fats and oils for a generation or more. Even common tallow candles were expensive and rare in ordinary households. Rapeseed was grown extensively in France and the Low Countries, but the crop was only large enough to supply a few industries, lighthouses, and homes of the wealthy with lubricants or illuminants. In large part it was this scarcity of oils that led Selligue and others during the 1840's to exploit schists and other mineral bitumens which, by British and subsequent American standards, were too poor in quality for commercial development.

* Expressed in index numbers (1862–100), the wholesale price of kerosene rose from 100 in 1862 to about 210 in 1865, whereas the index of general wholesale prices in the United States increased from 100 to approximately 177 during this same period.[29]

Although they represented a notable technological accomplishment, the mineral-oil industries in France, Switzerland, and several German states remained small and did little to relieve shortages. At the same time they were vigilant in their search for better raw materials. Thus when deposits of Rangoon petroleum in Burma became available for exports in the 1850's, the continental refiners competed actively, but unsuccessfully, with British interests for the concession. It remained for the Price Candle Company in England to begin regular commercial refining of illuminants, naphthas, paraffin, and lubricants from Rangoon petroleum in 1858. It was also during the late 1850's that James Young in Scotland, the world's largest manufacturer of coal oil, carried out a highly successful invasion of markets on the Continent with his recently developed paraffin illuminating oils and lamps. Although charging premium prices, Young exposed a potential demand for mineral illuminating oils in Europe that small local refiners, handicapped by inferior raw materials and high production costs, had only begun to exploit.

These general characteristics of the European market were common knowledge to many individuals both in Europe and in America. They were not surprised when a prominent French chemist, A. Gelee, after analyzing a sample of crude oil from the Drake well early in 1860, predicted a world-wide revolution in illumination and lubrication if the new substance could be had in quantity.[*] Interest in petroleum spread quickly to other parts of Europe, and during 1860 at least 40 casks of crude oil were shipped from the Oil Region to Britain and the Continent for chemical analysis and trial distillation.[30]

Certain of the features that were to characterize the foreign trade in petroleum over the succeeding decades began to emerge during 1861, when American exports of crude and illuminating oil totaled some 37,000 barrels. These shipments were distributed among 32 different ports scattered throughout the world including China and Latin America. But with the exception of Australia, which took 4,209 barrels, all but a small fraction of the shipments went to European ports. Over half the total was received at Liverpool, London, and Glasgow, the remainder going to Spain, France, the Low Countries, and Germany.[31]

It was in 1861 that the first shipload of petroleum crossed the Atlantic, a notable achievement in view of the dubious acceptance of petroleum as

[*] Captain Charles Townsend, brother of James Townsend, president of the Seneca Oil Company which owned the Drake well, left the sample at Havre late in 1859 for delivery to Gelee.

cargo. In November of that year, the 224-ton sailing ship, *Elizabeth Watts*, was chartered by Peter Wright & Sons of Philadelphia, for a shipment of some 3,000 barrels of crude to London. It took two weeks to load the cargo, while the shipper obtained a crew only by getting the men drunk and practically shanghaiing them.[32] The vessel's successful voyage, despite these difficulties, marked the beginning of a regular traffic in full-cargo shipments.

Despite many unresolved wartime uncertainties, American exports of petroleum increased during 1862 to approximately 277,000 barrels, of which about 95,000 barrels were illuminating oil. (See Table 13:2.) Although the number of shipping destinations rose to 47, located in almost every part of the world, at least four-fifths of these shipments went to Europe, where Britain remained the leading importer by purchasing more than half the total.[33] Considering the incomplete reorganization of the domestic refining industry at this time, the fact that about one-third of the total exports were illuminating oil was encouraging to American refiners.[34]

In large part this expansion of exports was in response to low prices of crude that averaged, in the Region, about $1.04 a barrel in 1862. There were, however, several other influences that served to stimulate petroleum exports at this time. One was the active role undertaken by American consuls in promoting petroleum illuminants. Dr. A. W. Crawford, for example, spent some months in Pittsburgh familiarizing himself with the qualities of petroleum before taking up his new assignment as consul at Antwerp in September, 1861. Failing at the outset to impress local merchants, he imported kerosene and a quantity of lamps at his own expense. By demonstration and the distribution of samples, he helped stimulate interest among marketers that soon moved Antwerp into the front ranks of importers of kerosene.[35] At the request of New York merchants, A. J. Stevens, consul to Leghorn, Italy, where petroleum was virtually unknown, performed a similar service in 1862, as did W. W. Murphy at Frankfort. These activities set a pattern that was subsequently followed on behalf of the new product by American consuls wherever they were stationed.

American lamp manufacturers also provided an important stimulus to foreign sales. Already following mass-production methods, they began to export large quantities of burners priced as low as 25¢ each, or $2.00 a dozen.[36] In 1863, the *London Ironmonger* reported the large British industry had been overwhelmed by the immense shipments of cheap American lamps pouring not only into England, but into France, Belgium, Germany, Holland and other countries on the Continent as well.[37] One dealer in Hamburg, for example, reported the importation of American lamps worth $15,000 in 1862 and $200,000 in 1863.[38] Even though Euro-

TABLE 13:2

Exports of crude and refined: 1862–66
(thousands of barrels, calendar-year basis)

Year	Total exports	Crude exports	Refined exports	Per cent refined
1862	277	182	95	34
1863	707	202	505	71
1864	777	233	544	70
1865	746	137	609	82
1866	1,686	270	1,416	84

Source: Appendix A.

pean manufacturers began to expand their production of petroleum burning lamps, American exports still totaled an impressive $439,320 in 1864 and $384,898 in 1865.[39]

Exports in 1862, however, as shown in Table 13:2, were but a preview of an expansion that lifted shipments abroad to 707,000 barrels in 1863 and to a level of almost 1.7 million barrels in 1866.

Cheap lamps and the sales promotion by American consuls played a part in this expansion, but they were only two of a number of favorable influences. Perhaps the decisive factor in building up foreign trade in petroleum during the war years was the combination of American currency inflation and the exemption of exports from the domestic tax on refined oils. Some measure of the extent to which this combination helped subsidize exports during this period is indicated in Table 13:3. In 1862, for example, the differential between the gold price per gallon of kerosene (paid by foreigners) and the domestic currency price was about 4¢. A year later the differential between the gold price and domestic prices (including the tax) was nearly 20¢. In 1864, the differential was over 42¢; in 1865, about 34¢; and in 1866 almost 32¢.

An important secondary effect of these benefits is also disclosed. The value of refined exports per gallon ran considerably higher than crude. Since favorable rates of exchange yielded an advantage roughly proportionate to the value of the products, they tended to favor exports of refined rather than crude.

Aside from these considerations, there were numerous other circumstances that encouraged export sales. Favorable export prices in terms of currency alone caused many refiners and marketers to sell abroad. A termination of their responsibility for further handling upon delivery at dockside furthered this preference, as did the location of the biggest domestic markets at the major export ports. Moreover, because of the va-

TABLE 13:3

Average wholesale prices for refined petroleum at New York: 1862–73 (cents per gallon)

Year	Export		Domestic	
	Currency (tax free)	Gold (tax free)	Currency (tax free)	Currency (taxed) *
1862	36.36	32.11	36.36	
1863	44.75	31.10		51.74
1864	65.00	32.00		74.61
1865	58.75	37.40		71.88
1866	42.50	30.08		62.00
1867	28.38	20.60		44.00
1868	29.50	21.16	29.50 †	41.00 †
1869	32.75	24.59	32.75	
1870	26.38	22.96	26.38	
1871	24.25	21.69	24.25	
1872	23.63	20.97	23.63	
1873	17.88	15.71	17.88	

Source: Folger, *Pa. Industrial Statistics,* 1892, B199–B204; Wesley C. Mitchell, *Gold, Prices, and Wages, Under the Greenback Standard* (Berkeley, Calif.: Univ. of Calif. Press, 1908), 6; S. S. Hayes, *Report . . . on Petroleum,* 24; *Report . . . on U. S. Commerce and Navigation for the Fiscal Year Ending June 30,* 1873, Appendix B xx–xxxiv
* From Sept. 1, 1862 to April 1, 1864 the tax was 10¢ per gallon; from April 1, 1864 to April 1, 1868, 20¢; from April 1 to July 1, 1868, 10¢.
† The average price for the first half of the year was 41¢. The average price for the year as a whole was 29½¢.

garies of ocean transport, foreign buyers tended to place a large portion of their orders far in advance of delivery, which helped reduce the risks of market changes for both refiners and marketers. Finally, foreign sales made it possible to avoid unfair competition with refiners and marketers who evaded all or part of the domestic tax.[40]

Two trends characteristic of the American export trade in petroleum for several decades to come had already emerged by the end of 1866. One was the predominant attraction of Europe as a market for American petroleum. The second was the increasing importance of exports of refined oil relative to crude.

Although American illuminating oil was being shipped to ports in Latin

America, Asia, Australia, and Africa, by early 1863 the predominance of the European market both for illuminating oil and crude was already firmly established. From mid-1863 to mid-1866, Europe's share of total American exports of illuminating oil ranged between 80–85 per cent, while its share of crude exports was about 94 per cent. (See Table 13:4.)

In large part the attraction of the European market was a reflection of that continent's advanced stage of economic development. A need for artificial illumination and familiarity with lamps, particularly in the more industrialized and urbanized areas of Britain and western Europe, made customers generally receptive to the use of petroleum illuminants. It was individuals in the industrial areas also who were attracted by the possibilities of establishing domestic refining capacity for processing imported crude.

These influences were reflected in the distribution of petroleum exports to the various countries in Europe. Britain, France, Belgium, Germany and the Netherlands, roughly in that order, accounted for all but a small proportion of the crude and refined oil shipped to Europe between 1863 and the end of 1866. (See Table 13:4.) The shift in the relative proportions of crude and refined-oil exports was closely associated with a determined, but only partially successful effort to establish petroleum-refining capacity in Europe, more specifically in England and France.

In France pressure to maintain or extend domestic refining capacity stemmed primarily from manufacturing and chemical interests, including French refiners who were quick to respond to the possibilities of using crude as a raw material. Early in 1863, three of the largest refineries, located near the port of Marseilles, had already begun to process crude imported from America.[41] The subsequent growth and survival of French petroleum refining, however, owed much to the import duties on refined oils of all types. Depending on whether the oils were shipped by foreign or French vessels, the duties varied from about 1.4¢ to 2.3¢ a gallon.[42] As shown in Table 13:4, the tariff did not exclude the importation of illuminating oil into France. But the continued imports of crude in excess of refined through 1866 suggests that French refiners enjoyed a considerable degree of protection.

In the British Isles there also was the strongest kind of pressure to develop petroleum refining capacity for both home consumption and export. Unlike France, none of this agitation stemmed from a desire on the part of coal and shale-oil refiners to convert to American crude. James Young, who virtually monopolized the British mineral-oil industry, for example, was confident of his ability to compete on a basis of quality with

TABLE 13:4

Exports of crude and illuminating oil to selected European countries: 1864-66 (40-gallon barrels)

Year ending June 30	Total world	Total Europe	% Europe	England	Germany	France	Belgium	Netherlands	Russia	Other Europe
1864 Crude	249,516	237,104	95	109,598	20,171	56,832	18,114	25,386	4,934	2,069
Illuminating Oil	291,189	238,433	82	88,167	15,864	34,169	56,004	12,853	8,667	22,709
1865 Crude	307,347	287,404	94	101,280	27,868	97,432	32,152	12,032	4,066	12,574
Illuminating Oil	292,363	233,017	80	76,631	26,877	41,770	63,927	7,551	4,536	11,725
1866 Crude	401,448	372,099	93	128,364	22,633	140,548	35,968	2,049	22,532	20,005
Illuminating Oil	837,747	711,540	85	230,853	110,907	44,328	182,206	18,521	40,677	84,048

Source: Folger, Pa. Industrial Statistics, 1892, 170B-172B; Reports . . . on U.S. Commerce and Navigation for the Years 1864-1866.

petroleum burning oils, whether imported from America or refined locally.

Unlike the French, British refiners were not favored by tariff protection. Britain, beginning in May, 1860, did impose a duty of about 2¢ a gallon on both crude and refined oils. But contrary to the interpretation by some historians, this duty was not a discriminatory tax against petroleum in favor of the domestic coal-oil industry.[43] It was a part of a revenue measure that applied to all bulk commodities and was repealed in June, 1863.[44]

Leadership in promoting petroleum-refining capacity in England was centered in a group of British industrialists and financiers, including the prominent London financial house of Peto & Betts. In 1861 this group took over financial control of the Atlantic & Great Western Railroad in the United States and almost immediately began extending its connections with the Oil Region.[45] The acquisition of the Atlantic & Great Western was, however, only part of a larger plan to link Pennsylvania crude production with refining capacity in Britain in order to make that country a major supplier of refined petroleum throughout Europe.

This plan was outlined in the prospectus of the Petroleum Trading Company, organized early in 1863 by Peto & Betts.[46] To insure adequate supplies of raw material for its refining capacity, the firm proposed to revolutionize the ocean transport of crude. Evaporation losses, freight, and insurance rates, and related costs were needlessly high, it was argued, only because of carriage in barrels in ships unsuited to the cargo. To solve this problem, the firm arranged for the construction of several iron ships, of about 400 tons each, fitted with compartments designed to carry crude oil in bulk.

Three vessels of this type, the *Ramsey, Atlantic,* and *Great Western,* began the bulk carriage of crude oil during 1863. In contrast to a number of later experiments these ships were technologically successful, although they by no means eliminated the hazards of fire and explosion that plagued subsequent attempts to ship in bulk. It was, in part, this problem that led to their diversion after a few voyages to traffic which was less dangerous and which, unlike petroleum, permitted pay-loads on return voyages.[47]

As shown in Table 13:4, the failure to introduce bulk shipments had no immediate impact on British imports of crude petroleum, which exceeded refined oil by considerable margins during 1864 and 1865. But in 1866, the balance shifted when crude imports were little more than half the volume of refined receipts. A year later they were less than one-tenth.

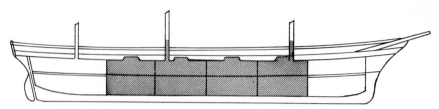

FIGURE 13:3. *Diagram of ocean going tanker* Atlantic *showing location of storage tanks and expansion cylinders.*

This change reflected the virtual collapse of the British petroleum refining capacity. In part this decline was the result of domestic competition. In 1864, James Young's patents covering the refining of mineral oils expired. Almost at the same time, there was a shift in the source of raw materials used, from Boghead coal to newly discovered and extensive deposits of cheap shales. The result was a boom in the domestic coal-oil industry that brought 38 new entries into refining in 1864 and 120 in 1865, and an expansion in output to 500,000 barrels annually.[48] Another factor that hastened the decline of the remaining British petroleum refining capacity was the failure of Peto & Betts in the financial panic that occurred in 1866.[49] Basically, however, the collapse of the British venture made clear the difficulties of establishing an extensive petroleum-refining industry abroad without the imposition of discriminatory duties. In the absence of an established system of maritime bulk shipments, the problems of ocean transport of crude in barrels, with accompanying heavy losses through leakage and hazards of fire, proved insurmountable.

In the absence of maritime bulk shipments, petroleum continued to move into foreign trade in slow, wooden sailing ships with modest carrying capacities. All crude cargoes were shipped in barrels or casks which took a long time to load and unload. Most of the refined was also shipped in barrels, but it was already standard practice to package illuminating oil in 5-gallon tins, two to a wooden case, particularly when routes or destinations involved exposure to tropical heats.

At the same time considerable progress had been made by the mid-1860's in the organization of ocean traffic in petroleum. A growing fleet of sailing vessels engaged in the trade relieved the scarcity of shipping space. There was also a marked improvement in scheduling shipments and in storage and handling facilities at major harbors like London and Liverpool, also Antwerp and Amsterdam. These changes were reflected in declining freight and insurance rates. Freight rates from Philadelphia to London, for example, declined from 8*s*. per barrel in 1863 (plus 5 per cent primage) to between 5*s*./6d. to 6*s*./3d. in 1865. British insurance companies during this period also reduced their rates on petroleum cargoes from America from £7/7*s*. to £2/10*s*.[50]

Along with improvements in the organization of transport there was an equally important development of marketing channels serving the export trade. Much of the initial promotion of petroleum exports had been undertaken by domestic refiner-marketers, such as Samuel Downer of Boston and Charles Lockhart of Pittsburgh or marketing firms like Horatio Eagle & Company of New York. But by the mid-1860's this responsibility had largely been taken over by foreign marketing firms, many of them with long experience in the trade in sperm, whale, lard, and other oils.

In some instances foreign buyers sent one of their own employees or a partner to the United States to act as their purchasing agent. More commonly they bought through the New York brokerage or wholesale firms, which emerged as specialists in handling foreign orders. These firms not only purchased the oil for their customers and had it inspected, but also arranged for shipping space and insurance coverage.[51]

The buying power of foreign purchasers was sufficient for them to bring pressure on refiners to deliver illuminating oil with a fire test of 115° F that became standard for the export trade in 1863. In 1867, exporters, under pressure from their customers, also adopted improved specifications for barrels and standardization of sizes within the range of 40–46 gallons.[52]

It was against this background that exports of petroleum continued to grow, relative to domestic sales, throughout the first half of the 1860's: by 1866 the predominance of the foreign markets was clearly established. This fact made the petroleum industry unique among American manufactures. In addition, its products ranked first among all exports of manufactured goods. In helping to fill the void in foreign trade left by the elimination of cotton and turpentine, petroleum exports strengthened the nation's balance of trade during the war years. In addition, the industry yielded nearly $7 million in tax revenues.[53]

Postwar foreign markets and marketing

Export sales of petroleum showed no signs of abating in the postwar years. As indicated in the following tabulation, they continued upward in an unbroken trend until a dip in 1872, occasioned by the Franco-Prussian

War, reaching a total of nearly 5.4 million barrels in 1873. The high ratio of refined to crude petroleum reached by the mid-1860's was also maintained.

TABLE 13:5

Exports of crude and refined: 1867–73 (thousands of barrels, calendar-year basis)

Year	Total exports	Crude exports	Refined exports	% Refined
1867	1,677	132	1,545	92
1868	2,473	194	2,279	92
1869	2,537	360	2,177	86
1870	2,872	289	2,583	90
1871	3,389	269	3,120	92
1872	3,210	390	2,820	88
1873	5,368	468	4,900	91

Source: Appendix A.

The disappearance of several of the earlier stimulants to foreign markets made this export performance all the more reassuring to the American industry. In April, 1868, the domestic duty on refined products was cut from 20¢ to 10¢ a gallon, and eliminated entirely two months later. Even more significant was the fading of the margin between the gold and currency prices as the domestic price level began to decline after the war. By 1869, the differential was about 8¢ a gallon; by 1873 it was a little over 2 cents. (See Table 13:3.)

There were, however, other factors that served to stimulate export sales. One was the drop in export prices for refined products, which after 1866 ranged between 20¢ and 25¢ a gallon through 1872. Another was the clear superiority of kerosene to alternative illuminants in the foreign markets plus a growing familiarity with the product, particularly in Europe.

Nothing seemed to retard demand, although the unsafe quality of much of the refined petroleum shipped abroad did act as a deterrent. American experience with the safety problem, for example, had its counterpart in England. As early as 1866 British scientists, including some who had formerly been leading proponents of petroleum, began agitation for regulatory legislation.[54] Reports by insurance companies and public agencies on fire casualties verified the mounting losses from unsafe kerosene. Several major disasters in transporting petroleum on land and by water in 1868 prompted steamship lines to ban petroleum cargoes on mail packets and several railroads to refuse to accept all cargoes of crude and refined.[55]

Parliament enacted the Petroleum Act of 1868, which ostensibly tightened storage regulations, provided a licensing system to control storage, established 100° F as the minimum flash-point of most oil sold in commerce, and laid down specifications for testing apparatus and procedure. But imprecise and incomplete definitions of standards and enforcement methods left so many loopholes that the legislation had little effect.[56] This was also largely true of the Act of 1871, which did little more than consolidate previous enactments.[57] These legislative attempts did, however, increase public awareness of safety problems and helped to stimulate the development of better testing equipment.

Unsafe kerosene also caused some concern at many points on the Continent, despite excellent regulation of sales and storage at Paris, Brussels, and several other cities and ports. Numerous markets on the Continent with a flash-point requirement of only 90° F or less regularly served as the dumping grounds for British merchants with inferior products that could not meet the 100°-F test standard in home markets.[58]

As shown in Table 13:6, Europe continued to be the main market for American petroleum, absorbing 90 per cent or more of refined exports after 1868, and almost all of the crude. Within this broad pattern there were some significant shifts in distribution. The permanent collapse of the English petroleum refining industry is plainly revealed in the negligible amount of crude shipped to the British Isles after 1866. British imports of crude over the following years were limited to the relatively modest needs of its chemical industry.

Britain continued to be a major importer of refined products, even though it was eclipsed as the number one foreign market after 1867 by Belgium and the Germanies. To some extent the relative decline of the British imports of refined was conditioned by competition from the domestic shale-oil industry.[59]

In 1867, France superseded Great Britain as the world's largest importer of crude. The continuation of the modest duties from 1.4¢ to 2.3¢ a gallon on kerosene imports while crude imports were duty-free was important in enabling French petroleum refiners to continue their operations.[60] These duties did not, however, shut out refined products, as evidenced by French imports from 1866 through 1869. In the first two years, refined imports actually exceeded those of crude; while in 1869 there was a differential of only about one-third in favor of crude imports.

The events that put French petroleum refiners in an almost impregnable position came as an aftermath of the Franco-Prussian War. In a wave of nationalism following the end of the war, the French National Assembly in 1871 enacted a new scale of duties that fell lightly on crude but which practically excluded imports of petroleum illuminants. This schedule amounted to duties of about 1.25¢ a gallon on crude petroleum, 14.8¢ on refined kerosene, and 18.5¢ on benzine and naphthas.[61] Early formation of the *Cartel des Dix*, a monopoly which refined 80 per cent of all petroleum sold in France through 1903, completed the creation of a protected domestic-refining industry.[62]

The effect of the new duties was to produce a radical readjustment in the ratio of crude and refined imports. For the fiscal year ending June 30, 1872, as shown in Table 13:6, crude imports soared to a record 233,000 barrels, as refined dropped to 53,000 barrels. A year later the volume of crude imported exceeded 300,000 barrels, while kerosene imports were reduced to a mere trickle of less than 18,000 barrels.

Belgium's position as the second most important market for refined petroleum after 1866 reflected the mushrooming of markets in the Lowlands and in many of the German States which were supplied through Antwerp. After 1868, crude imports into Antwerp were exceeded only by shipments to France and the combined German ports. A part of the crude went to a cluster of twenty small refineries in the vicinity of Antwerp; the remainder to small refineries in nearby German states.

Imports via Belgium, however, were only a small part of expanding shipments that lifted the German states after 1867 to the position of the number one importer of American-refined oil. The major factor behind this expansion was the growing industrialization of the German economy during the latter half of the nineteenth century. Accompanying this growth was an increasing urbanization and a corresponding increase in demand for artificial illumination. With animal and vegetable oils in short supply, German consumers were thus increasingly attracted to petroleum illuminants and German marketers were especially active in promoting the trade.

Toward the close of the decade there was also, for the first time, a significant expansion in the distribution of kerosene into European areas that were predominantly rural, deficient in communication and transportation and removed from the main centers of trade and commerce. Typical of

TABLE 13:6

Exports of crude and illuminating oil to selected European countries: 1867–73 (40-gallon barrels)

Year ending June 30	Total world	Total Europe	% Europe	England	Germany	France	Belgium	Netherlands	Russia	Other Europe
1867 Crude	183,606	176,357	96	35,142	5,796	113,982	14,794	1,562	..	5,081
Illuminating Oil	1,553,137	1,382,663	89	428,917	278,507	115,979	316,709	73,025	21,690	147,836
1868 Crude	250,741	248,579	99	60,529	22,720	96,930	49,346	3,439	..	15,615
Illuminating Oil	1,680,333	1,479,299	88	202,653	397,426	119,327	277,341	105,018	52,454	325,080
1869 Crude	322,612	321,804	99.7	9,550	71,065	154,022	62,526	9,355	2,725	12,561
Illuminating Oil	2,110,086	1,951,901	93	276,002	537,521	106,550	323,302	116,816	97,356	494,354
1870 Crude	248,876	248,769	99.9	4,714	25,442	170,697	36,218	..	..	11,698
Illuminating Oil	2,447,561	2,212,112	90	209,653	781,028	73,127	375,575	174,415	50,585	547,729
1871 Crude	236,520	228,381	97	28,197	30,459	88,178	59,376	3,248	2,631	16,292
Illuminating Oil	3,171,144	2,885,950	91	378,617	859,542	56,931	434,615	197,259	179,052	779,934
1872 Crude	338,994	330,594	98	5,435	40,011	233,197	48,032	..	..	3,919
Illuminating Oil	3,063,488	2,759,951	90	252,203	876,395	53,048	409,082	253,837	133,193	782,193
1873 Crude	460,985	441,698	96	16,168	71,188	308,008	26,691	..	14,253	5,390
Illuminating Oil	3,952,560	3,538,570	90	393,529	1,302,843	17,629	565,414	241,904	185,097	832,154

Source: Folger, Pa. Industrial Statistics, 1892, 170B-172B; Reports . . . on U.S. Commerce and Navigation for the Years 1867–1873.

such areas were Italy and Spain, which imported only small amounts of kerosene before 1868, but whose combined annual shipments totaled approximately 200,000 barrels by 1871. This movement, despite long winters of curtailed daylight, did not extend to the Scandinavian countries which remained the most retarded markets on the continent for kerosene lamp light. Imports of less than 8,000 barrels in 1867 and 1868 were followed by two years in which there were no imports.[63] It was not until the mid-1870's that Scandinavian imports began to approximate the volumes of such secondary markets as Italy and Spain.

Considering the vast distances and difficult transportation routes involved, perhaps the performance of kerosene demand in Russia outranks that of any economically backward area. As shown in Table 13:6, imports of refined in 1866, only four years after the first sample shipment reached Odessa, exceeded 40,000 barrels. Annual imports of a little over 50,000 barrels in 1868 and 1870 were punctuated by one of 97,000 barrels in 1869. When Russian imports climbed to 179,000 barrels in 1871, and 185,000 barrels in 1872, one of the most important secondary markets in Europe had been established.

Not revealed by these data were developments within Russia that had important implications for the future of the American petroleum industry. Since early in the century the petroleum deposits at Baku had been worked by the Meerzoeff family of merchants under a monopoly grant from the crown. None of the succeeding generations of Meerzoeff's had taken any effective measures to develop production, being content to sell small amounts of crude for delivery to nearby Middle-Eastern markets. Refining was begun on a small scale in 1858 but as late as 1873 output remained small.[64]

There was, however, a growing Russian interest in native petroleum, not as an illuminant, but as a substitute for coal. Most of the coal used in Russia had to be imported from England at almost prohibitive costs, while in the Caspian Sea District and over vast stretches along the Volga, wood was so scarce that cutting was restricted even to obtain industrial fuels. Backed by Russian naval authorities at Baku and operators of merchant marine fleets in the Caspian Sea, several Russian engineers began about 1865 to develop apparatus to burn petroleum under marine boilers. After many experiments and modifications of apparatus, two of the engineers succeeded in developing effective burners. In 1870, the first of these was installed on the steamship *Iran*. An additional half dozen merchant ships were similarly equipped in 1873, and conversions of the entire Caspian merchant marine of 40 vessels to oil burners soon followed.[65]

These successes with oil fuel provided both a market for native petroleum and a means to obtain transportation facilities necessary to develop an industry. While they made no immediate impression on the Russian illuminating oil market, they did lay the groundwork for the development of an important domestic refining industry.

Postwar domestic markets and marketing

Although overshadowed by the flow of refined products into world markets in the postwar years, there was nevertheless an impressive growth in domestic sales from about 770,000 barrels in 1867 to approximately 1.85 million in 1873. (See Table 13:7.)

A major factor in stimulating the domestic demand was the reduction of the domestic duty of 20¢ a gallon to 10¢ a gallon in April, 1868, and its final removal in July of that year. Wholesale prices that had averaged some 41¢ a gallon during the first half of 1868, dropped to less than 33¢ in 1869, averaged about 25¢ during the next three years, and fell to less than 18 cents in 1873. (See Table 13:3.)

Throughout the postwar years, domestic consumption of kerosene continued to be concentrated mainly in cities and towns and in the more heavily populated rural areas in New England and the Middle Atlantic states. Thus, while marketers, following the close of the Civil War immediately began laying plans to extend kerosene distribution to the South and west of the Mississippi, these areas did not make any significant contributions to domestic demand until late in the decade.

Among the factors that tended to delay an expansion of kerosene sales in the newly restored Southern markets were the difficulties associated with Reconstruction and a restoration of the South's rail facilities. There was a further problem of familiarizing many of the inhabitants with kerosene. Prior to the war, kerosene shipments had been extended to major Southern coastal cities on the Atlantic and Gulf, but for most plantation homes camphene was the accepted illuminant, while lamplight in any form had yet to reach the Negro population and many of the whites in the rural interior.

A quick resumption after the war of kerosene shipments by water to the coastal cities served to stimulate sales to a moderate extent. Jobbers and wholesalers at these ports also began pushing distribution to country stores in the interior, which began to multiply after the break-up of the planta- tion system. But it was not until the project of rehabilitating and enlarg- ing the Southern lines by some 8,000 miles neared completion early in the 1870's that the transport barriers were reduced sufficiently to permit mar- keters to expand sales in the interior regions of the South. Consumers were persuaded to accept the new illuminant in part through the efforts of drummers sent out to demonstrate kerosene lamps.

TABLE 13:7

Distribution of refined output: 1867–73
(thousands of barrels, calendar-year basis)

Year	Total refined output	Domestic consumption	Exported	% Exported
1867	2,418	773	1,545	64
1868	3,410	1,131	2,279	67
1869	3,267	1,089	2,177	67
1870	3,875	1,292	2,583	67
1871	4,050	930	3,120	77
1872	4,200	1,380	2,820	67
1873	6,750	1,850	4,900	73

Source: See Appendix A.

Distribution of petroleum illuminants proceeded somewhat more rap- idly in the Midwest with its established markets and more extensive trans- port facilities. Few of the older residents or recent immigrants in this area were unacquainted with the use of kerosene lamps. But as was also true of the South, it was not until the completion of a railway expansion of some 28,000 miles in 1873 that kerosene could be sold in any quantity in an area that remained predominantly rural.

The potentials of both the Southern and Midwestern markets were rec- ognized by a number of domestic marketing firms that began operating during the late 1860's. At Louisville, F. B. Carley in 1869 founded the firm of Chess, Carley & Company and began successfully pressing sales of barreled oil from his small refinery there by rail south into Tennessee, northern Mississippi, Alabama, and Georgia, and western sections of North and South Carolina.[66] Alexander and James McDonald, owners of Alexander McDonald & Company at Cincinnati, also began late in the 1860's to extend distribution beyond local markets into southern Ohio,

Indiana, and Illinois.[67] In St. Louis, Henry Clay Pierce soon began to follow the expansion of railroad lines into the Southwest, developing local
marketing outlets in territory that was still in frontier stages.[68] At Richmond, King & Troy, wholesale druggists who had pioneered coal-oil and
petroleum distribution between 1858 and 1861, resumed business after
the War and soon became the largest oil dealers in Virginia.[69]

What the availability of kerosene meant to western pioneer families is
brought out by Hamlin Garland in his children's classic, *Boy Life on the
Prairie*. Living on a newly "homesteaded" farm in northern Iowa late in
1868, Garland recalls that candles and a square candle-lantern were the
family's only source of artificial light during an unusually severe winter.
During the threshing season in the fall of 1869 a wonderful transformation
took place when the men came into the dining room for their evening
meal and found the table lighted with a kerosene lamp. A year later a
new kerosene lantern brought welcome illumination as the members of
the family performed long hours of chores and cared for the animals in
the darkness of early morning and late evening.[70]

The growing acceptance of kerosene by housewives generally was revealed in a book published in 1869, *American Woman's Home or Principles of Domestic Science*, written by Harriet Beecher Stowe (author of
Uncle Tom's Cabin) and her sister, Catherine Beecher. They pointed
out that while the best illuminant where available was gas, "good kerosene
gives a light which leaves little to be desired." For general use the newly
introduced "student reading lamp" with an Argand-type burner was recommended, but all households should be equipped with small, portable
lamps, and kitchen lamps with broad bottoms to make them less easy to
upset. Above all, housewives were cautioned against the use of poor oils
responsible for "those terrible explosions."[71]

This advice to avoid unsafe oils reflected the problem of kerosene
adulteration that became more acute as the expanding output was accompanied by a proportionate increase in naphthas and a temptation to
add it to illuminants. But beyond a series of state acts banning the use of
kerosene lamps on railroad trains and ships, legislation was little more effective in the postwar years than it had been earlier.[72] But as had also
been true of the war years, there was little to indicate that poor quality
kerosene was a serious handicap to its acceptance by domestic consumers
in the postwar years.

In view of the almost unbroken trend of expansion that marked the distribution of petroleum both at home and abroad since 1861, most industry

members early in the 1870's were convinced that the potential demand for their products was virtually unlimited. And considering the remarkable developments in drilling, refining technology, and transport during the decade, there seemed to be no serious obstacles in the way of continued expansion. But both the speed and nature of these developments had brought stresses within the industry that by the end of 1872 had reached critical proportions.

Part FIVE

Reaction to instability

Chapter 14

Refiners, railroads, and Rockefeller

In the years following the success of the Drake well, the American petroleum industry experienced a remarkable rate of growth. Drawing on the rich heritage of technology and marketing experience of the late 1840's and the 1850's, oilmen developed a complex organization that was supplying a substantial portion of the illuminating and lubricating needs of consumers in the major marketing areas of the world.

In the early 1870's, the petroleum industry was in many respects typical of American businesses at the time. Producers, refiners, and marketers all operated overwhelmingly on an individual or partnership basis. Little capital was required to enter any phase of the industry and competition was vigorous at all levels.

Experience also revealed unusual hazards of engaging in an industry based upon a mineral resource whose discovery and output followed an

irregular and capricious pattern. New oil pools appeared and were depleted with startling rapidity, affecting not only the fortunes of producers,
but those of the rest of the industry as well. Lack of balance between various segments of the industry appeared to be chronic; crude production, refinery capacity and throughput, and market demand were rarely
in equilibrium. First, production would outrun throughput by refineries;
then manufacturing capacity would exceed either current crude production or the rate at which refined products could be absorbed by the market. These more or less continuous maladjustments were reflected in wide
fluctuations in prices of crude and refined products.

Instability in the early 1870's

Early in the 1870's the lack of balance between production and facilities
for shipping, processing, and marketing petroleum was of particular concern to refiners. By the most conservative estimates, total refining capacity during 1871–72 of at least 12 million barrels annually (see Table 12:1)
was more than double refinery receipts of crude which amounted to
5.23 million barrels in 1871 and 5.66 million barrels in 1872.[1] At the same
time, total demand approximated crude production at $4.00 per barrel.
So long as this situation persisted, refiners' hopes for declines in the price
of crude were chimerical. Notwithstanding their charges about "obstinate"
producers refusing to make price concessions, it was clear that the refiners' troubles were due to excess capacity. And with the wave of refinery
suspensions that hit the industry late in 1871, the market painfully began
weeding out the marginal firms.

Not all firms, however, were equally affected by the wave of refinery
suspensions that hit the industry during 1871. Some by virtue of their size
and efficiency at least broke even if they did not operate at a profit. Survival, however, was not solely a function of processing costs. A company's
line of credit also helped determine its staying power and, if need be, its
ability to absorb losses over an extended length of time. Trade-marks or
reputation could further relieve some of the pressure on margins, although they could hardly provide market space for substantial excess refining capacity. A much more readily available prop to crumbling profit
margins was to trim transport costs which could account for as much as
20 per cent of the final product.*

* During the early 1870's, the New York export price of refined kerosene averaged
  around $10.00 per barrel. Open crude rates to Cleveland were approximately 40¢

As the success of Rockefeller demonstrated, this was a device which larger and more strategically located companies found easiest to utilize. There was some indication that the railroads were beginning to dislike the practice. A shipper given one concession was likely to ask for more, and although rebate contracts were secretly concluded, rumors about them led other shippers to seek similar aid. Nonetheless, it was virtually impossible to eliminate personal discrimination. It enabled the roads to compete for special freights without sparking a general rate war and each line was unwilling to risk losing large shipments to competitors by abandoning the practice unilaterally. From time to time, the railroads did agree collectively to abandon such concessions, but the secrecy of the rebate practically vitiated enforcement.

The significance of the railroads to the industry is obvious; but the other side of the coin—the importance of petroleum freight to the trunk lines—is more difficult to assess. On the one hand, much of the trunk lines' energies were absorbed in the struggle for through freight between the seaboard and the mid-west. Up to the mid-1870's, conflict over this traffic was largely a seasonal affair, sparked by the yearly re-opening of the Erie Canal after the winter, and centered on west-bound trade from New York to Buffalo and Chicago.* In competition with water routes, the railroads could not help but compete with one another, particularly since each trunk line was trying to consolidate its route and increase its market share in the long-haul traffic. After rates had been driven sometimes as low as 12¢ per hundredweight, a truce was generally called, agreements were signed, and railroad men declared they had learned the folly of their ways,[3] But as soon as the water routes re-opened, or one of the lines —usually the obstreperous Erie—sought to alter the balance of power, the cycle was repeated.

In this larger context, competition for petroleum was definitely a battle on a secondary, though not unimportant, front. Although oil rates were undoubtedly reduced along with many others during periods of general warfare, the rate wars themselves were touched off by broader considera-

---

per barrel; to New York between $1.50 and $1.90. Refined from Cleveland to the seaboard was carried at $1.25. Assuming 1.4 barrels of crude yielded one barrel of refined, total transport costs on a barrel of kerosene from Cleveland to New York were $1.80; on the crude equivalent from the Regions to New York refineries $2.30.

* This, of course, was not always the case. Sometimes the eastbound grain and live-stock freight were involved. In 1869 and 1870, for example, cattle were carried from Buffalo to New York at $1.00 per head, and wheat between the same two points was shipped at 1½¢ per bushel.[2]

tions, and would have occurred in any case. The opening of a new line to the Regions, the acquisition of a better connection to one of the refining points, the unilateral reduction of petroleum tariffs, occasioned skirmishes in the petroleum trade; but these did not serve as signals for over-all struggles to re-align the balance of power on other freight.

This did not mean, however, that the trunk lines completely subordinated all other considerations to the western trade. Petroleum revenues alone were sufficient to warrant a fair amount of attention by the oil carriers. In 1871 the *New York Commercial Bulletin* claimed that the Pennsylvania's profits on oil were second only to those on livestock; by 1872 the $40 million of petroleum products sent abroad (most of which was carried to ports by rail) ranked fourth among U. S. exports.[4]

Not all of this freight was necessarily profitable; in fact, by the end of 1871 the railroads were claiming that they had been losing money on oil for the past year. They probably overstated the case, but nonetheless there seemed little hope that the decline in oil rates would reverse itself. The Pennsylvania, aligning squarely with New York, declared open warfare on Cleveland, while Cleveland refiners relied on the Erie to support them against Scott; the Lake Shore, completing its branch from Cleveland to Oil City, threatened the Erie's crude traffic to the Ohio refineries; and the Baltimore & Ohio was pushing a branch into Pittsburgh, challenging Scott and practically everyone else. Thus in the final months of 1871, the stage seemed to be set for an outbreak of a rate war that would be costly to all of the oil carriers involved.

There was, however, an alternative, one that the trunk lines had attempted in similar situations involving livestock and passenger traffic. This was to enter into an agreement to pool shipments. Although such attempts had consistently failed, the railroads continued to resort to pools, largely because no other alternative, short of government regulation, was then available to end unprofitable rate wars and risky competitive building. Consequently, it was not surprising that late in 1871, rumors about a railroad pool in the oil trade began to circulate in the Oil Regions. The rumors were soon to be substantiated.

The South Improvement Company

Apparently the idea of forming a pool at this time came from Tom Scott of the Pennsylvania Railroad. He in turn enlisted the enthusiastic support of Peter H. Watson, associated with the Lake Shore–New York Central lines, who took over active promotion of the scheme.[5] From the outset it was recognized that to be successful the plan would have to include at least the major refining interests. Any alliance confined to the railroads alone would be continually threatened by the danger that one road might accede to refiners' demands for rate concessions. Railroad history was already strewn with fragments of rate agreements broken in this way; the strong ties of each trunk line with particular refining centers made the occurrence all the more likely in the oil trade.

The vehicle chosen by Watson and his associates to carry out the plan to organize the railroads and the oil interests was the South Improvement Company, one of a series of special companies chartered by the Pennsylvania Legislature between 1868 and 1871. These charters reflected the political strength of the Pennsylvania Railroad and especially its vice-president, Tom Scott, who virtually controlled the Pennsylvania Legislature at the time. Known as the "Tom Scott" companies, each was marked by broad powers granted to the incorporators, including the right to hold the securities of other corporations. This was an important feature as the holding company had not yet become accepted in the United States, being judged by the courts in violation of common law, while New York and Ohio barred them specifically by legislation.

Described by one observer as ". . . well hatched eggs from one den of vipers," [6] these charters were designed to help Scott and his associates consolidate their widespread railroad interests and to branch out into other activities. The Pennsylvania Company (1870), for example, was used to bring together the complex holdings of the Pennsylvania Railroad Company. The Overland Contract Company, chartered the same year, changed its name to the Southern Railway Security Company and, by the end of 1871, owned a major stock interest in eight southern railroads, which in turn owned or leased some 1,500 miles of track. The Domain Land Company (1871) was renamed the California and Texas Construction Company and attempted unsuccessfully the construction of a western railroad.

Despite unfavorable criticism in the press, Scott's control of the Pennsylvania Legislature was strong enough to insure the incorporation of other similar companies, including the South Improvement Company, in 1871. No use had yet been found for this charter, when it was purchased by the promoters of the South Improvement scheme.

Like the others, the South Improvement Company had broad powers "to construct and operate any work, or works, public or private, designed to include, increase, facilitate, or develop trade, travel, or the transportation of freight, livestock, passengers, or any traffic by land or water, from or to any part of the United States." In carrying out any or all of these objectives the Company could acquire and hold the securities of other corporations, including those chartered and operating outside the boundaries of the state of Pennsylvania.

Armed with a charter ideally suited for carrying out the proposed pool, Watson spent the latter months of 1871 sounding out the country's leading refiners, including Rockefeller. Rockefeller, who had already begun to think of a combination or association of refiners, was interested.* He took an active role in persuading other refining groups that they should come into the plan. How successful Watson and Rockefeller were in arousing refinery interest in the plan was brought out when the first official meeting of the South Improvement Company was held in Philadelphia, January 2, 1872. At this meeting two thousand shares of stock were divided among the companies represented as follows: W. G. Warden and O. F. Waring, Philadelphia and Pittsburgh oil brokers, 475 shares each; John D. Rockefeller, Henry M. Flagler, and O. H. Payne, of Cleveland, 180 shares each; William Rockefeller and J. A. Bostwick, 180 shares each; Peter H. Watson, who was elected president, 100 shares; and ten shares each to William Frew, W. P. Logan, John L. Logan, Charles Lockhart, and R. S. Waring, all of Pittsburgh.[8] Missing from the group were representatives from the Oil Regions and, with the exception of Bostwick, New York. It was agreed at this time that all refiners willing to accept the basic principles of the organization should be brought into the organization.

The nature of these principles was made clear by contracts signed with the Pennsylvania, New York Central, and Erie railroads within a few weeks. The South Improvement Company was to act as an "evener" of oil traffic over these lines with 45 per cent going to the Pennsylvania, and 27.5 per cent each to the other two. In return for an increase in freight rates on crude and refined, the railroads agreed to charge non-members the full rate but to rebate a portion of the amounts paid by members to the South Improvement Company. The proposed schedule of open rates and rebates is shown in Table 14:1.

For the railroads the proposed rates, even with the rebates, promised to bring in a larger return on the oil traffic than was currently being re-

---

* Commenting on an interview with Watson, Rockefeller wrote his wife on November 30, 1871, "There is a *new* view of the question just introduced and I don't know how it may turn though am hopeful—indeed the project *grows on me.*" [7]

alized. For the non-members these rates would have resulted in substantial increases, in some instances double, over what they had been paying. This differential alone could have eliminated all shippers outside the South Improvement Company. But there was a further provision that the

TABLE 14:1

Rates and rebates on oil shipments in Pennsylvania RR–
South Improvement Co. Contract: January 18, 1872

	Gross rate	Rebate	Net rate
On Crude Petroleum (per 45-gallon barrel)			
From any common point* to			
Cleveland	$0.80	$0.40	$0.40
Pittsburgh	.80	.40	.40
New York	2.56	1.06	1.50
Philadelphia	2.41	1.06	1.35
Baltimore	2.41	1.06	1.35
Boston	2.71	1.06	1.65
On Refined Oil, etc. (per 45-gallon barrel)			
From Pittsburgh to			
New York	2.00	.50	1.50
Philadelphia	1.85	.50	1.35
Baltimore	1.85	.50	1.35
From Cleveland to			
Boston	2.15	.50	1.65
New York	2.00	.50	1.50
Philadelphia	1.85	.50	1.35
Baltimore	1.85	.50	1.35
From any common point to			
New York	2.92	1.32	1.60
Philadelphia	2.77	1.32	1.45
Baltimore	2.77	1.32	1.45
Boston	3.07	1.32	1.75

* "Common points" were shipping points in the Regions. Rates from all common points to a given destination were equal.
Source: U.S., House of Representatives, Committee on Manufactures, *Report on Investigation of Trusts*, 50th Cong., 1 sess. House Report No. 3112 (Washington: Government Printing Office, 1888), 358–59.

South Improvement Company would also receive drawbacks, equivalent to the rebates, on all petroleum shipped by non-members. In other words, the South Improvement Company was to receive a substantial portion of the increase in freight rates. Allowances for rebates and drawbacks would bring into the South Improvement Company an annual sum estimated at $5–6 million. "This fund could be used in a variety of ways to kill any attempt by rival railroad managers, producers, refiners, speculators, or foreign export agents, to break down the contract." [9] Moreover, a daily report of all oil shipments, by amount, origin and destination, and names of shippers was to be sent to the South Improvement Company by the railroads. This would have provided members with a close check on the business of competitors. To insure full payment of drawbacks and rebates the railroads also agreed to open their books for inspection by the members of the South Improvement Company. As Allan Nevins stated: "Of all devices for the extinction of competitors, this was the cruelest and most deadly yet conceived by any group of American industrialists." [10]

At the organization meeting it was agreed that all refiners who wished to join were to be included in the South Improvement Company. The inclusion of the producers was another matter. Rockefeller doubted the feasibility of making them parties to the agreement, but Tom Scott insisted that the plan would not succeed if they were not invited to participate. A draft of a proposal to be made to the producers was drawn up which would have guaranteed minimum prices for crude in return for effective limitations on the flow of oil from the fields. But before it could be shown to the well owners in the Regions, news of the South Improvement Company leaked out. The wave of indignation that swept through the Regions made any such agreement extremely unlikely.

The event that touched off the reaction was the premature imposition of the new rate schedule on February 25, 1872, by an officer of the Jamestown & Franklin branch of the Lake Shore, who announced that the rate on crude to New York had been increased from 87¢ to $2.14 a barrel. The reaction in the Oil Regions was immediate. Two days later three thousand men gathered in the Titusville Opera House as "Well owners, refiners, oil brokers, shop keepers, bankers, all trooped in . . . [bearing] banners with defiant slogans: 'No Compromise'; 'Down with the Conspirators'; 'Don't Give Up the Ship.'" [11] This meeting marked the beginning of the "oil war" between the producers and refiners in the Regions and the interests promoting the South Improvement Company.

The *Titusville Morning Herald* viewed the South Improvement Company as a plot to eliminate the Titusville and Oil City refiners, located at

> . . . the only natural refining point in the country . . . our refiners have for years had to contend against draw backs granted by railroad companies to some of the largest refining houses in Cleveland and Pittsburgh and not withstanding this they are languishing . . . while Titusville refiners have been flourishing . . . now these [Cleveland and Pittsburgh] refiners are endeavoring to combine to wipe Titusville and Oil City refiners out of existence.[12]

As their frequent attempts to combine had indicated, the producers were not opposed to monopoly on principle. This fact, especially in view of their lack of success in forming an effective combination of producers, did not make them any less apprehensive over the prospect of being forced to sell their product to a gigantic monopoly buyer. Suggestions that the proposed rise in freight rates would be passed on to consumers were not convincing to spokesmen for the producers. Commented the *Titusville Morning Herald*, "Past statistics show this to be, if not a subterfuge, at least a fallacy." Experience had demonstrated that as freight rates advanced, price of oil in the Regions fell proportionately and *vice versa*. "Who ever heard of an advance in ocean freights raising the price of oil in Liverpool?" [13]

Aided by the refiners who had been uninvited or who had refused to join the South Improvement Company, the Petroleum Producers' Association was rejuvenated as the Petroleum Producers' Union. A campaign was laid out to impose an embargo on crude oil shipments to the South Improvement Company members, to bring public pressure to bear on the railroads, and to have the South Improvement Company charter revoked by the Pennsylvania Legislature.

Although not complete, the embargo was generally accepted with enthusiasm by the producers and was, on the whole, quite effective. Total shipments of crude dropped from 400,000 barrels in February to around 276,000 barrels in March. [14] Cleveland and Pittsburgh were especially affected, Cleveland receiving only 15,415 barrels in March compared to 73,000 barrels in February and Pittsburgh receiving 39,000 barrels compared to 92,000 barrels for the same months. [15] The *Oil City Derrick* reported on March 26 that five thousand employees were idle as a result of the Cleveland refineries closing down for lack of crude. Those who did sell to South Improvement Company members ran the risk of personal injury or damage to their property from indignant producers.

The railroads were quick to capitulate. Scott assured the representatives of the Regions that the Pennsylvania railroad "will not do anything which will interfere with the mutual and relative interests of producers and shippers." [16] Vanderbilt blamed his son for the part played by the New York Central. [17] Meeting in New York on March 25 with representatives of the "producers and refiners," the Erie, New York Central, Pennsylvania, Lake Shore, and Atlantic & Great Western, agreed to cancel the contracts with the South Improvement Company. A new rate schedule was agreed upon with the proclamation that henceforth all rates were to be public and equal, with no drawbacks, rebates, or other special concessions.[18] Interestingly enough, the rate schedules adopted at this time were almost identical to the *net rates* the railroads would have received under their contracts with the South Improvement Company.

On April 2, 1872, the Pennsylvania Legislature not only revoked the South Improvement Company charter, but went on to pass a free pipeline bill, conferring upon pipeline companies the right of eminent domain. The victory, however, was a hollow one; the Pennsylvania Railroad, which had been blocking the measure since it was first introduced in 1868, managed to restrict the bill to the eight oil-producing counties. Allegheny county, in which Pittsburgh lay, was excluded, and the Pennsylvania's hold on the crude traffic to the Iron City remained unchallenged.[19] The lack of an adequate free pipeline bill was to be a thorn in the side of the Regions for the next ten years, but for the present, the producers felt they had accomplished a signal victory. On April 9, after being reassured that the South Improvement Company contracts had in fact been abrogated, the Producers' Union announced the end of the embargo on crude shipments, and the "oil war" was officially ended.

"The Conquest of Cleveland"

While all refiners associated with the South Improvement Company plan came in for sharp criticism and suspicion from the Regions, Rockefeller and Standard were viewed with the greatest distrust. This was in part because Standard Oil was at the end of 1871 the largest single refiner in the industry, and in part because of its location in Cleveland, which had openly announced its intention of capturing the bulk of the nation's refining capacity. This feeling was intensified when it was learned that during the furor over the South Improvement Company proposal, Rockefeller had quietly absorbed the bulk of the refining capacity of Cleveland in addition to enlarging Standard's holdings in the New York City area.

It was an impressive group of firms that by April 1, 1872, had sold or merged their interests with Standard. Heading the list were five of the largest refiners in the Cleveland area: Clark, Payne & Company; Westlake, Hutchins & Company; Alexander, Scofield & Company; Hanna, Baslington & Company; and Clark, Schurmer & Company.[20] In addition nine other smaller Cleveland firms plus seven whose locations were "not noted in the records, but were probably all in Cleveland" were also acquired.[21]

The most important addition outside Cleveland came when Jabez A. Bostwick & Company of New York, prominent export house and owner of a refinery on Long Island and terminal facilities in New York harbor, decided
to join Standard. By acquiring practically the entire refining capacity of
Cleveland, approximately 12,000 barrels of crude daily, Standard now
controlled (with its New York refineries) over one-quarter of the total
daily capacity of the industry, estimated at about 46,000 barrels.[22]

FROM MC CLURE'S MAGAZINE, VOL. XX (1902)

John D. Rockefeller in 1872.

Like most contemporary businessmen, Rockefeller tried to keep his operations secret. But as news of his accomplishments leaked out there was
a growing feeling that only by using the South Improvement Company
plan as a threat could Rockefeller have absorbed such a large segment of
the refining industry in such a short period of time. Accounts of this episode in the history of Standard are sharply contradictory. According to
Ida Tarbell, Rockefeller, armed with the South Improvement Company
charter and contracts with the railroads, pointed out to the Cleveland refiners the futility of outsiders attempting to remain in operation. Without
indicating the possibility that all refiners might be invited to join the plan,
he allegedly suggested two alternatives to almost certain bankruptcy.
One was to exchange their holdings for Standard stock; the other was to

accept cash. In either case the value of the property was to be set by Rockefeller's appraisers.[23]

Support of this version of Rockefeller's tactics came from a number of individuals whose properties were acquired by Standard during this period. John Alexander, a partner in Alexander Scofield & Company, told a Congressional Committee in 1872, "There was pressure brought to bear on my mind and upon almost all citizens of Cleveland engaged in the oil business, to the effect that unless we went into the South Improvement Company, we were virtually killed as refiners." [24] Alexander did not directly charge Rockefeller with making threats, but Frank Rockefeller (younger brother of John and William), of the same firm, testified that Flagler and John Rockefeller had threatened Alexander, Scofield & Company with extinction, "If you don't sell your property, it will be valueless, because we have advantages with the railroads." [25] According to Robert Hanna, partner in Hanna, Baslington & Company, when he told Flagler and Rockefeller, "But we don't want to sell," Rockefeller's answer was, "You can never make money, in my judgement. You can't compete with Standard. We have all the large refineries now. If you refuse to sell, it will end in your being crushed." [26]

All three refiners declared that Standard had paid them much less than their properties were worth; in the case of Alexander, Scofield & Company the partners' valuation was $150,000; Standard's, $65,000; Hanna, Baslington's estimated their properties at $76,000; they received $45,000.[27]

According to the alternative version, Rockefeller was largely carrying out a plan he had envisioned prior to the incubus of the South Improvement Company, a plan that, with Standard as the nucleus, would eventually consolidate "all oil refining firms and corporations into one great organization." [28] This plan would have been extended to the Cleveland area in any event, given the distressed conditions of refining in that city in early 1872. It was largely the aftermath of almost two years of low returns or losses in refining, not the threat of the South Improvement Company, that was most effective. The complaints of pressure and of low evaluation of properties were primarily the reactions of individuals who resented the rise of Standard, or who, if they took cash, regretted their decision as the stock of Standard began to appreciate.

This was the general interpretation advanced by Rockefeller himself some forty years after the events.[29] As he recalled it, he participated in the South Improvement Company with considerable misgivings.

We acceded to it because he [Tom Scott] and the Philadelphia and Pittsburgh men, we hoped, would be helpful to us ultimately. We were willing to
go with them as far as the plan could be used; so that we would be in a position to say, 'Now try our plan.' Thus we would be in a much better position
to get their co-operation than if we had said 'No' from the start.[30]

He denied ever using the South Improvement Company as a threat. He
insisted that, far from being unwilling to sell, his competitors in the Cleveland area were eager to unload properties that had been losing money for
several years. As evidence of the lack of coercion he pointed to the number of individuals who came into Standard during these years and who became prominent members of the Standard management. "Some competitors, he admitted, did not regard such proposals favorably. 'With these our
relations continued entirely pleasantly, until at length one by one they
were pleased . . . to join their interest with ours.' "[31]

Between these widely divergent explanations several points emerge
which suggest both versions had an element of truth. For one thing, it
was almost inevitable that differences would arise over the question of the
valuation of the properties acquired. The owners, such as John Alexander
of Alexander, Scofield & Company, for example, talked largely in terms
of the "$50,000 [spent] on our works during the past year [and the]
$60,000 or $70,000 before that" and concluded that the properties were
worth at least 75 per cent of their cost.[32] Rockefeller and his associates,
however, put more emphasis on the "use-value" of the plant and equipment they proposed to take over and objected to paying high prices for
what they considered in many instances "obsolete equipment and doubtful good will."[33] In certain instances Standard's officers attached a high
value to good will, especially if it brought exceptional talent into the
management group. As an example, Clark, Payne, & Company recieved
four thousand shares of stock (par value $400,000) for properties ap-
praised at $251,110; the difference being paid for good will and an agree-
ment that Colonel Payne join the Standard hierarchy.[34]

Beyond the claims and counterclaims over the extent to which Rockefeller specifically used the South Improvement plan as a threat was the far
more important fact that such a plan existed. The refining interests of
Cleveland, familiar with the methods used by Rockefeller in developing
his company to the end of 1871, had little reason to think he would not
take full advantage of the opportunity to push his plans to completion.
The fact that such a possibility existed was sufficient to convince ex-

perienced operators of the devastating implications for those who remained outside of the rate and rebate structure agreed upon by the participants in the South Improvement Company.

It was not until after the South Improvement scheme collapsed that
Rockefeller's conquest of Cleveland refining interests was clearly revealed. Realization that the new Standard company was prepared to refine more oil than all the plants in the Regions, or those in New York, or
of Pittsburgh, Philadelphia, and Baltimore combined, prompted the
Cleveland Herald to speak of "the South Improvement Company *alias*
Standard Oil Company." [35] And in mid-April, 1872, the *New York Bulletin*
reported, "The trade here regards the Standard Oil Company as simply
taking the place of the South Improvement Company, and as being ready
at any moment to make the same attempt to control the trade as its progenitor did." [36] While this statement was phophetic regarding the future, it
exaggerated the role that Standard was prepared to assume immediately.
Rockefeller was willing to make one more attempt to bring stability to the
industry through co-operation with refiners and producers before launching his "own plan."

The National Refiners' Association

The proposal that attracted Rockefeller's attention came to be known
as the "Pittsburgh Plan." The South Improvement Company had scarcely
been interred by the Pennsylvania Legislature when a group of Pittsburgh
refiners, led by William G. Warden and Charles Lockhart, in April, 1872,
advanced a new scheme to unite the industry's refining interests and to
make peace with the producers. Their idea was quickly endorsed by
Rockefeller and the principal refiners in Philadelphia.

In its initial phase the Pittsburgh plan called for a loose but comprehensive organization of refiners under a central board. The board was to
handle the purchase of crude for all members, allocate refining quotas, fix
prices, negotiate uniform freight rates with the railroads, and distribute profits according to the value of the property of each participant.

To win support of the refiners in the Regions, representatives of the
group, including Rockefeller and Flagler, went to Titusville during the
second week in May, 1872. Their welcome was by no means enthusiastic.
The *Petroleum Centre Record* noted the arrival of Rockefeller, Flagler
and others "of South Improvement notoriety" and went on to say, "It is to
be hoped [the Regions'] refiners will not allow themselves to be soft-

soaped by the honeyed words of the monopolists and conspirators." [37] A few days later the promoters of the Pittsburgh plan appeared before open meetings in Titusville of the refiners where they outlined their proposals and indicated a willingness to co-operate with the producers.

The great majority of the local refining interests rejected the plan, but John D. Archbold and J. J. Vandergrift, among the ablest of the group and bitter opponents of the South Improvement Company, were won over. When they were joined by Charles Pratt and H. H. Rogers, leading independent refiners of New York, the promoters decided to proceed with their organization.

In August, 1872, the Petroleum Refiners' Association, better known as the National Refiners' Association, was formed with John D. Rockefeller, president; J. J. Vandergrift, vice-president; and Charles Pratt, treasurer. Open to any refiner who wished to join, the Association was not incorporated but operated under a "charter" or basic agreement that regulated the activities of its members. It was a representative body covering five districts, Cleveland, Pittsburgh, the Oil Regions, New York, and Philadelphia. Each district elected three members of a fifteen-man board of directors, responsible for general policies, including purchases of crude and sales of refined. In addition, each district elected a member of a Board of Agents whose job was to carry out the policies of the directors, including an allotment of crude to each of the districts to be refined. Allotments within the districts were made by local boards. No member was to purchase or sell except within the limits of his particular allotment. Members further pledged themselves to complete secrecy concerning the Association's activities and to refrain from speculating in oil.

The strength of the Association lay in the prominence of the membership, the large proportion of refining capacity represented and the bargaining position the group held opposite the railroads. Its most obvious weaknesses stemmed from any lack of financial penalties that might be imposed on recalcitrant members and the absence of control over outside refiners and the suppliers of crude. The reaction of the great majority of the producers to the formal organization of the National Refiners' Association was probably reflected by an anonymous correspondent of the *Oil City Derrick* who concluded that the producers would "be at the mercy of the combinations' agents," that there would be "no competition in the trade," and "that the agents or their employers [would] be able to dictate at any time the price of crude." [38] Producers began to think again in terms of reviving their own organization.

But there was a more immediate and pressing reason for reviving a producers' organization. After several years of exploration to the south of Tideout and Oil City, a number of strikes in late 1871 had indicated the presence of oil in commercial quantities in Clarion, Butler, and Armstrong counties. Throughout 1872, oilmen had flocked into the district signing up leases and sinking new wells. With the demise of the South Improvement Company, oil began to pour out of the southern fields in increasing volume.[*] The average price of crude in August, 1872, of $3.47 for a barrel was almost $1.00 less than the price a year earlier.

Within a few weeks of the meeting of the refiners' group, the Petroleum Producers' Association was reactivated. Its members set out to enlist the support of every man in the Regions to agree to stop drilling for six months, beginning in September, 1872. The theory was that with no new wells coming in, production would drop off sufficiently to restore higher prices for crude. Public sentiment, combined with occasional force, led to considerable success in carrying out this plan and prices that averaged $3.15 per barrel in September moved up to $4.17 the following month. But when the embargo was lifted late in October, production soared to new levels and prices again sagged to $3.29 in December.

Convinced that a more comprehensive plan was needed to control output, Captain William Hasson, leader of Producers' Association, proposed the formation of a Petroleum Producers' Agency, capitalized at $1,000,000, with stock to be owned by actual producers or their associates. This corporation was to purchase all available oil from the Producers' Association at $5.00 per barrel. Oil that could not be marketed at this price was to be stored while the Agency took additional steps to cut back production or, if necessary, establish its own refineries to maintain the prices. By early November, 1872, the stock of the Petroleum Agency had been subscribed.[40]

Rockefeller, on behalf of the Refiners' Association, immediately got in touch with the officers of the Producers' Agency, offering to buy oil at prices fixed by the latter group provided proper steps were taken to maintain them. Hasson objected to any such alliance but was outvoted by his executive committee and on December 19, 1872, an agreement was signed between the two groups.

Dubbed the "Treaty of Titusville," the compact called for the appointment of joint committees to see that its provisions were carried out. The

[*] Deliveries of crude from the Regions that had dropped to slightly over 276,000 barrels in March, 1872, rose to 428,500 barrels in April, and were over 510,000 barrels in May.[39]

producers were to sell only through their association while the refiners were to buy only from the producers and in such amounts as were mutually decided upon. A base price of $4.00 per barrel was established, with the provision that if refined oil sold in New York at 26¢ per gallon, no further amount was to be paid the Producers' group. But with each advance of 1¢ per gallon in the price of refined, refiners would add 25¢ per barrel to the price of crude until it reached $5.00.[41]

The weaknesses of this alliance between the producers and refiners for the purpose of regulating output and prices of crude soon became apparent. The Producers' group was unable to enforce their recommendation that new drilling be suspended for another six months. Instead the signing of the "Treaty" and an initial order by the refiners for 200,000 barrels of crude at $3.25 (which the joint committee felt was more realistic than $4.00) was the signal for new drilling and an increased output of crude that within two months brought the price as low as $2.00 per barrel. By mid-January, 1873, it was obvious that the Producers' Association would not be able to keep up its end of the agreement and the contract with the refiners was cancelled.

After noting that the pledge to stop drilling for six months "had tended to increase development by persons who wished to take advantage of the idleness of others," the *Oil City Derrick* wrote the epitaph to the Producers' Agency, "So we come to the end of this short-lived combination. It was wrong in principle, impossible in practice, and inconsistent with the record of the oil producers of Pennsylvania."[42]

Not only were the well owners back to the disorganized conditions of before, but production soared to new highs over the succeeding twelve months as the southern fields were further exploited. Output of crude that had averaged 21,000 barrels a day in 1872 rose to 33,000 barrels in 1873. Any hope that a balance would be restored between production and demand at prices for crude that approached $5.00 or even $4.00 per barrel faded rapidly. Even though inventories rose during the year from about 1.08 million barrels to 1.63 million barrels, prices of crude in the Regions that had averaged $3.75 per barrel in 1872 sagged to $1.80 in 1873.[43]

Meanwhile the members of the Refiners' Association had been working to strengthen their organization. Additional firms were persuaded to join the group and committees were appointed to work with refiners who remained outside. A plan for implementing the division of refining allotments and sales was adopted, including a schedule of fines for those who failed to pay their dues or exceeded their allotted sales quotas.

But these moves were insufficient to enable the Association to meet the objectives of its founders. With authority dispersed among the five districts it proved increasingly difficult to enforce the rules and regulations on quotas and prices. Even more serious, outside refiners took advantage of the situation by stepping up their output and dumped enough illuminating oil on the market to nullify the efforts of the Association to keep prices up. When a suggestion was turned down that outside refining capacity should be purchased, the organization rapidly deteriorated. In June, 1873, less a year from the date of its organization, the National Refiners' Association was dissolved by the mutual consent of its members.

Even with the breakup of the Refiners' Association, the refining interests generally seem to have been in a somewhat more favorable position in 1873 than in 1871 or 1872. For one thing, refining margins that had moved up in 1872 were apparently maintained throughout 1873.* Secondly, an increase of "refinery throughput" from about 5.9 million barrels of crude in 1872 to some 9.5 million barrels in 1873 gave some relief to the pressure of excess refining capacity that loomed so heavily over this branch of the industry during 1871 and 1872. But this relief was only moderate at best. Estimates of the industry's daily capacity of refined oil for 1873 put the figure at 47,000 barrels.[44] Average daily throughput for the year of approximately 26,000 barrels of crude (about 19,500 barrels of refined) suggests that the industry was still over extended and why "many refiners [during 1873] had to be shut down part of the time or else not operated to their full capacity." [45]

Railroad relations in transition

The South Improvement Company was bred by two factors over which the Oil Regions had no control. On the one hand was a refining industry

---

* This conclusion is based on the following tabulation showing average annual prices per gallon of crude in the Regions and export prices in New York of crude and illuminating oils:

Year	Crude in the Regions	Transport costs	Crude in New York	Refiner's margin	Illuminating oil in New York
1871	10.48*¢	7.65¢	18.13¢	6.12¢	24.25¢
1872	8.90	7.35	16.25	7.38	23.63
1873	4.27	6.36	10.63	7.62	18.25

*Amounts in cents per gallon throughout table.
Source: *Derrick's Hand-Book,* 1,711,783.

with excess capacity, selling more or less in a common east coast and export market, but whose manufacturing centers were so located that transport differentials between one refining center and another assumed prime importance. On the other, there was the over-extended railway network, ready to accept all freight at competitive points so long as rates covered operating expenses. Given these conditions, refiners would still seek rebates and railroads would still grant them despite assurances and contracts to the contrary.

Under these circumstances the aim of "equal and open rates to all" announced by the railroads following the collapse of the South Improvement Company represented wishful thinking. This becomes clear when the rate schedule announced in March, 1872, is examined more closely. Under this schedule, for example, refiners in the Regions were to pay $1.50 per barrel on all refined oil shipped to New York. While this same rate also applied to shipments of refined oil from Pittsburgh and Cleveland, refiners in the latter centers had to ship 1.4 barrels of crude from the Regions at 50¢ a barrel for each barrel of refined products (assuming 70 per cent yield of refined products). Their total freight bill would thus amount to $2.20 per barrel compared to the $1.50 charged refiners in the Regions.[46]

Refiners in Pittsburgh and Cleveland could hardly be expected to bear this disadvantage with equanimity. Even those who framed the schedule apparently had their doubts as to its workability. One of the signers—possibly with the benefit of hindsight to be sure—commented in 1879,

> At the time this arrangement was made, I do not think anybody who was acquainted with the business expected that it would last; . . . it was an impossible agreement; the immediate effect of it would have been to have utterly destroyed 55 per cent of the refining interest of the country; that is to say, Cleveland and Pittsburgh, which during the previous four years had shipped 55 per cent of all the oil out of the oil regions. . . .[47]

Furthermore, the new arrangements were not likely to stabilize relations among the trunk lines. The threat to the balance of power was too great. The New York Central and the Erie had built their export-oil traffic from Cleveland in the face of water competition during seven months of the year, and both had too large a stake in this trade to allow it to be injured seriously. To be sure Cleveland might have turned from exporting to developing the domestic market in the West, and the New York roads would presumably have garnered some of the increased west-bound traffic. This, however, promised to be a relatively slow process, during which the railroads would continually face stiff water competition over the Great Lakes from Cleveland. Moreover, the total volume of western shipments would have been relatively small at best.* The Central and the Erie had superior connections into New York City, an undoubted advantage if, as was possible, the new rates induced a great share of crude and refined oil to by-pass Cleveland and go direct to the seaboard. But it was doubtful whether such shipments would compensate for the loss of the Cleveland traffic, especially in competition with the Pennsylvania Railroad.

In contrast to the New York Central and the Erie, the Pennsylvania anticipated a substantial increase in its oil freight under the new rates.† It was in a much better position to accommodate an increase in refined shipments from the Regions than was either of its competitors. Through control of branches and connecting roads, the Pennsylvania was assured of excellent connections throughout the area; the New York roads reached the producing districts only at Oil City, and even here they depended upon Pennsylvania lines for delivery from the fields. Although its cumbersome connections into New York City were something of a disadvantage for the export trade, the Pennsylvania hoped to offset this by drawing export oil to Philadelphia at lower rates. And unlike the Central and Erie, the Pennsylvania had little, if any, Cleveland freight to lose.

But such considerations were only relevant if the announced rate structure could be enforced—a condition that Vanderbilt and the Erie management would hardly let be fulfilled. Thus when crude-oil dealer William Scheide wrote his employer that "there was no possibility of this arrangement continuing—the New York roads would not agree to it," he was only voicing what the railroads must have known even as they signed the contract.[49]

Scheide's letter was prophetic. Even for this period, when the lives of inclusive railroad agreements were counted in months, the March contract was surprisingly shortlived. Scheide himself, a member of the committee that had negotiated the schedule with the railroads, admitted that within a week after the embargo on the South Improvement Company

---

* Data on western shipments are virtually non-existent. However, exports accounted for approximately two-thirds of total production, in addition to which there was considerable domestic consumption on the urbanized eastern seaboard.

† When Colonel Potts, of the Pennsylvania's Empire line, heard the terms of the agreement, he commented, "I guess we will have to build some more cars right away."[48]

had been rescinded he approached Empire and received a "satisfactory" rebate.[50]

Standard Oil was not idle at this time. Under a one-year contract beginning April 1, 1872, Henry M. Flagler obtained preferential rates from the New York Central. During the seven and one-half months, from April to mid-November, when water competed with the railroads for Cleveland's freight, Standard's refined oil was to be carried to the seaboard at $1.25 per barrel; for the remaining four months, scheduled rates were raised to $1.40. In return, the railroad was guaranteed shipments of at least 100,000 barrels per month, but up to the end of the year generally carried considerably more.*

By the end of 1872, however, friction had developed between the New York Central and Standard. The Pennsylvania cut its rate on refined oil to New York to $1.05 per barrel on December 1. Standard was still shipping eastward over the New York roads exclusively and Flagler demanded that Vanderbilt meet the cut. Otherwise the large shipments that the road had enjoyed would be discontinued; Standard would be "compelled" to close its refineries, Flagler maintained, "for we could not afford to pay $1.25, when other people were only paying $1.05." †

While Vanderbilt had little desire to reduce a profitable rate, neither was he anxious to lose Standard's shipments of 4,000 barrels or more per day.‡ Somewhat grudgingly he agreed to meet the Pennsylvania rate. But a month later Vanderbilt restored the $1.25 rate. Standard paid the higher tariff for the remaining three months of the contract, but the New York Central received only the minimum shipments guaranteed in the agreement.

Before the termination of the contract, however, Rockefeller and Flagler patched up their differences with Vanderbilt. Standard continued to ship over the New York Central, apparently under an arrangement

---

* The smallest monthly shipment during this period, for example, was 108,000 barrels; the largest, more than 180,000 barrels.[51]

† Note that the railroad had apparently already granted Standard Oil a concession: under the original terms of the contract, rates after mid-November were to be $1.40 per barrel. Or perhaps, testifying seven years later, Flagler unintentionally confused the two rates.[52]

‡ Vanderbilt asked Flagler if he thought it just for the railroad to bear the entire reduction. Flagler certainly did, expressing in his argument what later characterized Standard's attitude on relations with the railroads: "I told him that if he should stand any part of it, he should stand it all. I said, it is a transportation fight and not a fight of the manufacturers. When it comes to competition of the manufacturers we would take care of ourselves." [53]

whereby the railroad bypassed its own terminal facilities in New York City and carried Standard's freight to the latter's yards at Hunter's Point, Brooklyn, at no extra charge.*

The Erie's lot at the same time was not a happy one. Its management was torn apart by the bitter struggle that ousted Gould and Fisk; its large capital stock was waterlogged; and while the Pennsylvania was shipping some 300,000 barrels per month and the New York Central was carrying Rockefeller's seaboard volume, the Erie averaged an oil freight of only 53,000 barrels monthly. Although the causes of the Erie's oil troubles were manifold, George R. Blanchard, who joined the road as general freight agent in 1872, laid most of them at the door of the Pennsylvania Railroad. The Pennsylvania granted large concessions to the Empire Transportation Company, which in turn controlled the Union Pipe Line, one of the largest in the Regions. Empire and Union, backed by these funds, granted large rebates to shippers in the Oil Regions. Blanchard righteously implied that such concessions were in violation of the March 25 contract with the producers and refiners.†

There was a good deal of truth in Blanchard's contentions. Empire did receive large concessions from the Pennsylvania—up to 30 per cent off the open rate. In part they were justifiable as commissions paid for soliciting business for the Pennsylvania lines; in part they constituted payment for the use of Empire's tank cars. Both types of compensation were accepted railroad procedure although the 20 per cent allowance for car service was almost double the mileage usually paid on foreign cars. But whether justified or not, these funds did enable the transportation company to rebate substantially.

There were, however, other important factors working against the Erie, part of them of the road's own making. The old broad gauge track, together with its connections into the Regions continued to be a serious problem. The Erie's broad gauge extended only to Oil City, from which production was rapidly receding, moving south to Parker's Landing. Oil had to be hauled fifty miles from the producing districts, over the Pennsylvania's Oil Creek & Alleghany River Railroad, to Oil City; there it was

---

* The terms of the agreement are obscure, but testimony of George R. Blanchard in 1879 suggests the terms indicated.[54]

† While president of the Erie, Gould had purchased control of the powerful Pennsylvania Transportation Company pipeline to help increase the road's oil freight. But shortly after, Tom Scott of the Pennsylvania Railroad bought into the company and presumably his influence prevented the pipeline from aiding the Erie in its competition against the Pennsylvania Railroad-Empire amalgamation.[55]

pumped into storage tanks and from these the broad-gauge tank cars were filled. This decidedly cumbersome arrangement was apparently all the more hampered by a lack of tank cars. Lack of funds undoubtedly prevented the Erie from alleviating the shortage, while shippers were loath to commit a heavy investment in tank cars that could be used only on one road.

There were additional troubles on the seaboard end of the line. The Erie's docks at Weehawken had, since 1869, been leased by Jabez A. Bostwick. A large shipper on his own account, Bostwick worked in close co-operation with Henry Harley, the Erie's oil agent under the Gould regime.* Both obtained extremely favorable rates to New York on their own shipments over the Erie, while Bostwick apparently gave his own and Harley's oil price concessions and handling preferences at the Weehawken terminal. When outsiders found they could not get terms matching Bostwick's, they started shipping over other lines, and the Erie's oil freight dwindled.[56]

Blanchard set out to remedy this situation when he joined the Erie as general freight agent in October, 1872. Negotiations were undertaken with a group of refiners in the Regions, and in March, 1873, an agreement was concluded under which, in return for a rate of $1.25 per barrel, the Erie and the Atlantic & Great Western were guaranteed the entire New York-bound output of these refiners for the remainder of the year.

The results of this agreement were sadly disappointing. Although the Erie's total oil freight averaged some 10,000 barrels per month more than during the previous year, there was apparently little increase in its refined-products traffic. And all but one of the refiners who had signed the agreement broke their contract and transferred their shipments to the Empire in November, 1873.† Instead almost the whole of the expansion came when Neyhart & Grandin, one of the largest crude-oil jobbing firms, broke with the Pennsylvania and entered into an exclusive contract with the Erie in June, 1873. Under the terms of this agreement Neyhart & Grandin built tank cars for the standard gauge track road between Parker's Landing and Oil City. In return for keeping these cars "running constantly," the jobbing firm received a net rate to New York of about 90¢ per

---

* It will be recalled that Harley was one of the original owners of the powerful Abbott and Harley pipeline, in which Gould bought a controlling interest in 1868. It was at this time that Gould appointed Harley to be the Erie's oil agent.
† The Erie thus found itself with only one small shipper of refined oil, G. Heye, and to keep his trade they had to give him a $1.10 rate.[57]

barrel compared to the "open rate" of $1.25.[58] With Neyhart & Grandin's business, the Erie was able to ship approximately 762,000 barrels in 1873, about 15 per cent of the combined shipments of petroleum (both crude and refined) to the east coast for that year.

While this volume of crude shipments was gratifying, the Erie management was eager to rebuild its tonnage of refined oil. Negotiations were opened with Standard for a substantial share of the latter's business, but no agreement was reached during 1873. Thus at the year's end the Erie was still hauling no more than a trickle of refined oil over its lines. This was hardly a situation that made for stability in the competitive relationships among the major trunk lines.

By contrast with both the New York Central and the Erie, the Pennsylvania was in a highly favorable position. From April, 1872, to the end of 1873 it handled over 60 per cent of the total petroleum traffic going to the eastern seaboard. The great bulk of this traffic was made up of crude and refined oil from the Regions, where Empire's widespread gathering lines and efficient purchasing system ably supplemented the network of the Pennsylvania branch lines that combed and surrounded the area. Furthermore, the Pennsylvania, unlike the New York roads, was not dependent upon Standard for business. It was supplying the transport needs of a large proportion of the independent interests that at the end of 1873 were still absorbing from 50 to 60 per cent of total shipments. Many of them were large, but many more were small, and none was so strategically located in respect to any one of the trunk lines as to command the allegiance or pursuit that Standard did from the New York Central and the Erie.

The Rockefeller plan

Of all the major refining interests operating in 1873, Standard was undoubtedly in the best position to take advantage of better refining margins and increased output. Throughout the latter part of 1872 and early 1873, Rockefeller and his associates devoted a large portion of their attention to consolidating and improving the company's holdings in the Cleveland area. This was no small task. Among the recently acquired properties were several plants specializing in tar manufacture, acid restoration, and the production of lubricating oils. These were brought under central management control and their output added to the company's line of refined products, hitherto confined to kerosene and naphtha. A decision was made to concentrate kerosene and naphtha refining to the half dozen of the largest and most efficient plants owned by Standard in Cleveland and to dismantle or scrap the rest. This still left Cleveland with a daily capacity close to 10,000 barrels of refined products.*

The management also took a number of other steps to improve Standard's position. In addition to obtaining "competitive" rates on refined shipments from the New York Central, Rockefeller and Flagler enlarged Standard's New York City holdings by acquiring the Devoe Manufacturing Company in January, 1873. This move strengthened the organization's position in the foreign sales as Devoe had been a successful exporter of "case oil" drawn from its refinery on Long Island and packaged in tins of its own manufacture. Rockefeller also took an initial step in strengthening Standard's marketing connections in 1873 by purchasing a half interest in Chess, Carley & Company of Louisville, Kentucky, already a leading wholesaler in the southeastern United States.

Judging from the share allotted to Standard, the company operated its Cleveland refineries during the life of the National Refiners' Association at about a third of their estimated daily capacity of 10,000 barrels of refined products.† Following the collapse of the Association, Standard stepped up its average production of refined oils at Cleveland to an estimated 6,250 barrels daily during the latter half of 1873.[61]

This rate of output meant that Standard was supplying about one-third of the industry's total output of refinery products from Cleveland, which added to the product of the New York plants probably raised the proportion to a figure between 30 and 40 per cent. Considering the fact that Standard's share of the market in January, 1871, was about 10 per cent, this was a remarkable advance.‡ But Rockefeller and his associates were fully aware that their position in the industry was by no means secure; the fact that Standard's Cleveland refineries were operating at

---

* Allan Nevins puts the figure at "easily 12,500 barrels of refined oil daily," but this statement is at variance with his earlier estimate of 10,000 barrels and his reference to the figure of 12,700 barrels of *crude* for the entire Cleveland area.[59]

† Between August 15, 1871, and April 30, 1873, the National Refiners' Association purchased some 3.56 million barrels of crude, of which 1.08 million barrels (30.6 per cent) went to Standard in Cleveland. Allocation to the company's refineries in New York added another 3.4 per cent to their proportion, but unless Standard bought a substantial amount of the 1.18 million barrels sold outside the Association, its average daily crude throughput would have been about 4,200 barrels or approximately 3,000 barrels of refined.[60]

‡ While Standard's profits are not available for 1873 there is little reason to think that net earnings were not well above the approximate $538,000 (nearly 22 per cent) on a $2.5 million capitalization reported for 1872.[62]

about two-thirds capacity only emphasized the overexpanded condition of the refining industry generally. An alliance between the Pennsylvania Railroad and refining groups in the Regions, Pittsburgh, Philadelphia and New York could well encourage a further expansion of capacity in those areas. Unless some effective means could be worked out to bring refiners under central control, the situation from Standard's point of view could easily deteriorate.

For Rockefeller, the collapse of the National Refiners' Association dispelled any lingering doubts he may have had about the effectiveness of loose combinations as a means of controlling refining capacity and "stabilizing" the industry.* He was all the more firmly convinced that what he later referred to as "our plan" should be put into operation. The extent to which "our plan" at this time envisioned a far-reaching integration of the industry under the aegis of Standard is open to question.† But it is clear that the immediate objective was to bring the great bulk of the industry's refining capacity under the ownership or control of Standard.

This in itself was a program to challenge the imagination. If successful, Standard would become for all practical purposes the sole purchaser of crude; its bargaining position opposite the pertroleum carriers would virtually insure preferential transportation rates, while marketers would have little choice but to turn to Standard for their supplies.

Such a program would inevitably encounter bitter opposition, but conditions in the industry generally worsened during 1873 and Rockefeller saw little reason why the tactics followed in the Cleveland area could not be extended with equal effectiveness to the other refining centers. His role in the National Refiners' Association had brought him into close contact with the leading refiners in the Regions, Pittsburgh, Philadelphia, and New York. He was confident that these individuals could be persuaded to join or sell out to Standard; their co-operation would open the way for the absorption of the rest.

---

* Years later Rockefeller stated, "It was apparent to me early in the organization of the Refiners' Association that among so many men untrained in business there were many who could not be relied upon to aid in solving a problem so difficult as the reformation my associates and I sought to bring about in this industry. . . . We proved that the producers' and refiners' associations were ropes of sand." [63]

† Referring to his concept of Standard's role in the industry, Rockefeller stated many years later, "The idea was mine . . . [and it] was persisted in, too, in spite of opposition of some who became fainthearted at the magnitude of the undertaking, as it constantly assumed larger proportions." [64]

Part SIX

Expansion, adjustment,
and integration:
1874–1884

Chapter 15

Production and marketing of crude

In April, 1873, when the collapse of the Petroleum Producers' Agency and the surging output of Clarion, Butler, and Armstrong counties in the Regions had dimmed any early prospect of cutting back production, the *Titusville Morning Herald* (April 30, 1873) took occasion to sound an optimistic note. Under the heading "Is Petroleum a Necessity?" the paper stated, "The production of petroleum has now become of such commercial and social importance to the world that if it were suddenly to cease no other known substance could supply its place, and such an event could not be looked upon in any other light than of a widespread calamity."

There was more than a germ of truth in the opinion expressed by the Regions' leading newspaper. Within the short space of thirteen years, kerosene had become firmly established as a leading illuminant, especially in the European and American markets, where it was rivaled only by artificial gas in urban communities. Successful deodorizing of heavy oils had removed one of the principal barriers to an extended employment of petroleum lubricants, while industrial uses gave promise of a widening market for the naphtha group of refinery products.

As the industry moved into the 1874–84 period, however, the important question was not whether a latent and possible unlimited demand existed for petroleum products, but how rapidly such a demand could be developed. For producers, the answer to this question was of major interest, particularly in view of their experience during 1872 and 1873, when an approximate 50 per cent increase in output was accompanied by a 50 per cent drop in the average price received for crude. If output continued to expand, there was a real threat that the market would be unable to absorb the greater output without a further drop in prices.

To a large extent this possibility was realized during the decade. As illustrated in Chart 15:1, production rose from 10.9 million barrels in 1874 to 24 million barrels in 1884. From 1874–78 approximately one barrel of every three produced was stored. In the first two years the average yearly price was in excess of $1.20; in 1876–77 prices doubled; in 1878 they fell back to a $1.20 per barrel level. Although not as "well off" aggregatively as in previous years, producers during this five-year period still fared better than they did after the Bradford field to the north was fully exploited. In 1884 Bradford reached its production pinnacle and the price and inventory picture darkened. By 1884 the average yearly price of crude in the fields stood at 83¢ a barrel. As crude poured from Bradford at a rate faster than the market could absorb, inventories rose precipately; between 1880 and 1884 every barrel of crude produced was represented by about 1½ barrels of inventory. (See Table 15:1.)

Excessive inventories and falling prices were usually the signals for "stopping the drill," and the literature of this period is rich with discussions of producers' schemes to restrain output. On the surface such actions might be interpreted as an indication that producers wanted to bolster the price of crude because they were losing money on their operations. This was in part the case. Declining crude prices did put pressure on many marginal producers, particularly those operating in older, more "settled" producing districts where the possibility of opening up new highly productive wells was severely limited.

The basic explanation of the failure of cutback movements, however, lay in the fiercely competitive structure that production had early as-

CHART 15:1. *Output, inventories, and prices of crude oil (per barrel),*
Appalachian fields, 1874–84. (*Sources:* Mineral Resources, 1891, *419;*
1895–96, *678, 687;* 1896–97, *176;* 1908, *351;* Pa. Industrial Statistics,
1892, *89.*)

sumed and retained for many decades to come.* As sources of crude physically widened and total output swelled, the inducement to become a producer or to expand into new fields made it virtually impossible to persuade the many thousands of producers to agree to any effective measures that would restrict production. And such behavior was not completely emotional and irrational. Within the competitive framework there was a wide range of cost-price conditions in which operations could be profitably carried on in both old and newer developing fields.

Drilling requirements and techniques had changed little since the mid 1860's. Average capital requirements to sink a well between 1874 and 1884 were still relatively low—$3,500 to $6,000. And it was more than possible to start wells with much less ready cash by utilizing used equipment. Second-hand boilers, engines and casing, in fact, were widely employed in the Regions.†

Once a well was drilled, operating costs were not excessively high. The major items were pumping costs for nonflowing wells and royalty payments. The range of pumping costs for the period was between $1.50–$5.00 per day, with an approximate average of $2.00.‡ Royalty payments were almost universally one-eighth of the net proceeds per well.§

* Data are not available to quantify the degree of competition; at best it must be deduced from the social, economic and legal factors. The number of firms must have been many thousands, possibly as high as 16,000, throughout most of the last quarter of the nineteenth century. There are available, however, some illustrative, even though sketchy statistics about firm size. In 1880, for example, the eight largest firms in the Oil Regions operated 600 wells upon 20,000 acres. On the average each of the eight firms operated 75 wells, and together, somewhat more than 3 per cent of all existing wells. This was probably an upper limit; seven years earlier, in 1873, the *Titusville Morning Herald* reported an oil company operating 38 wells as one of the largest producers in the Region.[1]

† In 1880, for example, about half of the wells drilled at Bradford were "supplied with engines and boilers from wells abandoned in the lower county," some of which dated from 1865. (Peckham, *Census Report*, 143–44.) Short-term credit from drilling contractors and equipment houses was also apparently easy to come by.[2]

‡ Pumping costs theoretically correspond to variable costs and as such should include wages, fuel, depreciation, maintenance and repair of pumping equipment, and royalty oil paid to lessor. Oil field practice, however, generally only considered wages and fuel as pumping costs proper. Engine and boiler costs, the largest cost item in the depreciation, maintenance and repair category, was generally included in drilling costs. This same practice has been followed in the following examples and estimates.[3]

§ By 1870 the one-eighth royalty payment had become well established. Cash bonuses were also widely used. Peckham cited bonuses of $100 to $500 per acre, but for rare choice sites as high as $1,000 and $5,000 per acre was actually paid. The one-eighth royalty has been used in the present calculations, but no attempt was made to estimate cash bonuses.[4]

Thus, most producers, provided they were able to market their output, could profitably operate their wells within a fairly flexible price range. From 1874–85 the average yearly price of crude per barrel ranged from a high of $2.58 to a low of 78¢; average for the period was approximately $1.25.* A well yielding 20 barrels per day, with drilling costs of $3,500, $2.00 per day pumping cost, and the usual one-eighth royalty would, at the 1884 average yearly price of 84¢, pay off its entire capital investment in just under 250 days. Thereafter it would continue to yield a daily net profit of $14.45. At the end of the first year it would have returned a net profit of nearly $1,600, or approximately 44 per cent on capital invested.

On the same cost and price assumptions, a 5-barrel per day pumper would return its original investment in just under 1,250 days. With seven years as the average life of a well during this period, wells of modest, even small, yield could be profitable.† Moreover, wells yielding 5–20 barrels per day were not at all uncommon, while some brought in a daily production running into hundreds or even thousands of barrels. According to a contemporary observer in one area of the Bradford field, producers could ordinarily expect to find a well "which will pay for itself in from five to eight months." [6] Even discounting such enthusiasm, a reading of the contemporary literature indicates that only a short search of any "oil town" was needed to find a successful and profitable producer.[7]

Conservation and the "Rule of Capture"

Although economic wastes did not seem to limit producer drilling activity, conservation had not escaped notice in the Regions. Geologists

---

* Average initial production was approximately twenty barrels and the average production of all wells about 3.5 barrels. The average life of a well for this period was estimated at seven years.[5]

† The examples above, of course, are not presented as "proof" of the profitability of petroleum production. Any such calculation would require much more detailed figures than are available. Indeed one notable omission is allowing for the risk of tapping dry holes. The ratio of dry holes fluctuated greatly over the twelve years from 1876–88, varying from as low as 3 per cent to as high as 25 per cent. The average rate for the period amounted to 12.5 per cent. To illustrate in somewhat more detail what the risk of dry holes meant, consider the case of the twenty-barrel well. Previous calculation established a net profit of 44 per cent. If nothing but twenty-barrel pumpers were drilled, and 3 per cent of the holes were dry, the average rate of return on all investment would be 42.7 per cent; with 25 per cent dry holes, it would drop to 33 per cent; and if the ratio of dry holes to wells drilled was the one-eighth that constituted the Pennsylvania average for this twelve-year period, return on investment would amount to 38.5 per cent.

from time to time severely warned the Regions about their disregard of conservation practices. But geologic pronouncements were not held in high regard by "practical" oilmen.

Many geologists agreed that sub-surface structure was the crucial factor for oil and gas accumulation. With considerable "scientific" justification many argued that the faster oil was extracted the more was gas pressure lost, allowing oil to run through the veins of the earth never to return. As early as 1874, when Butler, Clarion, and Armstrong were in their flush days, an Oil Region geologist, J. F. Carll cautioned producers: "We have reaped this fine harvest of mineral wealth in a most reckless and wasteful manner." [8] Ten years later the same writer questioned the propriety of indiscriminate exploitation. It was not a question of the complete exhaustion of the Region, he noted, rather the crucial issue was whether a "production of 70,000 . . . or even 60,000 barrels [per day] can be maintained over the next five years." The writer thought not. After years of having been "excessively and wastefully depleted" the established areas were declining; fruitless prospecting in recent years indicated that the most important productive pools were already discovered, with not even a "hint" of any major new sources on the horizon.

> From this time forward we have no reasonable ground for expecting . . . that these oil fields will continue to supply the world with cheap light as they have in the past . . . therefore it is wise to pause and consider how best to husband our remaining resources and make the most of them.[9]

A year earlier, Professor J. P. Lesley expressed even stronger concern. Speaking before the American Institute of Engineers, Lesley ominously prophesied the virtual extinction of the fields within the lifetime of young men then living. The flood of the preceding two decades, he believed, was but a "temporary and vanishing phenomenon . . . of which the people of Pennsylvania have foolishly taken so little advantage. . . ." [10] It could not be expected to continue in the future.

Lesley and Carll were Jeremiahs. The competitive struggle between producers was fierce, and the institutional setting was stacked on the side of rapid exploitation. Conservation was a latter-day affair.

From the beginning, petroleum production was enmeshed in a tangled legal web that forced immediate exploitation. Under the theories of property law applied to oil and gas, delay in development subjected a producer to two dangers: adjoining operators could drain his oil, and failure to drill could result in having his lease declared "forfeit," "terminated," or "abandoned" to the lessor. This was because property law concerning ownership of oil and gas said that surface owners over a common pool may take and keep all the oil and gas which can be taken from their wells, regardless of possible drainage from adjoining land or diminution of flow in neighboring wells.[11] This was the "Rule of Capture" and with only little modification remained the legal basis of production until the 1930's.

The fugacious character of oil also created unique problems for leasing arrangements. The interest of the lessor prompted quick and diligent drilling operations, but a lessee, holding many leases, with limited finances, drilling equipment, and skilled labor, conceivably would prefer a cautious drilling schedule for all his leaseholds.

Early leases attempted to protect the lessor and stipulated a given period of time in which a producer (lessee) had to begin drilling a well. But it remained for the courts to decide whether the lessee had pursued his obligation to drill with "due diligence" and if failure to do so constituted forfeiture of the lease. This position, in sympathy with the lessor's interest, tended "to force immediate exploitation," thereby neglecting the issue of conservation, and reinforcing the consequences brought about by the "Rule of Capture." (For a more complete presentation of the "Rule of Capture," see Appendix E.)

The legal framework gave credence to the contemporary desire to "get the oil out fast"; the cost-price profitability structure allowed rapid development with the risk of only minimum penalties from the market; together they gave production its competitive dress, its historic pattern of development, its problems, indeed its success.

This was the setting in which producers were to operate in the coming decades. Had a producer in Oil City or Titusville known what the future held, perhaps all would have been different. What is more likely, even with a "crystal ball," the producer of 1874—much like his counterpart ten, twenty, fifty years later—would have persisted as long as there was a scent of oil in the air. And in 1874 there was the smell of crude in innumerable towns, streams, villages, and districts. The only question was where the next "black gold" would be struck.

Production in the Lower Region

As shown in Chart 15:2 productivity of the Oil City–Tidioute regions —for the most part the sole producing district for the first decade of the industry's history—began a downward slump in 1871 which with only slight improvement in 1875–77, fell off until the district finally yielded less

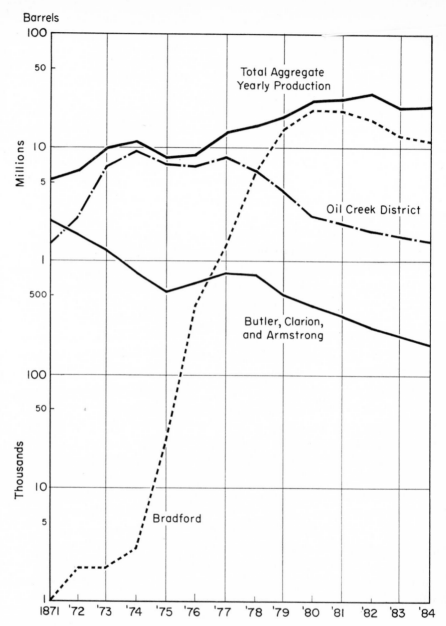

CHART 15:2. *Average annual production of crude from major Pennsylvania fields, 1871–84.* (*Source:* Pa. Industrial Statistics, 1892, 85B, 87B; Carll, *Geological Survey of Pa.,* 7th Report on the Oil & Gas Fields of W. Pa., for 1887, 1888, 40.)

than 200,000 barrels yearly. But drilling activity had early moved beyond the confines of this historic area, maintaining and even increasing national petroleum output, so that by 1874 a producing area had been outlined in Pennsylvania that extended like a giant inverted cornucopia.

With the base in Jefferson, Washington, and Allegheny * counties, the field stretched a hundred miles to the north and east, including all or parts of Butler, Venango, Warren, Crawford, Lawrence, Armstrong, Clarion, and McKean counties, with the tip terminating in Cattaragus county in New York State. Expansion during the following decade moved in two directions: one thrust in 1872 was south into the lower fields of Butler and Clarion counties. The other expansion moved northeasternly into Warren and McKean counties where the Bradford field was opened in 1877. In the excitement that permeated the Regions after one strike followed another, there was hardly a man without some idea of where oil could be found. He could usually find someone to agree with him, and both would reinforce their belief by putting the drill to the ground.

This was particularly apparent in the "Lower Regions," where one of the mainsprings in opening the new fields was C. S. Angell, who backed up his "belt theory" for locating oil deposits by drilling. By 1870 there was activity two miles southwest of Parker near Bear Creek, as well as around the mouth of the Clarion River. Eventually a tract of proven land was outlined 35 miles long and 3–5 miles in breadth. St. Petersburg, Antwerp, Turkey City, Petrolia, Parkers Landing, Karns City, Modoc, Millestown, and many other towns, villages, and cities flourished along this narrow strip. "The opening of every new section," a contemporary journal reported, "was the signal for the formation of a new city." [12] Petrolia and Parker's Landing became centers of the oil trade. Parkers' fame in particular spread far and wide.

The feverish excitement to drill soon made Butler, Clarion, and Armstrong counties the most important oil producing district in the state. In 1871 the district's output was 1.4 million barrels, or nearly 30 per cent of the entire U. S. production. [13] By August, 1872, Butler and Clarion flowed oil so rapidly and prodigiously that available tanks and pipeline terminals were unable to accommodate it. [14] In the same month the price of crude

* Allegeny (Alleghany) County—both spellings were used in contemporary literature—in New York is not to be confused with Allegheny County in Pennsylvania. The New York county was part of the Bradford field and became important about 1881–82. Allegheny County in Pennsylvania, in which Pittsburgh is located, was a part of the Southern field and became an important oil producing district in 1887.

was a dollar less than it had been in August of the previous year, and was approaching $3.00 per barrel.[15]

The producers looked upon these prices with great concern, indeed disdain. The Petroleum Producers Association, formed in the heat of battle over the South Improvement Company contract, was revived and through a Producer's Agency attempted to control production by forbidding drilling for six months in order to rally prices. In the famous "Treaty of Titusville," the Agency contracted with the Refiner's Association, formed during the summer of 1872, to sell oil at $5.00 per barrel throughout the fall of that year.

Although Rockefeller, in a conciliatory mood, bought several thousand barrels of crude during the life of the treaty, the small producers viewed the whole affair with alarm. Most of them felt the agreement aided mainly their larger competitors. Furthermore, forced idleness in the Regions was a vain illusion. Public opinion was not enough to curtail drilling, particularly in the new and virile Butler and Clarion fields, for no matter how persuasive the argument, it held little attachment for producers engaged in the exciting and profitable occupation of opening new territory. When in January, 1873, the alliance collapsed, it came as no surprise to those seasoned in Oil Region affairs.[16]

Meanwhile producers at Butler and Clarion went about the serious business of drilling. By the end of 1874, with over 1,000 wells drilled and completed, the district produced 9.1 million barrels of oil—83 per cent of the state's entire output. Existing transportation and storage facilities were pressed to handle this output and great effort was expended to push lines and build tankage. By 1874 the 2,000 odd miles of two-inch pipe that covered the Regions only two years previous had been more than doubled. Still, large amounts of crude were run into nearby creeks and farms before the situation was brought under control. This was no small-size boom.[17]

The year 1875 began with inventories of nearly 4 million barrels and prices just above $1.00, and still intensive drilling continued, although yields from new wells were not as bountiful as they had been only two years ago. Once-prolific wells had settled, and their yield was considerably reduced. All of these factors contributed to an approximate 20 per cent decline in output over the previous year. In the following two months, despite continued drilling activity, yields were smaller, dry holes more numerous, and output fell an additional 300,000 barrels. An increase in refinery demand and a general improvement in the crude market reduced inventories, and prices rallied to $2.56 per barrel, the highest they had been for some time.

Their heads loggy from the "much better" prices, producers proceeded again to pepper the lower Regions with the drill; new wells completed increased by nearly a thousand, and output in 1877 rose to 8.5 million barrels. Market conditions remained stable, however, and prices dipped slightly to $2.42, with only about 600,000 barrels of current output added to inventories. At the same time Bradford, on the brink of prolific production, helped raise the entire output of Pennsylvania to 13.1 million barrels, an increase of about 4 million barrels over 1876 for the entire state. This signaled the beginning of the end for the Butler and Clarion regions. After 1877 their productivity plummeted, and by 1884 output was annually less than a million barrels. (See Chart 15:2.)

The extent of Bradford's productive capacity was not yet known, and its weight not yet felt in the marketplace. Butler, Clarion, and Armstrong had for the time answered the question of whether or not commercial quantities of oil existed beyond the Tidioute–Oil City regions. Pennsylvania still reigned supreme as the country's premier oil-producing state. Here competition was tough, stiffened by the surging, plusating, energies of men intent on capturing a fortune. But when Bradford, like a quick and decisive stroke, purposed to more than take up the slack in production that set in at the Lower Regions in 1878–1879, a restless many quickly packed their belongings and headed north to McKean County.

Bradford

Three years after the Drake well there occurred the first attempt to find oil within the limits of the Bradford pool itself. Two wells had been drilled, but both stopped before reaching 1,000 feet and failed to tap the Bradford sand.[18]

In 1864 one Job Moses bought 10,000 acres of land in Cattarangus County, New York, about seven miles north of Bradford. Moses probably drilled six wells. The following year Jonathon Watson of Oil Creek also bored two holes.[19] In 1868 a small pumping well was struck near Bradford.

It was possibly because of a geological miscalculation that most of these initial efforts resulted in failure and that the field was not developed at an earlier date. Most of the wells hoped to find the "oil sands of the Venango group at about the same depth as along Oil Creek in Venango

County. . . ." [20] Actually the Venango oil sand group (600–800 feet depth) cropped out at Bradford and the paying Bradford sand occurred at greater depths (1,200 feet). In any event, as 1870 approached, the area had not excited the interest of producers throughout the state. Pleasantville and Shamburg in the Upper Oil Creek district were in the midst of wonderful production, and operators were apparently reluctant to move to other areas. For the time being, Bradford was a sleepy little hollow of fewer than 300 inhabitants. [21]

In 1871 three companies were organized at Bradford to prospect for oil in the Tuna Valley. Early in the summer of 1872 one of the companies struck a 20-barrel well. Ordinarily such a find would have made oilmen stand up and take notice; but Butler, Clarion, and Armstrong were at their height, and the event went unnoticed. During 1874–1875 several more 20-barrel wells dotted the scene. A 200-barrel producer was completed late in 1875, about three miles north of Bradford, and others of similar proportions were soon discovered. By December the Olean Pipe Line, with eleven miles of gathering lines, was in operation, and at least one entrepreneur was so firmly convinced of the area's potential as to begin buying Bradford crude for cash. [22]

A sure sign that a new place name was about to be added to oildom was a plea by the local newspaper for a drastic reduction in the number of "gin mills" and a "more favorable observance of law and order." [23] The action was well advised, for within a month after the new year an exodus of producers from the lower country began. These men "have faith in the future of the field," wrote a contemporary observer, "and are proceeding with developments as fast as conditions will permit." [24]

Drilling activity began to steam-roll. In mid-January, 1876, the *Cattaragus Republican* counted 18 producing wells and nearly 40 others drilling. Within a month there were 37 producing wells and over 100 drilling, of which only three were dusters. [25] The rush to Bradford began in earnest.

Inheriting all the growing pains of a new oil field, Bradford experienced wretched transportation and storage facilities, which made development, even residential settlement, an affair of extreme hardship. Population increased to more than 3,000, and producers still flooded the area from all over the state. In classic understatement, the *Bradford Era* observed in December, 1876:

> The need of more tankage and pipe-line facilities is becoming apparent . . .
> and the question of what to do with the oil is becoming the paramount one in
> the minds of producers. [26]

By December, 1877, with pipeline runs averaging 7,500 barrels daily, the need for more tankage and transportation facilities was strict necessity.

But onward went the drill and production soared. Kendall Creek was now "Mecca of the oilmen" and together with Tarport, Knapps Creek, and Quintiple Tract became added placenames to the oilman's stock-in-trade vocabulary. Anything that was large enough to set a rig on was potential oil land. There was oil in the air, the water, and the land. One contemporary reported the streams of McKean County as being "literally rivers of oil." [27] A more poetic, even health-conscious observer, fashioned from Bradford experience a line soon to be well known throughout oildom: "Oil, oil everywhere, and no untainted water to drink." [28]

By the end of 1877 all available storage facilities were filled to overflowing, and pipelines were strapped to handle the fields' prodigious output. The Empire Transportation Company had early thrown a network of pipelines in the field, but they were far from adequate. In 1878 Bradford produced 6.5 million barrels of crude—42 per cent of the nation's entire total. Three years later Bradford's output was 50 per cent greater than the entire production of the country in 1878. In 1880 there were approximately 7,500 producing wells; one year later there were 11,200, and by 1883 output totaled 23 million barrels, about 83 per cent of the nation's total. Such rapid development was unequaled for at least another quarter century, and in terms of total output, Bradford was one of the nine all-time largest producers in the United States.

By its very magnitude this strike shook the Pennsylvania Oil Regions in every direction. Prices fell, stocks accumulated, storage tanks overflowed covering the earth with black scum, fires were rampant, and wastage was more than any man could accurately count. One vague estimate placed the loss of crude through fire during 1880 at nearly half-million barrels, and in May of the same year possibly as many as 576 rigs were destroyed as a result of brush fires.[29] But oil-scented trees, farm land, and drinking water that tasted like petroleum were not obstacles to driving holes into the oil-bearing sand. Landowners insisted, and producers found the attraction far too irresistible.

"Immediate shipment"

Markets, transportation, and prices all felt the impact of this unrestrained production. Moreover, the producers shared a growing and well-grounded fear that despite the collapse of the South Improvement scheme, Standard was rapidly assuming a dominant role in the purchase, transportation, and refining of petroleum. Rockefeller and his associates made every effort to conduct their moves in these areas under the utmost secrecy. But if the extent of Standard's control over refining and its relations with the carriers remained somewhat obscure, the operations of two Standard affiliates between 1877 and 1879, the United Pipe Line Company and Jabez A. Bostwick & Company (the crude oil purchasing agent for Standard's refineries), gave the producers abundant evidence of the power of Standard to affect the marketing and pricing of crude.

Events that dramatized the position of Standard as the dominant purchaser and shipper of crude came in the Bradford field, where, late in 1877, Standard's United Pipeline was having difficulty in handling the rapidly increasing output. By December, United's pipeline runs were averaging nearly 9,000 barrels daily. Storage tanks were full, even though the firm was feverishly building or leasing new capacity.

It was against this background that United announced to Bradford producers that no further crude would be run from the fields through the company's gathering lines unless it had already been sold for immediate delivery. This announcement, tagged as the "immediate shipment" order, brought quick and violent protests from producers.* By refusing to run crude from the wells unless it was sold, United forced many producers into distress sales, primarily to Standard agents, at prices ranging between 2½ and 25¢ a barrel below the prices quoted in the oil exchanges, which by 1879 averaged below $1.00 per barrel.

Resentment over the effects of the immediate shipment order brought the Bradford producers into the recently organized Petroleum Producers' Union. The Union proposed two lines of action. One was to organize a shutdown movement which they felt would stem the Bradford crisis and in general benefit the producing interests of the Regions. The other was to lend support to various plans for extending an independent pipeline to the seaboard.[31] Neither effort was successful.

It is true that construction of trunk pipelines began in 1878 but it was nearly 1880 before they began to operate effectively.† Restricting output, which could have greatly aided the producers position, also languished. Stopping the drill in virgin and virile oil territory was virtually impossible.

---

* As the *Bradford Era*, tongue-in-cheek, wrote, "The immediate shipment screw which has been turned on the Bradford producers is the cause of a great deal of profanity."[30]

† See Chapter 17.

As a result, by the summer of 1878 Bradford's daily output rose to between 15,000 to 20,000 barrels—double the output of some six months earlier. United was adding between 2,000 and 5,000 barrels per day to its stocks. Despite increased storage capacity, tanks overflowed in June and again in July. [32]

While the producers were aware of the difficulties of providing storage facilities at Bradford, they nevertheless blamed the Rockefeller combination for not taking other steps to relieve the problem. More specifically they charged that there was sufficient equipment to move the oil from Bradford to other storage areas. Standard, they were convinced, was using the "immediate shipment" order as a means of obtaining cheap crude. [33]

Standard agents added to the resentment of producers when in August, 1878, they refused to buy crude at any price because of an alleged tank car shortage. Crowds gathered at Bradford, and violence seemed imminent, until tension was eased by an announcement that United Pipe and the railroads were to be prosecuted in the Pennsylvania courts. Ten days later, United rescinded its "immediate shipment" order. [34]

The experiences of 1877–1878 were but a preview of the reactions that followed the reintroduction of "immediate shipment" by United Pipe Line early in June, 1879. In what producers had by now come to fear most—a public notice—United announced at this time that it could only accept a fixed amount of crude daily from producer wells and that anything beyond this amount would be accepted only if the producer provided his own tankage. [35]

In defending its order, United argued that output had so far outrun available storage facilities that it could not receive any more crude in its storage tanks until the owners of the crude already in the tanks sold it and it was physically removed. Sale of the certificates representing the crude was insufficient to consider a tank as emptied. Under these circumstances United stipulated that until further notice only such crude that was destined for immediate shipment to some buyer's shipping point would be run through its pipeline facilities.

United's announcement led to an immediate agitation for the producers to unite and to bring pressure for a cancellation of the order. [36] Yet there was more to the situation than could be met by simple resistance. In the interval between the cancellation of the first "immediate shipment" order and the new announcement, output of crude had continued to rise at an increasing rate, and by June, 1879, had reached a daily average of 36,615 barrels, nearly twice the rate recorded in June, 1878. Storage facilities were full or overflowing, and, although new tankage was still being built as quickly as men and materials would permit, the question of adequate capacity was a serious one. United claimed to have 1.6 million barrels tankage of its own, an additional 2.3 million under leasing contracts, and 800,000 barrels under construction.[37] But by the end of July and throughout the next 60 days, stocks were added to at an average daily rate of 10,000 barrels, not counting what was being run into nearby creeks and farms. Total pipeline stocks by the end of July and the middle of August were between 6 million and 6.2 million barrels.[38]

The pipeline's plea that all tankage was full may have been justified. United's critics, however, claimed that there was plenty of empty tankage in the lower fields (Butler and Clarion) and that had the firm wished to connect Bradford production with this idle tankage the situation could have been greatly eased. The indictment was probably correct. Output in Butler, Clarion and Armstrong had decreased by nearly one half from the previous year to a paltry 1,166 barrels daily average.[39] Eventually United did connect this idle tankage with Bradford but not until late in 1879. In fact, the pipeline connecting the Lower Region was not started until the middle of September. It is quite possible that this delay may have been due to administrative difficulties within the company. But the fact that, despite a storage problem at Bradford that showed no sign of abating, United delayed some eighteen months before moving to utilize the southern tankage raised serious doubts about the motives of the Standard organization.[40]

Whether the crisis was precipitated by Standard is difficult to say. It is certain, however, that the Standard organization was quick to turn events in its favor. United's order, whatever its justification, presented producers without storage tanks of their own with three equally unattractive alternatives: to run their crude to waste; to sell to "scalpers"—producers with extra tankage who bought crude at 20¢ or more off the exchange price; or to sell at similar scalper's prices to Standard's purchasing agency, J. A. Bostwick & Company.[41] Since Standard by this time controlled between 85 and 90 per cent of the industry's refining capacity, most sellers had little choice but to sell to the Standard agency, however distasteful this may have been.*

Many tank owners (so-called "tankers") were placed in a similarly difficult position. Some 670 producers, with storage capacities that ranged from 1,000 barrels to less than 250 barrels, had leased their tanks to United

---

* Standard's acquisition of refining capacity is treated in Chapter 14.

Pipe Line in return for guaranteed pipeline accommodation on demand. One reason why they had done so was because it relieved them of a large portion of storage costs. More important they wanted the maneuverability that a permanent tankage dump close to trunk lines offered when their leaseholds were widely scattered.

The tanker's understanding (and this was the interpretation given the leasing contracts throughout the Regions up until this time) was that the leased facilities were theirs to use upon demand, constrained only by the sale of a certificate in order to have the tankage considered empty. Denied this interpretation by the United order demanding physical removal, the tankers claimed to have been deprived of their legal rights. Both parties met formally at least twice during July and August in an attempt to arbitrate their differences. But the pipeline held strong to its position, and, although individual suits in equity were started against the line, United was still able to force tankers to comply with its order.*

Under these circumstances, tankers faced essentially the same alternatives as the producer with no tankage. They might sell to independent scalpers, but the majority turned to J. A. Bostwick & Company, which was in a position to make an immediate shipment of all crude it purchased, thereby giving the tanker an equivalent amount of storage space in United's tankage once the crude was removed.[43]

The operations of Standard's purchasing company during this period were hardly calculated to generate good will among the producers. But it was the so-called "sealed offer" phase of the Bostwick company's buying policy that aroused the greatest wrath among the Region's press and producers. Introduced in the late spring of 1879, under this arrangement Bostwick & Company announced that it would accept only written offers from producers, stating how much crude they wished to sell and their lowest price. These offers were received during the forenoon, and at 12 o'clock, when United indicated the amount of storage available for Bostwick & Company, the firm's Bradford agent, Joseph Seep, would, starting with the lowest bids, equate the crude offers with the available shipping space. As the *Oil, Paint & Drug Reporter* observed, this plan "simply arrays the producers against one another and is the most effective method possible of 'bearing' the prices." [44]

Yielding to the protests of producers' and the Region's press, the "sealed offer" program was abandoned late in July. In its stead, Bostwick & Com-

---

* For whatever it was worth, the tanker's position was considered "sound" by Bostwick.[42]

pany thereafter simply posted a daily price for oil they would accept for immediate shipment.[45]

While it insured a uniform price, this change brought little improvement as far as the producers were concerned. Quotations ordinarily ran between 20¢–30¢ a barrel below the closing price on the Oil City Oil Exchange; in fact the latter quotations became little more than nominal. As far as most producers were concerned Standard was still using its bargaining position to obtain "cheap" crude for its refineries at the expense of the sellers.

Just how much crude was moved by Bostwick & Company under "immediate shipment" terms is impossible to determine. The *Titusville Morning Herald* reported throngs of producers hanging around the Bostwick office, apparently in an effort to intimidate any one from selling to the Standard buyer. But the *Herald* in the same issue reported a conversation that seemed to typify the attitude of many producers. "I am not only in debt," this anonymous producer said, "but want money in the worst kind of manner, and should consider myself very foolish to lose both my oil and my money too, at the dictation of anybody. . . . Working for a principle is all right but it don't feed the babies."[46] Apparently there were many who shared this view as daily shipments out of Bradford for July, August, and September averaged between 45,000–55,000 barrels, compared to daily pipeline runs that averaged for the same three months between 35,000–38,000 barrels.[47]

Throughout the period that United's "immediate shipment" order was in force, Jabez A. Bostwick not only denied any connection between his operations and United's action but laid the entire blame for the Bradford situation on United. In a letter to the *Bradford Era* early in August, 1879, Bostwick, a one-time director of the pipeline company, claimed he had resigned and no longer had any voice in United's policies. He went on to charge United with bad faith and claimed that if its management had shared his desire to relieve the producers, the "immediate shipment" plan would have worked to everyone's satisfaction.[48] In a subsequent letter Bostwick defended his pricing policy by stating,

> People who have tankage sufficient to take care of their oil have no necessity for selling shipment oil; but people who have no tankage during time of overproduction, as at present, are obliged to provide some way for moving the oil. Of course if we pay the same for shipment oil that we do for speculative oil, then everybody would want to sell shipment oil because there could be no possible question about that relieving their tankage.[49]

United, at least in public, made no attempt to rebut Bostwick's accusations. It did, however, vigorously defend its actions, and in a letter to the tankers' committee argued:

> There are some 9 million barrels of oil held in storage. The markets are clogged with refined. Oil is now being shipped faster than it is needed by the trade. Shipments must soon decline. What is to be done with the rapidly accumulating surplus? No company could guarantee to take care of it, consistent with remaining on a sound financial basis, because the fluctuations of the business are such that in a very short time a vast amount of the tankage erected at so enormous an expense would be practically useless. Besides, unlimited tankage encourages unlimited production, which is suicidal.[50]

Although it incorrectly stated the amount of stocks by nearly 3 million barrels, this and later statements certainly made clear that United did not consider its behavior at Bradford as improper as Bostwick had charged.

In December United cancelled the "immediate shipment" order, even though output continued to rise from the Bradford field.[51] In large part United was able to take this action because of the pipeline connection made in the fall of the year with its empty Lower Region storage tanks. But the situation was also relieved by the prodigious amount of tankage hastily constructed between August and December, variously estimated at from 2.5 million to 3 million barrels.[52]

Events in the summer of 1879 signalled the auspicious future that lay ahead of Bradford; output would rise until in 1883 this field would outdistance all previous producing areas and many yet to come; storage and pipeline equipment were to be taxed to their utmost for yet another two years; and crude oil prices at Bradford would also plummet still farther downward. But the days of "immediate shipment" never returned. Facilities, however tight, were always just sufficient to avert crises. By October, 1881, the United Line was building tankage capacity at a rate of 180,-000 barrels weekly—which occasioned one of the most outspoken trade journals to remark: "Pity that the farseeing Standard couldn't have made a similar calculation two years ago. . . ."[53]

There were ramifications of the "immediate shipment" experience which carried great importance for the future. For one thing, the Bradford crisis and Standard's ability to enforce its policies there represented a barometer of the strength and position which the Rockefeller organization had by this time come to wield. Bostwick's posted prices foreshadowed the decline in the role of the crude oil exchanges as Standard during the succeeding decade increasingly bought its oil directly from producers. But more than any other event since South Improvement Company days, "immediate shipment" solidified the antagonism of producers for the Standard Oil Company. From then, until Standard's dissolution in 1911, producers sought chinks in the combine's armor. Seaboard pipelines, independent refining schemes—any thing or plan they thought would free them from the Rockefeller yoke—always received a sympathetic hearing.

After Bradford

From its climax of 23 million barrels in 1883 the Bradford field gradually declined, and in 1888 production was little over 5 million barrels.* At its peak in 1880–1883 this phenomenal field not only furnished the bulk of the world's consumption but, at the same time, added 5 million barrels to inventories.[54]

This abundance of oil brought prices of all Pennsylvania crude down drastically. In 1877 the average price of Pennsylvania crude per barrel was $2.42. In 1882 it was under 80¢ and in July of that year reached the "depth below the depth" of 49½¢ per barrel.[55] Bradford was primarily responsible for these prices, but the low levels of 1882 were in no less measure caused by events outside that field.

In 1882 a prodigious oil strike in Cherry Grove Township, Warren County, caused the market to spiral further downward. Captain Peter Grace and George Dimick, experienced oilmen, were drilling a wildcat well at Cherry Grove. This was the famous "646," which probably more than any other single well up to that time, excited oil traders. It was shrouded in mystery; a fence surrounded the drilling site, and armed guards held back the inquisitive many. This only served to increase interest, and at least one rumor a day circulated in the Oil Regions. On March 29 came the report: "The mystery on 646 made a flow, the extent of which is unknown, but it exercises a depressing influence on the oil market." [56]

But "mystery" is not without danger and four days later: "The mystery on 646 still guarded and would-be intruders are warned away with the

---

* Bradford wells were destined to have a long life largely because of the success of secondary recovery methods, primarily water-flooding, early in the twentieth century. During the present period, however, the depletion of Bradford wells followed the normal pattern—large initial production dramatically tapering off after the second or third years, and then falling even further until output was nominal. (See Chart 15:3.)

Thousand Barrels

Wells cleaned out

CHART 15:3. *Annual average production from Bradford lease of Bird and Ball, 1878–91.* (*Source: Chas. R. Fettke,* The Bradford Oil Field, *294.*)

A Reminiscence of Three Dollar Oil

Seventy Cents Oil — Its Cause — And Effect

Contemporary impressions of the effect of price changes in crude on producers.

contents of shot guns whistling about their ears." [57] Within 24 hours still another report: "The mystery on 646 flows every three hours, say persons who have been near the well, the flows being of three minutes duration." [58]

With oil scouts dangling from every available tree and bush, the flow of news from "646" grew every day; business at the Exchange was conducted in a tense atmosphere of uncertainty, with "everybody trying to guess the 646 conundrum, and no one satisfied with the information obtained; the trade badly mixed on it." [59]

On May 18, "646"'s capacity became known: "First estimates place[d] its production at 100 barrels; then it increased to 150; at 3 P.M. estimated good for 200 barrels; 6 P.M., estimated good for 500 barrels; began flowing steadily. . . ." [60]

The success of "646" caused one of the most colorful furors yet seen in the Regions. On May 19 this wildcat was daily producing 1,000 barrels—and some said 2,500 barrels. The scramble for leases was on. Land prices soared, and by September 1, only four months after "646," there were 215 producing wells in the field. With the help of plenty of glycerine, daily output rose to 40,000 barrels; and within three months inventories were increased by nearly 1 million barrels. Prices dropped to a low of 49½¢.

The furor died down just as quickly when Cherry Grove "proved but a flash on the petroleum horizon." [61] By the end of October the yield had dropped to 9,000 barrels per day and the collapse of the pool was imminent. The average for 1883 was only 2,000, and for 1884 merely 400 barrels per day.[62] Thus ended one of the briefest and most colorful spurts in petroleum production history. Innumerable producers who had descended upon Cherry Grove, particularly Bradford operators, started the long trek back to where they came from, or to wherever else the call of the drill would next come.[63]

With the decline of the Bradford field and the almost complete extinction of Cherry Grove, production in Pennsylvania expanded once more rhythmically southward. In September and October, 1884, gushers were opened at Thorn Creek in Butler County. The Phillips, Christie, and Armstrong No. 2 wells, located within a few hundred feet of each other, again stirred producers. Each of these wells produced between 4,000 and 10,000 barrels per day! What made their performance even more spectacular was that at first they appeared to be dry. Armstrong No. 2, for example, was thought a "duster" until it was torpedoed. Then there occurred, one excited observer wrote, the grandest scene ever witnessed in oildom. When the shot took effect, and the barren rock, as if smitten by the rod of Moses, poured forth its torrent of oil, it was such a magnificent and awful spectacle that only a painters brush or a poets pencil could do it justice. . . . It was like the loosing of a thunderbolt. For a moment the cloud of gas hid the derrick from sight, and then as it cleared away a solid gold column . . . shot from the derrick floor 80 feet through the air. . . ."[64]

Within 24 hours after being shot it was estimated that Armstrong No. 2 produced 8,000–10,000 barrels.[65]

There was the usual cry to stop the drill, as the flood of crude pushed prices down. But nothing came of it and production mounted, stocks increased, and prices continued to fall. Coupled with the economic inertia the country was experiencing, 1884 was a discomforting year for Pennsylvania producers.[66]

There were many reflective oilmen who could not share the unbridled optimism of a contemporary journal which wrote, "The present outlook is that the Pennsylvania and New York fields will be able to produce oil in unlimited quantities for . . . years to come." [67] But those who had seen Bradford rush in like a lion and then slowly skid into oblivion wondered. One Oil Region editorialist saw only doom; pointing a warning finger at producers, he wrote, "The region has seen its zenith. Not able to deny the fact, oilmen approach it with reluctance. . . . The downlet career of the oil regions is an assured fact not to be stopped by denials." [68] A few days earlier the same commentator seriously encouraged the Regions to appraise the economic facts of life by developing new manufactures "to take over the place left by the business of refining petroleum and transporting petroleum." * This was a repeat of a warning sounded two years earlier when Bradford wells were being daily shot with nitroglycerin. Convinced that this could not go on forever and the field still live, the *Herald* wrote: "It would be a calamity if no new field is found in the immediate future. . . ." [70]

Strikes south of Butler and Armstrong momentarily allayed anxiety as to where the next drilling location could be found. But these wells were deeper, production more hazardous, and more costly to drill; they were not the "sure things" that Pennsylvania producers had become accus-

---

* The warning was not without justification. By 1880 only ten refining plants remained in Titusville and Oil City. Standard, through Acme Oil Company, had literally gobbled up the many refineries that had once dotted the area. By pruning, combining, and dismantling refineries, Standard reduced refining capacity in this area, one of the few districts in which this ever occurred.[69]

tomed to expect. Two "professional geologists"—who by this time must have become prophets of doom to the producing fraternity—saw no future at all in these strikes. They even went so far as to say:

> There are not at present any reasonable grounds for anticipating the discovery of new fields [in Pennsylvania] which will add enough to the declining products of the old to enable the output to keep pace with the shipments or consumption." [71]

Just what the future held for the Pennsylvania region was somewhat problematical. Sources of supply outside the district were known, and who could say that they would not soon be brought into commercial production? Indeed it was only a short hop across the Appalachian range into Ohio where the question was finally and affirmatively answered.

Chapter 16

Pipelines, railroads, and refineries

By the opening of the 1870's, it appeared that the major technical bottlenecks in petroleum transportation had been broken. A 2,000-mile web of gathering lines enabled any producer to run his crude from wells to railhead and river banks at one-tenth the rates formerly charged by teamster. For longer hauls the three oil-carrying railroad trunk lines had more or less adequately linked the fields and refining points, while a general adoption of the cylinder-type metal tank car increased savings in the transit of crude to refineries. While these advances had led to sharp reductions over the preceding decade, transportation costs at the beginning of 1874 were still among the industry's largest expense items. Time had not erased the rate differentials among various refining centers, nor brought an end to personal discrimination.

It was from attempts to overcome these differentials that the foundation was laid for the evolution of the long-distance pipeline. But until the practicability of this major innovation was fully demonstrated at the end of the decade, the transport needs of the industry were met by an expansion and multiplication of existing techniques—new railroads, more tank cars, and the spread of gathering lines.

As might be expected, in view of the tremendous expansion in the production of crude and the discoveries of new fields, critical shortages of pipelines and storage facilities did emerge. Yet most members of the industry were less concerned with the adequacy of transport facilities than they were over rate levels and rate structures. Producers and refiners alike were convinced that high rate levels would reduce their operating margins; and their competitive positions were markedly affected by the maintenance or imposition of discriminatory rate schedules.

These considerations lent particular significance to the revived agitation in 1874 by major trunk lines to form a pooling agreement to raise freight rates on all commodities, including crude petroleum. But more than railroad rates were involved. Early in 1874 the gathering lines also considered the possibility of a pooling agreement that would meet their problems.

Pipeline competition

The difficulties facing the gathering lines grew out of the movement of production into the Lower Regions. By 1872, pools in Butler, Clarion, and Armstrong counties produced some 2.5 million barrels, approximately one-third of Pennsylvania's total crude output. A year later output from these fields rose to 6.9 million barrels, about two-thirds of the Pennsylvania total.[1] Despite hurried, even frantic, efforts to push lines and build tankage, waste oil blackened the ground and scummed the creeks of the new districts before the situation was brought under control.[2]

The great demand for transportation provided many new firms with opportunities to enter the pipeline business. The Antwerp line, established the year before, began expanding its tankage to 90,000 barrels in February, 1873, and later in the year completed an extension from Clarion county to Oil City.[3] Another new company, somewhat pretentiously named the Grand Trunk Pipe, was organized in fall and began "dragging its slow length" over the thirty-odd miles from Foxburgh to Titusville in November, 1873.[4] With $200,000 subscribed by "prominent producers" around Parker's Landing, the Ocean Pipe Line Company pro-

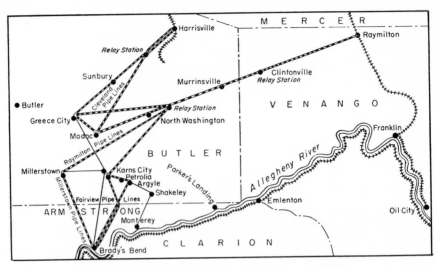

MAP 16:1. *United Pipe Line System, 1874.*

posed to run lines from the Allegheny Valley Railroad "all through the lower oil field." [5]

Alongside these fledglings, the larger established companies moved into the Butler-Clarion-Armstrong fields. The Empire Transportation Company, acting as the right hand of the Pennsylvania railroad, was apparently the first and most active. In April, 1872, it had acquired the Mutual, a company with 70–100 miles of line, and in the following September purchased the Union Pipe Line, adding another 125 miles to its holdings in the lower fields. In January, 1873, it began constructing a 60-mile extension into Butler. In October all of these lower division holdings were consolidated into the Union Pipe Line, a company whose 300 miles of line could run 26,400 barrels per day into tanks with an aggregate capacity of 300,000 barrels. [6] Since the pipelines were dealers as well as transporters of oil, these Empire acquisitions enabled the Pennsylvania to obtain a lion's share of the flush production.

Challenging Empire's power was the extensive Vandergrift and Foreman network. Long engaged in transportation and pipeline operations in the older fields, Vandergrift and Foreman began building a small seven-mile line in Butler as early as May, 1872; in September completion of the thirty-mile "Cleveland" lines, in which the Standard Oil Company also had a small interest, marked their second extension into this area; by fall of 1873 they had built and acquired a system that had a total capacity of 20,000 barrels per day. At this same time Vandergrift and Foreman also consolidated their holdings in the southern fields into a single organization, United Pipe.[7]

United was the second largest gathering system in the developing territory, but its importance was not due to sheer size alone. By supplying the Lake Shore and the Atlantic & Great Western with pipeline connections, it strengthened their positions, *vis-à-vis* the Pennsylvania: the New York roads had little track of their own into the new districts, and had heretofore depended upon the Pennsylvania-controlled Allegheny Valley railroad to deliver crude from pipemouth to their southernmost terminals around Oil City.[8] So it was understandable that William Vanderbilt (son of the Commodore) and Amasa Stone, of the Lake Shore, should each subscribe to one-sixth United's capital stock. Secondly, another third of the company's capital was taken by the Standard Oil Company, which contributed $233,333 in cash and its interest in the "Cleveland" lines.[9] This investment marked Standard's entrance into pipeline development in the lower fields on a major scale and subsequently provided Rockefeller with a powerful competitive tool.*

This expansion by old and new companies alike boosted total pipe mileage in the Regions from 2,000 miles in 1872 to 4,000 two years later.[10] It not only bridged the first temporary gap in pipe and tank facilities that immediately followed the opening of the new fields, but in many areas converted a deficit into a surplus of capacity.†

Early in 1874 there was an estimated half-million barrels of empty tankage in the lower districts—despite reports of large amounts being held for speculators—while in the Regions as a whole approximately one-third of total tank capacity went unused.[12] By mid-1874 one estimate placed pipeline capacity at more than twice that required by production.[13] But at the same time, local newspapers continued to report new construction as new pools came in.*

This apparent paradox—excess capacity on the one hand coupled with

* The reader will recall that, under a secret contract in 1868, Rockefeller had acquired a one-quarter interest in Gould's Allegheny Transportation Company (later known as the Pennsylvania Transportation Company), whose extensive lines were important in the northern fields.
† As early as May, 1873, the Ocean line, that grandiose scheme to run pipe throughout the lower districts, was taken over by Empire's Union Pipe Company; the cause, opined the *Titusville Morning Herald*, probably lay in "the recent large increase in pipe facilities rendering many new lines unnecessary." [11]
‡ In fall and winter of 1873, the Antwerp built from Clarion County to Oil City, and the Grand Trunk started edging towards Titusville. During the same period, Vandergrift and Foreman began laying pipes in the Millerstown field and the Grand Trunk finished its extensions from Parker to Millerstown.[14]

still further expansion on the other—points up the basic problem facing the gathering lines. A pipe could neither be ripped up nor reduced in diameter simply because the output of the wells it served had dropped by half. And it was common knowledge that wells that had come in big, as those in the lower field had, more often than not declined precipitously within a few months. As the fields expanded, this process tended to cumulate, for each new well subsequently became an old one whose output would use its pipeline connections to less than their full extent. The situation was aggravated by more gradual expansion of the southern fields after 1873. Companies with excess capacity in declining areas thus had less opportunity to bolster their profits by moving into newer pools, where transportation was at a premium. New construction, prompted by overestimation, or promoted to duplicate existing facilities for competitive reasons, added still more to excess capacity.

Sporadic rate-cutting and intermittent rebating had been fairly common among the gathering lines, but under the competitive pressures built up by 1874, these practices became particularly marked and widespread. To move oil for which they held title, the transportation companies reportedly sold it at less than cost; open rates of 30¢ per barrel were reduced to one-half and one-third that amount.[15] Some lines incurred absolute losses; others at least saw their profits reduced; and all faced the likelihood that rates would be slashed even further.*

This unprofitable prospect was brightened in summer, 1874, when Joseph Potts, of the Empire Transportation Company, proposed a scheme to restore rates and check competition. In this proposal he was joined by Henry Harley, whose Pennsylvania Transportation Company had an extensive gathering system in the upper Regions.

Their proposal covered more than pipeline traffic; it was an integral part of a broader plan to control competition for petroleum freight all the way from well to market. The trunk lines could not avoid becoming involved, and a brief review of their situation at the time will indicate why they were more than ready to try this type of scheme.

The railroad-pipeline pool of 1874

Although 1873 marked a high point in railroad investment, it also opened a dismal half-decade for trunk lines tapping the upper Mississippi

---

* There are very little data on pipeline costs at this time, but they probably averaged around 8¢–10¢ per barrel. A pipeline pool organized in 1874 fixed the rate structure so that any line not in the pool would have to cut its net rates to 8¢ per barrel to compete; presumably this figure approximated, or was slightly less than cost.[16]

Valley.[17] In September the failure of Jay Cooke & Company, financiers of the Northern Pacific railway, precipitated a panic that spread from Wall Street to the commercial industrial and agricultural sectors of the economy. Merchants forced to operate on narrowed margins and reduced volume, and farmers burdened with a heavy mortgage debt clamored for special railroad rates. At the same time, railroad facilities into the grain and meat belt of the Midwest had caught up with—in many cases exceeded—the needs of the area, providing the trunk lines with still more incentive to accede to the demands of their customers. Under these pressures the rate structure crumbled, and soon railroads were quoting rates as low as ten cents per hundredweight on grain from Chicago to New York.

A bad situation was made worse by the resurgence of the Baltimore & Ohio railroad and the belligerent attitude of its president, John Garrett. Chiefly engaged in carrying materiel during the Civil War, the B & O had been excluded from the through trade to the seaboard, but since the late 1860's had been skirmishing on the periphery of trunk-line operations, extending its connections into New York and the Ohio coal fields. In 1873 the road made its real bid for power as it started pushing toward Chicago. During the following year the B & O's president, determined "like another Samson . . . [to] pull down the temple of rates around the heads of these other trunk lines," [18] sniped at his competitors in a series of hit-and-run encounters.

Garrett's moves in regard to the petroleum trade were initially of little significance, but ominous in the highly charged atmosphere. The B & O had completed its Connellsville branch to Pittsburgh in June, 1871, breaking the Pennsylvania's monopoly of that city's trade. Despite expectations that a large oil traffic would thus be diverted to Baltimore, the B & O was hauling only 3,000 barrels per month to the Maryland port by early 1874.[19] But by this time a pipeline from the oil fields to Pittsburgh was being projected, and Garrett was reportedly negotiating to carry oil from the line to Baltimore.

More immediately disturbing were the maneuvers of the Erie-Atlantic & Great Western roads. Having lost virtually all of their refined oil freight to the New York Central-Lake Shore lines in late 1873, they opened negotiations with Standard for the "largest [shipments] made by any single oil refining organization" to that date. In April, 1874, both signed a contract guaranteeing the Erie and the Atlantic one-half of Standard's eastbound shipments in return for a promise to grant Standard rates as low as those which competitors might receive from any trunk line. How

Vanderbilt would react was uncertain. The New York Central and Lake Shore lines had been carrying the bulk of Rockefeller's eastbound oil for the past year. Although there were signs that Vanderbilt was beginning to doubt whether this freight was worth fighting for, he could easily take drastic action.*

At this juncture, Potts and Harley advanced their plan to organize both pipelines and railroads into one large pool. The timing was apt, for in an effort to restore order to the general freight structure, the Pennsylvania, the Erie, and the Central had arranged a meeting at Saratoga, New York, to devise an agreement; Garrett was invited to attend but refused, thus seriously limiting the effectiveness of the compact.[22]

The petroleum trade was not discussed at Saratoga, but Scott of the Pennsylvania, Blanchard of the Erie, and Vanderbilt did give its representatives a hearing when the triumvirate met a short time later in Long Branch, New Jersey. Potts suggested that the trunk lines divide their oil freight according to certain proportions. New York was to receive 62.5 per cent of total shipments to the three major Atlantic ports, the same percentage that it had in the year ending July 3, 1874. All three trunk lines were to share equally in this traffic, but the Pennsylvania, the only member with decent connections to Philadelphia and Baltimore, was the only road to be allowed to haul the remaining 37.5 per cent going to these ports.†

Potts also proposed a rate schedule that would literally equalize transportation costs to any given port on the Atlantic seaboard for all refiners; in other words, a Cleveland refiner would pay the same amount to ship his oil a total of 780 miles from well to New York as would his competitor in the Regions who was only 450–500 miles away. This meant that crude petroleum would be carried to Cleveland and Pittsburgh free of charge, while refiners in the Regions would lose any geographical advantages gained by the proximity to crude sources.‡

---

* On December 30, 1873, Amasa Stone, managing director of the Lake Shore, wrote William H. Vanderbilt, president of the Lake Shore, ". . . Referring to changing the contract with the Standard Oil Co., I can only say . . . that I do not see how we could possibly afford to make lower rates *even if it lost to us that trade.*" [Italics added.][21]

† Significantly, a rider provided that the Baltimore division could be revised "if the Baltimore and Ohio Railroad Company effects a connection in future with the Pennsylvania oil regions"—a qualification that, so far as the Pennsylvania was concerned, could cut both ways.[23]

‡ Because of the weight loss in refining, fourteen barrels of crude were considered equivalent to ten barrels of refined petroleum.

The rationale behind this rate structure is not altogether clear. Blanchard was undoubtedly correct when he later claimed that the transportation interests had not been alone in pushing the scheme. Refiners and shippers, both in the west and on the seaboard, he said, "urgently represented . . . it was in every way desirable," although the reasons that prompted the seaboard firms to press for the plan are obscure.[24] Aside from these outside pressures, the railroads themselves may have concluded that this rate structure represented a workable, though not ideal, compromise that would maintain the pool. All signed the pact, which was to become effective October 1, 1874.

On behalf of the gathering lines, Henry Harley argued that unless the pipelines were included in the pooling arrangements the railroad scheme would fail. If one pipe cut its rate, the railroad to which it pumped could gain a larger share, precipitating a rate war and quickly destroying the pool.

The railroads recognized the merits of Harley's argument. On September 4, 1874, nine major pipelines entered a joint agreement with the trunk roads covering rates. Open pipeage to railheads was fixed at 30¢ per barrel; the pipelines were to retain eight cents, placing the remainder in a "common fund" which was to be prorated to members of the pool.[25] The railroads lent a helping hand by adding 22¢ to their own rates, with the understanding that this sum would be refunded to shippers who used pipelines which maintained "agreed rates of pipeage."[26] Thus a shipper using a pipeline not in the compact would have to pay the full railroad rates; the only way an outside line could offset this was to cut its own pipeage to eight cents or less—that is, to levels approximating, if not less than, cost. Oil piped directly to refiners in the Regions was ineligible for the rebate.

Reaction to the proposed scheme was mixed. The *Titusville Morning Herald* stated, "We do not believe that such an outrageous and withal so bungling a system of charges will be submitted to by the people of the Oil Regions."[27] The reaction of the rival *Courier* was more vehement. It warned, "The people will not part with their sovereign rights, nor allow themselves to be ruled by King Pool."[28] Oddly enough, refiners along the Creek offered little organized resistance. They obtained some minor concessions with which they were apparently content, even though the fundamentals of the scheme were maintained.[29]

When the new schedules were put into effect, it was the producers who organized the opposition in the Regions. The new tariffs repre-

sented an estimated 45-cent advance over previous rates and the producers feared that the higher charges would further depress crude prices
below their current one dollar levels. Viewing the cheaper transportation
promised by a pipeline already pushing toward Pittsburgh, they charged
that the pipeline pool would discourage new ventures and completely
negate the already limited free pipeline bill of 1872.[30] A producers' meeting at Parker's Landing denounced the conspiracy, resolved to test the
plan's constitutionality and pledged another shutdown movement—if
enough signatures could be secured.[31]

Pittsburgh refiners, whose long-standing antagonism to Cleveland was
sharpened, were another source of protest. They joined the producers in
pledging support to the Pittsburgh pipeline, and petitioned the city council for rights to lay a product pipeline to the B & O, which was offering
to haul oil to the seaboard at rates 40–50 per cent below those charged
by the pooled roads.[32]

But these defiant gestures came to nothing. "Enthusiastic borers for
half-dollar oil," spurred by new discoveries in Butler, undermined the
shut-down movement.[33] Thus, despite some minor concessions on rate-
levels granted the Regions later in the year, petroleum transportation remained essentially unchanged until the dissolution of the pipeline pool
in early 1875.

Two factors, mutual suspicion among its members and attacks from the
outside, were largely responsible for the subsequent breakup of the pipeline agreement. Its immediate dissolution came when the Erie accused
the Pennsylvania of granting inordinate rebates to its Empire affiliate.
The Erie in turn was suspected, with justification, of rebating to shippers who used outside lines. As time passed more lines were admitted to
the pool. This made policing the compact more difficult and infringement
correspondingly became more common.[34]

These difficulties might have been resolved except for the actions of
the B & O, which throughout had sniped at all types of freight handled by
the Pennsylvania. Scott, reluctant to destroy the pooling agreements, at
first attempted to confine the fight with the B & O to local shipments, but
in mid-February, 1875, the Pennsylvania introduced severe cuts in rates
that brought on a short but vicious rate war among the Big Four. The
elaborate rate structures were an immediate casualty.

The railroad rate war affecting traffic between Chicago and New York
ended in June, 1875, when the Pennsylvania concluded an uneasy truce
with the B & O that provided for the maintenance of rates between the

two roads. But even before the truce had been arranged, Empire was able to revive the pipeline pool in March, 1875, by announcing a new rate schedule. Although charges were to be somewhat lower than under the 1874 scheme, the new schedule still retained the 22-cent pipeage rebate, "as under the late tariff." [35] References to the "associated lines" persisted in the press until the combination was dissolved in June, 1876.[36] The continuation of the pipeline pool plus special rebate contracts between the railroads and Standard prevented open warfare over the petroleum trade from breaking out until 1877.

The Columbia Conduit Company

While the railroads and the pipelines were forging their combination in the summer of 1874, a patent medicine manufacturer turned oil producer, began laying a long-distance crude pipeline from the Regions in the general direction of Pittsburgh. The idea of piping oil directly to refiners outside the Regions was of course not new at this time. A decade before, when Van Sycle proved the workability of gathering lines, there had been a discussion of a pipeline to the seaboard. Following the collapse of the South Improvement Company, Cleveland interests had urged the construction of a 150-mile line to tap the Regions.[37]

But the barriers to a practical extension of principles worked out in the gathering lines to longer distances were formidable and the Pittsburgh line was the first to get beyond the planning stage. While the success of this project was hardly overwhelming, its experience mirrored typical reactions to any major innovation and previewed the types of obstacles subsequent promoters of long-distance pipelines would encounter. Moreover, even before the line began to deliver oil from its Pittsburgh terminus in 1876, it became a factor in the already complicated relationships among producers, gathering lines, refiners, and the railroads.

The man behind the scheme was one "Doctor" David Hostetter, manufacturer of a wormwood and whiskey patent medicine that had endowed its owner with his title and a moderate fortune.[38] A shrewd and diversified businessman, he had participated in the "oil ring" of 1869, an attempted corner of the market led by Jay Gould, and later tried to monopolize Pittsburgh's gas industry. After heavy losses on investment in Butler production, he turned to the pipeline project.[39]

Somewhere along the line Hostetter acquired the charter of the Columbia Conduit Company, issued about a month after the free pipeline bill had been enacted in March, 1872. The Pennsylvania Railroad had managed to have Allegheny County, in which Pittsburgh was located, excluded from the eminent domain provisions of the free pipe bill itself; but for reasons not at all clear, the legislature had empowered Conduit to lay pipe from Butler to Sharpsburg, Allegheny County, ten or twelve miles from Pittsburgh.[40]

In June, 1874, after some $7,000 had been spent on surveys, Hostetter began laying a three-inch line over the 32-mile stretch to Sharpsburg. With booster stations every five miles, he hoped to deliver 4,000 barrels per day to Pittsburgh, at rates 25¢ less than the prevailing 55-cent rate from the wells. Line was being laid at the rate of a mile a day, and operations were scheduled to begin September 1.[41]

Just what Hostetter intended to do when the line was completed was not clear. He conferred with B & O officials on shipping crude to Baltimore; he was said to be in Philadelphia for the same purpose, while the *Pittsburgh Evening Leader* reported that Columbia would simply feed Pittsburgh refineries and no more.[42]

The rumors indicated one of the serious obstacles facing the line: there were too many irreconcilable industry interests. Producers might favor the line—and even their support was limited—but refiners in the Regions opposed advantages to Pittsburgh at their expense. Pittsburgh had smarted too long under Tom Scott to see crude oil diverted to seaboard refiners. Nor would Philadelphia countenance a loss of trade to Baltimore. Hostetter thus found it impossible to rally sufficient consolidated support.

Not the least of Conduit's difficulties arose from the fact that its route at several points crossed tracks and culverts of the Pennsylvania Railroad. The latter would hardly brook further interference with its Pittsburgh trade, especially when the B & O stood to gain and negotiations for the railroad pool had begun.

Conduit first clashed with the Pennsylvania soon after the first pipe was put down. Hostetter tried to cross under a culvert on land of the railroad's West Penn branch at Fairview, ten or twelve miles north of Pittsburgh. Hostetter claimed that since his charter carried the right of eminent domain, the railroad legally had to accept Conduit's bond for damages and allow passage. When the West Penn refused, the matter went to the Butler County courts, which handed down a decision in favor to the pipeline.

Conduit, which had in the meantime continued to build towards Pittsburgh, promptly ran a line under the culvert and the West Penn, with equal promptness, tore it up. More litigation carried the case up to the state Supreme Court where the original decision was reversed in November, 1874.[43]

Conduit's case was still moving through the courts when news of the railroad-pipeline pool broke, and Hostetter rallied at least moral support from the producers and Pittsburgh refiners. The latter, though they feared that Conduit might divert crude to Baltimore via the B & O, saw that the line could also obtain them cheaper transportation to tidewater. Since mid-September they had been petitioning the city council for rights to build a refined products pipeline to the B & O's Connellsville branch. With the triad of Conduit, the product line, and the railway, they hoped to land kerosene at the seaboard for about $1.40, as compared to the $1.96 charged by the pooled pipelines and the Pennsylvania Railroad.[*]

Pittsburgh oil was already being barged and trucked to the B & O when the city council approved the product line in October, but for some undisclosed reason, the project hung fire. The fact that the Pennsylvania Railroad intimated that the New York roads were already rebating to New York dealers, and hinted at rate concessions for Pittsburgh may have been among the reasons for delay.[45]

At the same time, Hostetter was hedging against an adverse decision on the Fairview case by attempting to complete connections elsewhere. Some miles above Fairview, Conduit had leased a stretch of land where the tracks of the West Penn crossed a stream. Hostetter felt that his lease of the land protected him even though the railroad claimed jurisdiction over both the right of way and the stream.[46] Consequently on the night of November 27 (a week after the Pennsylvania Supreme Court blocked the Fairview crossing), a force of Conduit men ran a pipe under the railroad. Hostetter left a "strong party" to cope with any attempts to tear up the line and appealed for an injunction to restrain the railroad. He started pumping oil to Pittsburgh,[47] but a week later the injunction was denied. The West Penn ripped up the line and literally fortified the crossing.[48]

At this point Hostetter abandoned further appeal to the courts and

* Hostetter was to charge 30¢ per barrel pipeage; the B & O offered to carry oil to Baltimore for 90¢; no estimates for piping refined oil are available, but probably costs did not exceed eight cents, the cost of barging the oil to the railroad connection. Assuming 1.4 barrels of crude yielded one barrel of kerosene, total transport costs by this arrangement amounted to approximately $1.40 (1.4 × .30 + .90 + .08). The Pennsylvania Railroad was charging a net rate of $1.54 to run refined oil to Philadelphia, to which must be added 42¢ pipeage (1.4 × .30). Under the pool, crude oil was of course carried to Pittsburgh free of charge.[44]

shifted his battle to the Legislature. In December, 1874, he proposed to submit to the winter session of that body a bill granting "to all parties who so desire" privileges to incorporate pipelines as common carriers, with the same rights of eminent domain, protections for crossings and connections as were afforded railroads.[49] By focusing on a *cause celebre* he apparently hoped to consolidate the laggard support proferred by the producers and Pittsburgh refiners, but paradoxically, he stimulated the opposition.

Against Hostetter and the producers was ranged almost every other oil interest in the state. Fearing that Conduit would increase Pittsburgh's trade, refiners in the Regions opposed the bill. One writer even had some good things to say about the railroad scheme for placing the Creek on the same basis as the Iron City.[50] The *Titusville Morning Herald*, heretofore fairly sympathetic to Conduit, added regionalist arguments, commonly employed in the era when the railroads were shifting relative economic advantages between different areas: the bill would enable Conduit to send crude to Baltimore via the B & O, "to the advantage of foreign capitalists who care nothing for our local prosperity." [51] The same charge was leveled by a not unimportant minority in Pittsburgh, some of whom, significantly, had recently been purchased by the Standard Oil Company. Hostetter's earlier negotiations with the B & O provided some basis for this contention, but most of the Pittsburgh refiners apparently discounted this threat, and the city's Chamber of Commerce worked for the bill's passage in the Legislature.[52]

The Baltimore threat so worried Philadelphia refiners that the city council passed resolutions opposing the free pipeline law. The Quaker City had sustained almost the entire 50,000 barrel drop in national exports for 1874, while exports from the Maryland port had increased 150 per cent over the previous year. In the face of this drop, assurances from Butler legislators that all of Pittsburgh's refined oil would be exported via Philadelphia [53] were less than convincing, especially since Baltimore dealers had guaranteed "every facility," including dredging the harbor, to Pittsburgh refiners.[54]

The Pennsylvania Railroad had ample grounds for opposing the free pipeline bill, and undoubtedly the railroad's opposition counted heavily in the State Legislature. Under the railroad pool, Scott's road hauled all petroleum destined for Philadelphia and Baltimore. If, as seemed likely, Conduit connected with the B & O, the Pennsylvania would either have its share cut, or see the pool disrupted.

The fight against the bill thus revealed inter-regional conflicts which Hostetter could not reconcile but which united opposition to him. On March 4, 1875, the proposed free pipeline law received its "quietus in the State Senate . . . by a vote of nearly two to one." [55] The lower chamber adjourned without taking any definite action on the measure.[56] Hostetter's buoyant optimism over his project dimmed. The pipeline, with small chance of crossing the right-of-way of the West Penn railroad, had siphoned some $400,000 out of his pockets. Hostetter was negotiating with Empire to purchase his property when in April, 1875, he accepted an offer from the partnership of D. McKelvy & Company to take a lease on Conduit.

While they had no firsthand experience with pipeline operations, the partners of D. McKelvy & Company—Byron D. Benson, Robert E. Hopkins, and David McKelvy—had been in the oil business for some years. They had joined forces in 1870 and out of operations that combined lumbering with the production of crude in the Regions had built up a moderate fortune for the firm over the succeeding five years.

Under the terms of the lease, Hostetter agreed to finance new construction and D. McKelvy & Company, in return for a quarter of the profits, was to provide working capital.[57]

The partners immediately came up with an ingenious suggestion to solve the problem of crossing the West Penn right-of-way. Rather than try to complete the connection under the West Penn, they proposed to take oil from one end of the broken connection, haul it across the tracks in wagons and redeposit it in the pipe on the other side of the railroad.

On the evening of June 29, 1875, an ungainly vehicle carrying 100 barrels of crude crossed an intersection of the railroad and a public road without the driver's "being mangled or shot at." Thereafter two wagons daily hauled 5,000 barrels over the crossing. Pipeline connections were made with the Citizens' and American refineries of Pittsburgh, each with approximately 1,000 barrels daily capacity.[58]

Refiners using the pipeline and the B & O were given a $1.15 through rate from the wells to Baltimore.[59] But in view of Garrett's truce with the Pennsylvania, concluded about this same time, it was doubtful whether the rate would continue to bring relief to the independents. Early in 1876, when both roads increased their open tariffs to $1.88 per barrel, the *Pittsburgh Chronicle* lamented, "Neither the B & O R.R. nor the Columbia Conduit Company have proved the Moses of Pittsburgh refiners, to lead them from their troubles. . . ." [60]

Conduit accordingly devised a cheaper route, this time via the Chesa-peake & Ohio Railroad. Pittsburgh refiners, after receiving their crude oil from the pipeline, barged their product 300 miles down the Ohio River to Huntington, West Virginia. Here the oil was picked up by the C & O and hauled to Richmond, to be exported directly or carried to New York by water. The route was almost three times longer than the distance to tide-water via the Pennsylvania Railroad, but it landed oil in New York for about 20¢ less than the trunk lines. If the cargo were exported directly from Richmond, the saving amounted to 50¢ per barrel.* The first ship-ment, consisting of 3,000 barrels of refined product, went over the C & O in May, 1876.[62]

The new route had serious drawbacks: water-levels around the Pitts-burgh docks, for example, dropped seriously during summer months, pre-venting vessels from being loaded during the very season that shipments were heaviest.[63] And despite increasingly strained relations with Garrett, especially after he was known to be favoring Standard Oil, Conduit con-tinued to use the B & O. Thus, by one means or another, the pipeline helped keep a dwindling group of independents alive for about a year and a half, with a tidy profit to itself. In the first ten months of 1876 Con-duit netted $110,000 and by 1877 had supplemented its main line to Pitts-burgh with 100 miles of gathering lines.[64]

Yet Conduit had only limited usefulness to the independents. Not only did it fail to effect any radical change in the market structure; it did not even maintain the status quo. By the end of 1874 several of the largest re-fineries had joined Standard, and within the succeeding three years other independents succumbed. As the number of Conduit's non-Standard cus-tomers dwindled, the two corporations increasingly followed live-and-let-live policies.

Producers in the Regions who had counted on the pipeline to provide lower transport costs and to break the hold of Standard in the crude mar-kets were particularly disappointed. But their hopes for relief were re-vived when Henry Harley, president of the Pennsylvania Transportation Company, proposed in 1876 to link his gathering lines to the eastern mar-kets by the construction of a long-distance pipeline to the seaboard at Bal-timore.

* The *Pittsburgh Chronicle* reported that total cost would run approximately $1.35, but this apparently did not include Conduit's pipeage to the refinery. The Pennsyl-vania and the B & O charged $1.88, but under the pipeline pool this presumably included a 31-cent pipeage rebate (22¢ × 1.4). Lewis Irwin later recalled obtaining a 90-cent rate from his works in Pittsburgh to Richmond.[61]

In proposing such a line Harley took little stock in an earlier engineering estimate that, while "entirely practical" from an engineering viewpoint, the cost of such a project would be over $4 million.[65] His engineer, General Hermann Haupt, a veteran of the Civil War and formerly in charge of the construction of the famous Hoosac tunnel, came up with an estimate of $1.25 million. He proposed a 300-mile, four-inch line with pumping stations with 100-horsepower engines spaced an average of 15 miles apart. Critics charged that this proposal greatly exaggerated the maximum pumping distance, but subsequent experience proved Haupt correct.[66]

In anticipation of building his line, Harley in 1875 obtained a broad Pennsylvania charter that reportedly granted him rights to lay pipe anywhere in the state. The acquisition of the Baltimore Transportation Company, a Maryland corporation, gave him equally broad powers in that state. With the placing of orders for some 8,000 tons of lap-welded pipe in June, 1876, Harley expected completion by the following December.[67]

Reactions to the undertaking were mixed. The *Pittsburgh Commercial* predicted that refining would be shifted to the seaboard, with "disastrous" effects on Pittsburgh's oil interests,[68] but the *Chronicle* was almost complacent:

> From the fact that the Standard folks are apparently doing nothing to oppose this line . . . oil men are not disposed to "take much stock" in the ultimate completion of the enterprise and are awaiting further developments.[69]

Further developments, however, failed to materialize when internal difficulties involving the management of the Pennsylvania Transportation Company brought the project to a halt. Pipeline companies commonly issued certificates for more crude oil than they had on hand, either to "loan" oil to their customers, or to speculate on the exchanges themselves. Legislation to abolish the practice and require monthly reports of stocks had been on the books since 1874, but cavalier methods of reporting had rendered the law generally ineffective.

In the spring of 1876 rumors began to spread through the exchanges that, contrary to the results shown by a recent gauging of its holdings, the Pennsylvania Transportation Company was short on oil. Another examination revealed a shortage of between 50,000 and 100,000 barrels. Harley, his lieutenants, and several gaugers were arrested, and the company went into receivership.[70]

Despite its failure, Harley's venture was indicative of how even Conduit's limited success has stimulated an interest in the possibilities of long-distance pipelines. Within three years this interest would lead to a practical demonstration of the feasibility of this revolutionary innovation in petroleum transportation. Of much more immediate concern to producers, refiners, and carriers alike, however, was the growing power of Standard in the area of transportation.

Standard's moves into transportation

Strong evidence of Standard's growing role in the field of transportation is revealed in the terms of the rail pool of 1874 and the pipeline agreements of 1874 and 1875. Clearly indicated was the fact that even at this relatively early date, the Standard Oil Company exercised considerable influence over petroleum transportation from well-head to refinery. It was also apparent that this influence, particularly as it affected the structure and functioning of the gathering lines, posed a potential threat to the preferred position of the Pennsylvania-Empire system in the area of petroleum transportation. Finally there was a growing awareness on the part of all interested parties that Standard was using its position to change the competitive structure of the industry.

In 1874 control over refining capacity in Cleveland and New York alone gave Standard a strong bargaining position, one that Rockefeller was able to exploit in his relations with the New York Central and the Erie. Moreover this position had been further buttressed by an extension of Standard's operations into gathering lines and terminal facilities. Rockefeller had taken a major step into the pipeline field in 1873 when, at the time Vandergrift and Foreman organized their pipeline system under the United Pipe Line Company, he acquired a one-third interest in the stock of the new concern. Association with United Pipe Line apparently paid a quick dividend, for a year later Vandergrift and Foreman sold their Imperial Refining Company, located near Oil City, to Standard. Whether this sale was a part of the arrangement for Standard to invest in United is not clear, but it did bring under Standard's control the largest single refinery in the Regions, with an estimated daily capacity of 1,400 barrels. In 1873 Rockefeller on his own account joined Jabez A. Bostwick in forming the American Transfer Company. They built a small gathering system in Clarion County which was taken over in 1874 by Standard of Ohio.

Since the late 1860's Standard had operated freight yards at Hunter's

Point, in Brooklyn, New York, where oil was unloaded and inspected. Through a lease taken by Bostwick in 1872, Standard obtained control of the New York Central's oil docks at Sixty-fifth Street, New York City.[71] Finally, as a part of the deal with the Erie Railroad in 1874 to handle at least half of Standard's east bound shipments, the latter secured a lease of the Erie's extensive oil terminal facilities at Weehawken, New Jersey.

These arrangements virtually insured that all petroleum and petroleum products shipped by any railroad except the Pennsylvania would be handled through terminal facilities belonging to or controlled by Standard. Not all shipments received at these various points, of course, were made by the Rockefeller concern, and there is no evidence of discrimination against outsiders. But for Standard there were two important advantages in the situation. The monetary return from operations, although not revealed, was undoubtedly attractive. Moreover, Rockefeller was able to keep track of competitors' freight. To be sure, the Erie's only other shipper of any significance at this time was the firm of Neyhart and Grandin, crude oil dealers, but the importance of their freight was not inconsiderable. By volume it amounted to approximately three-fifths of Standard's tonnage, and—even more important—most of it went to independent refiners on the seaboard.*

It seems quite possible that Standard was already strong enough to influence the terms of the railroad pool of 1874 that equalized transport rates to all refiners. Certain railway spokesmen, such as George Blanchard of the Erie, subsequently denied that the Standard Oil Company had had any hand in forming the pool beyond the hearing given all oil interests.[73] But Cleveland had more to gain from the rate structure than any other refining point, and Standard controlled virtually all of the city's refining capacity. The railroads must at least have had the company's Ohio position in mind when they committed themselves to rate equalization.

There is no doubt, however, that Standard played an important role in the division of profits that were provided in the pipeline pool of 1874. The United Pipe Line, in which Standard held a one-third interest, was allotted 29.5 per cent, the largest share granted to a single member. Standard's wholly owned American Transfer Company, with only 50 miles of lines was given a 7 per cent allocation. In comparison, Empire's Union

---

* Even with the Rockefeller contract, Erie's *share* in the total freight rose only slightly, from 17 to about 19 per cent; its absolute volume increased from 762,000 barrels in 1873 to 934,000 barrels in 1874. The New York Central carried the same amount.[72]

line, which represented the Pennsylvania railroad in the Regions, laid claim to a little over one-quarter of total disbursements.*

In explaining their pipeline investments, Standard officials in later years claimed that existing lines had not been able to tap new fields rapidly enough, had been generally inefficient, and had needed the "Standard touch" to assure the company of adequate regular crude supplies. But the market for transportation was hardly so inefficient as all this. A temporary shortage during periods of flush production was unavoidable, as Standard itself was to find when the Bradford fields came in. Otherwise, Empire and other pipelines had long proven their ability to adapt swiftly to changes in supply and maintain a modicum of order for their customers.

Two other considerations were probably more relevant. First, there were substantial profits to be made in pipeline transportation, even during low periods, if one was able to move with the fields. But of equal if not greater importance was the fact that control over gathering lines gave Standard a better bargaining position *vis-à-vis* the Pennsylvania Railroad. As of 1874 the Pennsylvania alone of the three major trunk lines remained relatively independent of Standard. Through the far-flung and aggressive operations of Empire, approximately three million barrels of oil (the largest shipped by any railroad) went eastward over the Pennsylvania's lines. Although a part of this total went to Standard refineries, at least two-thirds consisted of shipments for independents.[75] Already the Pennsylvania management was aware that its continued support to independent manufacturers was running counter to Rockefeller's ambitions; that Standard, allied with the Erie and the New York Central, was becoming a power to be reckoned with.[76]

Standard did not let its interest in the pipeline pools of 1874 and 1875 interfere with further steps to strengthen its position with the New York carriers. The Erie especially became more firmly tied to Standard when the latter agreed in 1874 to furnish tank cars, subsequently putting 520 broad-gauged cars on the Erie and the Atlantic & Great Western.† It was

* Although a member's share consisted of its proportion of payments out of the pool's "common fund," the pool's executive committee was instructed to parcel actual oil shipments among the lines in the same proportions as these payments were divided. The same body also assumed responsibility for assigning each member his pro-rata of any extensions required by new territories.[74]

† This move was of considerable importance to the Erie. It owned only 285 tank cars and, short of cash and credit, had little prospect of obtaining more. The Atlantic & Great Western was in even worse shape with creditors threatening to take over its cars.[77]

implicitly understood that, inasmuch as Standard had furnished the equipment, it would have a claim prior to those of competitors who used it.

Further, Standard received the customary one-cent per mile that the railroads paid on all "foreign" cars in service. Notwithstanding the fact that tank cars cost almost twice as much as ordinary box cars ($800 as compared to $425), Blanchard of the Erie thought the mileage would yield "more than bank rates of interest on the investment." *

Even before the delivery of the broad-gauged tank cars had been completed, the Atlantic & Great Western was brought into still closer association with Standard. With no track of its own to the southern fields, the Atlantic & Great Western depended on the Allegheny Valley Railroad, a standard-gauged line, to bring the bulk of its crude from pipe mouth north to Oil City. The inherent disadvantages of reloading shipments directly from narrow-gauged to broad-gauged cars were aggravated by the Atlantic & Great Western's shortage of rolling stock. All too often, from the standpoint of the Allegheny line, it was forced to hold its own loaded cars idle at the terminal point because Atlantic & Great Western cars were not available.

In late summer of 1875, this bottleneck was relieved when Standard's American Transfer Company agreed to take oil from the Allegheny Valley's cars and pipe it across the river at Oil City to storage tanks on the other side. The fact that it paid 10¢ a barrel for this service—some eight times the transfer charge of the Allegheny—gives some impression of the pressure put upon management of the Atlantic & Great Western. These charges brought the pipeline a net profit of 2¢ a barrel (20 per cent of its gross revenue) on a commodity with a high turnover.†

Further evidence of Standard's growing influence in transportation came in 1875, when the New York Central, the Erie, and the Pennsylvania agreed to give the Rockefeller combine a rebate of approximately

---

* Blanchard estimated an average tank car traveled about 2,500 miles per month. At one cent per mile, the car would then pay for itself in a little over two years. (Discounting at a compound annual rate of 10 per cent would lengthen the pay-off period to about three and one-half years.) The life of the car was considerably longer, abstracting from damage through loss and fire. Blanchard thought it would last "almost indefinitely." [78]

† Blanchard, however, did not regard the charges excessive. Perhaps the fact that American Transfer assumed risk for fire and loss while the oil was in storage outweighed many other considerations. If so, the Erie's attitude indicates the apprehension with which the railroads regarded oil freight, even at this late date. [79]

10 per cent of the open rates on petroleum shipments—a concession that apparently was continued through 1876.*

Whether this concession bore any close relationship to cost-savings derived from assured volume shipments is impossible to tell. In some cases, such as the lease of the Erie yards and the lease of the New York Central's terminal, it might be argued that the rebates partly constituted payment for such services as inspection and risk-bearing. But significantly, these were only subsidiary considerations in the eyes of the trunk lines. As A. J. Cassatt, vice president of the Pennsylvania Railroad, put it, "We found that they [Standard Oil] were getting very strong . . . and, if we wanted to retain our full share of the business and get fair rates on it, it would be necessary to make arrangements to protect ourselves." [81] Although other shippers obtained rebates, they could not hope to match the size being attained by Standard Oil, and correspondingly lacked that company's bargaining power.

Standard's acquisition of refineries

As the record shows, Rockefeller and his associates were unusually skillful in taking advantage of the difficulties and rivalries among the pipelines and railroads to strengthen their position in the field of petroleum transportation. In expanding into transportation, the organization took on few if any operations that were not profitable. But Rockefeller's fundamental objective at the end of 1873, the keystone to what he described as "our plan," was to bring the great bulk of the industry's refining capacity under Standard's domination. He and his fellow executives were primarily interested in using their expanding control over gathering lines, tank cars, and terminal facilities and stronger bargaining position opposite the railroads to bring pressure on the remaining independent refiners. Moreover, Rockefeller did not wait until Standard had become a dominant force in the field of petroleum transportation before launching his campaign, and by the end of 1876 had already made impressive strides toward achieving his goal.

In general, Rockefeller followed the tactics that had earlier worked so

---

* Professors Nevins and Johnson have interpreted these contracts to be another pool, in which the railroads paid Standard a 10 per cent rebate to act as "evener." Both cite the same testimony by Blanchard and Cassatt. There is little in this testimony to support this interpretation. Cassatt did admit that Standard at this time "was dividing its transportation between these different roads," but in context "dividing" did not connote a joint formal agreement to act as "evener." [80]

successfully to bring the Cleveland refiners into line. The first step was to persuade the leading refiners in a particular area to join the Standard organization. With these firms as the nucleus Standard could use its increasing power in the field of transportation to bring pressure on the remaining independents to sell or merge their interests with Standard.

Moving slowly in 1874, Rockefeller and Flagler found three major refining interests receptive to their suggestion of a merger. These included William Warden, head of the Atlantic Refining Company, the largest plant in Philadelphia; the partners of Lockhart, Frew & Company, which operated seven refineries in Pittsburgh (about half the city's total refining capacity); and Charles Pratt and H. H. Rogers, who headed up the relatively small but highly successful Charles Pratt & Company of Newtown, Long Island, New York.

There is no indication that Standard put any overt pressure on these prospective associates; it was hardly necessary. The idea of combination was not new to any of the members of the group. William Warden and the partners of Lockhart, Frew & Company had participated in the South Improvement scheme, while, along with Pratt, they had all been members of the National Refiners' Association.

By autumn, 1874, the details for a merger had been worked out. A valuation was placed on the respective properties which was applied to the purchase of stock of Standard of Ohio.[82] It was further agreed that their union with Standard was to remain an "absolute secret," that each was to continue operations under its respective firm name and same top management personnel, and that each was to absorb the other refiners in their respective areas as "rapidly as persuasion or other means could bring it about." * "Thus in what was a threefold merger the Standard obtained the largest refinery in the Philadelphia area, more than half the refining capacity of Pittsburgh, and the most widely known of the New York independents."[84]

Along with the refining properties came equally important management personnel. William Warden; the partners of Lockhart & Frew, Charles Lockhart, R. J. Waring, and William Frew; and Pratt and Rogers all became important executives in the Standard organization.

---

* It should be added that the association of the Philadelphia and Pittsburgh refiners with Standard did not long remain an "absolute secret." Early in 1875, for example, the *Pittsburgh Gazette* suggested that Frew's connection with the Cleveland interests accounted for his objection to a free-pipeline bill being considered by the Pennsylvania Legislature.[83]

In a move calculated to give Standard a closer contact with the remaining major refining interests, Rockefeller in the spring of 1875 took an active part in formation of the Central Refiners' Association, successor to the National Refiners' Association that had collapsed two years before. Like its predecessor, the Central Refiners' Association was to be open to all refiners who, in return for leasing their property to the Association, were given the right to subscribe to stock. An executive committee representing each of the five chief refining districts was appointed with power to control all purchases of crude and all sales of refined, to make all rate agreements with railroads and pipelines, and to divide profits among the stockholders. The president of the Association was John D. Rockefeller.[85]

There is little evidence to suggest that the Central Refiners' Association ever functioned effectively to meet its stated objectives. Apparently the major independents were interested and of course the refiners already associated with Standard became members. The proposal for the Central Refiners' Association to make all contracts with the railroads proved a particular stumbling block. Since the Pennsylvania Railroad was to be allotted all of Pittsburgh's shipments under the plan, refiners in that city bridled at the scheme, making it clear that they had no intention of entering the Association on these terms. The mere threat of an alternate route over Conduit and the B & O reinforced their demands that the executive committee keep its hands off transportation. The committee reluctantly agreed—a concession that made it virtually impossible for the cartel to perform its other functions.[86]

But if the Central Refiners' Association failed in these respects, it was nevertheless useful to Standard's management. As Ida Tarbell points out, it was "a most clever device," one that "furnished the secret partners of Mr. Rockefeller a plausible proposition with which to approach the firms . . . they wished to . . . control."[87] "In other words," as Rockefeller's biographer puts it, "if a refiner objected to an outright merger, he might be brought into the Association and then be gradually won over to the idea of a closer relationship."[88]

There were two refining areas where Standard was particularly eager to establish bridgeheads during 1875. One was the Regions, where the great bulk of the refining capacity was still in the hands of independents and where antagonism to Standard was the most intense. The other included the area in and around Marietta, Ohio, and Parkersburg, West Virginia, southwest of Pittsburgh.

Standard's interest in the West Virginia and southern Ohio refineries was not prompted so much by their capacity, which was relatively small, but by the threat posed by the B & O. In 1874, John Garrett, president of the B & O, had clearly indicated his intention to increase his railroad's share of the trunk-line traffic, including oil shipments, between the East Coast and the Midwest. It was not only Standard's interest in the railroad and pipeline pools of 1874 that was threatened. With a main line running from Baltimore to Chicago and St. Louis, the B & O was in an excellent position to strengthen independent refineries in the West Virginia–southern Ohio district by hauling their products to western and eastern markets. With a branch line to Pittsburgh and an agreement with Conduit to take deliveries of its crude, the B & O could also build up the position of the independent refiners in Pittsburgh or increase the shipments of crude, both from West Virginia and the Oil Regions, to the Baltimore refineries.

Standard's executives decided that the best way to meet the threat posed by the B & O and to expand their interests to the south and west of Pittsburgh was to ally themselves with a leading refiner in the area. Their choice fell on J. N. Camden of Parkersburg, part owner of the largest refinery in the town and a well-known and well-liked figure in the area. In May, 1875, Camden and his partner, Colonel W. P. Thompson, in return for an exchange of a portion of their interest for Standard of Ohio stock, agreed to head up a new company, the Camden Consolidated Oil Company, which succeeded the former concern, J. N. Camden & Company. The role of Camden Consolidated was clearly outlined. Without revealing the fact that the majority of its stock was held by Standard, J. N. Camden and Colonel Thompson were to absorb the bulk of the refining capacity in the territory served by the main line of the B & O.

Within a few months after the conclusion of these arrangements, Standard achieved an even more impressive victory in the Regions. Here the key figure was John D. Archbold, implacable foe of Rockefeller and Standard during the South Improvement Company episode and the manager of Porter, Morehouse & Company in Titusville, next to Standard's Imperial Oil, the largest refinery in the Regions. It took the better part of a year to convince Archbold that he should join forces with his former opponents. Possibly the fact that several of his allies in the fight against Standard, notably Vandergrift and Pratt, had come to an agreement with Rockefeller may have influenced his decision. Nor could he fail to have been impressed with Standard's role in the arrangements made under the railroad and pipeline pools of 1874 and 1875 that were generally unfavorable to refiners in the Regions.

Whatever the reasons, Archbold in the fall of 1875 came to terms with Standard. He became the head of a newly formed concern, the Acme Oil Company of New York, a wholly owned subsidiary of Standard of Ohio that was to operate in the Regions in essentially the same manner as Camden functioned in the areas south and west of Pittsburgh. Acme immediately purchased the property of Porter, Morehouse & Company and that of Bennett, Warner & Company, also in Titusville, the third largest refinery in the Regions. Archbold became one of the most active officers in the Standard organization.

With Standard affiliates in New York, Philadelphia, Pittsburgh, West Virginia, southern Ohio, and the Regions, Rockefeller's plans to absorb the bulk of the industry's refining capacity began to gain momentum. Archbold was particularly active. During 1876, on the behalf of Acme, he purchased seven Regions' refineries and leased another, along with six plants, either purchased or leased, in other parts of Pennsylvania and New York state.[89] In the Philadelphia area William Warden added early in 1876 the Franklin Oil Works to the holdings of Atlantic Refining; later that year he arranged for other leading refiners in the area such as W. L. Elkins, Waring, King & Company, Malcolm Lloyd, and the Harknesses to join forces with Atlantic.[90] Lockhart and Frew picked up a few small refineries in the Pittsburgh area in 1875 and added a major firm, that of August H. Tack, in 1876. In New York, Pratt and Rogers bought a half interest in James Donald & Company in 1876.[91]

It was the Camden Consolidated Oil Company, however, that became involved in the most complicated negotiations during this period. Indicative of the secrecy that was maintained over Camden Consolidated's ties with Standard was the contract made with the B & O in September, 1875. Under the impression that he was fighting both Standard and the Pennsylvania-Empire combination, Garrett gave Camden Consolidated preferential rates over the B & O on both crude and refined oil, with the added provision that Camden Consolidated was to buy 50,000 barrels of crude per month from Conduit and ship it either as crude or refined over the B & O.[92]

Part of this crude was sold to refiners in Pittsburgh; a part went to Parkersburg refineries, while the rest was shipped to Baltimore, where it was sold to refiners or was exported. During early 1876 Camden Consolidated built an oil wharf and warehouse in Baltimore to handle its export orders.[93]

Meanwhile Camden and Thompson had not been idle in the Parkersburg-Marietta area. But after an initial success in absorbing a number of small refineries during late 1875 and early 1876 they began to encounter difficulties with owners who demanded what Camden considered ex- horbitant prices for their properties. Rather than "pension" a "whole crew of broken down oil men," Camden hoped to "starve" them out. Late in 1876 he came up with a suggestion as to how this might be done. This was to cut off the supply of West Virginia crude—approximately 12,000 barrels per month—which would force the refiners in the area either to shut down or sell out. With Rockefeller's approval, contracts with the West Virginia producers were made that insured the delivery of the bulk of their output to Camden Consolidated.[94]

But the main objective of Camden and Thompson during 1876 was the Baltimore area. Late in that year the leading firm in the city, W. C. West & Son, was approached with the proposition to merge its property with the other Baltimore refineries into one large company that would in turn be absorbed by Camden. Negotiations hit a snag when the Wests, father and son, insisted on large salaries with the new company.[95] It was clear that pressure would have to be applied to the Baltimore area before arrangements satisfactory to Camden could be completed.

By the end of 1876 Standard's affiliates had made considerable progress in reducing the independent refining capacity in their respective areas. But there were still important independent interests in Pittsburgh, Phila- delphia, Baltimore, the Oil Regions, and New York, whereas New Eng- land remained almost entirely outside the Standard orbit.

Crude producers, unable to bring about any effective control over drill- ing activities, were quick to blame Rockefeller and Standard as a major factor in the low prices they received. To be sure, the details of Standard's relations with the pipelines and railroads were not widely known, and Standard's absorption of the major independents outside Cleveland and Pittsburgh remained a closely guarded secret. But producers were gener- ally aware of the growing power of Standard in the field of petroleum transportation, and they correctly assumed that the Central Refiners' As- sociation was dominated by Rockefeller.

As 1876 drew to a close, there were two developments that promised to give producers and refiners some relief from the alliances between Stand- ard and the railroads. One was a new plan for an independent long-dis- tance pipeline, announced almost immediately after Harley's failure. The other was a growing rift between Standard and the Pennsylvania-Empire combination, which flared into open conflict early in 1877.

Challenges to Standard

The new pipeline proposal appeared to have all the essential ingredients for success. Initiated by the Producers' Consolidated Land & Petroleum Company, a $2.5 million producing and refining corporation, the project had the financial backing of a syndicate of Boston, Union Pacific, and New York capitalists. Already the owner of the 100-mile Relief gathering pipeline in Butler county and a $500,000 refinery in the New York City area, Producers' Consolidated proposed to construct a trunk line to New York by way of Buffalo, where a new refinery was to be built.[96]

Bradford producers were particularly interested in the Buffalo line, less because of a lack of transportation and storage facilities than because of the price situation on Bradford crude. Empire had built its Olean pipeline in the newly discovered fields late in 1875, followed shortly by Producers' Consolidated's McKean line and a branch of Standard's American Transfer.[97] Except for sporadic shortages, these companies were as yet well able to keep pace with increasing production. Bradford crude, however, sold at a discount of 20–30 per cent below oil from the southern fields. The differential was in part due to short-run difficulties in refining northern crude, but the producers felt that the pipelines were "squeezing" and "cornering" their market, and hoped that another outlet would give better prices.[98]

By February, 1877, surveys to Buffalo were completed and by May a bill authorizing the Buffalo–New York City section had passed the Albany House.[99] But Producers' Consolidated never laid a mile of its projected line. Like Harley's project of the year before, it too succumbed from internal difficulties. Early in 1877 it was discovered that the president of Producers' Consolidated, Frederick Prentice, had defaulted on $500,000 in speculative loans against which he had illegitimately pledged company property. Prentice, a major figure in promoting the line, resigned and the new management evidenced little enthusiasm for the project. Some months later any hope that the proposed line would be of assistance to producers or independent refiners vanished completely when the Standard Oil Company purchased 65 per cent of Producers' Consolidated stock.[100]

As far as Standard was concerned, the threat posed by Producers' Consolidated during 1877 was of minor importance compared to the challenge offered by the Pennsylvania-Empire combination. The Pennsylvania's opposition to free pipeline bills, its involvement in the railroad pools, and its rebates to Standard had drawn the two organizations closer together. Nevertheless, while the New York Central and the Erie were becoming increasingly tied to Standard, the Pennsylvania-Empire alliance continued to draw the bulk of its traffic from independent shippers.

Open rivalry between the Pennsylvania and Standard had undoubtedly been checked by the pipeline pools of 1874 and 1875, but these had done little to hinder Standard's control over refining. This, more than Standard's growing power in the field of petroleum transportation, was what worried Joseph Potts, president of Empire. As he later stated,

> We reached the conclusion that there were three great divisions in the petroleum business—the production, the carriage of it, and the preparation of it for market. If any one party controlled absolutely any one of those three divisions, they practically would have a very fair show of controlling the others. We were particularly solicitous about the transportation, and we were a little afraid that the refiners might combine in a single institution . . . We therefore suggested to the Pennsylvania Road that we should do what we did not wish to do— . . . become interested in one or more refineries.[101]

With the backing of the Pennsylvania management, Potts took steps to insure the maintenance of refining capacity outside the Standard's orbit. Early in 1876 Empire acquired controlling interest in the Sone & Fleming refinery in New York, a works with about 2,000 barrels per day capacity. Standard Oil protested this move but took no definite action. When Empire bought the entire stock of the Philadelphia Refining Company, however, Rockefeller withdrew all shipments from the Pennsylvania. As John D. Archbold later rationalized:

> We objected to their going into the refining business on the ground that as a common carrier they had no right to do so . . . it was fair to presume that if they were themselves also manufacturers their facilities would be first used by themselves and to our exclusion.[102]

Archbold of course saw no objection to Standard's acting concurrently as a common carrier and refiner, supposedly because alternative means of transport were open to competitors. In the Empire case, alternative routes were inadequate, according to him, since the other lines lacked sufficient equipment to handle Standard's business. Yet Standard managed miraculously to send all its eastern shipments by competitive routes. The other roads, fearing that the Pennsylvania would gain an advantage over them, were more than willing allies.

The battle between Empire and Standard began quietly in March, 1877, about the same time that a general rate war on Chicago–New York freight broke out among the four trunk lines. The struggle became progressively worse. The Pennsylvania instructed Empire to buy and sell oil for "such prices as they [*sic*] could get," the difference constituting the railroad's freight charges. The general level of the Pennsylvania's rates during the contest is not known, but in at least one instance, the road paid eight cents per barrel to carry oil. The case was admittedly exceptional, but with the summer rates plunged "very low indeed," and showed every prospect of going lower.[103]

Rockefeller and his allies among the carriers also suffered losses in these markets. The Erie's crude rates to the seaboard dropped to 35¢ a barrel in spring and summer of 1877, less than one-third of their March tariffs. If the Erie and New York Central did not incur losses comparable to those experienced by the Pennsylvania, it was because they were in part absorbed by Standard as a part of the cost of meeting the challenge of Empire.

In a desperate attempt to maintain an outlet for crude, Empire started enlarging its Philadelphia plant to 6,000 barrels per day and concluded some eighteen or twenty contracts with such relatively well-known independent refiners as Bush & Denslow and Lombard, Ayres.[104] To insure adequate crude supplies, Empire and the Pennsylvania signed a three-year contract with a group of producers, assuring them rates as low as any competitors might obtain.[105]

Nonetheless, the Pennsylvania-Empire combination rapidly lost ground. Empire never succeeded in working its refineries to capacity, and even alliances with independent refiners and producers did not prevent the Pennsylvania's share of east-bound oil freight from falling from 52 to 30 per cent during the contest.[106]

Export figures for the first five months of 1877 reflected the shift. Philadelphia and Baltimore, the main termini of the Pennsylvania, exported only 29 per cent of the total, as compared to 45 per cent for the comparable period of 1876. New York, receiving the bulk of Standard's oil, sent 74,500,000 gallons abroad, or 32,500,000 gallons more than from January through May of the year before.[107]

The Pennsylvania management became increasingly aware of the difficulty of justifying to its stockholders the railroad's costly support of Empire. The latter was a separate corporate entity, and Pennsylvania shareholders had long resented Empire's earning large profits on services which, they alleged, the railroad itself could perform. Although an 1874 committee of railroad stockholders found Empire "a necessary product of the peculiar character of western traffic," [108] many still suspected that the fast freight company's primary purpose was to pump a major share of $400,000 in annual dividends into the pockets of Pennsylvania officials who held the bulk of Empire's equity. It now appeared that the railroad was sacrificing further dividends in a fruitless struggle. The situation was difficult at best, especially since the Pennsylvania had reduced its second quarterly dividend; events during the summer of 1877 were to make the railroad's position completely untenable from any point of view.

The severe depression of the 1870's, the rate wars of the preceding half-decade, large bonded debts, and the practice of paying high dividends on highly dilute stocks had all combined in varying degrees to strain the finances of all the trunk lines. In early summer they all reduced wages an average 10 per cent and in mid-July trainmen on the B & O and the Pennsylvania led a walk-out that spread to the Erie and the New York Central. Riots broke out on the B & O's main line, and the situation at Pittsburgh was particularly tense. Strikers and unemployed milled about the Pennsylvania's yards for two days while some 2,000 freight cars piled up along the tracks. On July 21, when the militia fired into the crowd to clear the tracks, mob violence erupted. The rioters set oil and coal cars aflame and ran them against the roundhouse where the troops had barricaded themselves. The militia escaped cremation, but the railroad buildings and equipment—the main objects of the riot—did not. A mile of freight cars stood ablaze; those that did not burn were pillaged; where the flames did not spread of themselves, the crowds carried them. By July 24, when the frenzy had dissipated itself, the Associated Press described the scene:

> Where lately stood magnificent buildings, teeming with life and business are seen naught but charred timbers and heaps of ashes. Of the hundreds of railroad cars . . . remain but wheels and axles which are broken and twisted in all possible shapes. Never was destruction more complete; never was the power of a mob more completely illustrated.[109]

This property damage of at least $2 million, added to previous losses, convinced the Pennsylvania management that it could no longer afford to combat Standard. The latter had no objection to Empire's maintaining its extensive transportation facilities, but demanded that it get out of the refining business. This was the very position that Joseph Potts, head of Empire, had fought to avoid. Rather than make such a concession he preferred to see his company dissolved.

This was the action the Pennsylvania management decided to follow. Exercising an option in its contract with Empire to buy the latter's assets, the railroad borrowed $1.65 million from Standard to purchase Empire's 4,000 general merchandise cars, and another $900,000 to buy its 1,500 tank cars. Empire's shops, piers, and depots were sold for $450,000 in railroad bonds. As a part of the deal Standard acquired Empire's refining properties for an outlay in cash of a little over $500,000.[110] For additional cash Standard's United Pipe Line Company bought Empire's 520 miles of gathering lines.* Except for two lake steamer lines, which continued to function as separate corporations, the Empire Transportation Company ceased to exist at the final stockholders' meeting held on October 17, 1877.

About the same time Standard took one further important step to strengthen its position in the transport field by purchasing the Columbia Conduit Company. Hostetter, who had counted on the B & O to maintain outlets with independents, became progressively irritated with the railroad's increasing cooperation with Standard. The B & O's breaches of contract with Conduit, according to Hostetter, became "so flagrant and outrageous as to have become too grievous to bear."[112] Moreover, in early summer Standard began to lay a competing pipeline to Pittsburgh; it was nearly completed.[113] Hostetter willingly accepted $1 million for his pipeline, which represented a $500,000 investment, and several refineries with an aggregate capacity of between 750 barrels and 1,500 barrels per day.[114] Hostetter in turn gave one-fourth of the payment to Benson, McKelvy & Hopkins for relinquishing their lease on the line.[115]

Standard's further acquisition of refining capacity

With the defeat of Empire, "Standard leaders recognized that the strategic hour had struck for pressing their campaign to absorb the 'outsiders' by lease, merger, or purchase."[116] The capitulation of the Pennsylvania affiliate made the railroads almost completely dependent on Standard for their petroleum traffic. The Rockefeller organization controlled every important gathering line to railheads. It operated under lease the New York oil terminal facilities of the New York Central and the Erie. By far the major share of the nation's tank cars were under its formal or informal control, and some three-quarters of all trunk line petroleum freight was shipped on Standard account.

---

* Standard either loaned or paid $4,000,000 in cash at the liquidation. John D. Rockefeller himself drove from bank to bank to round up short-term credit, telling bank presidents, "I need it all! . . . Give me what you have!"[111]

Standard's power was clearly recognized in the terms of the peace treaty that followed the end of the rate war of 1877. An agreement made in October of that year provided for a division of petroleum shipments to the east coast among the New York Central, the Erie, the Pennsylvania, and the B & O. In return for a "commission" of 10 per cent on the open rate for its own shipments, Standard acted as an "evener" by adjusting its shipments over the different roads. A similar pool of west-bound general merchandise freight out of New York had been established in June, 1877, but this was administered by a commissioner, chosen by the trunk lines, who was not concurrently a competitor of their customers.[117]

Standard lost no time in securing from the railroads further concessions that bore close resemblance to the rebate proposals under the South Improvement scheme. About the time of the Empire sale the New York Central and the Erie began paying "commissions" of 20–35¢ a barrel to Standard's pipeline subsidiary, The American Transfer Company, on all crude oil carried by the railroads. In other words, Standard received a rebate not only on its own shipments of crude but on every barrel shipped by independents as well.[118]

Early in 1878 the Pennsylvania, recognizing that American Transfer could easily divert traffic to other lines, granted the pipeline's demand for a similar "commission" of 22½¢ a barrel. In July, 1878, the trunk lines granted Standard an additional 15¢ per barrel rebate on its own shipments of crude oil.*

Standard further used its pipeline monopoly to help subsidize its refineries in their competition with independents. As compared to the open pipeage of 30¢ per barrel, Standard refineries paid Standard pipelines only 17½¢.[120] For the Rockefeller organization as a whole, this was merely an internal transfer; the refineries' increased earnings were offset by the pipelines' lower profits. But outside refiners, not being vertically integrated, were unable to gain any of the cost savings from pipeline transportation anywhere in their organizations. Thus Standard's plants could cut their prices, if need be, below cost to independents and still make money; moreover, the 17½¢ rate still gave the Standard pipeline an adequate margin, for pipe cost probably averaged no more than 8–10¢ per barrel. The sum total of these concessions meant that, instead of pay-

---

* Cassatt testified that Malcolm Lloyd, the Pennsylvania's only other crude shipper, also received the fifteen-cent rebate. He did not, of course, share in the 10 per cent "commission" paid Standard Oil or the 22½¢ paid American Transfer.[119]

ing combined pipe and rail open rates of $1.70 on crude from the southern fields to New York ($1.45 from Bradford), Standard's charges were only $1.06 (84¢ from Bradford).

At the same time, the Pennsylvania discovered that the New York roads, since May, 1878, had been giving Rockefeller's inland refineries a net rate of 80¢ per barrel on New York-bound kerosene. The cut had been made at Standard's urging when it was clear that independents were planning a new, cheaper route via the Erie Canal. The Pennsylvania, now similarly pressured, followed suit.*

It was against this background that "Rockefeller captains raised the tempo of their activities" [122] and "our plan" to acquire the bulk of the remaining independent refineries moved into its final stages. Settlement of the Empire dispute had brought an immediate addition to Standard's holdings of the New York works of Sone & Fleming and the Philadelphia Refining Company, while the purchase of Conduit contributed three, possibly five, more small refineries in the Pittsburgh area.[123]

In contrast to Philadelphia, where most of the refining capacity was already under Standard control, a small group of refiners in the New York City area continued to maintain their independence. Late in 1877 a renewed drive by Lockhart & Frew brought most of the remaining Pittsburgh refineries into the combination, including Flack & Elkins; McKee & Sons; E. I. Waring; Ferris & Emmons; Paine, Ablet & Company; and the R. S. Germania Refinery.[124]

Camden was just as successful in the Baltimore area. To bring pressure on the recalcitrant Wests, Camden Consolidated sold refined oil from its Parkersburg refineries at a loss in the Baltimore market. Several independent refineries were purchased and negotiations opened with Poultney & Moale, a firm which not only owned its own refinery but acted as a buyer of crude for a number of small Baltimore refineries.[125]

Late in 1877 the Wests finally came to terms and the Baltimore United Oil Company was formed, with Camden Consolidated holding over four-fifths of its stock in trust for Standard. In addition to some eleven refineries with a combined daily capacity of about 2,250 barrels, its properties included paraffin, gasoline, and barrel works, warehouses, and wharves. Camden proudly reported, "We have cleaned up every seed in which a refining interest could spring in Baltimore." [126]

---

* The tariff on petroleum refined in transit was $1.90. Of this, the railroads refunded the charge for hauling the crude-equivalent of a barrel of refined oil to the refinery. [121]

Meanwhile Archbold was active in the Regions. In 1877–78 he brought some 27 refineries, most of them small, under Standard control. He also found time to work with H. H. Rogers in New England organizing in late 1877 the Maverick Oil Company. In 1878 the Maverick Oil Company acquired a controlling interest in three leading Boston firms (Pierce & Canterbury, Stephen Jannay & Company, and T. H. Whittemore), plus a half interest in the Portland Kerosene Company.[127]

While all of the independent refining capacity did not succumb, the goal Rockefeller set for Standard in 1873 was largely achieved by the end of 1878. Standard either owned or had under lease over 90 per cent of the total refining investments in the United States.*

In building a horizontal combination at the refining level backed by a virtual monopoly over the transport and handling of crude, Rockefeller and his associates brought about a profound change in the competitive structure of the American petroleum industry. As the principal buyer and shipper of the output of the Regions, the Standard management was in a position to exert a major influence on prices and the volume of petroleum that moved from producers through the refineries and into distribution. This power in itself was sufficient to make Standard's operations highly profitable.

Standard executives were already considering the possibilities of extending the organization's activities into new areas. There were obvious advantages in complementing their horizontal combination with a vertical organization that would extend Standard's control even more effectively back into production and forward into marketing. But the Standard management was too preoccupied with more immediate and pressing problems to give much attention to working out a full-fledged expansion program.

Rapid acquisition of transport facilities and refining capacity over the preceding five years had left the Standard organization in a "confused, almost chaotic state," [129] badly in need of internal adjustment and reorganization. Concurrently, producers and remaining independent refiners launched an attack on Standard's position in transportation and refining.

---

* Total investment was estimated at about $36 million, of which Standard controlled approximately $33 million.[128]

Chapter 17

Tidewater, Standard, and long-distance pipelines

Empire had barely capitulated to Standard in September, 1877, when the Producers' Union, long floundering with ineffective shutdown movements, announced a new organizing program to fight the latest "crowning iniquity of the monopoly." [1] As it evolved, the program involved a three-pronged attack on Standard and its relationships with the railroads: first, a renewed attempt to organize an effective shutdown movement; second, an appeal to the United States Congress and the Pennsylvania Legislature and to the courts for legal action; third, participation in still another pipeline project that would link the Regions with Baltimore.

Legal attack

The shutdown movement failed completely. Although prices had declined steadily through 1877, the prolific Bradford fields offered a good chance of sizable returns, even with crude selling at $1.80 per barrel. By the time prices dropped to $1.00 per barrel in the summer of 1878, current output and stocks were so large that a shutdown offered only minor improvement.

Attempts by the Union to obtain relief through legal procedures were scarcely more successful. Skeptical of the possibility of gaining a favorable reaction from the Pennsylvania State Legislature, the producers decided initially to carry their cause to the United States Congress. The move was not unprecedented. In 1876 producers and independent refiners had helped support the unsuccessful attempt of Pittsburgh Congressman James Hopkins to outlaw rate discrimination in interstate commerce. Basing their work largely on Hopkins' measure, the Union's legal committee now drafted a modified, weaker version sponsored by Representative John Reagan of Texas. While railroad influence in the Senate managed to get the new proposal pigeonholed, the Reagan and Hopkins bills marked the first Congressional milestones in the road that ultimately led to the Interstate Commerce Act of 1887.

Turning to Harrisburg, the Union fared no better. The influence of the Pennsylvania Railroad was still strong enough in the Pennsylvania Legislature to block a proposal to outlaw railroad rate discrimination on intrastate shipments.[2] A free pipeline bill, introduced at the January, 1878, session aroused stiff opposition, much of it undoubtedly inspired by the Standard Oil Company. Confirmation of rumors that the producers planned to back the pipeline to Baltimore, however, gave a new edge to the free pipeline controversy. The Pittsburgh Chamber of Commerce again voiced its fears that refining would be completely shifted to the coast. Even though the producers offered to abandon the seaboard line if the law broke Standard's monopoly and some two hundred Pittsburgh firms announced their sympathy with the Union's cause and supported the bill, Allegheny County legislators were unconvinced.[3] The Philadelphia Commercial Exchange also opposed the measure, despite a Union promise to abandon the Baltimore line and terminate the first trunk pipeline laid under the bill in the Quaker City. After a stormy four-hour session on the floor of the legislature, a coalition of eastern Pennsylvania and Allegheny County senators succeeded in tabling the measure.[4]

Failure at Harrisburg brought the Union's legal committee home in May, 1878, heaping recriminations upon the heads of the state's "ignorant, corrupt and unprincipled" legislators.[5] An appeal to the governor, however, brought a promise of satisfaction. In August, 1878, the Commonwealth of Pennsylvania brought a *quo warranto* suit against Standard's United pipeline, seeking to revoke its charter for violating its duties as a common carrier. At the same time the state requested injunctions against the Pennsylvania, Atlantic & Great Western, and Lake Shore railroads and their connections, to prohibit discrimination in favor of the Standard Oil Company.

By this time events in the Regions had sharpened the edge of antagonism against the Standard Oil enclave. Faced with the burgeoning output of the Bradford fields, United Pipe invoked an immediate shipment order in December, 1877, and maintained it until August, 1878. Producers, forced to sell at distress prices, were quick to charge Standard with trying to obtain cheap crude at their expense.

If the producers expected to see their cause dramatically redressed in the courts, they were greatly disappointed. After months of delay, the suit against the Pennsylvania Railroad was finally brought to trial in January, 1879. It yielded rich testimony (particularly from the Pennsylvania's vice-president, A. J. Cassatt), which documented Standard Oil's hold over the railroads. But proceedings came to a halt when the attorney-general ordered a suspension of the case against the Pennsylvania until the prosecution had completed taking testimony in the suits against United Pipe Line, Lake Shore, and Atlantic & Great Western. This move was fatal to the producers' action, for the defendants in the other cases managed to keep themselves beyond the reach of the court. The Lake Shore, with its books and offices outside of Pennsylvania, was not disposed to co-operate. A legal technicality placed the Atlantic & Great Western beyond the jurisdiction of the state's Supreme Court, while United Pipe Line officials suddenly found it necessary to conduct an increasing proportion of their business in Olean, New York—except on Sundays, when Pennsylvania subpoenas could not be served on them.[6] After a year of litigation, the cases, though still not formally disposed of, had come to a dead end.

Although convinced that nothing could be expected from these suits, the Union's counsels believed that testimony in the Pennsylvania Railroad case had furnished enough evidence to charge various Standard Oil officials with conspiracy in restraint of trade under common law. A Clarion County grand jury agreed, as expected, and in April, 1879, indicted such illustrious members of the Standard Oil hierarchy as John D. and William Rockefeller, Jabez Bostwick, Daniel O'Day, William G. Warden, Charles Lockhart, and Henry M. Flagler.

Proceedings were again delayed. The Rockefellers and Bostwick, both nonresidents of Pennsylvania, had to be extradited from New York, but Governor Hoyt of Pennsylvania postponed applying for extradition. Hoyt knew that the producers, their funds running low, were considering a Standard proposal to settle out of court. He suspected that the conspiracy cases, if brought to trial, would be used "simply as a leverage for and against the parties to these negotiations" and refused to let the "highest processes of the commonwealth" become an instrument of intra-industry conflict. In addition, he was under pressure from officials of the Pennsylvania Railroad not to stir already troubled waters, but to let the matter die in the courts.*

When the proposed settlement fell through, a jurisdictional dispute arose. Because of hostility in Clarion County, Standard petitioned to have the case tried before the state Supreme Court. Arguments on the motion forestalled further action on the conspiracy until January, 1880, when the case was settled out of court. In the meantime, attempts to press *quo warranto* proceedings against United Pipe were effectively hamstrung because Standard officials pleaded that their testimony in the pipeline case might incriminate them in the impending conspiracy trial.[8]

The Hepburn investigation

Although the Union's litigation produced no concrete results, it and the famed Hepburn investigation, conducted by the New York state assembly during 1879, did much to penetrate the aura of secrecy that had surrounded Standard Oil's relations with the trunk lines. Prompted by an increasing resentment against railroad rate structures that discriminated against places and persons, financial manipulations such as stock watering, and other corporate abuses, the New York investigators called upon a distinguished group of railroad and shipping officials to testify.[9]

Among those who testified was George Blanchard, vice-president of the Erie, who gave the most detailed account of the nature of petroleum transportation since 1872. Supplemented by the testimony of various oilmen (among whom the Standard officials were the least informative),

* Cf. the following letter, dated November 21, 1878, from Tom Scott, president of the Pennsylvania, to State Senator Don Cameron: ". . . It would be wise for you to see the newly-elected Governor Hoyt and have a clear understanding with him to say nothing on the oil questions in his message but let the whole matter now in court remain there for its adjudication . . ."[7]

these hearings gave the public the nearest thing to a complete picture of the working of the Standard Oil combination available at that time.

Perhaps their most significant contribution was to give explicit documentation of the rebates granted Standard Oil and its pipelines over the years. They also illustrated how Standard used its pipelines to control competition in the refining segment of the industry. Rockefeller no doubt could have obtained railroad rebates without his pipeline network, but the gathering lines made doubly certain that the railroads would not similarly favor competing refiners.°

Witnesses further claimed that Standard used its command over tank cars to obstruct crude supplies to independents during the troubled spring and summer of 1878. Whether the combine needlessly tied up rolling stock to the exclusion of outside refiners is difficult to determine. There were charges that long lines of cars stood idle while independents clamored for them. But in the general confusion bottlenecks could occur, and the cars may merely have been waiting for engines to take them out, as the railroads alleged.

Standard's own demands, however, were large enough to utilize fully not only those cars which it directly owned but also those under railroad control, and it is certain that the roads refused to allocate the cars on a pro-rata basis. As George Blanchard explained, "If the Standard Company came in first, and they would use all our cars, they could have had them as the largest shippers we had; if the Standard wanted them they would have had that preference." [11]

How the Rockefeller organization exercised that preference was revealed by Josiah Lombard, one of the larger independents. When Lombard tried, in the hectic summer of 1878, to get one hundred cars of oil shipped out of the Regions, the Erie informed him that, "Charles Pratt & Company have ordered all the cars we have," adding that he could obtain cars only when Pratt had no further need for them.† In a similar attempt to obtain transportation over the New York Central, Lombard was told that the road had neither cars nor terminal facilities available at the

---

° A. J. Cassatt of the Pennsylvania Railroad recognized this power when he told the court: "They [American Transfer Company] could have opened those pipe lines to shipments at very low rates by water and to rival roads, and could have diverted a large quantity of oil from our lines if they had chosen to do so." [10]

† Most of the Standard cars hauled over the Erie were in Pratt's name. (Nevins, *John D. Rockefeller*, I, 562–63.) But apparently Pratt also had prior call on the 285 tank cars owned by the Erie itself.

time.[12] The Pennsylvania Railroad allowed outside refiners to put their own cars on the line. But it refused to carry them at rates as low as those granted to Rockefeller unless the independents guaranteed a like quantity of oil. This was manifestly impossible.[13]

The Hepburn Committee condemned these practices as violations of common-carrier obligations and sound railway economy. It found not a single instance since the Empire War "wherein it [the petroleum traffic] resulted to the advantage of these roads." [14] In an indignant if oversimplified analysis, Committee Counsel Simon Sterne attacked "shortsighted" and "imbecile" trunk line managers who allowed their roads to become utterly subservient to Standard Oil's demands. Disregarding any problems of surplus railroad capacity, Sterne maintained that competition in the petroleum industry would have meant competitive bidding for railroad services with correspondingly high rail rates and better dividends to railroad stockholders. The effect on the consumer bothered Sterne not at all. Even at high rates, he argued, oil was the cheapest illuminant and lubricant in the world. Moreover, the extra burden

> would largely have fallen upon the consumer in Europe . . . In that respect a freight charge upon petroleum differed from the freight charge upon grain, because as we had in petroleum the monopoly of the product the ultimate rate to the consumer was an arbitrary one—the price yielded being the amount fixed for the product, plus the freight charge and the profit thereon —instead of being, as in the case of grain, a price . . . determined by the competition . . . [with other countries].[15]

The argument was not new. For years oilmen had been claiming that the burden of higher prices could easily be passed on to foreign consumers. But for a free trader like Sterne, it was an astonishing piece of analysis. Carried to its logical conclusion, it placed him in a position quite close to such railroad officials as Cassatt, who later opposed the seaboard pipeline because it would lower prices and bring less money into the country.[16]

Whether in fact trunk line profits would have been greater without a Standard Oil Company, as the Committee alleged, is open to question. Since there is little to indicate that the volume of shipments would have been less, the answer turns basically on the probable level of the average rate over the years from 1870 to 1879. Quite likely net rates during periods when there were no rate wars would have been higher than they were. Few shippers could have received the rebates granted Standard Oil without guaranteeing Standard's volume. But it is equally possible that without Standard to act as an "evener" the number of rate wars over the petroleum trade would have been more numerous or more intense.*

Suppositions of what might have been aside, the Hepburn Committee sought to determine whether the petroleum trade in fact yielded the railroads any profits at all. The officials who testified seemed to have no very clear idea of the actual facts of the matter.† Few could say whether the petroleum trade had been profitable, although most agreed that they would rather have it than not. Generally, their answers implied that each road wanted to keep the traffic from going to competitors. Erie's George Blanchard was quite explicit. He was of the opinion that even during rate wars oil freights at least covered operating expenses.‡ Reviewing the Erie's relationships with Standard since 1874 he estimated that the railroad had averaged two to five times more revenue on petroleum than on other through freights, ". . . clearly the most profitable business that . . . the Erie Railway has done of all its through freight eastward." [19]

Independent evidence suggests that Blanchard was close to being correct. Franklin B. Gowen, president of the Reading, estimated that a barrel of crude could be hauled 400 miles from the Regions to New York for 35–45¢. According to annual reports of the New York roads, total cost (including fixed charges) per ton-mile, averaged over all freight, approximated seven mills.[20] Apportioning the latter figure to a 400-pound barrel of east-bound crude would yield a cost of 56¢. The lowest "peacetime" net rates that Standard had yet received were 88¢ from the lower fields and 66¢ from Bradford. Even though the railroads paid three-quarters of a cent per mile on round trips of Standard-owned cars, the trunk lines apparently received enough from these low rates to earn a profit ranging at least from 4¢ to 10¢ a barrel. Still this was partly offset

---

* There is also a question of whether costs without a Standard Oil Company would have been the same. Any attempt, however, to deduce this effect on profits is hopeless. Railroad managers were quick to point out that there were economies to be gained from assured, large volume shipments. But correspondingly, the large shipper received a rebate which bore no necessary relation to cost-savings.

† Cf. the comments of the Committee: "They [railroad officials] gave us the cost of their aggregate business but could not approximate even the cost in any of its various details; could not separate New York State from through business, nor give any comparison of the cost of through and local business." [17]

‡ Speaking of the trade during an oil rate war, he told the Committee: "I don't believe that, considering the percentage that [oil] business bears to the total that the [Erie Railway] company could economize in the expenses of the road to the extent that the loss of this entire business would reduce its earnings." [18]

by several factors. Tank cars, for example, were often returned empty. East-bound refined oil sent from Cleveland was carried at a loss because rates, following the principle established in 1874–75, still equalized total transportation cost to any given point on the coast regardless of the point of origin. Nonetheless, oil on the whole probably had a better earnings record than many other types of through freight. Grain, for example, was carried 1,000 miles from Chicago to New York for 15¢ and 20¢ per hundred-weight.* The trunk lines consequently had a profitable stake in the oil trade that they could be expected to fight to protect.

The general public was less concerned about Standard's effect on the railroads than about the latter's responsibility for Standard's monopoly. The testimony before the Pennsylvania courts and the Hepburn Committee was extensively reported in the trade journals and the daily papers.[22] In both sources the reaction was almost universally antagonistic. A *New York Graphic* cartoonist pictured the "Iron Hand" of Standard clutching the Regions' press and crushing the public.[23] According to the *Oil, Paint & Drug Reporter*, "There never has existed in the United States a corporation as soulless, so grasping, so utterly destitute of the sense of commercial responsibility and so damaging to the commercial prosperity of the country as is the Standard Oil Company." [24]

These sentiments were practically identical with those expressed by Simon Sterne, counsel for the Hepburn Committee. Despite the fact that the railroads had admitted making special rates on at least 50 per cent of their total traffic of all kinds, Sterne singled out the Standard Oil compacts as "the most shameless perversion of the duties of a common carrier to private ends . . . in the history of the world." Noting that railroad charges were important costs in many items carried over great distances of the United States, he asked, "What is to prevent the creation of like monopolies in any other product of our country?" [25] Amid these verbal and legal attacks, Standard remained silent. To reveal as little as possible had been the company's policy for the preceding seven years, and apparently Rockefeller saw nothing in the present situation to distinguish it from the past.

---

* Data for arriving at more than a rough approximation of transport costs are woefully inadequate. First, applying an over-all average to a particular class of freight probably distorts the true cost of the latter. Gowen's statement, based on a more intimate knowledge of the business, implies that the distortion may be upward. Secondly, the average itself is none too reliable. Railroads apparently allocated their expenses between passenger and freight traffic in proportion to the revenues from each—a highly questionable procedure.[21]

Trunk-line proposals

Shutdown movements and free pipeline bills had long been part of the producers' routine artillery in their campaigns against combinations among railroads and refiners. The appeal to the courts, though never before used formally by a producers' organization, had turbulent precedents in the Oil Regions. Even the Union's proposal to back a long-distance pipeline to tidewater found an antecedent in the producers' support of Hostetter's Columbia Conduit Company.

The Union was, in fact, joining a venture that Benson, McKelvy, and Hopkins, of Conduit fame, had started on their own account shortly after Harley's fiasco in 1877. They were joined by Harley's chief engineer, General Hermann Haupt, who spent 1877 in secretly purchasing a 230-mile right-of-way from Clarion to Baltimore. Mindful of Hostetter's difficulties, Haupt was confident that railroad crossings would present no problem. A recent decision of the Pennsylvania Supreme Court had confirmed the rights of pipelines to cross railways where the latter had only an easement and did not hold the property in fee. All of the line's crossings were planned to be made at such points.[26]

Opposition to the proposed seaboard line came from essentially the same sources that had opposed earlier efforts. The Pittsburgh and Regions' refining interests understandably objected to any move that would build up the seaboard ports at their expense. Philadelphia was similarly disturbed over the possibility that the line might terminate in Baltimore. With a secure, continuous right-of-way from the southern districts of the Oil Regions eastward across Pennsylvania, the promoters, however, went ahead with their plans. In May, 1878, workmen began clearing the way for the telegraph line that would parallel the pipeline. Byron Benson of the partnership became particularly active. He enlisted Philadelphia capital, apparently with the understanding that the line would not extend to Baltimore. Instead, arrangements were made with the Reading Railroad to pick up oil at Columbia and carry it to Philadelphia, some 60 miles distant.[27]

But the project stagnated. Despite reportedly liberal subscriptions from the Producers' Union, capital was still bearish.[28] Nor was the Reading in a position to give substantial aid to independents. It depended for its tank car requirements on the Green Line, the former Empire division now absorbed by Standard, and had been prorating its freight with the Pennsylvania since October, 1877.[29] But perhaps the most important factor was the absolute decline of the southern fields. Production in the Butler-Clarion-Armstrong districts dropped 1,750,000 barrels from 1877 to 1878. Benson, McKelvy, and Hopkins shifted their sights northward to Bradford, where even the "Standard touch" faltered in providing the flush production with adequate pipeline and storage facilities.

Even before McKelvy & Company's partners considered abandoning their southern route, the pressure on transportation facilities had led another group to project a pipeline from Bradford. Organized in spring, 1878, the Equitable Petroleum Company was financed mainly by Bradford producers, who were said to account for one-quarter of the field's output, with some help from independent refiners in New York City. One of its leading figures was Lewis Emery, Jr., a Bradford producer and refiner who had come in 1865 to the Regions from a southern Illinois merchandise and milling business. A flamboyant opponent of Standard Oil, Emery like many others attributed the combine's power almost completely to its control over transportation. This view was to lead him to play an increasingly active role in pipeline development both in the market place and on the floor of the Pennsylvania Assembly where he began a ten-year term in 1878.[30]

Rather than head directly to the seaboard, Equitable planned a half-million dollar pipe westward to Buffalo. This terminal point meant a relatively shorter line, with consequently less trouble over rights-of-way, and it enabled Equitable to tap two sources of support: seaboard independents could be cheaply serviced through the Erie Canal, and Buffalo refiners could reach eastern markets by the same route. By painting glowing pictures of the great prosperity that Bradford offered the city, Emery touched a responsive chord among Buffalo businessmen. The prospect of reviving its near-dormant refining industry gained the canal city's support for the New York free pipeline bill that became law in April, 1878.[31]

Although Equitable fought for the New York pipeline law, the victory was dubious. Governor Lucius Robinson of New York showed little enthusiasm for the measure. Openly skeptical of the bill's constitutionality, he allowed it to become law only in order to test the measure in the courts.[32] Nor did it seem likely that the pipeline would reach New York state for some time to come, and there was no such law in Pennsylvania. Equitable built a seven-mile common carrier pipeline to Coryville, Pennsylvania, on the Buffalo, New York & Philadelphia Railroad. The railroad agreed to haul the oil to Buffalo, and the pipeline company purchased a fleet of 20 canal boats to move the cargo eastward through the canal.[33]

On July 17, 1878, the Buffalo, New York & Philadelphia Railroad took the first twenty-five carloads of oil from Equitable's lines. Three weeks later the cargo completed its trip to New York.[34] The cost was estimated to be, at most, one-half the $1.15 all-rail route to the seaboard.[35]

Throughout the summer and fall, business prospered. Pipeline runs averaged 1,500–2,000 barrels per day, and the original two-inch line was extended somewhat westward by a four-inch connection to Rixford, Pennsylvania, near the state line. By September, Equitable had added at least 100,000 barrels of tankage at new termini.[36]

Winter, however, would close the canal, and Equitable had not devised an alternative route to the seaboard. In August there were hopes of prying the Erie or the New York Central away from the Standard orbit, but these negotiations failed.[37] Instead, the trunk lines cut their rates to Standard still further to combat the canal shipments.[38] Fortunately for Equitable, Benson, McKelvy, and Hopkins had completed their plans to lay a pipeline east from Bradford. In October, 1878, Equitable announced it would use this outlet, rather optimistically anticipating its completion before severe winter weather.[39]

Tidewater

The partnership's plans for a Bradford line were formalized with the organization of the Tidewater Pipe Company, Ltd., on November 22, 1878.[40] The name was somewhat misleading, for the company was not restricted to transportation, nor did it intend to pipe directly to tidewater. Rather, Benson, McKelvy, and Hopkins presented a modified proposal to Franklin B. Gowen, president of the Reading Railroad. They would span a 100-mile stretch from Coryville, at the southern tip of Bradford fields, to Williamsport, if the Reading would take the oil to Philadelphia and New York at reasonable rates (see Map 17:1). Gowen, then feuding with the Pennsylvania over anthracite freight, was interested, and on behalf of the Reading subscribed $250,000, about half of Tidewater's initial capital.[41] In return, Tidewater placed itself under $100,000 penalty not to extend its line farther east of Williamsport for eight years.[42]

The company itself was a hybrid, bred by the state's legal and political system. Incorporation would have required a special charter, hardly feasible with a legislature largely controlled by the Pennsylvania Railroad. Tidewater was organized instead under a Pennsylvania law passed in 1874 that permitted the formation of limited liability partnerships. There were 26 partners in the original association; they could sell their shares to others, and newcomers were added. But unlike corporate stocks, Tidewater's shares conferred no automatic vote in the company's management. That prerogative came only upon being elected to membership by the existing partners.* The individual subscriptions ran from $100,000 to $1,000, providing an initial capital of $515,000.

With Benson as president, Tidewater started vigorously. To free the Reading of dependence on Standard's tank cars, orders were placed for 200 bulk cars at the Reading Iron Works, which also joined National Tube in turning out 5,000 tons of six-inch wrought-iron pipe for Tidewater.[44] On February 22, 1879, the first 34 sections of pipe were laid in an almost straight-line route from Coryville southeast to Williamsport.[45] In April, with over two-thirds of the line connected, pipe was going down at the rate of two miles per day. Simultaneously, Tidewater acquired a valuable asset by absorbing Equitable's 2,500-barrel per day gathering system.[46]

This rate of progress was made despite numerous problems that taxed the ingenuity of Tidewater's management. With no power to acquire property through condemnation proceedings, the company had to lease its right-of-way, striking such bargains as it could with landowners along the way. Standard and the trunk lines took steps to block the project by leasing probable stretches before Tidewater bid, or by exploiting loopholes and defects in old titles.

To outwit these moves Tidewater's men ran fake surveys to mislead the opposition and spent long hours searching through musty volumes of old land records. Even when they seemed to have a tract secured, trouble still arose. A farmer in Lycoming County, for example, almost blocked the line by withdrawing his promise to lease until Tidewater's representative found a usable sixteen-foot wide strip next to the farmer's property. Another grant, already signed, was salvaged only after a wild night-ride by horseback over 17 miles of muddy road got it to the recorder's office just before a midnight deadline.[47]

* As one official explained, ". . . If the stock [sic] were readily transferable on the market, and any purchaser has the right of voting in the company's affairs, the independence would be imperiled as there would be nothing to prevent oppositionists from buying up a controlling interest." [43]

Sledging the pipe for "Benson's Folly."

Probably the most hectic event of this sort occurred when a five-foot snowfall prevented work from being completed on one section by the date stipulated in the lease. Denied an extension, all of the company's resources in the area were immediately shifted to laying pipe from either end of the property. Every team available was hired to haul pipe, often as far as 40 miles over unbroken, snowdrifted roads. Sometimes teamsters had to dump most of their loads in order to make headway, arriving with only a few sections. To make up the deficit, line already laid on nearby leases was ripped up. The job, seemingly impossible when first undertaken, was completed within seven hours of deadline.[48]

Tidewater had to cross under two railroads, but its precautions to iron-clad its right-of-way saved it from both the overt violence and the costly litigation that had helped undermine Columbia Conduit. Benson, Mc-Kelvy, and Hopkins had learned their lesson from Conduit's earlier misadventures. Thus when the Northern Central tore up a section of line laid under a railroad culvert, Tidewater obtained an injunction and proceeded undisturbed, with only two weeks' delay.[49]

The most formidable technical problem connected with the line was to devise pumps adequate to force the oil to Williamsport. Tidewater planned to use only two pumping stations, each equipped with $21,000 worth of machinery. One, at Coryville, propelled the oil 22 miles up a 69-

foot incline. The second, located seven or eight miles from the highest point on the line, lifted the oil another 480 feet to the summit, where gravitation carried it the remaining 70 miles to Williamsport, 2,500 feet below the summit.[50]

The unprecedented pressures involved required special cylinder jackets. When preliminary tests were made with ordinary castings, oil streamed through cylinder walls in rivulets. Conventional direct-acting pumps caused pipeline pressures to vary anywhere from 15 to 275 pounds to the square inch, frequently bursting lines. Tidewater's engineers reduced the variation to fifteen pounds per square inch by adapting the triplex pump of the Holly Water Works. Three pistons were used, each timed for its descent to begin just before the others started their upward strokes. The pumps also economized on fuel, for steam entered the cylinders only at the beginning of each stroke, then expanded to carry the piston the rest of the way. Tidewater's heavier machinery provided enough inertia to compensate for the quick change in pressure when the steam injection was cut off.[51]

Tidewater's 80 crewmen found that laying the line was hard work, even for men accustomed to heavy labor. The terrain was mountainous and the winter severe; the spring thaw brought mud. The 18-foot sections of pipe, each weighing 340 pounds, were the largest ever used to that date, and were clumsy to handle, especially since tools were, by modern standards, of the most rudimentary sort. Eleven-man gangs laid the sections and screwed the ends into four-inch long joints with specially designed tongs. Except for horses to haul pipe from rail depots, nonhuman power was almost completely absent.

But despite recurrent delays and after an estimated investment of $750,000, the line was completed on May 22, 1879. Nearly half a million barrels of oil were stored at Coryville. At the Williamsport terminus some 60,000 barrels of storage capacity stood waiting while the newly manufactured tank cars began accumulating at the siding of the Reading Railroad.[52]

To dispose of the 6,000 barrels that would flow daily through the line, Tidewater's representative made arrangements with the Logan & Farnsworth refinery in Philadelphia and the new Lombard & Ayres works at Bayonne, New Jersey. The company originally counted on the capacity of six more New York independents, but these had been absorbed by Standard Oil in the interim.[53] On the other hand, an unexpected market developed in Williamsport, where local interests were constructing the 1,000-

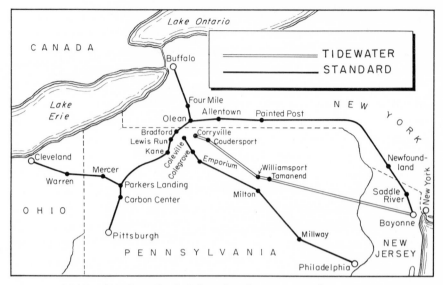

MAP 17:1. *Standard Oil and Tidewater pipelines, 1884.*

barrel per day plant of the recently organized Solar Refining Company.[54]

During the three-month construction period, skeptics named the line "Benson's folly." Although there were few who doubted that oil could be piped 100 miles, there were many who did not believe that it could be done with only two pumping stations.[55] The skeptical, the hopeful, and the curious gathered at the Coryville pumping station to see the 80-horsepower engines force the first oil into the pipe on the afternoon of May 28, 1879. It traveled slowly, at about half a mile an hour, reaching the relay station at Coudersport two days later. The summit was easily crested and, except for minor obstructions which were quickly removed, the oil rolled uneventfully towards its terminus.

On June 4 the tempo in Williamsport quickened. The oil, reported near the city, was expected momentarily and crowds began to gather around the receiving tanks. At 7:20 p.m. the area echoed with the loud, hollow sound of air being forced from the line by moving oil. Then the crude itself began to flow "in strong volume"—about 250 barrels an hour—and those near the tanks filled bottles of oil for souvenirs. Within a week the Reading started deliveries to the seaboard, via connections with the Jersey Central and the Pennsylvania. As the *Titusville Morning Herald* noted, a new era in petroleum transportation had been opened.[56]

Reaction to Tidewater

While there were few individuals associated with the oil industry who failed to recognize the long-run implications of Tidewater's success, it is true that Erie's vice-president, George Blanchard, was unconvinced. As he told the Hepburn Committee, "I believe that when this winter comes, the practical difficulties growing out of the season are [going] to be encountered [by Tidewater] . . . that interferences with the pipes are much more probable than they are with railroads by people along the line . . ." Blanchard did not deny that it was "practicable" to pipe crude, but he opined, "I don't believe it practicable . . . at anything approaching the present prices, and any interest on their investment." [57]

In one respect, his evaluation turned out correctly; in fact, "the difficulties growing out of the season" did not wait until winter. Buried only across cultivated land, Tidewater's pipe expanded out of all proportion under the hot summer sun, running "everywhere except in its proper position; telegraph poles and small trees that were fifteen to twenty feet distant . . . were pushed over by it, and in places it obstructed the highway." [58] Engineers feared that the line would snap, but fortunately no serious breaks occurred. This experience, however, led Tidewater to bury the entire length of the line—a $30,000 project that shortage of funds delayed until spring of 1880. [59]

Blanchard, however, was quite wrong in his estimates of the costs of transporting crude by the new trunk line. In combination with the Reading, Tidewater could reportedly land a barrel of oil at seaboard for 30¢, and General Haupt, the line's chief engineer, estimated that oil could be piped the entire distance to the east coast for 16⅔ cents a barrel. [60] Such rates were well below the levels at which the railroads could hope to compete. The New York Central's William Vanderbilt was no doubt voicing the opinion of many of his colleagues when he predicted that the railroads would be forced to surrender the crude oil traffic. "It came from the canal to us," he told the Hepburn Committee, "and is now going from us to other means of transportation." [61]

Vanderbilt's admission notwithstanding, it was soon clear that the railroads would not surrender their crude oil traffic without a fight. Nor was Standard willing to remain passive in the face of the threat Tidewater posed to its position in the transportation of oil.

The railroads launched their attack against the Tidewater-Reading combination in June, 1879, by cutting crude rates to New York to 15¢ and 20¢ a barrel, about half the cost of carriage. Standard's United pipeline co-operated in the war by cutting local gathering line pipeage charges to 5¢ a barrel.[62] The Reading was attacked by hostile connections over which it entered Philadelphia and New York; for use of a short but necessary stretch of line owned by the Pennsylvania Railroad, for example, it had to pay 15¢ per barrel.[63]

Paradoxically, Tidewater was also experiencing difficulty in building up its volume at the very time that flush Bradford production had led Standard's United pipeline to invoke the immediate shipment order of June, 1879.[64] Resentment against the order led some producers to switch from United to Tidewater, but the gain to the independent was minimal, and Tidewater seldom ran more than 3,000 barrels per day from producers' wells. An ever-present shortage of funds delayed expansion of the pipeline's storage facilities on both ends of the line, while Standard requisitioned stray tank cars and tied up the capacity of tank builders.[65] This helped to reduce Tidewater's shipments to final buyers to 1,700 barrels daily. Instead of the anticipated 10 to 13 per cent, Tidewater was handling only 2 or 3 per cent of total shipments out of the Regions during the latter part of 1879.*

The rate war continued to the end of 1879, at great cost to both sides. Gowen later claimed, probably with some exaggeration, that the Reading lost $500,000 on the oil trade during that period, while the combined losses to the Erie, New York Central, and Pennsylvania must have approximated $1.2 million.[66] By early 1880, however, the latter bowed to the inevitable, their decision undoubtedly colored by the fact that Standard was beginning to build trunk pipelines. In February the combatants negotiated a truce that increased seaboard crude rates to 60¢ from Bradford and allotted one-sixth of total coast-bound shipments to the Tidewater-Reading-Jersey Central group, enough to keep the pipeline working near capacity.[67] The new tariffs, reflecting the impact of Tidewater, were nearly as low as the net rates that Standard Oil had been paying during 1877–79, but they were still fairly profitable to the railroads and probably even more remunerative to Tidewater.

Probably by more than coincidence, the Tidewater agreement was announced at about the same time that accord on another front was at-

---

* Data estimated from monthly reports of pipeline runs and shipments published in the *Titusville Morning Herald,* August, 1879 through December, 1879.

tained. Delays in prosecuting the cases against United Pipe Line, Standard Oil officials, and the railroads led the Producers' Union to settle out of court in January, 1880. In return the railroads assured open rates to all, and equal rates to shippers *of like quantities,* with a vague promise that the differential favoring large shippers would be "reasonable." United Pipe Line on its part guaranteed adequate tankage, provided production was kept down to 65,000 barrels per day, while Standard agreed to pay the $40,000 litigation expenses which the Union had incurred.[68]

This was far less than had been hoped for, and the trade press took a generally skeptical attitude. "The most serious discriminations have been made by the railroad companies," the *Oil, Paint & Drug Reporter* somberly observed, "and they are by no means remedied by the agreement . . ."[69] Many producers agreed but, lacking any better alternative, voted to accept the compromise.[70]

Standard's trunk lines

The Hepburn Report and the Producers' Union charged that the Standard Oil Company had instigated the war against Tidewater, demanding protection against "injury by competition."[71] But Henry M. Flagler, in charge of Standard's transportation affairs, later claimed that his company was an unwilling partner, entering the hostilities, "like fools," only at the insistence of the railroads. According to his version, he had told the railroads, "Now, we can make a satisfactory arrangement with the Tidewater Pipe Line and avoid all this contest. It is not necessary to throw away any money."[72]

Rockefeller's actions about this time indicate that he too was somewhat reluctant to start a rate war. While United was cutting its pipeline rates, Rockefeller offered first to buy Benson's interest in Tidewater and, failing that, to guarantee the pipeline sales up to its capacity, an overture that was also rejected.[73]

These moves illustrate the temporarily entrapped position of the Standard Oil Company. Having built an elaborate rebate structure, Rockefeller was hesitant to antagonize the railroads needlessly. His lieutenants, however, had been predicting Tidewater's eventual success since its inception, and in March, 1879, Rockefeller went so far as to suggest a conference "regarding this pipeline question." Yet a week before Tidewater ran its first oil, he was still confident of the railroads' ability to compete success-

COURTESY THE OHIO OIL COMPANY

Laying an early pipe line.

fully.* The completion of Tidewater, however, quickly resolved any doubts in his mind concerning the future role of trunk pipelines in crude oil transportation. Standard immediately embarked on its own program of long-distance pipeline development.

The rate war against Tidewater was only four months old when Standard let its first contracts for a 100-mile pipeline to Cleveland.[75] Towards the end of 1879 some 40 crewmen and 80 wagon teams began laying five-inch pipe from Butler County, completing their work in March, 1880, at a cost variously estimated at $100,000 and $500,000.[76] The differential between the 12 to 20-cent pipeage costs and rail rates, which ran between 35¢ and 50¢ per barrel, was enough, according to *The New York Times*, to make Standard's position in the domestic market "well-nigh impregnable against all competition."[77]

* J. D. Rockefeller to L. Haldeman, May 22, 1879: "We entertain no doubt of the Railroad's ability of successfully competing for the transportation of oil and we do not want to invest any money in transportation enterprises in competition with the roads." A week earlier Rockefeller had written to Camden that he was "a little skeptical about their [Tidewater] doing it. They are quite likely to have some disappointments yet before consummating all their plans . . ."[74]

In 1880 the Rockefeller organization demonstrated its readiness to use the line to check competitors. Although intended to service only Standard's refineries, the Cleveland pipe also delivered oil, at a charge of 20¢ per barrel, to Scofield, Shurmer & Teagle, a Cleveland refinery with which Standard had signed a profit and market-sharing agreement. But when Scofield, Shurmer & Teagle sold more than its allotted 85,000 barrels per year, Standard cut off pipeline deliveries to the firm.[78]

Rockefeller, however, knew that the industry's major markets, and the areas where Tidewater was likely to foster independent capacity, were still along the eastern seaboard and abroad. Accordingly, in December, 1879, he offered the Erie railway $50,000 in cash (plus assurances that the road's oil traffic would be protected) for rights to lay pipe along the railroad right-of-way between the Bradford fields and Standard's works at Bayonne and Weehawken.[79] Agreement reached, construction began in February, 1880, for a line eastward from Olean, closely paralleling the Erie's route. Besides minimizing costs of right-of-way and haulage, this enabled Standard to operate parts of its line without waiting until the entire 400 mile-distance was spanned. Thus by June, 1880, oil was being piped about 100 miles from Olean to Painted Post, whence it was hauled further east by the Erie.[80] A few months later construction from the eastern terminus started in the opposite direction.[81]

While the New York line was progressing, Standard began laying another 80-mile conduit south from Bradford to Jersey Shore, Pennsylvania, on a branch of the Pennsylvania railroad.[82] Still a fourth line, this one paralleling the Buffalo & Southeastern railway to Buffalo, was built so quietly that relatively few people were aware it was flowing crude to the city by August, 1880.[83]

This extensive construction by Standard created more than a little apprehension in the Tidewater group. In addition, President Byron D. Benson feared that a new independent pipeline to Buffalo, plans for which were announced early in 1880,[84] would divert independent shipments from his company. He consequently offered in March and April of that year to negotiate with the Standard to raise pipeline rates and to prevent entry by other pipe companies, specifically mentioning the Buffalo line. But, despite earlier overtures to Tidewater, Rockefeller was now sufficiently satisfied with the status quo to table Benson's proposal for the time being.[85]

Rockefeller's refusal led several Tidewater stockholders, anxious to protect their transportation investment against Standard's further absorption of independent processors, to organize two quarter-million-dollar refineries of their own. One, the Ocean Oil Company, was located at Bayonne; the other, the Chester Oil Company, at Thurlow, Pennsylvania. With individual capacities of approximately 2,000 barrels per day, both plants began operating in spring, 1881.[86]

The anti-Standard *Oil, Paint & Drug Reporter* saw that more than this would be needed to maintain the independent refiners. Completion of Standard's Bayonne line, it said, would make Tidewater's extension to the coast an "absolute necessity." [87] Such a move, in the eyes of the *Reporter*, would dispense "retributive justice" on a grand scale. The Reading and the Jersey Central, which had invested in Tidewater, would share in the line's profits, while the Pennsylvania and the New York roads would be left without any crude oil freight.[88]

Reading's President Franklin B. Gowen was less sanguine. The depression of the 1870's, the rate wars, and Reading's unprofitable investments in coal mines had driven the road to bankruptcy. Under personal attack by a faction of his shareholders, Gowen was unwilling to risk further losses on Tidewater's account should a pipeline war with Standard develop.[89] At first he tried to reconcile the two pipelines. When this failed (because of disagreement over Tidewater's proposed share), he did not allow his position as a Tidewater director to interfere with his duties as president of the Reading. In March, 1881, the Reading signed a separate contract, guaranteeing Standard rates as low as its competitors (i.e., Tidewater) received. Standard, in return, extended the southern leg of its Bradford pipeline to the Reading's depot at Milton.[90]

The Jersey Central offered more active opposition to the eastern section of Standard's pipeline. This opposition stemmed essentially from its interest in Standard Oil's freight, which it picked up from the Pennsylvania Railroad at Allentown and hauled to Standard's yards at Weehawken. Reluctant to see itself displaced, the railroad tried to impede the pipeline's progress by refusing permission to cross a Central bridge in Bayonne. The attempt was unsuccessful. On the evening of September 22, 1880, the common council of Bayonne, claiming that the railroad had no control over the bridge, granted Standard permission to lay pipes through the streets of Bayonne. The mayor immediately signed the decree. Standard Oil confidence in the verdict was clear from the fact that it had 800 men waiting to begin construction at the moment authorization was granted. By the following morning they had completed their work; a week later the Jersey Central, faced with a *fait accompli,* agreed to leave the pipes unmolested.[91]

But a week later Standard Oil met dissension in its own ranks. The new irritant was the Pennsylvania Railroad, which nearly halved its Bradford rates to 33¢ per barrel in October, 1880.[92] Until the year's end it played no favorites, shipped at open rates to all comers, and even invited independents to build refineries along its tracks.[93] Although Standard initially availed itself of the low rates to reopen several Philadelphia refineries, it subsequently withdrew and the Pennsylvania's share of eastbound shipments dropped from approximately 27 per cent (guaranteed under the February agreement with Tidewater) to about half that proportion.[94] The railroad thereupon joined the National Storage Company, a $3.5 million organization headed by former Empire president Joseph Potts, in plans to build a refinery at Communipaw in the Bayonne area.[95]

The reasons for the Pennsylvania's aggressive stance are not clear; its action must have been somewhat puzzling to Standard's executives. Only a few months before, George B. Roberts, acting president of the railroad, had urged Rockefeller to divide the market with Tidewater before a "long and angry controversy" developed.[96] And the fact that Standard was piping oil from Bradford to the railroad at Jersey Shore had seemed to tie the Pennsylvania still more firmly to the Rockefeller combine. The most likely explanation is that the Pennsylvania feared that once Standard's pipe was extended to the seaboard the railroads would be devoid of crude oil freight. The Pennsylvania's efforts to prevent the line from being laid in Newark (allegedly because leakage would injure the oyster beds) lend plausibility to this conclusion.[97]

For Standard, the Pennsylvania's resistance was a major nuisance, though hardly a dangerous threat. That the railroad might seriously compete with the (completed) pipeline was out of the question. The Pennsylvania could, however, hamper material shipments and access to rights-of-way while the pipeline was being built. Its encouragement of the growth of independent refining capacity would make further absorption by Standard more costly, and Standard could still profitably use the railroad's co-operation on shipments of refined products. Consequently, Standard signed a compromise with the Pennsylvania on May 6, 1881. Under this agreement, the through rate from the mouth of the gathering line, via pipeline or rail, was fixed at 40¢, to be divided in agreed-upon proportions. If the rate fell below the 40-cent level, the Pennsylvania was to receive a portion of the 20-cent local gathering charges.[98]

The arrangement was secret, but outsiders suspected that some agreement had been reached for the Pennsylvania's proportion of eastward crude shipments rose from an average of 14 per cent over the preceding four months to about 38 per cent in March and April. Moreover, work on the National Storage Company's refinery at Communipaw was abandoned when Potts sold his company to Standard. The latter, in turn, incorporated National Storage into a massive, newly formed consolidation of Standard pipeline interests, known as the National Transit Company.[99]

At the time of its formation in 1881, National Transit was the largest single unit among Standard's holdings. Purchase of the entire stock of United Pipe Line gave National Transit control of a 3,000-mile gathering system with some 30–40 million barrels of storage capacity. This in turn fed into Standard pipelines completed to Buffalo and Cleveland and the line under construction from Olean to the Bayonne area. With the exception of Tidewater and a few microscopic independent gathering lines, National Transit controlled virtually every inch of pipe in and out of the Regions.*

Still National Transit had no influence over crude production, and new strikes in Bradford boosted output from 54,000 barrels per day in 1879 to 75,000 in 1881 and 82,000 in 1882.[101] Under the 1880 compromise with the producers, Standard obligated itself to provide storage facilities only if average daily crude output did not exceed 65,000 barrels. Nonetheless, United Pipe was reported to be building 180,000 barrels of iron tankage per week in the fall.[102] Recent pressure of public opinion may have made Standard more willing to accommodate; in addition, the action complemented Standard's policy of offering premiums to producers connected with Tidewater, which was having trouble in providing adequate tankage.[103]

Reduction of pipeline competition

The pressing needs of the Regions did not blind Standard to the fact that its main advantage in transportation lay in control of trunk pipelines

---

* National Transit was organized, not without poetic justice, under one of the old catch-all charters that the Pennsylvania Legislature had issued at Tom Scott's behest in 1871. Originally granted to the Overland Contract Company (later the Southern Railway Security Company), the charter carried the same broad powers as had that of the South Improvement Company. With it the Pennsylvania railroad obtained control over 1,500 miles of southern track, but failed to pay state taxes on the company's capital stock. The Commonwealth seized the charter in 1873, holding it until sale to Standard Oil in 1881 for $16,251.[100]

to the seaboard. Indeed, as premium payments to producers indicated, Standard's frenetic activity in the field was partly aimed at increasing this control, *vis-à-vis* Tidewater. At the same time, work continued on National Transit's line to the coast. By early May, 1881—about fourteen months after construction had been inauguarated—eleven pumping stations were forcing 7,000 barrels per day to Otisville, within 60 miles of New York.[104] The remaining section was completed early in the following year, the last grant across the Hudson River being signed in March, and a second line, starting about one hundred miles east of Olean, was well under way.[105]

At the same time, National Transit acquired another addition to its network—the Buffalo & Rock City Pipeline, the four-inch line to Buffalo that had troubled Byron D. Benson. The first company to be organized under the New York free pipeline act of 1878, the Buffalo & Rock City began operation in August, 1881.[106] Charging a rate of 10¢ per barrel, the line could deliver about 5,000 barrels per day. About 1,500 barrels per day were absorbed by the Atlas works, in which the Kalbfleisch Brothers, who built the pipeline, had an interest. The dozen other refiners in the city probably took some of their requirements from the line as well.[107]

The New York Times' Buffalo correspondent extravagantly hailed the new enterprise as the "most formidable victory ever achieved over the Standard Oil Company." But the independent's victory was extremely short-lived, for Standard completed its own line to Buffalo a year earlier. In January, 1882, the Buffalo & Rock City passed under Standard Oil control; United subsequently tore up its own line to the city and raised pipeage from the fields to 25¢ per barrel.[108] A few months later the Atlas refinery, suffering from Standard's competition, mismanagement, and internal dissension, was also sold to the Rockefeller combine.[109]

There remained only Tidewater, which continued in its curious status as a fairly vigorous, though somewhat reluctant, competitor. By early 1882 it was pumping from 5,000 to 8,000 barrels per day from Bradford, averaging about 11.5 per cent of total shipments from the Regions.[110] Its affiliated refineries, according to *The New York Times,* were "steadily assuming [an] important place in the oil trade." [111] The pipeline itself had been extended another 50 miles southeast to Tamanend, Pennsylvania, in 1881.[112]

Until 1881 this expansion was financed on a hand-to-mouth basis. Benson, McKelvy, and Hopkins pledged their producing properties as collateral and short-term loans, based on the company's oil in storage, were continually refunded with Philadelphia banks.[113] In June, Tidewater tapped a rather astonishing source of long-term funds by selling a minority block of voting stock to the Standard Oil Company, with guarantees that the latter would be represented on the board. The pipeline was pressed for capital, but it is doubtful that Tidewater would have agreed to sell without an underlying willingness to come to terms with Standard.[114] A year later a Standard offer to purchase the company for $5 million was rejected only because some of the directors insisted on more.[115]

In need of still more funds in the fall of 1882, Tidewater turned to a more independent source, the First National Bank of New York. A syndicate, headed by George F. Baker of the bank, took $825,000 of 6 per cent bonds at 90 and purchased $1.375 million in loan certificates, which apparently bore dividends equal to those paid on common stock as well as interest.[116]

This transaction was not completed without some difficulty. A minority of Tidewater's stockholders (led by Hascal L. Taylor and John Satterfield, who jointly controlled the largest producing company in the Regions) had unsuccessfully tried to convince the bank that the pipeline was near bankruptcy. Failing this, they took more drastic action at the company's sparsely attended annual meeting in January, 1883. Since the yearly reports had not been prepared in time, Tidewater's management intended to call the meeting to order and then formally postpone it. The minority, however, took the incumbent management by surprise, called the meeting to order and elected its own officers. There were irregularities in the appointment of tellers to count the vote and in the handling of proxies. In fact, when the matter was brought to court, the judge condemned the "whole proceeding from beginning to end . . . [as] farcical, its character fraudulent, and its results void." [117]

Minority spokesmen claimed that they were trying to protect their own interests against a corrupt management that cloaked its incompetence by blaming Standard Oil. As producers, the faction charged that Tidewater had not provided adequate tankage to handle their output. As stockholders, they had yet to receive a dividend, while Benson and his associates had voted themselves $20,000 in back salary, despite promises that they would take no payment until Tidewater reached the seaboard.[118] Tidewater's management, however, felt that the real purpose of the attempted coup was to gain control in order to sell out to Standard.[119]

In view of its own past record, the management's righteous indignation was somewhat misplaced. Although Taylor and Satterfield retained personal control of their Tidewater stock until 1884 (comprising about 30 per cent of the total), certain properties of their Union Oil Company were sold to Standard in September, 1882. And in November, 1881, one Standard executive wrote that the Taylor group clearly intended to obtain control of the pipeline in order to strike a bargain with Standard.[120]

A few months after these battles, a reform administration in Pennsylvania finally enacted an unqualified free pipeline law over the opposition of the Pennsylvania Railroad and Standard spokesmen in the Legislature. Two weeks later, the Pennsylvania Assembly again overrode Standard's objections and prohibited the merging of competing lines. This measure, said supporters, was necessary to prevent "monopolies" from absorbing new lines which might be built under the free pipeline act.[121]

Although the trade press welcomed the measure as a symbolic victory for the independents, it was generally skeptical of the bill's practical results. The arguments advanced are significant, for their insistence that trunk pipelines were profitable only as parts of integrated companies delineated a theme upon which the industry was to play variations for the next eighty years. The *Oil, Paint & Drug Reporter* wrote:

> . . . We do not imagine that capitalists will be found who will care to enter into competition with the National Transit Co., representing the Standard's pipe line system, or the Tidewater Pipe Line Co. These two companies have extended connections which will probably afford facilities for the transportation of all oil that there is an outlet for at any distance from the regions. Being large consumers themselves . . . they possess advantages which no line constructed merely for transportation purposes, can obtain.[122]

While this put the case a little too strongly, the prediction was in general correct. When competing lines were established, they sooner or later became divisions of integrated firms. For the present, however, this argument implied that if the free pipeline act was to foster more competition to Standard's line, it could do so only by enabling Tidewater to extend to the coast more easily.

There was no doubt that the free pipeline law could have helped Tidewater to this end. Although the company theoretically owned a right-of-way to Baltimore, many of the grants, conditional upon work being started within a specified period, had been allowed to lapse.[123] That the law would have had a significant affect on market structure is more dubious. Standard could (and probably would) have declared open war to prevent Tidewater from increasing its market share further.

But the need for such action did not arise. As already noted, Tidewater's directors had fended off attacks less because they valued their independence than because they wanted better terms from Standard. In October, 1883, only four months after the free pipe bill had been passed, they finally compromised. Prohibited by law from merging, the two companies signed a market-sharing agreement. Of the total volume of oil transported and refined by the two concerns, Tidewater was to handle a maximum of 11.5 per cent, roughly its proportion before the agreement.[124]

Thus, for the time being, any possibility of the pipeline act's affecting market structure was reduced to nil. As *The New York Times* editorialized, "So far as beneficial competition and the maintenance of low charges for transportation are concerned, the establishment of the pool is probably as satisfactory to the Standard and as unsatisfactory to the public as the complete absorption . . . of the Tidewater Line could have been." [125]

Impact of trunk pipelines on railcarriers and refiners

At the time of the agreement Tidewater had little more than half a line to the seaboard. It did not reach the coast until four years later. In contrast, Standard's two six-inch lines could pump about 20,000 barrels per day from the Bradford fields to Bayonne and Long Island, and a third line was under construction.[126]

As shown in Map 17:1, National Transit had also laid 280 miles of six-inch pipe from Colgrove, Bradford field, to Philadelphia; a five-inch branch connected with this line at Millway and ran 70 miles to Baltimore.[127] The completion of these lines in July, 1883, reflected the pull that trunk pipelines exerted upon refineries to locate at marketing points. Although no reliable cost estimates are available, it was undoubtedly cheaper to pipe crude over long distances than it was to send refined products by tank car. (It was generally believed that products would be injured in piping them over comparable distances.)

Thus there was some speculation that Standard intended to increase its refining activity in Philadelphia and Baltimore at the expense of its interior refineries, particularly those in Pittsburgh. This was in part borne out. The 284,000 barrels of crude shipped to Philadelphia in September, 1883, was more than triple the average of the preceding five months and nearly double the peak level attained in that period.[128] But Baltimore's crude receipts showed no increase, and the Baltimore line worked with so much excess capacity that the wisdom of the investment can be seriously questioned.*

The pipelines' pull to the market was somberly illustrated in November, 1883, when National Transit closed the great pump at Freeport on the pipeline to Pittsburgh. Pittsburgh's share of total crude shipments dropped from 14 per cent in 1879 to 3 per cent in 1883; monthly receipts of 50,000 barrels at the latter date were but one-third their 1879 levels. As Standard began dismantling the Star and Brilliant refineries, only one plant remained in operation.[130] The Iron City had lost out to the coastal refineries in the export trade and had yielded to Cleveland in the growing domestic market.

Cleveland, fed through Standard's first completed trunk line, was second only to New York in crude receipts. During the second and third quarters of 1883 an average of 460,000 barrels per month entered the lake port; its share of total crude shipments from the Regions rose from 18 per cent in 1879–80 to 27 per cent.[131] Like Pittsburgh, Cleveland had been shut out of the export markets by the trunk pipeline, but being closer to the major inland markets than was the Iron City, it could service many of them with cheaper transportation over the Great Lakes.†

With trunk pipelines entering every major refining center, the railroads more or less permanently lost the bulk of the crude oil trade. In 1888 they carried but 28 per cent of crude-oil shipments to points served by pipelines.[133] Standard took steps to soften the blow, partly to avoid troublesome rate wars, but mainly because the roads were still needed to haul refined products and supply fuel and material to pipeline stations. Thus, early in 1884, National Transit granted the Erie a direct subsidy amount-

---

* Baltimore's weekly refining capacity ran about 15,000 barrels per week; the line reportedly had a capacity of 10,000 barrels per day, although this was probably an exaggeration; Standard's six-inch lines to New York, with 10 pumping stations placed about 28 miles apart, could run 10,000 barrels per day; but the Baltimore line was only five inches in diameter and but one pumping station forced the oil through the entire 70 miles distance.[129]

† There is, however, one phenomenon that is inconsistent with the trend to locate at the market. Between 1879 and 1882 monthly crude shipments to "local points"— presumably along the Allegheny and Ohio rivers—rose from 4 to 11 per cent of the total. This rise during the same period as Pittsburgh's decline is inexplicable on the basis of information discovered. One possible explanation does suggest itself: as will be shown, independent refiners gained nothing from the trunk pipelines because pipeage rates were kept equal to rail rates. The next best thing was to locate in the field or near cheap water routes. This is consistent with a tendency on Standard Oil's part to follow an increasingly "live and let live" policy after 1880—provided that the independents did not gain too large a share of the market.[132]

ing to 2⅞¢ per barrel on all crude oil shipped to Standard's coastal "interests." This benefited the railway to the extent of $287,685 in two six-month periods.[134]

In August, 1884, National Transit substituted a new contract with the Pennsylvania Railroad for the one concluded while the New York pipeline was under construction. The railroad was now guaranteed 26 per cent of the combined total of crude and refined oil shipments to the Atlantic coast; if it carried less it would receive a portion of National Transit's revenue on the difference in quantity. In such instances the rate was to be divided on a sliding scale that guaranteed the pipeline a minimum of six cents and a maximum of ten cents per barrel pipeage from Olean to the Atlantic coast.[135] The railroad, in an extreme case, could carry absolutely no oil and still be paid for hauling one-quarter of the seaboard shipments. This interesting arrangement remained in effect until 1893, although the contract was not formally cancelled until 1905.[136]

On these contracts was based the entire rate structure to the seaboard for the next four years. Pipeage and rail rates from delivery points in the field were kept equal at 45¢ per barrel to New York (40¢ to Philadelphia); rates on eastbound refined oil from the Regions and Pittsburgh were 30 per cent more to adjust for the weight loss in refining.[137] The parity between railroad tariffs and pipeage, of course, deprived the nonintegrated refiner of the advantages of pipeline transportation while it filled the treasury of National Transit.

For if the rates were high enough to satisfy the railroads, they were more than satisfactory to the pipeline. That National Transit accepted a six-cent minimum in its contract with the Pennsylvania Railroad is *prima facie* evidence that at least the pipeline's out-of-pocket unit costs for piping oil from Olean to the seaboard did not greatly exceed that figure. In addition, gathering charges remained at 20¢ per barrel despite declines in cost. Between 1878 and 1888, for example, not only had the price of 3-inch pipe fallen from 50¢ to 10¢ per foot, but the pipe in the latter year leaked less.[138] One independent testified that in one six-month period the 20-cent rate had earned him $40,000 profit on his $70,000 investment in a gathering line.[139]

To the producers' charges that both gathering and trunk pipeage were exorbitant,[140] Standard Oil officials had two replies. Either they denied reliable knowledge of their costs or they maintained that seemingly high charges were needed to compensate for risks: wells, pools, and entire fields might play out before the cost of laying lines to them had been recovered.[141]

There was some truth to the claim so far as gathering lines were concerned. In 1882, for example, National Transit contracted for over 3 million barrels of tankage and laid pipes to handle an anticipated output of 10,000 barrels per day from the incoming Allegheny County district, only to have the field decline to one-twentieth that amount within a year.[142] But, even so the investment was not a total loss; unused pipe was transferred; oil from new areas was piped to empty tankage in other districts.

The validity of this argument when applied to the entire Pennsylvania field is more difficult to assess. Whatever might happen to individual pools in the field, the important quantity for the trunk pipeline was the aggregate volume of the field as a whole. Certainly when Standard began building its trunk line system, it seemed hardly likely that the output of the Pennsylvania Regions would drop seriously enough to imperil the earning power of the lines. It is true that by the late 1880's Standard was concerned over the incipient decline in Appalachian output, but apparently it did not regard either the economic or geologic exhaustion of the field as imminent. It was, instead, rather generally agreed in the councils of the combination that Standard's offering a higher price would induce greater production and exploratory effort.[143]

Along with the risk argument, Standard spokesmen often invoked a corollary proposition that pipeline and storage operations, if they were to prevent oil from running to waste, necessitated an organization as large as Standard. Appraising his company's efforts in this direction, John D. Rockefeller himself concluded that

> . . . It was a matter of great moment, of great value to the whole interest . . . that we should thus have put in the necessary capital not only to supply as needed those gathering lines but to supply them promptly, and not to wait a day or two or three, as might have been the case in this inferior service previous to our taking hold of the interest . . . And then we went on with the necessary larger pipes to get it [crude oil] away to where it would find its market . . . Conservative men in our connection never would have consented to make such large expenditures . . . We saved it [crude oil]. It could not have been done under the system that was previously in vogue. There was not the capital there; there was not the organization there. It was a godsend to the region that we moved . . . and moved boldly and continued so long as the necessity existed.[144]

Some might have considered Standard's "godsend" a mixed blessing and that Rockefeller exaggerated the superiority of his lines over the system "previously in vogue." The gap left by Standard's lag in entering the Macksburg, Ohio, field, for example, was filled by the independent Ohio

Transit Company (which in turn was absorbed by National Transit in 1885).[145] The "inferior service" of previous arrangements in large part arose during periods of sudden flush production in new fields—when no one could have laid lines and built tanks fast enough to prevent waste. But these delays were exceptional or unavoidable. Following the immediate shipment troubles of the late 1870's, Standard's pipelines were criticized on many counts—but few complaints arose about their ability to get pipes and tanks into the fields.

Nor is there much doubt that pipelines required large organizations. With costs (including materials) of laying 6-inch pipe running between $6,500 and $7,000 per mile, a 300-mile trunk line to the seaboard was necessarily a $2 million investment. Tidewater's problems in providing tankage indicated that the more a company could afford to invest in fixed capital equipment, the better its position was likely to be. And while factors that make for firms of large size do not necessarily lead to monopoly, available evidence strongly suggests that pipelines were, by their very nature, regional monopolies.

The technology of some industries is such that the larger a firm becomes, the more economies of scale it obtains, until one firm, by exhausting these economies, can force all others out of the market. The greatest source of pipeline economies is increasing line diameter, which (in the simplest case) increases surface area by the square inch, but increases volume by the cubic inch.* Until the mid-1880's the largest diameter in use was six inches, and the largest throughput about 10,000 barrels per day. This apparently approximated the optimum scale for this size line.† To carry greater amounts, separate lines were constructed. Although the capacity of existing lines could have been increased by adding booster stations, the increment would have been less than proportionate to the increase in pumping power. O'Day, for example, roughly estimated that doubling the number of stations on National Transit's pipes would have increased the latter's capacity by only 50 per cent.[147]

Crude oil output ran between 35,000–45,000 barrels per day, 5,000

---

* Throughput, however, would not rise by the cube of the dimension because friction would increase as oil came in contact with the (larger) inside diameter of the pipe. But the increase in friction would be less than proportionate to the increase in throughput since only the outer "layer" of oil would touch the pipe. Although stations on large diameter lines require more horsepower, pumping power also exhibits economies of scale.[146]

† "Optimum" in the sense that any greater outputs could be obtained only at higher unit costs.

to 8,000 of which went to local points with no need for trunk pipeline service. Nonetheless, the Regions were producing more than one pipeline could optimally handle.

Crude oil shipped by pipeline, however, was divided among the three major refining centers of New York, Philadelphia, and Cleveland. If the crude oil demand of any of these was no greater than the optimum capacity of a single pipe, then obviously only one line to that point would pay. This was clearly the case with Philadelphia: its crude receipts were approximately equal to the optimal rate of National Transit's line from Colegrove. In fact, since probably not all crude to the Quaker City went through the pipe, an expansion of throughput would not have raised unit costs and may have lowered them. New York's crude consumption of 25,000–30,000 barrels per day, even when reduced by 25 per cent to adjust for maximum possible shipments over the Pennsylvania Railroad, was clearly beyond the optimal range of one six-inch line.* Significantly, National Transit had two and a half lines from Bradford to the Bayonne-New York area. Although there were probably savings if one company owned both lines, it is conceivable that they could have been separately owned and operated.

It is still debatable whether separate ownership would have led to lower rates than a monopoly of both lines. With growing shipments to New York, two independent lines could have divided the market, by either tacit or open agreement, at rates which yielded quasi-monopoly profits and still kept equipment working near optimal capacity.† Much the same analysis applies to Cleveland, whose crude receipts of 16,000 barrels per day required two 5-inch lines at most. It was, therefore, almost inevitable that each refining point should be served either by a monopolistic pipeline, or by combinations of lines that would be less than freely competitive.

Standard spokesmen generally ignored the regional monopoly aspects of the problem. They implied that a national monopoly was more efficient

---

* From April through June, 1883, crude-oil shipments to New York averaged 27,500 barrels per day. Refineries in Bayonne were apparently counted in New York's crude receipts.[148]

† Alternatively, they may have followed the same pattern of rate wars and pooling agreements as the railroads did at competitive points. This would probably have lowered the average rate over time, but it seems doubtful that the situation could have persisted. For one thing, such lines did not have the non-competitive freight that railroads used to supplement their earnings. In any event, the competitive market processes were not likely to work in these situations.

than a series of independently owned lines. This was a debatable contention. There were few, if any, additional economies of scale open to a national amalgamation of pipelines. It is true that a company with lines to several refining points was less dependent upon the survival of any of them; but major refining areas had not changed swiftly or drastically in the past, and in the 1880's few oilmen anticipated that future geographical shifts would be so radical as to endanger lines built to "old" refining centers.

It can be argued that inevitable regional monopolies of this type are by their very nature public utilities. This idea, not entirely foreign even at a relatively early date, gained stronger support during the 1880's. The free pipeline bills, for example, were designed to endow the lines with privileges granted other public utility-type carriers. Had refining been structured competitively—particularly if no single processor had enough capacity to absorb a pipeline's optimal throughput—the common-carrier aspects of pipelines would have been even more obvious. But this phase was obscured by the fact that trunk lines developed for the most part as a part of an integrated monopoly. Independent refiners, fearing they would be refused accommodation during crucial periods, were reluctant to depend upon National Transit for transportation. Nor did the parity between rail rates and pipes offer the nonintegrated refiner any incentive to use pipeline transportation. Thus it was that Standard's trunk lines from the Regions "as a practical matter" neither carried outsiders' oil, nor were asked to do so.[149]

It would be too much to say that this situation precluded any possibility for competition to challenge Standard's dominant position in the refining segment of the industry. But it did mean that to occupy anything more than a peripheral position in the market any refiner or group of refiners must be prepared to operate on a scale sufficiently large to warrant the inclusion of trunk pipelines.

Chapter 18

The growth and reorganization
of refining capacity

To most oilmen accustomed to the competitive conditions of the 1860's, it appeared that Rockefeller's success in acquiring refining capacity and solidifying his control over transport facilities between 1872 and 1878 had made it virtually impossible for independent refiners to survive. The Tidewater innovation involving long-distance crude lines was still in the planning stage, and although Bradford had been discovered, its potential as a possible basis for independent operations had yet to be revealed.

Events of the next two years did little to improve the prospects for the emergence of refiners operating outside the Standard orbit. For a time it appeared that the development of Bradford as one of the truly prolific oil fields in the history of the industry might weaken Standard's control over storage and gathering line facilities. But once the circumstances fostering the "immediate shipment" orders of 1878–79 had been reduced to manageable proportions, this possibility largely disappeared. By moving swiftly to construct its own crude trunk lines, Standard was able to neutralize the threat of Tidewater to its position in the long-distance transport of crude.

These moves alone would not have assured Standard of maintaining its dominant position in the industry. It is true that the management's task was made somewhat easier by the fact that the basic refining technologies developed during the 1860's changed little between 1874 and 1884. But the need to provide additional refining capacity to handle the huge crude reserves uncovered at Bradford, the trend toward larger-scale refining establishments, and the growing importance of by-products, all presented a challenge to the industrial combination Rockefeller had built by the late 1870's. By meeting each of these challenges, Standard, by the mid-1880's, continued to hold its position as the most important single processor of crude oil. But to maintain this position, Rockefeller and his associates were forced to concede a more important role to independents than had been anticipated by the end of the 1870's.

The rise of independent refining capacity

A variety of factors was responsible for the growth of independent refining capacity after 1878. Access to crude supplies, although subject to many uncertainties, was still open to independents. Of major importance in attracting new capacity were the declining costs of the principal refining inputs. As shown in the following tabulation, between 1877 and 1884 the wholesale prices of sulphuric acid and caustic soda dropped

Year	Sulphuric acid [1] year-end price per 100 lbs.	Caustic soda [1] year-end price per 100 lbs.	Crude oil [2] average annual price per 42-gal. bbl.
1877	$2.25	$4.32	$2.42
1878	1.63	3.75	1.19
1879	1.63	4.38	.86
1880	1.63	3.54	.95
1881	1.63	3.53	.86
1882	1.50	3.34	.78
1883	1.50	3.25	1.06
1884	1.63	3.15	.84

[1] Anne Bezanson, *Wholesale Prices in Philadelphia: 1852–1896*, 2, 306.
[2] *Mineral Resources, 1891*, 419.

over one-third, while there was an almost two-thirds decline in the price of crude.

Although Standard, with the exception of Tidewater, owned and operated the major crude trunk lines after 1880, for refiners not located near wellheads there was the alternative, although generally more expensive, method of shipping by railroad tank car. Tidewater itself stimulated the construction of new plants at Thurlow, Philadelphia, and the New Jersey shore district, which had a combined daily crude charging capacity in 1882 of between 5,000 barrels and 6,000 barrels.[1] The independent line from Bradford to Buffalo, built under such duress and fanfare in 1880, also added to independent capacity. This included the Atlas works, which charged between 1,500 barrels and 2,000 barrels of crude daily, plus a cluster of smaller plants which also drew their crude from the line.[2]

Perhaps the most important addition of new independent refining capacity, however, came from plants built in the Oil Region, which were devoted to the manufacture of by-products, particularly lubricants, from the paraffin-rich Bradford crude. By all indications (such as spreading industrialization at home and abroad and advancing prices and limited supplies of animal and vegetable lubricants) the time was propitious for an expansion of petroleum lubricating oil manufacture.[3]

For the independent refiners, manufacture of lubricating oils offered particularly attractive opportunities in competition with the Rockefeller combination. It was possible even for relatively small refiners to differentiate their output and to cultivate market positions for branded products based on premium quality and customer service. Premium prices in turn fattened refinery margins which permitted greater flexibility in absorbing distribution costs, substantially reducing locational factors such as inhibited kerosene manufacture.

Illustrative of the kind of specialized refining capacity which had become well established by the mid-1880's was the Clark & Warren refinery at Corry. With a daily crude charging-capacity of about 300 barrels, the company produced a total of thirty-nine different grades of refined oils. These included a certain amount of kerosene, but the main emphasis was on specialties such as a 300° F fire-test illuminating oil "used on ocean liners and Pullman cars" and over two dozen varieties of lubricating oils having a fire test "ranging from 500 to 600 degrees Fahrenheit." These were "particularized and described in a neatly printed pamphlet" which went out to the trade.[4]

It is impossible to segregate, let alone enumerate, the types and amount of independent refining capacity which appeared between 1880 and 1884. One estimate for 1881, as shown in Table 18:2, put the total daily refining capacity of the industry at approximately 97,760 barrels with Standard controlling about 90 per cent (88,745 barrels) and independents the remainder (9,015 barrels).[5] The trade journals are littered with accounts of new plants built during the next few years. In 1884, F. Q. Barstow reported to Rockefeller that there were some 93 refineries operating outside the Standard orbit.[6] The best contemporary estimate suggests that by the mid-1880's, total daily refining capacity had grown to about 125,000 barrels of which 96,000 barrels (77 per cent) was owned by Standard and nearly 29,000 barrels (23 per cent) by outsiders.[7]

This approximate trebling of independent refining capacity within four years was, under the circumstances, quite impressive. A substantial portion—a little over 40 per cent—of the total increase of nearly 21,000 barrels per day was directly associated with the Tidewater and Buffalo pipelines. The remainder represented capacity built mainly in the Oil Regions, Pittsburgh, and Philadelphia.

A qualitative impression of the new independent capacity is also limited by imprecise and often inaccurate data. But judging from contemporary evidence, perhaps as much as one-half of the total independent capacity in 1884 owed its existence to a widening range of petroleum products. It confirmed the prediction made in 1881 by *Petroleum Age* that if independent activity was to increase over the coming years it would be fathered through by-product manufacture.[8]

Standard was by no means unaware of the potentials of by-product manufacture. But faced with the dual task of completing its own network of long-distance pipelines and consolidating and revamping the refining capacity already acquired, the management chose to devote the major portion of its refining resources to supplying the rapidly expanding and equally profitable demand for illuminating oil. Thus it was not until after the reorganization and streamlining of the administration under the famous trust agreement of 1882 that Rockefeller and his associates gave serious attention to a full-scale integration of by-product manufacture into their operations.

The Standard Oil trust

There were many reasons that dictated an administrative reorganization of the vast array of companies garnered under the Rockefeller alliance. Basically the need was for some legal instrument by which diverse properties could be effectively consolidated and an administrative policy and communications framework established. Delegated authority and efficient administration always characterized the Rockefeller companies, but now the immense scope of operations, and rising public criticism manifested in legislative and judicial attacks at one or another operating flanks, made administrative and legal reform imperative.*

As long as operations were confined within the state of Ohio there was no legal problem involved in expanding the holdings of Standard of Ohio. While the Cleveland concern continued as the nucleus of the organization over the succeeding years, under its charter it had no legal right to own plants in other states nor to hold stock in other companies.

These restrictions led the Standard management to an early use of the trustee device. When the Long Island Oil Company of New York was acquired in 1872, for example, its stock was transferred to Henry M. Flagler, secretary of Standard of Ohio, as trustee, not for Standard of Ohio, but for the stockholders of the Ohio corporation. This procedure, with Flagler, William Rockefeller, W. G. Warden, J. A. Bostwick and others functioning as trustees, set a pattern for the acquisition of the bulk of the companies brought under Standard control over the following half-dozen years.[9]

In addition to providing the Standard management with a flexible means of expanding its holdings outside Ohio, the trustee device had the further advantage of maintaining a "veil of secrecy" over Standard's acquisitions. By following a strict legal interpretation of the role of the trustees—that they were not owners and were acting simply on behalf of stockholders and not for the corporation—officers from Rockefeller on down could deny under oath that Standard of Ohio either owned or controlled particular properties. This was convincingly demonstrated during the first national investigation of the "trust problem" in 1888 when Standard officials repeatedly denied on the witness stand facts which only a strict legal interpretation of a trustee's duties allowed them to do.[10]

But these advantages had to be weighed against certain disadvantages which became increasingly apparent as Standard's holdings mushroomed after 1872. The death of any of the important trustees could in itself raise difficult legal problems. There was always the possibility, however remote, that disagreement among members of the Standard management might prompt a particular trustee to "make trouble" for his associates.

---

* The events and impulses, including prior administrative and policy experiments, are discussed at length in Hidy & Hidy, *Pioneering*, 40–75.

To avoid difficulties of this sort, an agreement was drawn up in 1879 under which all properties formerly held in the names of individuals were transferred to three trustees, selected from among the employees at the Cleveland office. They were to manage the stocks and other interests under their trusteeship for the exclusive benefit of the stockholders of Standard Oil of Ohio.[11] Under their original instructions, the trustees were supposed to divide the shares of the Standard-controlled companies among the shareholders of Standard of Ohio. This was never seriously attempted and their main task was to divide profits annually in proportion to the stock held by Standard of Ohio's shareholders.

While this move served to concentrate the legal holdings of the associated companies in the Cleveland office and to minimize the danger of a "conflict of interest" among the trustees, it still fell short of an adequate form of organization. The agreement made no provision for continuity of trustees in case of death or resignation nor for the transfer of certificates of ownership.

There was a further and perhaps more serious limitation arising from a Pennsylvania law which "forbade corporations of other states to acquire any realty in Pennsylvania either directly, or through any trustee or other device whatsoever."[12] Whether Standard of Ohio's method of holding properties was in violation of this law was open to question. A more immediate threat was posed in 1882 when a suit for $3,145,000 was brought by the Commonwealth of Pennsylvania against Standard for the recovery of taxes under a law passed in 1868 that apparently made the entire capital stock of corporations subject to tax regardless of the amount of business done in Pennsylvania. When the court on April 4, 1882, ruled that only property located in Pennsylvania was subject to taxation, Standard's tax bill was cut to $3,300.[13] But this experience helped convince the management that Standard of Ohio should "be divested of its out-of-state holdings as soon as practicable."[14]

There were several alternatives which might have been adopted. One was to form a holding company; the second was to form a joint-stock company; the third was to undertake "a careful development of the plan of holding stocks in trust, while the business was managed by elected representatives of the beneficiaries."[15]

The difficulty with the holding-company proposal was to secure a charter granting the necessary powers. Among the various states at this time only New York granted such charters, but these required a special act by the legislature which, in the judgment of Standard's legal advisors, would be difficult if not impossible to obtain. New York also provided for the formation of unincorporated joint-stock associations with most of the characteristics of a corporation. The principal disadvantage of the joint-stock association in the eyes of the Standard management was the loss of secrecy.

These considerations led to the decision to amplify the trustee device, and on January 2, 1882, the Standard Oil Trust Agreement was drawn up and signed by the forty-one stockholders of Standard of Ohio and the three trustees who had held the stock or equities in some forty specifically named companies associated with Standard.[16] As worked out by Standard's lawyers, with S. C. T. Dodd taking the leading role, the agreement provided solutions to all the problems troubling the management, and more.*

The value of the properties put into the Trust was set at $70 million against which 700,000 trust certificates (par value $100) were issued to the stockholders of Standard of Ohio. The agreement not only named the nine trustees who were to take over the administration of the Trust, but provided for one-third of the group to be elected each year by trust certificate-holders or their proxies. A wide range of powers and duties was conferred on the trustees. Of particular significance was Article 15, which read:

> It shall be the duty of said Trustees to exercise general supervision over the affairs of said Standard Oil Companies, and as far as practicable over the other Companies or Partnerships, any portion of whose stock is held in said trust. It shall be their duty as Stockholders of said Companies to elect as Directors and Officers thereof faithful and competent men. They may elect themselves to such position when they see fit so to do, and shall endeavor to have the affairs of said Companies managed and directed in the manner they may deem most conducive to the best interests of the holders of said Trust Certificates.[17]

Looking ahead to the possibility of organizing holdings more effectively on a state-wide basis, the agreement also provided for the formation of corporations bearing the name Standard Oil Company in New York and New Jersey as well as in other states and territories whenever deemed

---

* Indeed the trustee device as worked out by Standard marked an innovation in business organization which was quickly and widely adopted by other American industrialists who were looking for some method of forming inter-state combinations which were both legal and workable. Its association with large industrial organizations, including Standard, subsequently made the term "trust," at least in popular usage, synonymous with monopoly.

advisable. The Standard Oil Company of New Jersey and the Standard Oil Company of New York were incorporated in 1882, followed over the next few years by namesakes in Indiana, Iowa, Kentucky, and later California.

The formation of the Trust not only served to concentrate the legal holdings of Standard's subsidiary and associated companies. By its very simplicity the original trust agreement gave Rockefeller and his associates both the shield of legality and the administrative flexibility they needed to operate effectively what had become virtually global properties.

Responsibility for major policy decisions remained with John D. and William Rockefeller, H. M. Flagler, S. V. Harkness, and Oliver H. Payne, who together owned some four-sevenths of the outstanding shares of the Trust. Later joined by J. D. Archbold, Daniel O'Day, and the already established Jabez Bostwick, this group formed the top management nucleus of one of the greatest industrial machines of all time.*

For a decade or more prior to 1882 the top management group had increasingly delegated operating responsibility to the managers or heads of divisions. But in arriving at policy decisions the members had drawn heavily on the advice of the committees made up of personnel familiar with particular phases of the business. This pioneer practice which was continued and expanded after the formation of the Trust made the committee system one of the distinguishing features of the Standard management.†

With its legal and administrative network formalized, the management focused its full attention on the problems of readjusting and consolidating its complex of properties into an effective operating organization. One of the most important of these problems involved a reappraisal and reassortment of the Trust's holdings of refining properties.

Reorganization of Standard's refining capacity

In carrying out his "original plan," Rockefeller by the early 1880's had already brought about an extensive reorganization in the structure of the

* The 1882 agreement carried the Trust through one of its most productive periods. But in 1892 as a consequence of a court decision in Ohio, the enterprise was reorganized under twenty corporations.[18]
† Between 1887 and 1892, for example, the Trust maintained seven major committees responsible for advising on the following operations: case and can, cooperage, domestic trade, export trade, lubricating oil, manufacturing, and transportation.[19]

industry's refining capacity. Two changes between 1873 and 1880 were particularly striking: one, the reduction in the number of individual refineries; the other, the shift in capacity from the Regions to the major marketing or distributing centers for refined products.

In 1873, when Standard controlled about 10 per cent of total refining capacity, there were, as shown in Table 18:1, approximately 103 refineries

TABLE 18:1
Number and location of refineries: 1870–80

Area	1870 *	1873 †	1880 ‡
Kentucky	2	..	1
Portland, Maine	2	1	1
New Jersey	..	..	2
New York Metr. Area	16	15	18
Buffalo, New York	..	2	}3
Other New York	..	2	
Cleveland, Ohio	24	6	18
Marietta Region, Ohio	..	..	6
Pennsylvania Oil Regions		27	19
Pittsburgh		22	9
Philadelphia	}94	16	7
Erie		6	..
Other Pennsylvania		..	..
West Virginia	9	..	5
Massachusetts	8	2	2
Baltimore, Maryland	..	8	3
California	..	..	..
Colorado	..	..	..
All other	..	..	..
TOTAL:	155	103	89

* Theodore Childs Boyden, "The Location of Petroleum Refining in the United States" (unpublished Ph.D. thesis, Dept. of Economics, Harvard University, 1955), 120.
† Boyden, "Location of Petroleum Refining," 123; *Derrick's Hand-Book*, I, 780; J. T. Henry, *Early History*, 315–24.
‡ Peckham, *Census Report*, 186.

in existence, with the heaviest concentration in the Oil Regions, Pittsburgh, the New York metropolitan area, Philadelphia, and Cleveland and Erie, in that order. In 1880, when Standard's share of total capacity was about 90 per cent, the total number of refineries had been reduced to 89. Most of this decline occurred in the Regions, Pittsburgh, Philadelphia, and Erie where the number of refineries was reduced by about one-half. By contrast there was an addition of five plants in the New York–New Jersey area while the number in Cleveland grew from six to twelve.

While Standard had gone far by the early 1880's in reorganizing and consolidating the refining properties it had acquired so rapidly between 1873 and 1880, the management in 1882 still had to make a number of major policy decisions. These included the final disposition of the capacity acquired over the preceding decade, the total amount of refining capacity which should be maintained by the Trust, and how this capacity should be allocated among the principal refining centers. There was also the basic question of Standard's policy toward the growing number of independents.

It is evident that even prior to the trust agreement, Rockefeller and his associates had come to the conclusion that it was no longer economically feasible to control 90 per cent or more of domestic refining capacity. Noting the emergence of certain independent capacity, the New York *Shipping and Commercial List* early in 1882 observed, ". . . the general supposition now is that the Standard managers have come to look upon competition as inevitable." [20] According to Rockefeller's biographer, Allan Nevins, the Standard leaders felt by this time that their dominant position in the industry would not be endangered so long as they could "prevent outsiders from doing more than one-fifth of the business." [21]

To make sure that outsiders would not exceed this limit, Standard had moved early in 1882 to acquire the Buffalo and Rock City pipeline connecting Bradford with Buffalo. In the face of higher pipeage rates, the Atlas refinery and the much-publicized Solar works sold out to Standard a few months later. [22] The major step to prevent independent capacity from getting out of hand was, of course, the agreement reached with Tidewater in 1883, limiting the latter's share of the combined Standard-Tidewater shipments and refining of crude to 11.5 per cent. [23]

Experience with what it termed "blackmailing" activities may have served to dampen some of the Standard management's enthusiasm for acquiring outside refineries wholesale. The low crude prices which prevailed during and after the height of the Bradford boom, plus the fact that the initial capital costs of constructing a quasi-permanent refinery were quite minimal, prompted a number of individuals to build plants with the expectation they might be sold to Standard at a profit.

Rockefeller's personal reaction to "such adventurers" was one of righteous indignation. He much preferred whenever possible to let them "sweat it out" and serve as an example to other "blackmailers." But at times it was expedient to come to terms. Of one such group at least, Colonel Thompson expressed the hope that association with Standard would improve their business ethics. "H & J are small, mean sort of fellows," he wrote Rockefeller in April, 1879, "but I suppose we will all have to admit that we have frequently to make arrangements with that kind of people, and very often it changes and improves them. . . ."[24]

At most the number of "nuisance" plants bought by Standard could not have been large. With few exceptions the Trust after 1881 concentrated on acquisitions of plants which had a specific manufacturing value. Some were purchased to obtain patents or particular refining techniques; others

TABLE 18:2

Estimated refining capacities of major refining areas: 1873, 1881, 1884 (barrels per day)

	1873 *	1881 †		1884 ‡	
		Standard	Independent	Standard	Independent ¶
Cleveland	12,732	21,425	na	22,000	
Pittsburgh	8,990	15,035	1,730	na	
New York–New Jersey	9,790	34,300	2,571	43,000	
Philadelphia	2,061	11,000	4,457	13,000	
Baltimore	1,098	na	na	..	
Erie	1,168	na	na	na	
Boston	600	na	na	na	
Buffalo	450	na	na	1,600	
Oil Regions	9,231	6,985	275	na	
Others	450	na	na	na	
		88,745	9,015	96,000	28,868
Total U. S. Refining Capacity	46,570	97,760		124,868	

* J. T. Henry, *Early History*, 315–24.
† Emery Testimony, *1888 Trust Investigation*, 232–35.
‡ Hidy & Hidy, *Pioneering*, 100–101, 120. Estimates for Standard's capacity in Cleveland and Buffalo added.
¶ There is no breakdown available for the location of independent capacity in 1884. The best estimates suggest that at least a third was located in the Regions with the rest divided principally between Cleveland, Pittsburgh, and Philadelphia. Cf. *1888 Trust Investigation*, 438, 440.

for the quality, type, or brand names of their products.[25] Even so, the net addition to Standard's total daily refining capacity was not small, amounting to approximately 5,000 barrels.[26] This addition plus a moderate expansion of previously acquired plants served to increase Standard's total refining capacity, from just under 89,000 barrels per day in 1881 to approximately 96,000 barrels in 1884. (See Table 18:2.)

At the same time the management was adding to total capacity, it continued to concentrate a large portion of its holdings at the major marketing areas, and to reduce the number of its plants. Standard by 1881 had already concentrated over 90 per cent of its refining capacity at the major marketing centers. The locational pull of the export sale of illuminating oil is evident by the proportion of capacity, about 39 per cent, located in the New York–New Jersey area and to a lesser extent in the case of Philadelphia with approximately 12.5 per cent. Cleveland, with about a quarter of the total capacity, and Pittsburgh, with about 17 per cent, sup-

TABLE 18:3

Major Standard refining plants: number and location, 1882–86

Location	Number of plants	
	1882	1886
Major Refineries		
Baltimore	2	2
Buffalo	1	1
Cleveland	4	1
New York Metropolitan Area	14	10
Oil Regions	15	5
Philadelphia	6	2
Pittsburgh	11	1
TOTAL:	53	22
Lubricating Plants		
Inland Oil Regions	7	3
New York	2	1
Cleveland	1	1
Rochester	1	1
TOTAL:	11	6
Specialties		
New York	4	4
Oil Regions	1	1
TOTAL:	5	5

Source: Adapted from Hidy & Hidy, *Pioneering*, 102–5.

plied a portion of the export demand; but their principal outlets were the domestic markets of the West and South.

Three years later, daily capacity in the New York–New Jersey area of some 43,000 barrels accounted for almost 45 per cent of Standard's total refining capacity. An addition of about 2,000 barrels to Standard plants in Philadelphia maintained that city's share of total capacity at about 13.5 per cent. Operations at Cleveland, which was supplying the domestic trade almost exclusively, were only slightly larger in 1884 than in 1881. What facilities Standard still maintained at Pittsburgh in 1884 is not clear, but it seems likely that the Trust had already started closing down much of its capacity which left Pittsburgh with only one refinery in 1886. (See Table 18:3.) *

The trend toward a concentration of the Trust's expanding total refining capacity in fewer plants is clearly shown in Table 18:3. Between 1882 and 1886 the total number of Standard's "major" refining units—devoted primarily to the production of illuminating oil—was reduced from 53 to 22. Of the approximately 22 plants operating in 1885, three alone (the Cleveland, Philadelphia, and the Bayonne, New Jersey refineries) processed nearly two-fifths of Standard's total throughput of approximately 17.7 million barrels of crude in 1885. With an average daily charging capacity of 6,500 barrels each, these three plants were the largest refining units in the world.[27]

In addition to consolidating capacity in fewer plants, Standard also converted a number of its smaller refineries to the manufacture of specialties in order to tap the growing demand for petroleum by-products. Lubricating oil in particular loomed high in the Trust's future production plans. The lubricating plants as shown in Table 18:3 were concentrated in the Oil Regions, specifically in 1886 in Franklin, Pennsylvania, where Standard's Eclipse, Galena, and Signal lubricating oils were manufactured. With one exception the five specialty plants devoted to the output of paraffin wax and a variety of other by-products were located in New York.[28]

By the mid-1880's the Standard management had done a remarkable job in respect to its refining capacity. Recognizing the key importance of efficient refining operations in maintaining their dominant position in the industry, Rockefeller and his associates moved with characteristic speed and precision in reorganizing, consolidating and expanding the Trust's refining properties. By concentrating output in large units, they established manufacturing operations on an unparalleled basis of mass production. They set a pattern which enabled the Standard organization to capitalize on its growing scope of operations throughout the rest of the century in supplying the demand for refined products quickly, efficiently, and inexpensively.

Refining technology

In many ways the fact that the technology of petroleum refining changed little during 1874–84 simplified Standard's task of bringing its refining capacity under control. Moreover, the Standard management was fully alive to the possibility of having its accomplishments disturbed, if not upset, through technological change. To avoid such a contingency, Rockefeller and his associates systematically acquired operative or commercially potential patents affecting nearly all branches of the industry. Indeed, even by the early 1880's, Standard's control over basic patents was so extensive that it was virtually assured of participating in any new processes or techniques affecting petroleum.

In 1882, by one careful estimate, the newly formed Trust held at least 20 patents for general refining processes and apparatus, 28 pertaining to lubricant manufacture, in addition to those covering fractionating towers, deodorization techniques, and continuous distillation. Standard affiliates, including the appropriately named Combined Patents Can Company and the Atlantic Refining Company, also held close to 80 patents covering nearly everything in known contemporary refining operations including the manufacture of cans and shipping cases.[29]

Some of the most important patent rights covering the manufacture and processing of lubricants were acquired by Standard before 1882 through the purchase of a controlling stock interest in the Vacuum Oil Company of Rochester, New York, and the Chesebrough Manufacturing Company located in the New York City area. A three-quarter interest in the Vacuum concern, pioneer manufacturers of lubricating oils by means of vacuum steam distillation, was purchased by Standard representatives in 1879 for $200,000.[30] The Chesebrough Company, famous for its trade-marked "Vaseline" petroleum jelly and holder of several basic patents concerning filtration using bone dust or animal charcoal, became associated with the Rockefeller organization by a purchase of just over 50 per cent of its stock in 1881.[31]

One prize which did not fall under Standard's control until 1884, and then only after long, drawn out and bitter litigation in the courts, was Merrill's patent for deodorized or neutral oil. Discovered in 1867 and patented in 1869, Merrill's process immediately increased his sales of lubricants several fold, at premium prices, and it made lubricants a significant export for the first time. Before the ink of the patent was dry, Dr. S. D. Tweddle of Pittsburgh pressed into the heart of Merrill's territory in Boston with a deodorized "Topaz" lubricant, sold at cut prices. Sales literature gave dealers an extraordinary assurance from a man of very limited resources. It exonerated them of all responsibility for infringement. The supplier would assume all responsibility and costs of any litigation. Merrill soon learned the reason for this when he brought suit for infringement against a local dealer named Yeomans who was marketing "Topaz" oil. Standard Oil of Pittsburgh, owned by Charles Lockhart until he affiliated with Rockefeller in 1874, had assured Tweddle financial backing in the event of litigation over infringement.[32]

At a direct cost of $100,000 to Merrill, to say nothing of the loss of markets usurped by infringers, litigations dragged out for six years until he finally received a favorable judgment in the U. S. Supreme Court in 1876. Due to a technicality that his initial patent should have been drafted into two rather than one (one specifying neutral oil by whatever process made, the second as processed in his special still and apparatus for making it), the verdict awarded Merrill no damages and forced him to apply for two properly drafted reissues which had no retroactive force. Merrill was forced to bring suit again under the reissued patents, after which defense attorneys finally counseled Standard Oil to reverse tactics and take out a license.* Just how much Merrill collected in total damages is not clear. But in 1884, a year and a half before their expiration, he sold the patents to H. H. Rogers and John D. Archbold of Standard for $20,-000.[34]

Circumstances related to transportation of crude to his Boston refinery rather than subsequent difficulties in enforcing his patents may have influenced his decision. Merrill's tank cars holding about 80 barrels brought crude oil from the Regions to the Boston & Albany depot for a cost of about $40.00 a load. But to move the cars over the 2½-mile haul to his South Boston refinery via the Marginal Railroad Company he had to pay $10.00 a car, or about 80¢ a ton-mile. Critics at least were inclined to see a connection between these rates and the fact that J. A. Bostwick, a Standard executive, was president of the New York & New England Railroad, which owned the Marginal road. Whatever the circumstances, Merrill decided in 1887 to dismantle the historic refinery at Boston, selling his 80 or 90 tank cars to the Standard Trust.[35]

Standard's failure to develop certain of the processes acquired during these years is somewhat puzzling. This is particularly true of continuous distillation, a "near miss" in the 1860's, which had the same fate during the 1870's and 1880's, despite efforts by Samuel Van Syckle and others.[36] Van Syckle had gone so far as to solicit financial backing to refine crude from the Bradford field and with much fanfare the Solar works was built

---

* When the first of this new series of suits against Standard resulted in an award of $10,000 for infringement in a single refinery, the *Titusville Herald* reported that John D. Archbold offered Merrill $1.5 million for his patents.[33]

in Buffalo about 1880 to utilize his process.* Difficulties with handling the heavy Bradford crude by the continuous process apparently were an important factor in limiting the Solar plant's production to a few hundred barrels which were sold to the adjoining Atlas works. The company was temporarily revitalized financially by the Tidewater Pipeline Company, but shortly thereafter passed into the control of Standard.[37]

As subsequent American experience during the twentieth century revealed, the economies of continuous distillation, despite Van Syckle's initial claims, could not be achieved except by fairly large installations. Apparently reluctant to engage in any wholesale reconstruction of its existing capacity, Standard chose not to put the continuous process on a commercial basis.[38] This decision was a great disappointment to Van Syckle, famous in his own time for his inventive contributions to the petroleum industry, but who died in 1894, largely unrewarded for his efforts.[39]

Against this general background, the changes in refining techniques and equipment, both within and outside Standard, were relatively minor, modifying rather than revolutionizing practices developed by the early 1870's. In the construction of stills, steel had largely replaced iron bottoms. And, although stills of about 1,200 barrels crude-charging capacity were perhaps the most popular, as in the 1860's a great variety of stills, in size and shape, were continued in operation. These included the upright or cheese-box type, even though the wrought-iron horizontal stills, which could be heated more uniformly, were already gaining what would eventually prove a decisive foothold in refinery equipment and design. The vacuum still, following its introduction by the Vacuum Oil Company, continued to be used primarily for the preparation of high-grade lubricating oils.[40]

Condensers, agitators and pumps, the primary equipment of contemporary refineries, underwent only minor evolutionary steps. The ordinary iron piping, coiled in a tank of water, generally adopted by the 1870's, gave way in turn to long, straight iron pipes. Both mechanical and air agitators continued in use but with a strong trend in favor of this latter type. Latest design in pumps was the compact, powerful Worthington type.

Although basic refining techniques remained relatively stable, refiners ran into difficulties as they began processing Bradford crude for illuminating oil. The early kerosene from this type of crude, when used in conventional lamps, showed a distressing tendency to smoke, change color, and

---

* This plant should not be confused with the Solar Oil Company of Williamsport, Pennsylvania, nor the plant built later at Lima, Ohio.

to die out after only a portion of the oil in the reservoir of the lamp had been consumed.

As more Bradford crude was used in refineries during the late 1870's, there was a corresponding increase in complaints by consumers. The poor quality of American kerosene was the main subject for discussion at a meeting of the leading European petroleum distributors held at Bremen in the spring of 1879. Explanation at this time that the poor burning quality of the oil was attributable to the type of lamp wicks utilized failed to convince most of those attending the meeting.* In fact, this explanation was quite far from the mark.

A more adequate explanation stemmed from the difficulty encountered by refiners in obtaining their customary yields of illuminating oil from paraffin-rich Bradford crude without a deterioration in quality. In processing crudes from the older Pennsylvania fields, for example, fractional distillation was carried down to 38° B or 36° B before cracking to achieve yields of 70–75 per cent of illuminating oil. But with Bradford crude fractional distillation stopped at 46° B or even 50° B, resulting in a reduction in the yields of straight run kerosene of 15–20 per cent.

If subsequent cracking had been carried no further proportionately than with Lower Region crude, there would have been no deterioration in quality because the proportion of cracked distillates blended with fractionated, quality-controlling distillates would have remained about the same, although at a heavy sacrifice in yields. It was thus the pressure to expand yields, which at best were lower with paraffinic crudes, which led refiners to push the slow destructive distillation or cracking of Bradford crude at least to 36° B or 38° B. The result was a corresponding decrease in the proportion of quality-controlling fractional distillates in the illuminating oil, and a lowering of its burning qualities.

There were additional complications. Unless the cracked distillates from Bradford crude were chilled and the paraffin "expressed" or removed before blending, the kerosene would inevitably gum and clog lamp wicks. Even if the higher temperatures involved in cracking reduced the paraffin content somewhat, they tended to boost the gum-forming diolefins and otherwise impair the burning properties of the illuminating oil.†

---

* See Chapter 19 for details of this meeting.

† Two further considerations may be noted. With the Bradford type of crude, the burning qualities of the straight-run distillate would be excellent, but without removing or expressing the paraffin, its "cloud test" would be poor and, when used cold, it would not be sufficiently fluid to burn well. With the cracked distillate the same characteristic would prevail in greater degree, and without any compensating

In essence the difficulty with Bradford crude was not a lack of appropriate refining technology. But in order to maintain their accustomed yields of illuminating oil without a loss in quality, refiners were forced to exercise a much more precise control over both their distilling and treating operations. It is apparent from the subsequent decrease in complaints that the quality of illuminating oil from Bradford crude was generally satisfactory after 1880.

Working with the techniques developed during the 1860's, the leading by-product manufacturers had by the early 1870's not only set a pattern for the succeeding decade but in addition had already established a high reputation for the quality and variety of their products. This was particularly true in the field of lubricants. The Philadelphia Centennial Exposition of 1876, where American petroleum products received a preponderance of the citations and awards, did much to call public attention both at home and abroad to the significance of these developments and the new status of petroleum lubricants.

Joshua Merrill understandably received a grand award for his deodorized or neutral oils, but his other lubricants also received a special merit citation for both excellence in quality and variety. No less revealing were the awards to F. S. Pease of Buffalo, New York, one of America's leading pioneers in improving the art of compounding premium lubricants, for his exhibit of some fifty products ranging from heavy-cylinder oils to sewing-machine oils. Displaying a "complete series of animal, vegetable and mineral oils, and products," his exhibit was cited as "of special interest and importance . . . showing great progress in the production of refined burning oils and first-class lubricators." In the class of heavy reduced petroleum lubricants compounded with graphite, the solid lubricant of the Galena Oil Works at Franklin, Pennsylvania, received commendation for "special superiority of a dark petroleum lubricating oil." In a field closely related to processing of lubricants was the citation of the Chesebrough Manufacturing Company's "Vaseline, or petroleum jelly," for its "novelty, great value in pharmacy, unequaled purity, and purity of manufacture." [41]

While petroleum lubricants had assumed a key position in the total market by the mid-1870's, they still supplemented rather than replaced animal and vegetable oils in the production and blending of lubricants. Determining the appropriate combination to obtain a product with re-

---

improvement in burning qualities at ordinary temperatures. The increased formation of diolefins in cracking a paraffin crude also would have complicated treating problems, and markedly so when cracking was conducted at abnormal temperatures to distillates of 36° B.

quired lubricating properties remained more of an art than a science. With the rapidly expanding demand for an increasing variety of lubricants during 1874–1884 and a multiplication in the number of mineral-oil producers, difficulties were encountered in maintaining quality and development of uniform standards.

The basic problem of obtaining uniform testing standards stemmed from the fact that no single test, or combination of tests, was fully adequate to measure the various characteristics of the oils. Because of their relatively low boiling points, for example, it was imperative to apply a fire test to petroleum lubricants to determine the conditions under which they might be used with safety. Equally basic was the gravity test, which gave a rough measure of the "body" of the oil and the uses to which it might be put. This was also true of the cold test, which indicated the temperatures at which the oil would congeal. By the early 1870's experienced processors and certain purchasers of lubricants recognized that none of these tests, including the gravity measurement, gave a satisfactory indication of the viscosity, body, and related properties of oils, characteristics which were of primary importance for satisfactory lubricating qualities.

One further difficulty which continued to plague manufacturers and customers throughout the decade was the problem of relating these various tests to actual working conditions. In 1881, for example, the generally accepted method of obtaining a cold-test oil was to chill a sample of the oil in an ordinary 4-ounce test tube to determine the temperature at which it would flow freely from the tube.

Early in 1881 the New York Produce Exchange announced the discontinuation of this test, claiming it was unnecessarily severe. Instead the Exchange provided that oils had met the cold test satisfactorily if at a specified temperature they ran freely from the bunghole of a barrel, simulated for testing purposes in a vessel with an opening corresponding to the size of a bunghole.[42] Quick consumer reaction pointed up the inadequacy of the proposed test and forced its abandonment. As the *Oil, Paint & Drug Reporter* pointed out,

. . . consumers [object to the bunghole test] on the ground that it does not represent the actual quality of the oil. [They demand] that the cold test shall represent the temperature at which it will run from a can. It is found in practice that there are 10 or 15 degrees difference between a cup test and a weather test in a can, or an ordinary 4-oz. sample bottle.[43]

Despite these difficulties, there was a gradual improvement both in the quality of lubricants and testing techniques by the mid-1880's. In large part this resulted from greater experience on the part of producers and a more widespread use of the best practices available a decade earlier.

Much of the improvement in testing methods was based on the work of George Thurston, who by 1872 had developed a machine for testing friction. His demonstration, that at least half the power in an average mill was consumed by friction and that this loss could be cut by 50 per cent with better lubricants, impressed a small but growing group of industrial users among textile manufacturers. In certain instances studies undertaken by industrialists themselves served to publicize the value of testing, as when the New England Cotton Manufacturers' Association in 1878 adopted the evaporation test to reduce fire hazards.[44]

In using his friction machine, Thurston attempted to obtain a single, direct test of lubricating qualities under conditions simulating operating pressures, temperatures, and speeds. But he was fully aware that another index of the body of the oil—its viscosity, stickiness, or degree of cohesion between its molecules—varied widely at different temperatures. He accordingly supplemented his friction-machine tests with a modified Napier-type viscometer, an instrument which determined the viscosity or flow characteristics of oil in terms of the time it required for a sample to flow from the top to the bottom of a heated, inclined glass plane.[45] By the mid-1880's, improved viscometers had begun to replace friction machines as the most valuable single test of lubricating oils.*

Further improvements in quality of lubricants resulted when, during the early 1880's, several of the more progressive refiners began sending trained representatives into the field to observe and advise their customers on the proper choice of lubricants. This practice became increasingly widespread over the succeeding decades.

Refining costs and margins

The general organization of the various refining operations changed little from the pattern which had emerged by the early 1870's. Perhaps the only complete refinery, producing illuminating oil and all commercial by-products in a single establishment, was Downer's plant in Boston, although several general refineries in the Standard group, including Pratt's plant in New York and the Bayonne works, were turning out most by-

---

* Dr. C. Engler's viscometer developed for the German railways in 1884, and Boverton Redwood's model introduced a year later, remained standard in most foreign countries for several decades, as did Dr. Saybolt's instrument, introduced in 1878, for the United States.[46]

products from the light fractions and at least lubricants from the heavy fractions. The proportion of illuminating oil from crude inputs and the amount of cracked kerosene distillates were sharply reduced in this type of operation since a high proportion of the intermediate distillates were worked for neutral oil, light lubricants, and other by-products.

By contrast, the largest group of refineries were operated to maximize yields of light products and illuminating oils. Perhaps best designated as ordinary refineries, they extended slow destructive distillation as far as possible through the intermediate fractions to 36° B gravity, and either worked the residuum into lubricants and wax or sold it to specialty refiners. Specialty refiners, typically small in capacity, distilled residuum or selected heavy lubricating crudes with superheated steam, often in a partial vacuum. Their output included heavy cylinder and journal oils, sometimes light harness oils, and frequently one or more of the numerous brands of petrolatum such as vaseline, cosmoline, saxoline, and petrolina, which were employed as unguents in pharmaceutics and secondarily, but extensively, as lubricants. There were also a few specialty plants which limited their operations to the processing of reduced crude oil lubricants.

While refining technology and the general organization of capacity remained relatively stable, there was a marked downward trend in refining costs between 1874–84. Lowered prices of materials, such as sulphuric acid and caustic soda, were in part responsible, but of equal or greater significance were the economies of scale through increasing plant capacities.

It is true that individual refineries continued to vary widely in size, from plants with daily capacities of 40 barrels of crude to Standard's Bayonne works which by the mid-1880's had a daily charging capacity of 6,000–7,000 barrels. Despite these divergencies, most of the plants constructed during 1880 and 1884 generally ranged in daily crude-charging capacity from 1,500–2,000 barrels.[47]

Judging from contemporary evidence, the average or typical refining costs in 1880 for a plant with a daily charging capacity ranging between 1,500 and 2,000 barrels was approximately 2½¢ per gallon, or about half the generally accepted figure for the early 1870's of 5¢ a gallon.[48] By mid-1880's the average or typical cost for plants of this size may have been further reduced to 1½¢ per gallon.*

To the extent these costs were characteristically achieved by the best independent refiners, they remained well above the carefully prepared

---

* John Teagle, prominent Cleveland refiner, testifying in 1888, estimated the cost of refining 150,000 barrels a year at $70,000, or roughly between 1½–2¢ per gallon.[49]

estimates by Standard in 1884 which indicated a refining cost of .534¢ per gallon in all of its refining operations.[50] Thus contrary to the opinion of a number of contemporaries who denied any superior manufacturing advantages to Standard, it would appear that the Trust by virtue of its size, scope of operations, and integration, had achieved impressive economies of scale.*

This downward trend in manufacturing costs reflected improvements in scale and operating techniques of the utmost significance for the refin-

TABLE 18:4

Margins between New York wholesale prices of illuminating oil and delivered price of crude at New York: 1873–84

	1873	1874	1875	1876	1877	1878	1879	1880	1881	1882	1883	1884
Wholesale prices of illuminating oil	17.9	13.0	13.0	19.1	15.5	10.7	8.1	9.0	8.0	7.4	8.0	8.5
Less delivered cost of crude	10.6	6.0	6.5	10.5	9.1	7.2	3.6	6.5	6.6	6.6	7.1	7.0
Margin *	7.3	7.0	6.5	8.6	6.4	3.5	4.5	2.5	1.4	.8	.9	1.5

* Computed.
Source: *Pa. Industrial Statistics, 1892,* 200–201, 203–4.

ing segment of the American petroleum industry. Coupled with declining transport rates and lowering crude prices, the reduction made it possible for refiners to offset, in part at least, the effects of a reduction in margin between the delivered price of crude and the selling prices of refined products.

As shown in Table 18:4, the "margin" was substantially narrowed in respect to illuminating oil, but for at least two reasons it is not likely that total refining margins fell this far. First, most refiners were producing a larger proportion of marketable products from given inputs of crude. Second, as shown in the first tabulation on page 485, the dollar-value index of the principal by-products fell about 30 per cent between the mid-1870's and the mid-1880's, compared to an approximate drop of 44 per cent for illuminating oil.

To what extent refiners generally were actually able to operate profit-

---

* Among those who questioned Standard's claim to lower manufacturing costs was Lewis Emery, Jr.[51]

ably during the mid-1880's is not clear, but at least one contemporary source was of the opinion early in 1884 that most of them were making good profits.[52] This was certainly true of Standard, which not only had low refining costs but which through its control over trunk pipelines en-

Product/average value and index	1873–75	1878–80	1883–85
Illuminating oils			
Average value per barrel	$ 6.35	$3.35	$3.53
Index	100	53	56
Lubricants			
Average value per barrel	$12.52	$5.81	$8.74
Index	100	46	69
Naphtha-benzine-gasoline			
Average value per barrel	$ 4.16	$1.97	$2.85
Index	100	48	69

Source: Computed from Peckham, Census Report, 278; *Petroleum Age* (Sept., 1886) 1428.

joyed a substantial cost advantage on delivered prices of crude over those who shipped by tank car.

Whatever the actual situation, the possibility of profitable operations was sufficient to induce an expansion of refining capacity capable of processing the substantial increase in the volume of crude available between 1874 and 1884. The result, as shown in Table 18:5, was an impressive expansion in the total volume of refined products. The output of il-

TABLE 18:5

Refinery receipts of crude and the output of major refined products: 1873–85 * (*in thousands of 42-gallon barrels*)

	Yearly average 1873–75	Yearly average 1878–80	Yearly average 1883–85
Refinery receipts of crude	7,883.1	12,951.4	17,879.7
Output of refined †			
Illuminating oil	6,529.5	10,799.8	15,171.4
% of total	83	86	82
Naphtha-benzine-gasoline	894.7	1,482.2	2,442.0
% of total	10.5	11.5	13
Lubricating oils	225.8	376.1	884.2
% of total	2.5	2.5	4.5
Total refined ‡	7,650.0	12,656.1	18,497.6

* Estimated: See footnote, Appendix B.
† Does not include paraffin waxes.
‡ Differences between refinery receipts and total output of refined products accounted for chiefly by additions or subtractions from stocks of crude at refineries.

luminating oil, which continued to account for over 80 per cent of the total, grew nearly two and one-half times: from about 6.5 million barrels to over 15 million barrels. There was a more than corresponding increase in the naphtha-benzine-gasoline group of products—from almost 900,000 barrels to over 2.4 million barrels. The growth of lubricating oils was particularly impressive. Not only did the volume increase from approximately 225,000 barrels to over 880,000 barrels, but in 1883–85 lubricating oils averaged some 4.5 per cent of the total refinery output, compared to 2.5 per cent a decade earlier.

Against this background of rising illuminating oil and by-product manu-

TABLE 18:6

Annual operating rate of refining capacity: 1873–84
(42-gallon barrels)

	Barrels		
	1873	1881	1884
Rated capacity *	13,971,000	29,328,000	37,460,400
Less stand-by capacity †	3,492,750	7,332,000	9,365,100
Working capacity	10,478,250	21,996,000	28,095,300
Refinery receipts of crude ‡	9,156,175	19,321,635	21,761,129
Excess working capacity	1,322,075	2,674,365	6,234,171
Operating Rate of working capacity	88%	90%	78%

* See Table 18:2. Daily capacities were multiplied by 300 days to obtain an annual figure.
† Calculated at 25 per cent based on contemporary estimates. Cf. *US* v. *SONJ* (1906), *Transcript*, VI, 2697–98.
‡ Total pipeline shipments minus exports.

facture, the picture which emerges is one of noteworthy progress in mass-scale processing of crude. Within a decade, output and capacity had more than doubled and petroleum's position among the world's industrial giants seemed firmly established.

Yet by the mid-1880's there were indications of possible difficulties ahead. Attempts to measure capacity during this period are at best subject to a wide margin of error, which become even more impressionistic when applied to indicate excess capacity. But, judging from the calculations shown in Table 18:6, excess capacity (the rates of operating to working capacity), which had dropped from about 12 per cent in 1873 to 10 per cent in 1881, rose over the next three years to approximately 22 per cent.

It is true that the world's demand for petroleum products showed no signs of abating, as will be brought out in the following two chapters. But with flush production at Bradford beginning to subside, unless new sources of crude could be developed, there was a real danger that excess capacity would necessitate a sharp reorganization and readjustment of the refining segment of the industry.

Chapter 19

Foreign markets and marketing channels

With some three-fourths of the principal product of American refineries going into foreign trade by the early 1870's, industry members understandably were interested in events affecting demand for petroleum
products outside the United States. Any anxiety over the prospects of declining foreign demand was tempered by the wide acceptance kerosene
had already achieved in world markets and by the fact that the United
States was virtually the sole supplier of this valuable product.

The following extract from the Pennsylvania Geological Report of 1874
is typical of contemporary comments on the position of the American industry.

> Our refined Petroleum has penetrated to the most distant parts of the
> world, it brightens the lazy winter nights of Sweden, and Norway, and even
> Iceland, it is sent to Australia and New Zealand, to China, Japan, to Russia,

Germany, Austria, France and Great Britain, in the face of the fact that in everyone of the countries named, surface oil or bituminous shales, exist in quantity that would seem to be fatal to competition.

Although the question of whether ". . . the drill in other countries . . . would find oil . . . that some day may interest us," the authors antic-

TABLE 19:1

Production and distribution of crude and refined petroleum products: 1873–85 (thousands of 42-gallon barrels)

	Yearly average 1873–75	Yearly average 1878–80	Yearly average 1883–85
Crude oil			
Production	9,869.4	20,532.4	23,175.6
Exports	401.7	709.5	1,747.1
Percentage exported	4.1	3.5	7.5
Crude refined †	7,883.1	12,951.4	17,879.7
Illuminating oils			
Production	6,529.5	10,799.8	15,171.4
Exports	4,903.5	7,602.7	10,475.3
Domestic	1,626.0	3,197.1	4,696.1
Percentage exported	75.1	70.4	69.0
Naphtha-benzine-gasoline			
Production	894.7	1,482.2	2,442.0
Exports	277.1	381.5	363.3
Domestic	617.6	1,100.7	2,078.7
Percentage exported	31.0	25.7	14.9
Lubricating oils			
Production	225.8	376.1	884.2
Exports	27.2	89.7	278.4
Domestic	198.6	286.4	605.8
Percentage exported	12.0	23.8	31.5

Source: Table 18:5; Appendix D:1.

ipated no immediate threat to the dominant role of the United States in supplying the world's needs for petroleum products.[1]

The approximate division of sales between the domestic and foreign markets, shown in Table 19:1, indicates the continued importance of export sales for the American industry during the succeeding decade. The proportion of illuminating oil shipped abroad during these years is partic-

Barrels

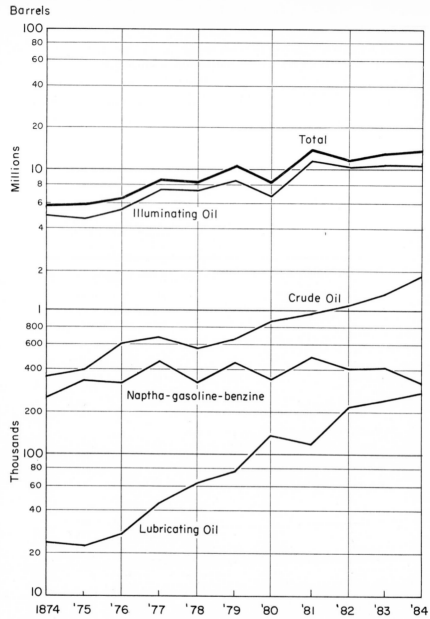

CHART 19:1. *U. S. exports of crude and refined petroleum (42 gallon barrel equivalents): 1874–84.* (*Source:* Mineral Resources, *1905, 883–84.*)

ularly striking, even though by 1884 about one-third of the output went
into the United States domestic market compared to less than 25 per cent
in 1874. A growing domestic demand also absorbed an increasing proportion of the naphtha-benzine-gasoline group of products. By contrast,
nearly one-third of American production of lubricating oils was being exported in 1884 compared to about 10 per cent ten years earlier. Finally,
it may be noted that although exports of crude had grown to over 1.7 million barrels by 1884, this amount was only about 7.5 per cent of the production from American fields in that year.

Changes in both the total and relative proportions of petroleum and
refined products sold abroad from 1874 through 1884 are revealed in the
annual data on exports shown in Chart 19:1. Except for a drop in 1881,
total exports moved up impressively, from approximately 5.6 million barrels to almost 13 million barrels. Illuminating oil remained the number one
export, although as a percentage of the total it dropped from about 88 per
cent in 1874 to approximately 80 per cent in 1884. The greater part of the
proportionate drop in illuminating oil was offset by exports of crude that
rose from 9 per cent of the total to over 14 per cent. Exports of the
naphtha-benzine-gasoline group of products reached a peak of about 4
per cent of total exports in 1881, only to fall to approximately 2.5 per cent
three years later. The most spectacular rate of increase was registered by
lubricating oils that grew from less than 0.5 per cent of the total to about
2.2 per cent by the end of the period. As indicated by the following tabulation, petroleum products continued to rank high among the commodity
exports of the United States during this period.

Average annual values of leading U. S. exports,
1781–1885

Years	Exports/millions of dollars			
	Cotton	Wheat	Meat	Petroleum
1871–75	$205.6	$ 82.2	$33.9	$36.9
1876–80	183.5	134.0	66.7	43.8
1881–85	218.8	157.6	69.3	47.8

Source: U. S. Dept. of Commerce, Bureau of the Census, *Statistical Abstract of the United States*,
1946 (Washington: Government Printing Office, 1946), 904–5.

The pattern of foreign demand

The changing pattern of distribution of illuminating oil exports among
the major foreign marketing areas is shown in Table 19:2. The distribu-

tion for 1874 suggests that although the authors of the Pennsylvania Geological Report were correct in stating that American kerosene "had penetrated the most distant parts of the world," it was at best casting only a
feeble glow outside Europe. Of the approximately 153,000 barrels shipped
to Asia (the second most important foreign marketing area), some 70,-
000 barrels went to the East Indies, another 50,000 to Asiatic Turkey,
while only about 2,000 barrels went to "light the lamps of China." *

TABLE 19:2

*Distribution of illuminating oil exported by major marketing areas
(thousands of 42-gallon barrels)*

	1874		1879		1884	
	Quantity	Per cent	Quantity	Per cent	Quantity	Per cent
TOTAL	5,171.9	100.0	7,894.9	100.0	9,895.6	100.0
Europe	4,629.4	89.5	6,340.5	80.3	6,931.7	70.0
North America	111.4	2.2	137.1	1.7	192.4	1.9
South America	109.6	2.1	204.5	2.6	275.4	2.8
Asia	152.5	2.9	986.8	12.5	2,142.0	21.6
Oceana	89.4	1.7	69.3	.9	112.5	1.1
Africa	63.7	1.2	143.6	1.8	184.1	1.9
Not Classified	16.0	.3	13.0	.2	57.4	.6

Source: see Appendix D:1.

The reasons for the strong European demand stemmed from the fact
that European experience with artificial illuminants began a half-century
or more before the birth of the modern petroleum industry. High costs of
illuminating materials had limited lamplight to the homes of the
wealthier classes, but as American-made kerosene became available at
relatively low prices, lamplight began to spread gradually among the
middle and lower income groups in the society.

In non-European foreign markets a number of obstacles prevented a
similar demand pattern from emerging before the mid-1870's. Only a
small percentage of the population was accustomed to the use of lamps,
while average real incomes were considerably lower than those received by most Europeans. Furthermore, a lack of fully developed distributing channels, longer distances from export ports in the United
States, and the need for special packaging, all tended to make costs and
prices of illuminating oil higher in these markets than in Europe.

* See Appendix D:5.

The subsequent growth in the proportion of illuminating oil exports to non-European markets suggests that the obstacles to a greater consumption of kerosene in other parts of the world began to disappear by the end of the decade. Reports from the various American consuls stationed in China give a graphic picture of how the use of kerosene was beginning to spread there by the mid-1880's. In 1883 the American vice-consul at the port of Newchwang, China, on the coast north of Shanghai, reported,

Ten years ago [kerosene] was only procurable at foreign stores, and being only imported in small quantities for the use of the foreign residents, was three times as dear as the oil expressed from beans in this province; the foreign lamps . . . in which . . . it could be burned were only procurable at foreign stores and at prices ranging from $3 upwards.

Owing, therefore, to the high price of the oil and lamps they were quite neglected by the Chinese, with the exception of a very few among the agents of southern [Shanghai] firms, and these only burned a little in their private rooms.

After a while these agents proceeded to procure their oil and lamps direct from Shanghai, a year's supply at a time. Hereupon some of the local merchants . . . from whom they hire rooms, . . . observing that the oil was cleaner than the native article and gave a brighter light and did not require such constant trimming, induced their lodgers to bring them a little from Shanghai, and this led to small quantities being imported and offered for sale in Chinese shops.

Later on, again, some Chinese opened stores, the business of which consisted chiefly in retailing miscellaneous articles of foreign manufacture, amongst which were lamps, and thus the quantity of lamps imported increased year by year and the price declined at the same time, so that now tin lamps for fixing on the walls and glass lamps to stand on desks or tables are retailed as low as 30 cents each, and this naturally facilitates the use of oil. . . . When first imported [kerosene] was retailed at 10 to 12 cents per catty [approximately one-fifth of a gallon]; three or four years ago it could be bought for 6 to 7 cents per catty; at present it is so cheap that [it] can be bought for . . . little more than 3 cents per catty. . . . It thus compares very favorably with bean oil, which is being retailed at about 5½ cents per catty, and has to be fetched in the buyer's own can or bottle. At present, then, the oil is used in this port by all the southern agents; it is also used at desks, in rooms for receiving visitors, and in hanging lamps in passages; by the grain and the principal retail dealers, so that the consumption is steadily and rapidly increasing among the trading class.

Private persons, generally speaking, have one or two lamps and some oil which they only use on great occasions, such as the New Year's festivities, or when entertaining friends; a few whose dwellings are larger than ordinary use this oil in their libraries and reception rooms, but like the smaller householders they do not burn it in the rooms in daily and general use, fearing that children or ignorant persons may get hold of it and thus give rise to some disaster.[2]

There was general agreement among the American consular representatives that ignorance of the properties and use of kerosene was the principal remaining barrier to sharply expanding sales of kerosene. This ignorance had been chiefly responsible for fires and explosions that had in turn prompted various governmental authorities to impose restrictions on its use; restrictions that in the case of Foo Chow in 1882, for example, prohibited either the sale or use of kerosene within the city limits. One of the most effective means of overcoming this obstacle was proposed by the American consul at Hankow, who in 1882 urged the preparation of a pamphlet to be printed in Chinese, dealing with the benefits and proper use of kerosene.[3] Such a pamphlet, he pointed out, could be circulated to the "better classes" and to approximately one-seventh of the coolies who were literate.*

Whatever the expansion of sales to other areas might portend for the future, imports that rose from about 4.6 million to just under 7 million barrels between 1874 and 1884 maintained Europe as the principal market for American kerosene by a wide margin. In general this increase was in response to a continuation of influences that accounted for the original significance of the European markets. A growth in the population from about 308 million to approximately 335 million and an extension and improvement of transport facilities, particularly on the Continent, were important factors.

Perhaps even more significant in expanding demand was the response of European customers to lower prices. As shown in Chart 19:2, New York export quotations for illuminating oil, which were closely reflected in England and the important port of Hamburg, Germany, rose sharply in 1876, only to drop to levels that throughout the early 1880's ran a third or more below the average price per gallon in 1874.

While the relative distribution of the growing volume of illuminating imports among old and new customers is impossible to determine, it may be noted that the average per capita consumption of kerosene in Europe

---

* The need for educating the public was not limited to the less developed areas of the world. In 1886 a distributor in Birmingham, England published an 80-page pamphlet entitled, "Petroleum: valuable hints to those who use it." The firm gave away 90,000 copies with the object to try to inform the public on the safe use of kerosene.[4]

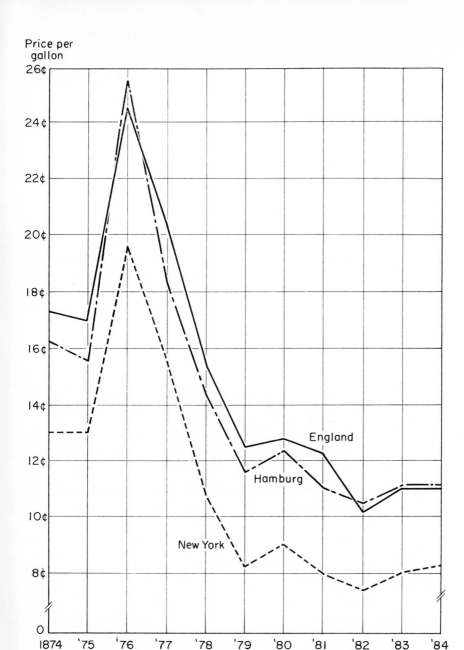

Price per
gallon

CHART 19:2. *U. S. wholesale prices of illuminating oil per gallon: Eng-*
land, Hamburg, and New York, 1874–84. (*Sources:* Industrial Statistics,
1892, B 203–4; *U. S. Congress, Senate, Committee on Finance,* Wholesale Prices, Wages, and Transportation, *Report by Mr. Nelson Aldrich,*
March 3, 1893, Part 1, 52d Cong., 2d Sess., Report 1394 (*Washington:*
Government Printing Office, 1893), 239; Aldrich, Committee on Fi-
nance, Wholesale Prices, Wages, and Transportation, *362.*)

grew from about two-thirds to nine-tenths of a gallon between 1874 and 1884. Used in the typical household lamps available at the time, this expansion represented an approximate increase in lamplight hours from 92 to 126.[5]

In contrast to illuminating oil, the pattern of foreign markets for American exports of crude, naphthas, and lubricating oil changed very little between 1874 and 1884. As indicated in Table 19:3, Europe was virtually the sole buyer of these products at both the beginning and end of the period. With few exceptions, the proportions of these products going to the various European countries also remained relatively stable. France, which had introduced differential tariff duties on crude and refined products early in the 1870's to encourage its domestic refining, purchased some 80 per cent of American crude exported to Europe in 1874; ten years later its share was about 60 per cent, with another 19 per cent going to Spain which had followed France's example of introducing high duties on refined oils.

TABLE 19:3

Proportions of total U. S. exports of crude, naphthas, and lubricating oils to Europe: 1874, 1879, 1884 (in thousands of 42-gallon barrels)

Year	Crude oil			Naphtha-benzine-gasoline			Lubricating oil		
	Total U. S. exports	Exports to Europe	Per cent of total Europe	Total U. S. exports	Exports to Europe	Per cent of total Europe	Total U. S. exports	Exports to Europe	Per cent of total Europe
1874	423.2	401.8	94.9	231.8	229.9	99.2	29.6	27.8	93.9
1879	616.1	573.0	93.0	358.4	354.7	99.0	59.2	55.2	93.2
1884	1,599.7	1,526.4	95.4	358.2	346.7	96.8	250.4	238.4	95.2

Source of data: Appendix D:1.

A demand for solvents and cleaning fluids in the United Kingdom, France, and Belgium absorbed about 86 per cent of the European imports of naphthas, benzine, and gasoline in 1874. With Germany replacing Belgium, the top three importers accounted for nearly 91 per cent in 1884. As might be expected from its advanced stage of industrialization, the United Kingdom purchased 83 per cent of the 28,000 barrels of lubricating oil exported to Europe in 1874. Ten years later the United Kingdom's share of the 240,000 barrels was about 51 per cent, with another 21 per cent going to its growing industrial rival, Germany.*

* See Appendix D:2.

Apparatus for filling tin cans with oil in the 1890's.

Marketing channels

Despite an approximate 100 per cent increase in volume, methods of shipping and handling petroleum products for overseas distribution remained essentially unchanged over the 1874–84 period. Virtually all of the exports of crude and refined products during these years were transported in sailing ships. Cargoes usually ran between 7,000 and 10,000 (50-gallon) barrels, although shipments up to 14,000 barrels were not unknown. Loading and unloading these vessels still took a month's time and the average voyage from New York to Bremen, for example, consumed forty-five days and the return trip thirty-five days.[6]

Except where climatic conditions en route or lack of adequate transportation facilities at their destination dictated the use of the square, 5-gallon tins packed two to a wooden crate, both crude and refined products were packaged in wooden barrels. Of the approximate 350 million gallons of illuminating oil exported in 1880, for example, about three-fourths were shipped in barrels for delivery to ports principally in western Europe. The remainder, packaged in some 8.65 million cases, went largely to ports in the eastern Mediterranean and the Far East.[7] Because the wood in the cases and the tin containers were highly prized by customers in these parts of the world there was no return of the packages.*

---

* According to J. J. F. Bandinell, American Vice-Consul at Newchwang, China, wooden crates were selling locally for 3–4¢ each for firewood, while the cans found a ready market among used metal dealers at 7¢.[8]

This was not the case, however, in respect to barrels which, once used for petroleum or petroleum products, were generally unsuited because of the danger of contamination for packaging most other liquids. But their re-use by American refineries led to an important return trade in "empties" that, according to estimates made in 1883, amounted to about four-fifths of the barrels exported.[9]

Except for minor modifications, the marketing channels serving the export trade in petroleum products also remained essentially unchanged throughout the 1874–84 period. Naturally, Standard's acquisition of the bulk of the domestic refining capacity during these years forced foreign buyers to obtain their supplies more and more from a single seller. As far as Standard was concerned, however, so long as the United States was virtually the sole supplier of crude and illuminating oil, there was little reason to assume a more active responsibility for the distribution and sale of petroleum products abroad.

Indeed, the major foreign marketing activity undertaken by Standard during these years involved specialty products, particularly lubricating oils, that were sold in competition with other types of lubricants. By the early 1870's a number of processors of lubricating oils and greases had already begun marketing their products under trade names either through sales offices or sales representatives located usually in New York City.

Late in the 1870's, Standard acquired a financial interest in Thompson, Bedford & Company of New York, which already marketed a line of branded lubricants. To handle the foreign distribution of lubricating oils and greases produced by plants associated with Standard, Thompson, Bedford & Company established sales representatives in the principal European markets.[10] Undoubtedly the most far-flung marketing operation that emerged during this period belonged to the Chesebrough Manufacturing Company that became affiliated with Standard in 1880. To distribute "Vaseline," Chesebrough had sales offices in Montreal, London, Paris, Barcelona, Hamburg, Rio de Janerio, and Buenos Aires.[11]

Yet Standard did not remain completely passive in respect to the foreign marketing of illuminating oil. In 1879, for example, Standard of Ohio bought a half interest in Meissner, Ackermann & Company that had important and widespread mercantile connections in western Europe and the Mediterranean area. Two years later, prompted by the high import duties on kerosene sold in Cuba, John D. Archbold on behalf of the Standard management joined with a Cuban capitalist to produce and market illuminating oil from a small refinery constructed in Havana. In 1882 the West India Refining Company was formed to take over the property and serve as a link between the partnership and the Standard Trust. "Such operations abroad were the exception in the early 1880's, however. Standard Oil usually sold its kerosene manufactured in the United States f.o.b. to merchants in the ports of the United States and there its responsibility and control ended." [12]

In large part this essentially passive role on the part of Standard in foreign distribution was made possible by the continued willingness of foreign import houses—many of them long established—with their knowledge of particular foreign markets, the rules and regulations affecting imports, and familiarity with language and customs, to take on the marketing functions. But by the mid-70's, the export trade began to gravitate into the hands of individuals or firms which increasingly specialized in petroleum products. This growing specialization was made possible by the increasing volume of petroleum exports, plus the fact that most of the shipments were between a few Atlantic-coast ports and a dozen or so major ports in western and southern Europe.

Located in the principal foreign marketing centers, the import firms arranged either through resident representatives or commission agents in the United States to handle their purchases, provide for shipping space and for marine insurance to cover the shipment of cargoes from dockside in the American ports to their destination abroad.

One such firm was C. T. Bowring & Company of London, Liverpool, and Cardiff, founded in 1812 as a shipping concern. [13] To handle a growing volume of American "exports of beef, pork, flour and beans," Bowring & Company established a New York branch in 1862 under the partnership of Bowring & Archibald. Appointed agents of Lloyd's of London about 1870, the New York partners became increasingly involved as brokers in providing marine insurance and cargo space for the growing petroleum trade. They also began forwarding large cargoes of barrelled oil on their own account, some of which was sold to distributors on the Continent.* But most of their shipments went to the British partners who built up extensive storage facilities for handling oil in the London area and were doing a large wholesale business in the United Kingdom. [15]

---

* An indication of the firm's position among the New York exporters is the fact that T. B. Bowring was elected to the five-man Executive Committee on Petroleum of the New York Produce Exchange for the years 1876–80. [14]

In virtually all cases the rules and regulations governing the handling and storage of petroleum products in the various import ports required special facilities.* In some instances these were supplied by local or state authorities; more commonly they were the responsibility of the shippers.† Whether the import houses paid rent or owned their facilities, they still had to have adequate storage space. The seasonal nature of the demand for illuminating oil required purchases in anticipation of future sales to distributors. To ensure adequate supplies as well as to hedge against abrupt price changes, European importers, prior to the advent of the tank steamer in the mid-1880's, customarily held sizeable inventories in storage. Some impression of the magnitude of these inventories during the late 1870's is revealed by the following tabulation of petroleum imports and stocks (in thousands of 50-gallon barrels) held at the end of the year in eleven principal European markets: ‡

Year	Imports	Stocks
1874	4,231.6	627.8
1875	3,000.9	509.3
1876	2,785.4	195.8
1877	4,096.2	716.1
1878	3,628.7	629.5
1879	3,962.3	825.5
1880	4,139.5	1,043.1

Source: Peckham, *Census Report*, 279.

In fact, as Boverton Redwood described the British market during the 1870's and early 1880's, "The speculative character of the trade at that time, and the enormous fluctuations in the price that prevailed, led to

* Safety regulations were imposed almost universally wherever petroleum products were handled and stored. In 1882, for example, the American consul at Batavia, Java, reported that all petroleum imports had to be placed in a private warehouse, built according to government specifications, that was located 100 meters from any other structure and surrounded by a deep ditch.[16]

† Between 1876 and 1877, for example, the Mersey Docks and Harbor Board of Liverpool had the Herculaneum Dock constructed especially to accommodate oil receipts and storage. Capable of storing 60,000 barrels, "with all needed facilities," the dock cost some 16,198 pounds sterling. (House of Commons, *Report of Select Committee on Petroleum* [1896], 56.) In this connection it is of interest to note that in 1876 the municipality of Venice, Italy, made an arrangement with the Italian government to operate public warehouses in the locality of Sacca Sessola for the storage of petroleum imports.[17]

‡ Hamburg, Rotterdam, Amsterdam, Bremen, Stettin, Dantzic, Antwerp, Konigsburg, St. Petersburg, Trieste, London.

importations very largely in excess of the consumption, and unnecessarily large stocks being aggregated in the centers of importation." Redwood was also struck by the trade practices whereby "almost every cargo, or parcel [of oil] imported being sold many times over." [18]

This flurry of activity reflected the character of British marketing channels during this period. Apparently few of the large import houses were as fully integrated as the New Patent Candle Company of Plymouth which marketed in the Cornwall, Devon areas of southwest England and the Channel Islands. Transporting oil directly from the United States this firm received its shipments at the "ordinary" landing wharves of the Great Western Railway docks at Plymouth. From there it was immediately shipped to various storage depots owned by the Candle Company, ranging in capacity from 3,000–5,000 barrels. From these depots the company in turn sold to domestic wholesalers or jobbers. For purchasers within a seven- or eight-mile radius of a storage depot, oil would be sent out by road wagon. For delivery to more distant points, shipments went by rail or, in the case of the islands, by coastal steamers. [19]

Most of the oil imported into Britain, however, seems to have been distributed through somewhat more conventional marketing channels. The large import houses typically took no responsibility for distribution beyond the ports of entry. In some instances they bought their oil in anticipation of holding it in their own or public storage facilities at the ports of entry, pending subsequent sale to domestic wholesalers at the going market price. In other cases they purchased on a basis of the orders of domestic wholesalers who might take immediate delivery for transfer to their own storage depots or who might prefer to pay storage charges to keep the oil at the receiving ports until it was sold. Before the oil reached the thousands of retail dealers scattered throughout Britain, however, it might be sold in smaller lots to intermediate-size wholesalers, who in turn sold in still smaller amounts to jobbers who supplied the retail trade.

This somewhat elaborate marketing system was not without its advantages. Speculation was spread among the various middlemen who could vary their inventories within the limits of the storage facilities available to them and according to their evaluation of future prices. Moreover, the multiplicity of small wholesalers and jobbers made it possible for retail dealers, chandlers, ironmongers, grocers, and other petty traders who handled oil as a side line to carry small stocks, seldom over 5–10 gal-

Petroleum dock at port of Hamburg.

lons. These small stocks in turn reflected the hand-to-mouth buying habits of retail customers who frequently bought their oil a pint at a time.[20]

Marketing channels and methods of distribution serving the major markets on the Continent followed a similar pattern. As pointed out by two German experts on marketing history,

> . . . In the early days of the trade, oil was sold for the German market f.o.b. New York to any purchaser. Although the trade was an open trade, the advantages of large operations and large capital made themselves felt, and certain big importing houses, such as Schutte and Riedemann, came to do a considerable portion of the business. They were simply importers, however, and sold their oil to the wholesale trade, and the oil was distributed throughout the Empire by the usual trade organization—a hierarchy of wholesalers—until it finally reached the retail shopkeeper and, through him, the consumer. . . . It was a trade which anyone could enter and in which anyone was free to buy any kind of oil and sell it where he pleased. There were, of course, trade relations of a more or less permanent character established by the parties in the business, but both the commercial practices and the fact that oil supplies could be obtained by anybody kept the trade open and free.[21]

Trade relations

Although the organization and functioning of the export part of the industry remained essentially unchanged, the 1874–84 decade was far from uneventful as far as foreign trade in petroleum products was concerned. An expansion of over 100 per cent in total quantities shipped abroad during these years affected the marketing structure. Price movements associated with this expansion evoked mixed reactions from sellers and buyers, while dissatisfaction with the quality and handling of petroleum exports reached near crisis proportions in 1879.

All contracts involving sales for the export markets followed well-established procedures that, were formalized early in the 1870's under the rules and regulations of the New York Produce Exchange. Among the forty-four rules in force in 1874, the following provisions may be noted. Petroleum inspectors were to be appointed by the Petroleum Committee of the Exchange, but buyers had the privilege of selecting from the list inspectors, whose principal duty was to certify that the terms of the contract between the buyers and seller had been fulfilled. Specifications were prescribed for crude and refined products to be delivered under sales contracts. In the case of illuminating oil, it was to be "standard white or better with a fire test of 110 degrees Fahrenheit or higher," and packaged in barrels "painted blue with white heads and well glued." [22] For both crude and illuminating oil it was assumed that a gallon was equivalent to 6½ pounds. No provision was made, however, for a routine inspection of the quality of refined oil. Inspectors apparently limited their activities to gauging shipments for volume.

However important these rules and regulations were in reducing misunderstanding between buyers and sellers and providing an orderly method of handling disputes, they did not put an end to complaints by foreign purchasers respecting the quality of products, their packaging, or the manner in which they were inspected. Given the organization of the export trade in which responsibility for enforcing standards was diffused, such complaints were almost inevitable.

Typically, complaints were voiced by the export houses who had a vested interest in maintaining standards of quality and packaging. Occupying an intermediate position in the process of distribution that began with the American refiners and ended with the distribution of the products to the final consumers, they were quick to voice their dissatisfaction with the quality of the product or methods of packaging or shipping.

Complaints of this sort might be brought to the attention of the representatives of the export houses on the five-man Petroleum Committee of the New York Produce Exchange.* An alternative was to bring pressure on the American refiners by calling public attention to the problem.

By way of illustration, the prominent brokerage firm of Meissner, Ackermann & Co. of New York issued a circular to the trade in 1874 pointing out the poor manner in which refined petroleum products were prepared for shipment abroad. More specifically, the refineries were charged with using green rather than seasoned wood in the preparation of barrels; of reusing old barrels without properly repairing them; and of either neglecting to seal them with glue or of using glue of a "wretched quality." As a result the losses from leakage, it was stated, ran as high as 40–50 per cent; losses which were borne by the European dealers. A "prominent European buyer" was quoted to the effect:

> You will therefore perceive we cannot recommence this business as long as your petroleum refiners continue to carry on shipments in this way by which they ruin the trade of the article. If you cannot give us safe guarantee that you will only ship barrels which are in fine shipping order all further orders of petroleum will be quite useless . . . anything derogatory to the product of your refineries is direct leverage in favor of the refining interests of Europe, which are only awaiting a favorable moment for their extension.[24]

According to the *Titusville Morning Herald*, a subsequent circular issued by Meissner, Ackermann & Company to its European customers pointed out that the members of the American industry were not entirely at fault; much of the loss of oil was the result of careless handling and storing barrels after they had reached their destination.[25]

There was little leakage between the time the barrels left the refineries and were placed aboard ships, but by the time the ships had reached port the hoops were loose as a result of the movement of the sea and temperature changes. The condition of the barrels became worse as they were taken from shipboard in slings, four or five at a time, rolled over rough ground which broke the inner coating of glue, stored in tiers four or five barrels high, and subjected to wide temperature changes for six months to a year. Meissner, Ackermann & Company, it was stated, advised its customers intending to hold oil any length of time to specify in their contracts with sellers that it be supplied in "carefully prepared extra glued barrels which are adapted for storage."[26]

---

* Between 1874 and 1883, for example, C. F. Ackermann of Meissner & Ackermann, Josiah Lombard of Lombard & Ayers, in addition to T. B. Bowring of Bowring & Archibald, all served one or more terms as members of the Petroleum Committee.[23]

Judging from the continuing complaints by foreign purchasers there was little improvement over the succeeding years. To these protests were added a growing criticism of the quality of illuminating oil shipped abroad as crude from the Bradford region began to move through the refineries. In view of Standard's growing control over the domestic refining industry it is not astonishing that these complaints were no longer directed against American refiners in general but specifically against Standard. There was increasing evidence that reactions might lead to further legislation in various European markets that would be highly unfavorable to American imports.

One such warning came from James R. Weaver, United States consul at Antwerp. In a report dated February 19, 1879, he outlined the current objections of Belgian consumers to American illuminating oil. The principal defect, he pointed out, was that it did not produce a clear, bright flame and after a half-hour of burning would begin to smoke and finally die out. "There was a question whether the difficulty came from imperfect refinement or the nature of the wicks used in connection with the lamps." In either event he felt compelled to advise American refiners that, "unless confidence in American illuminating oil . . . is redeemed it is likely that restrictions will be imposed upon the importation of refined petroleum." [27]

It was at a conference held at Bremen late in February, 1879, however, that the intensity of the European discontent with American illuminating oil was dramatically revealed. Some indication of the interest on the Continent in the petroleum trade at this time is indicated by the fact that the meeting drew delegates from the chambers of commerce located in the principal European commercial centers.* The immediate occasion for the gathering was to consider legislation proposed to the German Reichstag which provided severe penalties for persons who handled "unsafe" petroleum products. While indicating their disapproval of the proposed legislation, the principal business of the delegates at the meeting was to discuss the "petroleum question," more particularly the poor quality of the illuminating oil from the refineries controlled by the Standard group.

Reporting on the conference, Wilson King, American consul at Bremen, noted, "I regret to say that a bitter feeling towards Americans was somewhat snarlingly expressed by the spokesman of the Hamburg delegation

---

* These included Antwerp, Amsterdam, Berlin, Breslau, Christiania, Copenhagen, Danzig, Frankfort on Main, Hamburg, Koenigsberg, Lubeck, Mannheim, Nurnberg, Rostock, Rotterdam, Stettin, Trieste, Moscow, and Vienna.

and only very mildly resented by the gentlemen from Bremen and Breslau." [28] The delegates' complaints regarding the quality of American illuminating oil related by King were essentially the same as those made by the Belgians.

In anticipation of such criticism Standard sent F. W. Lockwood, prominent oil inspector from New York, to represent it at the meetings. According to Lockwood, the trouble with the illuminating oil came from the fact that much of it was refined from Bradford crude and contained a larger proportion of "heavy" parts than oil from the types of crude formerly processed. Used in the lamps with tightly woven wicks currently popular on the Continent, it was not a satisfactory illuminant as this type of wick took off the lighter portions of the oil first, leaving the heavy portion to be burned later. This accounted for the short period of illumination, followed by smoking and final extinction of the flame.

Lockwood demonstrated that with a loosely woven wick the "Bradford" illuminating oil would work satisfactorily. Taking up the question of safety, he argued that lamp explosions were brought about principally by the length of the wicks used, as ". . . petroleum oil flowing through so long a wick becomes heated and therefore evaporates from the wick before reaching the flame. Therefore the wick becomes charred and dried by this evaporation and the flame goes downward seeking moisture and may reach the oil in the lamp." [29]

To the suggestion that better refining methods would improve the quality of illuminating oil, Lockwood asserted that it was impossible for the American refiners to produce the former quality of illuminating oil from crude from the Bradford district. By this statement he clearly implied that the only solution to the "quality" problem was either to change the wick arrangements on the lamps already in use by consumers or possibly to replace them entirely with new lamps—a suggestion hardly popular with the representatives of the European petroleum trade gathered at the convention.

It was quite apparent that most of the delegates shared the views of the *Oil, Paint & Drug Reporter*, the highly articulate critic of Standard, which dismissed Lockwood's assertion of the impossibility of getting high grade illuminating oil from Bradford crude as "bosh." Said the editor, "The only trouble is that from a given quantity of Bradford oil the refined product of equal quality must be smaller than from the oils of the old fields. Those . . . refiners who recognize this difference . . . have no trouble in turning out a burning oil which satisfies the requirements of both foreign and home trade." The blame rested squarely on Rockefeller and his associates who ". . . in order to prevent losses from the unreasonably low prices which they have insisted upon maintaining for the purpose of crippling 'outside' refiners, have worked the Bradford oils less than was consistent with a good quality of refined. The result has been a gummy, slow-burning oil." The editorial ended with the comment that it was not realistic to expect the foreign consumers to buy new lamps when the problem could be solved by the use of better refining methods.[30]

Whatever the merits of this argument, the delegates of the Bremen conference concluded that "the complaints regarding the inferior quality of much of the petroleum recently received from America and especially of the different brands of the Standard Oil Company are fully justified." [31] A set of resolutions was adopted which demanded ". . . from the American refiners and especially from the Standard Oil Company" greater care in the refining of crude oil in order that "petroleum in the future be again as free as it formerly was from acids and heavy oils, [and] that inferior qualities may no longer be shipped to Europe and that the consumer may again receive the former customary good quality." They also wanted American petroleum inspectors ". . . to undertake the inspection of the petroleum as to its quality in the most careful manner and that certificates given by them regarding petroleum shall accurately describe the goods." They insisted that American petroleum exporters permit only officially appointed inspectors to give certificates. They further urged American refineries and produce exchanges to cooperate with the European exchanges for the purpose of deciding upon and introducing uniform testing methods by which the quality of petroleum could be judged.

The resolutions ended with a number of strongly worded specific recommendations designed to remedy the chronic problems involved in packaging and handling petroleum products for the export market. The New York Produce Exchange, it was suggested, should make it the "duty of the refiners to use only barrels of well-seasoned, air-dried, split (not sawed) white-oak stays and heads. They urged that barrels be more carefully glued and not filled until the glue is thoroughly dry. Finally, the Exchange was to impress on the American inspectors their obligation to "exercise a sharp and effective control, in the interest of the Continental petroleum importers, over the condition of the barrels and their gross and tare weight." [32]

In view of the fact that some 80 per cent of illuminating oil exports went to Europe in 1879, protests by the members of the Bremen conference could hardly be ignored. Standard's officers were quick to move into action. P. W. Lockwood and William Rockefeller wrote August Nibelthau, president of the Petroleum Congress of Berlin, assuring him that greater care would be taken by the American refiners in preparing and shipping petroleum for the European markets. They also reported that a special committee had been formed to work with the New York Produce Exchange in putting into effect the recommendations of the Bremen congress.[33]

After a series of open hearings, the committee headed by H. H. Rogers of Charles Pratt & Company, a part of the Standard combination, submitted its report early in June, 1879. On the question of the quality of American illuminating oils, it was stated that satisfying foreign buyers was not the sole responsibility of the refiners. Europeans were urged to adopt short, loosely woven wicks, which would eliminate trouble.

The remainder of the report followed the recommendations of the Bremen conference practically in their entirety. A number of new regulations and testing methods were suggested for improving inspection, including the use of the spark test. Illuminating oil was defined as "standard white or better, with a specific gravity not below 45° Baume." Barrels were to be made of well seasoned oak stays with six or eight iron hoops, depending on the gauge of the metal used. When filled they were to weigh between 360 pounds and 460 pounds.

The committee report was accepted, and the suggested rules, adopted by the New York Produce Exchange, went into effect on August 1, 1879.[34] The same set of regulations was adopted by the produce exchanges in the other important export ports in the United States including Boston, Philadelphia, and Baltimore.

While their imposition did not end complaints, the adoption of these regulations marked a milestone in the handling of petroleum exports. By their actions on the committee that formulated the new rules, the Standard executives showed an awareness of how better regulations of this type might improve relations with foreign buyers.

It was perhaps only natural, however, that Standard would refuse to accept full responsibility for improving the quality of illuminating oil going into exports. In part this attitude reflected internal difficulties in bringing refining operations of newly acquired plants under central control. Furthermore, since a negligible quantity of crude petroleum was produced outside the United States, buyers had little choice but to turn to Standard for their supplies. Thus pressure could be brought to bear on the European dealers to take the initiative in bringing about a change in lamps that would solve the problems associated with the Bradford crude. This attitude did little to soften the resentment of European buyers over the question of quality, but rather, made them eager to develop some alternative source of supply as a more effective means of bringing pressure on Standard.

For some five years following the Bremen conference, there were no significant changes in the conduct of foreign trade. On the horizon, however, there loomed one development that promised to bring about a major readjustment in the competitive structure of the export markets. The Russian petroleum industry was gaining momentum.

The rise of the Russian industry

In 1874 there was little to indicate that Russian oil would ever become a formidable competitor of the American industry, even in the Russian home market. To be sure, the striking of the first large flowing well in the Baku area in 1873 gave promise of large potential supplies of crude. And by 1874 the Russian industry was supplying about one-third of the domestic demand for kerosene compared to one-fifth two years earlier. (See Table 19:4.)

But this expansion had come in spite of obstacles which still threatened to limit any extensive future growth of the industry. Crude oil, for example, was transported at considerable expense from the wells to the refineries over distances ranging from six to nine miles in large, wooden barrels attached to two-wheeled, horse-drawn carts. Refinery techniques, judged by contemporary American standards, were primitive. Baku refineries generally practiced simple batch-distillation, with a separate firing and run for each cut. The major output was ostalki fuel oil with kerosene as a by-product. No lubricants were produced, and the naphtha, gasoline, and benzine fractions were thrown away.[35]

Undoubtedly the principal obstacle to future expansion was the difficulty of reaching the major marketing areas of Russia, such as Moscow or St. Petersburg, a thousand or more miles to the north. In the wood-scarce Baku region, barrels were expensive and often of poor quality. Once packaged, every barrel of refined oil had to be shipped by boat across the land-locked Caspian Sea and transferred to river craft and railroads for transport to the interior. As was true of the American experience, the leakage from the barrels was great, which prompted the government-

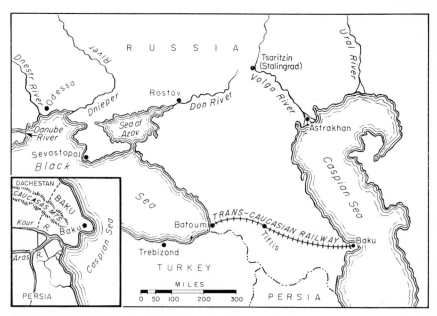

MAP 19:1. *Location of Russian petroleum industry. Inset shows detail of Aphsheron Penninsula.*

subsidized shipping and railroad lines to exact "heavy freights for conveying such inconvenient cargo to Russia." [36]

But if signs of momentous changes were lacking in 1874, there was abundant evidence of a major transformation of the Russian petroleum industry by the end of another ten years. As indicated in Chart 19:3, output of crude from the Russian fields grew from less than a half-million barrels in 1874 to nearly 9 million barrels in 1884. Much of this increased production came from the discovery of new flowing wells or "petroleum fountains" in the district. One well struck in 1875, for example, yielded 14,000 barrels of crude every twenty-four hours; another brought into production the following year flowed at a rate of 6,500 barrels a day for a period of three months. Perhaps the most spectacular well brought in during these years was the Drooja "fountain," which in the fall of 1883 began flowing at a daily rate of 43,000 barrels until it was capped some three months later. [37]

By 1884 the number of refineries in the Baku region had increased to nearly two hundred. While a large portion of their output continued to be made up of fuel oil, the volume of kerosene refined grew about as fast as crude production (see Chart 19:3), rising from about 144,000 barrels in

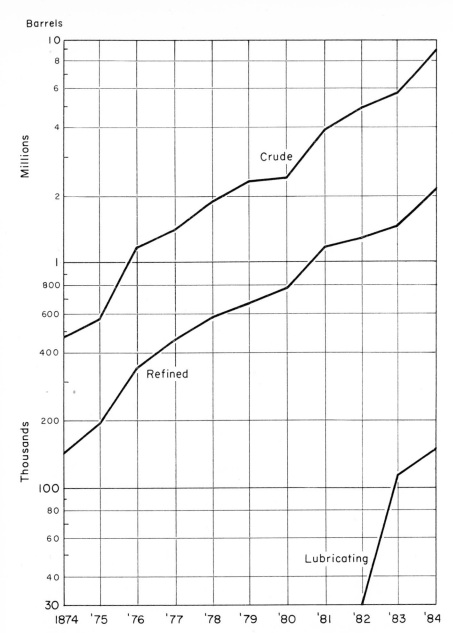

CHART 19:3. *Russian output of crude, refined, and lubricating oils (42 gallon barrel equivalents), 1874–84. (Sources: 1874–79—Marvin,* The Region of the Eternal Fire, *205, 207, 208, 1880–84—Mineral Resources, 1896, 885.)*

1874 to over 2 million barrels in 1884.* Adding lubricants to their production early in the 1880's the Russian refineries produced nearly 150,000 barrels of lubricating oil by 1884.

Some indication of the growing significance of the Russian refining industry in the domestic illumination market is revealed in Table 19:4. By 1876, Russian kerosene was already supplying more than half the total amount distributed in the home market. Four years later the proportion

TABLE 19:4

Russian domestic production and imports of illuminating oil: 1872–80 (thousands of 42-gallon barrels)

Year	Russian production	Imported from U. S.	Total
1872	39.0	175.9	214.9
1873	149.3	265.3	414.6
1874	143.8	248.0	391.8
1875	198.6	260.6	459.2
1876	347.9	261.5	609.4
1877	742.8	167.1	639.9
1878	594.4	195.4	789.8
1879	670.3	168.2	838.5
1880	914.0	142.0	1,056.0

Source: Marvin, *The Region of the Eternal Fire*, 300, 302.

had risen to almost 90 per cent. Within another four years the Russian industry had virtually completed its conquest of the domestic market and stood poised for an invasion of the illumination and lubricating markets of western Europe.†

How significant this competition for the export trade would be was a question as yet unanswered. As shown in Table 19:5, Russian outputs of crude, illuminating oil, and lubricants were still well below American production figures for 1884. But the fact that in terms of American outputs, the Russians had increased their volume of crude from 4 per cent to over 36 per cent, their illuminating oil from about 2 per cent to 14 per cent, and were producing about 16 per cent as much lubricating oil in

---

* Data on the outputs of fuel oil (in thousands of barrels) first available in 1880 were as follows: 1880, 697; 1881, 913; 1882, 1,768; 1883, 1,846; 1884, 2,868.[38]

† Comparable data on imports are not available for 1881–84, but American exports direct to Russia in the last year amounted to only about 12,500 barrels. See Appendix D:2.

1884, indicates how rapidly the Russian industry had been gaining on its American rival.

Perhaps of even greater importance for the future rivalry for the world petroleum markets were the series of innovations in refining, rail and water transportation, and marketing which had occurred in the Russian industry over the decade. Much of the credit for this transformation belongs to the Nobel brothers, who first entered the Baku area in 1875.

TABLE 19:5

Selected data on U. S. and Russian petroleum industries: 1874–84 (thousands of 42-gallon barrels)

	1874	1879	1884
Crude oil			
U. S. *	10,926.9	19,914.1	24,218.4
Russia †	475.3	2,254.5	8,841.0
Total	11,402.2	22,168.6	33,059.4
% Russia to U. S.	4.3	11.3	36.5
Illuminating oil			
U. S.‡	6,372.8	11,430.0	15,450.4
Russia †	143.8	670.3	2,161.0
Total	6,516.6	12,100.3	17,611.4
% Russia to U. S.	2.3	5.9	14.0
Lubricating oil			
U. S.‡	220.4	395.3	898.7
Russia †	..	..	147.0
Total	..	..	1,045.7
% Russia to U. S.	..	..	16.4

* *Mineral Resources, 1905, 883–84.*
† *1874 and 1879: Marvin, The Region of the Eternal Fire, 205, 207, 208, 300. 1884: Mineral Resources, 1896, 885.*
‡ *Estimated. See Appendix B.*

The Nobel brothers, Robert, Ludwig, and Alfred, were no strangers to the Russian scene. Their father, Immanuel Nobel, hired by the Russian government in 1838 to manufacture torpedoes which he had developed for military use, moved his family from Sweden to St. Petersburg in 1842.[39] During the Crimean War of 1854–56, the senior Nobel and his sons were kept busy turning out steam engines for the Russian navy. The conclusion of the war brought an end to government orders and the business of Nobel & Sons was forced into receivership in 1859.

For the next ten years each of the brothers followed separate careers. Alfred, the youngest, turned his attention to chemistry and after returning to Sweden in 1864 obtained his patent for dynamite in 1867. Ludwig, the middle brother, remained in St. Petersburg where he operated a profitable manufacturing establishment providing small arms and ammunition for the Russian government. Robert, the oldest, after indifferent success in selling petroleum products and later nitroglycerin in Finland, rejoined Ludwig in 1870.

COURTESY SHELL OIL COMPANY

Robert Nobel

It was on a journey to the Caucasus in 1874 in search of walnut wood for gun stocks that Robert became interested in the possibilities of Baku petroleum. With Ludwig supplying the capital, he returned to Baku the following year where he began the construction of a refinery that was in operation in 1876. Two years later the Nobel Brothers Naphtha Company was incorporated with an original capital of about $2.5 million, a part of which was contributed by Alfred.[40]

Robert Nobel had scarcely begun his refining operations when he rebelled at the costly method of hauling crude oil to the refineries in carts. Unable to enlist support from neighboring refineries, he built an 8-mile pipeline from the oil fields to Black Town, which repaid its cost, about $50,000, within a year.[41]

Seeking to integrate production with refining and transportation, the

Nobels next acquired oil properties and imported six drillers from the United States who were set "to work boring in the Pennsylvania fashion." Pennsylvania "know-how" in capping spouting wells paid immediate dividends, but many other modifications, such as the use of greatly enlarged tubing, had to be made to suit Baku conditions. Initially the Nobels suffered setbacks in obtaining their own production, but superior drilling techniques and organization soon brought them large supplies of inexpensive oil.[42]

COURTESY SHELL OIL COMPANY

Ludwig Nobel

When Robert Nobel retired because of ill health in 1879, the Nobel Brothers had completed the integration of their operations from production through refining. In that year their output of illuminating oil had risen to a respectable 8 per cent of the total production of the Baku refineries. (See Table 19:6.) It remained for Ludwig Nobel, who assumed active control after Robert's retirement, to extend the organization's operations into distribution and marketing.

As early as 1878, the Nobels had suggested to Caucasus & Mercury Company, which operated a fleet of government subsidized steamships on the Caspian, that they fit some of their ships with tanks to haul refined oil in bulk. Skeptical of the feasibility of the plan, the shipping company was uninterested. Ludwig accordingly turned to the Lindholmen–Motala ship building firm in Sweden, which designed and built the steam tanker *Zoroaster* that was delivered to the Nobels in 1879. Equipped with 21 vertical tanks capable of holding some 250 tons (about 1,550 42-gallon barrels), the *Zoroaster* was immediately put into service on the Caspian.* Her success as a bulk carrier led the Nobels to add to their fleet of tankers on the Caspian and by 1883 they had a dozen or more in the service. These were in turn supplemented by a fleet of smaller tank steamers and iron barges which carried oil from the entrance of the Volga river upstream about 100 miles to Tsaritzin (present-day Stalingrad), where the Griazi-Tsaritzin Railway Company had its southern terminus.[44]

Having eliminated the use of barrels for shipment of refined oil across the Caspian and on the Volga, Ludwig Nobel turned to the possibility of extending bulk transportation to rail shipments. When the directors of the Griazi-Tsaritzin Railway Company refused to supply specialized rolling stock, the Nobels began building their own iron railway tank cars. These tank cars were relatively small, each holding about 10 tons (approximately 2,600 gallons), but by 1883 there were over 1,500 Nobel Brothers' tank cars operating on the Russian railroads.[45]

To complete their distributing and marketing system, the Nobels also had to build extensive storage facilities which they located at twenty-six strategic marketing centers throughout European Russia, including Poland and Finland. In 1883 the largest of these, located at Orel in middle Russia, was equipped with storage reservoirs capable of holding 18 million gallons of kerosene at one time. Five other large depots, two at Moscow, and one each at St. Petersburg, Warsaw, and Saratoff, had a combined storage capacity of 10 million gallons, while twenty-one smaller stations had tanks capable of storing another 2.8 million gallons.

By the early 1880's, the Nobels had worked out an extensive distributing system well adapted to the seasonal demands of the Russian market. Throughout the summer months they kept their Caspian fleet busy hauling illuminating oil to the mouth of the Volga. There the oil was transferred to smaller tankers and barges for shipment to Tsaritzin, where it was in turn put into tank cars for shipment to the storage depots. During the fall and winter burning season, the tank cars were shifted from their

---

* The tanks were subsequently removed and the oil was carried in the "skin of the ship." Other features of the *Zoroaster's* design included the placement "of her engines amidships" and the use of fuel oil "to raise steam." However, "No arrangements appear to have been made for the expansion of the oil cargo." [43]

base at Tsaritzin to the storage depots where they were put to hauling oil by the "trainload to petroleum dealers [throughout] provincial Russia, who [brought] their own barrels to the railway stations and [carried] it away in this form to their own stores."[46]

At the same time the Nobels were pushing their operations back into production and forward into marketing, they maintained a careful control over their refining processes. From the outset, Robert Nobel had concentrated on the production of a high-grade illuminating oil. Continued research on improving refining techniques paid a big dividend when in 1883, apparently following a suggestion of Alfred Nobel, the Nobels began using the process of continuous distillation in their plants.[47] Introduced

TABLE 19:6

*Proportion of total Russian output of illu-
minating oil produced by Nobel Brothers:
1876–83 (thousands of 42-gallon barrels)*

Year	Total Russian production	Production by Nobel Brothers	% of total by Nobel Brothers
1876	347.9	.6	.2
1877	472.8	15.2	3.2
1878	594.4	27.7	4.7
1879	670.3	54.8	8.2
1880	914.0	146.2	16.0
1881	1,115.1	304.7	27.3
1882	1,230.9	438.7	35.6
1883	1,255.2	645.9	51.5

Source: Marvin, *The Region of the Eternal Fire*, 300.

some twenty years before its adoption in the United States, continuous distillation not only brought the Nobels savings in costs and improved yields, but enabled them to add lubricants to their refinery output.[48]

The introduction of continuous distillation was the final step in the series of developments which by the early 1880's made Nobel Brothers the outstanding firm in the Russian petroleum industry. Their expanding status was reflected in the organization's share of Russian output of illuminating oil which, as shown in Table 19:6, was over 50 per cent in 1883. Moreover, the Nobels had set an example which other firms interested in the industry were beginning to follow.

It was true that the majority of the nearly two hundred refineries operating at Baku in 1883 were small and still followed primitive methods of distillation, concentrating on the production of fuel oil, with kerosene as a by-product. But several companies by this time had followed the lead of the Nobels and were operating large refineries equipped to produce good quality kerosene and lubricants. In addition the Caucasus & Mercury Company had begun converting a part of its Caspian fleet to haul oil in bulk, while a newly formed Russian company had arranged with the Griazi-Tsaritzin Railway to distribute refined oil by tank cars to the Russian markets.[49]

The growth of the Russian industry during the 1870's created no alarm over the imminence of a competitive threat to American oil. Nor was this circumstance the result of a lack of continuity in American observers in the Baku vicinity, stemming all the way back to 1864, when Colonel Gowen drilled several wells near Kerth.[50] Most observers also had been competent, including professional oil men like Lewis Emery, Jr., H. M. W. Tweedle, and drillers that the Nobels themselves had imported.[51]

A few looked no further than the abundant contemporary manifestations of a badly organized industry: primitive refining techniques; a majority of wells dug rather than drilled; and enormous wastage of crude from uncapped flowing wells.[52] Most observers were more perceptive, but even they read into the lower kerosene yields of Baku crude a limitation that was illusory. They failed to realize that given the necessary incentives a huge surplus of crude oil within a few miles of refining capacity at Baku could easily be diverted to the production of large volumes of kerosene.

The year 1880 yielded a more definitive American report on the Russian industry than any of its numerous predecessors. It resulted from extended first-hand inspection of Baku operations by William Brough, an experienced Pennsylvania oilman from Franklin. Like other informed observers, he did not for a minute minimize vast distances from markets and inadequate transportation facilities as formidable barriers, but neither did he jump to conclusions on the superiority of American illuminating oils. Anticipating with remarkable insight later findings by Boverton Redwood in England and scientists in Germany and elsewhere on the Continent, he evaluated Russian kerosene as equal to American in color, odor, and fire-test.

One further consequential ingredient of the Russian scene caught Brough's objective eye. In January, 1880, the Russian government had granted a concession to build a railroad from Baku to Tiflis, which in turn was connected by rail with Poti, a port on the Black Sea. Here loomed "an outlet for Baku oil to the markets of Europe, and will bring it into direct competition with American oil in those markets." Furthermore, construction of the road was, "measured by Russian standards, progressing rapidly." [53]

The completion of the road in mid-1883 broke the major transportation barrier that had confined the Russian petroleum industry largely to the Russian home market. With the rail line in operation the distance from Baku to the nearest hold of European vessels was cut ". . . from 2,000 miles, intermittent water and rail transportation through western Russia to the Baltic, to 560 miles." [54]

While the newly completed railroad proved far from adequate to meet the traffic demands put upon it, shipments during the first six months operation included 14 million (imperial gallons) of petroleum products, 11.7 million barrels of which were illuminating oil. During the full year of 1884 total shipments over the road amounted to some 25.5 million gallons of which 20.2 million (equal to 553,000 forty-two-gallon barrels) were kerosene.[55] Compared with American exports of some 10.3 million barrels of illuminating oil in 1884, this was a mere trickle. But if this start did not portend a flood, it did mark the beginning of a flow of Russian petroleum products into the export trade that in less than a half dozen years reached respectable and, from Standard's point of view, alarming proportions.

Chapter 20

Domestic markets and marketing channels

While the export trade continued to absorb a major portion of the output of American refineries between 1874 and 1884, this decade marked the beginning of a trend that would eventually reverse the relative positions of the foreign and domestic markets. With total refinery output expanding sharply, this trend was reflected in an impressive growth in the volume of domestic sales during the decade.

Specifically, the shift in the proportion of illuminating oil retained for the home market, from about one quarter to just under a third, resulted in an increase in domestic consumption from about 1.6 million barrels to nearly 4.7 million barrels. In the mid-1870's, some 69 per cent of the total output of naphtha-benzine-gasoline—approximately 600,000 barrels—was consumed domestically. A decade later 85 per cent, or over two million barrels, was retained for the home trade. And even though the proportion of lubricating oil sold at home dropped from about 88 per cent to less than 69 per cent, the absolute volume distributed in the domestic market rose from just under 200,000 barrels to over 600,000 barrels.*

Growth in domestic demand

Basic to the expanding domestic market for petroleum products was an increase in the population of the United States from about 44 million in 1874 to approximately 55 million in 1884, coupled with an even faster growth in national income that raised real income per capita by approximately 50 per cent.† Such a population clearly offered an expanding potential demand for the output of the domestic refineries. But an approximate threefold increase in the volumes of illuminating oil, naphtha-benzine-gasoline, and lubricating oil sold domestically suggests other characteristics of the American market that were particularly favorable.

In respect to the naphtha-benzine-gasoline group of products, for example, a considerable portion of expanding sales during this period went to supply the demand for solvents. By far the larger proportion, however, was used in gas manufacturing and reflected the spread of the Gale-Rand processes of naphtha gas production and enrichment among American gas manufacturers.

In turn, the continued mechanization of industry, agriculture, and transportation, coupled with lower prices and an improved product, led to expanding domestic sales of petroleum lubricants. The approximately 600,000 barrels of petroleum greases and oils sold annually during 1883–1885, for example, went to lubricate a portion of the half-million agricultural implements (with moving parts) purchased by farmers in 1880; [2] the 24,353 locomotives, 16,644 passenger cars, nearly 821,000 freight cars utilized by the railroads in 1884; and the 12 million or more cotton spindles in operation. [3]

Kerosene's growing popularity in the American illuminating market during these years was the result of a complex of interrelated factors. Just as they had been important in stimulating foreign sales, falling prices unquestionably played a significant role in the growth of domestic consumption of illuminating oil that, measured in *per capita* terms, rose from about 1.5 gallons to approximately 3.6 gallons. It is true that domestic prices gen-

* See Table 19:1.
† The best estimates indicate that per capita income in constant dollars rose from $237 in 1869 to $309 in 1879, and to $383 in 1889. [1]

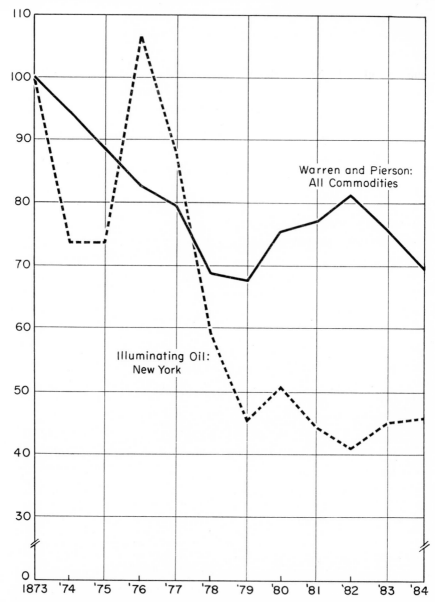

CHART 20:1. *General wholesale price index and New York illuminating oil wholesale price index (1873 = 100), 1873–84. (Sources:* Historical Statistics of U. S., 1879–1945, *231;* Derrick's Hand-Book, *I, 783.)*

erally followed a downward trend throughout this period. But as shown in Chart 20:1, except for 1876 and 1877 the wholesale price index of illuminating oil in the New York market, starting from a common base of 100 in 1873, remained well below the general wholesale price index.

It is noteworthy that this rather extraordinary growth took place in spite of a failure to impose strict legal controls over the quality of kerosene sold in the domestic market. According to the *Titusville Herald* in April, 1873, "One of the greatest drawbacks to a more universal introduction of kerosene at home is the dangerous character of the mixtures that have been sold under the name of safe kerosene for several years past." [4] The following year the same source called attention to the fact that although Europeans paid more for their refined oil they consumed much more than Americans. The reason was simply, ". . . that legislation in foreign countries has done there what legislation has failed to do here, and a perfectly safe article has been secured to our foreign consumers while we have had to use unsafe oil." [5]

Judging from complaints of foreign buyers during the late 1870's the *Herald* clearly overstated the case by describing the product sold abroad as "perfectly safe." Yet, however unsatisfactory foreign legislation was in controlling the quality of exported kerosene, the rules imposed by the produce exchanges alone were undoubtedly more effective in securing minimum standards of quality and safety of exported kerosene than were achieved in the United States under a tangle of laws and regulations imposed by Federal, state, and municipal governments.

The general inadequacy of legislation affecting the quality of kerosene sold in the United States in 1880 was revealed in a report to the Census of that year. According to the report, 17 out of 38 states had no special legislation on the subject, although a number of municipalities in such states had adopted ordinances providing for some test of safety. But the returns generally ". . . exhibited the confused condition of legislation regarding petroleum enacted by so many different legislative bodies more or less influenced by a great variety of opinions and interests." [6]

It was pointed out that some refiners, careful of the quality of their products, were eager to have safety standards applied to all illuminating oils. Others took the attitude that the responsibility should rest with the purchasers and that legislation, if any, should provide the lowest test possible. The same division of opinion applied to distributors and dealers, some of whom were willing to chance the danger of blending naphtha, benzine, or gasoline with the kerosene they marketed.

In November, 1881, the *Oil, Paint & Drug Reporter* took occasion to point out the inadequacy of existing measures to regulate the quality of kerosene. Basing its article on reports made before the *United Fire Underwriters of America,* the paper noted that, according to a survey made in 1876, between 5,000 and 6,000 persons lost their lives annually from fires caused by lamp explosions. Because little had been done during the intervening years to reduce these hazards, the editor was sharply critical of the underwriting group for failing to push more vigorously for the passage and enforcement of stricter legislation governing the sale of illuminating oil.[7]

Yet, by the early 1880's agitation over the quality of kerosene sold in the domestic market had already begun to subside. In part this was attributable to a gradual improvement in legislation and its enforcement. In part it reflected a greater concern by Rockefeller and his associates over the product that Standard was supplying in increasing amounts to the domestic markets. Finally, it is apparent that economic considerations played an important role in securing a safer illuminating oil for domestic consumers. As shown in the following tabulation, the margin between kerosene and naphtha prices narrowed after 1873 so that by mid-1880's there was little reason to use naphthas to adulterate illuminating oils; in fact, in 1881 the reverse was true.

Year	New York wholesale prices per gallon in barrels	
	Illuminating oils	Refined naphtha
1873	17.88¢	11.13¢
1874	13.00	9.00
1875	13.00	9.63
1876	19.13	11.38
1877	15.50	9.75
1878	10.75	7.13
1879	8.13	6.50
1880	9.00	7.75
1881	8.00	9.63
1882	7.38	6.25
1883	8.00	5.88
1884	8.13	7.50
1885	8.00	7.13

Source: *Pa. Industrial Statistics,* 1892, B201–4.

In the United States as in Europe, petroleum illuminants had been most readily accepted by customers living in industrial-urban communities. Not only was their desire for artificial illumination more intense

Plain and fancy kerosene containers.

than that of most rural inhabitants, but their needs could be supplied more readily and at a lower cost by existing transport and marketing facilities.

Yet there were few places in America at this time where kerosene was not known or used. It was already a familiar item in the inventories of most of the country stores that served a great portion of rural America, where it contributed to the "odoriferous inventory" of their entire stock, compounded out of the

> glaze on the calicoes and the starch in the checks, rotting cabbages and potatoes, spring-onion sets, cheese, neet's-foot oil, leather polish on new shoes, oil and wax on saddles, horse collars and buggy harness, kerosene, sardines, salmon, the stove, tobacco, the cat, the customers, asafetida, peppermint and wintergreen candy, and engine oil.[8]

But for some three-quarters of the population that lived in rural areas, particularly those located south of the Ohio and west of the Mississippi rivers, kerosene was still more of a luxury than a necessity.

It was the urban population located principally in the northeastern part of the United States, that grew from about 12 million to 18 million between 1874 and 1884 which continued to supply the major portion of its illuminating needs with kerosene. It is true that during this period individuals living in communities with populations of 2,500 or more could increasingly draw upon local gas manufacturing plants to illuminate their houses, factories, and public buildings.* Despite its initial advantages over competing light sources and subsequent improvements, gas met only a portion of the illuminating needs of urban customers, chiefly among commercial and industrial users and the private residences of the middle and upper income groups. Most urban residents continued to use kerosene whose "cheapness and simplicity of lighting . . . made it [particularly] desirable for many applications."[10]

Even as late as 1885, the *Oil, Paint & Drug Reporter* claimed that "by far the greater portion" of New York's million inhabitants "use oil for lighting purposes in preference to gas. Let any one travel on the 'L' roads when darkness sets in and observe the windows of every story as he passes," it was suggested. "He will see lamps, big and little and of various degrees of workmanship, burning instead of gas jets. Only on the ground floors, which are used for stores will he find gas used. . . . The working people, artificers and laborers who in Europe would burn candles in their rooms here use oil." The principal reason was that "a family can be supplied with lamp oil for $10 a year" whereas it was not uncommon for the gas bill of the more well-to-do householders to run that much per month.[11]

But kerosene was not only the "poor man's illuminant" in urban communities. Indeed industry spokesmen were proud of its appeal to the fashionable tastes of the more well-to-do classes and were quick to point out its advantages over its principal rival. In explaining what appeared to be a growing demand for expensive lamps, one observer stated,

> A greater finish can be given to a lamp than a gas fitting, and a more artistic design is attainable. Globes of cut glass are produced in curious yet beautiful shapes, in color of the softest, the purest azure, in varied tints and hues that appeal to the tastes even of the artist, . . . Everyone knows how much easier it is to read by the softened light of an oil lamp than by the glare of gas.[12]

To another observer in 1883,

> The increasing cost and elegance of lamps for burning kerosene oil is one of the remarkable features of our nineteenth century civilization, especially so when we consider that gas has been and electricity will be a strong rival to oil as an illuminator. The richest porcelain vases, mounted in fine brass work, supplied with costly burners, and the whole surmounted with decorated

---

* According to Census estimates, in 1890 only 350 of some 1,244 communities with 2,500 or more inhabitants in the United States did not have gas plants.[9]

shades of the most expensive sort, form a combination of handicraft which in many cases is disposed of at remarkable figures. One hundred and fifty dollars . . . is not an unusual price for a fine kerosene library lamp, and it must be confessed that the fashion, though expensive, is a very sensible one, especially if the lamp is used both as an ornament and an illuminator.[13]

While urban sales continued to expand, a significant change in the domestic market during 1874–84 came as an increasing amount of the industry's output of illuminting oil reached the primarily rural areas where the great majority of Americans lived. Some quantitative impression of the growth of non-urban dwellers in the United States is revealed by an expansion in the number of farms from about 3 million in 1874 to nearly 4¼ million in 1884 and an increase in the population living in towns of less than 2,500 inhabitants and in rural areas (including farm families) from approximately 32 million to nearly 38 million.*

This rural expansion would have had relatively little effect on illuminating oil sales except for other developments affecting the distribution of petroleum products. Of major importance was the expansion of the American railway network from about 72,000 miles in 1874 to over 125,-000 miles in 1884. Some four-fifths of this expansion occurred in the areas west of the Alleghenies and south of the Ohio River.†

This expansion of railroad mileage not only served to bring more adequate transport facilities to an increasing proportion of the population. It also brought about more intensive competition among the trunk lines for traffic that flared out in the series of rate wars during the 1870's and 1880's. As a result the general level of freight rates dropped substantially.‡

These developments did much to cut the costs of distribution to rural America, but their impact was both gradual and unevenly spread throughout the markets. To be sure, one careful student of marketing history noted, "A look at the stock of a good, lively country store at the time of the Philadelphia Centennial in 1876 would have been enough to convert any citizen to a belief in progress. Lamps and lamp chimneys, and the whole

* Reliable estimates for the number of farms and distribution of population are available for census years only. The figures quoted are rough approximations of the number of farms and the size of the rural population at points midway between the census reports, population figures for rural areas were as follows: 1870, 28,656; 1880, 36,026; 1890, 40,841. Data on the number of farms were as follows: 1870, 2,660; 1880, 4,000; 1890, 4,505.[14]
† Data on railroad mileage estimated from Interstate Commerce Commission reports for 1870, 1880, and 1890.[15]
‡ The best estimate suggests a drop in average freight rates of about two-thirds between 1870 and 1885.[16]

class of merchandise known as 'kerosene goods' would seem to be a marvel to eyes that had strained to see at night by means of a lighted rag, soaked in beef tallow and draped over the edge of a dish." [17]

Yet some five years after the Centennial, one contemporary trade journal, commenting on an estimated 20 per cent increase in total domestic consumption for 1881 over 1880, pointed out that because of exhorbitant rates for transportation and other causes, "much of the oil sold in the far western and southern states has retailed for 45¢–75¢ a gallon." [18] Purchases at these prices (the New York wholesale price for 1880 averaged about 8¢ per gallon) prompted the further comment that while "people will have it [kerosene] at any price within reason, or at a cost of 50 per cent or perhaps 100 per cent above any competing substance," it appeared that "the trade in petroleum in America was only in its infancy . . . a mere babe compared to what it will become when the causes which combine to prevent its growth shall have been ascertained and removed." [19]

Closely related to the growth in railway mileage were two other closely interrelated developments that gave promise of bringing greater maturity to the trade in petroleum. One was the gradual application of bulk handling by rail to refined products—principally illuminating oil—which further cut costs of distribution. The second was the emergence of large "independent" wholesale marketers. Waging vigorous marketing programs in their respective territories, these specialists eventually came to dominate petroleum distribution west of the Alleghenies and south of the Ohio River.

The beginnings of bulk shipment of refined oil

All of the objections to barreled shipments of crude (barrel weight, the tendency of contents to leak or evaporate, and the problems associated with their return for re-use) also applied to the shipment of refined products in barrels. With the growth in the domestic transport system and the geographic extension of the market, dissatisfaction with barreled shipments of refinery products mounted because these diseconomies tended to accumulate the farther the packages were shipped and the longer they were in transit.

Experiments with the prototype of the modern tank car, developed late in the 1860's for the transport of crude, soon demonstrated that these cars, whose 5,000–6,000-gallon capacity was nearly double that of the average

Iron boiler type tank car, 1880.

boxcar loaded with barrels, were readily adaptable to shipment of naphthas, kerosene, or lubricating oils. Indeed, refined products were less hazardous to ship by tank car since they contained none of the highly volatile and gaseous elements found in crude.

The introduction of bulk shipments by rail was the major step in the evolution of a system that ultimately enabled marketers to move refined oil from refineries to final consumers without special packaging. But even this first step evolved slowly over the following decade. The tank wagon, which extended bulk handling further to retailers or consumers, lagged even further behind.

While he may have overstated the case, one prominent marketer pointed up perhaps the major advantage of bulk shipments by rail. "The introduction of tank cars as a means of transporting refined products, particularly those of lighter gravity, . . . amounts in savings in evaporation alone to as much as 50 and 75 per cent of the total contents, dependent upon the distance to be transported." [20] Even if they are qualified, these percentages alone suggest a compelling reason to substitute tank cars for shipments in barrels.

Tank-car shipments offered another attraction in that, prior to restrictions imposed by the Interstate Commerce Commission in 1886, the railroads typically hauled bulk shipments in carload lots at lower rates per

gallon than similar lots of packaged goods. In part this concession reflected the fact that bulk shippers were saved the expense of paying
freight on the 65-pound weight on the barrel. But often the differential
was greater than would be warranted simply by this difference in weight.
In such cases, railroad officials justified the rate on the grounds that bulk
oil required less handling. This was partly true, since responsibility for unloading barrel shipments was ordinarily assumed by the consignee.* The
savings on this score were considerable. According to one railroad freight
agent, barreled oil shipped south prior to 1886 paid ton-mile rates from 20
to 25 per cent higher than tank car shipments.[21]

But before shippers could take advantage of the differential freight
rates they had to have tank cars. The railroads generally were unwilling
to assume the initiative in supplying tank cars, even though railroad operators differed widely in their evaluation of bulk transport. As the general freight agent of the Pennsylvania Railroad put it in 1888,

> The movement in barrels I have always considered preferable for two rea-
> sons: First, we could load barrels in a car that would carry a return carload.
> There are no back-loads for a tank car . . . Secondly, if there comes a colli-
> sion or fire, packages being separate, we are enabled to save some of the ton-
> nage. There is no hope of saving the contents of a tank car once it gets on fire.
> Barrels you can sometimes scatter and roll off and break up.[22]

An opposite point of view was expressed by a freight agent of the Illinois Central Railroad. In defending "lower rates on tank oil than upon
oil in barrels" he stated,

> [shipment by tank car] is considered a much safer plan to carry oil. There is
> less danger of fire, and there are other reasons . . . The barrelled oil we gen-
> erally load to about half the capacity that we do in tanks as to weight. We
> load 50 to 60 barrels to a car instead of 85 to 120; and when the [box] car is
> used for coal oil we either have to keep it exclusively in that service or allow
> it to remain on side tracks for a long time to get the odor out of it so that we
> can carry other freight . . . We cannot carry coal oil with other goods with-
> out damaging and destroying them in a great measure.[23]

Perhaps the major reason for the railroads' lack of enthusiasm arose
from financial considerations. Tank cars cost about $700, almost twice as
much as ordinary box cars; they were a relatively specialized type of
equipment, and with the financial stringency and intensified competition
of the 1870's, the roads apparently preferred to channel their limited funds
into more generalized types of capital equipment. They were therefore no

---

* To a much greater extent than the railroads liked to admit officially, the lower rates
  on tank-car freight constituted a rebate to the major users of this method—the
  Standard Oil Company and its affiliates.

more eager to supply bulk cars for kerosene than they had been earlier to provide similar equipment for crude oil. Since by the early 1870's it was standard practice to ship crude by tank car, the oil carriers generally allocated such cars as they did to fill the continually urgent demands from the Regions.*

This shortage was eased somewhat with the advent of the trunk pipeline in the early 1880's, but in the interim, shippers who wanted to move oil in bulk were largely thrown on their own resources. Some of the larger firms had tank-car fleets built for themselves. More commonly they purchased tanks costing about $250 each, which could be installed on railroad flatcars.

Whether refiners or marketers took the initiative in arranging for tank cars depended on particular circumstances, principally upon the sales volume and capital resources of the respective parties.† In any event, the added capital outlay, together with other expenditures required for bulk handling, tended to penalize smaller members of the industry. Although never completely paralyzed, they found it increasingly difficult to maintain their market positions. They were squeezed from behind by Standard Oil's absorption of independent refineries and from all sides by the growth of large marketing companies working in close alliance with the Standard organization.

Since the bulk system shifted packaging operations from refineries to intermediate distributing points, some provision had to be made at the latter to store oil during the interval between its receipt and subsequent re-sale. Unless he was part of a vertically integrated operation, the distributor commonly bore the expense of providing such storage facilities. Small marketers at times avoided the extra cost by running oil directly from tank cars into barrels, a practice that apparently began before much bulk storage was constructed, and which continued at least as late as

---

* The fact that most of the cars for through freight were technically provided by fast-freight companies, rather than the railroads themselves, does not substantially alter the foregoing argument. Existing evidence indicates that the only fast-freight line to supply tank cars in any considerable quantity was the Empire line, controlled by the Pennsylvania Railroad. The New York Central seems to have been generally wary about providing equipment to move and store oil; the Atlantic & Great Western, throughout most of the 1870's, was too poor to do much in this direction.

† When a distributor, F. B. Carley, set out to monopolize the southern trade in petroleum, he agreed to run his own cars over the Louisville & Nashville Railroad; on the other hand, when refiner George Rice found he could not compete against the low rates obtained on bulk freight by Standard Oil, he ordered ten tank cars from the Harrisburg Car Works. Apparently none of his customers was large enough to assume the investment.[24]

1906.[25] But oil handled in this manner was still subject to considerable evaporation and leakage during the time it was part of inventory.

More prevalent—and more significant—was the practice of providing storage tanks at receiving stations, which sprang up almost concurrently with the advent of bulk-product shipment. At these stations, the contents of the tank car could be pumped immediately into storage tanks, and be held in bulk until the time came to package it for subsequent distribution.

According to one refiner, bulk stations were "costly affairs as a usual thing," [26] but actually costs and size were relative matters, varying with the nature and extent of the station's market. Few matched Waters-Pierce's mammoth installation in St. Louis, estimated to be worth $250,000 in 1888. The Standard Oil plant in New Orleans, costing about $20,000, was probably a representative figure for stations in concentrated metropolitan areas that distributed kerosene to sub-depots in outlying districts. Perhaps the most prevalent type was the small terminal, like one constructed by Chess-Carley for $2,000 at Selma, Alabama.[27]

Larger depots, with storage tanks holding up to 1,000 barrels, generally handled a wide product line, ranging from kerosene and gasoline through lubricants and paraffin. Sometimes the lubricants were blended at the bulk station itself and sold under the marketer's brand. Barrels were often manufactured as well as repaired at the station, since the advent of tank-wagon delivery did not completely displace packaged sales. This was particularly true of stations which shipped in less than carload lots to smaller depots.

In contrast, small terminals, designed to handle kerosene and perhaps one or two other products, had but minimal equipment, such as a railroad siding, one to three storage tanks, with capacities ranging from 100 barrels to 500 barrels each, and a warehouse for storing and repairing packages. Many of them were units of great multistate marketing companies like Waters-Pierce or Chess-Carley; others were independently owned, although neither the proportions in each category nor the volume of business handled can be estimated on the basis of available statistics.*

---

* It is obvious that the amount of business could not have been very large, even by the standards of the time. An independent dealer, Harlow Dow, for example, sold only 100 barrels per week in 1885—apparently a peak year, judging from his statements—and to do this he had to ship to points as far as ninety miles away. The distance may have been exceptional, for stiff Standard Oil competition in his immediate territory perhaps sent Dow further afield than the usual 10–20-mile radius of operations, but the sales volume was not uncommon.[28]

Scattered evidence indicates that while tank cars and bulk stations were beginning to play an important role in the distribution of refined products by the early 1880's, their development had been slow and uneven up to that date. This gradual evolution in the application of bulk handling methods to refined products was in marked contrast to the relatively swift and dramatic changes in the handling of crude. Gathering lines replaced treamsters in a half-decade; within a few years after their introduction, tank cars were standard equipment for crude shipments, only to be replaced even more rapidly by trunk pipelines.

Differences in demography of the markets for crude and refined products largely account for slower response of the latter to the possibilities of bulk handling methods. In replacing a highly inefficient method of transporting crude, for example, the gathering lines found a ready market for their services. Since the lines tended to terminate at common shipping stations, large supplies were available for shipment to refineries. To be sure, it was necessary to invest in tank cars. But as refiners in turn customarily bought in large lots, the advent of the tank car rendered any special packaging between wells and refineries superfluous.

These same considerations accelerated the acceptance of the trunk pipelines. Gathering lines could marshall a large throughput at one end of the line, while concentration of refinery capacity in localized areas assured a demand at the other end sufficient to cover pipeage costs. In short, bulk transportation paid only when a sufficiently large amount could be brought to a given point for subsequent break-down into smaller lots.

While these conditions were not rare in the refined product market during the 1870's and early 1880's, they were far less typical. Even in the mid-1880's, relatively few bulk plants were established in the Middle Atlantic and New England states. With the majority of customers nearby, savings from the reduction in leakage or evaporation or differential freight rates hardly warranted any extensive use of the tank car and the transfer of storage and packaging functions from refineries to intermediate distributing points.

Markets located beyond the principal refining centers varied widely. Chicago by the late 1870's was handling several hundred thousand barrels annually, but many rural villages were not large enough to take even a full carload per shipment.

The distributing techniques utilized also reflected demographic differences. Bulk stations first appeared at two types of points: either large cities or smaller rural towns with good railroad connections. The first not

Pioneer distributor of gasoline.

only consumed large quantities of illuminants, but also served as entrepots for the surrounding area. The second served as distributing centers for a wide region—sometimes half a state.

Thus Standard of Ohio had established bulk stations in Chicago, Detroit, and other mid-western centers as early as 1875, yet in 1881 it was shipping only a little over half of its total sales of 55 million gallons in bulk.[29] In 1882 Standard Oil "interests," which accounted for some 80 per cent or more of industry sales, controlled about 130 marketing stations, scattered principally in areas south and west of the Atlantic seaboard. Their location in cities and larger towns suggests that many, if not the majority, of these were bulk stations. Nonetheless, Standard Vice-President John D. Archbold testified that, "In 1882 and up to that period . . . distribution [in the domestic market] . . . was largely made in barrels."[30]

But if the conquest of barrels was far from complete by the mid-1880's, bulk handling had already set a pattern which became increasingly significant over succeeding decades. Of more immediate significance was the part-bulk shipments by rail had played in changing the structure of domestic marketing channels.

Changes in domestic marketing channels

The slow spread of bulk handling methods into the eastern sections of the United States during the late 1870's and early 1880's had little impact on the marketing structure in these areas. Nor did Standard's increasing control over domestic refining capacity bring about any immediate change in distributing channels. All of the major units absorbed by Standard had established relations with the trade. In some instances they sold directly to retail outlets, but more commonly middlemen were used. Commission agents or regional jobbers acted as intermediaries in eastern markets, while large wholesale drug and grocery houses distributed to southern and western retailers.

The general policy of Rockefeller and his associates was to continue the marketing procedures of the companies they acquired. In part this decision reflected the preoccupation of Standard executives with the problems of bringing refining and transportation under control, a task which left little time to consider major shifts in marketing practices. But there were further considerations. By maintaining existing marketing relationships, much of the good will of the newly acquired companies could be retained, the value of their brand names could be exploited, and the fact that these companies were part of the Standard organization might be kept secret.

Within this general framework, Standard did move to bring dealers under contract and began to allocate certain major eastern marketing areas to particular refineries. By the late 1870's, for example, the middle-Atlantic states were covered by Standard units in New York, Philadelphia, and Pittsburgh. The Camden Consolidated Oil Company, formed in 1875 and operating chiefly out of Parkersburg, West Virginia, established marketing control over West Virginia, Maryland, and Virginia. Standard units in Boston, and Portland, Maine, supplied the New England area.[31]

As early as 1878, Colonel W. P. Thompson of the Camden Consolidated Oil Company suggested to Rockefeller that Standard take a more active role in the distribution of its products. Closer supervision of marketing by the company's top management at Cleveland, he argued, would both re-tain ". . . to ourselves the benefits we are entitled to in the home trade" and provide the company's agents better protection against "outside distributing agents."[32] This program, Colonel Thompson went on to say, could well be, "the foundation of the most magnificent commercial system and in my judgement the plan should be adopted for every portion of the U. S. territory." [33]

Despite Rockefeller's expressed enthusiasm for the plan, Standard was not quite ready in 1878 to go as far as Colonel Thompson suggested. The management, however, did take steps to protect its domestic marketing channels by acquiring a financial interest in certain leading wholesale houses that had emerged by the late 1870's. In fact, Rockefeller had made the first move in this direction as early as 1873 by purchasing a half interest in Chess, Carley & Company of Louisville, Kentucky. This purchase set a pattern for subsequent alliances. But beyond arranging for a geographic allocation of sales territories, Standard allowed its various associated marketing companies to continue under the active management of their founders.

The marketing companies that attracted the attention of the Standard management were all specializing in the marketing of petroleum and allied products. Located in the areas west and south of the principal refining centers, their growth was a major factor in the decline in the relative position of the large grocery and drug wholesaler in the domestic distribution of refinery output. They owed much of their success to the vigor with which they built bulk stations and pushed bulk shipments by rail. This was particularly true of Chess, Carley & Company and Waters-Pierce & Company, of St. Louis. Under their respective founders these companies not only demonstrated the advantages of bulk handling but followed competitive practices that were effective in securing a virtual monopoly of sales in their particular sales territories.

Chess, Carley & Company

F. B. Carley, the principal founder of the firm of Chess, Carley & Company, began operations in 1869 by distributing oil from a small refinery in Louisville.[34] Business soon outstripped his own refining capacity and Carley began buying from outside refineries in Pittsburgh and Cleveland. The company had already built up a considerable wholesaling business in oil and naval stores by 1873, when Rockefeller suggested a merger. Carley welcomed the suggestion and sold a half interest in the firm to Standard of Ohio in that year.[35] For the next thirteen years, however, he continued to manage Chess, Carley & Company with a free hand, giving up the business in 1886, when Standard bought out his interests and merged the company into Standard Oil Company of Kentucky.

Prior to 1873 Carley had apparently been successful in pushing the sales of barreled oil that was shipped by rail south into Tennessee, the northern sections of Mississippi, Alabama, and Georgia and the western parts of North and South Carolina. But with his supply lines guaranteed and added capital resources made available by his association with Standard, Carley, in 1873, began to give serious consideration to the possibility of capturing the petroleum market throughout the entire southeastern section of the United States.

The potential of this predominantly rural market was impressive. Hard hit by the Civil War and reconstruction, the South now showed signs of economic recovery. It had already attracted the attention of railroad builders who added some 10,000 miles of rails from 1874 to 1884. Its population of about 11.7 million in 1874 grew another 2.5 million over the same period. Yet, by the mid-1870's kerosene had made its way into few white homes, and not at all into the homes of the recently freed slaves.

But to capture this market, freight rates south from Louisville would need to be competitive with rates by water to gulf ports from east-coast refineries. Noting that the southern marketing centers were also located in the heart of the turpentine-producing area, Carley conceived the idea of using tank cars to transport illuminating oil south, where he could pick up return shipments of turpentine that could be marketed profitably in the northwestern section of the United States.*

Carley went to the Louisville & Nashville Railroad with his idea. Under an agreement with the railroad, he had a number of tanks made, ranging in capacity from 3,300–3,600 gallons each, which were put on flat cars belonging to the railroad. While the contents of these tanks weighed about 10 per cent more, the railroad agreed to accept 22,000 pounds as a basis for establishing rates. Coupled with the savings in the weight of the barrels, which were included in the rate on boxcar shipments, this concession meant a reduction in freight costs by tank car over boxcar of about 25 per cent.

This saving is brought out by the tabulation on the following page.

---

* As he put it, "we might have to lose money on the oil for two or three or four years but [the reduction] of leakage [on shipments south in barrels] and . . . of turpentine [on the return trip] . . . would furnish a margin that would enable us to supply Chicago and the Northwest with turpentine and still hold a rail route [south] on oil." [36]

Description of weight	Tank car	Boxcar
Assumed weight of contents	22,000 lbs.	22,000 lbs.
Weight of barrels	. . . . . .	3,850 *
Assumed net weight of contents	22,000 lbs.	18,150 lbs.
Assumed "overweight allowance"	2,200 †	. . . . . .
Actual net weight of contents	24,200 lbs.	18,150 lbs.

* 55 filled barrels at 400 pounds each = 22,000 lbs. Average weight of empty barrel is 70 lbs., times 55 (barrels) = 3850 lbs.
† 10 per cent.

This advantage was further increased when the Louisville & Nashville put tank car shipments in a sixth-class rate rather than a fifth-class rate that applied to barrels.[37] The railroad was willing to grant these concessions in return for guarantees by Carley of return loads on north-bound tank cars.

Carley lost little time putting his plan into action. The chief turpentine-producing center on the route of the L&N was the area around Mobile, Alabama, where for many years turpentine had been barreled and shipped by water to New Orleans for subsequent shipment up the Mississippi River. Carley proposed to the officers of the L&N, "If we put these tanks [tank cars of illuminating oil] into New Orleans and you will bring them back empty as far as Mobile, we can take that Mobile turpentine into the Northwest over the line of your road, comparatively none of which you are getting now."[38] In fact, he guaranteed that not over 10 per cent of the cars would return empty.

When the railroad agreed to this plan, Carley met with the turpentine producers in the Mobile area and contracted to buy their entire output. "We therefore had the exclusive handling except a very slight percentage against us of the return freight from Mobile North, and we made money on the turpentine most of the time." Moreover while Chess-Carley lost money on the "oil branch of the business in *some* [italics added] of the Southern markets, [he] succeeded in getting most of the turpentine business of the entire Northwest as the years went on."[39]

Carley followed the same pattern of purchases in other turpentine-producing centers, and soon added rosin and cotton-seed oil to his return shipments. Over the years he expanded his tank-car fleet and as other railroads started shipping petroleum products by tank car he began moving a part of his cargoes over other lines "so as to stand neutral between the different roads."[40] Even though he also shipped in barrels, he made every effort to keep the rates on barreled oil high in order to maintain his advantage of shipping by tank car. Even when rivals shipped in tank cars,

Carley concluded on a basis of his experience, "no outside refinery could have put that oil down there . . . [into his marketing area] without reference to other articles . . . they would have suffered a loss on their oil if they did not take back the turpentine." [41]

Carley used his bargaining position to bring pressure on southern jobbers to stop buying from east-coast refiners and to purchase oil, barreled at bulk stations, which Chess, Carley & Company established in the principal marketing centers. He offered the jobbing trade contracts under which they were guaranteed a fixed mark up per gallon in return for an agreement that they would not purchase from other suppliers and would maintain prices at levels set by Chess, Carley & Company. To insure compliance, he also introduced a system of rebates which were paid as long as prices were maintained at these levels. [42]

Not all jobbers, however, were eager to come under the Chess-Carley banner. Some preferred to buy from previously established sources of supply; some objected to buying from a "Standard Oil" company. In other instances Chess-Carley's policy of maintaining prices at relatively high levels attracted refiners and marketers outside the Standard orbit to start shipping oil to southern dealers. Some sent their oil over rail lines whose ties with Chess-Carley were not so close as were the L&N's. Others took advantage of water transportation, chiefly down the Mississippi River.

Carley responded to these challenges to his marketing position by engaging in ruthless competition. Letters to George Rice, independent refiner of Marietta, Ohio, who attempted to sell in Chess-Carley territory, give some impression of the tactics used against the customers of an outside supplier. On January 7, 1881, W. O. Peeples of Chattanooga wrote, "Chess, Carley & Company, on hearing that we had another car of oil on the road, stampeded and cut the price of oil that they had been selling at 21¢ to 17¢." [43] A. Haas & Brothers in a letter dated February 5, 1881, stated, "Yes, we are afraid of Chess, Carley & Company, but only to the extent they will sell for less money than you will allow us to sell. We are meeting but one difficulty; when we approach large buyers we get this answer 'we can't afford to buy from you; we don't like Chess, Carley & Company, but when they find you are getting the trade they will smash prices and Rice will drop you.' Whenever we can assure the trade that we will stay in the market you may count on your getting a big part of the trade." [44] On October 18, 1882, Gavin & Company of Memphis wrote, "The Standard Oil Company has been monopolizing the oil trade of this region for some time, Memphis particularly. One day oil is up to 20¢ and over, and when any person attempts to import here, other than the vassals of the Standard Oil Company it is put down to 7¢ per gallon. Can we make any permanent arrangement with you by which we can baffle such monopoly and not lose much? Can you furnish the oil in carloads—tank lots?" [45]

While Carley was prepared to cut prices on illuminating oil if necessary, he preferred wherever possible to compete on other items. "For instance," he explained, "I would go to New Orleans and Memphis and get the jobbers, instead of competing on oil and making that their leading article, to make their fight on sugar or lard or anything they chose and get them to stand neutral on oil." [46]

If these tactics failed, he resorted to other means. In dealing with a wholesale grocer, for example, he would not hesitate to sell groceries to the wholesaler's customers at cut-rate prices. In one instance at least he carried competition to the retail level by establishing a grocery store in Columbus, Ohio. [47] As he wrote to Wilkerson & Company of Nashville, Tennessee (a customer of George Rice), on December 10, 1880,

> it is with great reluctance that we undertake serious competition with anyone. *And certainly this competition will not be confined to coal oil or any one article, and will not be limited to any one year.* We always stand ready to make reasonable arrangements with anyone who chooses to appear in our line of business. And it will be unlike anything we have done heretofore if we permit anyone to force us into an arrangement which is not reasonable. Any loss, however great, is better to us than a record of this kind. [48]

While illuminating oil was its main business, Chess, Carley & Company sold a wide variety of refined petroleum and associated products including lubricating oils, greases, whale oil, lard oil, harness oil, and the like, often offering two or more grades in each category. Although Chess-Carley's association with Standard Oil was common knowledge, the company pushed its own brands; its premium kerosene bearing the name Kinslow and the other products generally some variation of the initials of the company, such as 3C Paraffine Oil, 3C Valvine Cylinder Oil, and 3C Machinery Oil—Golden. [49]

Carley's association with Standard also linked his sales tactics with those of Rockefeller and his associates. Carley stoutly denied any connection. "I was very fortunate in competing," he stated in 1888, "but I want to say that it was not by the direction of the Standard Oil Company or at their request. It was simply my way of doing business. I thought it was cheaper in the long run to make the price cheap and to be done with it, than to fritter away the time with a competitor in a little competition. I put the prices down to the bone." [50]

His independence is borne out by the reaction of Standard executives. Rockefeller, in 1881, complained of Carley's "want of balance." [51] In the same year Colonel W. V. Thompson suspected that independent oil was going into Chess-Carley territory with the connivance of Carley who had made a deal against Standard's interests.[52] The fact that Chess, Carley & Company was successful in maintaining a virtual monopoly of the petroleum in the southeastern section of the United States, however, postponed any serious attempt to bring the company under closer management control of Standard until 1886 when Carley's interests were bought out.

At least one trade journal, *The Petroleum Age,* not particularly favorable to Standard, indicated Carley's competition had not "only been conducted to the advantage of the retail dealers and the consumers but to the railroads and newspapers as well." Noting that the press of the "southwest country" took a lively interest in the contest, the editor went on to say, "It awakens the rural public and calls its attention to the matter." This was of particular importance for stimulating future sales as "this 'rural public' of the South and South-West has not yet passed its 'teens" in oil intelligence. A glass lamp is [still] a luxury kept on the 'spare room' table as an ornament, the light rendered the visiting neighbors being shed from the old fashioned . . . tallow candle or dip." [53]

Waters-Pierce & Company

Henry Clay Pierce, started in the oil business shortly after the Civil War by taking over the distribution of oil from the refinery of his father-in-law, John R. Finlay, the "owner of the first refinery built west of the Mississippi." [54] In 1871 at the age of twenty-two, he bought out his father-in-law and two years later joined with W. H. Waters, a St. Louis businessman, in forming Waters-Pierce & Company.

While refining operations were continued at the St. Louis plant until 1878, Pierce devoted his principal attention over the intervening years to making Waters-Pierce & Company the dominant marketer of petroleum products in the area south and west of St. Louis. An increasing amount of the organization's supply of refined products was purchased from outside refineries, chiefly from Standard's refinery at Cleveland.

At the outset, Pierce faced quite a different marketing situation than his contemporary, Carley. The latter was selling in a more thickly settled region that, even by 1874, was served by some 14,000 miles of railroads in addition to water transportation to ports on the south Atlantic and Gulf of Mexico. Carley's major problem, which he solved so effectively, was to change the buying habits of a well-entrenched group of jobbers and to shift the major distributing channels from water to rail.

By contrast, the region outside St. Louis that Pierce blocked out as his marketing territory, including southern Missouri, Arkansas, western Louisiana, Texas, and the Oklahoma Territory, was still in the frontier phase of settlement. Its population was widely scattered, while the railroads had barely made a start toward providing adequate transport facilities. But this area was attracting the forerunners of those who participated in what has been described as,

> the greatest movement of peoples in the history of the United States. Millions of farmers, held back for a generation by the forbidding features of the Great Plains, surged westward between 1870 and 1890. They filled Kansas and Nebraska, engulfed the level grasslands of Dakota, occupied the rolling foothills of Wyoming and Montana, and in a desperate bid for vanishing lands, elbowed Indians from the last native sanctuary in Oklahoma.[55]

Between 1874 and 1884 this migration helped bring about an increase in the population in the area between the Mississippi and the Rockies from about 7.5 million to over 11.5 million.*

Pierce apparently was quick to appreciate the fact that the area was on the verge of a period of rapid growth; that a further expansion of railroad mileage would not only bring transport facilities closer to the existing population but would attract further settlement.

His plan to capture this market had two main features. The first was to prevent rivals from gaining a foothold by extending the marketing facilities of Waters-Pierce & Company as rapidly as potential sales appeared to warrant, for as he put it, "I knew that otherwise I would not be able to control the trade." [57]

Thus, starting in 1869, Pierce "personally . . . accompanied shipments from St. Louis radiating within the [Waters-Pierce] sales territory, . . . located stations, attended personally to [their] construction, the selection of agents, and the bases of sales and the organization of the local business . . . as the railroads extended throughout the territory." [58]

Secondly, Pierce buttressed his marketing position by introducing and expanding bulk-handling methods to the distribution of refined products.

---

* Data on the regional growth of population between 1874 and 1884 are estimates based on census reports for 1870, 1880, and 1890.[56]

His initial interest in bulk shipments by rail was prompted by the savings on the transport of oil between the chief refining centers to the East and St. Louis. By the time barreled oil had covered the more than 550 miles by rail from Cleveland or the more than 600 miles from Pittsburgh, it had suffered a heavy loss from evaporation and leakage. Pierce's volume of shipments was sufficient to assure him favorable treatment by the railroads in respect to rates and the co-operation of Standard in making tank cars available.

But Pierce's interest in the elimination of wooden barrels went well beyond the use of tank cars. There seems little doubt that he was, as he claimed, "the first person to invent and have manufactured and put into use tank wagons for the distribution of oil by having horses haul the wagons; also the first iron barrels ever used in the country . . . long before these appliances were ever handled by anyone else." [59]

While these methods of distribution were apparently first introduced in the St. Louis area by the mid-1870's, Pierce pushed their use vigorously throughout the firm's marketing territory. Wherever the volume of sales justified, bulk stations were installed equipped with "warehouses, . . . storage tanks and other arrangements for the storage of oil; tank cars for the distribution of oil between St. Louis or other initial point of shipment stations; tank wagons in which to locally distribute the oil and iron barrels and smaller metal receptacles for the transportation and delivery of oil within the radius of, say, less than 100 miles from each distributing station." [60]

By 1878, Waters-Pierce & Company had developed a far-flung marketing organization for the distribution of petroleum products in southern Missouri and the four states to the south and west and had extended sales across the border into Mexico. Pierce, on the whole, had been successful in his plan to exclude other jobbers from establishing themselves in this area and most of the firm's sales were made to dealers or in outlying areas directly to retail customers. Like Chess-Carley, Waters-Pierce offered a full line of refined petroleum products including illuminating oil, greases, lubricants and the like, most of which carried the firm's trade-marks.

A description of the company's main bulk station in St. Louis in 1878 gives some impression of the scope of business at that time. Covering a city block, the plant's equipment and facilities included

iron storage tanks, brick buildings, machinery for the manufacture of tin cans for receiving, storing and shipping oil; the manufacture of iron barrels for similar purposes; the manufacture of wooden barrels; the manufacture of wooden

cases to receive and ship the tin cans; the manufacture of tank wagons for local distribution of oils and machinery for the manufacture of axel grease; for
the manufacture and compounding of all kinds of lubricating oils, and machinery, appliances, and facilities for receiving, storing, handling and manufacturing and shipping practically all articles the Waters-Pierce Company
handled to the extent that business was transacted in and from St. Louis.[61]

Increased purchases by Waters-Pierce of its refined products from
Standard between 1873 and 1878 had strengthened the ties between the
two concerns. In 1878, Rockefeller suggested to Pierce that these relations be formalized under an agreement whereby Standard would acquire
financial interest in Waters-Pierce & Company with the understanding
that Pierce was to continue "to control and direct its policies and operations." [62] Waters-Pierce & Company was, accordingly, reorganized as the
Waters-Pierce Oil Company. Standard acquired a 40 per cent interest in
its stock and, interestingly enough, Chess, Carley & Company a 20 per
cent interest.

Pierce was proud of the fact that throughout the 1880's and 1890's Waters-Pierce sold between 90 per cent and 98 per cent of the petroleum
products distributed in the company's marketing territory. This record, he
insisted, was due "entirely to the better facilities afforded by the Waters-
Pierce Oil Company to the consumer than competitors could furnish, and
to very close attention to the prompt delivery of the best quality and
fullest measure of oil." [63]

There was a considerable element of truth in his claims but those, like
George Rice, who attempted to sell in Pierce's territory were inclined to
put more emphasis on the competitive practices he used to stifle competition. A letter from a Pine Bluff, Arkansas, dealer dated January 11, 1882,
stated,

> While the merchants here would like to buy from some other than the
> Standard, they can not afford to take the risks of loss. We have just had an
> example of 100 barrels opposition oil which were brought here, which had
> the effect of bringing the Waters-Pierce Oil Company's oil down from 18¢
> to 13¢, less than cost of opposition oil, with a refusal on their part to sell to
> any one that bought from other than their company.[64]

Another Pine Bluff dealer wrote on May 26, 1882, referring to the Waters,
Pierce & Company, "Their agent was here several times again yesterday
and called on us. He wished to make a contract with us and four or five
other large dealers, and failing to do so in short time, he threatened opening a retail house here, put oil down from 5¢ and 10¢ retail." [65] And a

dealer in Hot Springs, Arkansas, wrote on January 16, 1882, that having stopped buying his oil from Waters-Pierce he expected to "have a coal-oil war here very soon. Their agent has made threats to some of our merchants that they must or shall buy oil from them and no one else, or, otherwise, they would come here and ruin them by fair means if they could, by underhand ways if necessary." [66]

The fact that Pierce held only 40 per cent in the stock of the Waters-Pierce Oil Company had little effect on his domination of the company. In many respects Standard found him more difficult to control than Carley. One executive warned Rockefeller in 1882, "I believe Pierce is one of the most unsafe men we have connected with us today." [67] Another recognized his brilliance but observed that he "wouldn't play ball with the crowd and he liked to pull fast ones . . . he was polite and cordial enough but when he got into a jam with people . . . they knew they were fighting a Tartar. He was the nastiest fighter you ever saw." [68] The saving fact, from Standard's point of view, was that, like Carley, Pierce made large profits. Referring to the net profit of Waters-Pierce in 1885 which came to nearly $340,000 Colonel Thompson observed to Rockefeller that this amount reflected "very extraordinary results and is evidence of great talent in handling the situation." [69]

Other Standard marketing affiliates

Two other important independent midwestern wholesalers caught Rockefeller's eye during the late 1870's: Alexander McDonald & Company of Cincinnati and Chase, Hanford & Company of Chicago. Alexander and James McDonald, owners of Alexander McDonald & Company, began late in the 1860's as Cincinnati oil dealers. Their operations spread and by 1878 they were selling to customers in the southern portions of Ohio, Indiana, and Illinois. [70] In that year they joined Standard of Ohio in forming the Consolidated Tank Line Company, which took over Alexander McDonald & Company under an agreement whereby the brothers retained management control and 50 per cent of the stock of the new company, with Standard holding the rest. [71]

Consolidated began immediately to push its operations across the Mississippi to cover the states to the north of Waters-Pierce.* In 1878 a bulk station was established at Hannibal, Missouri, to serve as a distribut-

---

* Standard's acquisition, in 1878, of an interest in Waters-Pierce, which it shared with Chess, Carley & Company, made this division of sales territories possible.

ing center for northern Missouri, Kansas, and the Missouri River points, including Kansas City and other centers to the northwest. To supply customers in Iowa and the South Dakota territory, Consolidated Tank Line Company the same year purchased the business of L. J. Drake of Keokuk, Iowa, and reorganized it as the Keokuk Oil Tank Line Company with Drake as manager. Drake had pioneered bulk handling in Iowa by establishing a bulk station at Keokuk in 1874 where he began supplying jobbers with barreled oil shipped in by tank car from Standard's refineries in Cleveland.

With these stations as a nucleus, Consolidated Tank Line began to expand its marketing facilities in its sales territories. In 1882, for example, a warehouse was established at Des Moines for the storage of oil barreled at Keokuk. A year later storage tanks and barreling facilities were added to receive tank shipments direct from Cleveland, and in 1884 the first tank wagon, handling 12 barrels of oil, began delivering to the local trade in Des Moines.[72]

Consolidated Tank Line Company appears to have followed Carley's example of concentrating on supplying the jobbing trade. But the McDonalds were as aggressive as Carley or Pierce in meeting competition. George Rice, for one, found it just as difficult to sell in their territory as to compete with Carley or Pierce. For example, his agent in St. Joseph, Missouri, wrote Rice on March 3, 1881,

> The tank lines [Consolidated Tank Lines Company] sent a man here—a special man—some days since; hence the cause of oil from January 28 shipment being sold so low. He went to three of the houses and got them to agree not to deal with me longer, as they only made 25¢ a barrel buying of me, and that he (the Tank) would make [a] jobbing rate at 17½¢ [a gallon for] . . . 150° [oil] and 13½¢ [for] 112° [oil], and give them $1.50 per barrel off from that, which would be 3¢ a gallon off, being 10½ and 14½ net to them.[73]

Standard too found the brothers difficult to control, Rockefeller in 1883 being suspicious that Consolidated Tank Line Company was "not selling oil at prices decided by the central office." [74]

P. C. Hanford & Company, formed in 1879, represented the merger of two firms: Chase, Hanford & Company, and Kenly & Jenkins, which began as oil brokers and dealers in the Chicago area in the mid-1860's.*

---

* Chase, or Hanford was variously listed in Chicago business directories between 1866 and 1873 as: oil brokers; oil manufacturers and dealers; oils and paints; oil, wholesale and machinery; and oils, manufacturers and wholesale dealers. Kenly & Jenkins listed themselves variously as oil brokers; oil and drug brokers; oil manufacturers and dealers; and oils, manufacturers and wholesale dealers.

Within a decade they had become the leading wholesalers in the Chicago area and were buying most of their petroleum products from Standard. From Standard's point of view, however, the Chicago marketing situation was "little short of chaotic" with "irresponsible" jobbers, including Kenly and Jenkins whom Standard executives considered "unreliable," all too willing to sell poor quality and highly explosive illuminating oil to unwitting purchasers.[75]

Early in 1878, Standard suggested a merger of the two firms to be headed and controlled by P. C. Hanford with the "unreliable" Kenly and Jenkins as salaried employees. Kenly and Jenkins "proved difficult," and it was not until May, 1879 that P. C. Hanford & Company was formed with Standard holding 51 per cent of the stock.[76] The sales territory blocked out for the new company included northern Illinois and Wisconsin.[77]

In contrast to Chess-Carley, Waters-Pierce, and the Consolidated Tank Line Company, who owned their own bulk stations, Hanford & Company sold oil barreled at stations installed and held by Standard of Ohio at Englewood (Chicago), Milwaukee, and Rock Springs, Wisconsin.[78] It followed the familiar pattern of selling to dealers in the urban centers of Chicago and Milwaukee and to jobbers for distribution in the smaller cities and towns. In part because of "the poor roads from the bulk station [at Englewood] to the center of the city," Hanford was apparently slow in pushing tank wagon deliveries and it was not until 1883 that they were introduced into Chicago.[79]

To round out its national coverage of wholesaling in the United States, Standard formed two new marketing companies in 1884. One was the Continental Oil Company of Iowa, which combined the bulk stations and marketing facilities of Ohio Standard and the formerly competitive Continental Oil & Transportation Company in the Rocky Mountain area.[80] The second was Standard Oil Company of Iowa, which took over from Ohio Standard the responsibility for distribution in the Arizona territory, Idaho, Nevada, the Pacific Coast states, western Canada, and Alaska.[81]

The 1874–84 period in review

Expanding domestic markets and changes in methods of distribution were but one phase of the rapid growth and development of the American petroleum industry from the mid-1870's to the mid-1880's. It was during these years that discoveries of new fields, many miles to the northeast and to the southwest of Oil Creek, lifted production of crude from just under 11 million barrels in 1874 to over 30 million barrels in 1883.*

This rapid expansion in output, particularly as Bradford reached its peak of production early in the 1880's, was reflected in the prices of crude. Quotations at wellheads, which averaged over $2.50 a barrel in 1876, fell sharply over the next three years, ranging after 1879, between $1.05 and 78¢ a barrel.† Many producers, not unexpectedly, found it difficult to adjust their operations to lower levels of crude prices. But attempts to stop the drill through organized shutdown movements were generally unsuccessful. The lure of new fields and the possibility of striking large producing wells made it difficult for any sustained efforts to restrict production to be effective.

Yet it was the pressure from new discoveries coupled with lower prices that provided the dynamic charge for the expansion of the industry. They furnished the major incentive for the spread of better drilling techniques, improvements in the gathering and transport of crude, for expanding and enlarging refinery installations, and the exploitation of new uses and markets for refined products.

In production, for example, the experience and judgement gained during the years following Drake's discovery, began to bear fruit after 1874. Wells could be dug faster, deeper, and for the most part more cheaply and efficiently than had been true earlier. In transport, gathering lines and storage facilities were—with the exception of the brief months of "immediate shipment" at Bradford—extended to new fields with sufficient speed to minimize the wastes commonly associated with initial "flush production."

The major innovation affecting the transport of crude (and without doubt one of the outstanding industrial achievements of the period) was of course, the development of the long-distance crude pipeline. As a means of moving crude quickly, efficiently, and more economically to refineries far removed from producing centers, crude trunk lines were a tribute to the inventive talent and experimental drive of imaginative oilmen.

Growth in processing facilities in response to an ever-increasing flow of crude was scarcely less impressive. Total refining capacity (measured in terms of annual working capacity) grew about threefold, from about 10.5 million barrels of crude in 1873 to just over 28 million barrels in 1884.‡

---

* See Chart 15:1, p. 373.
† See Chart 15:1, p. 373.
‡ See Table 18:6, p. 486.

A portion of the new capacity owed its existence to a quantitatively small, but growing demand for petroleum by-products, particularly lubricants.

In addition, there was a further trend toward larger refinery installations. In 1873, plants with a daily charging capacity of 2,000–3,000 barrels of crude were the exception. A decade later plants of this size were not uncommon, and there were a number of installations capable of processing 5,000–6,000 barrels of crude daily. Closely associated with larger refining installations were economies in processing which, along with lower prices of chemical treating agents, cut refining costs nearly in half during the period.

One further technological innovation marked the 1873–84 period. This was the introduction of bulk methods of handling refined products in the domestic markets, foreshadowing substantial future economies in distribution. But as yet these methods had not been extended to the two-thirds or more of the output of American refineries sold abroad.

Larger population, rising material incomes, and the spread of industrialization were basic to the expanding demand both at home and abroad for petroleum products during 1874–84. But it is doubtful whether in the absence of an approximate 40 per cent decline in prices during these years (made possible by cheaper crude, more efficient means of transport, and reduced processing costs), that this alone would have been sufficient to provide a market for the annual sale of refined products that grew from about 7.65 million barrels in 1873–75 to almost 18.5 million barrels in 1883–85.

No feature of the development of the American petroleum industry during 1874–84 was more striking than the changes in its competitive structure brought about by John D. Rockefeller and his associates. On the eve of the South Improvement Company scheme, Standard's activities were largely confined to the operation of its refineries in Cleveland and New York, representing about 10 per cent of the industry's total refining capacity. It was still relatively easy for anyone with moderate capital resources to enter any phase of the oil business in competition with many other individuals or firms engaged in the same type of operation. A decade later the situation was fundamentally different. There was no part of the oil business from the production of crude to the marketing of refined products free of the influence or dominant control of the Standard Trust.

Standard's basic position at this time rested upon ownership of between 85 and 90 per cent of the industry's expanded refining capacity and control over an equally high proportion of gathering lines, storage tanks, tank cars, and crude trunk lines which moved crude from wellheads to markets. Producers, who still numbered in the hundreds, had no choice but to sell all but a small portion of their crude to Standard agents. Foreign distributors likewise were compelled to purchase the bulk of their requirements from Standard's refineries, while the Trust's affiliation with leading American distributors limited the opportunities for independents to operate within the domestic market.

Standard's position in the industry by mid-1880's reflected Rockefeller's unusual ability to organize his financial and managerial resources to achieve his goals. Starting with the immediate objective of capturing the bulk of the industry's refining capacity, he was aware at the outset of the importance of transport costs in maintaining successful refining operations. He showed great skill in exploiting the rivalry between the trunk line railroads to obtain favorable rate concessions, meanwhile strengthening Standard's transport position by acquiring tank cars and gathering line facilities. Standard's growing control over transport and refining capacity did not go unchallenged, particularly by Empire and Tidewater. But defeat of the Pennsylvania Railroad's affiliate and rapid construction of long-distance crude pipelines after their feasibility had been demonstrated by Tidewater left Standard virtually without competitors in the transport and refining segments of the industry.

No less impressive was Standard's achievement early in the 1880's of streamlining its internal organization and consolidating the operations of the conglomerate of companies and plants acquired since the early 1870's. The administrative reorganization culminating in the famous Trust agreement of 1882 and the capitalization of its buildings at over $70 million, signalled the status of Standard, not only in petroleum, but as a giant among the world's industrial organizations.

By mid-1880's the members of American petroleum industry, including Standard, had reason to be somewhat concerned over their future prospects. There were growing signs that output of crude from the Appalachian fields, which had sparked developments of the preceeding decade, was beginning to taper off. It is true that oil had been discovered many places elsewhere in the United States, but the extent of these fields, not to mention the problems of getting the crude to market, presented many uncertainties. Equally disturbing was the emergence of the Russian industry, with its reportedly unlimited supplies of crude, as a competitor of the American industry in world markets.

For Standard, with its heavy investments in transport and refining facilities, the prospects of declining supplies of domestic crude and growing competition in foreign markets were particularly disturbing. These possibilities were the occasion for a re-examination of the management policies which had proved so successful in bringing Standard to its dominant position in the industry. More specifically, the main issues were: first, whether Standard should extend its operations into the production of crude; and second, whether it should take a more active role in the marketing of refinery products. The majority of Standard's executives were reluctant to do either. How they were forced to reconsider their position will be brought out in suceeding chapters.

Part SEVEN

Integration and reaction
to foreign competition:
1884–1899

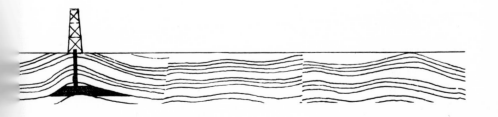

Chapter 21

The Appalachian fields and the rise of the
Pure Oil Company

$\mathbf{A}$ s the year 1884 came to a close, producers in the Pennsylvania Oil Regions could look back on nearly a quarter-century of uninterrupted development. To be sure, there had been periodic upsets and innumerable crises, but in general drilling had pushed forward with unending vigor. From the early beginnings around Oil Creek, petroleum territory now covered the northwestern corner of Pennsylvania and was already being extended further southwest. Soon production would spread over the entire western half of the state, from McKean County in the northwest-central section to Greene County in the extreme southwest tip of the state bordering West Virginia on the west and south.

By 1884, output of crude production had already been recorded in states beyond Pennsylvania, including Ohio and West Virginia, which had experienced an early but short-lived brush with petroleum production, and Colorado, California, Kentucky, and Tennessee. Already the seeds had been sown for future developments in California, Texas, and Oklahoma, but except for the famous Lima field, some time would elapse before these and other areas outside Pennsylvania would be developed on a commercial scale.

Within Pennsylvania, production moved steadily southward during 1885 and 1886 as the Butler and Armstrong districts were extended. Bradford, although past its peak, continued to supply the largest proportion of the Pennsylvania output, while in Butler, Warren, Clarion, and Venango counties there were widespread efforts under way to extend known producing districts. During 1885, for example, Thorn Creek and Baldridge in Butler County attracted particular attention; operations in the Clarendon and Tiona pools in Warren County were prosecuted vigorously. Cogley Run, in Clarion County, and Red Valley, in Venango County, were the new developments of the year, although neither assumed very great proportions. During 1886, Shannopin in Beaver County, Kane field in McKean and Elk counties, the Tarkill pool in Venango County, and the Grand Valley district in Warren County supplied new productive arenas. Beyond all doubts, however, it was the opening up of fields in Washington County, and later in Greene County, that brought the most significant additions to production in the Appalachian Regions since the great Bradford field.

The pioneer well in these fields was drilled, ostensibly for gas, by the Citizen's Fuel Company of Washington, Pennsylvania, on the Gant's farm in Washington County. Late in December, 1884, at a depth of 2,200 feet, a small showing of oil was encountered, and after two small flows during early 1885 the well was put on the pump. Encouraged with their first success, the Citizen's company selected another drilling site and in August, 1885, brought in the first full-fledged gusher in the Washington field. Early the following year the increase in exploratory activity engendered by these two wells began to materialize. In March, the Pew and Emerson "Manifold" well came in as a heavy producer, and in the following month the Thayer well was completed and initially gauged at 200 barrels daily.

From this point development was rapid. By May, 1886, daily output reached approximately 4,000 barrels; by June it had risen to 10,000, while by October daily production was pressing the 20,000-barrel mark.* It was

---

* Wells drilled and completed during 1886 numbered 254, of which 27 per cent (71), however, were dry holes. This percentage of dry holes was about twice as large as the over-all Pennsylvania rate.[1]

only a short time before production was extended from Washington into Greene County down the Ohio River and through the mountainous wilds of West Virginia.[2]

Output from the fields in Washington and Greene counties was sufficient to raise total production from the Appalachian fields from about 21.5 million barrels in 1885—the lowest total since 1879—to over 26 million barrels in 1886, and to just under 23 million barrels in 1887. But any satisfaction on the part of producers over the revival of lagging production was tempered by the fact that, although the number of wells drilled during 1886 and 1887 increased by nearly 50 per cent, the percentage of dry holes more than doubled. Moreover, the average yield per well-day, which fell to 2.8 barrels in 1885, rose only to 3.5 barrels in 1886, primarily on the strength of developments in Washington and Greene counties.[3]

In themselves, the higher proportion of dry holes and lower average yields per well were sufficient grounds for apprehension in the Regions. But of much more immediate concern to producers was the condition of the markets for crude. Despite the decline in production in 1885, crude inventories, after only a slight drop, leveled off at a figure of about 33.5 million barrels during 1886 and early 1887. Even more distrubing was the fact that the price of crude at the wells, which had averaged about 83¢ a barrel in 1884, declined to approximately 62¢ a barrel by mid-1887.[4]

In seeking a cause for their difficulties, the producers not unexpectedly found it in the operations of Standard's National Transit Company, which provided the gathering lines, storage tanks, and trunk-line facilities through which the great bulk of crude produced in the Regions moved on its way from the wells to the refineries.

The producers had no complaints about the procedures utilized by National Transit in handling oil through its facilities. Under practices dating back to the beginnings of bulk shipments of crude, oil was transferred from the producers' small settling tanks, a short distance from the wellheads, into National Transit's gathering lines and storage tanks located at the terminal points of the crude trunk lines. As soon as the oil was gauged, producers received a credit balance on the books of the pipeline company against which National Transit issued negotiable pipeline certificates in units of 1,000 barrels. Subject to instructions from the owners of pipeline certificates, National Transit held the oil in its tanks free for thirty days, thereafter charging a monthly fee for storage.

While National Transit's gathering line charges of 20¢ a barrel, regardless of the length of the haul, were considered higher than necessary to cover costs and risks, a crude inventory in excess of 30 million barrels focused the producers' main attack on the allegedly exhorbitant level of storage charges, 1¼¢ per barrel per month, and its effect on the price of crude. By the end of 1886 producers were generally convinced that a revision of the Pennsylvania laws affecting pipeline charges in general and storage rates in particular offered the most promising means of improving the price of crude.

The Billingsley bill

In focusing their attention on storage rates, the producers were far from clear in explaining just how lower charges, which were paid by the purchaser of oil, would be reflected in the price of crude at the wellheads. The *New York Tribune* summarized one of the more reasonable explanations: "The public, as a rule, prefers to operate for a rise, and yet there have been such disastrous experiences on that side that it has become a common saying: 'There's no use buying oil with the carrying charges eating you up.'" [5]

In other words, it was assumed that lower storage charges would raise prices, at least for a short time, in the oil exchanges by reducing the pressure to sell existing inventories of crude and by encouraging speculators to buy more oil to hold for future sale and delivery. But this line of argument failed to take into account Standard's role as the purchaser and refiner of all but a fraction of the crude produced in the Regions.

It was true that the Seep Purchasing Agency, which assumed responsibility for supplying Standard's refineries in 1884, obtained its crude either by acquiring pipeline certificates in the exchanges or by direct purchase from producers at prices fixed midway between the high and low quotations of the exchanges on the day of the purchase. [6] But the prices Standard was willing to pay for its crude reflected the management's estimate of the refined-product market and the rate of profits it was willing to accept. If prices in the exchanges rose above this level, the Seep firm, as the principal buyer of crude for immediate delivery, could bring pressure on the market simply by reducing its purchases.

Producers generally made little effort to spell out their economic reasoning. Thoroughly convinced that National Transit was getting rich at their expense, they were quite certain that if its earnings could somehow be reduced, their own incomes would automatically increase. This was the underlying assumption behind a bill introduced in the Pennsylvania Legislature in January, 1887, by Representative James Billingsley, of Washington County.

Up to this time only those pipelines which had used the free pipeline act to acquire land for their rights-of-way were required to act as common carriers. The Billingsley bill sought to extend this requirement to all pipelines, with some additional twists of its own. Under the proposed legislation all pipelines operating in Pennsylvania were to fulfill all requests to connect and to deliver oil to any point designated by the owner, regardless of distance involved or the volume of oil pumped.

From Standard's point of view this requirement was not particularly onerous, for National Transit had generally attempted to extend its gathering lines to all producing wells. But the proposal would have made it impossible to stop servicing marginal wells. Of major concern to the Standard management were the provisions of the bill providing for a close regulation of charges, particularly as there had been little or no systematic investigation for determining "reasonable" rates.

Specifically, under the bill, transit rates within any pipeline division (gathering lines or trunk lines) were not to exceed 10¢ per barrel. Storage charges were to be cut from the current figure of 1⅔¢ to just under a half cent per barrel per month, while deductions from the volume of stored oil for waste and sediment were to be limited to one-half of one per cent, rather than the prevailing three per cent per year.[7]

A bill designed to impose public regulation on pipelines could hardly have avoided controversy, particularly when its title began with the words, "An Act to punish . . . pipelines." But the debate soon degenerated to a melee of confused logic and personal villification.* Some independent refiners who favored cuts in storage rates, but did not accept the Billingsley bill *in toto*, were falsely accused of having sold their works to Standard.[9] The few small independent pipelines noted that many of the producers supporting the bill were nonetheless willing to switch from independent to National Transit lines when the latter offered a premium.[10]

Concomitant with these personal clashes ran a debate on more fundamental issues, independent of the personalities and motives of the participants. The *Titusville Morning Herald*, warning against "arraying the mandates of law against the laws that govern trade," compared the original bill to the "suicidal and short-lived granger legislation of the West."[11]

Standard Oil Counsel S. C. T. Dodd applied the same comparison to

* According to Patrick Boyle, pro-Standard editor of the *Oil City Derrick*, State Senator Lewis Emery supported the measure because Standard Oil had refused his offer to sell his refinery and "friendship" for $750,000. Emery indignantly denied the charge, although he did admit that he and his partners had named a price when Standard approached them in the fall of 1886, and had later refused a second Standard offer of a secret bonus arrangement.[8]

an amended version of the bill which, although it no longer required the lines to fulfill all requests to connect, he thought, reflected the failure of the original measure to allow reasonable charges. He concluded by contrasting the "simple and exact justice" desired by Standard with the "prejudice, passion, revenge" of "thousands . . . who would be benefited if they could obtain valuable service for nothing." [12] Spokesmen for the other side pointed out that laissez-faire presumed both parties to an exchange had adequate alternatives elsewhere—conditions hardly fulfilled in the pipeline case:

> The warehouse man, like the common carrier, especially when he enjoys a monopoly . . . does not meet his patrons upon equal footing, and for that reason the law does not allow him to stand upon his vantage ground and say to the public: 'You must pay me so much or go your way.'. . . he is held to 'exercise a public employment,'. . . and is held to certain duties to the public. Among those duties is to store whatever he holds himself out to store for a *reasonable* hire.[13]

But beyond mere assertion, however, neither side presented much evidence to justify its claims about the economic effects of the bill. Under the amended version, National Transit could have charged about 40¢ per barrel for piping from wellhead to Philadelphia.* This would have been about one-third less than prevailing rates (including 20¢ gathering charges); on the other hand, it would have been four times more than the pipeline's maximum revenue on oil shipped coastward for the Pennsylvania railroad under the contract of 1884. The latter, however, may not have reflected full unit costs.[15]

The proposed reduction in storage charges would have doubled the pay-out period of investment in a 35,000-barrel tank from two to four years.† This does not seem unreasonable, since the investment in much of the existing iron tankage had been recouped years before. Nor did it seem to impose impossible barriers to future construction; for even if particular producing areas ran dry before the end of the pay-out period, tanks could be connected to newer regions where storage facilities were short.‡

---

* Fifteen cents within a 50-mile radius from the receiving point, plus 6¢ for each 50 miles (or fraction) thereafter.[14]
† The reductions in storage charges were virtually the same in both the original and amended versions of the bill. Contemporary estimates placed the cost of constructing iron tankage at 25¢–30¢ per barrel.
‡ Perhaps the acme of mobility was attained in the late 1880's, when unused tanks were moved all the way from Bradford, New York, to the incoming Ohio and Indiana fields. See Chapter 22.

Almost everyone agreed, however, that the bill would apply only to oil produced in Pennsylvania and transported to refiners within the state.[16] This limitation would have created operating difficulties in separating inter-state from intra-state shipments, for the oil lost its identity once the pipeline accepted it. Even a Philadelphia refiner might receive oil produced in New York. But more than this, the limitation would have left about 75 per cent of refinery throughput beyond the jurisdiction of the act.* Only federal, rather than state, legislation could hope to accomplish the aims of the Billingsley bill.

Advocates of the bill never really came to grips with this problem. Instead, they reverted to the sectional argument which had become traditional in intra-industry conflict: namely, that if the lines running outside the state either failed to reduce rates or raised them, refining capacity would be driven to Pennsylvania.[18] But this was at best a remote possibility so long as the principal pipeline companies and refining capacity remained incorporated as a part of the Standard organization.

Soon after the bill had been introduced, a producers' committee went to Harrisburg to lobby and advise the Legislature on revisions.[19] About a month later, Colonel John J. Carter, a leading producer, organized a separate group seeking a compromise with National Transit. In part, Carter's negotiations were successful: the pipeline reduced storage charges to 25¢ per 1,000 barrels per day and lowered its deductions for waste and sediment from 3 per cent to the 2 per cent proposed in the amended version of the bill.

Although supporters of the bill hailed Standard's action as an admission that prevailing rates had been too high, they were far from satisfied with the concessions. Pipeage had not been cut, and National Transit's storage rates were still 60 per cent higher than the maximum that Billingsley proposed. Some thought that Carter, by dividing the producers, had unwittingly played into Standard's hands.[20]

In fact, Carter apparently only formalized a division of opinion that already existed among producers, many of whom favored a reduction in storage charges, but felt the proposed bill too radical and its probable effects too uncertain to command support. On the other hand, over 2,000 names appeared on a petition endorsing the measure, and Bradford pro-

---

* In 1890, Pennsylvania refineries processed 7.6 million barrels out of a total of 30.6 million barrels of crude throughput. Even this probably overstated the proportion that would have been covered by the Billingsley bill, for some of the oil used by Pennsylvania refineries came from New York.[17]

ducers (whose production constituted the bulk of existing stocks) passed resolutions rejecting Standard's compromise offer.[21]

In a pandemonium-ridden session of the Senate on April 28, the Billingsley bill, which had earlier passed the House, was defeated by a vote of 25 to 18. The *Philadelphia Times* attributed its defeat to the "wide difference of opinion among the oil producers themselves as to its effects," [22] but Emery and others attached more weight to bribes allegedly paid by Standard Oil. Although subsequent evidence indicates that money did indeed change hands, it also suggests that the money may have come from either or both sides.[23]

The Producers' Protective Association and the shutdown movement

Producers varied in their reaction to the failure of the Billingsley bill. Many, because of Standard's moderate concessions on pipeage and storage rates, were inclined to agree with the Philadelphia Times statement: "The oil men . . . have reason to believe they have gained a very substantial victory." [24] But the more militant producers gathered at Harrisburg viewed the outcome as a severe blow to their hopes. At meetings held at the state capital in the evening following the defeat of the bill, they once again urged a formal organization of producers as a means of opposing Standard.

Out of these meetings and similar gatherings held elsewhere in the Regions over the next few weeks, there emerged the Producers' Protective Association, which had as its chief objectives:

> To include in one organization of all producers of petroleum, and those who are engaged in industries incidental thereto, and known to be friendly to the producers' interests, in order that they may, by united action and all honorable means, protect and defend their industry against the aggressions of monopolistic transporters, refiners, buyers, and sellers of their products, in order that the producers may reap the just reward of their capital and labor, and to this end encourage and assist as far as possible the refining and marketing of their product and sale direct to the consumer by the producer.[25]

As the leaders of the movement traveled up and down the Regions soliciting membership, talk began to circulate concerning the possibility of a prolonged shutdown. A number of producers hoped that, although the new association was designed to set up refining and marketing outlets, it might also become the vehicle of such a movement. By the time the

Producers' Protective Association had adopted a constitution and established local assemblies in nearly every producing county in the Regions, this hope seemed fairly well grounded.

There were several large producers, however, among them T. W. Phillips, one of the original proponents of the Association, who objected to any shutdown agreement which did not include the largest buyer of crude oil, namely the Standard Oil Company. Unexpectedly the Standard management indicated its willingness to cooperate. At two meetings held at Saratoga and Niagra Falls, representatives of the Producers' Protective Association and Standard agreed on the terms of the now-famous contract signed by both parties on November 1, 1887.[26]

Various reasons, including a desire for better relations with producers and the public, have been advanced in explaining Standard's willingness to cooperate on a shutdown movement.* Whatever the public relations value of a more conciliatory attitude toward producers, it would appear that the controlling reason for entering the movement was Standard's concern over the long run domestic supplies of crude.[28]

Aware that "One of the greatest single threats to the security and longevity of the combination was the lack of assurance of the basic raw material, crude oil," [29] the implications of the growing number of dry holes and lower average yields per well in the Appalachian fields were not lost on the management of the Trust. To be sure, Standard had already moved into the Lima, Ohio, fields, which gave every indication of adding substantially to total United States production. But at best these fields could hardly offset any large drop in output from the Regions. Moreover, until a practical and economical method of refining the sulphur-laden Ohio product was developed, there was no assurance that it would supplement Appalachian crudes as a source of illuminating oil, lubricants, and other refinery by-products supplied by Standard to marketers at home and abroad.

Standard had another reason to be concerned about future supplies of domestic crude. By the late 1880's, the rapidly expanding Russian petroleum industry had already captured a substantial portion of the European

---

* John D. Rockefeller, in testifying before the New York State Trust Investigation, explained Standard's desire to align with producers in the following terms: "the inducement was for the purpose of accomplishing a harmonious feeling as between the interest of the Standard Oil Trust and the producers of petroleum . . . and we felt it to be in the interests of the American oil industry that a reasonable price should be had by the producer for the crude material, and we wanted to co-operate to that end." [27]

markets formerly supplied almost exclusively with American illuminating and lubricating oils and were extending their operations to the Far East.*

Thus for once in the long history of opposition between Standard and the producers there seemed to be an occasion for agreement. The Trust's interest in stimulating further exploration and drilling in the Appalachian fields by higher prices, even if it meant a temporary reduction in output, coincided with the immediate objectives of the producers for a better return for their crude.

At the meetings held to work out the details of the shutdown movement, Standard spokesmen emphasized the necessity of reducing the excessive inventories of crude. As David Kirk, one of the representatives of the Producers' Protective Association, put it:

> I might say that the Standard attributed all our troubles to existing stocks over supply; that we had a large stock on hand that was deteriorating in value—carried at vast expense that competed with us as producers, and until that was got out of the way we could not hope to get a fair price for our goods; and they convinced our committee and finally the committee convinced our assembly, that that was the trouble, and the best course was to get rid of that trouble first and take up the refining and marketing of our goods afterwards.[30]

Under the terms of the twelve-month contract signed by Standard Oil of New York and representatives of the Producers' Protective Association on November 1, 1887, the latter group was to encourage its membership to sign individual contracts binding them to stop new drilling or shooting of old wells and in general take such steps that would reduce average daily production some 17,500 barrels below the levels of about 60,000 barrels recorded during the preceding July and August. In return Standard agreed to set aside and hold 5 million barrels of "merchantable crude" at 62¢ a barrel to be sold or disposed of as the producers directed upon the successful completion of the contract.[31] Another provision of the original contract specified that 1 million barrels of the 5 million barrels was to be disposed of for the benefit of the workers thrown out of work by the shutdown; later Standard added another 1 million barrels for the same purpose.†

The reaction of the *Titusville Morning Herald*, which announced the signing of the contract with an emblazoned headline, "A New Era in Oil,

---

* See Chapter 24.
† The workers on their part quickly organized into what was called the Well-Driller's Union and it was through this agency that payments to idled workers was made.[32]

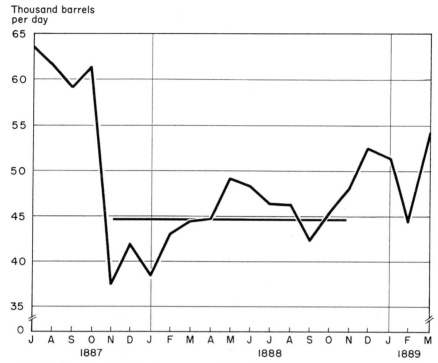

Thousand barrels
per day

CHART 21:1. *Monthly averages of daily production of crude from the Pennsylvania fields during the "1887–88 alliance." Straight line indicates the maximum daily output permitted under the agreement of November 1, 1887. (Sources:* Derrick's Hand-Book, I, 806; Pennsylvania Industrial Statistics, 1892, B85.)

The Great Shut-Down Movement Commences Today," was typical of reaction in the Regions.[33] Enthusiasm continued to run high throughout Pennsylvania oil country and after the contract had run a month it was generally acclaimed as a "shining success" and the "most rational alliance ever made" in the Oil Regions.[34] Another contemporary journal, more cautious in its appraisal, opined that the results of this "remarkable alliance had so far not been unsatisfactory." [35] But except for the usual summer-boom in drilling activity which even the most rigidly enforced shutdown could not fully curtail, the experience of one month under the agreement was to be repeated until the date of its termination on November 1, 1888.[36]

As shown in Chart 21:1, production was held within the prescribed maximum limits during eight of the twelve months covering the life of the contract. Considering previous attempts to cut back production, this was

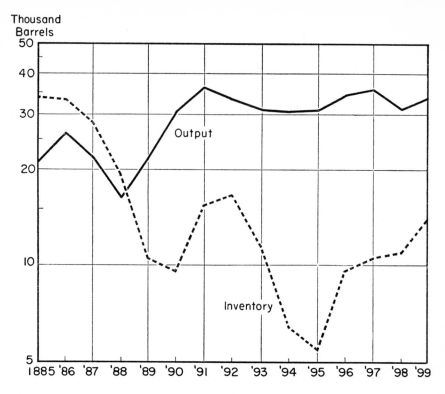

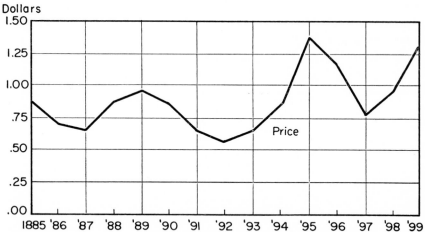

CHART 21:2. *Output, inventories, and prices of crude oil (per barrel), Appalachian fields, 1885–99. (Sources:* Mineral Resources, 1905, *815, 823, 852;* Derrick's Hand-Book, *I, 809–10.*)

a remarkable achievement, particularly in view of the developing fields in Washington and Greene counties. While these fields probably accounted for the summertime aberration from the prescribed maximum, the strength of the movement, which included nearly 85 per cent of the Regions' approximate 14,000 producers, was strong enough to resist the lure of a new and developing territory in its midst.[37]

It is also apparent from the data shown in Chart 21:2 that the objectives of reducing inventories of crude and raising the level of crude prices were also achieved. From a position of slightly over 32 million barrels in July, 1887, inventories declined steadily to a figure of under 20 million barrels in November, 1888. The movement of crude prices (pipeline certificates) was more erratic, rising slowly between November 1887 and the following March. By June these gains were almost wiped out as production moved up during the spring. But from the end of June until mid-September crude prices had increased on the average of 20¢, resting on the first of December, 1888, at approximately a 90¢ level.[38]

Just how much the members of the Producers' Protective Association and the workers received from the subsequent disposition of the 6 million barrels of crude credited to their accounts by Standard is difficult to determine. Rumors began circulating in the Regions at the close of the contract in November, 1888, that the Association might sell its oil at any time. Apparently about 500,000 barrels were marketed sometime during the next eight months, for on June 28, 1889, the Association's Executive Committee arranged to sell the final 3,500,000 barrels to Standard at 91½¢ a barrel.[39] A rough calculation, making allowance only for storage charges (25¢ per 1,000 barrels per month), would put the net return to the Producers' Protective Association from this sale at approximately $1,000,000.

What happened to the 2 million barrels set aside for the benefit of the workers is even more obscure. At least 1 million barrels were sold December 27, 1888, "at a net profit of 19¢ a barrel."[40] If the second 1 million barrels were sold on equally favorable terms, the total amount paid to the workers thrown out of work by the shutdown would have been in the neighborhood of $380,000. But whatever the amount, contemporary evidence indicates no dissatisfaction on the part of workers with their share.[41]

In large part the results Standard hoped to achieve by its participation in the shutdown movement were also realized. Its highly publicized cooperative venture with producers and the resulting rise in the price of crude won the Trust many friends both within and outside the Regions.[42]

More important, in terms of the management's longer-run objectives, within a month after the termination of the contract there was a sharp increase in drilling activity which resulted in over 6,000 new wells in 1889 and nearly 7,000 wells in 1890. With some variation this momentum continued throughout the 1890's with the annual number of new holes ranging between 2,600 and 7,500.[43]

The results of this activity were reflected in output from the Pennsylvania fields which, as shown in Chart 21:2, rose to over 30 million barrels in 1890 and averaged well over that figure throughout the remainder of the decade. At the turn of the century the Appalachian fields, despite a rapid growth in production in northwestern Ohio, were still supplying nearly 60 per cent of the total domestic output of crude.

Developments in the Ohio fields, however, were not without their effect on the fortunes of producers in the Regions. As will be noted in more detail in succeeding chapters, the major obstacles involved in refining Lima crude had been largely overcome by the early 1890's. As a result an increasing volume of the Ohio product began moving through the refineries in direct competition with Appalachian crude. This competition was heightened by the fact that while inventories of crude in the Regions after 1888 ranged only between one-half and one-fifth of current annual production, stocks in the Ohio field, already nearly 22 million barrels in 1890, were maintained well above current outputs for the following six years.

These developments, coupled with an increase in production from the Appalachian fields, had an immediate impact on the prices paid Region producers for their crude which, as shown in Chart 21:2, declined some 40 per cent between 1889 and 1892. It was not until market demand for refinery products began to catch up with the combined outputs of the Appalachian and Ohio fields that the price trend was reversed.

Declining prices of crude after 1889 soon brought the "honeymoon" between Standard and the producers to an end. Assured for the time being at least of large supplies of refinable Ohio crude, the management of the Trust lost much of its enthusiasm for stimulating output by maintaining a high level of crude prices. To the leaders of the Producers' Protective Association, it appeared that they should turn to the original purposes for which the organization had been founded. This was to compete with Standard by developing their own transport, refining, and marketing organization.

The Producers' Oil Company and The Producers' & Refiners' Oil Company

In deciding to move into transport, refining, and marketing the producers' group faced a formidable task. They could expect opposition from Standard at all levels, while the emergence of Russian oil added another potential competitor in the important export markets. There was the further problem of overcoming inertia and obtaining agreement on a plan acceptable to the membership of the association. As pointed out by the *Oil, Paint, & Drug Reporter* in the mid-1880's:

> The oil producers of today are thoroughly experienced in schemes in which they lose money. They no longer clasp hands, shed tears of sympathy and draw on their bank accounts to assist others in schemes to sell out to Standard. They have been too often betrayed.[44]

Finally the leaders had no definite program around which to mobilize support. When they subsequently did move, it was on a piecemeal basis.

Nevertheless, two events during 1890 served to convince the producers they should take some action. One was the decline in the price of crude from about $1.00 a barrel in January to 67¢ in December. The second came when Standard acquired control of the Union Oil Company. This purchase was particularly disturbing. Not only was Union one of the largest producers in the Regions, but its officers, H. L. Taylor and J. L. McKinney, had been members of the inner councils of the Producers' Protective Association.[45]

Early in 1891 the Producers' Protective Association began raising funds for a new company, and after about $500,000 had been subscribed, the articles of the Producers' Oil Company, Limited, were filed on June 9, 1891.[46] Producers' Oil, empowered to enter any segment of the industry, was formed under Pennsylvania's Limited Partnership Act. This act provided that each member was liable only to the extent of his equity and could participate in the company only by consent of a majority of the other shareholders. This form of organization was admittedly adopted to exclude possible Standard interference with the management and no one affiliated with the Trust was allowed to join the original subscribers.

By early 1892, Producers' Oil had laid a 15-mile gathering line from Coraopolis, a rail depot about ten miles northwest of Pittsburgh, to tap the newly-discovered McDonald fields on the Pennsylvania–West Virginia border. Fifty tank cars had been purchased and arrangements had been made with the Columbia Oil Company at Bayonne, New Jersey, to handle sales in the export market. But except for lower gathering charges, the producers found that, after paying the railroad charges on their tank cars, they made no more selling crude at the seaboard than in the Regions. For a time during 1892 they were even forced to sell a part of their crude for fuel oil in the Pittsburgh area.

It was to improve their crude marketing position that the management of Producers' Oil, during the summer of 1892, suggested an alliance with a group of independent refiners in the Regions. By the end of 1892 plans for a new company, the Producers' & Refiners' Oil Company, also a limited partnership, were completed. A part of the $250,000, subscribed jointly by Producers' Oil and the refiners, was used to extend the former's pipeline from Coraopolis north to Oil City and Titusville. Early in 1893 the refiners were receiving between 2,500 barrels and 3,000 barrels of crude daily over the line.[47]

Their participation in the Producers' & Refiners' Oil Company assured the refiners of 15¢ per barrel pipeage (as compared to National Transit's charge of 20¢),[48] a friendly source of crude, and the use of the fifty tank cars of the Producers' Oil Company. Of these, an assured supply of crude was probably the most important, since a number of the independents had previously relied on Standard's pipeline connections and even those owning leases seldom produced all of their refinery throughput. But sufficient throughput was not enough to maintain the independents in business and thus provide an assured crude oil market for the producing members of the alliance. Selling most of their product unbranded in the open market, the independent refiners were highly sensitive to any changes in costs or selling prices. Because of their location in the Regions they were particularly concerned about getting their refined products to markets as cheaply as possible.

The prospects for cheap transportation at this time were not encouraging. Following the passage of the Interstate Commerce Act of 1887, the Commission in September, 1887, had ordered the railroads to charge the same rates for barreled oil as they did for bulk oil, but allowed an added charge for the weight of the barrel.[49] The railroads accordingly established two sets of tariffs from the Regions to tidewater: 52¢ a barrel on tank car shipments and 66¢ on barreled cargo.[50]

The Regions' independents regarded this schedule as equivalent to a death sentence. Most of their east-bound oil went in barrels and even those who had their own tank cars used them principally in their western trade. The Pennsylvania was the only railroad with bulk cars among its rolling stock, but under an old contract these could only go to Standard-controlled yards in Jersey City.[51] Generally unable (or unwilling) to invest in more tanks of their own, the independents asked the Interstate Commerce Commission to rule that barrels should be carried free. The Commission so decreed in November, 1892, but the roads generally ignored the ruling and the refiners spent the next decade in the courts trying to get it enforced.[52]

The fifty tank cars owned by Producers' Oil were far from adequate to meet the transport needs of the refiners.* Under these circumstances a separate project, inaugurated by Lewis Emery, to build not only a crude pipeline, but also a refined products line, east from the Regions, aroused the interest of the Producers' & Refiners' Oil Company.

The United States Pipe Line Company

Following the defeat of the Billingsley bill in 1887, Lewis Emery had temporarily kept aloof from independent organizations. During the shutdown movement he refused to cut back production from the wells owned by his company, Logan, Emery & Weaver, unless either Standard or the Producers' Association guaranteed the firm's Philadelphia refinery sufficient crude to continue operations. Standard refused, but offered to buy the refining property which was sold to the Trust for $275,000.† Two years later, Emery built another refinery at Bradford, fed by a gathering line that by 1891 was delivering about 700 barrels per day.[55]

Although Emery shipped most of his refined oil to tidewater in tank cars, he was not satisfied with the margin that bulk rates left him. He knew that pipelines carried crude at still lower cost and concluded that they might do the same for kerosene.[56] The idea involved a gamble perhaps only second to that taken by Tidewater. Products pipelines had been used before, but only for short distances from refineries to railheads.

---

* The Mutual Oil Company at Titusville, for example, with but 450-barrel daily refining capacity alone kept thirty bulk cars running, as did the National Oil Works, with about the same capacity.[53]

† Interestingly enough, Emery had asked $750,000 for the works, including "good will" and the capitalized value of such locational features as a 350-foot frontage on the Delaware River. Prior to the shutdown, his opponents in the Billingsley fight had alleged that Emery's failure to obtain this amount in previous negotiations with Standard had led him to support the bill.[54]

The general opinion in the trade was that piping refined oil over several hundred miles would damage the product.[57]

Nonetheless by the summer of 1892 Emery had enlisted some ten other firms in the project. Among these were the Columbia Oil Company at Bayonne, the firm that had agreed to handle the exports of the Producers' Oil Company; Borne-Scrymser, a large refining and marketing concern at Elizabethport, New Jersey; and the Mutual, International, and National refineries, with a total of 1,500–2,000 barrels daily capacity in the Regions, already associated with the Producers' Oil & Refiners' Oil Com-

Lewis Emery

pany.[58] On September 20, 1892, the United States Pipe Line Company, capitalized at $600,000, was organized under the general incorporation laws of Pennsylvania.[59]

The plans of the new company called for the shipment of crude and refined oil from the Regions through separate, parallel 4-inch lines to Hancock, New York, about 100 miles from New York City; from there, a combined rail and water route was to complete the trip to tidewater. Excepting a few minor instances, the free pipeline act of 1883 saved the company serious right-of-way trouble in Pennsylvania and by December,

1892, the two lines were virtually complete except for a 150-foot span under an Erie Railway bridge near Hancock. (See Map 21:1.)

At Hancock, the Erie, allegedly at the instigation of Standard, posted armed guards to prevent the crossing of its tracks. Emery might have sought relief under New York's free pipeline act, but amendments in the early 1890's required, among other things, such close court supervision of condemnation proceedings that the work could have been tied up for months.[60] Consequently, he back-tracked some seventy miles to Athens, Pennsylvania, and projected his lines southeast to Wilkesbarre.[61] Following further skirmishes with railroad interests over rights-of-way, the lines reached Wilkesbarre in the spring of 1893. Each line could deliver up to 5,000 barrels daily to the Jersey Central Railroad, which had agreed to carry the oil over the remaining 170 miles to the coast.[62] Under an arrangement also completed about this time, Philip Poth, of Mannheim, Germany, and one of the leading marketers of petroleum products on the Continent, agreed to handle the bulk of United States' exports.

In June, 1893, the United States' pipeline delivered its first throughput of both crude and refined oil to Wilkesbarre.[63] The kerosene arrived undamaged by its journey and throughout the remainder of the year product shipments averaged approximately 2,000 barrels per day.[64] Both export kerosene and the higher-test water white (and, at least later, a third grade) traveled through the pipe with less than 50 feet of consecutive batches intermingling. At the terminal each grade was separated by a valve system that opened and closed lines to different tanks.[65] In proving that products could be piped over long distances, United States Pipe Line established a fact of major long-run significance. But the company's feat, even in the trade press, attracted relatively little attention, and no one was prompted to follow its example. Standard, with its refineries located at major marketing centers, had little need for a products line, while other independents lacked sufficiently concentrated markets to assure the throughput necessary for economical operations.

Although the United States Pipe Line Company encountered few serious technical problems in piping kerosene, it met with a number of legal and economic difficulties in effecting a consolidation with the Producers' Oil Company (including the Producers' & Refiners' Oil Company). The proposed consolidation had much to recommend it. The largest refiners in the Regions had subscribed both to the Producers' companies and to Emery's project, while the two pipeline systems complemented each other. But when Producers' Oil in 1893 tried to exchange its stock for shares in U. S. Pipe Line at par, the merger was blocked by Colonel John J. Carter, whose producing company was now Standard-controlled.

The point at issue was whether Carter, himself owner of 300 shares (of a total of 60,000) in Producers' Oil, could vote an additional 13,000 shares which National Transit had purchased and loaned to him. When the other shareholders, invoking the limited partnership provisions of their charter, ruled he could not, Carter appealed to the courts and held up the formal merger for several years.[66] The litigation, however, did not prevent the two independent companies from co-operating, without corporate ties, in the movement of crude and refined oil from the Regions to Wilkesbarre.*

A similar Standard attempt to acquire a voice in the management of United States Pipe Line was somewhat more successful. To prevent Standard from capturing the line through purchase of stock, the original subscribers had placed their stock in the hands of a trustee under an agreement that it could not be sold and only the trustee could vote it.[68] Despite this precaution, Standard by 1895 had acquired about one-third interest in the company, through the purchase of four of the largest refineries originally associated with the line.[69] A court order ruled that despite the original trust agreement, the ownership of the stock had to be recognized. As a result, various Standard executives served on the board. While not carrying sufficient power to formulate policies, these directorships did place Standard, as Archbold put it, "in the way of knowing what was being done." [70]

But United States Pipe Line and its affiliates faced a much more immediate and serious problem than court battles. As shown in Table 21:1, a rise in the price of crude, combined with a decline in the prices of illuminating oil, cut the average "margin" on export sales from 2.25¢ a gallon in 1892 to seven-tenths of a cent in 1894. The margins on premium water white oil, sold largely in the domestic market, were reduced during the same period from 4.1¢ to 2.8¢. To add to the difficulties Poth became embroiled in a price war with Standard's European marketing subsidiary.† While neither side would admit to having initiated the cuts, there was no mistaking the intensity of the fight against Poth.

---

* Carter in 1896 purchased these shares from National Transit, along with enough additional stock since acquired by that company, to give him control of Producers' Oil. Although Carter claimed that he intended to treat all parties impartially, the remaining members of Producers' Oil again refused to elect him to membership. This extended litigation until 1899 when the courts finally settled the issue by having Producers' Oil re-acquire Carter's stock at court-appraised value.[67]

† See Chapter 24.

The unusual, if not abnormal, relationship between crude and refined prices during these years was also quite understandably interpreted by the United States pipeline group as a tactic by Standard to drive them

TABLE 21:1

Margins between price of Appalachian crude and wholesale price of illuminating oils at New York: 1891–1900 (cents per gallon in bulk)

Year	Crude at wellhead	New York illuminating oil		Margins	
		Standard White Export	150F° Water White Domestic	Standard white	150F° Water white
1891	1.59	4.43	6.3	2.84	4.7
1892	1.32	3.57	5.4	2.25	4.1
1893	1.52	2.73	4.8	1.21	3.3
1894	1.99	2.69	4.8	.70	2.8
1895	3.18	4.86	6.8	1.68	3.6
1896	2.84	4.47	7.9	1.63	5.1
1897	1.87	3.41	6.5	1.54	4.6
1898	2.16	3.82	6.6	1.66	4.4
1899	3.10	5.48	7.7	2.38	4.6
1900	3.22	5.96	9.4	2.74	6.2

Source: *Report of the Commissioner of Corporations on the Petroleum Industry, Part 2, Prices & Profits* (Washington: Government Printing Office, 1907), 623.

out of business.* There was no question, given Standard's position in the industry, that the rise in crude and the fall in refined prices were the result of decisions made by the management of the Trust. But it seems highly improbable that the primary objective of these decisions was to eliminate the American independents. Their total sales were too small, relative to Standard's, to warrant an action which would involve a heavy loss on the latter's entire output. The more probable explanation is that the independents were caught in a larger struggle between Standard and members of the Russian industry for position in the markets of Europe and Asia.†

But whatever Standard's primary motivation, the effects on the independents was unmistakable. In an attempt to keep their refinery affiliates in operation, Producers' Oil in 1894 cut its gathering charges to 8¢ a barrel and United States Pipe Line began hauling refined oil for nothing.[72] Ac-

* This is the general interpretation both by contemporaries and several historians of Standard Oil.[71]
† See Chapter 24.

cording to one estimate they lost about $200,000 on their combined operations during 1893 and 1894.[73]

Late in 1894 the refiners appealed directly to Standard for relief. James W. Lee, counsel for United States Pipe Line, warned his former law partner, S. C. T. Dodd, that the independents' collapse would inevitably involve Standard in litigation and investigation, presumably on antitrust grounds. Dodd apparently agreed, but Standard's directors did not: "They thought they had matters their own way abroad," Dodd reportedly told Lee, "and wanted to keep them so." [74]

A committee of refiners which met with Standard's executives was no more successful in obtaining a promise of better prices. Standard did offer to buy their plants and pipelines at construction cost plus 12 per cent. With the exception of three refiners who sold out to the Trust in the spring of 1895,[75] this offer was unacceptable to the majority of the independents. They were willing to continue provided they could market their output at cost. With the promised co-operation and financial aid of the independent producers, who had gained from the improved prices of crude, they proposed to form a new company to take over responsibility for distribution and marketing.[76]

The Pure Oil Company

Plans for the proposed organization were announced at Bradford in January, 1895, amidst a fanfare of publicity. By fall sufficient capital had been pledged to form a new corporation, the Pure Oil Company, under the laws of New Jersey. As a precautionary move, arrangements were made to place 53 per cent of the original stock and half of all subsequent issues in a voting trust, to be administered by 15 trustees. They were charged with protecting and maintaining the "independent interests" against "monopoly in the business." [77]

For the time being at least, Pure Oil intended to ship via United States Pipe Line and the Jersey Central Railroad and to sell through Poth on the Continent. Just how this arrangement was to improve the profit position of its shareholders was not made clear. Fortunately for the promoters, this question did not have to be answered. A sharp rise in the price of crude in 1895, accompanied by an even sharper increase in the prices of illuminating oil, brought a substantial improvement in operating margins. (See Table 21:1.)

Independent refiners who had sold short interpreted the price increases as another Standard maneuver against them.[78] Their view gained some plausibility from the fact that in January, 1895, the Seep Agency began quoting its own prices for crude, announcing that it would no longer be guided by bids on the exchanges.[79] Actually, Seep's announcement was merely a formal recognition of Standard's influence on crude prices which had grown steadily over the preceding two decades.

The price movements of 1895 were basically due to a temporary shortage of supply. Appalachian production declined steadily from 35.8 million barrels in 1891 to 30.7 million in 1894; while in the latter year a cholera epidemic in Baku aggravated the shortage by virtually eliminating Russian supplies from world markets. Shipments of Lima oil to American consumers did not release enough Pennsylvania crude for export to halt the drain on inventories in the Appalachian fields. By 1895, stocks were less than half the level of 1893. (See Chart 21:2.)

Pure Oil in fact did not begin operating until the spring of 1896 because, as one of its officers explained later, "prices had changed so enormously" that the refiners did not require its services.[80] But the acquisition of Poth's interests by one of Standard's European marketing affiliates early in the year left the independents without a distributor on the Continent.* In April, 1896, Lewis Emery and David Kirk, president of Pure, hastened abroad and, quickly building tankage in Hamburg and Amsterdam, launched Pure's marketing operations in Europe.[83] A month earlier the company had begun tank wagon delivery in New York City.[84]

U. S. Pipe Line in the meantime had decided to strengthen Pure's position by extending its facilities from Wilkesbarre east to Bayonne. But upon entering New Jersey, which had no free pipeline legislation, the management once more encountered difficulties with the railroads. The principal contest arose over the legality of U. S. Pipe Line's action in running its pipes under the tracks of the Delaware, Lackawanna & Western railroad at Washington, New Jersey, in the fall of 1895. After several armed skirmishes the opposing parties agreed to let the issue be solved by the courts.[85] A favorable lower court decision in March, 1896, enabled U. S. Pipe Line to push construction to Hampton Junction where, during

---

* There are two versions of the Poth incident. According to the independents, Standard's representatives in Germany told Poth that the independents would soon fail and that "he had better get his money out . . . while he could." [81] In 1908 Archbold told a different story: Poth had distributed for Standard before assuming the business of Producers' Oil and U. S. Pipe Line. After the price wars of 1894–95, he reportedly approached DAPG, Standard's German subsidiary, to sell out.[82]

the next four years, the oil was picked up by the New Jersey Central and carried by rail over the remaining 50 miles to seaboard.[86]

Even though the lines had not reached the seaboard, the combined pipe and rail route via Hampton was of considerable benefit to the refiners affiliated with U. S. Pipe Line. It was about 12¢ per barrel cheaper than all-rail bulk shipments from the Regions; 26¢ less than shipping in barrels; and about equal to Standard's 45¢, all-pipeline rate.[87] Even a saving of 12¢ a barrel amounted to between 20 and 25 per cent of the gross margin between crude costs and the price of kerosene. The following tabulation indicates the approximate size of the margins, per 42-gallon barrel.

Year	Cost of crude including gathering charges	Kerosene export prices in bulk	Gross margin
1896	$1.39	$1.79	.40
1897	.98	1.43	.45
1898	1.11	1.60	.49

Source: *Derrick's Hand-Book*, II, 102, 108.

Meanwhile an appeal by the DL&W of the original decision had carried the case to the New Jersey Supreme Court, which in 1899 ruled against U. S. Pipe Line and ordered it to withdraw from the state.[88] This decision apparently came as no surprise to the pipeline. In anticipation of such a ruling, the management had already acquired land for a terminal at Marcus Hook, just below Philadelphia, on the Delaware River. The lines were pulled back to Wilkesbarre and relaid via Easton, Pennsylvania, to the new terminal where the first oil arrived in May, 1901.[89]

As early as 1897 a proposal to merge Producers' Oil and U. S. Pipe Line with Pure Oil formally recognized their community of interests. But opposition from some of the stockholders and Carter's litigation with Producers' Oil delayed consolidation until 1900. At that time Pure Oil acquired some 54 per cent of the stock of U. S. Pipe Line, 87 per cent of the shares of the Producers' & Refiners' Oil Company, and practically all of the stock of the Producers' Oil Company.[90]

Altogether Pure Oil's assets probably totalled between $2 million and $3 million. In addition to marketing facilities in Germany and the Netherlands, they included a controlling interest in two trunk lines spanning the 370 miles from Oil City to Marcus Hook and 500 miles of gathering lines in the McDonald, Bradford and Butler fields. Technically retaining their business identities, but operationally an integral part of Pure's activities

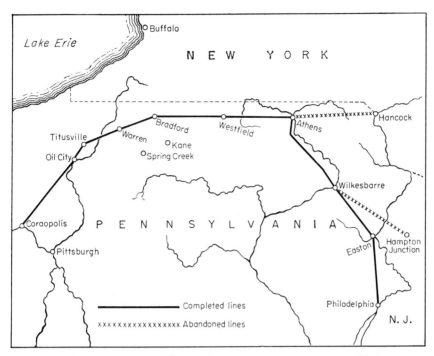

MAP 21:1. *Pure Oil's trunk pipe line system,* circa 1900.

were some 15 Regions refiners with daily capacity totaling 12,000 barrels.[91]

Within another four years Pure Oil became a fully integrated company. Because it had been forced at times to pay premiums ranging from one to nine cents a barrel to obtain crude, Pure Oil incorporated a subsidiary, the Pure Oil Producing Company, in 1902. Operating in southeastern Ohio and West Virginia, the new subsidiary contributed about 1,900 barrels daily, or approximately one-third of the total amount gathered by Pure's collecting system.[92]

Two years later Pure Oil completed its own refinery at Marcus Hook. Deliveries of crude which started at 600 barrels per day in 1904 were tripled by 1906.[93] The *Pennoil,* a tank steamer wholly owned by Pure, carried the company's products to Europe, where marketing facilities had been pushed further down the Rhine and some twenty Pure Oil barges plied the Dutch canals.[94]

Thus within a decade of its inception, the Pure Oil Company handled perhaps the largest volume of any independent in the United States. Certainly no other matched its combination of volume, facilities, and degree of integration. As one of the first fully integrated non-Standard firms, it foreshadowed the structure that major American competitors of Standard would assume for the next twenty years.[*]

Yet for all these achievements, Pure had only a slight effect on the basic structure and performance of the industry of its time. About 70 per cent of its products went abroad, principally to Germany and the Netherlands, where it supplied 15 and 25 per cent of total sales respectively. Although Standard was generally the price leader in these areas, Pure's market share was probably enough to place an upper limit on Standard's prices, particularly in local markets.[95] But such powerful competitors as the

Year	U. S. Pipe deliveries 42-gallon barrels	Pure sales in Germany and Holland 42-gallon barrels
1903	1,317,493	961,496
1904	1,372,462	1,043,564

Source: Petitioners' Exhibit 378, *US* v. *SONJ* (1906), *Transcript.* VIII, 904; *Report of the Commissioner of Corporations on the Petroleum Industry*, Part 3, Foreign Trade, 373, 509.

Nobels, Rothschilds and Deutschesbank (which had invested in Roumanian oil) were far more important determinants of Standard's price policy.

Pure's effect on American production, refining and transportation was perforce even smaller. Despite premium prices and drives to enlist widespread ownership, its pipeline runs indicate that it failed to rally a large following among the producers. Before the company entered production, its gathering lines collected about 5 per cent of total Appalachian runs, increasing their share to 8 per cent only after 1902. And Standard's coastbound shipments from Lima cut Pure's share in deliveries to the seaboard to about 4 per cent.

As shown in Table 21:2, the far greater proportion of Pure's seaboard deliveries consisted of kerosene. The 4-inch crude line could pump from 2,500 to 4,000 barrels per day; [†] yet until Pure completed its Marcus Hook refinery crude deliveries to tidewater were far below capacity. The 5-inch products line operated, on the whole, somewhat closer to a capacity

---

[*] By 1900 a few other independents, Sun in Ohio and the Pacific Coast Oil Company in California (later bought by Standard), may also have matched Pure's scale of integration, but their outputs were smaller and their markets primarily domestic.

[†] A. D. Wood in 1895 estimated that the United States pipeline could pump 5,000 barrels of crude and refined oil daily. It is not clear whether he was speaking of the combined or individual capacities of the crude and refined lines.[96]

approximating 3,000–4,000 barrels daily, and was probably the bulwark of Pure's competitive position abroad. Eight Regions refiners supplied virtually all of its cargo by the early 1900's.[97] These plants could have survived without the pipeline, since Standard maintained a sufficiently high

TABLE 21:2

Pipeline deliveries to Atlantic Coast: 1900–1905 (thousands of 42-gallon barrels)

Year	U. S. Pipe Crude	U. S. Pipe Refined	Standard Crude ‡	Tidewater Crude	Total
1900	247	717 †	29,882	2,721	32,567
1901	209	797 †	32,468	2,740	36,114
1902	244	981	31,741	3,184	36,150
1903	205	1,112	28,537	2,982	32,836
1904	394 *	978	27,793	2,737	31,902
1905	602 *	964	30,259	2,776	34,601

* Prior to 1904 (when Pure's refinery at Marcus Hook began operation), all of U.S. Pipe's crude deliveries terminated at Freemansburg, Pennsylvania whence they were shipped by rail to the Columbia Oil Company. The figures for 1904 and 1905 include 223,000 and 543,000 barrels of crude shipped to Marcus Hook.

† All deliveries of refined oil in 1900 went to Freemansburg; in 1901, approximately 207,000 barrels were shipped to this point, the rest directly to Marcus Hook. From 1902 to 1905, United States Pipe Line delivered all its refined cargo to Marcus Hook.

‡ Includes Lima crude shipped east through Standard's trunk lines.

official export price to enable other inland refiners to operate under an export agreement with it.[98] But the profits of Pure's affiliates would have been less, their independence threatened, and, by and large, their location uneconomical without the product line.

Tidewater, Crescent, and Standard

For the minority of producers who were still militantly anti-Standard, the Pure Oil Company filled the vacuum left by Tidewater. The latter, in their eyes, had forfeited its independence when it signed a market-sharing agreement with Standard's National Transit Company in 1883. This agreement was only the first tie that Tidewater, partly because of limited capital resources, developed with the Trust. In 1884 National Transit's treasurer purchased 31 per cent of Tidewater's stock from the dissident Taylor-Satterfield group.[99] More significantly, Standard began acting as Tidewater's broker in the late 1880's, handling from 50 to 75 per cent of its kerosene exports. Tidewater limited its direct foreign sales to case oil, deciding it could not afford to invest in tank steamers so long as others provided them on satisfactory terms.[100]

By 1887, when its line to Constable Hook, New Jersey was fully completed, Tidewater began to encounter a shortage of throughput.[101] In 1885, Tidewater gathered about 10 per cent of Appalachian crude production; by 1889 the figure had dropped to 5 per cent, principally because of the decline in the Bradford fields, Tidewater's main source of crude. Over the same period, it continued to ship about one-tenth of total deliveries from the Regions. The difference came from drawing down field stocks and purchasing supplementary throughput from Standard.[102] In May, 1890, National Transit placed these sales on a long-term basis, signing an eight-year contract to deliver between 2,000 barrels and 3,000 barrels daily to Tidewater's first station at Coryville.* As a result of these deliveries, which constituted from 30 to 40 per cent of its eastbound crude, Tidewater maintained its 10 per cent share of total shipments from the Regions throughout the 1890's.

But as Standard cemented relations with Tidewater in the late 1880's, discoveries in Washington and Greene counties stimulated competition from other carriers. O'Day complained of a "perfect flood of new competitive lines," [104] to refineries in the Regions and Pittsburgh, while the share of Appalachian production gathered by Standard declined from 88 per cent in 1886 to 77 per cent in 1889. (See Table 21:3.) Although this competition hardly threatened to inundate the Trust, it did lead to one major competitor, in addition to Pure Oil: a company headed by the 21-year-old William L. Mellon, of the Pittsburgh banking family.

Mellon entered the industry as a producer, leasing territory near Pittsburgh in the late 1880's and selling most of his output to Standard. His subsequent activities, far from being preconceived, resulted from a series of events and problems that gradually led him to expand operations. The first of these occurred in 1889, when Standard purchased the firm of Craig & Elkins, one of the largest of the independents. At its peak this company ran 85,000 barrels monthly through its pipeline to Pittsburgh and sold probably half that amount to refiners abroad, including Fenaille & Despaux, among the biggest in France.† Mellon learned that

---

* At the same time, the market-sharing agreement of 1883 was cancelled. In 1898 the supply contract with National Transit was renewed for another ten years.[103]

† The pipeline was known as the Western & Atlantic. The Globe refinery, Craig & Elkins' works at Pittsburgh, could handle perhaps 1,000 barrels daily, and some crude was also sold to other Pittsburgh independents.[105]

Fenaille & Despaux "disliked the Standard; or possibly certain Standard Oil people," and convinced the French firm to buy from him.[106] He simultaneously expanded his own production and was soon buying outsiders' crude as well. By August, 1891, Mellon's lines were collecting about 67,000 barrels per month, delivering about two-thirds that amount to Coraopolis

TABLE 21:3

Crude production, Standard pipeline runs, and Standard share in Appalachian fields: 1884–1900 (in thousands of 42-gallon barrels)

Year	Crude production	Standard runs *	Ratio of Standard runs to production
1884	23,956	20,256	84.5
1885	21,534	19,169	89.0
1886	26,550	23,340	87.9
1887	22,878	19,442	85.0
1888	16,941	13,933	82.2
1889	22,355	17,280	77.3
1890	30,073	23,997	79.8
1891	35,849	29,926	83.5
1892	33,432	26,975	80.7
1893	31,366	23,408	74.6
1894	30,738	23,925	77.7
1895	30,961	24,488	79.1
1896	33,972	29,464	86.7
1897	35,230	31,040	88.1
1898	31,717	27,773	87.6
1899	33,068	28,850	87.2
1900	36,295	31,985	88.1

Source: *Mineral Resources, 1905*, 823; Petitioners' Exhibit 373, *US v. SONJ* (1906), *Transcript*, VIII, 901.

* Excluding Tidewater's collections.

and Pittsburgh.[107] Selling primarily for export, he shipped crude to New York in 200 of his own tank cars.

Events in 1892 led Mellon to deepen his operations further. Early in the year, the Pennsylvania Railroad raised his rates to Tidewater. The increase, he wrote later, "was just enough to make our business profitless." [108] The Reading, however, agreed to take his oil at Carlisle, Pennsylvania, midway between Pittsburgh and the coast, and Mellon organized the Crescent pipeline to reach this point. But by the time the line reached

Carlisle, the Reading had a new president who refused to honor the contract. Anxious to avoid time-consuming litigation, Mellon pushed his 5-inch line directly to Marcus Hook, using Pennsylvania's free pipeline law to overcome the usual obstructions of the Pennsylvania railroad. On November 7, 1892, six months after starting east from the Regions, the Crescent pipeline delivered its first oil to the eastern terminals.[109]

With five pumping stations along its 270-mile route, the Crescent could deliver 8,000 barrels per day, then about 10 per cent of total pipeline deliveries in the Appalachian fields. Apparently it operated close to capacity. Mellon's lines were shipping about 250,000 barrels monthly out of the Regions in early 1893. Probably between 15,000 and 30,000 barrels of this amount went to the Bear Creek refinery, located near Franklin, Pennsylvania, in which Mellon had a half interest.[110] The rest went to the seaboard. This was more than Mellon could profitably export as crude, and there were few independents along the coast to take his oil. As a result, Mellon started to build his own refinery at Marcus Hook late in 1892.[111]

Standard was concerned about Mellon's growth, for his inroads into the French market aggravated the tense situation shaping there. In 1893 it was rumored that France might reduce the tariffs on Russian oil. More imminently, the Rothschilds of Paris, having invested heavily in the Russian petroleum industry, had close financial connections with the semi-cartelized French refiners. When Standard tried to negotiate a supply contract with the French syndicate, the latter attempted to win concessions by threatening to shift to the Crescent.[112]

The Russian threat also jeopardized Mellon's long-term position in his major market. Moreover, recent success in processing sulphur-bearing Ohio crude, comprising about one-third of American production, led Mellon to fear a long-term decline in oil prices. Both developments called for further investment if Crescent was to maintain its market share, and other interests were claiming Mellon's attentions and resources. Consequently in August, 1893, he accepted a Standard offer to buy the pipelines, some of the producing territory and the refineries.[113]

Since Pennsylvania law forbade the merging of competing pipelines, the Crescent's stock was personally bought by H. H. Rogers and two other Standard Oil representatives who paid Mellon $225,000 in cash with another $641,800 to come from future dividends.* Legally, the property re-

---

* Although the anti-merger law was generally not enforced, Standard was perhaps anxious to avoid an open breach. Ohio Standard had been successfully attacked in the Ohio courts and as a result Standard executives announced in 1892 that the Trust would be dissolved.

mained under the control of Mellon, who held title to the stock until the entire amount was paid.[114] Economically, Crescent ceased after 1893 to be an active competitor of Standard.

The Crescent transaction provided an ironic epilogue to earlier developments in the Pennsylvania Legislature. In June, 1893, that body had voted to repeal the pipeline anti-merger law. Governor Pattison, however, vetoed the measure contending that it would deprive the public of the benefits of competition among pipelines and shippers.[115]

In 1895 a second attempt at repeal was successful. Proponents of repeal maintained that the anti-merger law discouraged venture capital. When fields declined, they alleged, the fact that the lines serving them could not be sold entailed "enormous losses" upon their owners.[116] Although this argument failed to explain why a buyer should want to acquire a losing property in an exhausted pool, Governor Hastings partly adopted it in signing the bill:

> I am convinced . . . that the effect of the act of 1883 is directly the reverse of its ostensible object. Instead of encouraging completion and fostering the building of pipelines to compete with each other, the fact that when the property becomes unprofitable the owners are prohibited by law from selling it, must necessarily discourage investors in such enterprises.[117]

In view of the timing of both repeal bills, it is questionable whether their promoters shared Hastings' concern for maintaining competition. There is no proof that Standard was responsible for either measure, although its lobbyists kept close tab on relevant bills at Harrisburg. But the 1893 attempt antedated Standard's purchase of Crescent by three months, and nine months after the second was enacted the properties were formally transferred to National Transit. In the interim, Standard had allegedly offered to buy the gathering lines of the Producers' Oil Company.

In any event, the arguments of both Pattison and Hastings, taken at face value, indicate the confusion of much of the thought then underlying public policy toward pipelines. Neither explained just what they meant by the "benefits of competition." Presumably, they assumed that active bidding for customers would expand output and keep prices close to unit costs (including an unspecified "fair" profit). But this presumed a degree of entry that was not likely to obtain in trunk pipelines.

Thus, new entries and legislative agitation posed no real threat to Standard's control over Appalachian pipelines throughout the 1884–1900 period. How well it succeeded is roughly indicated in Table 21:3. In the mid-1880's, its gathering system collected 87–89 per cent of crude output. The decline in the latter part of the decade was in large part due to the entry of small lines of which O'Day complained. The even lower proportions from 1893 to 1895, are quite misleading since Standard did not officially admit its control of the Mellon system, nor did it count the lines' throughput as its own, until after their formal transfer to National Transit in 1895. Thus the data for 1896–1900, which include the Crescent's runs, give a more realistic impression of Standard's market share after 1892.

Geographically, the major extensions of Standard's Appalachian network went to the South and the West (see Map 22:2). In June, 1886, the Southwest Pennsylvania pipeline ran its first oil from the Washington fields in southern Pennsylvania to Standard's system in Butler county.[118] When West Virginia's production jumped to 500,000 barrels in 1889, four times its annual average since 1882, Standard entered the field with two new companies, the Eureka and the Southern pipelines. Together they began channeling West Virginia oil to the coast in February, 1891.[119] The Eureka collected crude from the wells and delivered it through 100 miles of 6-inch pipe to the Pennsylvania border; from there the Southern pipeline carried the oil 170 miles further east to Millway, where it joined Standard's already existing lines to Philadelphia and Baltimore.[120]

Standard's control rested upon two basic factors: the managerial skill and absolute size of Standard's pipeline operations and the large market share of the Standard organization as a whole. The first kept producers generally well satisfied with Standard's service, if not always with the prices it charged. Even Lewis Emery admitted, "they [Standard] have been as faithful people as ever were in the world to take care of the product."[121] The second, by depriving independent lines of potential outlets, heightened the entry barrier.

Conversely, the cost-advantage of pipelines was important in maintaining Standard's position in other segments of the industry. From 1888 to 1900, a refiner shipping by rail from the Regions to the coast paid 55 cents per barrel in bulk, 66¢ for barreled cargo. Given prevailing throughputs, pipelines in the late 1880's could pump a barrel of crude to tidewater for 21¢–25¢; a decade later, the figure had declined to 11¢ or 13¢.*

---

* Costs for the 1890's were taken from *Report of the Commissioner of Corporations on the Petroleum Industry*, Part I, *Position of Standard Oil*, 231. The Bureau used data for 1903 and 1904, which were approximately the same as in the late 1890's. Depreciation, representing about 40 per cent of total costs, was estimated at 5 and 10 per cent of the reproduction costs of fixed plant; interest, at 10 percent of this base. These data included Tidewater.

The figure for the late 1880's is much less reliable having been constructed from a variety of sources. Throughput of the seaboard lines was estimated at 11.7 million

Theoretically, an independent refiner on the seaboard could save 10¢–15¢ per barrel by switching from bulk rail shipment to Standard's trunk lines.[125] New York and Pennsylvania law had, by the early 1880's, declared pipelines to be common carriers, and Archbold claimed that Standard's lines had been so operated.[126] But he also admitted Standard's common carrier operations were of "very little practical significance" for hardly any independent oil had been tendered.[127]

Although the reasons for this are not altogether clear, a few explanations are apparent. The amount of independent capacity at the refining termini of trunk lines was comparatively small—a phenomenon that was itself the result of Standard's early gaining control of the lines. Moreover, until 1888 the non-Standard refineries at the seaboard had little to gain from using Standard's facilities. The Tidewater group, one of the largest, had its own line, while National Transit's pipeage was not cheaper than rail rates. Thirdly, independents outside the Regions apparently preferred to buy their crude from near-by Standard refineries, rather than purchase it in the field. From 1885 through 1890, Standard's refineries sold an average of 2.2 million barrels of crude annually to outsiders, almost all of the Trust's total outside crude sales.[128] This practice at least saved the independents the expense of maintaining their own purchasing agents in the field, and probably increased in importance as transactions on the exchanges diminished. Finally, the independents may have been suspect of Standard's intentions to operate as a common carrier, despite Archbold's statements to the contrary.*

---

barrels per year: between 1886 and 1889, the Trust's refineries annually consumed 19.5 million barrels of Pennsylvania-grade crude; it was assumed that this was pro-rated among the various refining centers according to their respective capacities. Of a total of 110,770 barrels daily capacity, Standard's tidewater refineries in New York, New Jersey, Baltimore and Philadelphia had about 68,460.[122]

Operating costs were estimated at 7¢ per barrel, the minimum that National Transit agreed to accept in 1884 for piping crude to New York on behalf of the Pennsylvania Railroad.

Depreciation on plant investment was again computed at 5 and 10 per cent; interest, at 10 per cent. Plant investment itself was estimated at $7,000 per mile of 6-inch line, a figure commonly cited at this time.[123]

In the late 1880's, Standard operated approximately 1,600 miles of such line in the Appalachian fields. Since there are no published mileage figures for Standard's Appalachian lines during this period, the lines known to be operated in 1889 were assumed to have the same mileage as in 1904, when data were published. Tidewater was omitted from this estimate.[124]

* A poll of independent refiners in the Appalachian fields in 1906, for example, revealed that they did not tender oil to Standard because they believed the tender would be refused.[129]

But if Standard was to insure its position in the industry it had to do more than maintain its control over the transport and handling of Appalachian crude. Important new discoveries in northwestern Ohio in the mid-1880's challenged the Trust to utilize its capital resources and experience to extend facilities and establish similar control in the new territory.

Chapter 22

Standard and the Ohio-Indiana fields

Since the earliest days of the industry the Mecca district, which spread across the border of western Pennsylvania into Ohio and the Macksburg fields in southeastern Ohio, had contributed modest amounts of crude to total domestic production.* While a portion of the Mecca crude was a "heavy" oil particularly valuable as a source of lubricants, most of the output from these fields was technically well suited for general refining operations and was usually classified among the Pennsylvania-type crudes.

Although production in the Mecca and Macksburg areas continued to expand, it was the development of the "Lima fields" during the mid-1880's which made Ohio a major crude producer. Centered in the northwestern part of the state, the first discoveries were made near the towns of Lima in

---

* Total production from these fields between 1860 and 1875 has been estimated at about 200,000 barrels, while for the 1860–84 period as a whole, output never exceeded 50,000 barrels in any one year.[1]

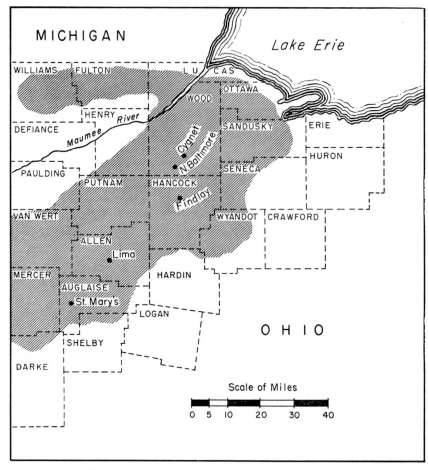

MAP 22:1. *Ohio oil producing territory in the 1890's.*

Allen County and Findlay in Hancock County. Developments subsequently spread northeast and southwest across the neighboring counties
of Mercer, Wyandot, Van Wert, Seneca, and Lucas. (See Map 22:1.)

Indications of both petroleum and gas had long been noted in northwestern Ohio, but on a basis of experience in Pennsylvania, most observers
discounted the possibility of discovering any important accumulations of
oil. The flat, drift-covered Ohio terrain, with its underlying limestone and
shale formations was in sharp contrast with the hills and sandstones of the
eastern producing centers. Thus when paying quantities of oil were discovered more or less as a fortuitous by-product of drilling for gas, they
came as a complete geological surprise.[2]

Explorations for gas began on an intensive scale in northwestern Ohio early in the 1880's, and in 1884 the first well to strike the vast gas reservoirs of the Trenton limestone was drilled near Findlay, Ohio. With promises of low-cost building sites and bargain rates for fuel, the area soon attracted a number of industrial establishments. Evidencing a flair for advertisement coupled with a complete disregard for resource conservation, a number of wells were ignited by their owners, lighting up the sky in a vast panorama.[3]

Because it clogged the gas lines, the small amount of oil encountered in a number of the early gas wells was considered a nuisance. But this attitude began to change early in 1885 when a well near Lima, Ohio, after being unsuccessfully drilled for gas, was shot with a charge of "rack-rock" (a mixture of chlorate of potash and di-nitrobenzole) and started to flow at a rate of about 18 barrels per day.[4]

While the news that Lima had "struck oil" aroused considerable interest in petroleum circles, it remained for two wells, drilled specifically for petroleum in the fall of 1885, to reveal the possibility of paying quantities of oil in northwestern Ohio. The first of these wells, the Citizens' well at Lima, was completed in November and, after a pump had been installed, yielded 1,450 barrels of oil during December, thereafter leveling off to a daily output of about 26 barrels.[5] The second well, completed about the same time, was the Matthias well, drilled near Findlay, Ohio, by the Findlay Gas Company. After reaching a depth of 1,321 feet, the well was torpedoed with a charge of 100 quarts of nitroglycerin. Following an initial output of 300 barrels per day, it settled down as a flowing well, yielding a daily production of about 35 barrels.[6]

Many experienced oilmen in Pennsylvania were inclined to discount the importance of the Ohio strikes. Samples from the pioneer wells indicated that the heavy, black Ohio crude had a high sulphur content, which not only gave it the name "skunk oil," but which stubbornly resisted removal by known refining methods.* Late in 1886 the United States Geological Survey noted that while some of the illuminating oil currently refined from Ohio crude was ". . . fully equal to the best oil of any fields . . . [most Ohio] refined oil that is in the markets cannot . . . endure the test; it crusts the wick and clouds the chimney of the lamp."[8]

Even if no satisfactory refining process could be discovered the Sur-

* Similar types of crude, discovered in Canada, were being currently processed there by Imperial Oil, but compared to Pennsylvania grade crude, they cost ten cents more per barrel to refine and yielded almost one-fourth less kerosene of inferior quality.[7]

vey nevertheless, in what must have pleased Ohio operators and attracted the interest of oilmen generally, concluded on an optimistic note. Pointing out that Ohio crude was already being used as a fuel, it prophesied,

> If all other and higher uses of petroleum are dropped entirely out of the account it is still evident that an enormous stock of fossil power, vastly greater than all that can be furnished by the newly discovered natural gas fields in this part of the country, is made available to us in the Trenton limestone oil.[9]

Whatever the uncertainties about the future uses of Lima crude, the tremendous potentials of the Ohio fields could not long be denied. With the discovery of the wells at Lima and Findlay, the stage was once again set in the history of the industry for the oil towns to burgeon and blossom overnight as the lure of discovery attracted hundreds of restless men into Ohio, bringing with them the knowledge acquired in the oil nurseries of the East.

Characteristics of the Lima fields

With only slight modifications, principally due to differences in geological structure, the theories, traditions, and techniques based on production experience in Pennsylvania could readily be applied to the Ohio-Indiana fields. In contrast to the anti-clinal folds and structures of the eastern regions, Ohio geologists found that at Lima the "terrace structure prevail[ed] and broad tracts of . . . level oil rock [were] revealed by the drill." [10]

The oil-producing formations were a magnesian limestone composition and contained from 24–39 per cent of carbonate of magnesia. The formations lay beneath gray and greenish-blue shales at surface depths varying between 1200 and 1500 feet. In the productive portion of the field, the limestone occured as a flat-lying terrace with well-marked boundaries of steeper descent on the east, west, and north. Within the terrace, slight rolls, domes, or hollows could be found.[11] The subdivisions of oil-bearing strata in turn consisted of relatively small and compact territories with few barren spots within each subdivision.*

It was not until a fairly extensive amount of test drilling had been undertaken that these characteristics came to be recognized. Once the boundaries of the producing regions were more clearly established, how-

---

* The area as defined in late 1887 was as follows: the Lima field, 11 miles by 3 miles; the Findlay field, 5 miles by 2 miles; and the North Baltimore field, 6 miles by 2 miles.[12]

ever, the physical characteristics of the Lima fields encouraged rapid exploitation.

Because of the compact nature of the oil-bearing strata, for example, the number of dry holes to the total number of wells completed dropped from about 25 per cent in 1886 to less than 5 per cent by 1889, and averaged between 5 per cent and 15 per cent over the succeeding decade. Moreover, the costs of bringing the Ohio wells into production were relatively low, averaging between $2,000 and $2,500 per well compared to Pennsylvania where by the late 1880's total expenses per well typically ranged between $3,200 and $5,300 with some wells costing as much as $8,000.[13]

In large part the cost advantage of the Lima fields was due to the fact that whereas the oil bearing strata in northwestern Ohio were seldom found at more than 1500 feet below the surface, in Pennsylvania at this time over 80 per cent of the wells were drilled to depths ranging between 1500 and 3000 feet.*

Also in contrast to the rock formations frequently encountered in the Regions, Ohio drillers found it relatively easy to penetrate the limestone and shale that overlay the producing "sands." According to one report, for example, it was seldom necessary to use jars to dislodge drill bits, while one set of tools might be employed in putting down a dozen wells.[14] One further comparative advantage of the Lima fields may be noted. In the Regions it was not uncommon to extend casing as far as 1600 ft–2000 ft into the well.[15] Ohio operators, on the other hand, were ordinarily able to exclude water and prevent cave-ins by limiting the amount of casing to about 400 feet.[16]

Another characteristic which distinguished the Ohio-Indiana fields from the Regions was the consideration given to the proper spacing of wells. In large part this attention to conservation reflected the policies of the large producing companies, many owning or having under lease many thousands of acres, which emerged soon after exploration and production got under way. It was, for example, a part of the announced policy of the Ohio Oil Company, which by 1887 became the largest of the producing companies, to "avoid the pursuit of that ancient folly of protecting the lines, and to limit the drilling to a large number of acres to the well." [17]

Among contemporaries there was considerable difference of opinion regarding the proper amount of acreage to assign to a well. Some argued that the amount should be fifty acres. In actual practice it appears that

* See Appendix F.

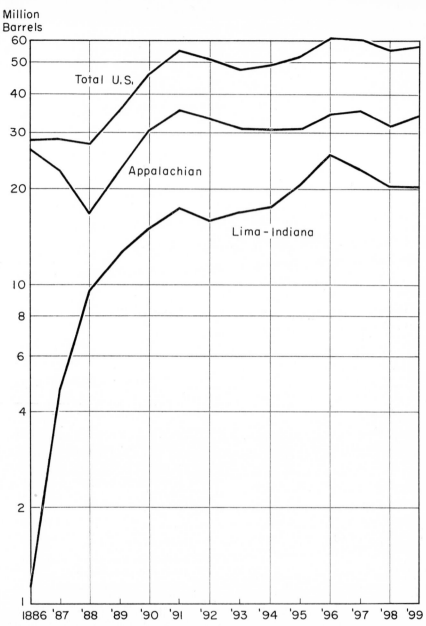

CHART 22:1. *Crude oil production, total U. S. and Appalachian and Lima-Indiana fields, 1886–99.* (*Source:* Petroleum Facts and Figures (*9th ed.*); *New York: American Petroleum Institute, 1951, 152.*)

10–20 acres was the most frequently used spacing arrangement, while most operators believed that in "territory where they were not obliged to guard against rival interests, economy would be consulted [sic] by giving to each well at least 20 acres." [18]

By following these practices the Ohio state geologist, Edward Orton, estimated that on the average a good territory would produce from 2,000–4,500 barrels per acre over a producing span of at least three years or longer.[19]

For a new field, the output of most of the wells struck in Ohio early in 1886 was modest, seldom exceeding 150 barrels per day. But late in 1886 and early 1887 the discovery of several large wells, one producing 1,500 barrels per day and a number of others which reached 1,000 barrels per day, gave an added stimulus to drilling.[20]

The generally favorable conditions affecting the costs of bringing new wells into production were reflected in the growing output from northwestern Ohio. As shown in Chart 22:1, the annual volume of Lima crude expanded rapidly from about one million barrels in 1886 to over 12 million barrels in 1889, about one-third of the nation's total output of crude for that year. Throughout the succeeding decade Ohio-Indiana continued to hold a position in production second only to the Appalachian fields, reaching a peak of just under 25.3 million barrels in 1896, when Lima crude accounted for over 40 per cent of total domestic output.

Although annual output did not fall below 20 million barrels during the next three years, there were, as shown in Table 22:1, ample signs of a future decline in production from the Ohio-Indiana fields. One important indication was the sharp increase in the number of abandoned wells after 1894. Even more revealing was the approximate 93 per cent decline in the average daily production per well over the twelve years starting in 1888. It was this trend which prompted Edward Orton to observe in 1897 that Lima began to die at the moment it began to live.[21]

Well before the Ohio-Indiana fields reached their peak production in the mid-1890's, however, they had already written a new chapter in the history of the American petroleum industry. This was not alone because they added an important new source of domestic oil. Initial difficulties with refining Lima crude led to its utilization as a fuel on a scale that insured a permanent place for fuel-oil in the future demand for the industry's products. To obtain a satisfactory method of refining Ohio oil involved methods of research on a scale that foreshadowed the kinds of experiments carried on by present-day refiners. Finally, the Ohio-Indiana fields supplied Standard with the incentive and opportunity not only to enlarge its transport, refining, and marketing activities, but to move into production on an extensive scale as well.

TABLE 22:1

Drilling activity in northwest Ohio: 1886–99

Year	New wells completed	Dry holes	New wells producing	Abandoned wells *	Total wells producing *	Average daily output per well
1886	228	59	169	15	155	18.8
1887	351	39	312	32	435	29.3
1888	531	27	504	0	939	28.3
1889	701	34	667	0	1606	20.7
1890	1969	193	1776	20	3362	12.2
1891	1575	250	1325	−2 †	4689	10.1
1892	1446	183	1263	−93 †	6045	6.9
1893	1569	203	1366	190	7221	5.2
1894	2472	384	2088	70	11076	4.0
1895	4489	564	3925	394	12770	3.4
1896	4458	550	3908	493	16185	3.5
1897	2486	384	2102	1572	16715	3.1
1898	2394	270	2124	774	18065	2.5
1899	5556	395	3164	1236	19993	2.2

Source: Arnold & Kemnitzer, *Petroleum in U.S.,* 245.
* As of December 31.
† Abandoned wells returned to production.

But before Lima crude could make any significant impression on the industry, there was an immediate problem to be solved. This was the problem of providing adequate facilities for the gathering and storage of crude. It was Standard, logically enough, which made its entry into Ohio early in 1886 to provide these facilities.

Transport and storage facilities

From the outset the Ohio oil fields were well situated with respect to good railroad connections to major markets. The Lake Erie & Western, running southwest from Cleveland, cut the area diagonally and touched major production points at Findlay, Lima, and St. Mary's. The Chicago & Atlantic and the Pittsburgh, Fort Wayne & Chicago both converged at Lima, putting the town within six hours' shipping time of Chicago. Over the Dayton & Michigan it was possible to reach Toledo three hours after leaving Lima, while the Toledo & Ohio Central divided the rich North Baltimore fields in Wood County. In addition, the Miami & Erie canal, connecting Toledo with Cincinnati, provided Lima and St. Mary's with a somewhat slower, but still practicable outlet.[22]

Even with good rail connections it was still necessary to provide local gathering facilities and storage capacity. The technique of piping oil from wellheads was quickly copied from the Pennsylvania fields. Shortly after the striking of the Citizens' Well, in November, 1885, the Edwards Oil Burner Company of Chicago laid a short line to the Lima Strawboard Company which used the oil as fuel.[23] The original throughput at most could not have exceeded more than 25 barrels per day, but Edwards, eying the drilling activity in the territory, anticipated a much larger scale. In February, 1886, when daily output from the Lima fields approximated 300 barrels, he suggested to Standard of Ohio that it purchase the light ends from a refinery he proposed to build at Lima to process crude for fuel oil, which he planned to market himself.[24]

Although Standard was not quite ready to deal with Edwards, the management had already decided that it should take an active role in the Ohio fields. In March, 1886, less than six months after the discovery of the pioneer wells at Lima and Findlay, National Transit extended its operations into Ohio by providing $100,000 and personnel to organize the Buckeye Pipe Line Company. In April, Buckeye bought out the Edwards line (though Edwards continued work on his refinery), and in June ran the first oil into its tanks.[25] The Seep purchasing agency followed Buckeye into the new fields, and began buying all the oil run by the pipeline at prices ranging from 35¢–40¢ per barrel. In addition, Standard began construction at Lima of a refinery of its own, the Solar Refining Company.

Standard's quick move into the Lima area was basically a reflection of the management's concern over future supplies from the Appalachian fields. If output from Ohio-Indiana was to provide a hedge against a possible decline from Pennsylvania, the Trust's executives wanted to make sure of their control over the bulk of production. Benjamin Brewster, writing to Rockefeller in June, 1886, gave a precise evaluation of Standard's position at the time, "The matter of Ohio oil is, it seems to me, important and *on its manufacturing value* hinges our policy. We cannot afford to allow anyone else to handle it, neither can we afford to load ourselves with it, at a price above its value. The question is, How to utilize it & where?" The answer to the question, he felt, lay with the Standard organization itself. "We can afford to spend time & money . . . to answer [it] intelligently." [26]

Within a year, mounting inventories of crude in Buckeye storage tanks made it evident that the problem of finding a use for Ohio oil was to be a real test of the organization's financial and technical resources. Nowhere in the Ohio-Indiana operations was the immediate need for adequate financial resources more apparent than in the case of pipelines and storage facilities. It was at least debatable, for example, whether National Transit operated at this time more efficiently than had non-integrated lines in Pennsylvania. But it is highly unlikely that any pipeline or group of pipelines not integrated with a large organization would have assumed the risks on a scale that could have matched Buckeye's first year record in the physical handling of Ohio oil.

As a new pipeline company Buckeye faced the old problems of extending its connections and storage capacities as rapidly as output increased. By the end of September, 1886, Buckeye had already filled four 35,000-barrel tanks and was building three more, including one in the still relatively modest Findlay area. A month later, pipeline runs averaged 4,000 barrels daily, triple their September levels, but in Lima producers were complaining that their oil was running to waste.[27] The lack of tankage, however, did not deter them from pressing development still further. In 1887, new wells near Lima flowed 150 barrels per day and the opening of the prolific Wood County pool, in northern Ohio, produced some of the largest wells known to that time.[28] As a result, Buckeye's runs were averaging over 10,000 barrels per day by summer of that year. The pipeline had built one tank farm just south of Lima and was constructing another near North Baltimore, the center of the Wood County development, as its principal receiving depots. Together they provided almost 2 million barrels of storage capacity; it was sorely needed, for the 2.7 million barrels collected during the first year of the pipeline's operations were offset by deliveries to refineries of only 350,000 barrels.[29]

Mounting stocks were, of course, not unusual in the first flush of a new field. But never in the history of pipelines had the market absorbed so small a proportion of current output. Even in Bradford's most prolific days, deliveries to refiners had equalled from 50–60 per cent of pipeline runs from wells.[30]

In terms of absolute additions to stocks, the amounts at Bradford were, of course, almost three times greater than at Lima, but this merely reinforces the argument: despite the smaller absolute quantities involved, a much smaller proportion of Lima crude entered the market.*

In part, the failure to move a larger portion of Lima crude on to refineries was attributable to the existence of large inventories of Pennsylvania oil. But a much more significant factor was the difficulty encountered by Standard in developing a method of refining sulphur crude.

This failure had not come for want of trying, but, "Attempts at Standard Oil plants in Cleveland, Oil City, and elsewhere to manufacture satisfactory products from Lima crude during the first six months of 1886 failed miserably." [31]

It is true that a number of small refineries were beginning to manufacture small amounts of inferior kerosene from Ohio crude. But earlier experience with Bradford crude had convinced the Standard Oil managers that to hold their major markets they "had to manufacture not only acceptable kerosene but specific qualities of kerosene to which customers had become accustomed." [32]

Buckeye as virtually the only outlet available to producers in the Ohio fields continued to acquire crude. But to slow down the rate of production as well as to reduce the cost of acquiring additional inventories the Standard management decided to reduce the posted price of crude. Thus starting in the summer of 1886 the Seep agency initiated the first of a series of price cuts that by mid-1887 brought the quotations on Lima crude from 40¢ a barrel to 15¢ a barrel, which became the ruling price for the next two and one-half years. [33]

Producers in the fields in and around Lima were quick to react to the pressure of lower prices. At a mass meeting held at Lima late in June, 1887, they appointed a committee to consult with Standard about the future prospects for crude. Specifically the committee members were instructed to find out what progress Standard was making in developing a satisfactory method of refining sour crude. [34] Daniel O'Day, head of National Transit, as spokesman for Standard replied that despite their continuation,

---

* In the interests of accuracy, it should be noted that the ratio of deliveries to runs only approximates the proportion of current output absorbed by refineries. Pipeline runs in any one month fall short of current crude output to the extent that producers increase the oil that they hold in tanks at the wells. Deliveries indicate transfer from pipeline tanks to refiners, but they take no account of changes in crude oil inventories at the refineries.

. . . the expensive experiments conducted by it [Standard] in seeking to obtain from Lima crude an illuminating oil that will fairly compete in the open market with Pennsylvania oil have resulted in complete failure . . . He further averred that the only use [Standard] had been able to find for the oil was for fuel.[35]

O'Day went on to point out that within a year of entering the Ohio fields, the Trust had already spent more than $2,000,000. Over $750,000 of this amount had gone for the purchase of oil alone, to say nothing of expenditures on gathering lines and storage tanks, while another $500,000 had been spent on the Solar refinery at Lima.* At least for publication O'Day termed these outlays a "great mistake." Not only was there no immediate prospect for earnings, but in virtually guaranteeing pipeline accommodations, Buckeye had only encouraged producers to sink more wells.[37]

Until some important use could be developed for Lima crude, the management suggested to producers that they should inaugurate a shutdown movement similar to the one currently operating in Pennsylvania. Otherwise, Standard warned, Buckeye could not continue to buy and store the expanding output of crude no matter how low the price.[38]

Standard, in fact, was already investigating the possibilities of the fuel-oil markets as outlets for Lima crude. Experts had been put to studying Russian oil-burning equipment, the best of the day, and in June, 1887, Major C. W. Owston, former superintendent of the George P. Smith farm in Franklin, Pennsylvania, had been put in charge of Buckeye's fuel-oil department.[39] In return for assurances by Standard that it would use the time to develop the fuel oil markets, a group of producers in the Lima district did agree early in July to stop new drilling for nine months.[40]

Even though Buckeye in some instances refused to supply additional tankage, against the attraction of new fields the shutdown movement had at best only a moderate influence on new drilling. This was particularly true in the areas around North Baltimore and Findlay where, during 1887, prolific wells and low drilling costs yielded good profits even with oil bringing only 15¢ a barrel.[41] Output from the Ohio-Indiana fields, as shown in Table 22:2, totaled some 4.7 million barrels in 1887, compared to 1.1 million barrels for 1886. Moreover, of the combined output for 1886–1887 of 5.8 million barrels, approximately 4.5 million barrels were in

---

* Some attempt had been made to economize on storage costs by bringing old tanks from Pennsylvania and rebuilding them at Lima, but poor workmanship soon caused them to deteriorate, and apparently new tankage had to be constructed.[36]

storage at the end of 1887, most of it in tanks owned by Buckeye.

By this time, however, the Standard management was much less concerned over the accumulation of inventories than had been the case even six months earlier. Even though experiments carried on in the various

TABLE 22:2

Production of crude oil in the Ohio-Indiana fields, shipments, year-end inventories and average annual prices of Lima crude: 1886– 99 (in millions of 42-gallon barrels)

Year	Output	Shipments	Inventories	Price per barrel
1886	1.1	...	...	...
1887	4.7	.8	4.5	$0.23
1888	9.7	3.1	10.2	.15
1889	12.2	5.8	14.8	.15
1890	15.1	6.2	21.6	.30
1891	17.5	12.2	22.6	.31
1892	15.9	16.5	19.1	.37 *
1893	16.0	14.7	19.0	.47
1894	17.3	14.5	20.2	.48
1895	20.2	16.8	21.5	.70
1896	25.3	20.0	23.3	.66
1897	22.8	20.1	22.8	.49
1898	20.3	24.4	15.2	.61
1899	20.2	24.6	10.5	.89

Sources: Report of the Commissioner of Corporations on the Petroleum Industry, Part 2, Prices & Profits, 103; Mineral Resources, 1892, 634, 641; Mineral Resources, 1894, 356–57; Mineral Resources, 1895–96, 678, 687; Report of the Census, Mines and Quarries (1902), 735.
* Average for North and South Lima combined.

Standard refineries had as yet failed to develop a satisfactory method of refining Lima crude, a rapidly expanding demand for fuel oil promised to provide significant marketing outlets for Ohio oil.

Early marketing of Lima oil

Standard was by no means the pioneer in the marketing of Lima crude for use as fuel oil. Since the middle of 1886 small operators, such as the Edwards Oil Burner Company, had been shipping Ohio oil to industrial customers as far away as Detroit. But it remained for Major Owston with his sales force of fifteen to thirty men to develop the market potential.

The enthusiasm of the head of Buckeye's fuel-oil department for his work was reflected in a circular which he issued in August, 1887, outlining the virtues of Lima crude as a fuel. As distinguished from coke, coal, or wood, Major Owston wrote, oil gave better control over the rate and magnitude of the fire, a more uniform heat unaffected by fluctuations in the weather, a cleaner fire less damaging to boilers and flues, economy in boiler capacity, and lower labor costs in tending the fires. Ohio crude was recommended for use in locomotives, ocean steamers, in the manufacture of glass, steel, crockery, sewer pipe, brick, and many other industrial products. Moreover, Owston concluded:

> The use of oil as a fuel and heat producer is no longer an experiment. Improved forms of burners and methods of applying are being rapidly introduced, and events in the past few months have demonstrated its entire practicability and economy over the ordinary methods.[42]

Owston's boast that oil for fuel was no longer in the experimental stages was well founded. Even before Standard was fully embarked on its marketing program there were already many satisfied customers. In October, 1887, *The Iron Trade Review* of Cleveland published a list of comments of manufacturers on results that they had achieved by using petroleum. "Oil fuel has increased our production 20 per cent," a representative of a Lima paper mill said, "and at considerably less cost than coal." "In the making of brick," P. L. Sword, a Cleveland manufacturer stated, "the use of oil as fuel means a saving in the cost of fuel of 15 per cent, in the cost of labor, 40 per cent, and a considerable increase in the value of the product." All agreed that oil as fuel was here to stay. As a representative of the Detroit Steel & Spring Works put it, "Oil, we are convinced, is the fuel of our business for the future." [43]

By December, 1887, there was sufficient oil moving to markets via tank car to warrant the construction of a short trunk line spanning the 50-mile distance between Lima and the Cygnet, North Baltimore, and Findlay fields to the northeast. Buckeye, which up to this time had confined its operations largely to supplying gathering lines and storage facilities located at nearby railroad shipping points, found farmers along the route quite willing to grant rights-of-way for 20¢ a rod. By January, 1888, the 6-inch line was rapidly moving to completion.

There were obvious attractions for terminating the line at Lima. It was there that Buckeye's major tank farm, capable of storing 1,225,000 barrels of crude, was located. Lima was also the site of Standard's Solar refinery where light ends (naphtha and gasoline) were "skimmed" from the crude before shipment to the fuel-oil markets.*

Construction had hardly begun on the Cygent-Lima line when rumors began to circulate that Buckeye was also planning a pipeline from Lima to Chicago. Standard executives were non-committal, probably to keep down costs of acquiring lands, but by March, 1888, the *Oil, Paint & Drug Reporter* was already discussing the project as a certainty, noting,

> The importance of this line cannot be overestimated. It connects the already overburdened and most extensive oil field of the world with the great western manufacturing center, . . . places fuel oil at the doors of thousands of . . . industries, furnishes an outlet for the . . . 4,000,000 barrels already stored in this field and makes room for the thousands of barrels that are daily coming to the surface and going to waste . . . for lack of surface facilities.[45]

Actual construction, which began in April, generally followed the route of the Chicago & Atlantic Railroad between Lima and Chicago. The railroad had much to gain and little to lose by granting the right-of-way to Buckeye. Although fuel oil shipments by tank car out of Lima were running about 7,000 barrels per day during the first six months of 1888, these were split among the various lines running through Lima.[46] Under its agreement with Buckeye, the Chicago & Atlantic not only obtained revenue from hauling materials for the pipeline but was guaranteed one-third of Standard's westbound shipments of refined products.[47]

Compared to the lines built earlier in Pennsylvania and New York, the 205-mile Chicago line was relatively easy to construct. In part this was because of the level terrain, the highest elevation encountered being only 10 feet above Lima. As a consequence only one pump, located at Lima, was originally needed for the entire 8-inch line, although it was reportedly the largest oil pump built to that date.[48]

But the most striking difference between the construction of the new line and those built in the early 1880's involved the substitution of machinery for human and animal power. A traction engine laid the pipe on the surface parallel to the railroad tracks; a steam-powered "screw machine," first employed on the Cygnet-Lima line, joined the threaded sections of the pipe; and a steam-powered ditching machine dug a 3-foot trench. Manual labor was principally confined to removing stumps and

---

* Apparently Standard attempted to make the fuel oil more attractive to customers by reducing its sulphur content. In December, 1886, one contemporary account described the Solar refinery as ". . . merely a big deodorizing establishment, using the Dufur process, which eliminates the bad odor at a cost of about 1½¢ per barrel."[44]

boulders which the ditcher would not handle and in placing the pipe in the trench.[49]

After four months of construction activity and an estimated outlay of $2 million, Buckeye was ready to pump oil to Chicago. The first of the 65,000 barrels needed to fill the pipeline left Lima on August 8, 1888, and entered receiving tanks on the outskirts of Chicago twelve days later. A week later the first tank cars were loaded at the western terminal to deliver the oil to industrial users.[50]

The Chicago press generally welcomed the completion of the line as a sign of industrial progress and an indication that the "fuel of the future" would be supplied in ample quantity. There was some disappointment over the fact that with the pipeline in operation the price of crude delivered at Chicago remained at 60¢ a barrel.[51] Major Owston in defending the price stated, "Having this pipe . . . in contemplation long before work was done on it we reduced the price of oil to a pipeline basis from the start." [52]

Even at 60¢ a barrel, consumption of Ohio crude as a fuel oil rose rapidly and by fall of 1888 daily shipments over the pipeline were running close to the 8,000-barrel capacity for which it was originally designed. Daily consumption by industrial plants in the Chicago area amounted to about 3,500 barrels, while the remainder went by tank car or barge to manufacturing establishments in 217 cities in twenty states and territories.[53] In addition to shipments from Chicago these markets were also supplied with another 3,500 barrels daily, which went directly by tank cars from Lima. All in all slightly over 3 million barrels of crude were distributed from Lima during 1888.[54]

Although Major Owston had proved himself a good prophet, distribution of 3 million barrels of fuel oil absorbed less than a third of the output from the Lima-Indiana fields, which amounted to almost 9.7 million barrels during 1888. As a result, Buckeye's inventories of crude by the end of the year were in excess of 9.8 million barrels, almost 6 million barrels larger than the amount held by the pipeline company twelve months earlier.

For the Standard management this accumulation of inventories was not an occasion for alarm. There was every indication that the fuel-oil markets for Ohio crude would continue to expand. It is true that under the system of internal accounting used by the Trust, Buckeye's fuel-oil marketing division operated at a loss, but returns on Standard's entire Lima-

Indiana operations yielded profits which during 1891, for example, amounted to approximately 13.5 per cent on the total investment.[55]

There was, however, a much more important reason for any lack of concern by Standard's executives with mounting inventories of crude. In October, 1888, Rockefeller received word that after almost two years of experimentation at Standard's refineries at Lima and Cleveland, ". . . we have succeeded in producing a merchantable oil [from Lima crude]." [56]

Actually, it took further experimentation extending over the next few years to perfect the new refinery process. But the prospect of obtaining high yields of good quality kerosene from Lima crude was sufficient for Standard to expand its operations based on Ohio-Indiana oil, including substantial purchases of producing properties.

Standard's moves into production

For Rockefeller and his associates the move into production represented a change in basic policy. For some years various members of the management group, many of whom had personally acquired an interest in producing companies, had strongly urged such a move but without success.

The prevailing management policy was expressed by Charles Pratt, who wrote Rockefeller in 1885, "Our business is that of manufacturers and it is my judgment an unfortunate thing for any manufacturer or merchant to allow his mind to have the care and friction which attends speculative ventures." [57]

By the mid-1880's the Trust did in fact own a number of producing properties, but these had been acquired more or less incidentally in connection with deals made primarily for other reasons. Such was the case, for example, when Standard purchased the Producers' Consolidated Land & Petroleum Company in 1877, which had planned an independent pipeline from connecting the Regions to Buffalo.* As late as 1888 the combined output from the various properties owned by the Trust did not exceed 200 barrels per day at a time when total daily production for the United States was averaging nearly 76,000 barrels.[58]

The management had made an exception to its rule of not going into production in respect to natural gas. Early in the 1880's, Daniel O'Day was

---

* Similarly Standard bought producing properties in West Virginia in 1885 to forestall the Baltimore & Ohio Railroad's threat of entering the business of refining.

authorized to spend $100,000 on a gas pipeline which supplied fuel to Standard's crude trunk-line pumping stations in the Regions. In 1883, National Transit's executives were further authorized to expand their operations in natural gas not only as pipeline operators but into production as well. Within three years National Transit had either organized or bought a controlling interest in nine natural gas companies operating in Pennsylvania, New York, and eastern Ohio representing a total investment of over $7 million. The financial success of this venture did much to convince the top management of the Trust that profits could be made in the production as well as in the distribution and marketing of a raw material.[59]

The development of a practical method of refining Lima crude offered the Standard management an unusually attractive opportunity to enter into production. Expanding output and the large number of producers who continued to operate with oil selling at 15¢ a barrel at the wellheads suggested that profits could be made even at this price.* With Pennsylvania crude selling at nearly six times this figure, there was every reason to expect the value of Lima crude to increase once the refining methods had been perfected.

Under these circumstances Standard, late in 1888, began to acquire extensive producing properties in the Lima-Indiana fields. In January, 1889, National Transit bought the entire interest in the Amazon Oil Company, which owned or had under lease nearly 6,000 acres in Wood and Sandusky counties with 32 producing wells and a monthly production of 75,000 barrels. Commented the Pittsburgh *Commercial Gazette,* "The Standard by this deal has obtained the bulk of Ohio oil lands. The only large tract not yet secured by the corporation adjoins the one just sold . . . owned by the Ohio Oil & Gas Trust. Should the big corporation secure this, it would be the complete master of the Buckeye state oil fields." [62]

Standard apparently did not buy the Ohio Oil & Gas Trust but in April, 1889, it acquired an even greater prize, the Ohio Oil Company, which in the fall of 1887 had been formed to consolidate the interests of a number of smaller companies and individual producers. Operating in Allen and Auglaize counties to the south of Lima, the Ohio company's

---

* There was much evidence to support this contention. The Ohio Oil Company, according to its own account, was largely built "on 15-cent oil." [60] The U. S. Geological Survey observed that there was every indication that many producers during "the era of 15 cent oil" had more than their share of success.[61]

assets as of January 1, 1889, were reported as including ". . . 20,911 acres of land in fee [or leased] . . . with 146⅓ wells thereon and all machinery, tools, supplies, etc. including 10 dwelling houses, 10 rigs up and undrilled, and 30,000 barrels oil in tanks" valued at nearly $900,000.[63]

Following the purchase of what Edward Orton, the state geologist of Ohio, described as ". . . the most sagacious or more fortunate of the [large] independent producers," Standard began to buy the holdings of "scores" of small companies and individual producers in all parts of the fields. These were supplemented by leases in promising districts as yet undeveloped and by the purchase of many thousands of acres in fee simple.[64]

By 1889, Standard was producing approximately two-fifths of the oil from the Ohio-Indiana fields, and by 1891 had boosted its percentage to well over half.[65] Once having decided to move into production the management began late in 1889 to acquire developed gas and oil properties in New York, Pennsylvania, West Virginia, and Kentucky. These purchases brought the proportion of oil produced by Standard in the Appalachian fields to nearly 14 per cent by 1891. In 1891, Standard, with an investment of some $22 million in producing properties, thus accounted for over a quarter of the total output of domestic crude.[66]

Reaction in the Pennsylvania Oil Regions to Standard's entry into production was varied. Many were inclined to view the move with suspicion and were curious about what the Trust expected to gain by it. Others greeted the new situation as signalling a new chapter in oil history when presumably Standard would be more sympathetic with the problems of the producers.

Among the most vocal objectors to Standard's move were the members of the oil exchanges. Already disturbed by the growing practice of the Trust to buy crude directly from producers, they correctly assumed that the volume of pipeline certificates would be further reduced as Standard began producing its own crude.*

Not all the members of the exchanges took this position. One anonymous correspondent of the *Bradford Star* pointed out to his colleagues,

> So long as the Standard does not rob the owners or steal the land, but pays full equivalent, and both parties to the deals are satisfied, it is difficult to understand and by what right we who have not any oil territory to sell do any grumbling. If the Standard robs the producer in no other way than by pay-

---

* For a discussion of the events leading to the final demise of the oil exchanges, see the following chapter.

ing him a good price for his property, can he complain? And if there is any crime, in the transaction which is the greater criminal, the fellow who buys or the fellow who sells and pockets the cash? [67]

Many of the Ohio producers who sold out to the Trust during 1889 and early 1890 were particularly suspicious that they had been duped. They alleged that by keeping the price of crude low even after it was clear that a practical refining method had been developed, the Standard management had bought their properties at bargain rates.

Commenting on these purchases, for example, Edward Orton noted late in 1889,

> . . . let it be remembered [that they] were made on a basis of oil at fifteen cents a barrel, the price to which the product of the Trenton limestone had been forced by the Standard control . . . [Since] this price has no relation to the intrinsic value of the oil . . . it is easy to see that the Standard Oil Company is likely to gather more wealth from Trenton limestone oil than all it has accumulated in the eastern fields.[68]

Standard's general defense of its policies regarding both the purchase of Lima crude and producing properties was that they were undertaken before it was clear that a practical method of refining sulphur crude was available. As John D. Archbold put it, ". . . it was another case of the farsightedness of the Standard Oil interests with reference to the possibility of the trade. We were willing to take that chance." [69]

The Whiting refinery

Whatever the uncertainties respecting the practicality of processing Lima crude, Standard had scarcely launched its program of acquiring producing properties when the management began planning an expansion of refining capacity. By early 1889 a decision had been made to enlarge the Solar plant at Lima. But because of Chicago's advantageous position relative to markets, it was decided instead to locate the major addition to capacity in a new refinery there.

The proposed location had several attractions over the alternative of concentrating all of the new capacity at Lima. It was undoubtedly cheaper to pipe crude through Buckeye's trunk line to Chicago than to ship refined oil the equivalent distance west by rail. And while Lima's rail connections were good they could not match those of Chicago, which was served by a dozen or so trunk lines to carry refined oil to markets throughout the West and Southwest.

The original plan called for the new plant to be located at the South Chicago terminus of the pipeline from Lima, but high land costs and taxes, trouble with insurance underwriters, and opposition from local residents led to a shift to Whiting, Indiana, some seventeen miles to the south. The disadvantage was not serious. Construction materials could be delivered, and once the works were built refined products could be shipped out by either the Pittsburgh, Fort Wayne & Chicago, the Lake Shore, or the Baltimore & Ohio. In addition, Standard in 1890 obtained control over the Chicago & Calumet Terminal Railroad, a belt line from Whiting to the trunk lines entering Chicago. The project was supervised by the Trust until a new subsidiary, Standard of Indiana, was incorporated in June, 1889.

If Standard did not succeed in concealing its activities it was not for want of trying. Dummy representatives purchased the land and when actual construction began an 8-foot fence surrounded the site. Some two weeks before construction was started on May 5, 1889, the *Chicago Tribune* quoted a statement allegedly from a Trust official that a "mammoth refinery" was to be built at Whiting and within a week after workmen began clearing the ground, the newspaper announced Standard's plans as a fact.[70] Yet, according to one reporter, as late as September, 1889, W. F. Cowan, in charge of construction, admitted only "after a careful period of reflection" that a refinery was being built, but no more.[71]

For most outsiders the construction of the Whiting plant was evidence that Standard, despite its evasiveness, had solved the major problems involved in refining sulphur crude. A further indication of Standard's anticipated utilization of Lima oil at Whiting came in June, 1890, when Buckeye completed a second eight-inch pipeline from Lima to Chicago, thus increasing its daily pumping capacity from about 8,000 barrels to 16,000 barrels.[72]

Meanwhile work on the new refinery moved slowly and it was not until September, 1890, some sixteen months after construction had started, that Whiting's stills received their first charge of Lima crude.[73]

Some time prior to the completion of the Whiting plant Standard began to reconsider its fuel oil marketing policies. Indeed, in April, 1889, the *Chicago Tribune* noted that the Trust was no longer willing to renew its long-term fuel oil contracts, suggesting that the management expected to use crude for higher-valued refinery products.[74]

This was undoubtedly true, although it did not mean that Standard had lost interest in fuel-oil sales. The Solar plant began refining kerosene by the Frasch process some time in 1889, but until the Whiting plant was completed and in full operation, Standard at best could refine only a fraction of the output of Ohio crude.* Furthermore, the refinery yields of naphtha, gasoline, and kerosene from Lima crude initially ran about 30 per cent (later stepped up to 45 per cent) leaving a residual which could either be sold as fuel oil or further processed to obtain lubricating oils, greases, and wax, depending on the relative marketing demands.

In fact, it appears that for at least a year or more after operations began at Whiting in the fall of 1890, the refinery did not push distillation beyond the kerosene range. Rockefeller, who took a dim view of utilizing crude for fuel purposes, wrote a sharp letter to Archbold late in 1891 when he learned that Whiting was only producing 60 barrels of by-product paraffin wax per day:

> With wax at say fifteen dollars a barrel and fuel oil between fifty and sixty cents, it seems we ought not let our valuable paraffine plant be idle in order to supply fuel oil. . . . My view is *we* should distill all the production of Lima oil, run our paraffine plant to full capacity, and supply tar, and ben-zine, if necessary for fuel.[76]

So long as the price of crude at the wells remained, as it did during 1889 and early 1890, at 15¢ a barrel, its use as fuel oil continued to grow, absorbing most of the 5.8 million barrels shipped in 1889.† And if Stand-ard was unwilling to enter into long-term contracts with customers, there were many independent dealers who would.‡

By early 1890 demand had grown to the point where independent buy-ers began paying 2½¢ a barrel more for crude than Standard's posted price of 15¢ a barrel. Standard's response on March 6, 1890, was to raise its price, for the first time in thirty months, to 20¢, only to be matched by in-dependents with the result that by May, Standard was paying 37½¢ a bar-rel for its purchased crude.[79] One immediate effect of the price rise was to stimulate production with the result that, despite a subsequent reduction to 30¢ a barrel in November, output for 1890 rose to 15.1 million barrels.

Apparently the higher prices for crude checked the growing demand

---

* The "consumption" of Lima crude by Standard refineries amounted to 1.1 million barrels in 1888; 1.63 million barrels in 1889; and 1.66 million barrels in 1890.[75]

† Throughout most of this period the price of fuel oil delivered f.o.b. Chicago appar-ently remained at about 60 cents a barrel and included the following cost items: oil at wellheads 15 cents; Buckeye's gathering line charges 20¢; "open freight" rate via railroad tank car, 23.6 cents; total 58.6 cents.[77]

‡ Most of Standard's contracts appear to have been made for one year, while those of independents were on a two-year basis.[78]

for fuel oil since, as shown in Table 22:2, shipments during 1890 of 6.2 million were only 400,000 barrels larger than the amount recorded in 1889. Crude inventories as a consequence reached a new high of 21.5 million barrels at the end of 1890.

For the fuel-oil dealers, particularly those who were under contract to deliver oil at prices based on 15¢-a-barrel crude, the rise in the cost of crude was most unwelcome. Their suspicions that the price changes were part of a plan by Standard to drive them out of business were further aroused during 1891 when the Chicago & Atlantic Railroad, apparently as a part of a general increase in freight rates initiated by the Central Freight Association, raised the Lima-to-Chicago rates on tank car shipments from 23.6¢ a barrel to about 30¢.[80] As the representative of the independent fuel-oil dealers of Chicago, C. D. Chamberlain, later testified, "I felt that inasmuch as water competition is recognized by the railroads . . . as a reason for reducing the rate . . . we were entitled to a lower rather than a higher rate, as independents, not having a pipeline, to compete with the [Buckeye] pipeline to Chicago." The fact that the Central Freight Association refused to consider the argument for lower rates Chamberlain attributed to Standard's influence for "a railroad man told me that Mr. [Howard] Page of the Standard Oil Company was present at the CFA councils deliberating the change."[81]

Independent refiners, on their part, were convinced that Standard's actions were an attempt to make ". . . the price of Lima crude sufficiently high so that competing refineries could not use it to advantage against Standard refineries."[82]

Whatever the original reasons or motivations, the growing refinery demand for Lima crude was reflected in subsequent increase in prices, which, as shown in Table 22:2, moved from about 30¢ a barrel in 1891 to 48¢ in 1894, thereafter ranging between 49¢ and 89¢ a barrel.

It was against this background of rising costs that residual fuel oils were increasingly substituted for unrefined crude in the fuel oil markets. This shift was reflected in Standard's production of fuel oil which, over Rockefeller's objections, rose from about 780,000 barrels in 1890 to nearly 6.9 million barrels by 1897.[83]

Despite earlier success in refining Lima crude at the Solar plant, it was not until the completion of the Whiting plant in September, 1890, that Ohio oil began to assume importance as a source of kerosene. With an original still charging capacity of 10,000 barrels a day—subsequently enlarged to about 36,000 barrels—the Indiana refinery absorbed about

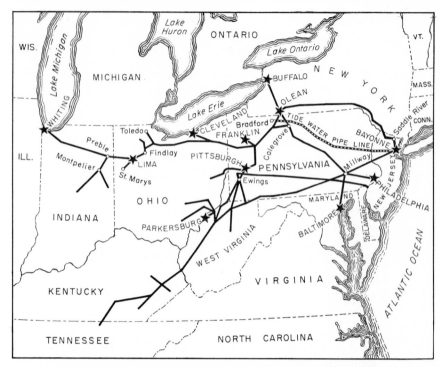

MAP 22:2 *Standard's trunk pipe line system* circa *1900.*

40 per cent of Buckeye's total shipments during the 1890's. And at least toward the end of the decade Whiting accounted for more than one-fifth of the combined throughput of Standard's refineries.[84]

The precise distribution of the remainder is not known but Standard lost little time in making Lima crude available to its major eastern refineries. Early in 1888 Buckeye began laying an 8-inch line east from Cygnet, Ohio, to Mantua, Ohio, where it joined National Transit's western extension connecting Bear Creek, Pennsylvania, with Cleveland.[85] In 1890 Buckeye began delivering Lima crude over this line to Standard's Cleveland refineries which were in the process of being thoroughly overhauled and which operated exclusively on sulphur crude throughout the rest of the decade.[86]

With the subsequent extension of the Cygnet-Mantua line to Colegrove, Pennsylvania, where it connected with National Transit's eastern network, Standard's plants at Olean, New York, were put to refining Lima crude in 1893. (See Map 22:2.)

Until the mid-1890's the illuminating oil refined from Lima crude went almost exclusively into domestic consumption. This was in part due to the geographic loacation of the fields relative to the markets west of the Alleghenies. It also reflected a ruling by the New York Produce Exchange passed in 1888, with Standard's support, which declared illuminating oil refined from sulphur crude was not acceptable for foreign trade. In 1895, Standard was sufficiently confident of the quality of its Lima refined oil to press for a reversal of the ruling which came in September of that year.[87] With this change, Lima crude began moving in increasing volume over the Buckeye-National Transit lines to Standard's refineries in Philadelphia and Bayonne, New Jersey.

It was Standard's financial and managerial resources which enabled it to assume the calculated financial risk of moving quickly into the Ohio-Indiana fields. By providing badly needed gathering lines and storage facilities the Trust undoubtedly prevented a wastage of crude such as had characterized earlier discoveries of flush fields. The fact that this risk paid off handsomely does not detract from Standard's more lasting contributions to the future of the American petroleum industry which arose from its operations in Ohio and Indiana. One was the intensive development of the fuel oil market which made it a permanent and major outlet for petroleum products in the years that lay ahead. Second was the perfection of a method of refining sulphur crude which could readily be applied to new discoveries of this type of oil elsewhere in the United States.

For Standard the emergence of the Ohio-Indiana oil fields could not have come at a more opportune time. Faced with the prospect of declining yields in the Appalachian area and rising competition outside the United States from the Russians, the expanding output of relatively cheap Lima crude made it possible for the Trust not only to maintain but to expand its extensive refining, transport, and marketing facilities.

Indeed, based on Lima crude, the American petroleum industry measured by processing and transport facilities and the volume of products manufactured and distributed, had emerged as one of the world's major industries. But by the turn of the century the incipient decline of the Ohio-Indiana fields made it clear that new sources of production would have to be found if this expansion was to be supported and maintained.

Chapter 23

Refining and refining capacity

The near-record accomplishments in the production and transportation of petroleum which characterized the 1884–99 period were no less matched in the refining segment of the industry. As shown in Table 23:1 refinery throughput of crude expanded some threefold, from about 18 million barrels annually during 1883–85 to 52 million barrels in 1899. This expanding throughput was reflected in a corresponding increase in output of refined products. There was, however, a shift in the relative proportions of particular refined products. Illuminating oil, which made up over four-fifths of the total output in 1883–85, contributed only about two-thirds at the end of the period. By contrast, output of lubricants grew from about 4 per cent to 9 per cent, while fuel oils, commercially insignificant prior to the early 1880's, amounted to some 11 per cent of the total by 1899.*

* The topped Lima crudes originally utilized as fuel oil are not included in this tabulation. The best estimate suggests that perhaps 4.5 million barrels of the Ohio product were sold as fuel in 1889.[1]

Back of this expanding output and increasing relative importance of by-products were several significant changes in the organization and structure of refining. Like other divisions of the industry, refining was dominated by Standard, which had more than maintained its share of expand-

TABLE 23:1

Refinery receipts of crude and output of major refined products: 1883–99 (millions of 42-gallon barrels)

	1883–85	1889	1894	1899
Refinery receipts of crude	21.4	30.7	47.8	52.0
Output of refined				
Illuminating oils	15.2	20.2	29.5	30.0
% of total	82.2	73.2	66.6	60.4
Naphtha-benzine-gasoline	2.4	3.9	6.1	6.7
% of total	13.0	14.1	13.8	13.5
Lubricating oils *	.9	1.8	3.3	4.1
% of total	4.9	6.5	7.4	8.2
Residuum		1.4	1.4	.7
% of total		5.1	3.2	1.4
Paraffin wax		.3	.6	.9
% of total		1.1	1.4	1.8
Fuel oil			3.4	7.3
% of total			7.7	14.7
Total refined production †	18.5	27.6	44.3	49.7

* Includes paraffin oils, neutral, filtered, cylinder, spindle, and other reduced oils.
† Differences between refinery receipts and total output of refined products accounted for chiefly by additions or subtractions from stocks of crude at refineries.
Sources: 1883–85: Table 19:1; 1889: *Eleventh Census of Manufacturing Industries: 1890*, 359–73; 1894: Extrapolated from 1890 and 1900 Census figures; 1899: *Twelfth Census of Manufactures: 1900*, 688.

ing capacity. Although forced to concede a major portion of the by-product manufacture to independents, Standard did move to exploit the market potentials created by increasing industrialization and urbanization in America and western Europe.

It was still possible to operate specialty plants on a small scale during this period. But (see Table 23:2) the general trend was toward plants with larger charging capacities and greater average investment. In large part the trend was a function of expanding markets, growing integration, and specialization in manufacturing. It further reflected the greater recovery of refined products from given outputs of crude which made it more economical to utilize a greater variety and number of stills and more elaborate treating equipment, which could be operated without a proportionate increase in the labor force.

Among the most important features of this decade and a half were the accomplishments associated with sulphur crude. For the first time in the industry's history, the Trenton Limestone deposits of Ohio and Indiana yielded a crude oil unrefinable by known methods and processes. Solving

TABLE 23:2

Comparative summary of the U. S. refining industry: 1880–99

	1880	1889	1899
No. of establishments or firms	86 *	94 ‡	67 ‡
No. of refining plants	89	104	75
"Capital" †	$27,325,746	$77,416,296	$ 95,327,892
Wage earners, average no. §	9,869	11,403	12,199
Value of products §	$43,705,218	$85,001,198	$123,929,384

* No statement as to idle firms.

† The Census' definition is as follows: "These figures represent, not the capital stock of the corporations engaged in the industry, but the value of the plants, together with the live or working capital." *Twelfth Census of Manufactures: 1900*, 683.

‡ Exclusive of two idle firms in 1899 and seven in 1889.

§ Taken from Dept. of Commerce & Labor, Bureau of the Census, *Manufactures, 1905*, Part 4, Special Reports on Selected Industries (Washington, Government Printing Office, 1908), 567.

this problem constituted a major technological achievement. Many processes were developed to solve the problem of refining sulphur crude, but the most commercially successful solution came from Herman Frasch, working under a "crash" research program sponsored by Standard.

The Frasch process

German-born Herman Frasch migrated as a young man to the United States in 1868 where he established a laboratory in Philadelphia. He spent the next several years developing processes of working petroleum on which he received a number of patents. Early in the 1880's, he was hired by the Imperial Oil Company of Canada to develop methods of refining kerosene from sulphurous Ontario crudes, similar to those found in northwestern Ohio. In July, 1886, some four months after moving into the Ohio fields, Standard hired Frasch and assigned him to the Cleveland refinery to apply his experience to Lima crudes.[2]

The limitations of kerosene from Ohio types of crude were already well known. As described by a contemporary, "When the crude was refined . . . , the sulphur carried along into [the] refined products, and when burned in lamps or used in other ways, produced very disagreeable effects [particularly an unpleasant odor] in the room in which the lamp was burning, smudging the chimney and making a very bad wick crust." [3]

Frasch's basic concept of how to remove the sulphur from the Lima crudes was simple. It involved the uniting of the sulphur with a solution of metallic oxide, forming an insoluble precipitate which could then be separated from the oil. In his patent application dated February 21, 1887, Frasch indicated that oxides of lead, bismuth, cadmium, mercury, iron, copper, or silver might be used. [4] But before the principles involved in the patent could be put on a commercial basis much further experimentation had to be made. There was the question as to whether the sulphur should be precipitated in the original distillation or in the redistillation of the kerosene fraction. Details regarding the operation and temperatures of the agitators also had to be worked out. A further problem involved the choice of the most efficient and economical type of oxide and the development of methods of recovering oxides for reuse.

Although further experimentation was needed to perfect the process, by fall, 1888, Frasch had developed the essentials for commercial operations. The first step, or primary distillation, was unchanged. But following primary distillation, the kerosene stock was run to a cheese box "sweetening still" (ranging in capacity between 1,000 barrels and 1,200 barrels) and mixed with an oxide compound consisting of two parts copper oxide, one part lead oxide, and two parts iron oxide. During redistillation a mechanical agitator moved constantly within the still to insure effective mixing of the oxides with the oil. Following this step, the product was chemically treated with sulphuric acid and caustic soda.

To recover the oxides, the precipitated sludge (sulphur dioxide and heavy petroleum fractions) from the "sweetening still" was subjected to filter pressing to remove the oil, followed by burning to remove any remaining organic materials, leaving only the metallic base of the original oxides. To prepare them for reuse, these were put into roasting ovens and exposed to air to reform oxides. [5]

The development of the Frasch process validated for Standard the management's earlier decision to become heavily engaged in the production and transport of Lima crude. Even before the process was fully worked out to produce illuminating oil comparable in quality to kerosene from Appalachian crude, construction was started in 1889 on the new refinery at Whiting.

Although it owned the Frasch patents, Standard was not the only entrant into refining Lima crudes. Attracted by the flush production from the Ohio fields a number of "Western independents" began refining operations in the area, using alternative methods of removing sulphur from crude. Among the more successful was the Paragon Refining Company of Toledo, Ohio. Paragon utilized what became known as the Pitt process, discovered by a high school chemistry teacher at Buffalo. Essentially "it removed the sulphur compounds by absorption into iron filings when the crude was fired to a vapor." [6] Other processes, of which little is known but which were apparently successful, were used by the Craig Oil Company, the Peerless Works, and later the Sun Oil Company, all of which had plants in Ohio.[7]

Considering the concern of many oilmen over the future of crude reserves, these technological achievements were enthusiastically received.[8] It was generally recognized, however, that the Ohio product was still inferior to most Appalachian crudes as a suitable raw material for refining.

Under ordinary distillation procedures, Lima and Appalachian crudes yielded roughly the same proportion of the lighter distilled fractions and residue stocks: on the average, ordinary distillation through to the kerosene fractions amounted to 50 to 55 per cent. Refining loss was approximately 6 to 7 per cent for both grades of crude, and the remaining stocks from which lubricants, waxes, and other by-products could be further distilled were approximately the same; between 40 and 45 per cent. This rough equivalence, however, masks quite substantial differences in the quality and refining flexibility of the two crudes. Cracking Appalachian crude increased yields of the lighter fractions to as high as 70 to 75 per cent, while it was virtually impossible to crack Lima crude and obtain a commercially satisfactory product.[*]

After the light fractions were cut, the residue—normally used for lubricants and paraffin wax stocks—differed similarly. In contrast to Appalachian stocks, under the most careful attention and the best operating conditions it was impossible to produce a lubricating oil from Lima oil which did not retain some sulphur. With even small sulphur content such a lubricant, when used for example against metals, would react to form a corrosive sulphide whose action would be further aggravated by the oil's inability to withstand high heats.

[*] The difficulty stemmed largely from the highly unpleasant odor of cracked Lima crude. Referred to by refiners as "stink," it was caused by a high proportion of mercaptans in the distillate. Deodorization costs would have been prohibitive with virtually no assurance that the odor could be completely removed and that yields of the kerosene fractions would have been sufficiently increased to warrant the extra expense of treating and deodorization.

In view of these technical and commercial restraints upon Lima crude the product yields were limited to illuminating oils (40 to 45 per cent) and the remainder (less refining loss) primarily to distilled fuel oil. At the Whiting refinery, for example, throughout the 1890's illuminating oil, distilled fuel oil, and naphtha-gasoline were the major products produced, in that order. Only a small amount of lubricants, principally greases, and insignificant quantities of paraffin wax were manufactured.[9]

The limitations of Lima crude tended to elevate the status of most Pennsylvania grade oils to a premium position as a source of lubricants and waxes. And with much of Standard's refining operations, almost of necessity, centered on the Ohio product, smaller eastern independents found it increasingly advantageous to limit their distillation of Appalachian crudes to the lighter illuminating oil fractions in order to maximize yields of lubricating stocks. By concentrating on the production of by-products, particularly lubricating oils, independents found they could build a secure market position through brand names, high-quality products, and superior service.

But to maintain operations the independents faced the problem of obtaining crude supplies at a time when opportunities for purchasing oil in the open market were becoming more restricted. This trend was reflected in the continued decline in the importance of the oil exchanges.

Integration and the decline of the oil exchanges

At the time of the discovery of the Lima fields it was still possible for nonintegrated refiners to obtain crude through the oil exchanges. Even Standard's refineries, which absorbed the great bulk of the oil produced at this time, acquired all but a small portion of their crude by purchasing pipeline certificates in the exchanges.[10]

In 1884 the Trust, in order to centralize its purchases of crude, organized the Joseph Seep Agency. At the same time Seep announced that his prices would be an average of the daily high and low quotations for certificates on the "leading oil exchanges."[11]

Since Standard was already the largest purchaser, transporter, and refiner of crude, these moves had little immediate effect on the functioning of the exchanges. But as Seep gradually began to supplement his purchases of pipeline certificates by buying direct from producers or Standard's pipeline subsidiaries, the volume of crude traded in the exchanges was correspondingly reduced. Transactions became increasingly speculative during the late 1880's, with many purchases and sales of certificates but with little actual oil changing hands. Standard, as well as independents, increasingly began to by-pass the exchanges by purchasing their crude direct from producers. By the early 1890's the exchanges had largely lost their marketing functions. As the *Derrick's Hand-Book* explained,

> Speculation in oil . . . began to decline after 1887. One exchange after another was abandoned, and in 1894 only one, that of Oil City, pretended to do any business. And here the transactions had come to be confined within narrow limits, and the sales and resales of oil had shrunk to very small proportions. For the entire year 1894 the total sales of oil reported . . . were but a fractional part of the business transacted during a single day in 1882 or 1883.[12]

The growing unimportance of the exchanges in facilitating the marketing of crude from the new Ohio fields matched their position in the eastern region. The Oil City Exchange, early in June, 1890, passed a resolution requesting that Buckeye list its certificates and "give the trade the opportunity of dealing in Lima oil." [13] Generally supported by the other exchanges, Buckeye agreed to a listing on a trial basis.[14] Within a few weeks exchanges were established at Lima and Findlay and trading in Ohio oil was begun at the eastern exchanges. During the first full trading day on August 18, 1890, close to 500,000 barrels were cleared at Findlay and Lima, and slightly more than that amount at Oil City.[15] But within a week sales were almost nonexistent and less than a month after opening the Ohio exchanges were closed. Soon after, Buckeye withdrew its authorization from the eastern exchanges and Lima crude again flowed through its accustomed marketing channels.

In January, 1895, Seep publicly closed the era of the oil exchanges in his famous circular. Dated January 23, it came as no surprise to most oilmen:

> The small amount of dealing in certificates on the exchanges renders the transactions there no longer a reliable indication of the value of the product. This necessitates a change in my custom of buying credit balances. Hereafter in all such purchases the price paid will be as high as the markets of the world will justify, but will not necessarily be the price bid on the exchanges for certificate oil. Daily quotations will be furnished you. . . .[16]

With the disappearance of the oil exchanges it was no longer possible to buy crude in an open market, and independent refiners, if they were to survive, were forced to develop their own sources. The independent refiners who managed to find success in Ohio, for example, were all vertically integrated, having their own production and transportation facilities and usually a working agreement with a marketing company for the distribution of refined products. Most far-sighted eastern independents had in some measure become vertically integrated late in the 1880's.[17] Thus by the early 1890's it was no longer possible to enter the refining business on any sizable basis simply by building a refining plant. Arrangements for crude supply and transportation were as much a component of entry cost as the plant itself. Combined they made the barriers to entry greater than at any previous period in the history of the industry.[18]

Plant scale, costs, and gross margins

Even though independent refiners might be completely integrated, few could hope to match the scale of operations of companies affiliated with Standard. In 1885 the 8,000 barrels per day charging capacity of the Bayonne plant alone was nearly equal to that of the 21 major independents in the Region, Pittsburgh, and Cleveland, that formed the Independent Petroleum Refiners' Bureau.[19] Ten years later the capacity of the Whiting works was greater than all the independent capacity combined, and nearly three times as large as Pure Oil, the major eastern competitor of Standard.[20]

While the plant scale was disproportionately in favor of Standard installations, this disparity, as shown in the following tabulation, was not reflected in Standard's level of processing costs, which rose slightly after 1895.

| | Refining Costs | |
Year	Cents per barrel	Cents per gallon
1890	17.33	0.41
1895	16.31	.39
1900	19.61	.47

Source: *US v. SONJ* (1906), *Brief of Facts and Argument for Petitioners*, I, 199.

But this increase is rather misleading. Cost estimates for 1890 and 1895 undoubtedly include some Appalachian crude processed in Standard refineries; by 1900 all the major plants of the Trust were running Lima crude, which in large part accounted for the slight rise in costs.* More-

---

* Before the improvements and final commercialization of the Frasch process the cost of processing Lima crude was generally considered from ¼¢ to ½¢ higher per gallon than working Appalachian crude.[21]

over, the largest plants of the Trust, particularly Bayonne and Whiting, may well have achieved processing costs even lower than the estimates shown in the tabulation.

No comparable summary of processing costs can be made either for the Ohio independents or Standard's eastern competitors. The best estimates place the processing costs for refiners of Appalachian crude in plants of about 3,000 barrels to 4,000 barrels daily charging capacity at ½ cent per gallon.[22] For smaller installations possibly ⅛¢ more per gallon was required.[23]

Thus while Standard's refineries were still attaining economies of scale, by the mid-1890's their advantage over independents was only nominal for two reasons: one, the growing use of Lima crude in Standard plants; the other, the economies of scale attained by major independents through their own vertical integration.

For most eastern independents, however, the general level of processing costs was much less important than the refining margins for by-products. As indicated in Table 23:3, gross refining margins on illuminating oil averaged about 5¢ a gallon between 1890 and 1899. By contrast, the margins for four representative grades of lubricating oils typically ran about twice this figure.*

Refinery output

The growing demand for lubricants and industrial fuels was largely a function of the tremendous strides made in the industrialization process not only in the United States but also in most of the Western world. The growth of lubricating oil production in the United States rode into importance on this wave of industrialization. From 1889 until mid-1892 demand for mineral lubricants was nothing short of phenomenal. The Standard Oil Company alone manufactured over 900,000 barrels of all types of lubricants in 1890, and probably over a million barrels for each of the following two years. In 1884–85 comparable output totals were not two-fifths of this amount.[25]

Trade was affected by the world-wide depression of the early 1890's, when exports, which had absorbed about 30 per cent of total domestic output of lubricants, fell off by about 10 per cent.[26] Nevertheless, pessi-

---

* Margins on premium grade oil were substantially higher. A neutral lubricating oil of 34 B gravity, for example, sold at wholesale for about 20¢ a gallon throughout most of the decade.[24]

TABLE 23:3

Gross margins, lubricating and illuminating oils: 1890–99 (all figures in cents per gallon)

	Wholesale price *	Appalachian crude at field †	Average open tank-car rate to seaboard ‡	Processing cost	Gross margin
1890					
Illuminating oil	9.95	2.10	1.50	.50	5.85
Lubricating oil	15.20	2.10	1.50	.50	11.10
1895					
Illuminating oil	9.32	3.20	1.50	.50	4.12
Lubricating oil	14.20	3.20	1.50	.50	9.00
1899					
Illuminating oil	10.15	3.10	1.50	.50	5.05
Lubricating oil	14.50	3.10	1.50	.50	9.40

* New York job lots, 150F° water-white. *Report of the Commissioner of Corporations on the Petroleum Industry*, Part 2, *Prices and Profits*, 229. Simple average of wholesale New York prices of paraffin, neutral, black, and cylinder oils. *OP&D Reporter* (Dec. 3, 1890, July 20, 1896); *Report of the Commissioner of Corporations on the Petroleum Industry*, Part 2, *Prices and Profits*, 933–34.
† Computed from *Derrick's Hand-Book*, II, 102.
‡ Explanation of the rate structure is handled explicitly in Chapter XXV.

mism amongst manufacturers was far from widespread. As E. T. Bedford, of Standard, pointed out in April, 1893,

> The sale of mineral lubricating oils is greater today than at any time in the history of the business. The increased demand is undoubtedly due to the improvement in quality, which induces buyers to give them the preference over all other lubricants. With . . . well recognized advantages it is not strange that they should have supplanted all other lubricating oils.[27]

J. E. Borne, of Borne-Scymser, another Standard affiliate, was even more optimistic. "The outlook," Borne said, "is about as good as it ever was. . . . The number of industries using mineral lubricating oils is increasing, and consequently the volume of business in this line of trade is steadily growing. . . . While many dealers report a dull trade, I cannot say that that has been our experience. We find the demand fully up to the average." [28]

A long editorial in the *Oil, Paint & Drug Reporter* (December, 1895) summarized the past gains and future prospects for mineral oil lubricants. Referring to an approximate 20 per cent increase in exports during the preceeding 18 months, the *Reporter* stated,

> Considering the general improvement of trade this year, it was natural to look for some expansion in the distribution of mineral lubricants, but the fact that the sale of this class of goods has increased disproportionately to the de-

mand for competing articles of animal or vegetable origin emphasizes the view that mineral lubricants are, year after year, becoming more firmly established in popularity.[29]

This confidence in the future prospects for sales of lubricating oils proved well grounded. As shown in Table 23:1, the production of lubricants by the American refineries, which had increased from about 900,000 barrels in 1883–85 to 2 million barrels in 1889, rose to 4.1 million barrels in 1899.

Any estimates of the distribution of lubricating oil production between Standard and non-Standard refiners are limited to observations for a few scattered years. In certain classes of lubricants it is clear that Standard produced the greater share. By one account, for example, Standard's Galena-Signal subsidiary supplied approximately 90 per cent of the market for railroad oils during the 1890's.[30] Considering all grades of lubricants, however, Standard's share in 1890 did not exceed 40 per cent. Seven years later, when the total output of lubricating oil was approximately 4.1 million barrels, Standard's share remained at about 40 per cent.[31]

While independents contributed nearly 60 per cent of the lubricants produced in the 1890's, their share of the other petroleum by-products was considerably less. In the manufacture of petrolatum, for example, there were a few independents, but the bulk of the business was taken by the Chesebrough works, subsidiary of the Standard Oil Company.

Because crude paraffin wax is a joint product of lubricating oil processing, it is not surprising that the independent's share of the total domestic production of waxes more nearly reflected their contribution to lubricating oil manufacture. As indicated in the following tabulation, output of paraffin wax grew eight-fold from 1888 to 1899. Standard's share of the total in 1892, the first year for which information is available, was nearly two-thirds; in 1899 half the wax came from independent refineries.

Domestic output of paraffin wax

Year	Total in millions of lbs	Standard's share of total percentage
1888 *	53.0	...
1892	114.9 †	65
1899 *	383.0 †	50

* Estimated from data in Table 22:1.
† Defendents' Exhibits, No. 283, *US* v. *SONJ* (1906), Transcript, XIX, 682.

In contrast to lubricants and waxes processed almost exclusively from Appalachian crudes, distilled fuel oils, the third largest product of American refineries by the late 1890's, was almost solely a by-product of Lima crude. Standard, which had dominated the early sale of topped Lima crude as a fuel oil, also dominated the subsequent production of the distilled product.

The only important competition faced by Standard in the fuel oil markets came from independent refiners of Ohio oil. With an aggregate daily crude refining capacity of between 7,000 barrels to 10,000 barrels, the independents were reasonably successful in selling considerable amounts

TABLE 23:4

Comparison of Standard's pipeline runs to total production, Lima-Indiana fields: 1890–99 (in millions of 42-gallon barrels)

Year	Total prod.	Standard pipeline runs *	Ratio of runs to production
1890	15.1	11.9	78.8
1891	17.4	14.5	83.3
1892	15.9	13.6	85.5
1893	15.9	14.4	90.5
1894	17.3	16.1	93.1
1895	20.1	18.4	91.0
1896	25.2	22.2	88.0
1897	22.8	19.7	85.9
1898	20.3	17.1	84.4
1899	20.2	17.2	85.1

* Buckeye and Indiana Pipeline.
Sources: *Mineral Resources, 1896–97,* 753, 808–9; *Report of the Commissioner of Corporations on the Petroleum Industry, Part 1, Position of the Standard Oil Company,* 145; C. N. Payne testimony, *US* v. *SONJ* (1906) *Transcript,* 1, 357–58. All figures rounded from raw data.

of fuel oil (and kerosene). In fact, because Standard's eastern refineries by the mid-1890's were processing substantial amounts of Lima crude, the independents may well have supplied a quarter or more of the fuel markets in the South and Midwest. But their share of the total domestic distribution of fuel oils, at most, could not have exceeded 15 per cent. As indicated in Table 23:4, Standard's share of pipeline runs of Lima crude averaged over 85 per cent throughout the 1890's.

Against this widening range of products from the heavier petroleum fractions—whose significance can hardly be overestimated—illuminating oil and (almost of necessity) its joint product, naphtha, continued by volume to be the major products of American refineries. The distribution of illuminating oil production between Standard and independent refiners is a further demonstration of the importance of lubricants and waxes for the success of the independents. In 1890, for example, of an estimated total output of illuminating oil of nearly 22 million barrels, the Trust's share was some 17.2 million barrels, or about 80 per cent. In 1897, when total illuminating oil output had expanded to about 33–35 million barrels, just over 26 million barrels were produced from Standard refineries.[32]

Thus, while independents were contributing about 60 per cent of the lubricants and nearly 50 per cent of the wax produced during the late 1890's, Standard continued by a substantial margin as the predominant American refiner. Its share of total output of refined products—from illuminating oil to paraffin waxes and fuel oils—was estimated by one Standard Oil official as follows: 1894, 81.4 per cent; 1895, 81.8 per cent; 1896, 82.1 per cent; 1897, 82.4 per cent; 1898, 83.7 per cent.[33]

Growth and location of refining capacity

One of the most important results of Frasch's demonstration of the commercial feasibility of processing Lima crude was its effect on refining capacity. Apparently the threat of crude exhaustion and the uncertainties associated with the shut-down movement of 1887–88 dampened enthusiasm for new entrants into refining, for in 1888 the operating rate of refineries was just over 80 per cent of total working capacity (see Table 23:5).

With the "conquering of Lima crude," doubts about the future prospects for crude reserves were largely dissipated and refining entered a new era of expansion, consolidation, and improvements. But despite a growth in plant and equipment of almost 70 per cent in less than ten years (in terms of daily crude charging capacity, roughly from 140,000 barrels to 220,000 barrels between 1888 and 1895–97), the industry, as shown in Table 23:5, operated during the late 1890's closer to its "working" and rated capacity than at any time in its history.

The growth in total refining capacity, as indicated in Table 23:6, was quite evenly distributed between Standard and the independents. Most of Standard's expansion in turn was accounted for by the construction of new plants and enlarging of existing works. The principal exceptions were the acquisition between 1887 and 1891 of the plants of Logan, Emery &

TABLE 23:5

Annual operating rate of refining capacity: 1884, 1888, 1895–97 (42-gallon barrels)

	1884	1888	1895–1897
Rated Capacity *	37,460,400	42,088,200	64,743,000
Less stand-by capacity †	9,365,100	10,522,050	13,685,750
Working capacity	28,095,300	31,566,150	51,057,250
Refinery receipts of crude ‡	21,761,129	26,601,936	47,889,022
Excess working capacity	6,334,171	4,964,214	3,168,228
Operating rate of working capacity	78%	85%	94%

* Daily capacities, taken from Table 23:6, were multiplied by 300 days to obtain an annual figure.
† The same method of calculation was used as in Table 18:6, although a 25 per cent deduction for stand-by capacity for 1895–97 may be somewhat high, considering the greater utilization of by-product manufacture by this time.
‡ Total pipeline shipments minus exports. Taken from *Derrick's Hand-Book*, I, 811, II, 144; *Mineral Resources, 1905*, 883.

TABLE 23:6

Estimated capacities of major refining areas (42-gallon barrels per day)

	1884		1888		1895–97	
	Standard	Independent	Standard	Independent	Standard	Independent
Cleveland	22,000 ‖	.....	22,000	3,310	20,000	3,310
Pittsburgh	.....	.....	.....	4,103	.....	5,000
New York–						
New Jersey	43,000	.....	49,266	3,500 ‡	55,000	7,000 §
Philadelphia	13,000	.....	17,200	3,500 ‡	45,000	3,500
Baltimore	.....	.....	2,000	610	3,000	.....
Whiting, Ind.	.....	.....	.....	.....	36,000	.....
Toledo-Lima	.....	.....	.....	.....	7,000	7,000
Buffalo	1,600 ‖	.....	3,000	1,500	4,500	.....
Oil Region	.....	.....	12,000	9,000	7,000	12,000 §
Other *	.....	.....	.....	4,000	2,500	.....
Std. acquisitions (from Jan. 1, 1888)	.....	.....	2,000 †	.....	3,000	.....
Total	96,000	28,868	110,771	29,523	183,000	37,810
Total U. S.	124,868		140,294		220,810	
% Standard	78%		79%		82%	

* Includes Western and Marietta, Ohio, plants.
† See letter of S. C. T. Dodd to Henry Bacon of May 23, 1888, in *1888 Trust Investigation*, 394.
‡ Includes Tidewater group, in which Standard held minority interest.
§ Includes Tidewater and Pure Oil Co. affiliated refineries.
‖ See footnote above, Table 18:2.
Sources: These estimates were built from the relatively firm estimate of 1884 in Hidy & Hidy, *Pioneering*, 100–101, 120, and statements submitted by Archbold, *Industrial Commission Report*, 1, 541–42, and earlier in *1888 Trust Investigation*, 438–40. Also C. B. Matthews testimony, *1888 Trust Investigation*, 424; Table 18:2; *OP&D Reporter* (March 4, 1891, March 1, 1897, Oct. 26, 1891); *Petroleum Age* (May, 1885), 1009 (Sept., 1886), 1462 (May, 1887), 1651–55 (July, 1887), 1682.

Weaver, Malcolm Lloyd, the Chester Oil Company (a former Tidewater affiliate) and the Globe Refining Company. These were all dismantled and the equipment moved to Philadelphia to increase the capacity of Standard's Atlantic Refining Company.[34]

Subsequent acquisitions included the two plants of the Bear Creek Refining Company at Marcus Hook and Pittsburgh, and the Mutual, Union and International Oil companies all in the Region. The purchase of the Bear Creek properties was part of an arrangement with the Mellon interests to strengthen Standard's export position.[*] In respect to the other plants, the principal motive was to acquire stock in the newly formed Pure Oil Company.

Of much greater importance in increasing Standard's refining capacity, however, was the building of new and the expansion of old existing plants. The Whiting works, for example, which by 1895–96 was capable of running at its planned capacity of 36,000 barrels of crude daily, was constructed early in the 1890's. Similarly the Solar works, with a daily capacity of 7,000 barrels built at Lima in 1886, was also new. The Philadelphia and eastern seaboard plants were all expanded; some by adding new buildings and installations, others by the addition of equipment from dismantled refineries. The Bayonne plant, for example, was by 1895–97 nearly two-thirds the size of Whiting, the largest of all Standard Oil plants. By 1900, Bayonne's capacity was nearly equal to Whiting's.

Accompanying this growth in Standard's total refining capacity was a further trend in the concentration of its operations geographically. Plants in New York–New Jersey, Philadelphia, Whiting and Cleveland, in that order, accounted for 160,000 barrels of Standard's total daily capacity of 183,000 barrels in 1895–97. Two-thirds of this amount, in turn, was contained at Whiting, Bayonne, Atlantic Refining's Philadelphia unit, the Long Island works, and Sone and Fleming in Brooklyn.[35]

The relative importance of the Trust's control over output from the Ohio fields is indicated by the fact that, of the approximate 46 million barrels of oil processed by Standard refineries in 1899, about one-half was Lima crude. By this time the three major refining Standard plants—Bayonne, Whiting, and Atlantic Refining—were processing Lima oil almost exclusively. Bayonne alone, for example, consumed some 6.7 million barrels of Ohio crude in 1899, about 90 per cent of its total requirements, and nearly 40 per cent of total shipments of Lima crude for that year.[36] An additional amount, close to 11 million barrels, went to Whiting.

[*] See Chapter 24.

In meeting the challenge of Lima crude and by moving rapidly to insure control over output, storage, and shipment of crude from the Ohio fields, Rockefeller and his associates made certain their continued dominance of the refining segment of the American industry. That Standard's control fell short of complete monopoly is indicated by the success of independents, particularly those specializing in the production of lubricants and waxes. Even in illuminating and fuel oils the integrated Ohio independents competed on a limited basis with Standard in the domestic market.

But if management had little reason to be concerned about its share of domestic sales, this was not true of markets outside the United States, where Standard faced vigorous competition from the Russians and, by the late 1890's, from oil produced in the Far East.

Chapter 24

Foreign marketing and the rise of
competition

For most of the twenty-five years following the Drake well, the American industry, virtually the world's only source of an exportable surplus of petroleum, had sold by far the larger portion of the output of its refineries outside the United States. As the first quarter-century of the industry's history drew to a close there was little reason to assume that demand for petroleum products would not continue to expand both at home and abroad. But when oil began moving over the Baku–Batoum railway during the 1880's, there was a growing realization that the emergence of the Russian industry posed a threat to the hitherto unchallenged position of the United States in the world's markets for petroleum.

Initial reactions of American observers both as to the significance and imminence of this threat were varied. According to the *Scientific Ameri-*

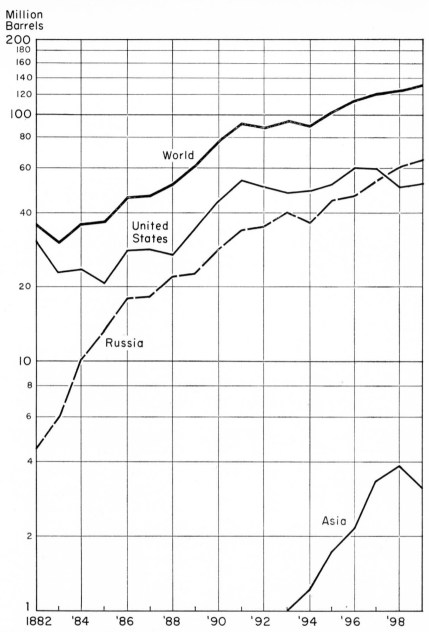

CHART 24:1. *World production of crude (42 gallon barrels) and output of major producing areas, 1882–99. (Source:* Mineral Resources, *1929, 470–71.*)

can in May, 1884, new management and improved transport facilities had not only "brought the Baku oil region out of obscurity," but Russian oil was already "a very formidable rival to American petroleum" in the European markets.[1] *Bradstreet's,* on the other hand, recognized that the oil wells at Baku were enormously productive, but, according to its own "reliable informants," refining and transport facilities were still "wretchedly inadequate" and because of shortsighted commercial policy of Russia "promised to remain so for some time to come." [2]

Subsequent developments soon put an end to any remaining questions whether the Russian industry was to be a formidable competitor in the world trade for petroleum. As shown in Chart 24:1, output of crude from the Russian fields (about 4.5 million barrels compared to United States' production of over 30 million barrels in 1882) expanded to 23 million barrels by 1888, equivalent to more than four-fifths of the American output for that year. Ten years later Russia had passed the United States as the world's leading producer of crude. Moreover, by 1898 production in "Asia"—the Dutch East Indies—was adding almost another 4 million barrels to total production.

Largely because of the quality of crude, Russian operators from the outset converted a smaller percentage of their refinery throughputs into illuminating and lubricating oils. The continuation of this trend, as shown in Table 24:1, kept the Russian production of these two leading petroleum export commodities well below American output. But against a background of rapidly expanding crude supplies between 1884 and 1899, Russian refiners, in terms of a combined output of the two industries, increased their proportion of illuminating oil from about 14 per cent to 30 per cent, and their lubricating oils from almost 16 per cent to just over 24 per cent. While it is impossible to indicate what proportion of the total output of the Russian refining industry was sold abroad during these years, data on the exports of illuminating oil show that by 1899 Russian oil was supplying over one-third of the world's markets for kerosene.*

Table 24:1 also gives some impression of the relative importance of export sales of illuminating oil for both the American and Russian industries during the 1884–99 perid. Following a trend already apparent a decade earlier, the American home market absorbed an increasing proportion of the kerosene output of the domestic refineries. Despite this trend, not

---

* Actually both the Russian and American shares of the total export market was somewhat lower than these figures suggest because of the rise in production of crude and refinery output in the Dutch East Indies. See p. 673.

only did exports increase, but in 1899 nearly three-fifths of the industry's total production of illuminating oil was still being sold outside the United States. By 1889 the Russians were marketing over half their kerosene abroad, while ten years later they were selling over two-thirds of their total production in the export markets.

TABLE 24:1

Selected data on U. S. and Russian petroleum industries: 1884–99 (in millions of 42-gallon barrels)

	1884				1889			
	U. S. *	Russia †	Total	% Russia of total	U. S. ‡	Russia †	Total	% Russia of total
Crude Production	24.22	8.84	33.06	26.7%	35.16	20.14	55.30	36.4%
Illuminating Oil	15.45	2.57	18.02	14.3%	20.19	7.14	27.33	26.1%
Exports	10.33	.30	10.63	2.8%	13.14	3.75	16.89	22.2%
% Exported	66.86%	11.67%	58.99%		65.08%	52.52%	61.80%	
Lubricating Oil	.90	.17	1.07	15.9%	1.83	.39	2.22	17.6%

	1894				1899			
	U. S.*	Russia †	Total	% Russia of total	U. S. §	Russia ‖	Total	% Russia of total
Crude Production	49.35	30.38	79.73	38.1	57.07	61.67	118.74	51.9
Illuminating Oil	29.46	8.32	37.78	22.0	29.95	12.88	42.83	30.1
Exports ¶	17.39	5.41	22.80	23.7	17.25	8.79	26.04	33.8
% Exported	59.03	65.02	60.35		57.60	68.25	60.80	
Lubricating Oil	3.28	.63	3.91	16.1	4.06	1.30	5.36	24.3

* Estimated.
† *Mineral Resources, 1896,* 885.
‡ *Eleventh Census of Manufacturing Industries: 1890.*
§ *Twelfth Census of Manufactures: 1900.*
‖ *Petroleum Review* (June 23, 1906), 405.
¶ U. S. Exports: Appendix D:1; Russian Exports: *Report of the Commissioner of Corporations on the Petroleum Industry,* Part 3, *Foreign Trade,* 198, 218.

Comparable information is not available for the exports of other petroleum products from Russia, which undoubtedly expanded sharply between 1884 and 1899. This competition did not, however, prevent a substantial increase in exports of American crude, lubricating oils, and the naphtha-benzine-gasoline group of products. (See pages 645, 646, 652.)

As Chart 24:2 shows, the annual kerosene exports from the United States also continued to expand during the greater part of the period, reaching a peak of about 19 million barrels in 1897, dropping off to a little over 17 million barrels in 1899. Throughout the first ten years there was an extension of the downward trend in export prices that began in the 1870's.

Million
Barrels

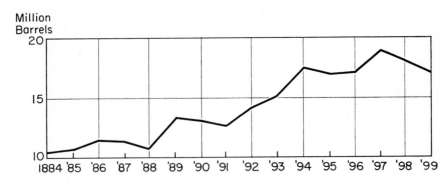

Cents per
Gallon

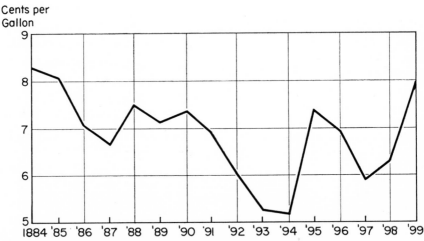

CHART 24:2. *U. S. exports of illuminating oil (in 42 gallon barrel equivalents) and New York export prices (per gallon), 1884–99. (Sources:* Mineral Resources, 1905, 884; Report of the Commissioner of Corporations on the Petroleum Industry, *Part 2, Prices and Profits, 622–23.)*

With the abrupt drop in Russian production (see Chart 24:1), owing to a cholera epidemic that swept through the Caucasus in 1894, prices rose sharply a year later. Following the restoration of Russian production, prices again fell off during 1896 and 1897, only to rise again when American production declined in 1898.

These data suggest two important characteristics of the world's markets for kerosene during this period. One was the relative inelasticity of demand at any particular time, as indicated by the sensitivity of refined prices to short run variations in outputs of crude. The second, of overwhelming significance, was the continued growth in demand at impressive rates, with the result that the world's total consumption of illuminating oil in 1899 was nearly two and one-half times as large as it had been fifteen years earlier.

In part this expanding demand was the result of consumer response to lower prices, particularly between 1884 and 1894. But demand was also affected by the intensive international competition for market shares that emerged during the mid-1880's. It was primarily a battle between industrial giants, with Standard, on the one hand, attempting to maintain its predominant position in the export markets against the inroads of the Nobels and the Rothschilds, on the other. The major area of competition was Europe, where the American industry in 1884 was selling over 70 per cent of its total exports of illuminating oil and over 95 per cent of its exports of crude, naphthas, and lubricating oil. But competition was no less intense in other foreign marketing areas, particularly Asia, the second largest outlet for illuminating oil. An important consequence of this rivalry was a radical readjustment of the entire competitive fabric of the world's petroleum trade, which was in turn reflected in a major revolution in marketing channels and methods of distribution.

The Baku to Batoum railway

Before Russian marketers could sell any substantial portion of their output abroad it was necessary to provide more adequate transport and handling facilities than were available at the time the state-owned and operated railway was completed from Baku to Batoum in 1883. Almost from the day the railroad was opened shippers complained about the shortage of tank cars, which as late as 1886 still numbered only a few hundred. This situation was relieved somewhat in 1887, when the state finally gave permission to private shippers to operate their own cars on the line; the Nobels alone immediately put 465 tank cars into service.[3]

A peak elevation of 3,000 feet on the railroad presented another traffic bottleneck, since only a half dozen cars could at one time be hauled the critical distance of 78 miles over the road's single track. In 1889, after considerable discussion, the Nobels installed a pipeline covering this distance of the road, which did much to relieve congestion.[4]

Meanwhile the two contractors who had completed the Trans-Caucasian Railway, A. Bunge and Serge Palshkofski, took the initiative in providing harbor and port facilities at Batoum. Forming the Batoum Petroleum & Trading Company in 1883, they set out to develop the petroleum trade on the Black Sea "as the Nobels had done in the Volga-Caspian area." [5] Within two years the promoters ran short of funds, but were able to complete their plans when in 1885 they obtained a loan of 2 million rubles from Baron Alphonse de Rothschild, head of the Paris Rothschilds. Under the loan arrangement, they agreed to turn over all their "refined oil for export . . . to the Rothschilds for disposal in foreign markets." [6]

The Rothschilds followed this loan to the Batoum Petroleum & Trading Company with other investments which soon placed them alongside the

Baron Alphonse de Rothschild

Nobels as the principal leaders in the Russian petroleum industry. In the process they acquired extensive production and refining facilities, but their main interest was in the marketing phase of the business. By the late 1880's the Rothschilds were selling oil for distribution in the major markets of western Europe and the Near and Far East.

Already shipping oil to eastern Germany, Austria-Hungary, and Finland via the Volga River and the Russian railway lines, the Nobels were quick to follow the Rothschilds in building their marketing facilities based at the Black Sea port. By 1885 they had added western Germany, England, Italy, Belgium, Denmark and "Scandinavia" to their market areas.[7]

The initial impact of Russian competition in markets formerly supplied almost exclusively by the United States was most noticeable in the countries bordering on the eastern Mediterranean and in southern Europe. As early as mid-1885 many German marketers, who drew their supplies from America, were voicing their concern over the possible loss of their business in southern Germany and Switzerland to Russian imports by way of Fiume, Trieste, and Genoa. According to one spokesman, "If Caucasian oil is to be prevented from having a considerable sale by this means it can only be done by reducing the tariffs [railroad rates] for petroleum from the North Sea ports." [8] By the end of 1885 it was further noted that Russian oil was already being shipped in tank cars from St. Petersburg to Lubeck and would "very soon be dispatched up the Rhine via Rotterdam." [9] But the event in 1885 that was to have the greatest impact on the marketing structure and methods of distribution in western Europe was the introduction by the Nobels of the ocean-going steam tanker.

Ocean-going tank steamers

While a few sailing ships fitted with iron tanks had begun hauling crude across the Atlantic during the 1860's, several factors long prevented bulk transportation from replacing barrels or tins, except on a limited scale. Most of the design problems encountered by pioneer barge and tank car builders also applied to marine bulk handling. It was difficult to prevent the movement of oil from throwing the ship out of ballast.* Even more important in restricting bulk transport were the dangers of fire and explosion. These hazards, serious enough with barrels which inevitably leaked a portion of their contents into the holds of the vessels, were more acute with tanks, for escaping oil collected at the sides and below the tanks, where it generated explosive gases.

Quite aside from technical limitations was the fact that few marketers during the 1860's and 1870's handled a sufficient volume to warrant the cost of installing the special equipment necessary to handle oil in bulk. In the absence of appropriate harbor facilities at the export and import ports, most shipowners engaged in petroleum trade had even less reason to convert their general cargo vessels into special bulk carriers, particu-

---

* According to one ship captain whose vessel had been wrecked while carrying oil in bulk in the 1870's, "The difficulty was that the oil seemed to move quicker than water, and in rough weather, when the vessel was pitched forward, the oil would rush down and force the vessel into the waves much the same as improperly stored bulk grain does some times in stormy weather." [10]

larly when such a conversion made it more difficult to obtain cargoes for return voyages to the United States.

Apparently one of the earliest opportunities for substantial shipments in bulk was offered by the French refining industry during the late 1870's. Concentrated along the River Seine between Le Havre and Rouen, where they had installed large tanks to receive and store their oil, French refiners increased their imports of American crude from about 300,000 barrels in 1874 to 450,000 barrels in 1879.*

Attracted by the possibilities of this traffic, a group of Norwegian ship-owners put four vessels, specially designed for bulk shipments, to hauling crude from Philadelphia in 1879.[11] Three of these, the wooden sailing ships, the *Jan Mayn*, the *Nordkyn*, and the *Lindernoes*, were equipped with a series of tanks, "in a manner superior to the bulk of the vessels [pre-viously] engaged in the trade." † While little is known of her design, it was the fourth vessel, the steamship, *Stat*, which foreshadowed "a new era in marine transport when she left Philadelphia for Rouen on October 18, 1879, with the first cargo of petroleum ever loaded in a tank steamer bound east." [13]

But the economic attractions of this traffic were not sufficient to offset the hazards involved. Following successful maiden voyages all four ships continued to haul crude from Philadelphia until late 1880, when some French merchants fitted the sailing ship, *Fanny*, with tanks and sent her to Philadelphia for a load of crude bound for Le Havre. The disappear-ance of the *Fanny* at sea, presumably by an explosion, prompted the Nor-wegian operators to withdraw from the trade the *Stat* and two of the three sailing vessels.[14]

There were no important new entries into ocean-going tank service un-til 1885, when four ships were put into operation. A variety of factors brought about this renewed interest in marine bulk transport: improve-ments in ship design, a growing volume of exports, difficulties encount-ered by the Russians in shipping their oil in barrels from Batoum, and a change in German import duties.

By the early 1880's naval architects had come to recognize that a well-

---

* See Appendix D:2.
† These ships were fitted with two large tanks extending the length of the ship and sealed from the top deck by a layer of felt and an inner ceiling of wood. The tanks were subdivided into compartments by lateral bulkheads, while "wash boards," running longitudinally, were "intended to prevent the rapid rush of the oil from side to side which would otherwise take place as the vessel rolled, in the event of the tanks becoming only partially filled." [12]

designed tank ship had to meet the following requirements. First, to avoid danger of leakage or the possibility of bursting tanks, provision had to be made for the expansion and contraction of oil resulting from changes in temperature. Second, in providing for expansion and contraction, it was also important to design the storage tanks in such a way that the movement of oil would not throw the ship out of ballast. This danger could be reduced by installing some automatic method of keeping the tanks full at all times. Further stability could be achieved by using a large number of individual tanks or by inserting logitudinal and lateral "wash" bulkheads in larger tanks, which would serve to brake the movement of oil. Third, for obvious reasons, some means had to be provided for the escape of ex-

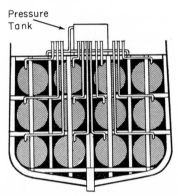

FIGURE 24:1. *Cross-section diagram of* Crusader *showing detail of storage tank installations.*

plosive gases given off by the oil in transit. Finally, particularly in the case of steam tankers, special precautions had to be taken to prevent oil from passing into the boiler room.[15]

A number of these improved design features were incorporated in the two sailing tankers introduced in the trans-Atlantic trade in 1885. The first of these, the converted wooden ship, *Crusader* (643 net tons), for example, was fitted with a multiple system of boiler type tanks, forty-seven in all, patented by her owner, L. V. Sone of New York. (See Figure 24:1.) Each tank was connected to two separate pipes. One pipe was used to fill and discharge the contents. The other pipe, extending well above deck, not only served to dissipate any accumulations of gas, but was also connected with a storage tank on deck which kept the main tanks full and the oil under constant pressure.[16]

The *Crusader's* design was too complex, and the use of boiler-type tanks was too wasteful of shipping space to be widely adopted. But the *Crusader* achieved one major distinction by sailing from Philadelphia to London in the fall of 1885 with the first cargo of kerosene to be shipped in bulk across the Atlantic.[17]

Improved design was no doubt a factor in the decision of the prominent German importer, Wilhelm Riedemann, to use his converted sailing ship, the *Andromeda*, for transporting crude from New York to Bremen that same year. A much larger vessel than the *Crusader*, the *Andromeda* (1,871 net tons) incorporated essentially the same design features as the American built ship. Each of her 72 separate steel tanks had a pipe for loading and unloading purposes, and each had a pipe to carry off gas accumulations, which was in turn connected with an auxiliary tank that maintained the storage tanks full of oil and under constant pressure.[18] Riedemann had another and perhaps more compelling reason to engage in bulk shipments. In 1885 the Reichstag, in response to pressure by German coopers, had placed a tax of $2.50 on each petroleum barrel imported into Germany.[19]

Whatever the motivation, the performances of both the *Andromeda* and the *Crusader* were, according to contemporary observers, well "beyond the theoretical [expectations] of those who have [long] agitated for an improvement in shipping facilities." [20] Their initial success did much to stimulate interest in American shipping circles regarding the possibilities of bulk shipments across the Atlantic.

Of much greater significance in marking a new era in marine bulk transportation were the two steam tankers put to carrying refined oil from Batoum to western Europe in 1885. Although the development of the tanker fleet on the Caspian following the introduction of the *Zoroaster* by the Nobels in 1879 had attracted considerable attention, there was a widespread feeling among shippers in Europe and the United States that steam tankers could not survive the severe weather encountered on the Atlantic.

But as the American consul at Antwerp noted in 1886:

> The importers of Russian petroleum have . . . experienced great difficulty and expense in procuring suitable barrels for the transportation of their oil; they were compelled to purchase empty American barrels and send them to Batoum, where they were refilled and reshipped by steam to Antwerp. In addition to the heavy expense thus incurred, the barrels were often not properly repaired and glued at Batoum, so that a large percentage of

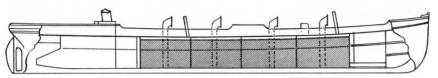

FIGURE 24:2. *Diagram of steam tanker* Sviet *showing location of storage tanks.*

the cargoes arrived here in bad condition and much loss was sustained from leakage, and from injury done to the oil from the glue which became detached from the barrels.[21]

These difficulties led the Nobels to pioneer the introduction of ocean going steam tankers for the transport of refined oil. Working through their Belgian marketing representative, Henri Rieth, of Antwerp, the Nobels made arrangements with the British shipbuilding firm of R. Craggs & Sons to convert the steam cargo vessel, the *Fergusons*, into a tanker. Immediately after delivery in the fall of 1885, the *Fergusons* hauled her first cargo of refined oil from Batoum to Antwerp and continued in bulk petroleum service until 1889, when she was destroyed by an explosion at Rouen.[22]

The second ship, the *Sviet*, (see Figure 24:2), ordered by the Nobels through their shipping subsidiary, the Black Sea Steam Navigation & Trading Company, was built by Lindholmen-Motala, the Swedish firm that had constructed the *Zoroaster*. Specially designed as an ocean-going tank steamer, the *Sviet* also left Batoum in the fall of 1885 for London, where her arrival "with more than a half million gallons of Russian [refined] oil in bulk . . . after a safe and easy passage across the Bay of Biscay, reduced the [skeptics of steam tankers] to silence." [22]

Aside from the installation of special cofferdams to prevent oil from reaching the boiler rooms, the internal construction of the *Fergusons* and the *Sviet* was quite similar to the design of the *Andromeda*.[23] These ships, along with the *Crusader*, all shared one design feature that made for an unnecessary fire hazard. This was the construction of tanks within the hull leaving narrow spaces at the sides and below the tanks where escaping oil could accumulate and generate explosive gas. Moreover, it was exceedingly difficult to clear these spaces and to make repairs.

This particular problem had been met on several ships put into the Caspian service between 1880 and 1885, which used the "skin of the ship" as a part of the storage compartments. Credit for introducing the first ocean-going steam tanker with this type of construction, however, must go to

Wilhelm Riedemann, whose ship, the *Gluckauf* (see Figure 24:3), began hauling refined oil from New York to Bremen in the fall of 1886. Built by the English firm of Armstrong, Mitchell & Company, the *Gluckauf* (2,307 gross tons) was 300 feet long, with a 37-foot beam and a 24-foot depth. Equipped with triple-expansion steam engines, she had a cruising speed of about 11 knots.* The *Gluckauf* had no sooner arrived at London when Riedemann ordered additional ships and by 1888 he had at least five steam tankers operating across the Atlantic.[26]

The introduction of the *Gluckauf* removed most of the remaining doubts about the feasibility of ocean-going steam tankers. Like most important innovations, they were introduced over the protests of vested interests. The use of pumps to load the *Gluckauf* for her maiden voyage from New York in 1886, for example, brought a violent reaction from dockyard workers skilled in handling petroleum products in barrels. Represented by the Knights of Labor, they had to be restrained by the police before the cargo was put on board.[27] Owners of sailing ships by 1887 were much concerned over the growing use of tankers, which was idling an increasing portion of their ships. Reportedly, Standard's own initial reaction was to look askance at the possibility that the disuse of barrels would throw the company's extensive cooperage establishments out of work.[28]

These obstacles and objections gradually gave way to the savings in money and time offered by the new steam tankers. According to estimates made in 1888, the port fees, inspection charges, loading, and other costs for a tanker carrying 20,000 barrels of petroleum products were $2,300 less than shipping the same amount in barrels. Compared to a maximum number of three round-trips across the Atlantic in one year for sailing

---

* Further details of the *Gluckauf's* construction were as follows: "The hull of the vessel was divided into compartments by transverse bulkheads, these compartments being again divided longitudinally by a central bulkhead, which extended through all the oil spaces. A pump-room was provided aft, which extended right across the vessel, and formed a cofferdam or safety-space, dividing the stokehold from the oil compartments; between the main and upper decks a narrowed trunk was arranged, the trunkway extending the whole length of the cargo-tanks. The transverse and longitudinal bulkheads being carried up to level of the upper deck, this arrangement allowed for expansion of oil from any rise in temperature, and by keeping the main portion of the tank below the main deck full, reduced the danger of movement of oil from rolling and pitching of the vessel, also providing spaces along the whole length of vessel on each side of the trunkway for the accommodation of light oil cargoes. The bulkheads were designed to stand the pressure of the full head of oil when one tank was full and the next one empty. A double bottom was fitted, for water-ballast only, under the machinery-space, and in the forehold. The machinery and boilers were placed aft, no tunnel for shafting being required. . . ."[25]

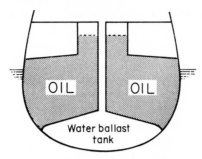

FIGURE 24:3A. *Cross-section diagram of* Gluckhauf *showing detail of storage tank installations.*

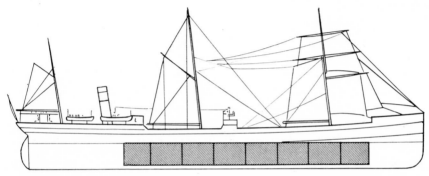

FIGURE 24:3B. *Tanker* Gluckhauf *showing position of storage tanks.*

ships, steam tankers by 1892 were averaging seven round-trips.[29] Their more frequent service, coupled with the fact that steam tankers, unlike the sailing ships, could operate year-round, reduced the pressure to build up inventories of oil abroad in anticipation of peak seasonal demands during the winter months. These savings were sufficient to stimulate the construction of 145 new steam tankers (353,000 gross tons) for the petroleum trade between 1885 and 1898.[30] By mid-1890's it was estimated that well over 90 per cent of both crude and refined oil (not shipped in cases) was being exported from the United States in tankers.[31]

Impact of Russian competition

Standard's executives were quite naturally concerned by the rise of Russian competition during the mid-1880's. Controlling some 90 per cent of the American industry, which in turn sold a high proportion of its refining output abroad, Standard could hardly afford to let the export markets go by default. Indeed, from the time William Rockefeller moved to New York in 1866, members of Standard had been fully aware of the signifi-cance of foreign demand for their products. It was on the continued im-portance of the export market that they had constructed trunk pipelines to their refining capacity concentrated on the Atlantic seaboard.

One of the first reactions of the management was to make a special ef-fort to obtain information about competitors and foreign market condi-tions. Reports began to flow into 26 Broadway from all over the world, in-cluding those from American consuls located at strategic points, some of whom were put on the Standard payroll. Company representatives were also sent abroad to get first-hand information and to strengthen the com-pany's connections with foreign marketers.

To meet growing competition from the Russians in the European mar-kets for lubricants, Standard's domestic affiliates specializing in the sale of lubricants moved to expand their operations abroad. Thompson & Bed-ford, for example, which had earlier established a sales office in Man-chester, England, appointed an experienced German oil broker, Leo Op-penheim, as its representative for Northern and Central Europe, while James G. Macgowan was sent from the United States to Paris, where he assumed responsibility for the company's sales in France, Switzerland, and the Mediterranean area. The Vacuum Oil Company, which in 1885 maintained a Liverpool sales office with four clerks and nineteen traveling salesmen, began to set up additional offices on the Continent during the early 1890's.[32]

European trade was generally divided between the two companies along the lines followed in the American domestic market. Thompson & Bedford dealt mainly with jobbers, while Vacuum sold directly to consum-ers. In part this division was to insure an adequate coverage of markets; and in part—at least according to critics—it was also an attempt to main-tain the fiction that Vacuum was an independent company, competing with Standard's Thompson & Bedford.[33]

The selling activities of both companies reflected the special character of the lubricating trade. As in the United States, European consumers were primarily industrial units, divided into two broad categories: trans-portation and manufacturing. The products supplied were extremely varied, with private or special brands or blends being of particular importance in attracting and maintaining customer loyalty. Vacuum was especially active in supplying lubricants suited to the special requirements of its customers, and in 1895 they established the first of a series of European compounding plants and lubricating oil refineries in England in order to service its customers more effectively.[34]

Standard's principal competitor for the European lubricating-oil markets during the 1880's was the firm of A. André Fils, which had begun distributing the Russian product as early as 1877. In 1885 the firm was appointed European marketing agent for the Nobels and, shortly thereafter, began shipping lubricating oil in tank steamers to receiving depots at Dunkirk, Point St. Louis-de-Rhone, Antwerp, Hamburg, and Trieste.[35]

In the early 1890's, A. André Fils joined the leading Russian refiners in the formation of the Industrial Supply Company to distribute lubricants in the European markets.* Some indication of the growing acceptance of Russian lubricants is indicated by the fact that total exports—mostly to Europe—which amounted to a little over 81,000 (42-gallon) barrels in 1884 grew to almost 2 million barrels by 1900.[37]

Differences in the character of the lubricants produced from Russian and American crudes brought about a broad division of markets between the Russian and American firms. Wherever an oil of high viscosity at comparatively low temperatures was needed, the Russian product was generally superior to the American oil.[38] The Industrial Supply Company, as a consequence, was most successful in marketing lubricants for heavy machinery and for railroad use particularly on the Continent. American oils, which maintained their viscosity at higher temperatures, were by contrast generally preferred as lubricants for cylinders, spindles, and general factory use.

Although by the turn of the century the Russians were apparently sell-

U. S. exports of lubricating oil (in thousands of 42-gallon barrels)

			European distribution		
Year	Total U. S. exports	Exports to Europe	United Kingdom	Germany Netherlands Belgium	France
1884	250.4	238.4	121.9	91.9	19.1
1889	599.2	552.2	338.1	163.2	42.6
1894	956.9	843.0	468.3	259.9	72.6
1899	1,605.3	1,279.1	627.5	409.3	154.8

Source: Appendix D:2.

* The members, in addition to A. André Fils, included; the Nobel Brothers; Schibaieff, Ltd.; Bnito (Rothschilds); Petroleum Works Albrecht & Co.; and the Lubricating Oil Import Co.[36]

ing more lubricating oils abroad than were American refineries, the Standard management had good reason to be pleased with the success of its own marketers. As shown in the tabulation on page 645, total exports of American lubricating oil to Europe, most of which was marketed by Standard, grew over five and one-half times between 1884 and 1899.

Standard also took steps during the 1880's to insure a continued sale of crude in the important French refining market, where the influence of the Rothschilds was growing. In order to make sure the French would not turn to Russia for their supplies, Standard negotiated in 1887 a six-year contract with a leading French refiner, A. & K. Deutch & Company, calling for the delivery of some 500,000 barrels of crude annually. Under this contract, which served as a basis for further negotiations in 1893, Standard in turn agreed to sell no kerosene for export to France and to operate the French concern's forwarding plant in the United States located on the Delaware River.[39]

Standard's early success in meeting the Russian challenge in the foreign marketing of lubricants and crude was not matched in respect to illuminating oil.* European customers generally throughout the mid-1880's con-

U. S. exports of naphthas (in thousands of 42-gallon barrels)

Year	Total U. S. exports	Exports to Europe	European distribution		
			Germany	France	United Kingdom
1884	358.2	346.7	39.6	147.2	127.0
1889	335.7	331.9	77.6	116.7	112.0
1894	370.4	362.9	101.9	89.6	162.7
1899	387.0	374.1	112.3	36.1	180.6

Source: Appendix D:2.

tinued to be critical of the quality of American kerosene.

In 1884 F. W. Lockwood was again sent abroad to investigate and to answer complaints about the quality of Standard's illuminating oil, this time in the United Kingdom. To accompany Lockwood on his journey, Standard employed Boverton Redwood, Britain's best-known petroleum

---

* Maintaining exports of the naphtha group of products posed no particular problem during the 1884–99 period. For one thing, Russian crude yielded low percentages of the lighter distillates. Secondly, the domestic market absorbed all but a small portion of the increasing naphtha output of the American refineries during these years with the result that, as shown in the tabulation on this page, exports remained relatively stable.

expert. Lockwood's explanation of the poor burning qualities of the American imports was essentially the same as presented at the Bremen conference five years earlier, and he again urged the adoption of better wicks as a solution to the problem. In this instance Standard made sure that better wicks would be available. Under a subsidy arrangement the largest wickmaker in England agreed to manufacture weekly 10,000 pounds of wicks designed to handle Standard's illuminating oil and at the same time to fit the types of lamps currently popular in Britain.[40]

These moves did not stop British complaints about the quality of American illuminating oil. Britons were particularly indignant late in 1885, when it was reported that the United States inspectors had stopped recording on their inspection certificates the results of tests with English wicks and lamps. Of special interest was Boverton Redwood's suggestion to English importers ". . . not to entirely stop the importation of American oil but to rely more largely upon the Russian product if the irregularities are continued."[41] This recommendation had particular force as the first shipments of illuminating oil from the Nobels had come into the British market in that year.[42]

The Standard management did not take these criticisms lightly. The company's manufacturing committee made a special effort beginning in 1885 to improve the quality of illuminating oil that went into the export trade and to make certain that it met the specifications of the various brands.* At the same time, the management gave serious attention, somewhat belatedly, to the shipping qualities of the barrels used in the export trade. In 1885 a decision was made to abandon the use of second-hand barrels for water white kerosene and naphtha. Two years later it was further decided to "double glue" all new barrels containing products for the export markets.[44]

Integration in foreign marketing

It soon became obvious to the Standard management that these measures alone could not prevent the Russians from capturing a large share of the markets for petroleum illuminants in western Europe. The effect of this competition was particularly noticeable in Britain, where the Rus-

---

* This concern over quality appears to have prompted Standard representatives to back the prohibition imposed by the New York Produce Exchange against shipping illuminating oil refined from Lima crude in 1887, a ban that was maintained until 1895.[43]

sians increased their share of the total imports of illuminating oil from less than 2 per cent in 1884 to nearly 30 per cent by 1888.[45]

Moreover, there was ample evidence that the leading members of the Russian industry were determined to maintain and expand their position in the United Kingdom. In the mid-1880's, the Rothschilds joined Frederick Lane, of the prominent British shipping firm of Lane & MacAndrew, in forming The Kerosene Company, Limited, with exclusive rights to import and sell Rothschild's kerosene and The Tank Storage & Carriage Company which was to store and distribute the product. The Nobels followed by appointing Bessler, Waechter & Company, a leading English marketing firm, as their exclusive agent for Great Britain.[46]

Faced by this rising tide of competition, the Standard management decided to take more drastic action by forming its own marketing affiliates for foreign sales and distribution. In arriving at this decision Standard proposed to engage in all phases of marketing, from refineries to final consumers, and to take full advantage of the potentials of bulk distribution wherever possible.

Standard took the first step toward implementing its new policy in Britain. In April, 1888, less than a month after the formation of the Kerosene Company by the Rothschilds, the Standard Trust organized a wholly owned subsidiary, the Anglo-American Petroleum Company, Limited. Capitalized at about $2.5 million, Anglo-American immediately put in orders for steam tankers, barges, tank cars, and tank wagons and began acquiring and building storage and bulk handling facilities at the major ports serving the British Isles.[47]

Within two years after the establishment of Anglo-American, Standard turned to extending its control over marketing on the Continent. Instead of forming new organizations in western Europe, the management decided to affiliate or absorb existing companies, in part because language and legal barriers affecting the petroleum trade "could be most easily surmounted by citizens of the different countries."[48] Perhaps a more significant reason was the fact that, unlike Britain, a number of the large import firms, particularly those operating in Germany, Holland, and Belgium, had already developed extensive bulk handling and marketing facilities.

Following this plan of action, Standard in February, 1890, persuaded Wilhelm Riedemann and A. K. Schütte & Sohn of Bremen to join in the formation of the German corporation, the Deutsch-Amerikanische Petroleum-Gessellschaft (DAPG), capitalized at about $2.15 million. Riedemann, an importer of American kerosene since 1865 and a pioneer steam tanker operator, had worked closely with A. K. Schütte & Sohn, who had developed an extensive marketing organization for the distribution of petroleum products in Germany, Holland, and Belgium. Their contributions to the new company not only included Riedemann's five steam tankers, but storage tanks, warehouses, barreling plants, tank barges, tank cars, and an experienced labor force.[49]

According to one contemporary American account, Schütte & Sohn joined DAPG because they were "powerless to resist." This, it was stated, came from the fact that Standard had also made arrangements with August Sanders & Company and G. H. J. Siemers & Company, two large Hamburg importers. If Schütte & Sohn had not agreed to join DAPG, Standard threatened to divert all shipments for the German, Dutch, and Belgian markets from Bremen to Hamburg.[50]

Whatever the nature of this pressure, DAPG, with Standard supplying an initial capital investment of about $800,000 and with the two Schüttes and Riedemann serving as managing directors, began expanding and enlarging its marketing and distributing system in the area extending from the Rhine to east Prussia.

To cover the Continental market in the Low Countries and west of the Rhine, Standard, in co-operation with three of the leading Dutch and Belgian oil merchants, Frederick Speth & Company, of Antwerp, and Horstmann & Company and Hermann Stursberg & Company, both of Rotterdam, in 1891 formed the American Petroleum Corporation, a Dutch corporation with a capitalization of about $2 million. To match Standard's investment of $1 million in the new company, the European members contributed extensive storage facilities at the principal Belgian and Dutch ports, tank cars, a steel sailing ship, seven ocean-going steam tankers, an iron steamer, and a tank ship which operated on the Rhine.[51]

An arrangement between DAPG and the American Petroleum Company provided for a division of marketing territories. DAPG withdrew from Holland and Belgium, while American Petroleum was no longer to sell in northern Germany and ports of the lower Rhine.[52]

To round out its European holdings, Standard, also in 1891, turned to Italy, subscribing about $500,000 to a new corporation, the Società Italo-Americana del Petrolio, which took over the business of Walter & Company, reputed to control the distribution of most of the petroleum products sold in the Italian markets.[53] During the same year the Trust acquired a minority interest in the Det Danske Petroleums-Aktienselskab, a Danish importing firm that had been formed in 1888 by a number of leading Scandanavian petroleum marketers. To meet competition in the Swiss markets, DAPG joined with Italo-Americana about 1893 in acquiring stock in two leading Swiss marketing companies.[54]

Throughout the early 1890's there was much speculation and uncertainty about the way the members of the Russian industry would react to Standard's development of its foreign marketing affiliates. One response was a series of attempts by the Russians to build up stronger marketing organizations of their own. In 1892, for example, the Nobels agreed to act as the marketing representative for six of the larger refiners (not including the Rothschilds) in the Baku area.[55]

The possibility of an even more effective marketing cartel seemed likely a year later when, at a conference called by the Russian government at Petrograd, representatives of some 60 per cent of the Russian refining capacity agreed to have the Nobels and the Rothschilds serve jointly as their marketing agents. It was quite apparent that the primary purpose of the association was not to compete with Standard, for at this same meeting the Nobels and the Rothschilds were also empowered to enter into a marketing agreement with the American rival.[56]

Even prior to 1893 there were strong rumors of an impending or actual "understanding" between the Russians and Standard. In 1891, for example, German sources called attention to "the formation of a ring by the two great controllers of the international petroleum business (the Standard Oil Company of New York and the Paris Rothschild house) for a division of the markets of the world and the fixing of a selling price for this article, so necessary for the use of the masses." [57]

Two years later, *Bradstreet's* reported a discussion between Standard and the Russian "oil kings," the Nobels and Rothschilds, of "a scheme for parcelling out between them the whole of the refined oil markets of the world." The plan, it was stated, was for Russia to withdraw from the German and English markets in return for Standard's agreement to pull out of Asia.[58]

Standard officials did in fact confer with members of the Russian industry on a number of occasions but no effective agreement providing for a global division of marketing territories was reached. The principal obstacle appears to have been the refusal of the Russian minister of finance to support a plan which would compel all Russian refining interests to join one organization for the purpose of selling in the export market. From Standard's point of view, there were obvious weaknesses in any agreement that left a part of the Russian industry free to compete wherever they wished.[59]

The Russians may well have been influenced in turn by the fact that Standard itself at this time lacked complete control over the exports of the American industry. Early in the 1890's, the Mellon interests were already operating the Crescent pipeline and had begun construction of a refinery at Marcus Hook.[60] A joint ownership (in 1890) with T. C. Bowring & Company of the Bear Creek Refining Company, which in turn owned the Bear Creek Shipping Company, Limited, enabled the Mellons to ship and distribute their refined products to the British markets. By 1893 they were negotiating with French refiners for sales of crude.[61]

During this same period the Producers Oil Company (reorganized as the Pure Oil Company in 1895) also began shipping refined oil to the German markets under an informal arrangement with Philip Poth of Mannheim. This association provided the Producers Oil Company with the services of an exceptionally able marketer, one who was determined to remain independent of Standard. Poth had pioneered the introduction of bulk shipments on the Rhine and of tank cars in Baden. He had tank installations at Rotterdam, Amsterdam, Vlissingen, Mülheim, Mainz, Mannheim, and Strasbourg. He was also closely associated with the shipping and import firm of Rassow, Jung & Company of Bremen.[62]

In the absence of any effective agreement providing for an international division of the world's petroleum trade, Standard moved in a variety of ways to strengthen and consolidate its position in particular markets. France was one important area that demanded immediate attention from the management early in 1893. The occasion was the approaching termination of Standard's 1887 contract with A. & K. Deutch & Company, which had guaranteed the company about half the French market for crude.

In view of France's growing political friendship with Russia and the continued influence of the Rothschilds among French official circles, it appeared that the French refineries as a group might turn to Russia for their crude. Moreover, because the members of the French industry were closely associated through interlocking investments with refineries in Spain, this decision would no doubt be extended to the Spanish markets.

In view of the French differential duties on crude and refined products, Standard's initial intention was to enter the refining business in France and start selling in French markets. The aroused French refining interests countered by announcing that if Standard followed this plan they would buy all of their crude either from Russia or from the Mellons. They further indicated they would make the fight against Standard a political issue with the French government.[63]

To avoid open conflict, both parties began discussion of an arrangement whereby the French refineries would buy crude from Standard provided the latter would not enter actively into refining. As negotiations dragged on during late 1893, Standard strengthened its bargaining position in two ways. To cut off the possibility of crude purchases from independents in the United States (as well as to reduce competition in the British markets), Standard made a deal with the Mellons whereby Anglo-American took over their interest in the Bear Creek Shipping Company and the Trust acquired their pipelines and the uncompleted refinery at Marcus Hook.[64] About the same time Standard formed a French corporation, Bedford et Compagnie, which immediately purchased a refinery located in the west of France and began construction of a new refinery at Rouen in northern France.

These steps, coupled with the current rumors that Standard and the Russian industry might come to an agreement on a division of world markets brought negotiations to a conclusion. A six-year contract was signed whereby the French refiners agreed to purchase the bulk of their crude from Standard while the latter agreed not to sell refined products (except lubricants and wax), including the output of its two refineries in France, to anyone outside the French group. The contracts with the French were followed by similar agreements with refining interests in Spain.[65]

The relative importance of the French and Spanish demand for American exports of crude oil is shown in the following tabulation.

U. S. exports of crude oil (in thousands of 42-gallon barrels)

Year	Total U. S. exports of crude	Exports to Europe	European distribution		
			France	Spain	All others
1884	1,599.7	1,526.4	919.5	290.3	316.6
1889	1,737.8	1,606.2	1,284.7	150.6	170.9
1894	2,903.0	2,535.7	2,010.4	361.3	164.0
1899	2,692.6	2,420.0	1,991.2	231.5	197.3

Source: Appendix D:2.

Although the rumors current during the negotiations with the French of an impending arrangement between the Russians and Standard exaggerated its scope, Standard did come to an understanding with the Rothschilds which substantially reduced competition in the illuminating markets both in Britain and on the Continent. Under a contract signed in 1893, Anglo-American took over Rothschilds two British marketing concerns, the Kerosene Company and the Tank Storage & Carriage Company. Shortly thereafter the Rothschilds leased their extensive storage and handling facilities at Antwerp to DAPG and the American Petroleum Company. As a part of the agreements, the Standard affiliates assumed the long-term contracts of their newly acquired (or leased) properties, obligating them to purchase and distribute various quotas of kerosene from the Rothschilds.

Three years later, under a similar arrangement with the Nobels, DAPG and the American Petroleum Company acquired a controlling interest in the former's long time marketing affiliate, Henri Rieth & Company of Antwerp.[66]

In 1896, DAPG scored an additional triumph by finally acquiring the property of Philip Poth, its most formidable competitor in the German markets. Throughout the early 1890's, Poth had continued to sell oil supplied him by the Producers Oil Company. But early in 1896, possibly under the mistaken impression that the Producers' companies had sold out to Standard, and that he was cut off from his supplies, Poth sold his interests to DAPG. The following year DAPG also bought out Rassow, Jung & Company, which was merged with the holdings acquired from Poth into a new company, Mannheim-Bremer Petroleum, AG.[67]

The spread of bulk distribution

Standard's success in absorbing or coming to an agreement with its principal competitors did not reflect the full extent of the organization's penetration and control over the European petroleum markets. By the late 1890's, Standard's marketing affiliates received most if not all of their oil by tank steamer. Wherever practical each company had pushed bulk distribution within its respective marketing territory. In varying degrees each had also moved closer to the trade, either by entering into long-term contracts with existing jobbing firms or by selling direct to retail outlets and in some instances to final consumers.

In the Scandinavian countries, for example, where most of the oil received was barreled for distribution, Standard's affiliate, Det Danske, sold almost exclusively to large wholesalers, who in turn supplied jobbers selling to retail outlets. Italo-Americana also sold through wholesalers and jobbers, but to supply customers in the mountainous areas of Switzerland or the hot, sparsely settled parts of Italy and Sicily, the company packaged most of its illuminating oil in cans, manufactured (after 1895) at its own plants.[68]

Having already inaugurated bulk distribution prior to their association with Standard, the European managers of DAPG and American Petroleum needed no urging from the parent company to expand the system. By the mid-1890's they had extended barge and tank car deliveries to receiving and storage depots serving local markets throughout Germany and the Low Countries. While some oil was delivered by tank wagon or in iron barrels to retail stores and even direct to householders in a few of the larger cities, throughout the markets served by DAPG and American Petroleum the great majority of dealers preferred to receive their oil in barrels.

Although neither company, with few exceptions, went into retailing, both DAPG and American Petroleum followed an aggressive policy of reducing competition and extending their control over the wholesaling functions by buying out or acquiring a financial interest in major independent import and shipping companies.*

Of the various Standard affiliates, it was Anglo-American which undoubtedly brought about the most radical changes in existing marketing channels and methods of distributing. At the time of Anglo-American's incorporation in 1888 the Russians had already captured nearly a third of the British market for imported illuminating oil. By shipping in tank steamers they had also reduced some of the pressure to build up extensive domestic inventories of oil in anticipation of a peak seasonal demand. But for the most part the Russian invasion scarcely disturbed the existing domestic marketing structure, with its large number of wholesalers, jobbers, and dealers. Nor did it affect the traditional methods of storing and distributing oil in barrels.

---

* Shortly after its formation in 1890, for example, DAPG reduced competition and added to its bulk handling facilities by purchasing several prominent German importing and marketing firms, including G. H. J. Siemers & Company of Hamburg and the August Sanders & Company and the American Petroleum & Storage Company of Stettin.[69]

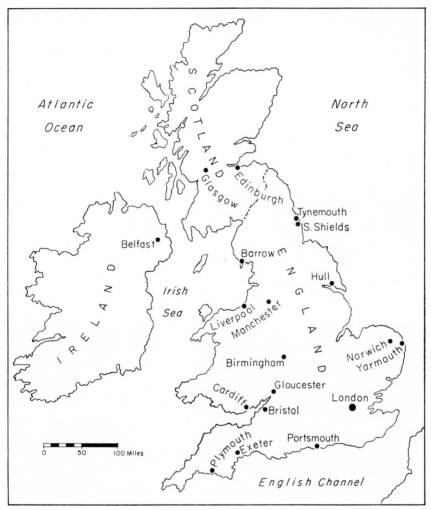

MAP 24:1. *Anglo-American Oil Company, major distributing centers, 1895.*

By contrast, Anglo-American by the mid-1890's had developed a system of bulk distribution in the British markets that was even more extensive than those of DAPG and American Petroleum on the Continent. By this time Anglo-American tank steamers were delivering oil to receiving depots owned by the company at ten or more major ports ringing the British Isles. (See Map 24:1.) From the receiving depots the oil was in turn shipped to Anglo-American bulk stations scattered throughout Britain.*

For certain markets oil was barreled at the receiving stations for subsequent distribution. As a representative of Anglo-American pointed out in 1897, there were,

> . . . districts where it is impossible to reach the dealers with our bulk system. Take Wales, for instance. It would be almost impossible to carry out the bulk system of distribution in a hilly country like that. Take Scotland. The same conditions exist there; and we find that in many of the centers in England certain districts where they continue to deal and will continue to deal, I believe, in barreled oil. . . .[71]

But where ever possible Anglo-American—by frequent deliveries and by selling storage equipment at cost—encouraged retail customers to accept tank wagon delivery.† In urban areas where street hawkers were quite common, this group obtained supplies at Anglo-American bulk stations to cover their markets.

The development of bulk distribution by Anglo-American, DAPG, and American Petroleum aroused considerable hostility among British and Continental marketers. German observers, for example, were quick to grasp the implications of DAPG's operations for the future marketing structure of middle Europe. Referring to the newly formed Standard affiliate, one German newspaper pointed out in June, 1891:

> The German firms which have merged their identity in that of this company attend only to the business of forwarding to the interior. They sell everywhere to inland wholesale dealers and the larger retail dealers in the required quantities. For this purpose the entire country is divided among the companies into fixed districts, so that for other merchants hardly standing room is left. Petroleum tanks have even been erected at various places in the interior, *e.g.*, at Riesa, Duisburg, Mannheim, and it is feared that next the retail dealer will be found superfluous if the company gives the small trade as a local monopoly to a branch. . . . One group of business men after another is thus made superfluous and pushed aside.[73]

In Britain, Anglo-American's policy of operating its own bulk storage and distributing facilities brought similar complaints. In commenting on

---

* In 1897, according to an Anglo-American representative, the company owned 207 bulk stations, 290 railway tank cars, 413 road tank wagons, 1 steam tug, 24 river barges, 671 horses and 65 vans and carts; it employed 1,859 workers and had about 100,000 customers on its books.[70]

† In the period 1892–1896, Anglo-American sold the following equipment to retail dealers: 11,895 tank cabinets, 10,537 circular tanks, and 5,189 galvanized painted tanks. Storage equipment was also purchased from other suppliers while many dealers were iron mongers and tinsmiths who made their own receptacles.[72]

a proposed law, backed by Anglo-American, which would have made it mandatory to store more than 60 gallons of illuminating oil in tanks, a representative of one British wholesaling firm explained:

> The immediate result of this would be that the Anglo-American Company, who have adopted the tank wagon method of distribution, would enjoy a monopoly of the oil trade, and all other firms, many of whom, like ourselves, have distributed the oil in barrels for the last 30 or 40 years . . . would have to retire from the business . . . because we should not be able to compete on its own lines with a great firm of almost unlimited capital. . . .[74]

Local dealers, many of whom combined a small jobbing trade with retail sales, were particularly vocal in condemning Anglo-American's system of tank wagon delivery. As the secretary of the Tallow Chandler's and Oil Dealer's Association of Liverpool put it in 1897,

> In the first instance, when the Anglo-American put their tanks on the ground they gave us their word that no less a quantity than 20 gallons would be delivered, but when they found that the retail dealers of Liverpool would not embrace the new system of tank wagon delivery, but preferred to take it in the old style of barrels, they in the words of their Liverpool manager, were forced to administer a stab in our backs; go really behind us and secure that trade which legitimately belonged to the Liverpool chandler doing his small wholesale business, and that is why we are objecting to the delivery of anything less than 10 gallons of oil.[75]

In large part these complaints about the introduction of bulk handling methods were the inevitable reactions to a technological innovation that had proved itself more economical than distribution in barrels. Even more disturbing to critics both in Britain and on the Continent was the fact that by utilizing bulk distribution to extend control over their respective marketing areas, Anglo-American, DAPG, and American Petroleum left little room for independent operators to function. It was further apparent that only by adopting a similar policy could other suppliers of refined products hope to offer any effective competition to Standard's European affiliates.

European markets and competition in the late 1890's

These moves toward integration and technical advances in methods of distribution in Europe were all part of a larger struggle for the world's petroleum markets that began on an intensive basis with the Russian invasion in the mid-1880's. They were the logical result of attempts by members of both the Russian and American industries to increase export sales in the face of rapidly expanding domestic crude supplies.

The early success of the Russians following the introduction of the tank steamers clearly revealed the basic weakness of Standard's position in the European markets. So long as Standard's control over distribution extended no further than the Atlantic seaboard, the management was in no position to bargain effectively with the Russians in the formation of an international marketing agreement. Nor did the possibility of driving out the Russians by sharp reductions in export prices appeal to Standard's executives. At best such a policy would have resulted in a costly price

TABLE 24:2

United Kingdom: imports and consumption of kerosene: 1884–99

	Imports *				Per capita consumption †	New York export price ‡
	(In thousands of Imperial barrels *)					
Year	U. S.	Russia	Total	% U. S.	In gallons	cents/gallon
1884	927.9	17.1	945.0	98.2	1.3	8.29
1885	1,367.7	70.1	1,437.8	95.1	2.0	8.09
1886	1,363.8	46.8	1,410.6	96.7	1.9	7.11
1887	1,444.4	188.5	1,632.9	88.5	2.2	6.73
1888	1,286.1	549.1	1,835.2	70.1	2.3	7.49
1889	1,355.6	771.2	2,126.8	63.7	2.7	7.12
1890	1,357.1	787.5	2,144.6	63.3	2.7	7.31
1891	1,647.8	830.9	2,478.7	66.5	3.2	6.93
1892	1,711.1	807.6	2,518.7	67.9	3.4	6.07
1893	2,209.6	743.1	2,952.7	74.8	3.8	5.23
1894	2,736.2	578.1	3,314.3	82.6	4.2	5.19
1895	2,730.0	602.9	3,332.9	81.9	3.7	7.36
1896	2,992.7	634.0	3,626.7	82.5	4.0	6.97
1897	2,755.5	494.3	3,249.8	84.8	3.5	5.91
1898	2,843.5	915.3	3,758.8	75.6	4.0	6.32
1899	2,701.8	1,340.4	4,042.2	66.8	4.3	7.98

* Report of the Commissioner of Corporations on the Petroleum Industry, Part 3, Foreign Trade, 440.
† House of Commons, Report of Select Committee on Petroleum (1896), Appendix 29, p. 739.
‡ Report of the Commissioner of Corporations on the Petroleum Industry, Part 2, Prices and Profits, 622–23.
§ Imperial barrel equals 43.2 American gallons.

war, with no assurance of any permanent improvement in Standard's marketing position. Under these circumstances the management decided to develop its own European marketing affiliates and to capitalize on the opportunities offered by the steam tankers to extend the system of bulk distribution as the primary method of attracting customers.

Standard's success in extending its marketing activities in turn prompted its principal competitors to reorganize their own marketing operations along similar lines. Thus the future role of the large integrated oil companies in the structure of the European petroleum markets was foreshadowed when the Pure Oil Company, following the purchase of Poth's interests by DAPG in 1896, sent Lewis Emery, Jr. to Europe to duplicate Standard's marketing organization on the Continent. Within two years Pure Oil had its own fleet of tank steamers, receiving and storage facilities at Rotterdam and Hamburg, tank cars and barges, and bulk stations for the efficient bulk marketing of its kerosene in Germany and the Low Countries.[75]

In 1897 the Rothschilds also decided to re-enter the British market. Again joining forces with Frederick Lane of Lane & McAndrew, they formed the Anglo-Caucasian Oil Company, Ltd., and immediately inaugurated an extensive system of bulk distribution including tank wagon deliveries, modeled after Anglo-American's methods of distribution.[72]

The broad results of this interplay of forces are perhaps most strikingly illustrated in the case of the United Kingdom. In 1884, as shown in Table 24:2, the United States supplied all but a tiny fraction of the illuminating oil imported into the United Kingdom. Russian development of the tank steamer was rewarded when by 1889 the proportion of Russian oil to total imports rose to over one-third. Standard's reaction to this challenge by forming Anglo-American was in turn vindicated by a growth in the share of American illuminating oil to a peak of almost 85 per cent by 1896. But with the re-entry of the Rothschilds as active marketers in 1897, the proportion of American imports was again reduced to about two-thirds by 1899.

This struggle for market shares occurred against a background of expanding demand by British consumers for imported illuminating oil, which was reflected in a more than fourfold increase in total sales between 1884 and 1899. To the conservative London *Economist* it appeared in 1890 that, "nothing . . . has happened to mar the even progress of the industry and the consumption of an article which has grown to be as much of a necessity as daily bread and coal, or water itself." [78]

But competition among the principal sellers in the British market undoubtedly played an important role in expanding total sales. This competition was particularly effective in stimulating demand when, as was the situation between 1884 and 1894, it was accompanied by a substantial decline in prices. The rapidity with which British customers turned to kerosene during the decade is indicated by the growth in per capita consumption from about 1.3 gallons annually in 1884 to approximately 4.2 gallons in 1894.

In the face of higher prices after 1894, total sales remained comparatively stable for some three years as per capita consumption varied between 3.5 gallons and 4 gallons. Although sales again moved up sharply following the re-entry of the Rothschilds (and a temporary dip in prices in 1897) per capita consumption in 1899 was barely above the 4.2-gallon figure reached six years earlier.

Higher prices alone do not account for the failure of per capita consumption of kerosene to increase more rapidly after 1894. It was during these years that the petroleum product began to meet increasing competition from alternate illuminants, notably gas and electricity, foreshadowing the future decline of kerosene as a source of light for the more urbanized markets of the world.

TABLE 24:3

European imports of U. S. and Russian illuminating oil (thousands of 42-gallon barrels)

Year	United Kingdom *			Continent			Total		
	U.S.	Russia	% Russia	U.S. †	Russia ‡	% Russia	U.S. †	Russia ‡	% Russia
1884	955	17	1.7	5,976	183	3.0	6,931	200	2.8
1889	1,396	794	36.3	6,464	1,081	14.3	7,860	1,875	19.3
1894	2,818	595	17.4	8,854	1,903	17.7	11,672	2,498	17.6
1899	2,782	1,380	33.2	10,018	2,367	19.1	12,800	3,747	22.6

* Source: Table 24:2.
† Source: Appendix D:2.
‡ Estimated for 1884 and 1889; for 1894 and 1899 see D. Ghambashidze, "The Russian Petroleum Industry and Its Prospects," *Journal of the Institute of Petroleum Technologists,* IV (1917), 172.

The absence of adequate data on the geographic distribution of Russian exports makes it difficult to measure Standard's success in maintaining its position in the major marketing areas on the Continent during these years. Estimates of the division of total imports of Russian and American illuminating oil between the United Kingdom and the rest of Europe, shown for selected years in Table 24:3, suggest that while the members of the Russian industry had recaptured a third of the British market by 1899, they were still supplying less than 20 per cent of the Continental markets. Not all of the American oil distributed on the Continent, of course, was supplied by Standard, as by the end of the 1890's Pure Oil's sales in Germany and the Low Countries were running about one million barrels annually.[79]

Yet on the whole, the Standard management had good reason to be pleased with its success in meeting the challenge posed by the Russian entry into the European illumination markets in the mid-1880's. Much of the approximate 40 per cent increase in American exports of kerosene to Europe between 1889 and 1899 stemmed directly from the vigorous activities of the Trust's foreign marketing affiliates. But Standard's executives could hardly afford to be complacent about their accomplishments. Their major competitors had been quick to adopt similar marketing organizations to sell in Europe. Nor had they been equally successful in defending their position against Russian competition outside Europe, particularly in the Far East.

New competition for world markets

The influences on Standard to change its export sales policies outside Europe varied widely during the 1884–89 period. In Latin America, for example, the management had little occasion to depart from the established practice of selling refined oil through brokers representing importers in the major distributing centers in the area.* The markets were small, widely diffused and local transport facilities insufficiently developed to warrant the introduction of bulk shipments. In Cuba, where a tariff discriminated against refined oil, Standard had operated a refinery near Morro Castle since 1881. Trade flourished and it purchased a rival refinery in 1887. Three years later the Trust built a refinery in Puerto Rico. For similar reasons, Standard decided in 1896 to construct a small refinery at Rio de Janeiro, but a change in the Brazilian tariff caused the abandonment of this project before the plant was completed.[80]

With the exception of Peru, where sufficient oil was discovered in the early 1890's to supply the domestic market for refined oil plus a small surplus of crude for export, Standard remained virtually the sole supplier of petroleum products for Latin America.† During the 1890's the management did make a minor adjustment in its policy of selling through brokers by appointing its own marketing representatives in the larger South American ports.[82] But this change at most had little effect on the sales of

* As will be noted in the following chapter, Mexico and Canada were included in the Trust's active program of integrating its domestic marketing operations during this period.
† In 1893 a shipment of crude oil was sent from Peru to California. It was less expensive to ship it from there than from the East coast. There was no tariff against this crude, as Peru did not place a duty on American oil.[81]

kerosene to Latin America, which grew from about 365,000 barrels in 1884 to over 900,000 barrels by 1899 (see Appendix D:4).

If Standard had little concern with maintaining its marketing position in nearby Latin America, the situation was quite different in the Middle East, where Russian oil had its first impact on American exports. The competitive disadvantage faced by American illuminating oil in the areas bordering on the eastern Mediterranean was pointed out as early as 1886 by the American consul at the Lebanese port of Beirut.

When Russian oil was first shipped to Beirut in 1885, American kerosene, because of its superior lighting qualities and lower prices, had long been preferred over locally produced olive oil as an illuminant. The Russian oil was initially inferior, but improvement in quality and its relative cheapness, caused the American oil to lose ground rapidly. The major price advantage came from shipping costs, which ran about ten cents a case from Batoum to Beirut, compared to twenty-three cents from American ports.[83]

Standard took action in this area through A. & H. Sefelder, agents for Meissner & Ackerman in the Eastern Mediterranean. The Sefelders were given information on prices and marketing policies. When freight rates increased in 1887, they were allowed to purchase oil from Russian suppliers, and at times engaged in pools with the Russian competitors.

These efforts were hardly enough to stem the flow of Russian oil into the Middle East, and in 1893 it was reported that, except for Italy, American oil had been virtually shut out of the ports in the Eastern Mediterranean.[84] Although Standard, through its European marketing affiliates, continued in its efforts to maintain sales in the Middle East, it was forced to concede by far the larger share of these markets to its Russian competitors. A rough measure of the extent of this concession is indicated by the decline in exports of American illuminating oil to the major marketing areas in the Eastern Mediterranean (the Ottoman Empire and Greece), shown in the following tabulation (in thousands of 42-gallon barrels).

Year	1884	1889	1894	1899
Quantity	329.9	84.1	100.6	62.4

Source: U. S., Treasury Dept., *Annual Report and Statements of the Chief of the Bureau of Statistics on the Foreign Commerce and Navigation of the U. S. for the Fiscal Year Ended June 30, 1884*, 234–37; . . . *Year Ending June 30, 1889*, 292–95; U. S., Bureau of Statistics, Dept. of Commerce and Labor, *The Foreign Commerce and Navigation of the U. S. for the Year Ending June 30, 1904*, II, *Imports and Exports of Merchandise, by Articles and Countries, 1894–1904*, 606–15.

Of the major foreign marketing areas outside western Europe, however, the Standard management was most concerned with maintaining its position in the Orient. Already absorbing some 2.1 million barrels of American kerosene in 1884, over 20 per cent of total exports for that year, the Asiatic countries offered a huge and relatively underdeveloped outlet for illuminating oil.

But this was a market that also appealed to the distributors of the Russian product. The Rothschilds early built a large plant at Batoum to pack-

One of several trade marks of Rothschild's Bnito.

age illuminating oil in tins and cases. The first Russian kerosene reached the west coast of India in 1886, when the ubiquitous Frederick Lane of Lane & McAndrew began shipping oil consigned to merchants in Bombay. By the early 1890's, Russian cased oil was competing with the American product in every major marketing area throughout the Orient.*

This growing volume of Russian shipments had little immediate effect on the sales of American illuminating oil to the Far East which, as shown

---

* Between November, 1886 and December, 1893, over 90 million cases of Russian kerosene were exported to the Far Eastern markets of which Lane & McAndrew shipped nearly 65 million cases.[85]

in Table 24:4, grew from 2.1 million barrels in 1884 to almost 4.6 million barrels in 1894. Under the circumstances, Standard for some half dozen years continued to sell cased oil bound for the Orient to merchants or their American representatives, f.o.b. New York or Philadelphia. In 1890 the management decided to extend its marketing control beyond the borders of the United States by channelling all Far Eastern sales through Anglo-American, which in turn distributed oil on consignment to local merchants located in the leading ports east of Suez.[86]

Two events during the early 1890's changed the character of the Far

Sir Marcus Samuel

Eastern trade and forced Standard to re-examine its marketing tactics in that growing market. One was the shipment of refined oil in bulk, through the Suez Canal, a revolutionary action initiated by Marcus Samuel. The second was the completion of the refinery of the Royal-Dutch at Pang-kalan Brandan in Northern Sumatra.

Marcus Samuel and his brother, Samuel Samuel, headed the English firm of M. Samuel & Company, established by their father in 1830 to trade with the Orient. Operating as brokers, agents, and importers and export-ers, the firm built up an extensive business sending out manufactured articles of every type to the East and securing return cargoes of such items as jute, tea, rice, and copra. While the firm had offices in a number of the principal Oriental ports, its special interest was in Japan, where it represented several shipping companies running to the Empire.

At the time of the elder Samuel's death in 1874, the Japanese were already launched in their program of westernization. By the late 1880's it was evident to the brothers Samuel that Japan was no longer a promising market for traditionally imported manufactured products.[87]

Looking for a more promising trading article, they were impressed with the possibilities of kerosene, a product not produced locally, but one that had found increasing acceptance by Japanese consumers since its introduction by American shippers in 1870.[88] By 1889, American exports to Japan, almost all supplied by Standard, amounted to just over 780,000 barrels (see Table 24:4).

The brothers were convinced that even though Standard was well established, there was an opportunity for a competitor to gain a foothold in the rapidly expanding Japanese market. Marcus Samuel accordingly visited Russia to see if some arrangement could be made with Standard's existing rivals. Arriving at Batoum in 1890 he found the management of Rothschild's Société Commerciale et Industrielle de Naphte Caspienne et de la Mer Noire—more commonly referred to as "Bnito"—receptive to his ideas.

Essentially what Marcus Samuel proposed was to extend the bulk system of distribution already under way in western Europe to the Far East, including tanker shipments to receiving depots at the principal coastal cities, the use of tank wagons or barges to forward the oil to interior bulk stations, and its final delivery to retail dealers in drums which would be returned when empty to the bulk stations.

This was a bold and imaginative plan, one that depended on the solution of a number of problems for success. Not the least of these was the fact that while the Suez Canal Authority at this time allowed merchant steamers loaded with cased oil to pass through the canal, it barred the passage of petroleum tank steamers. There was also the question of obtaining return cargoes for tankers—a necessity which Samuel, as an experienced shipper, realized was important in keeping operating costs down.

But Samuel had no intention of sending tankers on the long journey around the southern tip of Africa, being confident that he could get the

Canal authorities to relax their regulations. He was also confident that he could solve the problem of securing cargoes suitable for the return voyages.

The Bnito management was sufficiently impressed to enter into a ten-year contract in December, 1891, under which Samuel was to accept established quotas of illuminating oil, on consignment at Batoum, which was to be marketed "east of Suez." In return he agreed to pay Bnito the market price, f.o.b. Batoum, plus a 2.5 per cent commission for acting as supplier. One feature of the contract, which especially appealed to Samuel, was the provision that the oil did not have to be paid for until sufficient time had elapsed for it to reach the market and be sold. Under this arrangement he could carry on his shipping and marketing operations without investing any capital in the cargoes.[89]

Samuel did not, however, conclude his agreement with Bnito until he had taken steps to build up a marketing organization and until he was reasonably sure the Suez Canal Authority would allow petroleum tank steamers to pass through the canal.

To insure marketing outlets he formed the Tank Syndicate, an association of prominent shipping merchants located in the major ports in the Orient.* Members of the Tank Syndicate were guaranteed supplies of kerosene in return for building tank installations for receiving the oil in bulk and arranging for return cargoes.

Samuel's petition in 1891 on behalf of the Tank Syndicate to the Suez officials to relax their regulations prohibiting the passage of tankers through the canal brought a storm of protests. British shipping firms, which owned most of the general cargo steamers carrying cased oil from Batoum via Suez, as well as a large portion of the sailing vessels handling American cased oil sent from Philadelphia via the Cape of Good Hope to the Far East, were particularly concerned. They were joined in their protests by tin-plate manufacturers in Wales who feared a decline in demand for their product.[91] Much emphasis was put on the dangers of allowing bulk transport of oil through the Canal, which had earlier caused the Canal authorities, with good reason, to ban the passage of tankers.

In arriving at their decision, however, the Canal Authority directors took into account the great improvement in tanker design since the mid-1880's. Fortunately, for the Tank Syndicate's cause, the French directors

* By 1899 these included merchant houses in Bombay, Calcutta, Madras, Kurrachee, Columbo, Straits Settlement, Java, Saigon, Bangkok, China, and Japan.[90]

of the Canal Authority, already friendly to Russia, were in favor of changing the regulations, while Lord Salisbury, the British Prime Minister, was
interested in improving relations between Russia and Great Britain. The
result was a new set of regulations, issued in January, 1892, whereby tank
steamers of specified construction were granted the right-of-way through
the Canal beginning in July of that year.*

Meanwhile Samuel, anticipating a favorable decision by the Canal
authorities, had arranged with the English shipbuilding firm of William
Gray & Sons to help in the financing and to build the *Murex*, first of the
Tank Syndicate's tanker fleet. Launched in May, 1892, the *Murex*, a 4200-
ton vessel, began a new chapter in maritime bulk transportation and revolutionized Eastern marketing when she passed through the Suez some
two months later with a load of kerosene bound for the Far East.

For a time it appeared that Samuel's whole marketing plan might
founder on the attempt to extend bulk distribution to the retail dealers. As
anticipated by more experienced Far Eastern petroleum marketers, this
method of distribution was a complete failure, largely because ". . . almost everywhere in the East the empty tin had become an indispensable
material in the domestic economy of the people. It was used for transporting vegetable oils and dried fish, for carrying water, for roofing and for
turning out numerous articles which in Europe [were] made of iron, copper or tin." [93]

Contrary to the expectations of the traders in case oil, members of the
Tank Syndicate soon arrived at a satisfactory solution to their problem by
establishing canning factories at the loading ports, followed later by
plants at major inland centers. Even if tin cost a little more than the
price delivered to Batoum, Philadelphia, or New York, there was the offsetting factor of cheap labor available to fabricate and can the illuminant.
Local canning had the additional advantage of leaving the cans fresh and
shiney at the market, making them more desirable than tins, which rusted
during the long voyage from Batoum or eastern United States.

With the solution of the problem of local distribution, the victory of the
tanker system in shipping oil to the Orient was assured. The Tank Syndicate began expanding its operations by building additional receiving depots and canning factories and by adding to its tanker fleet.[94] While the

---

* It was also pointed out by contemporary observers that the English directors took
account of the fact that the Tank Syndicate was composed of British shipping and
marketing firms.[92]

Nobels and the Rothschilds continued to ship small amounts of cased oil via Suez, neither concern developed a tanker fleet or maintained tank installations in the Far Eastern markets.[95]

There is little question that the members of the Tank Syndicate handled the bulk of Russian shipments of illuminating oil to the Orient which, as shown in the following tabulation (in millions of 42-gallon barrels), grew over 50 per cent between 1893 and 1900.

Year	1893	1894	1895	1896	1897	1898	1899	1900
Quantity	3.50	2.87	3.62	4.12	4.25	4.12	4.87	5.37

Source: Ghambashidze, *Journal of the Institute of Petroleum Technologists*, IV (1917), 172.

The Tank Syndicate, led by Samuel, was not the only threat to Standard's position in the Far East. By the early 1890's Royal Dutch, drawing on local supplies of crude, had emerged also as a competitor for the Asiatic illuminating-oil markets.

Royal Dutch had its origin in 1880, when Aeilko Jans Zijlker, manager of the East Sumatra Tobacco Company, took some petroleum skimmed from a nearby pond to Batavia to be tested. The analysis indicated that a first quality illuminating oil could be produced from the crude with yields of about 60 per cent. Obtaining a concession to explore and produce petroleum in northern Sumatra from the Sultan of Langkat (who under Dutch rule had control over the area where the petroleum seepages were found), Zijlker set out to raise sufficient capital to start exploratory drilling. Finally in July, 1884, under the direction of the Netherland's Department of Mines, which maintained a water-drilling operation in Sumatra, the first test well was started. The original well produced no oil, but the second, Telaga Tunggall No. 1, drilled to a depth of about 72 feet by June, 1885, brought a yield of a little more than 5 barrels a day.[96]

This volume was hardly sufficient to form the basis of any extensive operation, and when further drilling failed to bring in any more producing wells, Zijlker, who had exhausted his slender capital resources, was about to give up on the project when reports by two government engineers indicated the probable existence of extensive oil deposits in the area and the possibility of profits to be made refining and selling illuminating oil in the Asian markets.[97]

Encouraged by these reports, plus the fact that Telaga Tunggall No. 1 had unexpectedly increased its output to about 164 barrels a day, Zijlker left Sumatra late in 1889 for The Netherlands, where he hoped to raise

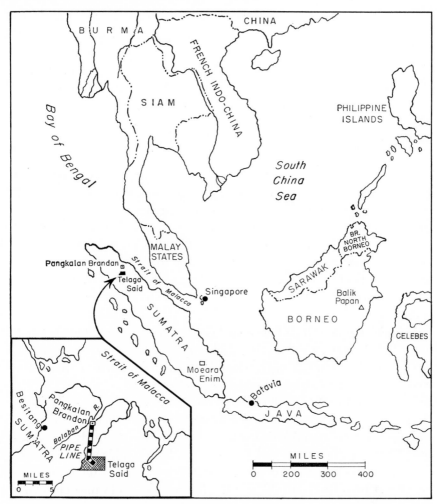

MAP 24:2. *Dutch East Indies oil fields and major markets, 1890's. Inset showing location of Royal Dutch installations. (Adapted from map, p. 27, Beaton, Enterprise in Oil.)*

sufficient funds to start commercial operations. Financial respectability for the project was gained when Dr. N. P. van den Berg, for 16 years director of the Java Bank and about to become Governor of The Netherlands Bank, agreed to serve as chairman of the proposed operating company.

By early spring of 1890 a prospectus was drawn up, and a distinguished board of directors was named. When the 1,300 shares of the new company, capitalized at about $520,000, were offered for sale, they were over-subscribed about four and one-half times. It remained only for King William III to sign the final document in May, 1890, to bring the Naamlooze Vennootschap Koninklijke Nederlandsche Maatschappij tot Exploitatie van Petroleumbronnen in Nederlandsche-Indië, or Royal Dutch, into legal existence. Zijlker, who had been paid $68,000 in cash and 200 shares for his concession, did not live to see the subsequent development of his project. He left Holland late in 1890 for Sumatra, where he died before the end of the year.[98]

Meanwhile the directors of Royal Dutch had appointed as president J. A. de Gelder, a young naval engineer with several years experience in the Indies. De Gelder immediately began ordering materials for the construction of a railroad, a pipeline, harbor facilities, and a refinery and packing plant to be located at Pangkalan Brandan, on the Babalan River near the Telaga Said concession. Two young graduates of the Delft engineering school were sent to the United States to observe drilling operations and to purchase drilling equipment and to hire drillers and refinery operators.

Although equipment and personnel began to arrive in the Langkat area in the fall of 1890, a combination of difficulties delayed any extensive developments for nearly a year, when J. B. A. Kessler arrived from The Netherlands to take charge. Kessler, with eleven years experience in the Indies as a partner in a leading Dutch trading firm, was well qualified for the job. Under his direction work was started on new wells, the railroad and pipeline were completed, and the refinery was made ready for operation. Finally on the morning of February 28, 1892, oil began to move under natural gas pressure from the well at Telaga Said to the refinery at Pangkalan Brandan.[99] At full capacity, the refinery could turn out about 1600 five-gallon cans a day.

Under the trade name "Crown Oil," the product was introduced to the commercial world by Kessler at Penang, important business center of the Straits Settlement. Tests carried on before a group of merchants, local officials and the press indicated that the quality of the Royal Dutch product compared favorably with both Russian and American oils. The response was enthusiastic, and Kessler contracted with Dutch firms at Penang, Medan, and Singapore to act as agents in distributing the oil to merchants in their respective areas.[100] After arranging through his friend H. W. A. Deterding, subagent of The Netherlands Trading Company, to advance funds against illuminating oil held in storage, Kessler returned to The

Refinery of Royal Dutch, Pangkalan Brandan, Sumatra, 1892.

Distilling section of the Pangkalan Brandan refinery, 1893–96, showing roofs over the stills which provided protection against heavy tropical rains.

Netherlands, turning over responsibility for operations to a Scot, James Waddell.[101]

Waddell, however, found it difficult to market even the modest output of the refinery. When prospects for a pickup in sales looked no better during late 1892, he wrote the directors in The Netherlands, urging them to sell out either to the Tank Syndicate or to the Rothschilds. A minority of the directors, including de Gelder, were inclined to follow Waddell's suggestion, but when the majority of the board disagreed, de Gelder resigned, being replaced as managing director by Kessler in November, 1892. The board authorized Kessler to negotiate with Samuel for a long-term contract to market Royal Dutch's output of illuminating oil, but negotiations came to a halt when Samuel insisted on a guarantee of supplies in excess of the output currently being produced in the Indies.

J. B. August Kessler

Meanwhile Waddell had also resigned and in May, 1893, Kessler returned to the Indies, where he found Waddell had allowed the plant and equipment to deteriorate. Once again Kessler was able to improve operations, so that by the end of 1894 new wells and better utilization of refining equipment had increased output of kerosene about 90,000 cases per month. Among the more obvious indications of Kessler's success was the declaration of an 8 per cent dividend to Royal Dutch shareholders at the end of the year.[102]

Despite this improvement Kessler became increasingly dissatisfied with the system of distributing oil through merchants who also handled competing brands. He was convinced that Royal Dutch should follow the example of the Tank Syndicate and begin shipping in bulk to its own tank installations and canning plants in the principal markets. To undertake this development, Kessler in 1896 persuaded Deterding to leave The

Netherlands Trading Society and take charge of sales for Royal Dutch. Deterding began immediately to reorganize the company's marketing structure and methods of distribution. Employees of Royal Dutch replaced the agent firms formerly selling "Crown" oil. Later the "Tiger" brand was introduced to compete with Russian oil. Tank sites were chosen at various locations in China, the Straits Settlements, and India. Facilities

H. W. A. Deterding

in the latter country were provided by the firm of Ogilvy, Gillanders & Company, which had previously been importing Russian oil. Tank steamers were ordered or chartered with the result that by the end of 1897 Royal Dutch was well equipped to distribute and market the output from the refinery at Pangkalan Brandan, which grew from 1,333,000 cases (320,000 barrels) in 1895, to 1,850,000 cases (440,000 barrels) in 1896, and to 4,565,000 cases (1,085,000 barrels) by 1897.[103] This growth was reflected in dividends on operations in 1895 of 44 per cent; 46.5 per cent for 1896; and 52 per cent for 1897.[104]

Royal Dutch's success did not go unnoticed by its two principal rivals in the Asiatic markets, and both Samuel and Standard approached Kessler with proposals for co-operation. Samuel was prompted to reopen negotiations with Kessler, chiefly because of rumors during the mid-1890's of the possible nationalization of the industry at Baku by the Russian government. The prospect of having the Tank Syndicate's heavily mortgaged fleet and extensive storage facilities made idle by such action made him particularly anxious to develop alternative sources of supply. Kessler was interested in Samuel's proposal made in the summer of 1896 to assume responsibility for marketing Royal Dutch's output of illuminating oil. Negotiations again broke down this time owing to Samuel's insistence that the oil be marketed under the brand names of the Tank Syndicate.[105]

Samuel turned to the development of oil properties which he had purchased on behalf of M. Samuel & Company in Borneo in that year and which he proposed to connect with a refinery to be located at Balik Papan. In order to insure the continued cooperation of the members of the Tank Syndicate, he organized The "Shell" Transport & Trading Company in 1897, named after the name of kerosene sold by M. Samuel & Company.* The new corporation issued stock in return for the oil properties of M. Samuel & Company, and the tanker fleet, receiving depots, and canning plants of the members of the Tank Syndicate.[106]

Standard's first approach to Royal Dutch came in 1895, when W. H. Libby suggested the possibility of an agreement between the two companies. Back of this move was the growing realization on the part of Standard's management of their geographic disadvantage in supplying the Asian markets from the east coast of the United States. Some 5,207 miles separated Suez from the Atlantic coast ports, whereas ships leaving Batoum had only to travel 1,464 miles to reach the canal. Both Standard and Shell in turn faced a handicap compared with Royal Dutch, although Shell's disadvantage was offset considerably by the ability of its members to obtain return cargoes.†

---

* The following companies were the original participants in the formation of the "Shell" Transport and Trading Company, Ltd.:

Name	Location
Borneo Company	Straits Settlement
Speidel & Company	Saigon
Kerr, Bolton & Company	Phillipines
Arnold Karberg & Company	Hong Kong
M. Samuel & Company	Japan
Best & Company	Madras
Graham & Company	Bremen
A. Runge & Company	London
Lane & Macandrew	London

Source: *U. S. Consular Reports*, LVII, No. 212 (1898), 51.

† Samuel was proud of the fact that the Tank Syndicate had been able to clean the tankers sufficiently to ship general cargoes on their return voyages to Europe.[107]

Libby's proposal in 1895 that Standard acquire a majority of the shares of Royal Dutch had little appeal to the directors of the latter company, who wished to maintain their corporate independence. It was agreed, however, that Standard should send representatives to Sumatra to survey the properties with the possibility of coming up with a more acceptable proposal.[108]

Early in 1897 two Standard representatives, C. F. Lufkin and John H. Fertig, journeyed to Asia where they made a careful survey of Royal Dutch's facilities. Both returned convinced that if Standard was to compete on favorable terms in the Orient it should acquire local producing and refining properties, and both recommended negotiating with Royal Dutch. In Fertig's judgment, "neither Russian nor American competitors with their kerosene carried from a distance could meet the new force of Royal Dutch in the oil industry in the East."[109]

TABLE 24:4

U. S. exports of illuminating oil to major Far Eastern markets: 1884–99 (in thousands of 42-gallon barrels)

Year	China	Japan	East Indies	British Australasia *	Total
1884	315.2	428.7	1,253.7	102.5	2,100.1
1889	394.5	780.7	1,625.5	179.8	2,980.5
1894	1,363.5	887.4	2,045.4	281.5	4,577.8
1899	970.9	778.7	844.8	342.8	2,937.2

* India, Australia, and New Zealand.
Source: Appendix D:5 and D:6.

On a basis of these recommendations, Libby reopened negotiations with Kessler in the summer of 1897. His proposition in essence was that the stock of Royal Dutch be increased from three to twelve million guilders and that Standard take over the entire block of additional stock at par. This offer was not attractive to the directors of the Netherlands company for two reasons. One was a strong desire to maintain the corporate identity of Royal Dutch, as a "national" enterprise which would be lost under the proposed division of stock. Secondly, they were not impressed with Standard's offer to take over the new stock at par, when shares of Royal Dutch were currently selling in the market at some eight times that figure.[110]

Failing to reach an agreement with Royal Dutch, Standard turned to the possibility of acquiring oil properties from other firms springing up in the Dutch East Indies. The most promising of these was the Moeara Enim Petroleum Company, which had a concession located on the southern end of Sumatra. A tentative agreement was reached in 1898 providing for the formation of a new corporation (to be jointly owned by Moeara Enim and Standard), which was to take over the oil properties and provide for refining and marketing facilities. But when news of the proposal reached Holland there were loud outcries in the press against allowing Standard to gain a foothold in Dutch territory. A letter from The Netherland's Minister of Commerce to the Moeara Enim directors, expressing some doubt whether the government would authorize the transfer of their concession to the new company, apparently brought an end to negotiations. Shortly thereafter Moeara Enim contracted with Shell to market its product.[111]

With the failure to come to an agreement either with Royal Dutch or Moeara Enim, Standard was forced into other tactics in an attempt to hold its dominant role in the Far Eastern markets. Special pamphlets outlining best methods of using and caring for kerosene lamps were prepared in various languages and distributed widely among potential customers. Arrangements were made with manufacturers to produce small lamps and wicks which were sold at low prices. In 1898, Standard even arranged to purchase Russian kerosene, which was marketed as a second grade along with the high-grade Devoe brand, to compete with the oils of Shell and Royal Dutch.[112]

On the whole these efforts were ineffectual in maintaining Standard's marketing position in Asia, where by 1899, sales of American illuminating oil were running well below the approximate 4.8 million barrels distributed by marketers of Russian oil. But more than Standard's relative marketing position was involved. In contrast to Europe, where total sales of American kerosene reached their peak at the end of the 1884–99 period, the volume of American exports to the Far East (see Table 24:4), declined from a level of almost 4.6 million barrels in 1894 to just under 3 million barrels by 1899. And although Royal Dutch's sales of illuminating oil amounted to only about 500,000 barrels in 1899, there was no reason to assume that its refinery output would long remain under the volume of over one million barrels recorded for 1897.[113]

By the turn of the century the implications of these developments were abundantly clear to the members of the Standard management. They faced the alternatives of either allowing their position to deteriorate still further in the markets of the Far East, or of building a fully integrated business in the Orient based on local supplies of crude.

Chapter 25

Domestic markets and marketing channels

$\mathbf{A}$s competition for sales outside the United States grew more intensive
after the mid-1880's, the domestic market assumed an increasing significance for the American petroleum industry. Between 1884 and 1899, as
shown in Table 25:1, the proportion of illuminating oil distributed in the
home market rose from about one-third to over two-fifths, while the proportion of naphthas retained for domestic consumption increased from
approximately 87 per cent to almost 94 per cent. Only in the case of lubricating oils was there a decrease in the share of the total output sold in the
United States from about two-thirds in 1884 to three-fifths in 1899.

With total output more than doubling between 1884 and 1894, the result was an impressive expansion in the volumes of refined products distributed domestically: illuminating oil from about 5.1 million to 12.1 million barrels; naphthas from 2.2 million to 5.7 million barrels; lubricating

oil from 600,000 to 2.3 million barrels; and fuel oils and residuum from less than 500,000 to over 3.5 million barrels.

Between 1894 and 1899, changes in both the output and the domestic distribution of refined products were closely related to the growing pro-

TABLE 25:1

Domestic distribution of refinery outputs: 1884–99 (thousands of 42-gallon barrels)

	1884	1889	1894	1899
Illuminating oil				
Production	15,450.4	20,191.2	29,457.5	29,953.8
Domestic	5,120.6	7,053.8	12,067.8	12,702.3
Percentage	33.1	34.9	41.0	42.1
Naphtha and Gasoline				
Production	2,485.1	3,915.6	6,113.9	6,682.5
Domestic	2,159.5	3,582.6	5,743.5	6,256.2
Percentage	86.9	91.5	93.9	93.6
Lubricating oils				
Production	898.7	1,834.5	3,288.4	4,056.6
Domestic	613.3	1,170.1	2,331.5	2,405.9
Percentage	68.2	63.8	70.9	59.3
Fuel and Residuum				
Production	482.9 *	1,195.0	3,522.0 *	8,239.1
Domestic	482.9	1,195.0	3,522.0	8,239.1
% of total	100.0	100.0	100.0	100.0

* Estimated.
Sources: Table 24:1; Report of the Commissioner of Corporations on the Petroleum Industry, Part 2, Prices & Profits, 231.

portion of Ohio oil processed by domestic refiners. Lima crude, with its relatively low yields of lighter fractions (compared to Appalachian crude), became increasingly important as a source of illuminating oils and naphthas. But the larger quantities of illuminating oil and naphthas from Ohio oil were largely offset by lower yields from Appalachian crudes, whose refiners found it more profitable to run to lubricants instead of cracking the heavier distillates to maximize yields of illuminating oil and naphthas. Thus, while total domestic crude production rose from about 53 million barrels in 1894 to 57 million barrels in 1899, outputs of illuminating oil and naphthas remained relatively stable. To maintain these outputs, however, the industry produced, as a necessary by-product of Lima crude, increasing quantities of heavy distillates unsuited for lubricants, which were converted into fuel oils and residuum.

With the aggregate outputs of illuminating oil and naphthas remaining almost constant and the proportion exported relatively unchanged, the quantities distributed domestically were likewise unaffected. On the other hand the domestic economy was able to absorb more than 8 million barrels of fuel oil and residuum in 1899, which was double the quantity consumed only five years before. By contrast, all but a small portion of the increased output of lubricating oils between 1894 and 1899 were sold abroad, where Standard was unusually successful in competing with foreign suppliers of lubricants.

Changes in demand

Both foreign competition and changes in proportions of refinery outputs had an influence on the relative importance of the home market during 1884–99. But of equal significance in affecting the domestic demand for petroleum products throughout this period was the continuation of the major demographic and economic trends characteristic of the American economy during the latter half of the nineteenth century. Population, for example, grew from approximately 55 million in 1884 to nearly 75 million in 1899. It was accompanied by an even faster rise in national income that raised average annual incomes per capita from about $150 to over $205.[1]

In 1899, some 47 million Americans (60 per cent) still lived in towns with less than 2,500 inhabitants or on farms, which expanded in number from about 4.3 million in 1884 to 5.7 million in 1899.[2] But the growing importance of industry and commerce in the economy was reflected in the expansion of the urban population from approximately 18 million (about one-third of the total) in 1884 to over 29 million (nearly 40 per cent of the total) in 1899.[3]

Further evidence of the changing significance of the industrial segment of the economy between 1879 and 1899 was the rise in the number of manufacturing establishments from about 253,000 to 512,000; an increase in the average industrial labor force from 2.72 million to 5.3 million; and a rise in the "value added by manufacture" from $1.9 billion to nearly $5.6 billion.[4]

An additional measure of the expanding American economy was the growth in transport facilities. Railroad mileage increased from about 125,-000 in 1884 to over 187,000 in 1899; locomotives from just under 25,000 to nearly 37,000; and freight cars from approximately 800,000 to almost 1.3 million. Supplementing the railroad network was an expansion in "surfaced highways" from approximately 99,000 miles in 1884 to almost 125,-000 miles in 1899.[5]

In general these characteristics of the American economy all contributed to an expanding potential demand for petroleum products. The rapid rise in population and incomes increased the market for artificial il-

TABLE 25:2

Prices of refined products: 1884–99 (cents per gallon)

Year	Illuminating oil *	Crude naphtha †	Lubricating oil ‡	Gas oil §
1884	8.3	6.0		
1885	8.1	7.1	12.7	—
1886	7.1	7.5	11.8	—
1887	6.8	7.7	10.4	—
1888	7.5	7.3	11.6	—
1889	7.1	8.0	11.3	—
1890	7.3	8.2	14.2	—
1891	6.9	6.9	15.0	—
1892	6.1	5.2	15.2	—
1893	5.2	5.6	14.0	—
1894	5.2	5.8	15.0	—
1895	7.4	8.2	12.2	—
1896	7.0	7.6	12.2	—
1897	5.9	5.9	12.7	1.9
1898	6.3	6.0	11.3	1.5
1899	8.0	10.0	12.7	2.5

* Report of the Commissioner of Corporations on the Petroleum Industry, Part 2, Prices & Profits, 229; the illuminating oil is 150° WW in barrels, job lot, New York, taken from the OP&D Reporter.
† Report of the Commissioner of Corporations on the Petroleum Industry, Part 2, Prices & Profits, 259.
‡ OP&D Reporter, 1885–97; Report of the Commissioner of Corporations on the Petroleum Industry, Part 2, Prices & Profits, 933, 934. The lub-oil price is the unweighted average of paraffin (25 gravity), neutral 34 (bloomless), and black. The figures for 1898 and 1899 were obtained from the Thompson & Bedford, Boston and Providence circulars, with the purchaser receiving a 5 per cent discount on the posted price.
§ Report of the Commissioner of Corporations on the Petroleum Industry, Part 2, Prices & Profits, 282.

lumination in factories, hotels, office buildings, and homes. The growth and continued mechanization of industry, agriculture, and transport called for greater utilization of lubricants. Manufacturing and the industrial establishments that grew in number and size offered additional outlets for solvents and industrial fuels.

Not all the elements in the situation between 1884 and 1899, however, were equally favorable to increased sales of petroleum products. Success in tapping the demand for particular products varied with the ability of marketers to compete on a cost and service basis with substitutes.

Illuminating oil, for example, continued to supply a substantial portion of the growing demand for illumination from the mid-1880's to the mid-1890's, when domestic sales of kerosene expanded sufficiently to raise the average annual consumption from about 3.9 gallons in 1884 to 7.4 gallons in 1894. In large part this growth was in response to prices, which as shown in Table 25:2 dropped nearly 40 per cent during the decade. With the reversal of the long-run downward trend in prices after 1894, however, an increasing number of customers turned to gas or electricity for their illumination. By 1899 the average per capita consumption of kerosene had dropped to 7.1 gallons.

Yet it is doubtful whether higher kerosene prices did more than accelerate the shift to alternative sources of illumination. By the early 1890's a series of changes in both industries had made manufactured gas and electricity increasingly formidable competitors to lamp light in the rapidly expanding urban communities.

The adoption of two innovations were of particular importance in the evolution of the manufactured gas industry during this period. One was the process of producing a "carburetted" water gas by the use of a "superheater," patented by Thaddeus Lowe in 1875.[6] The principal advantage of Lowe's process came from the successful utilization of heavy petroleum fractions to produce a water gas with excellent heating and illuminating qualities on a large scale.[7] Its gradual spread after the mid-1880's cut manufacturing costs substantially as fuel or gas oils were substituted for the more expensive naphthas for gas enrichment.

The competitive position of gas was further strengthened in the field of illumination when Carl Auer von Welsbach, working in the R. W. Bunsen laboratory in Germany, developed a practical gas mantle, which was patented in 1885. Made of cotton fabric impregnated with rare earths, the mantle when placed near a gas flame "became incandescent and gave off a steady white light. This light was not only vastly superior in quality to that from an open gas flame but much more economical."[8]

In 1890 the American rights to the Welsbach mantle were purchased by the United Gas Improvement Company of Philadelphia. Ten years later there was an estimated ten million in use in the United States, approximately half the number of incandescent electric lights.[9]

Electricity made its debut as an illuminant in the form of arc lights which, by the early 1880's, had achieved considerable acceptance for street lighting and other types of open air illumination. It was not until the development of the incandescent light bulb and circuits that permitted the low voltage transmission of current, however, that electricity became competitive with gas or kerosene for interior illumination.

The pioneer innovator in this development was Thomas A. Edison. By 1879 he had perfected a vacuum light bulb, which received an impressive demonstration in December of that year at his laboratory in Menlo Park, New Jersey. Some sixty lamps were strung on poles in the laboratory grounds and a few were installed in nearby private homes. The first commercial use was on the steamship *Columbia* in 1880, followed a year later by an installation in a lithographing shop in New York City.[10]

The characteristics of the incandescent electric lamp clearly placed it above kerosene or gas as a form of artificial illumination. The steady, soft light produced without an open flame was one of the chief reasons for its acceptance as a desired alternative. In addition, electricity filled a previously untapped market for use in flour mills, chemical plants, libraries, and even petroleum refineries, where it was too dangerous to utilize open-flame illuminants.[11]

The ultimate success of the incandescent system was related to the low-cost distribution of electricity from central generating plants. Under Edison's system, which used direct current, power stations had to be placed close to consumers to avoid excessive power leakage in transmission. It remained for George Westinghouse to develop the alternating current system capable of long distance power transmission with safety. Borrowing ideas from gas transmission and technology from Europe, he first demonstrated the possibilities of using alternating current in April, 1886, at the Westinghouse laboratories at Great Barrington.[12]

Although kerosene remained popular throughout the 1884–99 period, these developments in gas and electricity foreshadowed its future subordinate position in the domestic markets for illumination.

These developments in illumination also affected the demand for other petroleum products. This effect was most evident with respect to gas manufacture, long a major outlet for naphthas. The gas industry continued to draw on the petroleum industry for materials to enrich its products. But with the spread of the Lowe process, gas manufacturers increasingly substituted the cheaper gas oils for naphthas for this purpose. See Table 25:2. Thus it was the producers of gas oil rather than naphthas who supplied the bulk of the approximate 4.6 million barrels of oil purchased by the gas industry in 1899.[13]

The decline in the demand for naphthas as a gas enricher was in part offset by an expanding industrial demand for solvents. But to dispose of the rapidly increasing output of the light fractions, industry members, particularly Standard, began a vigorous drive early in the 1890's to extend the use of naphthas as a fuel.

Gasoline stoves were already widely used in the Midwest and South where alternative fuel supplies were not readily available. To tap the eastern markets, Standard encouraged manufacturers to produce improved gasoline stoves and to send out salesmen to instruct customers in their use. Standard also persuaded eastern insurance companies to follow the example of western firms in not charging higher fire insurance rates on buildings using vapor stoves. Coupled with an extensive advertizing campaign extolling the virtues of the gasoline stove—its coolness in the summer, its safety, and its economy—these efforts were sufficient to lift prices of naphthas after 1892, as shown in Table 25:2, to levels that for some years were higher than illuminating oil.*

Industry members found it somewhat easier to develop outlets for the heavy fuel or "reduced oil" which, along with gas oil, was produced as a by-product of the greater volume of Lima crude processed by refiners. Experience in selling "topped" Lima crude in the Midwest during the late 1880's had indicated a strong potential demand for fuel oils among industrial users. This experience had also demonstrated that many industrial consumers were attracted by fuel oil, even at prices that made it more expensive as a source of heat than coal.

In part this attraction reflected the lower operating expenses of utilizing fuel oils; fewer workers, greater economy in storage, and lower upkeep of furnaces. It also reflected the advantages of liquid fuel in making possible rapid adjustments in the amounts of heat generated. It was by emphasizing these advantages that marketers were able to distribute an estimated 3.5 million barrels of fuel oils in the industrial markets of the Midwest and the East by 1899.

The growth in the demand for petroleum lubricants was primarily a function of the expanding industrial and transportation sectors of the American economy. Increases in the number and size of railroad locomotives and rolling stock and faster running times brought a need for improvements in the quality of lubricants as well as for larger quantities.

---

* By the summer of 1899, gasoline was in such short supply that Standard refineries were storing kerosene in order to run enough crude to obtain needed light fractions.[14]

Similarly, a near doubling in the number of active cotton spindles from 10.7 million to 19.5 million during the last two decades of the century and the increase in industrial mechanization, measured by an expansion in horsepower available from steam engine installations from 11.6 million in 1879 to 38.4 million in 1899, were indicative of the growing industrial market for lubricants.[15] Of lesser importance was a growing consumer demand for lubricants, to keep bicycles, sewing machines and other household devices in order.

These forces were strong enough, as shown in Table 25:2, to hold the prices of lubricants relatively stable in the face of an approximate fourfold increase in domestic sales. Even though prices remained comparatively stable, improvements in the quality of products resulted in a lowering of the costs of lubrication.

Competition and marketing channels for by-products

The marketing channels through which the major by-products of the American refineries flowed reflected both the nature of the products and the character of demand. These two factors, along with Standard's control over refining capacity, were further mirrored in the nature of competition in these markets during 1884–99.

In concentrating its refining capacity on kerosene, the Trust of necessity became the major supplier of the domestic outputs of naphthas and fuel oils. With the bulk of these products sold to large industrial users, there was little need for an elaborate marketing apparatus. Most sales contracts, whether by Standard or independents, were made by refiners directly with consumers, and provided for tank-car delivery into the storage tanks of the purchasers.* The principal variation from this pattern involved the sale of naphthas to small industrial users and householders where the distribution followed essentially the same marketing channels as kerosene.

Standard's principal competition in the domestic markets for naphthas came from prominent East coast independents, such as Pure Oil and Tidewater, in addition to several Midwestern refiners of Lima crude. The latter also competed with Standard for the fuel-oil and gas-oil markets in the Midwest and the South.

A new element of competition appeared in these markets during the

---

* Among the more prominent independents following this pattern of marketing during the late 1890's was the Sun Oil Company of Toledo, which regularly supplied gas companies in Montreal and various towns in Iowa.[16]

1890's when The People's Gas Light & Coke Company of Chicago and the Indianapolis Gas Company began to integrate toward their supply. In 1890 the two gas companies assured themselves of adequate sources of gas oil at reasonable prices by forming a jointly held subsidiary, The Manhattan Oil Company, with its own pipeline facilities and a "refinery." Subsequently the Manhattan concern added other gas companies to its customers.[17]

The three lubricating oil markets (railroads, industrial, and consumer) continued throughout 1884–99 to be marked by diversity, both in the degree of competition and in the organization of marketing channels. By the mid-1890's, for example, Standard's subsidiaries, the Galena Oil Works and the Signal Oil Works, produced virtually all the lubricants used by the American railroads.

A part of the railroad demand was supplied through independent marketing and merchandising firms, such as Spear, Gregory & Company of Boston, which sold Standard lubricants on a commission basis.[18] A number of railroads in southwestern United States bought their lubricants from Standard's marketing affiliate, the Waters-Pierce Oil Company. The bulk of the railroad market, however, was supplied directly by the Galena and Signal companies.

Much of Standard's success in dominating the market for railroad lubricants rested on the high quality of the products manufactured by its lubricating plants, plus their aggressive marketing policies. Beginning in the mid-1880's, for example, Galena Oil began selling to railroads under annual contracts that guaranteed carriers a 10 per cent saving of the previous year's costs of lubricating its equipment.[19] Early in the 1890's these contracts were modified by guaranteeing the railroads a specified average cost of lubricating its equipment per one thousand miles. If costs ran higher than the amounts specified, Galena supplied extra lubricants free; if lower, the benefit went to the railroad.[20]

There is reason to believe that not all of Standard's success in dominating the railroad demand for lubricants was a result of the quality of its products or the services rendered customers. Independents who had previously supplied a portion of the railroad market testified that, beginning in the early 1890's, the railroads consistently refused to consider their bids on lubricating-oil contracts. The reason, they argued, was that Standard used its position as the largest shipper of petroleum products to bargain with the railroads for exclusive contracts covering the latters' lubricating needs.[21]

Compared to the railroad field, independents found it much easier to operate profitably in the market for industrial lubricants. With almost every technological and engineering advance calling for a new or different type of product, there were many opportunities for both producers and marketers outside the Standard orbit to cater to the highly diversified needs of industrial customers.

Once established, suppliers in the industrial markets were reasonably well insulated from intensive price competition, so long as they stood ready to service the specialized lubricating needs of their customers. Distribution of industrial lubricants was marked by a considerable variation in marketing channels. Refiners, for example, might sell directly to industrial users. Alternatively they might sell through sales agents or commission houses. Finally they might dispose of lubricating stocks in the "open markets," or produce exchanges, to wholesalers or blender marketers for distribution to the trade.

How far Standard supplied the domestic market for industrial lubricants is impossible to determine. There is little question, however, that the Trust distributed much more than the estimated 40 per cent of the total output of lubricating oils produced by Standard refineries.* Standard's specialized lubricating oil affiliates, including Thompson & Bedford and the Vacuum Oil Company and its general marketing companies, such as Waters-Pierce, competed actively with the independents in the industrial markets for lubricants. They supplemented the Trust's production of these types of lubricants by buying in the open market or directly from independent refiners.

The importance attached to the industrial demand by the Standard management is indicated by the marketing organization maintained by the Vacuum Oil Company. In addition to carrying a large number of grades of oil in stock, Vacuum in the 1890's had sales offices and blending facilities in every major industrial center in the United States.† Vacuum's sales force was said to be larger than the total number of salesmen employed by all of the other Standard companies.[23]

It was Standard's large and efficient sales organization that enabled it to dominate the market for lubricants sold to non-industrial consumers. These products, prepackaged and branded, were typically purchased by customers at local hardware or grocery stores. Sold as "convenience" items, their distribution required no specialized or technical knowledge on the part of retailers.

---

* See above, Chapter 23.
† Thompson and Bedford carried 47 standard lubricating oils in open stock.[22]

Unless their product had a particular brand appeal, independents found it difficult to compete in this market with Standard. The latter, whose representatives called regularly on the trade to promote the sale of illuminating oil and naphthas, could market lubricating shelf goods at very little added cost.

These developments in the sale and distribution of by-products provided independent refiners and marketers with opportunities to operate profitably on a limited scale. For the members of the Standard management, however, the growth of independents presented an unwelcome potential challenge to their position in the American markets. Already facing strong competition in the export trade, their response was not only to assume a more active role in the distribution of by-products. They also took steps to consolidate and extend their control over the important domestic market for illuminating oil.

Standard's marketing reorganization

Through affiliation with leading wholesale firms and the formation of others, Standard's share of the domestic market for illuminating oil by the mid-1880's was at least proportionate to its control over refining capacity, variously estimated at between 80 and 85 per cent. Among the independent refiners and marketers who emerged during the succeeding fifteen years there was no one with the stature of the Nobels or the Rothschilds to challenge Standard's position in the domestic market for illuminating oil.

As pointed out by his biographer, however, Rockefeller and his associates regarded all "competitors with a jaundiced eye." From their standpoint, they had by mid-1880's replaced "savage competition" with unified control, which made the business highly profitable and attractive to outsiders. They felt that, "If independents were to spring up on every side, they would soon disorganize the markets again" to the detriment of the Trust with its large fixed investment.[24] It was in part to avoid this possibility that the management not only took a more active role in the marketing of by-products, but moved to strengthen Standard's position in the distribution of illuminating oil as well. Even without the possibility of more active competition, its top executives, by mid-1880's, had reason to be dissatisfied with Standard's existing marketing structure.

Within the Standard hierarchy there was a conflict of interests between the marketing units, particularly those not owned 100 per cent by the

Trust. The marketers wanted large discounts on their purchases, while the refining units objected to any shift of "profits" to the marketing companies. Moreover Standard's executives found it difficult to exert any central control over the marketing procedures and policies of their major affiliates. They could never be sure that Carley, Pierce, Hanford, or the McDonald's would follow directives or suggestions originating at 26 Broadway. They were also impatient with the failure of a number of the Trust's marketers to push bulk-handling methods more vigorously. Finally, not all wholesalers were affiliated with Standard, and between the wholesalers and retailers generally there were scores of jobbers over wh·ch the Trust had no direct control. Among this group there were many who, from Standard's point of view, were all too ready to buy from independent refiners when the opportunity arose.

While there was a growing conviction on the part of its executives that the Trust's marketing organization and policies should be strengthened and extended, there was some uncertainty as to how far and how fast they should move toward integration. Rockefeller in 1885 expressed his opinion that there should be closer supervision of the stations selling Standard products. About the same time, F. Q. Barstow advocated a further extension of bulk stations, "without disturbing relations with good [jobbing] customers," while W. G. Warden urged a further extension of bulk stations to be administered from New York.[25]

With the formation of the Domestic Trading Committee in 1886, a more definite program began to emerge; one designed to reduce competition, to "get closer to the trade" by selling directly to retailers, and to extend the system of bulk distribution, including tank-wagon deliveries, to Standard's products sold in the domestic markets.[26]

To insure a closer and more effective central control over marketing practices and policies, the Domestic Trade Committee took steps to buy out the minority interests in a number of the principal marketing affiliates associated with Standard. In 1886, F. D. Carley sold his interest in Chess, Carley & Company, which was subsequently absorbed in the newly incorporated Standard Oil Company of Kentucky, owned 100 per cent by the Trust. In 1890, P. C. Hanford decided to retire and the Standard Oil Company of Illinois was incorporated to take over the business of the P. C. Hanford Oil Company. Two years later the McDonalds gave up their interests in the Consolidated Tank Line Company, which was merged with Standard of Kentucky. An attempt to buy out Henry Clay Pierce failed because of a disagreement over price and he retained his 40 per cent stock

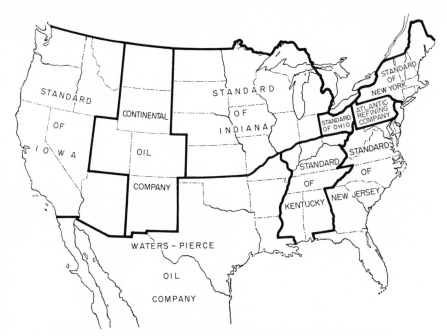

MAP 25:1. *Standard's marketing territories: 1892–99.*

interest and management control over the Waters-Pierce Oil Company.[27]

A further reorganization came in 1887 with the formation of Standard Oil of Minnesota, which took over Ohio Standard's marketing facilities in Minnesota and North Dakota. The following year, Standard's Continental Oil Company acquired a minority interest in the United Oil Company, a producing and refining corporation in Florence, Colorado.[28] In 1892, as a part of the reorganization of Standard interests following the dissolution of the Trust, the number of domestic marketing companies was reduced and their respective sales territories enlarged and consolidated into the nine major divisions shown on Map 25:1.

Prior to 1890, Standard's sales in Canada were handled through branch offices of its domestic affiliates, in competition with Canadian refiners and American independents, such as Scofield, Schurmer & Teagle. Beginning in the 1890's, Standard began purchasing Canadian companies and in 1898 acquired the Imperial Oil Company, a fully integrated firm that controlled some 25 per cent of Canadian production.[29]

With the consolidation of its other holdings under the Imperial Oil Company, Standard not only owned nearly all the Canadian refining capacity but held a completely integrated and nation-wide network for the distribution of petroleum products. By 1899, Imperial's share of the total Canadian market was estimated at over 60 per cent. The remainder was supplied by independent marketers who imported their refined oil from the United States.[30]

Standard continued to supply virtually the entire Mexican market for petroleum products through Waters-Pierce. Sometime in the 1880's, Waters-Pierce, which had long been a distributor of refined products south of the border, built a refinery at Tampico. In 1897 these facilities were expanded and a canning plant added.[31] Two years later, exports of crude from the United States to Mexico amounted to approximately 1.9 million barrels.* It is perhaps indicative of the position of Waters-Pierce in the Mexican market that during 1899 it realized a "profit" of 7.7¢ per gallon from sales of illuminating oil, approximately three times the average return from sales in the United States domestic market.[32]

Spread of bulk distribution

Meanwhile, under the prodding of the Domestic Trading Committee there was a marked expansion in the number of tank and bulk stations owned and operated by Standard marketing companies. Compared to the 130 stations reported six years earlier, the number by 1888 had risen to 313 stations, located principally in the upper Mississippi Valley. Judging from later figures, the number of stations grew at an accelerated rate over the succeeding decade and by the end of 1899 may well have been close to 3,000.† While a portion of these stations were added through purchase, the overwhelming majority were constructed by Standard.[34]

This expansion suggests that by 1899 Standard was well launched in providing what one of its veteran marketers subsequently described as "good [marketing] equipment." This was to "establish stations so near together that the entire country is covered from them . . . [adopting] as a general rule a radius of twelve miles for a station." The facilities used depended "on the size of the city or town," ranging from two 400-barrel storage tanks, a small warehouse, a stable, and horses and wagons for a small country station, to the elaborate layout of a bulk station like the one Waters-Pierce was operating in St. Louis.[35]

---

* See Appendix D:3.
† No precise data are available for the intervening years, but Standard reported its holdings at 3,573 stations in 1906.[33]

Early tank wagon introduced in the 1890's.

The extension of bulk and tank stations opened the way for a further spread of tank-wagon deliveries. Rockefeller was personally much interested in extending Standard's marketing operations in this fashion. As early as 1883 he had urged the employment of tank wagons in Cincinnati, Pittsburgh, and Chicago. Later in the same year he opined that Standard was not pushing "the tank-wagon business" in Western cities as it should. "I believe it is one of our best means for getting and holding the trade." [36]

Rockefeller's comments revealed a clear understanding of both the limitations and the advantages of tank-wagon deliveries. Wherever the number of customers was large enough within a territory that could be served by a horse and wagon in a day or possibly two days, this method of delivery offered further economies. Even where these economies were not shared with retailers, the latter found tank-wagon delivery safer and more convenient than barrels. They no longer had to handle bulky barrels or be concerned with their disposal when empty. Instead of buying a barrel of oil by weight, which was sold by the gallon, retailers could order the quantity desired and pay for the number of gallons received. Metal storage tanks at the retail establishments virtually eliminated losses through evaporation and leakage, reduced unpleasant odors and possible danger to surrounding merchandise, and cut down the risk of fire.

Tank-wagon delivery was an especially effective device for controlling sales in a particular community. Essentially a system of local transportation, one tank wagon could provide efficient delivery service for a given town or section of a city. Operating expenses and interest and depreciation on investment increased per unit of product handled, when a territory that could be covered by a single tank station and tank wagon was served by two or more in competition.[37] Thus when a single firm had already established facilities for delivering oil within a given territory, other concerns were hesitant to enter the field. This was especially true of smaller towns or communities.

The use of tank wagons for delivery between bulk stations and retail dealers, however, spread slowly. Reporting to the Census as of December 31, 1889, Standard listed only 252 tank wagons, along with 992 horses and 11 mules for its entire marketing system.[38] A breakdown of the Trust's delivery of refined oil by tank wagon, shown in the following tabulation, indicates that it was not until 1897 that more than half of Standard's sales of illuminating oil was distributed in this fashion.

Percentage of Standard's total sales of refined oil delivered by tank wagon

Year	1892	1893	1894	1895	1896	1897	1898	1899
Per cent	33	34	36	39	46	52	56	59

Source: Statement No. 407–S, Petroleum Statistics: 1906, 68 (Archives of Socony-Mobil Oil Company).

While tank-wagon deliveries spread slowly, there was a marked improvement in the type of packages used to distribute petroleum products from bulk or tank stations. By the early 1890's, the wooden barrel was largely restricted to shipments to customers in remote areas where it was expected the barrel would not be returned.[39] Elsewhere, wooden barrels had been replaced generally by iron containers, including the so-called 10-gallon "milk can," iron barrels holding between 50 and 54 gallons, and iron drums that varied in capacity between 100 and 110 gallons. The particular type of container utilized depended on a variety of circumstances. As the capacity of tank wagons grew larger (ranging, for example, between 300 and 1,000 gallons), it was frequently more convenient to use a can wagon holding between 25 and 35 cans for delivery, especially in country districts.[40]

No independent distributors of illuminating oil could hope to compete with Standard, except on a limited scale, without adopting bulk distribution. This was a possibility, however, open only to those firms with sufficient capital resources to supply their own tank cars, and in most instances to build their own bulk stations. It was for this reason that Standard's principal competition so far as bulk handling was concerned came almost entirely from large independents, such as the Pure Oil Company, Scofield, Schurmer & Teagle, the Sun Oil Company, and Crew-Levick.[41] The small independents were forced to distribute their illuminating oil in barrels, either in nearby local markets or sparsely populated areas where it was uneconomical to establish bulk plants.

Standard's acquisition of independent marketers

Simultaneously with the extension of bulk and tank stations and tank-wagon deliveries, Standard began increasingly to bypass the jobbing trade and sell directly to retailers. In certain markets, such as those served by the Waters-Pierce Oil Company and one or two of Standard's eastern marketing organizations, this move involved no major change. Elsewhere Standard's marketing companies varied their procedures according to the competitive situation in particular areas. Sometimes it was possible simply to enlarge the selling force and handling facilities of the Trust's existing marketing affiliates and begin selling directly to retail outlets. In other instances it was considered more advisable to buy up local or small regional jobbers and either merge their facilities or maintain them as "independent" companies selling in the same market with companies known to be a part of the Standard organization.

Reviewing this phase of the organization's marketing development, John G. Milburn, attorney for New Jersey Standard, stated,

> When in the '90's the Standard adopted the policy of reaching the retail merchants themselves, through the establishment of stations all over this country, the jobber had necessarily to a very large extent go out of business. . . . When the Standard reached the retailer the jobbers' profit was eliminated; and a great many of their plants . . . were taken by Standard . . . There are fifty or sixty of those little marketing concerns, and some you would not describe as little, that were acquired under those conditions.[42]

An examination of Table 25:3 gives some impression of the types of marketing companies acquired by various Standard units between 1886 and 1899. With few exceptions, such as the Dixie Oil Company of New Orleans, the Commercial Oil Company of Atlanta, and the Paragon Oil Company of Columbus, Ohio, they were all small concerns that sold to retail outlets either in a local or state-wide market.

Standard attempted to keep its acquisition of marketing companies a secret. According to the critics of Standard's marketing policies, these "bogus companies," as they were sometimes called, were maintained principally to compete with bona fide independents. As one employee of a "secret" Standard company explained it, "Any trade that the Standard Oil Company was enjoying was to be left alone. We were to work for the class

TABLE 25:3

Principal marketing companies acquired by Standard: 1886–99

Date	Name and location
1886	United Refiners Export Co., Perth Amboy, New Jersey
1887	Monarch Oil Co., Cincinnati, Ohio Globe Oil Co., Minneapolis, Minnesota
1888	West Michigan Oil Co., Michigan Chester Oil Co., Minnesota
1889	Mehlen's Family Oil Co., Ltd., Long Island City, N. Y. White's Golden Lubricator Co., Cincinnati, Ohio
1890	Underhay & Co., Boston, Massachusetts Des Moines Oil Tank Co., Des Moines, Iowa Blodgett-Moore Co., Augusta, Georgia Electric Light Oil Co., Philadelphia, Pennsylvania New Jersey Oil Co., Newark, New Jersey
1891	L. D. Mix Oil Co., Cleveland, Ohio
1892	Excelsior Oil Co., Wichita, Kansas Sutton Bros., Kansas City, Missouri
1895	Oakdale Oil Works, Philadelphia, Pennsylvania Nicolai Bros., Washington, D. C.
1897	New American Oil Co., Mansfield, Ohio
1898	Commercial Oil Co., Atlanta, Georgia Peoples Oil Co., Atlanta, Georgia
1899	Cassetty Oil Co., Nashville, Tennessee Protection Oil Co., Cincinnati, Ohio Paragon Oil Co., Columbus, Ohio Dixie Oil Co., Savannah, Georgia Southern Oil Co., Richmond, Virginia

Source: *Standard Oil Co. of New Jersey et al. v. United States of America,* Supreme Court of the U. S., October term, 1909, *Brief for Appellants,* II, *Facts,* 256–59.

of trade that the Standard Oil could not get or had difficulty in getting." This trade was described as "dissatisfied trade." Dissatisfaction might arise from a variety of causes but, "a great many, especially country merchants, did not want to buy from Standard . . . because it was a monopoly; and

preferred to buy from independents on that account." In general the "bogus company" was to go after the business of the independent companies
"as hard as they could." [43]

Insofar as Standard marketers followed this practice, it gave them an
effective and precise method of competition by enabling them "to carry
on a competitive warfare against the independent dealers and cut the
prices without involving [Standard's] prices directly, and without reducing [Standard's] prices on the larger part of [its] product sold in the same
locality." [44]

While Standard spokesmen admitted—ordinarily with considerable
reluctance—the operation of "secret" companies, they disputed vigorously
the reasons advanced by critics for following this practice.[45] Their general
defense was that "Nearly all the . . . companies or firms whose business
was thus purchased . . . [were] operated merely for the purpose of getting the benefits of brands and good will that went with the purchase and
in order to turn over the business to the purchaser to the best advantage." With few exceptions, it was stated, these companies sold at the
same prices as the regular "Standard" companies operating in their respective markets.[46]

As a general rule price competition was avoided by both Standard and
the independents. "Price wars" normally resulted when a new firm attempted to enter a market or an existing independent began to expand
its sales. For example, when Pure Oil began distributing by tank wagon
in the New York metropolitan market in 1896 the price of illuminating
oil was cut from 9.5¢ a gallon to 5.5¢.[47] Independents generally agreed,
however, that so long as they did not attempt to increase their share of
particular markets beyond 15 or 20 per cent, they were reasonably sure of
avoiding price competition. Otherwise, as one small marketer testified, an
independent found that Standard's prices "will be cut so low that he will
either lose his business or go broke on low prices." [48]

Standard's methods in checking on the business done by competitive
marketers were universally condemned by independents. This information was obtained through an elaborate and detailed system of reporting
instituted by the Trust's Domestic Trading Committee to administer its
far-flung marketing operations. Inspectors were sent into the field to
check on the condition of equipment, the efficiency of the labor force, and
the accuracy of the reports by local managers on the business of competitors.[49]

According to critics, the local managers were a part of Standard's spy system which would have done credit to any national intelligence group. Independents particularly resented the methods utilized by Standard representatives which ran the gamut from simple observation to bribing of railroad officials and clerks in competitors' offices.[50]

Refinery location and railroad rates

In addition to integrating its marketing operation and extending bulk handling, Standard's over-all domestic competitive strategy included two other important features. One was to locate its refining capacity close to major marketing and distributing centers. The second was to carry on extensive negotiations with the railroads for favorable rates on shipments of refined products.

The general location of Standard's refineries relative to those operated by independents is shown on Map 25:2. The independents were situated in metropolitan New York, western Pennsylvania, and Ohio. Standard not only maintained refining capacity in these areas, but also in Baltimore, Parkersburg (West Virginia), Whiting, and Olean (in upper New York state). From its plants at New York, Philadelphia, and Baltimore, Standard was in a strategic position to ship refined products by bulk steamer to major Atlantic coastal entrepots for inland distribution. For example, Standard could supply the eastern New England market through Boston at total transport costs ranging between 14¢ and 16¢ per hundredweight. Independents supplying this same market by rail, paid rates that varied between 28.5¢ and 34.5¢ per hundredweight.[51]

Where Standard's refinery locations gave them no inherent advantage in reaching major marketing areas, the management worked to obtain preferential treatment from the railroads. Negotiations with the railroads after 1887, however, had to take into account the effects of the Interstate Commerce Act, passed in that year, on relations between shippers and carriers. Designed to eliminate the abuses of preceding decades, the legislation specifically prohibited rebates, pooling arrangements, unreasonable rates, and personal or place discrimination. It made mandatory the filing of all rates on interstate shipments with the Interstate Commerce Commission, which was given responsibility for enforcing the act.[52] Shipments in intrastate commerce, however, were not subject to the federal restraints. Moreover, there was considerable doubt about the meaning, inclusiveness, and constitutionality of the provisions of the act.

It was under these circumstances that the traffic managers of the vari-

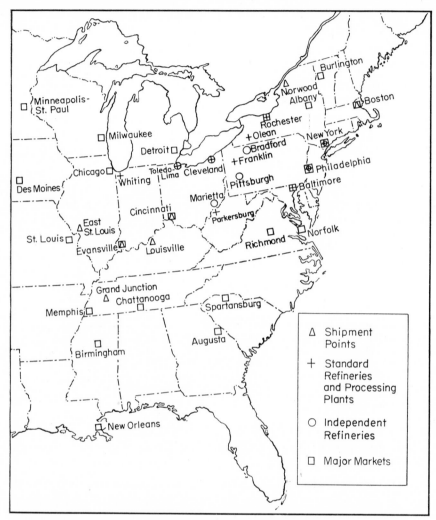

MAP 25:2. *U. S. major refining, shipping and marketing centers in the 1890's.*

ous Standard marketing agencies ". . . labored assiduously to get products classified in such a way as to subject them to the lowest possible railroad tariffs, took advantage of intra-state schedules, and adopted other means deemed legal by Standard Oil counsel in keeping transport costs to a minimum." [53]

One area where Standard was able to obtain preferential treatment from the railroads was in the markets served by the Olean refinery. For example, the open rate on refined petroleum, filed with the Interstate Commerce Commission, was 33.6¢ per barrel on shipments from Olean to Rochester, New York, 106 miles to the northeast. The open rate from the nearest independent refinery at Bradford, 167 miles from Rochester, was 46.4¢. Thus on a basis of open rates, Standard had an advantage of 12.8¢ a barrel. In reality, however, oil was shipped from Olean to Rochester for 9¢ a barrel under a special, unpublished, intrastate rate granted by the Pennsylvania Railroad. As a result, independents traveling a 58 per cent longer distance paid a rate that was 500 per cent more than the Stand-

TABLE 25:4

*Comparison of distances and freight rates on refined petro-leum products from Olean, New York, and Warren, Pennsylvania, to selected Vermont markets: 1906 (rate in cents per hundredweight) ** *

Destination	From Oleon via Norwood		From Warren		Rate advantage of Olean over Warren
	miles	rate	miles	rate	
Burlington	451	15.34	523	33.00	17.66
Rutland	518	16.02	456	23.50	7.48
Bellows Falls	571	16.89	509	23.50	6.61

* These rates were either identical or similar to the rate structure during the late 1890's.
Source: Bureau of Corporations, *Report of the Commissioner of Corporations on the Transportation of Petroleum*, 112.

ard. Under these circumstances it is not surprising that shipments by independents "into Rochester were very infrequent." [54]

This intrastate rate in conjunction with other special rates also gave Standard a significant shipping advantage over competitors in supplying markets in Vermont. The posted rate of the New York Central Railroad from Olean to Burlington, Rutland, and Bellows Falls, Vermont, via Norwood, New York, was 40.5¢ per hundredweight. Standard's rates, however, as shown in Table 25:4, ranged from 15.34¢ to 16.89¢ per hundredweight. Independents shipping from Warren, and in some instances over a shorter distance, paid a differential that ranged from 6.61¢ to 17.66¢ per hundredweight.

Standard was also successful in obtaining advantageous shipping rates from its Whiting refinery to the important markets in the southern and southwestern regions of the United States. Judging from the open or published rates, filed with the Interstate Commerce Commission, shown in

Table 25:5, Standard's Whiting refinery was at a disadvantage in supplying the southwestern market through the East St. Louis gateway. Indeed, any independent contemplating a refinery location to supply this market would have been dissuaded by these rates from establishing a plant in the Chicago area.

Actually, Standard, beginning in the early 1890's, shipped refined products from Chicago to East St. Louis over the Chicago & Eastern Illinois, the Chicago, Burlington & Quincy, and the Chicago & Alton railroads at

TABLE 25:5

Comparison of distances and freight rates on refined petroleum products to East St. Louis from selected refining centers: 1891–1900 (rate in cents per hundredweight)

From	Distance	Open rate	Actual rate	Whiting's rate advantage
Pittsburgh	613	22.5	22.5	16.5
Cleveland	522	17.5	17.5	11.5
Toledo	433	15.0	15.0	9.0
Whiting	281–345 *	18.0	6.0	
		10.0 C&A		

* Variation in mileage based on alternative routes.
Source: Bureau of Corporations, *Report of the Commissioner of Corporations on the Transportation of Petroleum*, 319, 322.

rates of 6.0¢ and 6.25¢ per hundredweight. Knowledge of these rates was limited to Standard and the carriers.* The 6.25¢ rate of the Chicago & Eastern Illinois, for example, applied only to shipments between Dalton, Illinois, an obscure junction in Chicago switching area, and East St. Louis. Moreover, all shipments by Standard over this road and the others were "blind billed" (the rate was not specified on the bills-of-lading, settlements being made between the general freight agents of the railroads and Standard's traffic managers). As a result of these arrangements Standard had an actual advantage over competitors in shipping to East St. Louis that, as shown in Table 25:5, varied between 9.0¢ and 16.5¢ per hundredweight.

Standard was also able to secure a substantial rate advantage in reaching the marketing areas south of the Ohio and east of the Mississippi rivers. Under schedules filed during the 1890's with the Interstate Com-

---

* The Chicago & Eastern Illinois rate of 6.25¢ was filed with the Interstate Commerce Commission in 1894, but the 6.0¢ rate of the other two roads was not.[55]

merce Commission by the Chicago and Ohio River Committee of the Central Freight Association, rates on petroleum products from Cleveland and Toledo to Cincinnati were set at 10.0¢ and 10.6¢ (respectively) per hundredweight, while the rate from Whiting to Evansville was 12¢ (reduced to 11¢ between 1896 and 1900).[56]

Until the mid-1890's Standard shipped via Evansville at the 12-cent rate plus local rates from Evansville to various southern markets. After

TABLE 25:6

Comparison of distances and freight rates on refined petroleum products from Whiting and Toledo to selected Southern markets: 1904 (rate in cents per hundredweight) *

Destination	From Whiting via Evansville		From Whiting via Grand Junction		From Toledo		Rate advantage of Grand Junction over Evansville	Rate advantage of Grand Junction over Toledo
	Miles	Rate	Miles	Rate	Miles	Rate		
Birmingham, Ala.	651	44.0	820	29.5	692	47.5	14.5	18.0
Florence, Ala.	570	41.0	692	26.6	638	44.5	14.4	17.9
Selma, Ala.	754	44.0	923	37.0	795	74.5	7.0	10.5
Spartansburg, S. C.	771	60.5	1,195	52.1	701	60.0	8.4	7.9
Atlanta, Ga.	733	44.0	1,003	33.2	687	47.5	10.8	14.3
Macon, Ga.	821	44.0	1,019	35.2	775	47.5	8.8	12.3
Augusta, Ga.	904	44.0	1,391	36.5	835	47.5	7.5	11.0
Chattanooga, Tenn.	595	39.5	849	25.9	549	43.0	13.6	17.1
Knoxville, Tenn.	572	43.5	960	34.8	502	43.0	8.7	8.2
Harriman, Tenn.	529	45.0	1,011	31.3	470	42.0	13.7	10.7

* These rates were either identical or similar to the rate structure during the late 1890's.
Source: Bureau of Corporations, *Report of the Commissioner of Corporations on the Transportation of Petroleum*, 253, 255.

1896, however, Standard benefited from a special through rate, filed in that year with the Interstate Commerce Commission by the Illinois Central, of 13¢ a hundredweight from Dalton, Illinois, to Grand Junction, Tennessee, some 250 miles to the south of Evansville. This rate, coupled with local tariffs from Grand Junction, enabled Standard, as indicated in Table 25:6, to reach various southern distributing centers at lower shipping costs than independents.

In conjunction with its refinery locations, the general effect of these differential freight rates was to confine the active competition of independents to areas where freight charges were a relatively small part of total costs of distribution. In these areas (western New York and Pennsylvania,

Ohio, Indiana, lower Michigan, and Illinois), independents competed vigorously with one another as well as with nearby Standard refineries. It was in these markets, for example, that the extension of bulk-handling methods were of particular significance in determining market shares.

Refinery location and differential freight rates were thus an integral part of the management's over-all marketing strategy. Coupled with extension into crude production and expansion of transport and refining facilities, it enabled Standard to maintain its dominant position in the domestic industry. This was accomplished at a time when control over the domestic markets was of particular importance in enabling Standard to compete more effectively with powerful rivals in the export trade. Whether Standard's apparent impregnable position in the American industry could be maintained in the years ahead was open to question. The imminent development of large new producing areas in the United States would, in itself, offer a major challenge to the Standard management. Equally challenging was the growing public hostility toward big business that focused particularly on the practices of Standard Oil.

Chapter 26

The public and the trusts

Having built their organization behind a curtain of secrecy in the 1870's, Rockefeller and his aides wanted nothing so much as to be let alone after they organized the Trust. Their hopes were disappointed, for the years from 1880 to 1900 saw Standard's operations gradually exposed not only to more, but increasingly hostile, publicity. This was inevitable, once state and federal governments began heeding the growing public reaction to railway abuses and the concern, evidenced somewhat later, over industrial combination. Since Standard had benefited as much as any from railroad discrimination and was the first to employ the trust device, it could not hope to escape investigation and attack, even though it understandably tried hard to do so.

It was perhaps accidental that the public obtained its first reasonably comprehensive picture of Standard's operations from the Hepburn investigation of New York railroads in 1879. But it was also symbolic of the course that attack would follow. From the early 1870's until the mid-'80's railroad abuses constituted the principle targets of reform groups. While the roads were not viewed as the founts from which all evils flowed, they were at least cited as representative of the worst elements in modern business practice. According to these arguments—and they were not entirely false—the trunk lines watered stock; they corrupted public office; their geographic discrimination distorted competition in the industries which they served. Standard was not unexpectedly taken as the prime, but not the only, example of the last point; in fact, for nearly twenty years after the Hepburn investigation, outside observers and independents within the petroleum industry gave rebates an inordinate share of the responsibility for Standard's monopoly.

While New York groups were trying to get the Albany Legislature to enact the Hepburn Committee's recommendations for railroad legislation, a journalist in the Midwest was digesting the Committee's findings into a virulent attack upon Standard Oil. As financial and railroad editor of the then liberal *Chicago Tribune,* Henry Demarest Lloyd was already beginning to use that paper's columns to inveigh against the corporate iniquity of the trunk lines, Standard Oil, and "the coal and iron and salt Pashas" whose methods redistributed "the wealth of the masses into the pockets of the monopolists." [1] In March, 1881, he reached an international audience when the *Atlantic Monthly* published his history of Standard, "The Story of a Great Monopoly." The *Atlantic* had to run seven printings of that issue and the *London Railway News* distributed free reprints of the article to English investors. [2]

Even within the limits of the facts then known, Lloyd's article contained some quite substantive inaccuracies, exaggerations, and distortions. But these very characteristics helped make it an effectively written polemic for the author's main theme: "monopolies" were to be curbed through federal regulation of the railroads. State legislatures were inadequate, for they themselves were pawns of the trunk lines and the combinations which the railroads fostered. The conclusions were not new, but no one had argued the points so colorfully and forcefully for wide circulation.

Standard did not reply to Lloyd's charges. The Trust's executives, like most corporation leaders of the time, firmly believed that their business was no one else's concern; that so long as their activities were legal, they were answerable to none but their stockholders. If Standard here misjudged the trend of public opinion, the mistake is understandable. Although railroad criticism was quite strong and widespread in the early 1880's, the attack on pools and trusts in manufacturing and marketing was only beginning. In this context, Lloyd and others like him were still minor voices in 1881. There were consequently ample grounds for believing that if left unanswered, their attacks would soon be forgotten.

Moreover, while some of Lloyd's charges could have been refuted, the answers might have led to embarrassing questions on other points—as Standard learned a year later. In 1882 the New York Senate investigated speculation in various markets and several Standard officials volunteered to testify, hoping to reply to allegations made against the Trust. But when asked about profits and other particulars, they took cover in evasion and wrangles over the propriety of the Committee's questions. As Professor Nevins points out, Standard's reticence did not necessarily connote a sense of guilt. Rockefeller, like Carnegie, was one of the most self-righteous executives in American corporate history. But the Trust's managers realized that facts and interpretations which they might advance in justification would hardly be regarded sympathetically by the opposition.[3]

Somewhat more attention was paid to criticism in the trade press and Regions' dailies which had long been hostile to the Standard. In 1881, Ohio Standard loaned money to the new manager of the *Oil City Derrick*, a personal friend of a Trust director. Four years later the paper was purchased by Patrick Boyle, a friend of Daniel O'Day. While Boyle apparently had no financial ties with the Trust until 1904, his ideological commitment was soon evident.[4] A similar change occurred in the policy of the influential *Oil, Paint & Drug Reporter* after it merged in 1883 with a rival trade journal allegedly financed by Standard. Formerly highly critical of the Trust, the *Reporter's* editorials thereafter paralleled the defensive argument advanced by Standard's executives.[5] Criticism in other Regions' newspapers subsided during the 1880's, but it is doubtful that Standard's interference was responsible in all cases.

Standard's only public reply to attacks in the general press before the late 1880's was written by J. N. Camden, formerly in charge of the company's Maryland and West Virginia operations and since 1881, United States Senator from West Virginia. Camden, answering a critical article that had appeared in the *North American Review*, gave a preview of the general line of defense that the Trust would adopt in the future. Standard was pictured as the agency that had stabilized the destructive forces of cutthroat competition in the industry. This it had accomplished through superior efficiency, in part resulting from "a consolidation of capital adequate to any possible business emergency." The less successful refiners were ipso facto the less efficient—an anticipation of the Social Darwinism that Dodd would later state more explicitly—and their failure had led them to make common cause with politicians in attacking the Trust.[6]

The Trust, however, was soon called upon to deliver more than Camden's broad assertions and denials. Agitation for railroad regulation was approaching its climax by the mid-1880's. When the Cullom Committee in 1885 began hearings on proposals to regulate interstate commerce, the rebates paid Standard were both attacked and defended in the testimony.[7] No new information resulted, but the hearings again called adverse attention to the Trust, as they did to the general growth of corporate power. Citing the railroads as the most conspicuous example of the problem, the Cullom Committee concluded,

> No general question of governmental policy occupies at this time so prominent a place in the thoughts of the people as that of controlling the steady growth and extending influence of corporate power and of regulating its relations to the public.[8]

When Congress passed the Interstate Commerce Act in 1887, *The New York Times* reviewed the history of the Standard, cottonseed oil and beef combines to demonstrate that "the notorious abuse of power in unjust [railroad rate] discrimination" was the "clearest and strongest argument" in support of the Act.[9] These were extreme positions, but they indicated that much of the reformist energy built up for railway legislation was beginning to turn to the broader issue of trust-busting.

The growing concern over "combinations" at this time in part reflected a sharpened awareness, probably due to railroad reform, of an older problem. Price-fixing associations and pooling arrangements had been operative in many industries since the Civil War. While most of these were local in character, the anthracite roads, segments of the metal trades and the salt producers had organized fairly effective regional pools by the middle of the 1870's.

The 1880's saw other industries adopt similar restrictive devices on both the regional and national levels. From 1881 to 1887, when they organized the Whiskey Trust, distillers repeatedly used national pools to try to limit output and dump surplus spirits on the export market. The first of a succession of collusive arrangements among the large meat packers was established in 1885. A year before, the cottonseed-oil industry became the first to follow Standard's example when about 90 per cent of national crushing capacity was placed in a trust. In 1887 about 70 per cent of sugar refining capacity came under trust control, and by 1889, members of the lead trust reportedly produced about nine-tenths of that industry's output.*

These were only the beginnings of a movement toward consolidation and market control that would reach its peak in the following decade. Public reaction against them can easily be exaggerated. The 1888 platforms of both major parties contained anti-trust planks, although it is questionable whether they represented the opinion of a significant part of the electorate. In the same year President Cleveland, who could hardly be classified as a radical, warned Congress that "Corporations . . . are fast becoming the people's masters," but made no recommendations as to how they might be made the people's servants.[10] Of the fourteen states which had by 1890 enacted constitutional or statutory provisions against various restraints of trade, at least seven had done so since 1887.[11] These laws, however, were rarely enforced. On the other hand, officials thought that public opinion was strong enough to warrant more than formal pronouncements. John Reagan, a Democrat from Texas and one of the principal champions of railroad regulation, and John Sherman, a Republican from Ohio, each introduced anti-trust bills in the U. S. Senate in August, 1888. The House had already instructed its Committee on Manufactures to investigate trusts in January of that year. The New York Senate undertook a similar investigation at the same time, while the Empire State's attorney general charged that the North River Sugar Refining Company had violated its corporate charter by participating in the Sugar Trust.[12]

These events, each of which would affect Standard, helped convince the Trust's executives that public sentiment was growing too strong to be ignored as it once had. While Congress was debating the Sherman Act in 1890, Rockefeller himself granted a long interview to the *New York World*, one of the more hostile New York papers. He replied to specific al-

---

* Many factors are responsible for these restrictive arrangements. Most of them were apparently attempts to eliminate allegedly cutthroat competition that arose when capacity (built during temporary booms) exceeded the quantity that the market would profitably absorb after the boom tapered off. The reasons for the demand movements and for the failure of capacity to adjust to them varied from industry to industry. One frequently cited generalization claims that economies of scale were already precluding the possibility of many firms' competing in the same market. But except for rather vague remarks—often based upon the testimony of the industrialists themselves—few attempts have been made to document this explanation.

legations against the Trust, but would not comment on the trust question generally. Two years earlier, however, S. C. T. Dodd had written a broader defense of combinations in his book, *Combinations: Their Uses and Abuses.* He somewhat obscured the issue by identifying the right to form industrial consolidations with the right of free association: "Every partnership is a combination. Every corporation is a combination. Destroy the right to combine and business on a large scale becomes at once impossible." Increase in scale, he argued, was an inherent part of the competitive struggle in which the fittest survived. Since the struggle could end in the elimination of competition, Dodd admitted that combines should be regulated. He denied, however, that this was a necessary result. Thus Standard was indeed a combination, but strong competitors precluded its monopolizing the market. There was obviously room to differ with Dodd on what constituted strong competition.

The immediate impetus leading Dodd to write his book was the New York Senate investigation, probing the Standard, cottonseed-oil, and sugar trusts, along with less formal arrangements in other industries. Standard executives on the witness stand admitted to refining about three-quarters of American crude output, a market share that, they claimed, was attained solely through superior ability and economies of scale. Rockefeller estimated that the Trust's yearly profits averaged about 10 per cent on capital stock (the actual figure was closer to 15 per cent), and produced, for its first public appearance, the Trust Agreement of 1882. Although both he and Archbold pictured the Trust as merely a convenient fiduciary device, they did admit under further questioning that the Trustees exercised at least indirect control of the constituent companies by electing their officers and directors.[13]

The investigation was, in the words of Professor Nevins, "a loose, hasty, and inconsequential affair." [14] The committee majority concluded that the "main purpose, management and effect" of all the combinations studied was to "annihilate" competition and fix buying and selling prices. Yet they were not convinced that explicit state anti-trust legislation was needed. Distinguishing "good" from "bad" combinations—a doctrine that common law had already expressed in rather obscure form—they preferred to wait and see if common law could not restrain the latter.[15]

Two months after they were called by the New York Senate, Standard executives were summoned before the Congressional committee investigating trusts. In general, they tried to follow the advice of Paul Babcock, president of Jersey Standard: "I think this anti-trust fever is a craze which we should meet in a very dignified way and *parry every question* with answers which while perfectly truthful are evasive of *bottom* facts." [16] The choice of Franklin B. Gowen as the Committee's counsel probably strengthened Standard's determination to use this tactic. As president of the Reading, Gowen had sided with Tidewater in the latter's early battles with Standard. Despite his subsequent contracts with the oil combine, he remained hostile. That evasion succeeded in large part was indicated by one committee member's remark to Rockefeller: ". . . the testimony goes to show that you are doing an immense business, and there seems to be nobody that knows anything positively about it!" [17]

Nonetheless the hearings added substantially to available information on the Trust. Much of this came from independent refiners and marketers like George Rice and Harlow Dow, who outlined in detail Standard's harassment. But F. D. Carley, Standard's principal marketer in the South, also readily described how he eliminated competition in that area. In addition, testimony concerning rebates and drawbacks, first taken in the Producers' Union cases of the late 1870's, was read into the record. A defense of its position necessarily involved the Trust in some disclosure.[18] Neither could Gowen's persistent questions be entirely evaded by vague generalities and mere assertion. The main burden of the testimony was the familiar theme of railroad discrimination. Emery summed up the independents' position: "I believe wealth has been derived not from legitimate returns of the business, but from discriminations and drawbacks given by the railroads." [19] Gowen's only explicit recommendation at the hearings was that federal courts be empowered to compel the railroads to grant equal rates to all, although at one point he hinted that pipelines might be made subject to the Interstate Commerce Act.[20] Standard admitted receiving rebates, pointing out that the practice had been general. But it again denied that its market position resulted from anything else than "the application of better methods and of better business principles than have been brought against us. . . ." Archbold, in one extreme statement, even termed the rebate system "a vexation. It was not one that we upheld or profited by, in my judgment to any degree whatever." [21]

The Committee's report took no stand on this or any other issue. It presented no analysis of causes nor recommendations for policy. It did not even summarize the evidence. Its main point was that, since Trusts, as distinct from their constituent companies, never did any business, they probably could not be prosecuted even if Congress passed laws against combinations in restraint of trade.[22]

"The Bosses of the Senate." Cartoon by Joseph Keppler, Puck, *January 23, 1889.*

This conclusion, however, was soon to be contested in the courts of New York and Ohio. In 1888, the attorney general of New York sought to vacate the charter of the North River Sugar Refining Company, a New York corporation whose stock was held by the Sugar Trust. The state based its petition on two main points: first, the company's affairs were not controlled by its own directors, as specified in the charter, but by the managers of the Trust; secondly, the company had become party to an unlawful monopoly "injurious to trade and commerce." [23] In 1889 the lower court agreed with both points and the appeals court reaffirmed the decision, though on somewhat narrower grounds.[24] The Trust Agreement was dissolved in 1891, but the constituent companies were soon merged into the American Sugar Refining Company.

In May, 1890, Ohio's Republican attorney general, David K. Watson, inaugurated quo warranto proceedings to revoke Standard of Ohio's charter for identical reasons.* Standard's reply was a general denial of the allegations: the Trust was in no way a combination "injurious to trade

* Professor Nevins notes that Watson "undoubtedly read of the New York suits, and doubtless heard of the Western State officials who were riding like knight-errants into the lists." [25]

and commerce"; the management of Ohio Standard was in the hands of its board of directors, elected by the stockholders; while certain stockholders had entered the Trust Agreement solely as individuals, the company as such had not. The New York courts, however, had already struck down this legal distinction in the Sugar case—a fact that probably led Standard to invoke the five-year statute of limitations as additional grounds for defense.[26]

Perhaps because many lawyers thought that he had a strong case,[27] Watson was pressured to abandon it. Standard limited its battle to the courts, but Mark Hanna, leader of the Ohio Republicans, called the attorney-general's attention to the benefits that Standard had bestowed upon Cleveland, and the opposition of that city's businessmen to the suit. He also warned that, for Watson's political future, "the identification of your office with litigation of this character is a great mistake." [28]

Nonetheless Watson persisted. In March, 1892, the state Supreme Court ordered Ohio Standard to sever its connections with the Trust, but did not vacate its charter because of the statute of limitations. Not only was the company deemed guilty of illegally having surrendered control of its affairs to the Trustees, but it was also convicted of the broader charge of having entered a combination to monopolize trade. While admitting that Standard may have improved the quality and lowered the price of petroleum products, the Court noted that

> such is not one of the usual or general results of a monopoly; and it is the policy of the law to regard, not what may, but what usually happens. Experience has shown that it is not wise to trust human cupidity where it has the opportunity to aggrandize itself at the expense of others.[29]

Three weeks after the decision, S. C. T. Dodd announced that the Trust would be dissolved. The decision, he said, was not an admission of guilt but a concession to public opinion. Implicitly assuming that Standard's market share was necessary to obtain its economies of scale, he maintained that Standard had only sought to lower costs and consumer prices. But the public, "unwisely" aroused, failed to distinguish between combinations for this purpose and those designed to restrict output.[30] Moreover, New York threatened to prosecute Standard companies incorporated in the Empire State if they continued in the Trust.

The dissolution involved considerable problems. The Trust, by the end of 1891, held varying proportions of the equity of 92 separate companies, and some 1,600 holders of Trust certificates had a share in each of them.

Since only the certificates had been traded for the last 10 years, there was no market valuation for the shares of the individual companies.

To simplify the matter, Standard reduced the number of constituent firms to twenty by having major units purchase the assets or the stock of others. Shares in each of these twenty companies were then to be pro-rated among owners of certificates according to the number of certificates they owned. Thus Rockefeller owned about 26.4 per cent of the total certificates outstanding. He exchanged these for 26.4 per cent of the number of shares held by the Trust in each company. At the other extreme, the owner of only one certificate was to receive but a fractional share, called scrip, in each of the twenty firms, and these paid no dividends.[31]

This was probably the simplest practical method of distributing the stock, but, as the press had correctly predicted,[32] it did not substantively alter the operations of the Standard Oil group. The eight original trustees and nine other Standard Oil executives held over half of the Trust certificates. Upon exchanging these for stock in 1892, they acquired working, and in most cases majority, control of the constituent companies.* Moreover, the presidents and directors of the latter were chiefly men who had been on the Trust's executive committee. The small certificate holders also had an interest in maintaining the status quo. Rather than exchange their holdings for scrip, they retained their certificates and continued to receive dividends from the combined profits of all Standard companies. Nor did Standard encourage them to surrender their shares in the Trust. As a result, about 49 per cent of the certificates remained outstanding until November, 1897.[33]

By the time that the Ohio cases were closed, the nation's anti-trust fervor had temporarily abated. Various states continued prosecutions, but on the national level, concern over combinations steadily diminished until the last two years of the decade.[34] The high McKinley tariff of 1890 made protection the principal issue in the presidential campaign of 1892. The panic of 1893 only made Cleveland's innately conservative administration all the more unwilling to risk impairing business confidence by making trusts a major issue. Recovery, labor unrest, tariffs and bimetallism absorbed the president's major attentions. By the presidential election of 1896, free silver had become the over-riding political issue. Even the Populists, in whom reformers like Henry Demarest Lloyd had placed

---

* They did not obtain majority control in all cases because the Trust, in some instances, had not held a sufficiently high proportion of the total shares.

their hopes for broad social reform, adopted this panacea to the virtual exclusion of other considerations.*

Nor did the federal attorney generals show much inclination to enforce the Sherman Act. District attorneys were generally left to take the initiative in instituting proceedings; Washington's communiques to them generally stressed the difficulties of enforcement and the need to proceed with caution. In 1895 the U. S. Supreme Court practically vitiated the measure by ruling that a monopoly of manufacture could not be construed as interstate commerce. Attorney General Olney, formerly counsel for the Whiskey Trust, hardly concealed his pleasure.[36]

These events, along with the Trust's formal compliance with the Ohio decision, gave Standard a respite from legal action until 1897. The attacks in the press on Trusts in general and Standard in particular also decreased during these years.[37] There was, however, one notable exception: in 1894 Harper's published Henry Demarest Lloyd's *Wealth against Commonwealth.*

Lloyd devoted the bulk of his 500 pages to Standard Oil's activities. After consulting state and federal investigations, court testimony and contemporary newspapers, he arranged his findings into an exaggerated, if sincere, indictment of corporate greed as epitomized in Standard. While still stressing railroad rebates as the foundations of Standard's control, he placed them in a rather unusual context:

> [Railroad] freight agents and general managers presented to the monopoly, out of the freight earnings of the oil business, the money with which to build the pipe lines that would destroy that branch of the business of the roads. . . . Here [in pipelines] its control has been the most complete; and here the reduction of price has been least.[38]

If this oversimplified the development of Standard's pipeline system, it nonetheless showed a greater awareness of its importance than had other popular tracts. Moreover, it reflected a point of view that the independents had only begun to appreciate. On the other hand, the exaggerations and distortions which had characterized his article of 1881 were also present in the larger work. Lloyd recounted, for example, the myth that Rockefeller had forced a widow into poverty by paying only a pittance for her refinery. Actually, the reverse was true, and Lloyd had documents at hand to prove this. The ex parte statements of other Standard opponents

---

* The Populist platform of 1896 did have an anti-monopoly plank, but it stressed public ownership of public utilities and land held by "other corporations in excess of their actual needs," rather than industrial combination.[35]

were so uncritically accepted that the *Nation,* hardly a spokesman for corporate interests, wrote, "A dog would not be hung upon such evidence. . . . Were we not satisfied from evidence *aliunde* that the managers of the Standard Oil Company had violated both law and justice . . . we should be inclined to acquit them after reading this screed." [39]

Other journals were more favorable. The *Outlook* termed it "the most powerful book on economics" since *Progress and Poverty;* the *Chicago Evening Post,* "a masterly and successful attempt to illustrate the movement of business feudalism." Many readers encouraged the author to have a cheap edition published. [40]

Yet the impact of Lloyd's work is difficult to assess. He provided continuity between the first wave of anti-trust sentiment, ending around 1892, and the second, that gave rise to the muckrakers. Like the latter, he helped significantly to mold the climate of opinion, and this was his primary objective. "The ignorance of the people is the real capital of monopoly," he wrote, "if they know, they will care." But knowledge and caring were not enough to reform: the energies aroused had to be channeled into a vehicle for effective action. The Progressives in part provided such a vehicle for the muckrakers in the early twentieth century, but in 1894 there were none to fulfill this function for Lloyd. As noted above, the Republicans and Democrats were concerned with other issues, and the Populists, who might have built up sufficient pressure to force the major parties to action, were already beginning to hitch their wagon to a silver star. This, as much as its traditional policy of secrecy, perhaps explained Standard's silence to Lloyd's charges.

Significantly, Standard next skirmished with a state, rather than the federal government. In November, 1897, Frank S. Monnett, attorney general of Ohio, started contempt proceedings against Ohio Standard, alleging that the latter had failed to comply with the court order to sever connections with the Trust. He based his case upon evidence brought him by George Rice, an independent refiner who had fought Standard in the market, on the witness stand and in the press.

Rice held seven shares in the Standard Oil Trust—all purchased after the announced dissolution—upon which he had regularly been paid dividends. Although never requested to exchange these for scrip, Rice did submit one share for cancellation on his own initiative. S. C. T. Dodd tried to discourage him but Rice persisted. He finally received fractional shares —some as small as 50/9725—in each company. More importantly, he proved to his own satisfaction what he probably had hoped to prove from the start: namely, that Standard was evading the court's order of 1892. He also convinced the crusading Monnett who, in addition to preparing suits against the Tobacco Trust, Western Union and several insurance companies, was drafting an anti-trust law that the legislature passed in April, 1898.

Monnett set himself a difficult task. The trustees had formally surrendered their certificates. That they still controlled the companies in the Standard group was technically merely the result of the distribution of Trust certificates before the dissolution. As for the remainder, Standard claimed that small shareholders could not be forced to surrender their certificates for non-paying scrip. Archbold called attention to proverbial "widows and orphans and hospitals," whom this procedure would have ruined.* This was not entirely an adequate explanation, for within three months after Monnett started proceedings, 100,000 certificates, about 21 per cent of the total then outstanding, were rather suddenly cancelled.[42] But the evidence, if weighty, was not conclusive.

Monnett hoped to find more definite proof of contempt in the distribution of Ohio Standard's dividends. Rockefeller admitted on the witness stand that about 13,600 shares of Ohio Standard were still represented by Trust certificates. If the company paid dividends on this equity to the trustees, it was directly violating the court's order. Ohio Standard denied it had done so; since 1892, it claimed, all its profits had gone to modernizing the decrepit Cleveland plant or into surplus. When Monnett wanted proof from the company's ledgers, Standard executives refused to produce the books. The attorney general, perhaps aware of the shortcomings of his evidence in the contempt suit, had indicted Ohio Standard, Buckeye, Ohio Oil and Solar Refining for violating the state's anti-trust act. Opening their books for the contempt case, Standard argued, might provide the state with evidence against them for the anti-trust suits. They continued to refuse even in the face of a court order. Allegations arose that the books had been burned, but the charge, which Standard denied, was never proved. At any rate, Monnett failed to produce sufficient evidence to convince the court and in 1900–1901 the state Supreme Court dismissed both the anti-trust and contempt cases.[43]

While Monnett was attacking Standard companies in Ohio, Waters-Pierce was being prosecuted under Texas' anti-trust law of 1889. The com-

---

* In 1897, a shareholder would have had to own 194½ shares in the Trust in order to obtain one full share of each of the 20 constituent companies. The cost of acquiring this equity was about $66,000.[41]

pany was charged with using a wide array of unfair competitive practices in its attempt to monopolize the state's oil business—among others, resale price maintenance, refusal to sell to potential competitors, coercion of dealers, and operation of bogus independent companies. In 1900, the U. S. Supreme Court affirming the state court's ruling, ordered Waters-Pierce to forfeit its permit to do business in Texas. Although technically more successful than Monnett's suits, the economic result of this action was negligible. The company was reorganized with identical capitalization and stockholders, resumed business in the state under a new permit, and, according to a subsequent conviction in 1907, continued to use the same tactics.[44]

At the same time they were trying to convince the public that their organization was really a collection of independent companies, Standard's top executives were also taking steps to consolidate its internal corporate structure. This move, begun in 1897, was designed to take advantage of New Jersey legislation passed in 1884 that allowed corporations chartered in the state to hold securities of other corporations. Despite misgivings on the part of Standard's chief counsel, S. C. T. Dodd, about the constitutionality of the New Jersey laws, the management decided to make the Standard Oil Company of New Jersey the top, or parent, holding company for its properties.* In June 1899, New Jersey Standard increased its capitalization to $110 million and began exchanging its shares for those of Standard affiliates and subsidiaries. By the end of the year, New Jersey Standard had a majority control of all but three of the 41 affiliated companies.[45]

Administratively, the Standard properties continued to be run after 1900 by the same relatively close-knit group as before. But the adopting of the holding company device reduced the possibility that the death of any of the principal shareholders might disturb or disrupt the continuity and functioning of the Standard Oil organization.

By this time the trust problem was again beginning to attract nationwide interest. The return of prosperity brought with it the first of the

---

* It should be noted, however, that there were also reasons to believe that at least the holding-company form would not be challenged. Chief among these reasons was the Supreme Court decision in the E. C. Knight case concluding that the Sherman Act did not apply to restraints of trade in manufacture. That few corporations shared Dodd's fears was demonstrated by the popularity of the holding-company device. During 1899–1901, an average of 2,000 corporations were chartered annually in New Jersey. Many of these—including the United States Steel Corporation—were holding companies.

great merger movements during 1898–1902. Statistics on the number of new consolidations during these years must be severely qualified because of the limitations of the basic data, but all indicate the same general trend. According to one fairly comprehensive series, there were 24 consolidations with authorized capital of $1 million or more formed in 1898. This was 50 per cent greater than the number in the highest previous year and the combined capitalization of $550 million was four times greater. In 1899, the number had increased to 105 whose aggregate capital amounted to $2.2 billion.[46] For obvious reasons, Standard could not hope to avoid attention.

The specific vehicle that again put Standard in the national spotlight was the Industrial Commission, a bipartisam committee of the U. S. Senate and House appointed to "collate information . . . and recommend legislation to meet the problems presented by labor, agriculture and capital." The leading figure behind the Commission was Representative Thomas W. Phillips, a trustee for U. S. Pipe Line and shareholder in the Pure Oil Company. Elected to the House in 1894, Phillips had repeatedly introduced bills to establish the commission since that date. That he finally succeeded in 1899 was partly due to the renewed concern over trusts,* although the Commission's reports also reflected the broader economic and social problems besetting the turbulent '90's: Chicago labor disputes, labor-management relations, immigration, agriculture and the marketing of farm products. To digest and analyze 19 volumes of testimony, the Commission employed a staff of academic experts, one of the first used in government investigation. Among them were John R. Commons on labor, William Z. Ripley on transportation and Jeremiah Jenks on trusts.

Along with a mass of factual material, the Commission's investigation of Standard Oil revealed what the contending parties thought the basic issues were. For Standard, they had changed little since 1888: large size (apparently identified with large market share) had been necessary to bring order out of chaos, improving efficiency and lowering prices to the consumer. Rockefeller not only adopted this view as a defense, but urged it as a positive course to be followed by other industries and encouraged by the government:

> There are other American products besides oil for which the markets of the
> world can be opened, and legislators will be blind to our best industrial in-

---

* E. Dana Durand, secretary of the Commission, later wrote that "The rise of the trusts was probably the chief ground which led to establishment . . ."[47]

terests if they unduly hinder by legislation the combination of persons and capital requisite for the attainment of so desirable an end.[48]

He joined S. C. T. Dodd in recommending that federal charters be granted to facilitate this process, providing that some sort of regulation be employed to prevent abuses of power.

The independents' testimony indicated that they had learned little over the preceding decade. Pure Oil trustee James W. Lee confused size and market share by proposing that a $1 million upper limit be placed on corporate capital, a recommendation that subsequently embarrassed him after Pure increased its authorized capital to $10 million in 1900.[49] But the independents' main proposals still focused on railway discrimination.[50] Archbold furnished a sheaf of letters from railway executives denying that Standard either sought or received rebates since the Interstate Commerce Act.[51] The independents, however, alleged that Standard had not only obtained direct rebates since 1887 (which was probably false) but also that it had gained the same effect through special concessions and rate combinations (which was true).* Several, disillusioned with the slow and costly procedures of the Interstate Commerce Commission and the courts, recommended government ownership of railways. Even Lewis Emery apparently failed to appreciate the significance of his trunk pipeline. Government ownership of the roads, he contended, "will do away with all difficulties that exist today, because the prime movers of all this trouble are the railroads."[52] Railroad discrimination on product shipments continued to play an important, though somewhat diminished, role in Standard's operations and provided the independents with a comfortable scapegoat. But the latter's near obsession with this problem made them unwilling or unable to appraise their own positions realistically. The economies of trunk pipelines and their effect on refinery location, for example, were barely considered, except in complaints that Standard's pipeline charges were excessive; and even then the independents implied that equal rail rates to all could offset this advantage. Similarly, their proposals to outlaw persistent selling below cost showed little awareness of the problems posed by different cost structures and economies of scale.

The Commission's final report, published in 1902, avoided both extremes. It granted that in many industries costs fell as output increased, but noted,

> Production on a large scale does not always imply monopoly. . . . Much
> weight must be given to the argument that, while production on a large scale

* See Chapter 25.

should be encouraged, no added gain to society comes through monopoly, and that every effort should be made to prevent monopoly through combination.[53]

After generally advocating thorough enforcement of existing anti-trust laws, the report specifically emphasized the need for publicity. The larger corporations, it recommended, should be required to publish reasonably detailed balance sheets and profit statements "to encourage competition when profits become excessive." This was to be supplemented by the investigations of a permanent federal agency on corporations and combinations.[54]

These proposals were moderate and the summaries of the evidence on combinations impartial. After reading an advance copy sent him by Senator Boise Penrose, Archbold wrote, "We think the report is so fair that we will not undertake to suggest any changes." * Yet for all its objectivity, the Commission probably aroused antagonism against Standard as much as it allayed it. The hearings uncovered or revived much damaging evidence and were given wide attention in the daily press. Even on points which Standard could convincingly rebut, people were likely to accept the independents' side of the story, particularly since the newly developing mass-circulation magazines were beginning to present generally unflattering portraits of the modern corporation.

In November, 1902 (the same year in which the Commission's final report appeared), *McClure's,* the largest and most influential of these periodicals, ran the first of Ida M. Tarbell's articles on the Standard Oil Company. Originally, S. S. McClure conceived the study as a part of a series that the magazine had been running on the "greatest American business achievements." The author shared McClure's hope that the Standard articles would be "received as a legitimate historical study." She combed court testimony and government hearings; she interviewed such prominent independents as Lewis Emery; S. C. T. Dodd, H. H. Rogers and Daniel O'Day gave her information, an indication of the more flexible public relations policy Standard had recently adopted to have its side of the story more fully presented.

---

* Archbold had, however, made suggestions before this. An earlier proposal required corporations to publish balance sheets, lists of shareholders and their equity, and statement of receipts, expenditures, and profits. Archbold wrote Penrose, his personal friend on the Commission, to try to block this recommendation. It should be noted that other corporations also pressured the Commission, and among the Standard hierarchy, Archbold was generally more aggressive on these matters than the older executives.[55]

Announcement

M
c
C
L
U
R
E
'S

F
O
R

1
9
0
3

McCLURE'S MAGAZINE has just closed the most successful and prosperous year in its history. It has been permanently enlarged to make room for new and broader editorial plans. There will be timely and important articles, character sketches of great men, reports of all that is going on in the world of science, exploration, politics, arts, and letters. In fiction, McCLURE'S stories of real life—of heroism, love, and adventure—will be unsurpassed as heretofore.

The leading feature for the ensuing year is *Miss Tarbell's History of*

The Standard Oil Company

This series of articles will furnish an Object-Lesson in the

PARAMOUNT PUBLIC QUESTION OF TO-DAY—THE QUESTION OF TRUSTS

Here we have a vivid illustration of how a trust is made, of what it costs to replace competition by a single-headed combination. This story of how the Standard Oil Co. grew from an insignificant beginning to a Practical Monopoly of one of the chief products of this country will constitute the most important, the most illuminating, and most interesting discussion of the Trust Problem which a magazine has ever laid before the reading public. It tells the amazing story of the attempt in 1872 of a few men, aided by the leading officials of the Standard Oil Co. and the *Railroad Kings of the day*—Vanderbilt, Gould, and Scott—to seize the entire oil industry,—an attempt frustrated by a popular uprising in the Oil Regions. It tells the story of how later the Standard was able to secure special freight rates which enabled it to become the most nearly perfect monopoly of the 19th century. It rehearses the tragedies which the rise of the Standard necessitated. It describes the ability, the energy, the perseverance employed to build up *THE MOST PERFECT BUSINESS ORGANIZATION* which the world knows.

This history is a great Human Drama—the story of the conflict of the two great Commercial Principles of the Day—Competition and Combination. Miss Tarbell, whose Life of Lincoln is well remembered for its bringing to light unknown and important facts, and for the sympathetic insight into the character of the great President, tells this history in the same way, without partisan passion and entirely from documents. *The sole aim is to record what actually happened—to tell how the Standard Oil Co. achieved its great position.* It will be profusely illustrated with drawings, portraits, maps, documents, etc.

4

Announcement by McClure's Magazine of forthcoming publication of Ida Tarbell's History of the Standard Oil Company.

The result, appearing intermittently in *McClure's* over the next three years, was considerably more balanced than any popular account published to that time. The chapter on the "Legitimate Greatness of the Standard Oil Company" gave due credit to the ability of its executives, its enlightened labor policy, the efficiency of its pipelines, its close study of the needs of the market, and the "real service" of its refineries in developing the Frasch process. If the other details in Miss Tarbell's portrait of Standard were unflattering, the reason lay as much in the subject as in the artist.

Yet the work was neither the objective study at which its author had aimed nor the fair presentation for which Standard had hoped. Inaccuracies and distortions detrimental to the company could easily have been avoided. Some of these were serious enough: there was, for example, no reason for repeating Lloyd's untrue story that Rockefeller had forced a widow into penury. Others were matters of degree: Rockefeller was not the prime mover behind the South Improvement Company, as claimed, but he undeniably embraced its purposes; although Standard's activities were too widespread for Rockefeller to have been as personally responsible for them as she said, many of her allegations were applicable to the group as a whole.

The main factor preventing *The History of the Standard Oil Company* from being received as a "legitimate historical study," however, was the underlying tone of the whole work. For Miss Tarbell was basically a moralist and in her book the producers and independents were on the side of the angels. They were taken to represent an ethically superior type of economic order with equal opportunity for all to compete under vaguely defined rules of "fair play." Standard was condemned because it had foreclosed opportunity and because its profits had been gained through exploiting the consumer and/or producer; but most of all because it had violated the rules of the game, and had caused others to violate them, indeed, even to admire the violations.

Miss Tarbell cast her recommendations for public policy in, by then, familiar terms, emphasizing the independents' needs for cheaper transportation to the virtual exclusion of everything else. Sherman acts and laws forbidding "unfair competition," she maintained, were academic until the transportation question was resolved. Standard's pipelines should be forced to operate as common carriers at equitable rates, and railroads forbidden to establish tariffs favoring points where Standard located refineries. She did not, however, indicate how these proscriptions could be enforced. The independents' experience with the Interstate Commerce Commission provided little grounds for hope via this route. Miss Tarbell recognized this, but, primarily interested in informing and arousing the public, she provided no adequate substitute beyond moral exhortation: "We, the people of the United States, and nobody else, must cure whatever is wrong in the industrial situation." [56]

Whatever weaknesses this moral tone imparted to the book, it was nonetheless the main source of the book's popularity. The general public (as opposed to particular groups whose interests had been injured) was beginning to reject the economic and political morality of the Gilded Age, and the Roosevelt administration, after fairly conservative pronouncements in its early years, was by 1902 ably dramatizing the issue in prosecuting the Northern Securities Company and the Beef Trust. Thus, when the *McClure's* articles appeared in book form in 1904, they were generally quite favorably received. The *Nation* issued the only significant dissent, criticizing the work for treating "grave problems sensationally" and for implying that Standard's competitive practices were exceptional. This review was probably the one that Standard had reprinted and distributed.[57]

Standard, having begun in the late 1880's to abandon the practice of merely ignoring criticism, did not remain passive under renewed attack. In the late 1890's, it hired the Jennings Advertising Agency to place favorable notices in newspapers in Ohio and western Pennsylvania, then focal points of strong criticism. Publishers were paid 3¢ a line to print articles which appeared as straight news or reader interest items. The practice was not unusual at the time and was adopted to substitute for Standard's previous policy of subsidizing or extending loans to various papers and periodicals.

Standard also supplied information to authors and distributed copies of several books in quantities ranging from 100–5,000 copies in the hope of getting "its side of the story" told. Thus H. H. Rogers granted interviews to Miss Tarbell, and Daniel O'Day and S. C. T. Dodd personally prepared material for her. After the publication of her Standard history and several articles which unfairly attacked Rockefeller personally they understandably regretted their efforts. In other cases, however, the policy produced results more favorable to the company. The most notable of these were two articles written by a young Harvard economist, Gilbert Holland Montague.

Montague had originally embarked upon his study of Standard as a research topic for a Harvard economics seminar. Relying upon published investigations, he apparently reached his conclusions independently, although Dodd read portions of the manuscript and corrected some factual errors. The articles, originally published in a Harvard periodical, the *Quarterly Journal of Economics,* presented a considerably more balanced, if more favorable, interpretation of Standard's growth than did Miss Tarbell.

Montague concluded that Standard had obtained its pre-eminent position in 1872 "by the sheer superiority of its organization and—so far as is known—quite unaided by unusual discrimination in rates." [58] After that, given the railroads' readiness to favor large shippers, its monopolization of the industry was "inevitable." If Standard had not become the most powerful firm in the industry, another would have. Later, its control of pipelines maintained its market position.

Montague accepted too readily the myth that pipelines prior to Standard had been unable to service the fields adequately; but he correctly noted that through its control of long distance conduits the Rockefeller combine could locate its refineries at strategic points where inter-railroad competition or alternative water routes kept freight rates low. [59] Although he thought that the question of Standard's effect on refined prices "will probably never be settled," he guardedly admitted that despite the growth of the by-product market, Standard's margins "only slightly decreased." [60] In general, he granted Standard's "tremendous" power but concluded that "it is only such power as naturally accrues to so large an aggregation of capital." Although monopoly had been "inevitable" in the past, competition and the threat of competition, through some means that Montague did not specify, would in the future prevent "the abuse of this great power." [61]

Standard found the articles congenial enough to warrant financing their publication in book form. Some 5,000 copies were printed; many were sold to the public, while others were distributed free to Standard executives. But the over-all effect could hardly offset that of Tarbell and a presidential administration that, for public consumption at least, had declared its hostility to "moneyed interests." The newly established Bureau of Corporations would soon publish investigations of Standard whose tone reflected this hostility. And in 1906, the Standard interests would themselves feel the impact of the Sherman Act.

Part EIGHT

Epilogue

Chapter 27

Summary and conclusion

The record of the American petroleum industry at the close of the nineteenth century was strikingly impressive. From an output of a few barrels at the time of the Drake well, production of petroleum had grown to nearly 60 million barrels annually by the turn of the century. Total output over this forty-year span amounted to almost one billion barrels of crude. This expanding volume of American oil went largely to supply the demand for an inexpensive illuminant that, almost from the very beginnings, had been world-wide. Indeed, few products associated with America have had so extensive an influence as kerosene on the daily living habits of so large a proportion of the world's population.

Long before 1900, however, demands for other petroleum products had augmented the pervasive influence of illuminating oil, and by late 1890's some 200 by-products accounted for at least half the value of the industry's total sales. In the domestic markets particularly, naphthas supplied medicine with local anesthetics, industry with solvents, and fuel for stoves and internal combustion engines. Petroleum wax, in addition to its use in the preparation of pharmaceuticals, was the principal material for the manufacture of candles. Heavy oils were of primary importance for the gas industry, besides adding to the supply of industrial fuels. Of even greater significance was the petroleum industry's contribution of a supply of high-quality lubricants sufficient to support the rapid industrial expansion in America and Europe during the last quarter of the century.

External forces associated with general economic growth played an important part in the rise of the petroleum industry in the nineteenth century. Yet this record could not have been achieved in the absence of contributions generated from within the industry that affected all stages of its operations from the production of crude to the distribution of the finished product. It is noteworthy that many of these contributions were developed within the first ten years of the industry's history.

It was during this period, for example, that the basic production technology for the rest of the century was laid. This was achieved when oil-men complemented methods of salt-well boring with such indispensable innovations for effective drilling and pumping as tubing, torpedoing, and flexible power hookups.

The foundations of nineteenth-century refining technology were likewise completed during the first decade. Coal-oil refining techniques for all types of simple distillation and treating were soon transferred to petroleum and expanded to general practice, as were the methods of destructive distillation, or cracking, of heavy fractions into illuminating oils. Petroleum refiners also evolved on a small commercial scale the entire range of by-products produced throughout the century. Finally, and perhaps the most difficult accomplishment of all, they succeeded by the early 1870's in carrying both ordinary and destructive distillation of illuminating oils from small to large-scale stills. Only one further major innovation was subsequently introduced late in the 1880's when Herman Frasch solved the problem of processing Lima crude by developing a practical method for precipitating and removing the high sulphur content of Ohio oil.

The early technical achievements in the development of methods for refining illuminating oils and petroleum by-products deserve a place high among all contributions to industrial progress. It is perhaps less to the industry's credit that, despite experiments in the United States indicating their potential feasibility, fractionating towers and continuous distillation were not put into commercial operation before the close of the century. Just how much the suppression of these innovations was due to economic factors and how much to a reluctance to introduce new methods of refining is, however, difficult to determine.

No innovation with its roots in the decade of the 1860's had a more significant impact on the entire petroleum industry than the development of bulk methods of transport. Gathering lines linked with railroad tank cars at the outset broke a bottleneck that seriously threatened the movement of crude from the Regions. In the subsequent evolution of bulk shipments, tank cars were increasingly used in the transport of refined products, once the practicability of the long-distance trunk pipelines for the movement of crude had been demonstrated. Credit for initiating and developing the overland phase of one of the notable industrial achievements of the nineteenth century goes to a small group of American entrepreneurs. It remained for the Russians, however, to extend bulk shipments of petroleum and its products throughout the markets of the world when they pioneered the first successful ocean-going tank steamers late in the 1880's.

The contribution of these major innovations to the more efficient organization of production, refining, and transportation was reflected in varying degrees in substantial cost-savings for the industry. Although methods of locating petroleum reserves remained more empirical than scientific, at least some of the cost-reducing possibilities of the advances in production techniques were realized in lower unit production costs. In refining it was the development of the larger stills that marked the critical break-through to lower processing costs. The major cost-reducing influences of the nineteenth century, however, were undoubtedly the long-distance crude trunk lines that made possible the rapid and relatively inexpensive transmission of crude, and tank cars and tank steamers that performed the same service for refined.

There is no question of the industry's achievements in improving its technical efficiency by 1900. But any objective appraisal of the industry's performance must take social costs into account as well as the extent to which cost savings were reflected in the prices of finished products to consumers.

A major social cost associated with the development of the industry occurred in the production of crude. From the outset, production was marked by the fierce and almost indiscriminate drilling of newly discovered fields which brought excessive wastage both above and below ground. Concern over the irresponsible manner of producing petroleum was largely confined to a few contemporary geologists. But their advice on proper well-spacing, the control of gas pressure, the danger from flooding, and other preventives of premature depletion and unnecessary exhaustion of fields went largely unheeded. Moreover, the weight of legal precedent tended to encourage rapid exploitation. Yet had there been any latent desire among the more prominent members of the industry to rationalize production, both legal means and the requisite geological and technological knowledge were available.

It is also evident that there were human as well as social costs involved in the use of petroleum illuminants. These were reflected in the number of explosions and fires associated with the use of kerosene lamps and the resulting casualties and property damage. In part these disasters were the result of ignorance or carelessness in handling the petroleum illuminant. But they were also a reflection of the inability or unwillingness of refiners initially to apply careful methods of separating the volatile naphtha fractions from illuminating oils in their distilling operations. It was only with the growing value of naphthas and better refining operations that the excessive dangers in using kerosene began to disappear after the mid-1880's.

It is virtually impossible to measure the extent to which consumers benefited by cost savings achieved by the industry. The decline in the wholesale price of kerosene, at a faster rate than the general wholesale price level, from approximately 45¢ a gallon (without tax) in 1863 to about 6¢ by the mid-1890's, suggests that at least a part of these savings were passed on to consumers.

It was argued by many contemporaries, however, that savings to consumers would have been even greater if the industry had been more competitively structured. In contrast to the early history of the industry, which was marked by a multiplicity of producers, refiners, and sellers of oil, Standard's domination by the mid-1880's made petroleum one of the most concentrated of American industries.

Although highly critical of the methods used by John D. Rockefeller and his associates in building up Standard, few contemporaries questioned the efficiency of the organization as it had evolved by 1900. There was much in the record to substantiate this evaluation. It is true that with the exception of the Frasch process Standard took little initiative in the promotion of technical innovations. But the management was quick to recognize the potentials of most major developments and aggressive in extending their use. This policy was well illustrated by the rapid incorporation of crude trunk lines into Standard's domestic operations once the feasibility of this method of transport had been demonstrated. Similarly the management integrated tank steamers into Standard's overseas marketing operations.

Standard spokesmen credited the performance of their organization to its size, scope of operations, and management efficiency, which were in turn reflected in its record of earnings. Critics, however, argued that Standard's profits were more of a reflection of its monopoly position in the industry than its efficiency. They further argued that in the absence of such monopoly power in the industry, not only would other oil companies have been able to operate as efficiently as Standard, but many more of the benefits of technological advances would have been passed on to consumers in the form of lower prices.

The merits of this line of argument are difficult to assess. It appears quite unlikely that only Rockefeller and his associates had the qualifications necessary to create an efficient organization at this stage of the industry's development. Moreover, a careful examination of the evidence available leads to the conclusion that Standard's initial position in the industry was based primarily on its control over transportation: first, through concessions from the railroads; and subsequently, through ownership of the major trunk pipeline facilities. While the management conducted few, if any, of Standard's operations at a loss, it also appears that a substantial portion of the organization's income was derived from its pipeline and transport facilities.

Given the structure of the American railroad industry and the social and economic environment of the nineteenth century, it was perhaps inevitable that the carriers would grant special rates to favored shippers. But had the railroads been able or willing to treat all shippers alike and, had trunk pipelines been able to develop as common carriers under effective governmental supervision, it is at least questionable whether Standard could ever have controlled such a large share of the oil business. It is further reasonable to suppose that with greater freedom of entry, other entrepreneurs would have built up organizations sufficiently large at the refining, and possibly the marketing level to achieve economies of scale. Whether such a development would have set the stage for co-operation among a relatively few large operators with no important reductions in prices, or alternatively have led to intensive price competition that was beneficial to consumers, is impossible to state. But there is no question that the probability of more vigorous competition under such an industry structure would have been much greater than it actually was under the aegis of Standard.

But whatever the cause of Standard's growth as a giant corporation, the future did portend important threats to maintaining this position. By the late 1890's there was increasingly vigorous competition abroad. At home mounting public indignation against monopolies in general, and Standard in particular, also presented possible legislative obstacles to the company's future role in the petroleum industry.

Perhaps the greatest threat to Standard's immediate position involved its ability to control any new producing areas, for under the stimulus of flush production, there was always the strong possibility of new competition arising. Sources of domestic production outside the eastern oil regions were known to exist and in California, and to a lesser extent in Texas, there were indications by the close of the century that significant crude reserves might be discovered. In the event that these locations developed into commercial producing regions, the Standard organization, to maintain its position, would have to mobilize rapidly its financial and managerial resources. Subsequent events would reveal whether or not these uncertainties could be neutralized to Standard's advantage.

Twentieth-century prelude

At the turn of the century, the American petroleum industry was still largely oriented to the production and distribution of illuminating oil and lubricants. Few of its leaders had yet grasped the implications of the development of the internal combustion engine as an outlet for the lighter fractions, nor had they fully recognized the potentials of extending the use of petroleum fuels to ships, locomotives, and even home furnaces.

Yet, in many respects, some forty years experience had left the industry reasonably well prepared to shift from a primary emphasis on the production of illuminants to supplying the expanding needs of transportation and industry for flexible and highly convenient sources of energy. There were, for example, no primary obstacles to the quick extension of bulk transport facilities to any important new sources of crude production. Refining techniques were fully adequate to supply a satisfactory fuel for the early gasoline engines. Marketing channels for the distribution of industrial fuels were sufficiently developed to meet the requirements of a much larger and more diversified demand.

But while the oil industry in 1900 was prepared to meet the problems of the immediate future, no phase of its operations was to remain unaffected by developments over the succeeding half century. The industry as it emerged by the late 1950's reflected the ability of its members to meet the challenges of the dynamic economic and social changes associated with the twentieth century.

Acknowledgments

The authors are deeply appreciative of the contributions of the many individuals and institutions to the preparation of this volume. We are particularly obligated to the members of the Petroleum History Committee appointed by the American Petroleum Institute, who, under the chairmanship of Stanton K. Smith, co-operated with Northwestern University in making this study possible. Dr. John W. Frey, who served as secretary of the Petroleum History Committee, was most helpful in supplying source materials and offering constructive suggestions at all stages in the preparation of the manuscript. James E. Moss, director of the A.P.I.'s Division of Transportation in Washington, D. C., generously lent us many items from his personal collection of documents covering early phases of the industry's history. William B. Harper, director of economic information for the A.P.I., also supplied us with valuable statistical data.

We are heavily indebted to the directors and staff members of various public, company, and university libraries for their willingness to lend us materials and documents dealing with the history of petroleum. Those who were especially co-operative included Miss Virginia Smythe, of the A.P.I. library in New York City; Mrs. Carol M. Cronkhite at the New York Public Library; M. P. Doss, of the Texas Company; William Wright, of the University of Pennsylvania Library; James Stevenson, director, and T. J. Garvin, curator of the Drake Museum at Titusville; Miss Bess Glenn and John E. Maddox, of the National Archives; John P. McGowan, librarian of the Technological Institute, Northwestern University; and Miss Marjorie Carpenter, Miss Aleen Baker, and Miss Mary L. Hilton, staff members of the Deering Library, Northwestern University.

We received many additional bibliographical suggestions and valuable materials from Brewster Jennings, former chairman of the board of the Socony-Vacuum Oil Company; James E. Boudreau, director, and Christopher Vogel, his associate, of the public relations department of the Ethyl Corporation; Philip D. Croll, public relations department, Sun Oil Company; Mrs. Margery M. Porter, assistant secretary, Standard Oil Company, New Jersey; K. G. McKenzie, formerly with the Texas Company; Kendall F. Beaton, member of the Shell Oil Company's public relations staff; Professor R. J. Forbes and W. F. W. DeVos, of the Royal Dutch–Shell Group at the Hague; and C. W. Bowring III, of Bowring and Company, New York City.

Our special thanks go to William B. Plummer, petro-chemical consultant, New York City, and Dr. Edmund L. d'Ouville and his associates of the Standard Oil Company, Indiana, for their generous assistance in the preparation of the technical chapters included in the volume.

Among those who read all or parts of the manuscript and contributed many helpful suggestions were Professor G. Herberton Evans, The Johns Hopkins University; Professor Thomas C. Cochran, University of Pennsylvania; Professor Robert W. Clower, Northwestern University; Professor Stuart Bruchey, Michigan State University; and Dr. Frank M. Surface, Standard Oil Company, New Jersey.

We owe a further debt to our former research associates, Professor Rudolph C. Blitz and Herbert D. Werner for their contributions. We are also indebted to Dr. Thomas L. Bushell, Dr. Phillip S. Landis, Mrs. Janet Petterson, and Mrs. Hazel W. Hall for their work as research assistants on the project.

Finally we wish to express our gratitude to Mrs. Kathleen S. Hines, who served so effectively as executive secretary of the project; to Mrs. Velma D. Veneziano, who successfully combined the duties of a research assistant and secretary; to Mrs. Margaret P. Praxl for taking such good care of the project library; and to Miss Miriam McCormack for her excellent services as contoura operator and typist.

THE AUTHORS

Appendixes

Refining operations, 1860–1873

The basic idea for the following tabulation of refinery throughputs and production for 1860–73 were drawn from a wide variety of sources. Wherever possible, the methods described in Appendix B were applied, but because of the lack of continuous or comparable statistical series, the results are at best approximate estimates of the growth in petroleum refining during these years and the relative importance of foreign and domestic markets for the industry. Their principal value is to indicate the major trends which have been carefully checked against the statements of the most reliable contemporary observers and members of the industry.

APPENDIX TABLE A

Refinery outputs and throughputs, 1860–1873
(calendar years/thousands of barrels)

	1860	1861	1862	1863	1864	1865	1866
Production *	500	2,114	3,056	2,611	2,116	2,498	3,600
Storage and waste †			−2,306	−984	+227	−111	−120
Net shipments ‡			750	1,802	2,343	2,387	3,480
Crude exports §			−182	−202	−233	−137	−270
Refinery throughput			568	1,425	2,110	2,250	3,210
Total refinery output ‖			335	855	1,266	1,336	2,049
Distribution of refinery output:							
Domestic markets ¶			240	350	722	727	633
Foreign markets §			95	505	544	609	1,416

(table cont. on next page)

(table cont. from previous page)

	1867	1868	1869	1870	1871	1872	1873
Production *	3,347	3,646	4,215	5,261	5,205	6,293	9,894
Storage and waste †	−495	+1,448	+845	+428	+460	−393	−394
Net shipments ‡	3,842	5,094	5,060	5,689	5,665	5,900	9,500
Crude exports §	−132	−194	−360	−289	−269	−390	−468
Refinery throughput	3,710	4,900	4,700	5,400	5,396	5,510	9,032
Total refinery output ‖	2,418	3,410	3,267	3,875	4,050	4,200	6,750
Distribution of refinery output:							
Domestic markets ¶	773	1,131	1,089	1,292	930	1,380	1,850
Foreign markets §	1,545	2,279	2,177	2,583	3,120	2,820	4,900

* *Mineral Resources, 1905*, 818.
† Estimated.
‡ Estimated for 1863–70; 1871–73 net shipments from *Derrick's Hand-Book*, I, 807–9.
§ *Shipping and Commercial List and New York Price Current,* 1864–1873; *Atlas of the Oil Region,* unnumbered page; Cone and Johns, *Petrolia,* 617; *32nd Annual Report of the Philadelphia Board of Trade, for the Year 1864* (1865), 36; *Mineral Resources, 1905,* 883.
‖ Estimated: Factored at 60 per cent yield of refined, 1863, 1864; at 65 per cent, 1865–67; at 70 per cent, 1868–70; at 75 per cent, 1871–73.
¶ Estimated: Internal Revenue Reports of Domestic Tax Collections plus 20 per cent in 1863, 15 per cent in 1864, 10 per cent in 1865, 5 per cent in 1866; estimated as one-half of exports 1867–70; 1871–72, net shipments from *Derrick's Hand-Book,* I, 807–9, less crude export (*Mineral Resources, 1905,* 883), factored at 75 per cent to total refined output, less refined export, *Mineral Resources, 1905,* 807–9.

Methods of estimating
refining operations, 1874–1899

The first decennial census covering the refinery output of the American petroleum industry was made in 1880. To derive roughly comparable estimates covering outputs and domestic distributions for other than census years, the following methods were utilized. Approximate figures for the amount of crude processed by domestic refineries were obtained by subtracting crude exports (*Mineral Resources, 1905*) from shipments or pipeline deliveries (*Derrick's Hand-Book,* I) from the Oil Region.

Output ratios of refined to crude, available in census reports for 1880, 1889, and 1899, were interpolated and applied to refinery throughputs of crude to derive the production of various refined products for census years. The amount and proportion of these outputs distributed in the domestic markets were in turn obtained by subtracting export data (*U. S. Commerce and Navigation Reports*), available on an annual basis, from the totals.

The following table shows the percentages which were applied to the volume of crude processed to obtain estimates of the outputs of refined products for the years indicated.

APPENDIX TABLE B

Year	Illuminating oils	Naptha-benzine-gasoline	Lubricating oils
1874	75 17	10.30	2.60
1879	75.17	10.30	2.60
1884	70.51	11.54	4.29
1889 *	65.85	12.77	5.98
1894	61.72	12.81	6.89
1899 *	57.59	12.85	7.80

* Census years.

U. S. exports of
illuminating oil, 1864–1873

The data for the following table were drawn originally from reports by various agencies of the Federal government, principally the Bureau of Statistics on the Commerce and Navigation of the United States. Although subject to frequent changes in methods of reporting and categories utilized, the material presented in the table below gives a reasonably accurate impression of the growth of export sales and the major marketing areas.

APPENDIX TABLE C

U. S. exports of illuminating oil to major geographic areas, 1864–73 (year ending June 30/40-gallon barrels)

Geographic area	1864	1865	1866	1867	1868
Europe	238,433	233,017	711,540	1,382,663	1,479,299
North America	16,369	24,728	41,915	42,187	51,797
South America	17,435	12,575	41,776	49,317	50,027
Asia and Oceana	11,996	20,900	31,128	69,307	91,523
Africa	6,956	1,063	2,277	9,512	7,563
All others	...	80	111	151	124
Total	291,189	292,363	837,747	1,553,137	1,680,333

Geographic area	1869	1870	1871	1872	1873
Europe	1,951,901	2,212,112	2,885,950	2,759,951	3,538,570
North America	65,420	91,518	84,061	87,218	88,810
South America	38,660	71,300	89,158	90,137	117,056
Asia and Oceana	47,578	58,292	83,142	104,401	167,935
Africa	1,132	5,353	6,277	6,059	31,547
All others	5,395	8,986	22,556	15,722	8,642
Total	2,110,086	2,447,561	3,171,144	3,063,488	3,952,560

Source: Folger, Pa. *Industrial Statistics,* 1892, 182B–184B.

U. S. exports of petroleum
and petroleum products, 1874–1899

The data presented in this appendix indicate the quantities and values of the exports of petroleum and petroleum products from the United States at five-year intervals between 1874 and 1899.* Because of frequent changes in the methods of reporting and categories utilized in the original governmental documents in which these figures were reported, it was necessary to regroup much of the data in order to present comparable time series. To facilitate comparisons, TABLES D:1–D:7 were set up to show (1) exports by major continental areas, and (2) exports to the principal countries in each of the following continental divisions: Europe; North America; South America; Oceana, including British Australasia (India, Australia, and New Zealand); and Africa.

---

* The sources for the tables in Appendix D are as follows:

1874—*Annual Report of the Chief of the Bureau of Statistics on the Commerce and Navigation of the United States for the Fiscal Year Ended June 30, 1874* (Washington: Government Printing Office, 1875), 212, 213.

1879—Treasury Dept., *Annual Statement of the Chief of the Bureau of Statistics on the Commerce & Navigation of the United States for the Fiscal Year Ended June 30, 1879. Foreign Commerce* (Wash.: G.P.O., 1880), 170–73.

1884—Treasury Dept., *Annual Report and Statements of the Chief of the Bureau of Statistics on the Commerce & Navigation of the United States for the Fiscal Year Ended June 30, 1884. Foreign Commerce* (Wash.: G.O.P., 1884), 234–37.

1889—Treasury Dept., *Annual Report & Statements of the Chief of the Bureau of Statistics on the Foreign Commerce & Navigation, Immigration, & Tonnage of the United States for the Fiscal Year Ending June 30, 1889* (Wash.: G.O.P., 1890), 292–95.

1894, 1899—Bureau of Statistics, Dept. of Commerce & Labor, *The Foreign Commerce & Navigation of the United States for the Year Ending June 30, 1904. Vol. 2, Imports and Exports of Merchandise, by Articles and Countries, 1894–1904* (Wash., G.O.P., 1904), 606–615.

APPENDIX TABLE D:1

U. S. exports of domestic petroleum and petroleum products to major geographic areas (quantities and values in thousands by five year intervals, 1874–99)

Year ending June 30	Product	Total		Europe		North America		South America	
		Gallons	Dollars	Gallons	Dollars	Gallons	Dollars	Gallons	Dollars
1874	Crude oil	17,776.4	2,107.8	16,875.7	1,966.2	899.0	141.3	1.8	.2
	Naphthas	9,737.5	1,050.2	9,655.2	1,018.6	59.8	25.6	14.1	3.8
	Illuminating oil	217,220.5	37,620.8	194,434.0	32,564.7	4,680.0	1,074.2	4,601.6	1,021.2
	Lubricating oil	1,244.3	404.2	1,166.1	366.9	38.3	20.2	27.6	13.0
	Residuum	43.5*	142.3	36.2*	124.9	5.9*	12.5	1.5*	4.9
1879	Crude oil	25,874.5	2,180.4	24,067.9	1,978.5	1,805.6	201.8	.9	.1
	Naphthas	15,054.4	1,258.8	14,895.9	1,235.0	32.6	4.3	106.8	16.5
	Illuminating oil	331,586.4	35,999.9	266,299.4	27,433.0	5,757.8	766.4	8,590.7	1,113.8
	Lubricating oil	2,487.7	655.5	2,319.3	588.5	117.4	46.3	38.3	14.8
	Residuum	78.7*	210.7	77.9*	209.6	.8*	1.1	—	—
1884	Crude oil	67,186.3	5,303.0	64,106.9	5,011.5	3,079.4	291.4	—	—
	Naphthas	15,045.4	1,072.7	14,560.6	982.0	186.3	29.0	199.5	32.6
	Illuminating oil	415,615.7	38,195.3	291,130.3	25,058.1	8,080.5	976.3	11,568.2	1,258.8
	Lubricating oil	10,515.5	2,179.6	10,010.8	1,996.8	225.1	81.2	130.2	52.5
	Residuum	126.1*	352.7	125.6*	349.8	.2*	2.3	—	—
1889	Crude oil	72,987.4	5,083.1	67,462.1	4,626.2	5,525.3	456.9	—	—
	Naphthas	14,100.1	1,155.7	13,939.7	1,120.7	31.3	4.9	65.5	12.3
	Illuminating oil	502,257.5	39,285.3	330,144.5	22,653.4	11,914.4	1,256.7	22,099.4	2,294.8
	Lubricating oil	25,166.9	4,292.0	23,190.5	3,761.3	496.6	157.4	719.8	216.2
	Residuum	40.1*	96.5	16.6*	48.5	23.4*	47.4	.2*	.6
1894	Crude oil	121,926.3	4,415.9	106,498.3	3,626.2	15,426.0	789.6	2.0	.1
	Naphthas	15,555.8	944.0	15,242.2	894.8	173.6	21.7	79.8	12.7
	Illuminating oil	730,368.6	30,676.2	490,252.3	15,224.3	14,182.8	1,065.6	23,341.8	1,698.0
	Lubricating oil	40,190.6	5,449.0	35,406.6	4,458.7	1,725.7	373.4	1,509.7	325.6
	Residuum	5.0*	14.7	2.1*	5.0	2.5*	8.0	.2*	.8
1899	Crude oil	113,088.1	5,202.9	101,640.5	4,600.9	11,447.6	602.0	137.7	20.8
	Naphthas	16,252.8	1,170.3	15,710.2	1,095.3	267.7	32.4		
	Illuminating oil	722,279.5	41,087.0	536,943.3	26,065.0	17,067.0	1,337.5	32,306.6	2,930.3
	Lubricating oil	67,424.4	7,943.2	53,722.3	6,078.6	1,966.0	328.5	2,899.3	424.3
	Residuum	730.2*	869.8	724.2*	858.4	5.3*	8.4	.7*	2.9

(table con't. on next page)

APPENDIX TABLE D:1 (cont.)

Year ending June 30	Product	Asia		Oceana		Africa		All other ports, countries, and seas	
		Gallons	Dollars	Gallons	Dollars	Gallons	Dollars	Gallons	Dollars
1874	Crude oil	—	—	—	—	—	—	—	—
	Naphthas	7.2	2.0	—	—	1.0	.2	—	—
	Illuminating oil	6,406.2	1,417.4	3,754.1	828.6	2,674.5	578.9	670.2	135.7
	Lubricating oil	.7	.3	—	—	11.6	3.8	—	—
	Residuum	—	—	—	—	—	—	—	—
1879	Crude oil	—	—	—	—	—	—	—	—
	Naphthas	—	—	1.8	.7	17.3	2.3	—	—
	Illuminating oil	41,446.2	5,347.1	2,911.7	490.8	6,033.0	776.8	547.8	72.0
	Lubricating oil	—	—	2.8	1.2	9.7	4.4	.3	.2
	Residuum	—	—	—	—	—	—	—	—
1884	Crude oil	—	—	—	—	—	—	—	—
	Naphthas	—	—	84.8	26.0	12.4	2.6	1.7	.3
	Illuminating oil	89,966.1	9,215.0	4,725.9	628.3	7,732.0	799.9	2,412.8	258.9
	Lubricating oil	.5	.3	145.7	47.9	3.1	1.0	—	—
	Residuum	—	—	.2*	.6	—	—	—	—
1889	Crude oil	—	—	—	—	—	—	—	—
	Naphthas	(a)	(a)	56.3	15.2	7.4	2.5	—	—
	Illuminating oil	120,162.8	11,073.2	9,576.1	1,163.5	6,384.8	656.2	1,975.7	187.5
	Lubricating oil	434.3	55.8	305.4	92.0	20.4	9.2	—	—
	Residuum	—	—	—	—	—	—	—	—
1894	Crude oil	—	—	—	—	—	—	—	—
	Naphthas	—	—	57.1	13.8	3.1	1.0	—	—
	Illuminating oil	182,239.5	11,067.8	13,302.6	1,103.1	7,049.5	517.4	—	—
	Lubricating oil	623.3	118.6	809.8	128.1	115.4	44.5	—	—
	Residuum	.3*	.9	—	—	—	—	—	—
1899	Crude oil	1.0	.2	—	—	—	—	—	—
	Naphthas	—	—	119.1	18.4	17.0	3.2	—	—
	Illuminating oil	109,120.9	8,126.0	15,469.6	1,558.2	11,372.2	1,070.0	—	—
	Lubricating oil	5,607.6	568.8	2,129.8	317.9	1,099.4	225.1	—	—
	Residuum	—	—	(a)*	(a)	(a)*	.1	—	—

(a) Quantities or values so small they do not show up when figures are given in thousands (less than 50).
* Barrels

743

U. S. exports of domestic petroleum and petroleum products to major European countries
Quantities and values in thousands by five year intervals, 1874–99

Year ending June 30	Product	Total Europe		Belgium		Denmark		France		Germany	
		Gallons	Dollars	Gallons	Dollars	Gallons	Dollars	Gallons	Dollars	Gallons	Dollars
1874	Crude oil	16,875.7	1,966.2	488.6	49.0	—	—	13,436.2	1,598.2	1,818.1	197.2
	Naphthas	9,655.2	1,018.6	1,000.8	101.2	—	—	2,650.0	285.6	884.9	76.2
	Illuminating oil	194,434.0	32,564.7	34,101.8	5,506.5	6,532.4	1,050.1	1,971.7	329.4	72,398.8	11,730.9
	Lubricating oil	1,166.1	366.9	41.3	11.0	2.6	1.0	—	—	40.9	15.1
	Residuum	36.2 *	124.9	.7 *	3.5	—	—	—	—	.5 *	1.9
1879	Crude oil	24,067.9	1,978.5	—	—	—	—	19,411.8	1,602.7	3,224.5	257.4
	Naphthas	14,895.9	1,235.0	1,397.0	127.9	—	—	4,258.9	349.0	1,554.0	116.6
	Illuminating oil	266,299.4	27,433.0	41,445.8	4,099.9	10,838.4	1,082.1	1,656.2	193.0	102,422.3	10,168.4
	Lubricating oil	2,319.3	588.5	200.3	53.6	46.9	13.1	5.4 *	15.4	458.9	98.2
	Residuum	77.9 *	209.6	3.0 *	8.9	—	—	—	—	—	—
1884	Crude oil	64,106.9	5,011.5	—	—	—	—	38,617.3	2,943.8	2,949.3	216.5
	Naphthas	14,560.6	982.0	751.3	57.8	—	—	6,181.8	390.3	1,661.8	106.6
	Illuminating oil	291,130.3	25,058.1	42,732.1	3,561.2	11,432.7	937.5	1,864.8	159.3	107,703.7	8,975.4
	Lubricating oil	10,010.8	1,996.8	1,118.1	219.3	40.9	10.6	800.5	186.7	2,095.2	392.0
	Residuum	125.6 *	349.8	4.5 *	12.5	—	—	33.7 *	97.3	—	—
1889	Crude oil	67,462.1	4,626.2	—	—	—	—	53,955.7	3,666.3	3,064.4	169.5
	Naphthas	13,939.7	1,120.7	309.0	25.9	—	—	4,900.7	415.2	3,257.9	241.4
	Illuminating oil	330,144.5	22,653.4	38,800.5	2,608.3	4,137.3	301.9	2,602.1	191.0	138,518.4	8,458.9
	Lubricating oil	23,190.5	3,761.3	1,765.0	257.4	60.3	11.2	1,787.4	358.9	3,252.9	564.2
	Residuum	16.6 *	48.5	—	—	(a) *	(a)	6.1 *	20.7	—	—
1894	Crude oil	106,498.3	3,626.2	—	—	—	—	84,435.0	2,958.2	4,877.6	134.6
	Naphthas	15,242.2	894.8	—	—	2.6	.2	3,764.6	227.8	4,278.8	233.9
	Illuminating oil	490,252.3	15,224.3	36,313.0	986.5	9,290.3	296.7	11,812.0	532.9	86,388.8	2,460.5
	Lubricating oil	35,406.6	4,458.7	2,931.2	333.8	74.3	12.7	3,050.5	511.6	5,637.5	678.5
	Residuum	2.1 *	5.0	—	—	—	—	—	—	—	—
1899	Crude oil	101,640.5	4,600.9	—	—	8.0	.9	83,630.5	3,832.8	3,485.4	139.5
	Naphthas	15,710.2	1,095.3	—	—	—	—	1,517.8	113.4	4,716.3	287.4
	Illuminating oil	536,943.3	26,065.0	40,715.7	2,001.1	17,548.1	865.6	3,994.9	256.1	115,124.6	5,240.4
	Lubricating oil	53,722.3	6,078.6	4,625.8	500.8	366.0	52.2	6,500.1	748.7	8,233.9	1,015.0
	Residuum	724.2 *	858.4	12.4 *	15.6	.1 *	—	—	—	—	.3

(table con't. on next page)

Year ending June 30	Product	Italy		Netherlands		Sweden & Norway		United Kingdom		Portugal	
		Gallons	Dollars	Gallons	Dollars	Gallons	Dollars	Gallons	Dollars	Gallons	Dollars
1874	Crude oil	—	—	—	—	425.4	48.4	490.8	49.5	—	—
	Naphthas	5.0	.7	—	—	253.2	27.7	4,675.3	509.4	5.0	.9
	Illuminating oil	9,190.0	1,745.6	13,901.2	2,385.9	1,412.7	218.4	26,079.0	4,353.4	1,297.6	210.3
	Lubricating oil	—	—	115.5	34.7	.2	.1	965.3	304.9	—	—
	Residuum	—	—	—	—	(a)*	.1	32.3*	110.9	—	—
1879	Crude oil	—	—	—	—	25.3	1.8	151.9	13.3	—	—
	Naphthas	—	—	—	—	576.8	45.5	7,043.1	590.3	7.2	1.0
	Illuminating oil	16,565.5	1,936.2	17,564.3	1,753.6	6,528.2	659.2	34,751.4	3,637.6	2,290.3	230.6
	Lubricating oil	—	—	556.5	146.0	—	—	1,009.4	264.6	—	—
	Residuum	—	—	—	—	—	—	69.5*	185.3	—	—
1884	Crude oil	—	—	—	—	151.2	10.9	—	—	—	—
	Naphthas	—	—	—	—	422.5	29.5	5,335.2	379.0	9.5	.9
	Illuminating oil	19,818.2	1,995.9	24,445.1	2,053.9	6,493.7	546.4	48,781.8	4,323.6	2,675.9	231.8
	Lubricating oil	53.0	11.3	647.7	129.3	4.3	2.3	5,118.4	1,019.6	—	—
	Residuum	7.7*	23.8	—	—	—	—	79.7*	216.2	—	—
1889	Crude oil	—	—	—	—	170.8	12.4	8.7	.5	—	—
	Naphthas	—	—	—	—	739.7	57.0	4,702.7	378.7	1.4	.2
	Illuminating oil	18,825.6	1,727.5	41,158.2	2,716.9	10,212.6	768.5	65,368.6	4,971.1	4,186.8	348.4
	Lubricating oil	283.3	45.2	1,838.2	263.1	(a)	(a)	14,198.7	2,259.4	.2	.1
	Residuum	—	—	—	—	—	—	10.4*	27.8	—	—
1894	Crude oil	—	—	—	—	185.4	11.7	6,834.8	410.7	—	—
	Naphthas	—	—	—	—	336.5	20.2	274,555.0	8,132.4	3.0	.4
	Illuminating oil	22,945.0	1,010.6	31,868.2	894.8	9,848.1	487.0	19,668.8	2,421.3	4,231.4	235.7
	Lubricating oil	1,356.3	177.2	2,346.9	286.7	20.9	2.8	2.1*	5.0	79.1	7.8
	Residuum	—	—	—	—	—	—	.3	(a)	—	—
1899	Crude oil	—	—	2,409.0	96.8	1,395.1	51.1	7,584.5	572.8	—	—
	Naphthas	—	—	1,477.0	85.1	404.0	35.4	178,796.5	8,563.5	2.6	.3
	Illuminating oil	19,750.2	948.5	138,188.3	6,732.2	17,345.4	1,037.2	26,353.1	2,887.5	2,692.5	195.2
	Lubricating oil	1,921.1	259.1	4,332.7	453.9	250.4	32.3	711.7*	842.5	317.5	37.2
	Residuum	—	—	—	—	—	—	—	—	—	—

(table con't. on next page)

Year Ending June 30	Product	Spain		Austria-Hungary		Russia		Other Europe	
		Gallons	Dollars	Gallons	Dollars	Gallons	Dollars	Gallons	Dollars
1874	Crude oil	—	—	—	—	216.5	23.8	(a)	(a)
	Naphthas	90.4	7.7	—	—	90.7	9.1	—	—
	Illuminating oil	5,544.1	1,010.0	3,774.9	630.4	7,022.7	1,194.8	11,207.1	2,198.9
	Lubricating oil	—	8.4	—	—	.2	.1	—	—
	Residuum	2.6*	103.4	—	—	—	—	—	—
1879	Crude oil	1,254.5	—	—	—	3,141.9	336.7	58.9	4.7
	Naphthas	—	—	—	—	47.2	13.0	—	—
	Illuminating oil	6,630.8	747.1	12,612.8	1,326.1	—	—	9,851.4	1,262.6
	Lubricating oil	—	—	—	—	—	—	—	—
	Residuum	—	—	—	—	—	—	—	—
1884	Crude oil	—	—	10,197.2	738.4	11.7	.9	186.9	17.1
	Naphthas	—	—	—	—	—	—	—	—
	Illuminating oil	12,191.8	1,102.0	12,839.7	1,051.3	529.2	42.8	11,054.6	1,107.2
	Lubricating oil	758.9	71.7	93.0	16.2	—	—	—	—
	Residuum	39.7	9.5	—	—	—	—	—	—
1889	Crude oil	6,327.1	535.6	3,935.4	241.9	28.2	2.3	5,307.9	484.3
	Naphthas	—	—	—	—	—	—	.1	(a)
	Illuminating oil	183.2	17.2	723.7	50.9	119.7	8.4	—	—
	Lubricating oil	3.2	1.3	1.2	.3	—	—	—	—
	Residuum	—	—	—	—	—	—	—	—
1894	Crude oil	15,176.0	454.7	1,824.3	67.0	21.6	1.5	.5	.1
	Naphthas	—	—	—	—	—	—	—	—
	Illuminating oil	1.8	.2	—	—	99.8	5.4	2,899.0	181.7
	Lubricating oil	154.2	17.1	41.5	4.6	44.5	4.6	.8	.1
	Residuum	—	—	—	—	—	—	—	—
1899	Crude oil	9,723.4	441.6	996.8	39.0	—	—	—	—
	Naphthas	—	—	—	—	—	—	—	—
	Illuminating oil	—	—	—	—	—	—	2,787.1	225.1
	Lubricating oil	105.8	11.0	641.7	68.4	71.2	11.6	3.0	.9
	Residuum	—	—	—	—	—	—	—	—

(a) Quantities or values so small they do not show up when figures are given in thousands (less than 50).   * Barrels.   Source: See page 739.

U. S. exports of domestic petroleum and petroleum products to major North American countries
Quantities and values in thousands by five year intervals, 1874–99

Year Ending June 30	Product	Total North America		Bermuda		British Honduras		Canada		Central American States **	
		Gallons	Dollars	Gallons	Dollars	Gallons	Dollars	Gallons	Dollars	Gallons	Dollars
1874	Crude oil	899.0	141.3	—	—	(Included in West Indies)		12.1	9.7	.1	(a)
	Naphthas	59.8	25.6	—	—			24.5	15.7	.1	(a)
	Illuminating oil	4,680.0	1,074.2	—	—			540.4	183.8	21.2	5.8
	Lubricating oil	38.3	20.2	—	—			9.6	3.2	(a)	(a)
	Residuum	5.9*	12.5	—	—			5.8*	12.2	—	—
1879	Crude oil	1,805.6	201.8	—	—	—	—	—	—	—	—
	Naphthas	82.6	4.3	—	—	—	—	4.9	.9	5.5	.9
	Illuminating oil	5,757.8	766.4	—	—	38.5	5.2	901.2	132.1	109.1	16.2
	Lubricating oil	117.4	46.3	—	—	1.3	.2	50.0	17.8	.7	.4
	Residuum	.8*	1.1	—	—	—	—	.8*	1.1	(a)*	(a)
1884	Crude oil	3,079.4	291.4	—	—	—	—	6.0	.7	—	—
	Naphthas	186.3	29.0	—	—	—	—	63.8	7.9	57.7	11.3
	Illuminating oil	8,080.5	976.3	—	—	61.0	7.0	2,760.6	319.7	344.0	45.4
	Lubricating oil	225.1	81.2	—	—	.4	.1	142.3	43.7	7.4	3.7
	Residuum	.2*	2.3	—	—	—	—	.2*	2.1	(a)*	(a)
1889	Crude oil	5,525.3	456.9	—	—	—	—	—	—	.7	.1
	Naphthas	31.3	4.9	—	—	.1	(a)	11.8	1.5	4.2	1.0
	Illuminating oil	11,914.4	1,256.7	—	—	93.7	10.2	4,838.8	479.5	509.7	72.4
	Lubricating oil	496.6	157.4	—	—	.2	.1	267.4	79.7	22.2	8.4
	Residuum	23.4*	47.4	—	—	(a)*	(a)	.7*	2.8	(a)*	.1
1894	Crude oil	15,426.0	789.6	13.8	2.6	—	—	19.4	1.2	—	—
	Naphthas	173.6	21.7	141.9	13.9	—	—	85.3	8.5	1.0	.2
	Illuminating oil	14,182.8	1,065.6	.2	.1	127.9	8.8	7,639.3	487.9	1,050.5	113.5
	Lubricating oil	1,725.7	373.4	—	—	(a)*	(a)	942.6	149.6	34.2	11.1
	Residuum	2.5*	8.0	—	—	(a)*	.1	2.2*	7.0	(a)*	.2
1899	Crude oil	11,447.6	602.0	22.3	2.9	.5	.1	20.5	1.4	.8	.1
	Naphthas	267.7	32.4	286.0	25.0	211.0	21.2	154.5	12.7	864.3	97.5
	Illuminating oil	17,067.0	1,337.5	.8	.2	1.8	.3	9,163.3	564.4	45.6	12.1
	Lubricating oil	1,966.0	328.5	(a)*	.1	(a)*	(a)	881.9	124.1	.1*	.4
	Residuum	5.3*	8.4	—	—	—	—	4.2*	5.7	—	—

(table con't. on next page)

APPENDIX TABLE D:3 (cont.)

Year Ending June 30	Product	Mexico		Miquelon, Langley, etc.		Newfoundland & Labrador		West Indies	
		Gallons	Dollars	Gallons	Dollars	Gallons	Dollars	Gallons	Dollars
1874	Crude oil	—	—	—	—	—	—	886.8	131.6
	Naphthas	.6	.2	—	—	—	—	34.6	9.7
	Illuminating oil	623.6	164.2	15.2	3.5	158.7	31.5	3,320.9	685.4
	Lubricating oil	.6	.1	—	—	—	—	28.1	16.9
	Residuum	—	—	—	—	—	—	.1*	.3
1879	Crude oil	—	.6	—	—	—	—	1,805.6	201.8
	Naphthas	7.6	.6	—	—	.2	(a)	14.4	1.9
	Illuminating oil	936.5	152.4	23.1	3.1	384.0	38.4	3,365.4	419.0
	Lubricating oil	2.7	1.6	—	—	.3	.1	62.4	26.2
	Residuum	—	—	—	—	—	—	—	—
1884	Crude oil	15.8	1.9	—	—	—	—	3,057.6	288.8
	Naphthas	35.8	4.5	—	—	—	—	29.0	5.3
	Illuminating oil	1,444.0	197.6	37.7	3.7	563.0	52.8	2,870.3	350.2
	Lubricating oil	21.0	8.3	—	—	4.8	1.6	49.2	23.8
	Residuum	(a)*	(a)	—	—	—	—	(a)*	.1
1889	Crude oil	1,881.4	160.0	—	—	—	—	3,643.2	296.8
	Naphthas	7.0	.7	.3	(a)	.3	(a)	7.6	1.6
	Illuminating oil	2,123.5	226.2	44.1	4.1	408.4	34.2	3,896.2	430.2
	Lubricating oil	77.3	21.4	.1	.1	19.7	5.8	109.7	41.9
	Residuum	(a)*	.2	—	—	—	—	22.6*	44.3
1894	Crude oil	8,026.2	337.9	—	—	—	—	7,380.5	450.5
	Naphthas	2.5	.6	—	—	3.7	.3	67.2	9.5
	Illuminating oil	388.9	64.8	50.4	3.6	609.1	40.6	4,174.9	332.6
	Lubricating oil	318.8	81.3	(a)*	(a)	12.7	4.0	417.1	127.3
	Residuum	(a)*	(a)	—	—	—	—	.2	.8
1899	Crude oil	7,969.9	395.4	—	—	—	—	3,457.2	205.2
	Naphthas	73.4	14.2	—	—	.3	(a)	15.9	2.5
	Illuminating oil	581.2	73.3	40.0	3.6	698.3	57.9	5,222.8	494.7
	Lubricating oil	605.2	104.0	—	—	14.0	3.5	416.7	84.2
	Residuum	.2*	.7	—	—	—	—	.8*	1.5

(a) Quantities or values so small they do not show up when figures are given in thousands (less than 50).   * Barrels.   ** Costa Rica, Guatemala, Honduras, Nicaragua, Panama, Salvador.

Source: See page 739.

APPENDIX TABLE D:4

U. S. exports of domestic petroleum and petroleum products to major South American countries

Quantities and values in thousands by five year intervals, 1874–99

Year Ending June 30	Product	Total South America		Argentina		Bolivia		Brazil	
		Gallons	Dollars	Gallons	Dollars	Gallons	Dollars	Gallons	Dollars
1874	Crude oil	1.8	.2	—	—	—	—	1.8	.2
	Naphthas	14.1	3.8	11.4	3.4	—	—	—	—
	Illuminating oil	4,601.6	1,021.2	657.5	139.6	—	—	2,075.9	467.9
	Lubricating oil	27.6	13.0	.2	.1	—	—	1.1	1.3
	Residuum	1.5*	4.9	—	—	—	—	—	—
1879	Crude oil	.9	.1	—	—	—	—	.9	.1
	Naphthas	106.8	16.5	6.5	1.5	—	—	84.4	12.2
	Illuminating oil	8,590.7	1,113.8	943.5	117.2	—	—	4,222.1	558.5
	Lubricating oil	38.3	14.8	.4	.1	—	—	25.0	9.6
	Residuum	—	—	—	—	—	—	—	—
1884	Crude oil	—	—	—	—	—	—	—	—
	Naphthas	199.5	32.6	9.3	2.2	—	—	151.0	23.2
	Illuminating oil	11,568.2	1,258.8	2,111.6	240.3	—	—	5,034.1	532.8
	Lubricating oil	130.2	52.5	24.5	13.4	—	—	38.2	11.8
	Residuum	—	—	—	—	—	—	—	—
1889	Crude oil	—	—	—	—	—	—	—	—
	Naphthas	65.5	12.3	41.0	7.3	—	—	15.5	2.4
	Illuminating oil	22,099.4	2,294.8	5,952.4	619.9	—	—	8,834.3	890.0
	Lubricating oil	719.8	216.2	431.3	99.6	—	—	113.9	45.4
	Residuum	.2*	.6	—	—	—	—	(a)*	(a)
1894	Crude oil	2.0	.1	—	—	—	—	—	—
	Naphthas	79.8	12.7	36.6	5.6	—	—	39.4	6.3
	Illuminating oil	23,341.8	1,698.0	3,162.8	247.8	2.8	.2	12,154.7	839.7
	Lubricating oil	1,509.7	325.6	504.1	95.0	—	—	432.5	98.5
	Residuum	.2*	.8	—	—	—	—	—	—
1899	Crude oil	—	—	—	—	—	—	—	—
	Naphthas	137.7	20.8	86.0	12.8	.3	(a)	26.6	4.5
	Illuminating oil	32,306.6	2,930.3	6,483.3	665.7	—	—	16,239.1	1,397.4
	Lubricating oil	2,899.3	424.3	1,396.0	177.6	—	—	671.5	97.2
	Residuum	.7*	2.9	(a)*	.1	—	—	.4*	2.1

(table con't. on next page)

Year Ending June 30	Product	Chile		Colombia		Ecuador		Guianas		Paraguay	
		Gallons	Dollars	Gallons	Dollars	Gallons	Dollars	Gallons	Dollars	Gallons	Dollars
1874	Crude oil	—	—	—	—	—	—	—	—	—	—
	Naphthas	—	—	—	—	—	—	—	—	—	—
	Illuminating oil	733.1	168.7	186.6	42.4	—	—	171.3	36.4	—	—
	Lubricating oil	7.0	5.0	17.0	5.3	—	—	1.5*	4.9	—	—
	Residuum	—	—	—	—	—	—	—	—	—	—
1879	Crude oil	—	—	—	—	—	—	—	—	—	—
	Naphthas	1.6	.3	2.7	.4	—	—	—	—	—	—
	Illuminating oil	832.5	107.1	192.4	27.6	—	—	775.2	97.2	—	—
	Lubricating oil	4.0	1.5	3.8	1.6	—	—	.2	.1	—	—
	Residuum	—	—	—	—	—	—	—	—	—	—
1884	Crude oil	—	—	—	—	—	—	—	—	—	—
	Naphthas	3.8	1.2	(a)	(a)	—	—	—	—	—	—
	Illuminating oil	953.1	104.7	402.7	50.5	—	—	491.2	55.9	—	—
	Lubricating oil	6.0	2.9	40.1	14.6	—	—	4.6	1.0	—	—
	Residuum	—	—	—	—	—	—	—	—	—	—
1889	Crude oil	—	—	—	—	—	—	—	—	—	—
	Naphthas	3.0	.9	.2	(a)	—	—	—	—	—	—
	Illuminating oil	2,035.9	209.0	573.3	72.6	67.0	8.3	651.3	67.5	—	—
	Lubricating oil	66.8	25.9	41.1	16.4	7.1	3.3	6.4	2.5	—	—
	Residuum	—	—	.1*	.4	—	—	—	—	—	—
1894	Crude oil	—	—	—	—	—	—	—	—	—	—
	Naphthas	.2	(a)	.6	.1	—	—	—	—	—	—
	Illuminating oil	1,919.9	133.6	735.2	61.3	259.1	20.2	936.9	63.8	—	—
	Lubricating oil	435.7	90.3	34.7	9.4	11.2	3.1	9.9	4.5	—	—
	Residuum	—	—	.1*	.6	.1*	.2	—	—	—	—
1899	Crude oil	—	—	—	—	—	—	—	—	—	—
	Naphthas	5.4	1.0	15.6	1.7	—	—	.1	(a)	—	—
	Illuminating oil	3,685.8	314.3	1,187.4	116.1	354.3	34.4	896.2	77.3	—	—
	Lubricating oil	619.3	100.5	58.0	11.8	10.1	2.5	9.9	2.2	.5	.1
	Residuum	(a)*	.1	.1*	.3	(a)*	.1	(a)*	(a)	—	—

(*table con't. on next page*)

APPENDIX TABLE D:4 (cont.)

Year Ending June 30	Product	Peru		Uruguay		Venezuela		All Other	
		Gallons	Dollars	Gallons	Dollars	Gallons	Dollars	Gallons	Dollars
1874	Crude oil	—	—	—	—	—	—	—	—
	Naphthas	.4	.1	1.5	.1	.8	.2	—	—
	Illuminating oil	257.2	55.1	325.3	69.4	194.7	41.7	—	—
	Lubricating oil	2.3	1.3	(a)	(a)	—	—	—	—
	Residuum	—	—	—	—	—	—	—	—
1879	Crude oil	—	—	—	—	—	—	—	—
	Naphthas	1.6	.2	9.6	1.8	.4	.1	26.5	3.2
	Illuminating oil	601.1	76.6	636.8	81.1	360.6	45.3	1.5	.7
	Lubricating oil	2.7	.9	.2	.1	.5	.2	—	—
	Residuum	—	—	—	—	—	—	—	—
1884	Crude oil	—	—	—	—	—	—	—	—
	Naphthas	2.0	.7	8.2	2.5	25.2	2.8	—	—
	Illuminating oil	295.9	30.2	1,507.7	161.1	721.3	77.6	50.7	5.7
	Lubricating oil	8.7	3.5	5.3	3.5	2.2	1.3	.6	.5
	Residuum	—	—	—	—	—	—	—	—
1889	Crude oil	—	—	—	—	—	—	—	—
	Naphthas	—	—	5.7	1.7	.1	(a)	—	—
	Illuminating oil	425.1	46.9	2,523.8	267.2	1,036.3	113.4	—	—
	Lubricating oil	21.5	10.2	15.1	4.6	16.6	8.3	—	—
	Residuum	(a)*	(a)	—	—	(a)*	.2	—	—
1894	Crude oil	—	—	2.0	.1	—	(a)	—	—
	Naphthas	.8	.2	2.1	.5	.2	—	—	—
	Illuminating oil	387.6	32.8	2,520.6	201.8	1,262.3	96.8	—	—
	Lubricating oil	44.2	13.7	24.9	5.8	12.5	5.3	—	—
	Residuum	—	—	—	—	(a)*	.1	—	—
1899	Crude oil	—	—	—	—	—	(a)	—	—
	Naphthas	.4	.1	3.5	.6	.1	—	—	—
	Illuminating oil	322.0	33.8	1,760.5	172.8	1,327.7	118.4	—	—
	Lubricating oil	70.3	17.1	45.8	9.4	18.0	5.9	—	—
	Residuum	—	—	—	—	.1*	.3	—	—

(a) Quantities or values so small they do not show up when figures are given in thousands (less than 50).   *Barrels.   Source: See page 739.

751

APPENDIX TABLE D:5

U. S. exports of domestic petroleum and petroleum products to major Asiatic countries and ports
Quantities and values in thousands by five year intervals, 1874–99

Year ending June 30	Product	Total Asia Gallons	Total Asia Dollars	Aden Gallons	Aden Dollars	Chinese Empire Gallons	Chinese Empire Dollars	China Gallons	China Dollars	East Indies Gallons	East Indies Dollars	Hongkong Gallons	Hongkong Dollars
1874	Crude oil	—	—	—	—	—	—	—	—	—	—	—	—
	Naphthas	7.2	2.0	—	—	—	—	—	—	—	—	—	—
	Illuminating oil	6,406.2	1,417.4	—	—	—	—	827.5	196.0	2,961.1	653.9	—	—
	Lubricating oil	.7	.3	—	—	—	—	.7	.3	—	—	.1	.1
	Residuum	—	—	—	—	—	—	—	—	—	—	—	—
1879	Crude oil	—	—	—	—	—	—	—	—	—	—	—	—
	Naphthas	—	—	—	—	—	—	—	—	—	—	—	—
	Illuminating oil	41,446.2	5,347.1	—	—	—	—	5,443.0	690.4	17,659.6	2,315.9	390.0	45.9
	Lubricating oil	—	—	—	—	—	—	—	—	—	—	—	—
	Residuum	—	—	—	—	—	—	—	—	—	—	—	—
1884	Crude oil	—	—	—	—	—	—	—	—	—	—	—	—
	Naphthas	—	—	—	—	—	—	—	—	—	—	—	—
	Illuminating oil	89,966.1	9,215.0	—	—	—	—	8,383.8	835.9	52,655.0	5,420.4	4,856.3	505.4
	Lubricating oil	.5	.3	—	—	—	—	.2	.2	—	—	.3	.1
	Residuum	—	—	—	—	—	—	—	—	—	—	—	—
1889	Crude oil	—	—	—	—	—	—	—	—	—	—	—	—
	Naphthas	(a)	(a)	—	—	—	—	—	—	—	—	—	—
	Illuminating oil	120,162.8	11,073.2	—	—	—	—	9,849.0	907.5	68,271.8	6,224.8	6,720.8	640.6
	Lubricating oil	434.3	55.8	—	—	—	—	2.8	1.1	355.9	37.8	.9	.4
	Residuum	—	—	—	—	—	—	—	—	—	—	—	—
1894	Crude oil	—	—	—	—	—	—	—	—	—	—	—	—
	Naphthas	—	—	—	—	—	—	—	—	—	—	—	—
	Illuminating oil	182,239.5	11,067.8	—	—	40,377.3	2,435.8	—	—	85,907.6	5,292.5	16,888.8	1,019.7
	Lubricating oil	623.3	118.6	—	—	9.8	2.8	—	—	540.1	94.2	13.1	4.4
	Residuum	.3*	.9	—	—	—	—	—	—	.3*	.9	—	—
1899	Crude oil	—	—	—	—	1.0	.2	—	—	—	—	—	—
	Naphthas	1.0	.2	67.0	6.2	—	—	—	—	—	—	—	—
	Illuminating oil	109,120.9	8,126.0	—	—	22,683.4	1,791.1	—	—	35,481.3	2,597.6	18,095.3	1,380.8
	Lubricating oil	5,607.6	568.8	—	—	185.4	25.3	—	—	4,420.7	404.1	103.1	18.6
	Residuum	—	—	—	—	—	—	—	—	—	—	—	—

(table con't. on next page)

752

APPENDIX TABLE D:5 (*cont.*)

Year ending June 30	Product	Japan Gallons	Japan Dollars	Korea Gallons	Korea Dollars	Russia (Asiatic) Gallons	Russia (Asiatic) Dollars	Turkey in Asia Gallons	Turkey in Asia Dollars	Siam Gallons	Siam Dollars	Other Asia Gallons	Other Asia Dollars
1874	Crude oil	—	—	—	—	—	—	—	—	—	—	—	—
	Naphthas	7.2	2.0	—	—	6.3	1.8	—	—	—	—	26.0	4.9
	Illuminating oil	526.2	120.1	—	—	—	—	2,059.1	440.6	—	—	—	—
	Lubricating oil	—	—	—	—	—	—	—	—	—	—	—	—
	Residuum	—	—	—	—	—	—	—	—	—	—	—	—
1879	Crude oil	—	—	—	—	—	—	—	—	—	—	—	—
	Naphthas	—	—	—	—	4.1	.7	—	—	—	—	—	—
	Illuminating oil	15,295.6	1,959.6	—	—	—	—	2,653.9	334.6	—	—	—	—
	Lubricating oil	—	—	—	—	—	—	—	—	—	—	—	—
	Residuum	—	—	—	—	—	—	—	—	—	—	—	—
1884	Crude oil	—	—	—	—	—	—	—	—	—	—	—	—
	Naphthas	—	—	—	—	35.3	6.8	—	—	—	—	—	—
	Illuminating oil	18,005.4	1,849.5	—	—	—	—	4,401.4	428.0	—	—	1,629.0	169.0
	Lubricating oil	—	—	—	—	—	—	—	—	—	—	—	—
	Residuum	—	—	—	—	—	—	—	—	—	—	—	—
1889	Crude oil	(a)	(a)	—	—	—	—	—	—	—	—	—	—
	Naphthas	32,791.1	3,069.8	—	—	2.5	.6	248.2	20.5	—	—	2,280.2	209.4
	Illuminating oil	74.7	16.5	—	—	—	—	—	—	—	—	—	—
	Lubricating oil	—	—	—	—	—	—	—	—	—	—	—	—
	Residuum	—	—	—	—	—	—	—	—	—	—	—	—
1894	Crude oil	—	—	—	—	—	—	—	—	—	—	—	—
	Naphthas	—	—	—	—	2.4	.4	—	—	—	—	—	—
	Illuminating oil	37,272.5	2,209.1	—	—	—	—	—	—	—	—	1,791.0	110.3
	Lubricating oil	60.3	17.2	—	—	—	—	—	—	—	—	—	—
	Residuum	—	—	—	—	—	—	—	—	—	—	—	—
1899	Crude oil	—	—	—	—	—	—	—	—	—	—	—	—
	Naphthas	—	—	—	—	8.7	1.4	—	—	—	—	—	—
	Illuminating oil	32,705.2	2,341.9	—	—	1.3	1.2	(a)	(a)	—	—	80.0	7.1
	Lubricating oil	897.1	119.6	—	—	—	—	—	—	—	—	—	—
	Residuum	—	—	—	—	—	—	—	—	—	—	—	—

(a) Quantities or values so small they do not show up when figures are given in thousands (less than 50).   *Barrels.   Source: See page 739.

U.S. exports of domestic petroleum and petroleum products to Oceana
Quantities and values by five year intervals, 1874–99

Year ending June 30	Product	Total Oceana		British Australasia		British Oceana		French Oceana	
		Gallons	Dollars	Gallons	Dollars	Gallons	Dollars	Gallons	Dollars
1874	Crude oil	—	—	—	—	—	—	—	—
	Naphthas	—	—	—	—	—	—	—	—
	Illuminating oil	3,754.1	828.6	3,693.0	811.4	—	—	—	—
	Lubricating oil	—	—	—	—	—	—	—	—
	Residuum	—	—	—	—	—	—	—	—
1879	Crude oil	—	—	—	—	—	—	—	—
	Naphthas	1.8	.7	—	—	—	—	—	—
	Illum nating oil	2,911.7	490.8	2,791.8	469.0	—	—	—	—
	Lubricating oil	2.8	1.2	2.6	1.0	—	—	—	—
	Residuum	—	—	—	—	—	—	—	—
1884	Crude oil	—	—	—	—	—	—	—	—
	Naphthas	84.8	26.0	55.8	18.5	—	—	—	—
	Illuminating oil	4,725.9	628.3	4,306.4	576.9	—	—	—	—
	Lubricating oil	145.7	47.9	136.9	42.6	—	—	—	—
	Residuum	.2*	.6	.2*	.6	—	—	—	—
1889	Crude oil	—	—	—	—	—	—	—	—
	Naphthas	56.3	15.2	53.1	14.1	—	—	—	—
	Illuminating oil	9,576.1	1,163.5	7,551.2	958.0	—	—	—	—
	Lubricating oil	305.4	92.0	288.2	84.5	—	—	—	—
	Residuum	—	—	—	—	—	—	—	—
1894	Crude oil	—	—	—	—	—	—	—	—
	Naphthas	57.1	13.8	54.9	13.2	—	—	—	—
	Illuminating oil	13,302.6	1,103.1	11,821.9	973.3	—	—	33.6	6.3
	Lubricating oil	809.8	128.1	774.8	115.9	—	—	—	—
	Residuum	—	—	—	—	—	—	—	—
1899	Crude oil	—	—	—	—	—	—	—	—
	Naphthas	119.1	18.4	118.0	18.3	—	—	—	—
	Illuminating oil	15,469.6	1,558.2	14,396.8	1,422.7	—	—	15.3	2.5
	Lubricating oil	2,129.8	317.9	2,029.7	288.8	—	—	(a)	(a)
	Residuum	(a)*	(a)	—	—	—	—	—	—

(table con't. on next page)

APPENDIX TABLE D:6 (cont.)

Year ending June 30	Product	German Oceana		Hawaii		Philippine Islands		All Other Oceana	
		Gallons	Dollars	Gallons	Dollars	Gallons	Dollars	Gallons	Dollars
1874	Crude oil	—	—	—	—	—	—	—	—
	Naphthas	—	—	—	—	—	—	—	—
	Illuminating oil	—	—	61.1	17.2	—	—	—	—
	Lubricating oil	—	—	—	—	—	—	—	—
	Residuum	—	—	—	—	—	—	—	—
1879	Crude oil	—	—	—	—	—	—	—	—
	Naphthas	—	—	1.8	.7	—	—	—	—
	Illuminating oil	—	—	119.9	21.8	—	—	—	—
	Lubricating oil	—	—	.3	.2	—	—	—	—
	Residuum	—	—	—	—	—	—	—	—
1884	Crude oil	—	—	—	—	—	—	—	—
	Naphthas	—	—	29.1	7.5	—	—	—	—
	Illuminating oil	—	—	419.5	51.4	—	—	—	—
	Lubricating oil	—	—	8.8	5.3	—	—	—	—
	Residuum	—	—	—	—	—	—	—	—
1889	Crude oil	—	—	—	—	—	—	—	—
	Naphthas	—	—	3.2	1.1	—	—	—	—
	Illuminating oil	—	—	380.6	50.8	1,644.3	154.7	—	—
	Lubricating oil	—	—	17.2	7.5	—	—	—	—
	Residuum	—	—	—	—	—	—	—	—
1894	Crude oil	—	—	—	—	—	—	—	—
	Naphthas	—	—	2.2	.6	—	—	—	—
	Illuminating oil	—	—	507.6	66.5	610.4	35.3	329.2	21.7
	Lubricating oil	—	—	34.7	12.1	.3	.2	—	—
	Residuum	—	—	—	—	—	—	—	—
1899	Crude oil	—	—	—	—	—	—	—	—
	Naphthas	—	—	1.1	.1	—	—	—	—
	Illuminating oil	6.9	1.0	1,049.2	131.9	.2	(a)	1.4	.3
	Lubricating oil	—	—	99.9	29.1	—	—	—	—
	Residuum	—	—	(a)*	(a)	—	—	—	—

(a) Quantities or values so small they do not show up when figures are given in thousands (less than 50).   * Barrels.   Source: See page 739.

755

U. S. exports of domestic petroleum and petroleum products to major African countries
Quantities and values in thousands by five year intervals, 1874–99

Year ending June 30	Product	Total Africa Gallons	Total Africa Dollars	British Africa Gallons	British Africa Dollars	Canary Islands Gallons	Canary Islands Dollars	French Africa Gallons	French Africa Dollars	German Africa Gallons	German Africa Dollars
1874	Crude oil	—	—	—	—	—	—	—	—	—	—
	Naphthas	1.0	.2	—	—	—	—	1.0	.2	—	—
	Illuminating oil	2,674.5	578.9	780.7	178.5	—	—	81.4	16.6	—	—
	Lubricating oil	11.6	3.8	11.6	3.8	—	—	—	—	—	—
	Residuum	—	—	—	—	—	—	—	—	—	—
1879	Crude oil	—	—	—	—	—	—	—	—	—	—
	Naphthas	17.3	2.3	1.3	.3	—	—	16.0	2.0	—	—
	Illuminating oil	6,033.0	776.8	867.2	128.6	—	—	1,082.7	138.0	—	—
	Lubricating oil	9.7	4.4	9.7	4.4	—	—	—	—	—	—
	Residuum	—	—	—	—	—	—	—	—	—	—
1884	Crude oil	—	—	—	—	—	—	—	—	—	—
	Naphthas	12.4	2.6	7.6	2.0	—	—	4.7	.6	—	—
	Illuminating oil	7,732.0	799.9	1,659.3	203.7	—	—	1,657.0	161.5	—	—
	Lubricating oil	3.1	1.0	3.1	1.0	—	—	—	—	—	—
	Residuum	—	—	—	—	—	—	—	—	—	—
1889	Crude oil	—	—	—	—	—	—	—	—	—	—
	Naphthas	7.4	2.5	7.4	2.5	—	—	—	—	—	—
	Illuminating oil	6,384.8	656.2	2,117.1	277.4	—	—	1,560.6	143.5	—	—
	Lubricating oil	20.4	9.2	16.2	7.2	—	—	—	—	—	—
	Residuum	—	—	—	—	—	—	—	—	—	—
1894	Crude oil	—	—	—	—	—	—	—	—	—	—
	Naphthas	3.1	1.0	3.1	1.0	—	—	—	—	—	—
	Illuminating oil	7,049.5	517.4	2,874.8	262.7	513.8	31.5	844.0	52.6	.1	(a)
	Lubricating oil	115.4	44.5	115.2	44.4	.1	(a)	—	—	—	—
	Residuum	—	—	—	—	—	—	—	—	—	—
1899	Crude oil	—	—	—	—	—	—	—	—	—	—
	Naphthas	17.0	3.2	15.8	3.0	—	—	—	—	—	—
	Illuminating oil	11,372.2	1,070.0	7,540.8	756.9	1,004.8	77.4	1,102.8	88.4	—	—
	Lubricating oil	1,099.4	225.1	866.1	181.4	.2	.1	6.0	.8	—	—
	Residuum	(a)*	.1	(a)*	.1	—	—	—	—	—	—

(table con't. on next page)

APPENDIX TABLE D:7 (cont.)

Year ending June 30	Product	Liberia Gallons	Liberia Dollars	Madagascar Gallons	Madagascar Dollars	Portuguese Africa Gallons	Portuguese Africa Dollars	Spanish Africa Gallons	Spanish Africa Dollars	Turkey in Africa Egypt Gallons	Turkey in Africa Egypt Dollars	All Other Africa Gallons	All Other Africa Dollars
1874	Crude oil	—	—	—	—	—	—	—	—	—	—	—	—
	Naphthas	—	—	—	—	—	—	—	—	—	—	—	—
	Illuminating oil	12.9	2.8	—	—	—	—	—	—	1,584.0	332.4	84.0	19.1
	Lubricating oil	.1	(a)	—	—	—	—	131.5	29.5	—	—	—	—
	Residuum	—	—	—	—	—	—	—	—	—	—	—	—
1879	Crude oil	—	—	—	—	—	—	—	—	—	—	—	—
	Naphthas	—	—	—	—	—	—	—	—	—	—	—	—
	Illuminating oil	16.0	1.9	—	—	4.2	.7	320.1	42.6	2,891.8	352.3	851.0	112.7
	Lubricating oil	—	—	—	—	—	—	—	—	—	—	—	—
	Residuum	—	—	—	—	—	—	—	—	—	—	—	—
1884	Crude oil	—	—	—	—	—	—	—	—	—	—	—	—
	Naphthas	.1	(a)	—	—	—	—	—	—	—	—	—	—
	Illuminating oil	48.8	4.9	—	—	229.0	24.2	83.5	8.9	3,528.8	346.4	525.6	50.3
	Lubricating oil	—	—	—	—	—	—	—	—	—	—	—	—
	Residuum	—	—	—	—	—	—	—	—	—	—	—	—
1889	Crude oil	—	—	—	—	—	—	—	—	—	—	—	—
	Naphthas	—	—	—	—	—	—	—	—	—	—	—	—
	Illuminating oil	28.8	2.9	—	—	2.0	.1	370.9	35.7	1,312.3	108.0	993.2	88.6
	Lubricating oil	—	—	—	—	—	—	(a)	(a)	4.1	2.0	—	—
	Residuum	—	—	—	—	—	—	—	—	—	—	—	—
1894	Crude oil	—	—	—	—	—	—	—	—	—	—	—	—
	Naphthas	—	—	—	—	—	—	—	—	—	—	—	—
	Illuminating oil	10.1	.6	—	—	23.7	1.8	—	—	2,038.1	125.4	745.0	42.8
	Lubricating oil	(a)	(a)	—	—	.1	(a)	—	—	—	—	—	—
	Residuum	—	—	—	—	—	—	—	—	—	—	—	—
1899	Crude oil	—	—	—	—	—	—	—	—	—	—	—	—
	Naphthas	—	—	—	—	1.0	.2	—	—	—	—	.2	(a)
	Illuminating oil	—	—	—	—	627.8	59.6	—	—	313.6	24.2	782.5	63.4
	Lubricating oil	—	—	—	—	91.3	19.2	—	—	21.4	4.8	114.5	18.7
	Residuum	—	—	—	—	—	—	—	—	—	—	—	—

a) Quantities or values so small they do not show up when figures are given in thousands (less than 50). * Barrels. Source: See page 739.

757

The legal framework
of crude-oil production

The two most important legal aspects affecting the conditions of crude-oil production during the nineteenth century were the "Rule of Capture" and the judicial construction of contemporary leasing provisions. Both legalities had similar influences and results on petroleum production: namely, to facilitate the rapid exploitation of crude reserves, but with strong safeguards for individual property rights. This appendix proposes to examine the major judicial decisions and, where possible, to indicate alternative lines of development which contemporaries might have applied.

The "Rule of Capture"

The position of the surface landowner in English mining law by the end of the eighteenth century was practically supreme. The legal maxim which governed crude-oil production in the United States during the nineteenth century, however, while similarly related to the primacy of the surface owner, took its direct ancestry from English doctrine dealing with percolating waters. *Acton* v. *Blundell,* decided in 1843, is generally cited as the leading authority.[1]

The suit arose because a percolating water well, dug in 1821 and used to operate a cotton mill, was rendered insufficient in 1837 by a coal pit sunk on neighboring ground. The owner of the well sued and based his claim upon the established English law of surface streams. This doctrine held that

each proprietor of the land has a right to the advantage of the stream flowing in its natural course over his land . . . not inconsistent with a similar right in the proprietors of the land above or below.[2]

Accordingly, neither party along the stream had the right to diminish or otherwise injure the quality of the water without the permission of other affected parties. But in the Blundell Case the court held that the surface-stream doctrine was inapplicable where water was drawn from wells.

Water which feeds a well, the court wrote, does so from neighboring soil, but not "openly in the sight of the neighboring proprietor, but through the hidden veins of the earth. . . ." It would be impossible for any proprietor to know "what portion of the water is taken from beneath his soil; how much he gives originally," or how much he transmits or receives until the well is sunk. Until this is done there cannot properly be said to be any flow of water at all.[3]

If, by analogy, the law of streams was applied to wells, the opinion continued, serious consequences were bound to occur. Should sinking a well give the owner an indefeasible right to the water, it would prevent a neighbor from making use of a spring or even from draining his land. Mining proper would likewise be affected:

[A] well may be sunk to supply a cottage, or a drinking-place for cattle; whilst the owner of the adjoining land may be prevented from winning metals and minerals of inestimable value . . . there is no limit of space within which the claim of right to an underground spring can be confined: in the present case the nearest coal-pit is at the distance of half a mile from the well: it is obvious the law must equally apply if there is an interval of many miles.[4]

The court concluded that the proper principle to apply was to give the "owner of the soil all that lies beneath his surface," and that he should be quite free to dig at his will and pleasure.

The ruling in *Acton* v. *Blundell* received its first application to petroleum deposits in the Pennsylvania state courts. In a case decided in 1875, the court invoked the "Rule of Capture" but recognized the power of the legislature to enforce a different one. "No doubt," the deciding judge wrote, "many thousands of dollars have been expended in oil and gas territory that would not have . . . if some rule had existed by which it could have been drilled."[5]

Other early cases took a slightly different approach to the rule by equating the characteristics of oil and gas to the law of wild animals or solid minerals. Substantively there was little difference between this approach to the rule and that stemming from percolating waters.[6] Probably because of the dissimilarity of precedent in early applications of the rule, it was not until the turn of the century that it acquired specific legal meaning. Three state supreme court cases are generally cited as authority.[7] In these cases the right to reduce to possession all the below-ground minerals as well as the right to prevent drainage by drilling off-set wells received explicit judicial sanction.

Leasing provisions

The strength of the "Rule of Capture" was significantly increased by judicial interpretation of contemporary leasing provisions. Indeed, such action might be called the practical working of the rule, since most oil land was operated under leases. Of the total of 1.56 million acres of oil property in 1890, for example, approximately four-fifths was leased.[8]

Landowners, producers, and especially the courts early became aware that the fugacious character of oil created unique problems for leasing arrangements. The interest of the lessor prompted quick and diligent drilling operations.[9] But a lessee, holding many leases, with limited finances, drilling equipment, and skilled labor, conceivably would prefer a cautious drilling schedule for all his leaseholds. Under such circumstances a conflict of interest was inevitable.

Early leases, attempting to protect the lessor, frequently stipulated a given period of time in which a producer (lessee) had to begin drilling a well. Test, producing, and offset wells were treated at considerable length in early leases. It remained for the courts to decide whether the lessee had pursued his obligation to drill with "due diligence" and if failure to do so constituted forfeiture of the lease.[10]

With the growing complexity of the technology and economics of drilling, entrepreneurs yearned for a more flexible operational framework. Rather than be bound "to immediately develop each tract of land or . . . suffer the loss of the lease by forfeiture," producers sought a compromise provision in leasing arrangements.[11]

The delay rental, prominent after 1874, appeared to afford the producers wider latitude in decision-making. Delay rentals were not rentals as the term is commonly used, since no payment was made for the use of land but rather in lieu of drilling operations.[12] Delay rentals, however, did not really solve producers' problems in the way they hoped. On the contrary, the device created a set of new problems.

One of the earliest cases of a delay rental was *Brown* v. *Vandergrift*, decided in 1875.[13] The protested lease stipulated that the lessee should have sole right to the oil for twenty years, that operations should commence within sixty days after the effective date of the lease and continue "with due diligence." If operations were suspended at any time for twenty days it was further agreed that the lease would be considered forfeited. Finally, the lease also stated that the lessee could elect to pay a $30.00 per month delay rental in the event that he wished to postpone drilling operations. The lessee paid the delay rental for four months, omitted payment for eleven months, and then tendered the amount for that time.

The court really went beyond the issues under immediate consideration. It held that the covenant of forfeiture was merely modified, not abrogated, by the delay-rental-payment clause, and the *lessor* could refuse the tender and insist on forfeiture. After noting the peculiarities of oil production, the court stated what amounted to a rabid defense of the lessor's position:

> it was necessary to guard the rights of the landowner, as well as the public interest, by numerous covenants, some of the most stringent kind, to prevent their lands from being burdened by unexecuted and profitless leases. . . .[14]

Ironically, the same court, in language smacking of complete ignorance of the problems involved, conceived a strong lessor's position as a deterrent to overdrilling:

> Without these guards the lands would be thatched over with oil leases by subletting, and a farm riddled with holes and bristled with derricks. . . .[15]

Courts invariably decided that immediate and exhaustive exploitation was "intended," even if omitted or ambiguously stated in the terms of a lease. In *Munroe* v. *Armstrong*, decided in 1880, the court remarked: "Perhaps in no other business is prompt performance of contracts so essential to the rights of the parties." The court went on to say:

> Forfeiture for non-development or delay is essential to private and public interests in relation to the use and alienation of property . . . [The] lease is of no value till developed. . . .[16]

Many courts also held the lessee's right to explore and produce inchoate (incomplete) until drilling was actually started. And even if there was an optional-drilling clause, the lessor had the privilege of canceling the lease at any time before drilling commenced.[17]

The extent to which the bench's absolute treatment of property rights protected the position of the landowner-lessor and encouraged development of individual leases is vividly illustrated in a Pennsylvania case of 1889. In *Galey* v. *Kellerman,* the lessor sued for recovery of delay rental after the lessee failed to drill or pay. The court held that the lessor had the option of either accepting forfeiture if the lessee failed to drill or pay, or waive it and decide to collect back rentals. This decision resulted in many suits for back rentals.

The severity with which the courts treated drill or pay leases undoubtedly prompted the later change in wording ("unless" was substituted for the "or" clause). Adopted in the early 1890's, the "unless" clause since then has not undergone any drastic modification. An early version read:

> this lease shall become null and void, and all rights hereunder shall cease . . . unless a well shall be completed within one month from the date hereof, *or unless* [italics added] the lessee shall pay at the rate of [a fixed sum of money] in advance, monthly, until the well is completed. . . .[18]

Prior to the 1880's most leases were for a definite time period, anywhere from ten to ninety-nine years. During the 1880's, however, this leasing provision was importantly modified. A short-period lease was commonly used (a few months to ten years), but the lease remained operative as long as oil was produced in paying quantities.[19] The long-term leases, based to some extent on a mistaken analogy of oil to solid minerals, created conditions which made some time modification a necessity.

> In the absence of a clause giving him [the lessee] option of renewing the lease or a continuance thereof for the productive life of the land, he was placed at a great disadvantage in negotiating for a renewal of the lease on the original terms, for the lessor could rightly insist on a renewal, if he were willing to grant it at all, on a basis of the value of the land for oil and gas purposes as determined by the conditions at the end of the first term. But the definite term lease was not unfavorable to the lessee alone. The leases usually contained a clause providing that the lessee might remove all machinery and equipment from the land. The courts held that all machinery, as well as the casings of the wells, were trade fixtures and removable by the lessee within the term. Therefore, while the lessor had the lessee at a disadvantage in contracting for renewal, the lessee might in turn remove all fixtures and well casings, and leave the property in such a condition that the lessor would have to grant a second lease of the premises on terms approximately the same as if the wells had not been drilled.[20]

Once adopted, the indefinite-term lease constituted an improvement for the parties immediately concerned.

Acceptance of the Rule: legal authority

The widespread acceptance of the "Rule of Capture," which was extended well into the twentieth century, and its reinforcement by court interpretations of leases, is generally credited to the weight of legal precedent: rules once established are difficult to upset. There were contemporary factors, however, which could have mitigated the use of the rule.

Early in the industry's history there had been two unsuccessful attempts to legislate a well-plugging law in Pennsylvania. Significantly, the bills were introduced by members of the legislature from oil-producing Venango County. It was not until 1878, however, that these earlier efforts bore fruit as a well-plugging law was adopted, specifically aimed at shutting off water from oil-bearing rock and preventing oil and gas from spoiling fresh water. For the most part this legislation was accepted by oil producers and the result was the elimination of one out of a considerable number of unwanted production practices. Other oil-producing states followed Pennsylvania's example on plugging legislation, but as Summers has observed:

> This type of piece-meal legislation did not prohibit all of the acts which might result in waste of oil or gas or injury to the oil and gas foundation. Furthermore, no administrative body was charged with the enforcement of this legislation.[21]

More consequential legislation of production methods was enacted during the early 1890's in the oil-producing state of Indiana. The first law, passed in 1891, prohibited the burning of gas in flambeau lights at the well. Two years later the Indiana legislature passed a more specific conservation measure which explicitly forbade the wasteful escape of gas into the open air at the wellhead. From a conservation viewpoint, the statute of 1893 was more significant and far-reaching: the first law sought to prevent wasteful *uses*, while the second law aimed at eliminating wasteful *methods* of production.

Contemporary courts upheld both statutes, but the 1893 act travelled through state and federal jurisdictions in the case of *Ohio Oil Co.* v. *State of Ohio* and finally reached the United States Supreme Court. The Indiana legislature intended the statute of 1893 to act as a fire-safety measure. In the state supreme court, however, this was ignored and the statute was upheld as a public welfare safeguard. When the case came before the United States Supreme Court the welfare aspect of waste prevention was again upheld, but upon the grounds of "correlative rights." The court refused to examine the wisdom and discretion of the statute but rather was concerned with whether the legislature had the power to enact it. Justice White said that the legislature could equalize "correlative rights" in a common pool and at the same time prevent wastage of a resource.[22]

Surface owners draw from a common source of supply of oil and gas and are equally privileged to take it. But an unlimited exercise of such privilege by one "may result in an undue proportion being attributed to one of the possessors of the right, to the detriment of others, or by waste by one or more to the annihilation of rights of the remainder." [23]

Justice White continued:

> Hence it is that the legislative power from the peculiar nature of the right and the objects upon which it is to be exerted, can be mainifested for the purpose of protecting all the collective owners, by securing a just distribution to arise from the enjoyment, by them, of their privilege to reduce to possession, and to reach the like-end by preventing waste.[24]

This language appears clear enough: correlative rights in a common oil and gas pool are subject to abuse in the absence of some regulation. Strangely enough it perplexed later jurists. Two eminent legalists writing in the 1930's stated the net effect of thirty years of judicial opinion concerning the Ohio case:

> Foremost among the doubts left by this case [was] whether legislation designed to regulate production to protect the correlative rights of the surface owners of a pool without the element of waste [was] constitutional.[25]

The "excessive waste" the Ohio case dealt with was the non-use of gas once it was above ground. How much or how fast gas should be allowed to escape in the first place was not spelled out. This meant, or so later courts interpreted it, that the "rule of capture" had not been retired or abandoned, but rather would be allowed to operate side by side with "correlative rights." To determine which one of the two concepts should have priority in a specific case, some courts before 1930 used the norm that the landowner should "take whatever of the oil can be captured so long as waste, as defined by statute is not committed." [26]

Acceptance of the Rule: scientific knowledge

The second most generally cited reason for the widespread application of the "Rule of Capture" was that contemporary producers were unaware of the technical implications of current production methods and consequently saw nothing wrong with them. To a certain extent this was true. On the other hand, there were contemporaries who demonstrated an understanding of the engineering problems in exploiting crude-oil deposits. As early as 1863, a popular journal observed that some of the most valuable wells ceased flowing because oil-producing districts were honeycombed with wells, dissipating the gas necessary to force oil to the surface.[27] Two years later the same journal reiterated concern for wasteful production:

> They [oil derricks] stand along the headwaters of Cherry Run like the masts of vessels. . . . It is no difficult matter to . . . count from 50 to 100 within half a mile. . . . The mutual effect upon each other of wells so closely sunk, cannot be calculated in advance. . . . The theory usually entertained is that the force of the gas being spent in two directions, it is unable to give direct effect to either . . .[28]

The eminent Pennsylvania geologist, H. F. Wrigley, was even more specific:

> . . . we know that a sand-rock [oil formation], if kept free from the surface water and pierced by only a moderate number of holes will last eight or ten years, [however,] the average life of wells has not practically reached three years. We do not exhaust our beds of sand-rock, but destroy them. We pluck the apple, so to speak, by rooting the tree.
>   It is this feature of oil production, which will decide for us within the next ten years, whether we shall still lead in this commodity or remain only an example for some wise Commonwealth to profit by. Had it been possible from the start to regulate drilling, it can hardly be questioned that one-third of the wells that have been drilled would have brought as great a return as we have had from them all thus far, and at one-third of the cost of producing.[29]

Moreover, even though "oil science" was not held in high esteem by oilmen, some notable advances were made during the 1880's. Dr. Israel Charles White, a geologist of West Virginia, lent prominence to the anticlinal theory by successfully locating oil fields in advance of the drill. White projected the important oil and gas field near Washington, Pennsylvania in 1884, the Grapeville gas field in Westmoreland County, the Belle Vernon field on the Monongahela River, the Taylortown field of

Washington County, and the Mannington field in West Virginia. Along with other contemporary "oil scientists," White by the 1880's already had established the essential importance of subterranean structure in oil and gas accumulation.[30]

Regulation of production in retrospect

Perhaps it is not too extravagant to speculate briefly on the possible fate of a state law around 1900 if it had attempted to regulate production more economically. The Indiana and Pennsylvania statutes did regulate some aspects of the industry. Modification of the "Rule of Capture" by "correlative rights" could be interpreted as an attempt at regulation. Moreover, in the analogous problems of drainage and irrigation which "like gas energy [are] not a respector of boundary lines," [31] courts as early as 1885 enforced virtually unit operations. Landowners were even compelled to permit construction of ditches on their land. The judicial language in some of these cases is striking:

> [It is] a just and constitutional exercise of the power of the legislature to establish regulations by which adjoining lands held by the various owners in severalty, and in the improvement of which all have a common interest, but which, by reason of the peculiar natural condition of the whole tract cannot be improved or enjoyed by any of them without the concurrence of all, may be reclaimed and made useful to all at their joint expense.[32]

It is not unreasonable to argue that a hypothetical state law limiting the number of wells in a pool stood a good chance of being upheld by the courts had they desired to expand the range of acceptable precedents. Both the knowledge and the legal apparatus were available to insure more orderly techniques of production. But as far as producers were concerned, the rapidly expanding markets for petroleum products and the possibility of quickly opening up new fields created an exciting environment, one in which they were not likely to champion legislation circumscribing individual property rights for the sake of orderly development.

Drilling costs

$\mathbf{T}$he following tables are intended to give an impression, rather than a firm estimate, of the range of contemporary drilling costs. In evaluating the tabular data on costs, it must be kept in mind that the last quarter of the nineteenth century was generally one of falling prices. On the basis of any price index (general or wholesale), prices dropped drastically from the middle 1860's until the beginning of the 1870's and then continued to fall more gradually with only small fluctuations until 1896. The drop from 1864 to 1870 was about equal to the decline between 1870 and 1896. The Synder-Tucker general price index (1910-1914 = 100), for example, fell from 129 in 1864 to 102 in 1870 and then to 71 in 1896.[1] When viewed against the broader background of generally falling prices, the gradual decline in drilling costs which emerges from the following estimates will appear much less impressive.

Supplementary notes

Ohio-Indiana. A paucity of data makes it impossible to give any comparable tabular estimate of production costs in the Ohio fields. The best estimate suggests that total costs of bringing wells into production typically ranged between $2,000 and $2,500. Before 1890 most Ohio wells were below 2000 feet and, with the exception of some areas in Indiana, this depth was not frequently exceeded throughout the 1890's.

Well Depths. It is impossible to establish any precise average well depth in a given year. An "educated guess" is that during 1869-72 the mode was somewhere between 900 and 1200 feet. In 1874, wells of 2000 feet depth were considered exceptional.[2] After 1890, production spread southwest in Pennsylvania and into West Virginia, and wells with depths of 3000 feet to as much as 3600 feet were commonly drilled.

APPENDIX TABLE F:1

Cost of drilling, Pennsylvania Fields: 1877–78, 1886–88
1877–1878

	Equipment	1000' well	1400' well	1600' well	2000' well
1	Pipe, casing, tubing, belting, packer, sucker rods, etc.	$ 861	$ 975	$1,012	$1,116
2	Derrick and rig	400	400	450	450
3	Engine and boiler	600	600	600	600
4	Drilling (contractor's price: $1/ft)	1,000	1,400	1,600	2,000
5	Torpedoes	100	100	100	100
6	100-bbl. storage tank	60	60	60	60
	Total	$3,021	$3,535	$3,822	$4,326

1886–1888

	Equipment	1000' well	1400' well	1600' well	2000' well
1	Pipe, casing, tubing, belting, packer, sucker rods, etc.	$532	$605	$641	$714
2	Derrick and rig	275–325	275–325	275–325	275–325
3	Engine and boiler	375–600	375–600	375–600	375–600
4	Drilling (contractor's price: 60¢/ft)	600	840	960	1,200
5	Torpedoes	120–180	120–180	120–180	120–180
6	Tank	100	100	100	100
	Total	$1,902–2,337	$2,215–2,759	$2,371–2,806	$2,684–3,119
	Average	$2,119	$2,483	$2,589	$2,951

Sources: *Mineral Resources*, 1888, 316–17; *Engineering and Mining Journal*, XLVI (Sept. 1, 1888), 179; *Oil and Gas Journal*, XXXIII (Aug. 23, 1934), 59.

APPENDIX TABLE F:2

Cost of drilling, Appalachian Fields: 1897–98

		Cost of well per foot			
	Equipment	1600-ft	2400-ft	2800-ft	3200-ft
1	Pipe, casing, tubing, belting, packer, sucker rods, etc.	$ 426	$ 555	$ 840	$1,440
2	Derrick and rig	300	300	300	500
3	Engine and boiler	600	600	600	700
4	Drilling (contractor's price: 45¢/ft)	720	980	1,260	1,920 *
5	Torpedoes	100	100	100	100
6	Storage tank	60	60	60	60
	Total	$2,206	$2,595	$3,160	$4,720

* Contractor's prices figured at 60¢ per foot.
Source: *Derrick's Hand-Book*, II, 177–183.

APPENDIX TABLE F:3

Estimated total drilling cost of individual wells in different Pennsylvania oil regions, 1865–89

Year	Area	Depth/ft	Total cost *	Cost/ft
1865	Venango County	600	$4,650	$7.58
1868–69	Tallman Farm	860	4,800	5.60
1871	Parker's Landing	1050	5,200	5.00
1876–77	Bradford †	1600–2000	4,000	2.00–2.50
1880	Bradford †	2000	2,632	1.30
1889	Washington County	2400	5,940	2.47

* Total cost includes: derrick, boiler and engine, casing, tubing, belting, sucker rods, etc., driller's costs (contract drilling price), tank, and in some cases torpedoes.

† Per foot cost figures for the Bradford field are from Peckham, *Census Report*, 142–48. They are considerably lower because of differences in allocating casing, boiler, and engine costs, and partly because of a decline in prices.

Sources: Cone & Johns, *Petrolia*, 463; *Derrick's Hand-Book*, I, 153, 278; *Oil and Gas Journal*, XXXIII (Aug. 23, 1934), 59, 59.1.

APPENDIX TABLE F:4

Drilling costs, Bradford and White Sands Fields: 1883–1887

Field/cost breakdown	Total cost
White Sands Cost of drilling, completing and putting to pumping 6,888 producing wells including bonuses paid for leases viz., $5,300 each.	$36,506,400
Cost of drilling 1,564 dry holes, including estimated cost of bonuses and refusals for leases, less value of machinery, etc., moved away at $3,200 each.	$ 5,004,800
Bradford-Allegany Field, 1887 Cost of drilling and completing 190 wells including bonuses paid for leases at $3,350 each.	$ 674,500

Source: *1888 Trust Investigation*, 747, 748.

APPENDIX TABLE F:5

Percentage distribution of 231 wells drilled in Pennsylvania: 1886–88

Depth class/ft	No. of wells	Distribution/%
less than 1000	23	9%
1000–1499	17	8
1500–1999 *	87	37
2000–2499	73	32
2500–2999	20	9
3000 and deeper	11	5

* Average depth: 1900 feet.

Source: This sample distribution was drawn from records in Carll, *Geological Survey of Penn., 7th Report on the Oil & Gas Fields of W. Penn.,* for 1887, 1888, 147–323.

APPENDIX TABLE F:6

Dry holes as a percentage of total wells drilled: Appalachian and Lima-Indiana Fields, 1876–1900

Year	Dry holes/per cent of total wells	
	Appalachian	Lima-Indiana
1876	11	
1877	17	
1878	11	
1879	5	
1880	3	
1881	5	
1882	5	
1883	9	
1884	11	
1885	13	
1886	17	27
1887	26	12
1888	25	6
1889	17	11
1890	17	13
1891	20	16
1892	25	15
1893	22	15
1894	23	15
1895	22	13
1896	24	13
1897	26	16
1898	26	12
1899	22	11
1900	24	12

Source: Computed from Arnold and Kemnitzer, *Petroleum in the U. S.* 70, 231.

Endnotes

Chapter 1

1 Thomas A. Gale, *The Wonder of the 19th Century! Rock Oil, in Pennsylvania and Elsewhere* (Erie, Pa.: Sloan and Griffeth, 1860).

2 Much of the material for this section on petroleum in antiquity has been drawn from R. J. Forbes, "Fifteen Centuries of Bitumen (A.D. 300–1860)," *Bitumen* (Jan.–June, 1937) [mimeographed translation from the German, 51 pp.]; R. J. Forbes, "Petroleum and Bitumen in Antiquity," *Ambix,* the Journal of the Society for the Study of Early Chemistry and Alchemy, London, II (1938), 68–92.

3 Forbes, *Ambix,* II (1938), 86; Reginald Campbell Thompson, *Assyrian Medical Texts* (Oxford: Oxford University Press, 1923), *passim.*

4 Forbes, *Ambix,* II (1938), 83; Forbes, *Bitumen* (Jan.–June, 1937), 3.

5 S. F. Peckham, *Production, Technology, and Uses of Petroleum and its Products,* 47th Cong., 2nd Sess., H.R. Misc. Doc. 42, Part 10 (Washington: U. S. Government Printing Office, 1884), 3, hereafter cited as Peckham, *Census Report;* Ralph Arnold and William J. Kemnitzer, *Petroleum in the United States and Possessions* (New York: Harper & Brothers, 1931), xvii, hereafter cited as Arnold & Kemnitzer, *Petroleum in U. S.*

6 Forbes, *Bitumen* (Jan.–June, 1937), 4.

7 Forbes, *Bitumen* (Jan.–June, 1937), 6, 9b; "Alchemy," *Encyclopedia Britannica* (11th ed.), I, 519–22.

8 Forbes, *Bitumen* (Jan.–June, 1937), 9b.

9 Forbes, *Bitumen* (Jan.–June, 1937), 7, 8, citing Auguste F. Mehren, *Manuel de la Cosmographie du Moyen Age* (Kopenhagen: C. A. Reitzel, 1874).

10 Forbes, *Bitumen* (Jan.–June, 1937), 8, 9.

11 Forbes, *Bitumen* (Jan.–June, 1937), 1–20; Marco Polo, *The Travels* ("Everyman's Library"; London: J. M. Dent & Co., [1926]), Book I, ch. iv & xi.

12 Forbes, *Bitumen* (Jan.–June, 1937), 14n; G. F. Oviedo y Valdes, *Historia general y naturel de los Indias* (Toledo, 1526; modern edition, Madrid, 1853).

13 Forbes, *Bitumen* (Jan.–June, 1937), 8n; Nicholas Monardes, *Joyfull Newes out of the Newe Founde Worlde,* trans. John Frampton, I (Stephen Gaselee ed., "The Tudor Translations," 2nd series, IX; London: Constable and Co. Ltd., 1925), 19.

14 Sir Walter Raleigh, "The Discovery of the Large, Rich, and Beautiful Empire of Guiana," in *Hakluyt's Voyages,* VII ("Everyman's Library"; London: J. M. Dent & Co., n.d.), 282.

15 Fray Bernardino de Sahagun, *Histoire General des Chose de la Nouvelle Espagnus* (Histoire general de Nueva Espana; Trad. franc. de Jourdanet et Simeon, Paris, 1880), 630, 631; Forbes, *Bitumen* (Jan.–June, 1937), 17, 18.

16 Forbes, *Bitumen* (Jan.–June, 1937), 18. For "phantastic" accounts uncritically reported, see E. G. Woodruff, "Natural Asphalts in Use Long Before They were Refined from Crude Oil," *Oil & Gas Journal* (March 28, 1935), 32, 33.

17 Forbes, *Bitumen* (Jan.–June, 1937), 19, 20; Monardes, *Joyfull Newes out of the Newe Founde Worlde,* II, 24.

18 *The Derrick's Hand-Book of Petroleum: A Complete Chronological and Statistical Review of Petroleum Developments from 1859 to 1898,* I (Oil City, Pa.: Derrick Publishing Company, 1898), 6, hereafter cited as *Derrick's Hand-Book.*

19 *Derrick's Hand-Book,* I, 6.

20 *General Map of the Middle British Colonies in America,* published in 1755. Max Savelle, *Seeds of Liberty: The Genesis of the American Mind* (New York: Alfred A. Knopf, 1948), 133–34.

21 Benjamin Franklin, *Autobiographical Writings,* ed. Carl Van Doren (2 vols.; New York: Viking Press, 1945), I, 298–306.

22 Benjamin Vincent, *Haydn's Dictionary of Dates* (New York: G. P. Putnam's Sons, 1889), 647.

23 *Derrick's Hand-Book,* I, 6, 7.
24 *Derrick's Hand-Book,* I, 7.
25 *Derrick's Hand-Book,* I, 7.
26 J. R. Dodd, *West Virginia, Farms and Forests, Mines and Oil-Wells* (Philadelphia: J. B. Lippincott Co., 1865), 244.
27 Dr. S. P. Hildreth, "Observations on the Saliferous Rock Formation, in the Valley of the Ohio," *The American Journal of Science and Arts,* XXIV (July, 1833), 64; Joseph C. G. Kennedy, *Preliminary Report on the Eighth Census, 1860,* House Executive Document, No. 116, 37 Cong., 2 sess. (Washington: Government Printing Office, 1862), 71, also attributes the name "genesee oil" to early discoveries of oil by the Seneca Indians near the head of the Genesee River, in New York.
28 W. H. Whitmore, *The Results of the Destructive Distillation of Bituminous Substances: A Report Presented to the Annual Meeting of the American Pharmaceutical Association at New York, September 10, 1860* (Boston: Henry W. Dutton & Son, 1860), 4, hereafter cited as Whitmore, *Destructive Distillation.*
29 Fortescue Cuming, *Sketches of a Tour to the Western Country, Through the States of Ohio and Kentucky; a Voyage Down the Ohio and Mississippi Rivers, and a Trip Through the Mississippi Territory and Part of West Florida* (Pittsburgh: Cramer, Spear, & Eichbaum, 1810), 84, hereafter cited as Cuming, *Sketches.*
30 Benjamin Silliman, "Notice of a Fountain of Petroleum, called the Oil Spring," *The American Journal of Science and Arts,* XXIII (Jan., 1833), 98.
31 Cited by Sherman Day, *Historical Collections of the State of Pennsylvania* (Philadelphia, 1843), 250.
32 Whitmore, *Destructive Distillation,* 4.
33 Cuming, *Sketches,* 13; William J. Buck, "Early Accounts of Petroleum in the United States," *Engineering and Mining Journal,* XXI (June 17, 1876), 588.
34 Buck, *Engineering & Mining Journal,* XXI (June 17, 1876), 588.
35 Ella Lonn, *Salt as a Factor in the Con-*

federacy (New York: Walter Neale, 1933), 14–17.
36 J. P. Hale, "Salt," *Resources and Industries of West Virginia,* ed. Professor M. F. Maury (Charleston, W. Va., 1876), 7–9, hereafter cited as Hale, "Salt," *Resources & Industries of W. Va.*
37 Hale, "Salt," *Resources & Industries of W. Va.,* 9–10.
38 Samuel W. Tait, Jr., *The Wildcatters: An Informal History of Oil-Hunting in America* (Princeton, N. J.: Princeton Univ. Press, 1946), 49; Hale, "Salt," *Resources & Industries of W. Va.,* 11, 12, 14.
39 Hale, "Salt," *Resources & Industries of W. Va.,* 13; Charles A. Whiteshot, *The Oil-Well Driller* (Mannington, W. Va.: By the author, 1905), 66.
40 Hale, "Salt," *Resources & Industries of W. Va.,* 15; Paul H. Giddens, *The Birth of the Oil Industry* (New York: The Macmillan Company, 1938), 6; Tait, *The Wildcatters,* 52.
41 Hildreth, *American Journal of Science & Arts,* XXIV (July, 1833), 64.
42 Hale, "Salt," *Resources & Industries of W. Va.,* 12.
43 Cf. Thomas F. Gordon, *Gazetteer of the State of New York* (Philadelphia: By the author, 1836), 356.
44 Hildreth, *American Journal of Science & Arts,* XXIV (July, 1833), 54, 55; Lonn, *Salt as a Factor in the Confederacy,* 20.
45 Hale, "Salt," *Resources & Industries of W. Va.,* 15; Tait, *The Wildcatters,* 49–50.
46 Cf. Peckham, *Census Report,* 8.
47 C. L. S. Mathews, *Burkesville* (Kentucky) *Courier,* Oct. 11, 1876.
48 James B. Sayers, "Pennsylvania, 1858–1948," *Conservation of Oil and Gas: A Legal History,* 1948, ed. Blakely M. Murphy (Chicago: Section of Mineral Law, American Bar Association, 1949), 425–26.
49 Giddens, *The Birth of the Oil Industry,* 24; *Derrick's Hand-Book,* I, 1034.
50 *Derrick's Hand-Book,* I, 1034.
51 J. T. Henry, *The Early and Later History of Petroleum, with Authentic Facts in Regard to its Development in Western Pennsylvania* (Philadelphia: James B. Rodgers Company, 1873),

56, 57, hereafter cited as J. T. Henry, *Early History*.
52 *Derrick's Hand-Book*, I, 1034.
53 *Kier v. Peterson*, 41 Pa. 357 (1861); *Lewis Peterson v. Samuel Kier*, 2 Pittsburgh 191, at 193 (1860); 154 January Term, 1859, Court of Common Pleas, No. 2, *Allegheny County, Pa., Appearance Docket 33*, Paper Books, Allegheny County Law Library, Pa. Supreme Court, 1860, Vol. 4, cited in Sayers, *Conservation of Oil & Gas*, ed. Murphy, 425–26.
54 *Derrick's Hand-Book*, I, 1034; J. T.

Henry, *Early History*, 58.
55 Cf. Giddens, *The Birth of the Oil Industry*, insert, 24–25.
56 J. T. Henry, *Early History*, 83.
57 Interview with George H. Bissell, in S. S. Hayes, *Report of the United States Revenue Commission on Petroleum as a Source of National Revenue*, Special Report No. 7, 39th Cong., 1st Sess., House Ex. Doc. 51, 1866, 4–5, hereafter cited as S. S. Hayes, *Report . . . on Petroleum*.
58 S. S. Hayes, *Report . . . on Petroleum*, 4.

Chapter 2

1 F. R. Moulton and J. J. Schiffers, *The Autobiography of Science* (Garden City, N. Y.: Doubleday and Co., 1945), 228.
2 H. Butterfield, *The Origins of Modern Science: 1300–1800* (London: C. Bells and Sons, 1950), 185; William Cecil Dampier, *A History of Science and Its Relations with Philosophy and Religion* (New York: The Macmillan Co., 1949), 183, hereafter cited as Dampier, *History of Science*.
3 M. E. Weeks, *Discovery of the Elements* (6th ed.; Easton, Pa.: Journal of Chemical Education, 1956), 227.
4 Dampier, *History of Science*, 184.
5 A. R. Hall, *The Scientific Revolution 1500–1800: The Formation of the Modern Scientific Attitude* (Boston: The Beacon Press, 1956), I, 335–36, hereafter cited as Hall, *The Scientific Revolution*.
6 *Scientific American*, II (Feb. 20, 1847), 174; *Scientific American*, VIII (Dec. 18, 1852), 112.
7 Walter S. Tower, *A History of the American Whale Fishery* (Philadelphia: University of Pennsylvania, 1907), 26–28, 106–108, hereafter cited as Tower, *American Whale Fishery*.
8 "The Spermaceti Works of the Last Century," *Hunt's Merchants' Magazine and Commercial Review*, XXXII (March, 1855), 385–87.
9 Quincy C. A. Norton, "Lamp," *The Encyclopedia Americana* (New York:

Americana Corporation, 1942), XVI, 680–81.
10 Norton, "Lamp," *Encyclopedia Americana*, XVI, 679.
11 "New Lamps for Old Ones," *The Living Age* (April 18, 1863), 106.
12 R. J. Forbes, *Short History of the Art of Distillation: From the Beginnings up to the Death of Cellier Blumenthal* (Leiden: E. J. Brill, 1948), 224–25.
13 William Haynes, *The American Chemical Industry: Background and Beginnings* (6 vols.; New York: D. Van Nostrand Co., Inc., 1954), I, 129.
14 Haynes, *American Chemical Industry*, I, 129–30.
15 *Oil, Paint, and Drug Reporter* (April 26, 1862), 816, hereafter cited as *OP&D Reporter*.
16 J. Leander Bishop, *A History of American Manufactures from 1608 to 1860* (Philadelphia: Edward Young & Co., 1868), II, 351; "Burning Fluids," *Scientific American*, XI (Aug. 2, 1856), 374.
17 William Hamilton, "Report on the Carcel, or Mechanical Lamp," *Journal of the Franklin Institute*, XXXV (Jan., 1843), 105–109; "Safety Fluid Lamp—Chemical Cause of Explosion," *Scientific American*, VIII (Dec. 18, 1852), 112.
18 Haynes, *American Chemical Industry*, I, 147–50.
19 *Scientific American*, VI (Aug. 23, 1851), 386; J. B. D. DeBow, *The In-*

dustrial Resources of the South and Western States (New Orleans: Office of DeBow's Review, 1852), I, 377–79.

20 Hamilton, *Journal of the Franklin Institute,* XXXV (Jan., 1843), 105–109.

21 Haynes, *American Chemical Industry,* I, 128.

22 Thomas Antisell, *The Manufacture of Photogenic or Hydro-Carbon Oils, from Coal and Other Bituminous Substances, Capable of Supplying Burning Fluids* (New York: D. Appleton & Co., 1859), 10.

23 M. Selligue, "Applied Chemistry: The Use of Oil Extracted for the Manufacture of Illuminating Gas," *Comptes Rendues des Séances de l'Académie des Sciences,* IV (Jan.–June, 1837), 969–70.

24 Messrs. Thenard, d'Ancet and Dumas, "Report on Selligue's Paper on New Processes for Manufacturing Illuminating Gases," *Comptes Rendues des Séances de l'Académie des Sciences,* X (Jan.–June, 1840), 861–65.

25 Frank H. Storer, "Review of Dr. Antisell's Work on Photogenic Oils," *American Journal of Science and Arts,* XXX (Sept., 1860), 259–61.

26 Cf. Copy of patent issued to James Young, October 7, 1850, reprinted in *Scientific American,* XIV (Feb. 12, 1859), 185.

27 H. R. J. Conacher, "History of the Scottish Oil-Shale Industry," Geological Survey, Scotland, *The Oil-Shales of the Lothians* (London: His Majesty's Stationery Office, 1927), 241.

28 Haynes, *American Chemical Industry,* I, 129–130.

29 "Gas Light Companies in the United States," *American Gas-Light Journal,* IV (June 15, 1863), 373.

30 Cf. "Gas Light in Factories," *Scientific American,* VII (Dec. 6, 1851), 93.

31 Frederick L. Collins, *Consolidated Gas Company of New York, A History* (New York: The Consolidated Gas Company, 1934), 38, 39, 41–45.

32 Haynes, *American Chemical Industry,* I, 129.

33 "Annual Meeting of the Gas-Light Association," *American Gas-Light Journal,* I (Dec. 15, 1860), 186–87.

34 *American Gas-Light Journal,* I (Dec. 15, 1860), 186–87; "New York Gas," *Scientific American,* V (Sept. 29, 1849), 13.

35 Louis Stotz and Alexander Jamison, *History of the Gas Industry* (New York: Stettiner Bros., 1938), 241, 242; Collins, *Consolidated Gas Company of New York,* 61.

36 *American Gas-Light Journal,* IV (June 15, 1863), 370–74; "Hints to Gas Consumers," *Scientific American,* II (Feb. 20, 1847), 174.

37 "Give us Cheap Gas," *Scientific American,* VIII (Dec. 11, 1852), 101.

38 "Tar Oils," *Scientific American,* XIV (Dec. 18, 1858), 118; Antisell, *Manufacture of Photogenic or Hydro-Carbon Oils,* 9, 10.

39 C. B. Mansfield, "On the Application of Certain Liquid Hydro-Carbons to Artificial Illumination," A Paper Read at a Meeting of the Institution of Civil Engineers in London, April 17, 1850, cited in *Annual of Scientific Discovery,* I (1850), 191–92.

40 Mansfield, *Annual of Scientific Discovery,* I (1850), 191–92.

41 "Standard Gas, Carburetting, etc.," *Scientific American,* XVIII (Oct. 2, 1866), 104.

42 "Spirit Gas Lamps," *Scientific American,* V (Jan. 26, 1850), 145; "Patent Safety Spirit Lamp," *Scientific American,* VI (Oct. 5, 1850), 24; "Safety Lamp," *Scientific American,* VIII (Nov. 27, 1852), 82.

43 "Gas Burner Lamps," *Scientific American,* III (July 2, 1860), 6, 7; "Lamps," *Appleton's Dictionary of Machines, Mechanics, Engine-Work and Engineering,* II (New York: D. Appleton and Company, 1867), 180–81.

Chapter 3

1 President's Annual Address, *Bulletin of the Natural History Society of New Brunswick,* IX (1890), 34.

2 G. W. Gesner, "Dr. Abraham Gesner, A Biographical Sketch," *Bulletin of the Natural History Society of New*

Brunswick, XIV (1896), 5, 7; Kendall Beaton, "Dr. Gesner's Kerosene: The Start of American Oil Refining," *The Business History Review,* XXIX (March, 1955), 33.

3 Beaton, *The Business History Review,* XXIX (March, 1955), 33–35; Abraham Gesner, *A Practical Treatise on Coal, Petroleum and Other Distilled Oils* (New York: Balliere Brothers, 1861), 8, 9, 74, hereafter cited as A. Gesner, *Treatise* (1861).

4 Obituary, Earl of Dundonald, *Scientific American,* III (Nov. 24, 1860), 346–47; Lord Thomas Cochrane, 11th Earl of Dundonald, and H. R. Foxbourne, *The Life of Thomas, Lord Cochrane, 10th Earl of Dundonald, G.C.B.* (2 vols.; London: R. Bentley, 1869), II, 307 ff.; see also Lord Thomas Cochrane, 10th Earl of Dundonald, *Autobiography of a Seaman* (2 vols.; London: R. Bentley, 1860–61).

5 U. S. Patent No. 7052, "Manufacture of Illuminating Gas from Bitumen," issued Jan. 29, 1850; "New Kind of Gas," *Scientific American,* V (Feb. 9, 1850), 164.

6 A. Gesner, article addressed to the Academy of Natural Sciences, *Scientific American,* V (Feb. 16, 1850), 172.

7 Anon. (probably G. W. Gesner or John Frederick Gesner), "Kerosene—The Origin of the Name, The History of a Great Industry Years Past, and the Possibility of its Revival," *The Engineering and Mining Journal* (Feb. 9, 1884), 99–100.

8 A. Gesner, *Scientific American,* V (Feb. 16, 1850), 172.

9 "Kerosene Gas—Nova Scotia Going-A-Head of 'Old Mother,'" *Scientific American,* VI (Dec. 7, 1850), 89. A search of the patent literature has failed to reveal any issue from this application, but it probably was for a patent on the machine to carburet atmospheric air with kerosene gas.

10 Beaton, *The Business History Review,* XXIX (March, 1955), 36, 37.

11 A. Gesner, *Treatise* (1861), 22; G. F. Matthew, "Abraham Genser, A Review of His Scientific Work," *Bulletin of the Natural History Society of New Brunswick,* XV (1897), 48.

12 Matthew, *Bulletin of the Natural History Society of New Brunswick,* XV (1897), 48; Storer, *American Journal of Science,* XXX (Sept., 1860).

13 Beaton, *The Business History Review,* XXIX (March, 1955), 37–39.

14 Horatio Eagle, Assigner of Patents, *Project for the Formation of a Company to Work Under the Combined Patent Rights (For the State of New York) of Dr. Abraham Gesner, of Halifax, N. S., And the Right Hon. The Earl of Dundonald of Middlesex, England* (New York: By the author, circa 1853), 1–8, pamphlet in the Bella C. Landauer Collection, New York Historical Society, New York, N. Y., hereafter cited as Eagle, *Prospectus.*

15 Beaton, *The Business History Review,* XXIX (March, 1955), 39.

16 U. S. Patents, Nos. 11,203, 11,204, 11,205.

17 Eagle, *Prospectus,* 2.

18 *Scientific American,* VIII (Dec. 11, 1852), 101.

19 Circular letter, "Kerosene Oils, Distilled from Coals. Not Explosive—Secured by Letters Patent," in Bella C. Landauer Collection, New York Historical Society.

20 Advertisement, "Kerosene or Coal Oil," *American Railroad Journal,* XXIX (March 1, 1856), 141.

21 *Engineering and Mining Journal* (Feb. 9, 1884), 99.

22 Albert Norton Leet, *Petroleum Distillation and Modes of Treating Hydrocarbons* (New York: Oil, Paint & Drug Publishing Co., 1884), 107, hereafter cited as Leet, *Petroleum Distillation;* A. C. F. [A. C. Ferris], "Petroleum Reminiscences and Early Oil Refining," *OP&D Reporter* (March 20, 1889), 40.

23 Merrill's "Reminiscences," *Derrick's Hand-Book,* I, 883.

24 G. W. Gesner, *Bulletin of the Natural History Society of New Brunswick,* XIV (1896), 7–9.

25 U. S. Patent No. 9,630, isssued March 29, 1853.

26 Merrill's "Reminiscences," *Derrick's Hand-Book,* I, 890.

27 Joshua Merrill's obituary, *Boston Evening Transcript* (Jan. 15, 1904), 3.

28 Merrill's "Reminiscences," *Derrick's Hand-Book*, I, 880, 881, 889, 890.

29 Testimony of William Atwood, *Merril v. Yeomans*, cited by Peckham, *Census Report*, 10.

30 Merrill's "Reminiscences," *Derrick's Hand-Book*, I, 881–82.

31 Merrill's "Reminiscences," *Derrick's Hand-Book*, I, 881.

32 Edwin M. Bailey, "The Dawn of Petroleum Refining," *Institute of Petroleum Review*, II (1948), 359.

33 Iltyd I. Redwood, *A Practical Treatise on Mineral Oils and Their By-Products* (London: E. & F. N. Spon, Ltd., 1897), 8, 21, 22, hereafter cited as I. I. Redwood, *Mineral Oils & Their By-Products*.

34 Merrill's "Reminiscences," *Derrick's Hand-Book*, I, 881.

35 A. Gesner, *Treatise* (1861), 106, 108, 111–18.

36 See Schieffelin Bros. Circular, reprinted in Gale, *Wonder of the 19th Century*, appendix.

37 See A. Gesner, *Treatise* (1861), 20–34; for the highest still temperatures normally applied in obtaining the burning oil portion of the run, see U. S. Patent No. 28,448 issued to Luther Atwood, May 29, 1860.

38 See U. S. Patent No. 28,448 issued to Luther Atwood, May 29, 1860.

39 S. Dana Hayes, "On the History and Manufacture of Petroleum Products: A Memoir Communicated to the Society of Arts, Massachusetts Institute of Technology, March 14, 1872," reprinted in the *American Chemist*, II (May, 1872), 401.

40 A. Gesner, *Treatise* (1861), 21, 23, 24, 34.

41 S. D. Hayes, *American Chemist*, II (May, 1872), 401; Merrill's "Reminiscences," *Derrick's Hand-Book*, I, 882–85.

42 S. D. Hayes, *American Chemist*, II (May, 1872), 401; Merrill's "Reminiscences," *Derrick's Hand-Book*, I, 885; U. S. Patent No. 28,448 issued to Luther Atwood May 29, 1860. For a more extended discussion of "cracking" see Chapter 9.

43 Arnold R. Daum, "The Illumination Revolution and the Rise of the Petroleum Industry, 1850–1863" (unpublished Ph.D. dissertation, Dept. of History, Columbia University, 1957), 393–427.

44 Boston Kerosene Oil Company Circular, Nov. 10, 1858, W. H. L. Smith Papers, Pennsylvania Historical Society; Merrill's "Reminiscences," *Derrick's Hand-Book*, I, 882.

45 Boston Kerosene Oil Co. Circular, Nov. 10, 1858, 2; "Sale of the Kerosene Oil-Works," *American Gas-Light Journal*, I (June 1, 1860), 250.

46 Merrill's "Reminiscences," *Derrick's Hand-Book*, I, 884.

47 Merrill's "Reminiscences," *Derrick's Hand-Book*, I, 884.

48 (June 7, 1856), 312.

49 Charles M. Wetherill, "On the Relative Costs of Illumination in Lafayette, Ind.," *American Gas-Light Journal*, I (June, 1860), 246, 247.

50 "The Manufacture of Coal Oil," *Scientific American*, XIII (April 17, 1858), 254.

51 Bishop, *History of American Manufactures*, III, 311.

52 "Coal Oil Manufacture," *Scientific American*, II (Jan. 2, 1860), 3.

53 *Scientific American*, II (Jan. 2, 1860), 3.

54 *Annual Statement of the Trade and Commerce of Cincinnati: For the Commercial Year Ending August 31, 1860* (Cincinnati: n.p., 1860), 29–30.

55 S. Kussart, *Allegheny Valley* (Pittsburgh: Burgum Printing Co., 1938), 203.

56 Cf. "Young's Paraffin Oil Patent," *Scientific American*, XII (Feb. 25, 1865), 135; Antisell, *The Manufacture of Photogenic or Hydro-Carbon Oils*, 15.

57 *Scientific American*, XI (Aug. 2, 1856), 374.

58 "Horrors of Burning Fluid," *Scientific American*, II (Jan. 21, 1860), 51; "Burning Fluid, Camphene," *Medical News*, XVII (Nov., 1859), 237.

59 "The Manufacture of Coal Oil—The First Patent," *Scientific American*, XIV (Feb. 19, 1859), 186.

60 *Scientific American*, II (Jan. 2, 1860), 3.

61 Cf. Daum, "The Illumination Revolution and the Rise of the Petroleum Industry," chapter xi.

62 *Scientific American*, XI (June 14, 1856), 374.

63 Merrill's "Reminiscences," *Derrick's Hand-Book*, I, 883; U. S. Patents No. 16,981, issued April 1857, 23,160, issued March 1858, and 29,260, issued May 1859.

64 "Coal Oil and Its Manufactures," *American Gas-Light Journal*, II (Dec. 1, 1860), 168; *Scientific American*, II (Jan. 2, 1860), 3.

65 Cf. A. C. F., *OP&D Reporter* (March 20, 1889), 40; A. C. F., "Early Petroleum Refining," *OP&D Reporter* (May 1, 1889), 9–10; *OP&D Reporter* (June 19, 1889), 27–30; Leet, *Petroleum Distillation*, 15–16, 19–21; Daum, "The Illumination Revolution and the Rise of the Petroleum Industry," 149–51, 152–58, 210–24.

66 Wetherill, *American Gas-Light Journal*, I (May 1, 1860 and June 1, 1860), 230–31, 245–47.

67 Antisell, *Manufacture of Photogenic or Hydro-Carbon Oils*, 133–35.

68 Merrill's "Reminiscences," *Derrick's Hand-Book*, I, 884; "The Coal Oil Business," *The Repository* (New London, Connecticut), III, (Oct. 18, 1860), 207, 208; "Coal Oil," *American Gas-Light Journal*, II (Jan. 15, 1861), 215. On manufacturing costs, see Report of Phil T. Ruggles, Receiver, April 20, 1860, cited in "Sale of the Works of the North American Kerosene Gas-Light Company," *American Gas-Light Journal*, I (May 1, 1860), 225.

69 Daum, "The Illumination Revolution and the Rise of the Petroleum Industry," chapters vi and vii.

Chapter 4

1 *Venango Spectator* (Franklin, Pennsylvania), May 30, 1855, in Giddens, *Birth of the Oil Industry*, 15.

2 Statement by Dr. F. B. Brewer, *Titusville Morning Herald*, Jan. 28, 1881, in Paul H. Giddens (comp. and ed.), *Pennsylvania Petroleum, 1750–1872: A Documentary History* (Titusville, Pennsylvania: Pennsylvania Historical and Museum Commission, 1947), 45–46, hereafter cited as Giddens, *Pa. Petroleum*.

3 Interview with E. L. Drake, *Philadelphia Times*, reprinted in *Titusville Morning Herald*, Sept. 11, 1879.

4 *Titusville Morning Herald*, Jan. 28, 1881, cited in Giddens, *Birth of the Oil Industry*, 31.

5 *Titusville Morning Herald*, Jan. 28, 1881, cited in Giddens, *Birth of the Oil Industry*, 31.

6 George H. Bissell's account, Feb. 1866, in S. S. Hayes, *Report . . . on Petroleum*, 4.

7 Giddens, *Birth of the Oil Industry*, 32.

8 J. T. Henry, *Early History*, 65–67.

9 Giddens, *Pa. Petroleum*, 35–36.

10 B. Silliman, *The American Journal of Science and Arts*, XXIII (Jan., 1833), 100–101.

11 Giddens, *Birth of the Oil Industry*, 37.

12 Eveleth to Dr. Brewer, Feb. 8, 1855, Brewer Papers, reprinted in Giddens, *Birth of the Oil Industry*, 38.

13 Ebenezer Brewer to Dr. Brewer, March 23, 1855, Brewer Papers.

14 Eveleth and Bissell to Dr. Brewer, Nov. 6, 1854, Brewer Papers.

15 Charles Richmond, Jr., to Dr. Brewer, April 20, 1855, Sheldon to Dr. Brewer, April 23, 1855, Brewer Papers.

16 Eagle, *Prospectus*; B. Silliman, Jr., *Report on the Rock Oil, or Petroleum, From Venango Co., Pennsylvania, with Special Reference to its Use for Illumination and other Purposes* (New Haven, Conn.: J. H. Benham's Steam Power Press, 1855), hereafter cited as B. Silliman, Jr., *Report on Rock Oil*. Gesner outlined many uses for tar, pitch, and related products ignored by Silliman: hydraulic cement, road-building material, composition for roofing, etc.

17 B. Silliman, Jr., *Report on Rock Oil*, 8, 12.

18　B. Silliman, Jr., *Report on Rock Oil*, 11.

19　B. Silliman, Jr., *Report on Rock Oil*, 14.

20　B. Silliman, Jr., *Report on Rock Oil*, 9, 10, 20.

21　Billings & Marsh, New London, Connecticut, Advertisement, *Scientific American*, II (June 16, 1860), 399.

22　B. Silliman, Jr., *Report on Rock Oil*, 8, 13.

23　B. Silliman, Jr., *Report on Rock Oil*, 9.

24　J. T. Henry, *Early History*, 75–78.

25　Giddens, *Birth of the Oil Industry*, 41.

26　Giddens, *Birth of the Oil Industry*, 42–43.

27　Eveleth to Dr. Brewer, June 25, 1855, Brewer Papers.

28　Giddens, *Birth of the Oil Industry*, 45.

29　Brewer to Bissell, Oct. 1855, cited in J. T. Henry, *Early History*, 80.

30　Brewer to Bissell, Oct. 1855, cited in J. T. Henry, *Early History*, 80.

31　J. T. Henry, *Early History*, 84–85.

32　J. T. Henry, *Early History*, 85; Townsend to Brewer, Watson & Company, Jan. 8, 1858, Townsend Papers.

33　J. T. Henry, *Early History*, 84–85. Personal recollections of George H. Bissell reported by S. S. Hayes in 1866 make no mention of the drugstore incident in 1856. They mention only that in 1858 the partners heard that Samuel Kier was getting his rock oil medicinal from a Tarentum salt well, and that they decided to sink an artesian well for petroleum, hiring Drake for the purpose. This version denies Drake's claim to having originated the idea of artesian boring for petroleum.

34　Townsend Collection #413, Drake Museum, 19, reprinted in Giddens, *Pa. Petroleum*, 60. The account is undated, supposedly supervised by James M. Townsend in its preparation.

35　Giddens, *Pa. Petroleum*, 57.

36　Giddens, *Pa. Petroleum*, 57; J. T. Henry, *Early History*, 87.

37　Giddens, *Pa. Petroleum*, 55–56; Giddens, *Birth of the Oil Industry*, 48–49.

38　Giddens, *Birth of the Oil Industry*, 49; J. T. Henry, *Early History*, 86.

39　J. T. Henry, *Early History*, 87.

40　Colonel E. L. Drake, MS, in Drake Museum, reprinted in Giddens, *Pa. Petroleum*, 66; interview with Drake, *Titusville Morning Herald*, July 27, 1866; interview with Drake, *Philadelphia Times*, Sept. 11, 1879.

41　Lease of the Pennsylvania Rock Oil Company of Pennsylvania to Drake and Bowditch, cited in Giddens, *Birth of the Oil Industry*, 52. See also Eveleth & Bissell to Townsend, Dec. 30, 1857, Townsend to Messrs. Brewer and Watson, Jan. 8, 1858, Townsend to Bissell, Jan. 14, 1858, Townsend Papers, reprinted in Giddens, *Pa. Petroleum*, 135–39.

42　Minute Book of the Seneca Oil Company, April 1, 1858. This entry along with other selected entries form the Minute Book, Townsend Collection #236, Drake Museum, are reproduced in Giddens, *Pa. Petroleum*, 139–54. See also Giddens, *Birth of the Oil Industry*, 52–53; J. T. Henry, *Early History*, 87–89.

43　Giddens, *Birth of the Oil Industry*, 53–54.

44　Giddens, *Birth of the Oil Industry*, 54; Drake's account, MS [circa 1870] in Drake Museum, 5.

45　*Titusville Morning Herald*, Oct. 23, 1868, cited in Giddens, *Birth of the Oil Industry*, 55.

46　Giddens, *Birth of the Oil Industry*, 56.

47　Seneca Oil Company account with Colonel Drake from March, 1858 to February, 1859, Townsend Papers, Drake Museum.

48　Giddens, *Birth of the Oil Industry*, 56–57; Interview with Drake, *Titusville Morning Herald*, July 27, 1866; Drake's account, MS [circa 1870] in Drake Museum; interview with Drake, *Philadelphia Times*, Sept. 1879.

49　Drake account, MS [circa 1870] in Drake Museum.

50　Giddens, *Birth of the Oil Industry*, 57.

51　Giddens, *Birth of the Oil Industry*, 57. Drake claimed that he advocated the idea of driving the iron tube down. See accounts based on interviews with Drake, *Titusville Morning Herald*, July 27, 1866; *Titusville Morning Herald*, Sept. 11, 1879, quoted in Giddens, *Pa. Petroleum*, 63, 70. William "Uncle Billy" Smith claimed that he originated the idea. See interview with William A. Smith, *Titusville Morning Herald*, Jan. 15, 1880, quoted in Giddens, *Pa. Petroleum*, 72–80.

52 Interview with Colonel E. L. Drake, *Philadelphia Times*, reprinted in *Titusville Morning Herald*, Sept. 11, 1879, quoted in Giddens, *Pa. Petroleum*, 71.

53 Edwin C. Bell, *History of Petroleum* (Titusville, Pa.: The Bugle Print, 1900), 21, cited in Giddens, *Birth of the Oil Industry*, 57.

54 Gale, *Wonder of the 19th Century*, 66–67.

55 Report of the agent of the Seneca Oil Co. for the quarter ending Aug. 31, 1859, Townsend Papers; Giddens,

Birth of the Oil Industry, 58.

56 James M. Townsend's account, Townsend Collection #413, Drake Museum, quoted in Giddens, *Pa. Petroleum*, 57.

57 Giddens, *Birth of the Oil Industry*, 59; interview with Samuel B. Smith, *Oil City Derrick*, Aug. 27, 1909.

58 Giddens, *Birth of the Oil Industry*, 59; interview with Samuel B. Smith, *Oil City Derrick*, Aug. 27, 1909.

59 Giddens, *Birth of the Oil Industry*, 69–70; James Dodd Henry, *History and Romance of the Petroleum Industry* (3 vols.; London: Bradbury, Agnew and Co., 1914), I, 153–55, hereafter cited as J. D. Henry, *History & Romance*.

60 See Giddens, *The Birth of the Oil Industry*, 65–68.

Chapter 5

1 J. T. Henry, *Early History*, 222.

2 George Rogers Taylor and Irene D. Neu, *The American Railroad Network, 1861–1890* (Cambridge, Massachusetts: Harvard University Press, 1956), 27–29.

3 Rolland Harper Maybee, *Railroad Competition and the Oil Trade, 1855–1873* (Mount Pleasant, Michigan: The Extension Press, 1940), 183, hereafter cited as Maybee, *Railroad Competition*.

4 Taylor & Neu, *The American Railroad Network*, 32, 38.

5 J. T. Henry, *Early History*, 92, 95–98, 108, 109.

6 Andrew Cone and Walter R. Johns, *Petrolia: A Brief History of the Pennsylvania Petroleum Region, Its Development, Growth, Resources, Etc., from 1859 to 1869* (New York: D. Appleton & Company, 1870), 71, 93, hereafter cited as Cone & Johns, *Petrolia*.

7 *Derrick's Hand-Book*, I, 149.

8 J. T. Henry, *Early History*, 97–98, 108–9. S. S. Hayes, *Report . . . on Petroleum*, 5; Cone & Johns, *Petrolia*, 221–22.

9 J. T. Henry, *Early History*, 349, 350.

10 William Kirkpatrick's account, circa

1912, in J. D. Henry, *History and Romance*, I, 159–60.

11 John J. McLaurin, *Sketches in Crude Oil* (Harrisburg: By the author, 1896), 100.

12 Victor S. Clark, *History of Manufactures in the United States: 1860–1915* (Washington, D. C.: The Carnegie Institution of Washington, 1928), II, 124.

13 *Derrick's Hand-Book*, I, 954, 990–91.

14 Tait, *The Wildcatters*, 74–77; "Petroleum—Its Sources—Various Theories," *Scientific American*, VII (July 19, 1862), 37; "Petroleum—Report of a Visit to the New Petroleum Region of Southwestern Pa.," *American Gas-Light Journal*, VI (Nov. 16, 1864), 145–46.

15 S. H. Stowell, "Petroleum," U. S., Department of the Interior, U. S. Geological Survey, *Mineral Resources of the United States* (Washington: Government Printing Office, 1883), 191–92, hereafter cited as *Mineral Resources*, and the specific year.

16 J. T. Henry, *Early History*, 248.

17 McLaurin, *Sketches in Crude Oil*, 121.

18 William Wright, *The Oil Regions of*

Pennsylvania (New York: Harper & Brothers, 1865), 62.

19 *Oil City Register*, Sept. 7, 1865, in Giddens, *Pa. Petroleum*, 297–98.

20 J. H. A. Bone, *Petroleum and Petroleum Wells* (Philadelphia: J. B. Lippincott Company, 1865), 33–46, reprinted in Giddens, *Pa. Petroleum*, 261–67; J. R. G. Hazard, Titusville, dispatch to *New York Daily Tribune*, reprinted in *Titusville Morning Herald*, Aug. 11, 1866, and in Giddens, *Pa. Petroleum*, (344–53), 344.

21 J. T. Henry, *Early History*, 95.

22 Giddens, *The Birth of the Oil Industry*, 70, 71.

23 *Derrick's Hand-Book*, I, 19.

24 H. C. Folger, Jr., "Petroleum, Its Production and Products," Commonwealth of Pennsylvania, *Annual Report of the Secretary of Internal Affairs*, XX, *Industrial Statistics, 1892*, Part 3 (Harrisburg: Edwin K. Meyers, State Printer, 1893), B. 29, hereafter cited as Folger, *Pa. Industrial Statistics, 1892*.

25 *Derrick's Hand-Book*, II, 201–3, 207.

26 *Derrick's Hand-Book*, II, 195–98, 201, 202; Gale, *The Wonder of the 19th Century*, 23.

27 *Derrick's Hand-Book*, II, 195–200, 202. As early as September 17, 1861, Brewer, Watson & Co., Chas. Hyde, Samuel Grandin, L. D. Wetmore, E. T. F. Valentine and L. L. Lowry paid $10,000 for the lease of the undivided half of 416 acres in Deerfield township from William W. Wallace, in addition to one-eighth royalty. *Derrick's Hand-Book*, II, 198.

28 J. T. Henry, *Early History*, 103–10.

29 J. T. Henry, *Early History*, 101, 224.

30 Rev. S. J. M. Eaton, *Petroleum: A History of the Oil Region of Venango County, Pennsylvania* (Philadelphia: J. P. Skelly & Co., 1866), 107, 108.

31 Anonymous, *Boston Traveler*, quoted in Edmund Morris, *Derrick and Drill* (New York: James Miller, 1865), 35, 36; J. T. Henry, *Early History*, 248–49; Eaton, *Petroleum*, 106, 107.

32 Eaton, *Petroleum*, 107, 108.

33 Eaton, *Petroleum*, 103–5, 112–14.

34 Eaton, *Petroleum*, 174, 175; J. T. Henry, *Early History*, 244, 245.

35 Gale, *Wonder of the 19th Century*, 31.

36 Gale, *Wonder of the 19th Century*, 25.

37 Cone & Johns, *Petrolia*, 123.

38 J. T. Henry, *Early History*, 219.

39 Eaton, *Petroleum*, 129–33; McLaurin, *Sketches in Crude Oil*, 332.

40 Gale, *Wonder of the 19th Century*, 32, 33.

41 Anonymous commentator, Sept. 1860, quoted in Morris, *Derrick & Drill*, 39.

42 *Derrick's Hand-Book*, I, 18–20; Bone, *Petroleum & Petroleum Wells*, 47–49.

43 Gale, *Wonder of the 19th Century*, 70, 71; *Derrick's Hand-Book*, I, 18.

44 Giddens, *Birth of the Oil Industry*, 72; *Warren Daily Mail*, Aug. 18, 1860.

45 *Derrick's Hand-Book*, I, 19.

46 *Derrick's Hand-Book*, I, 23.

47 Bone, *Petroleum & Petroleum Wells*, 15; Eaton, *Petroleum*, 178.

48 Gale, *Wonder of the 19th Century*, 72, 73.

49 C. H. Shattock, "Coal Oil in West Virginia," *Report of the Commissioner of Agriculture for the Year 1863* (Washington: G.P.O., 1863), 525–29.

50 *Derrick's Hand-Book*, I, 23.

51 T. S. Scoville, quoted in Morris, *Derrick & Drill*, 42.

52 Folger, *Pa. Industrial Statistics, 1892*, 86B, *Derrick's Hand-Book*, I, 17, reports 1860 production as 200,000 bbls., as does Giddens, *Birth of the Oil Industry*, 75.

53 *Derrick's Hand-Book*, I, 706.

54 Hayes, *Report . . . on Petroleum*, 5, 6.

55 Giddens, *Birth of the Oil Industry*, 83; J. T. Henry, *Early History*, 277.

56 Folger, *Pa. Industrial Statistics, 1892*, B17, 86B. This source overstates the wastage as 10 million barrels, or about five times the estimated output in 1861. Since the highest estimate for the peak rate of production, prevailing only in the fall quarter, is 8000 barrels daily, such wastage could not have been possible even assuming the peak rate of daily production throughout the entire year. Even 10 million gallons wastage (2,-500,000 barrels) strains plausability.

57 *Derrick's Hand-Book,* I, 706, 707.
58 Report of Philo T. Ruggles, Receiver, April 20, 1860, cited in *American Gas-Light Journal,* I (May 1, 1860), 225.
59 William H. Abbott, Centennial edition of Meadville (Pa.) *Daily Tribune Republican* (May 12, 1888), 164–65.
60 Gale, *Wonder of the 19th Century,* 40.
61 Joshua Merrill to W. H. L. Smith, Boston, Jan. 6, 1862. W. H. L. Smith Papers.
62 Giddens, *Pa. Petroleum,* 227.
63 Cone & Johns, *Petrolia,* 98–100; Morris, *Derrick & Drill,* 17, 22, 140, 141; J. R. G. Hazard, *New York Daily Tribune,* reprinted in *Titusville Morning Herald,* Aug. 8, 1868.
64 E. L. Drake to F. A. Townsend, Titusville, August 13, 1861, Townsend Collection #275; R. D. Leonard to Asahel Pierpont, Titusville, March 26, 1860, Townsend Collection #291.
65 *Venango Spectator,* Jan. 15, 1862.
66 R. D. Leonard to Asahel Pierpont, Titusville, March 26, 1860, Townsend Collection #291.
67 See Morris, *Derrick & Drill,* 34–39.
68 Maybee, *Railroad Competition,* 242.
69 Edwin C. Bell, in *Oil City Derrick,* quoted in *The Oil and Gas Journal,* XV (Jan. 4, 1917), 23.
70 McLaurin, *Sketches in Crude Oil,* 138.
71 William H. Abbott, account of first oil refineries, Centennial edition, *Daily Tribune Republican,* Meadville, Pennsylvania (May 12, 1888), 164–65.
72 A. C. Ferris to William A. Ives, New York, March 12, 1860, Townsend Collection #282.
73 *Scientific American,* LXVI (Feb. 27, 1862), 134.
74 *Oil City Register,* Sept. 27, 1866; *Pittsburgh Mining and Manufacturing Journal,* 1866, cited in Cone & Johns, *Petrolia,* 618.
75 *Twenty-Ninth Annual Report of the Philadelphia Board of Trade, for the Year 1861* (Philadelphia: King & Baird, Printers, 1862), 83.
76 *Twenty-Ninth Annual Report of the*

Philadelphia Board of Trade, for the Year 1861, 83–84.
77 *Scientific American,* II (Jan. 2, 1860), 3.
78 *The Repository,* III (Oct. 18, 1860), 207–8.
79 *Annual Statement of the Commerce of Cincinnati . . . for the Year 1860,* 29.
80 *Derrick's Hand-Book,* I, 781.
81 "Petroleum Oils Wanted," *American Gas-Light Journal,* II (Dec. 1, 1860), 168.
82 Samuel Downer to W. H. L. Smith, New York, Oct. 2, 1860, W. H. L. Smith Papers.
83 *Pittsburgh Post,* in the *American Gas-Light Journal,* II (Jan. 15, 1861), 215.
84 A. Gesner, *Treatise* (1861), 128–29; Kennedy, *Preliminary Report on the Eighth Census, 1860,* 78.
85 *Derrick's Hand-Book,* I, 19, 20, 780; *Pittsburgh Post,* in *American Gas-Light Journal,* II (Jan. 15, 1861); Giddens, *Birth of the Oil Industry,* 93; Kennedy, *Preliminary Report on the Eighth Census, 1860,* 78.
86 A. Gesner, *Treatise* (1861), 107, 121.
87 *Oil City Register,* Sept. 27, 1866; *Pittsburgh Mining and Manufacturing Journal,* 1866, cited in Cone & Johns, *Petrolia,* 618.
88 *Twenty-Ninth Annual Report of the Philadelphia Board of Trade, for the Year 1861,* 83.
89 *United States Railroad and Mining Register,* Jan. 14, 1865.
90 Samuel Downer to W. H. L. Smith, New York, Oct. 2, 1860, W. H. L. Smith Papers.
91 M. C. Greene to W. H. L. Smith, Boston, Dec. 30, 1861, W. H. L. Smith Papers.
92 Samuel Downer to W. H. L. Smith, Boston, Jan. 7, 1862, W. H. L. Smith Papers.
93 J. T. Henry, *Early History,* 439.
94 Gale, *Wonder of the 19th Century,* 40.
95 Gale, *Wonder of the 19th Century,* 40.
96 *The Pittsburgh Post,* December 18, 1861; *Derrick's Hand-Book,* I, 991.
97 *Cleveland Leader,* January 5, and 8, 1866; A. C. Ferris, "Petroleum Refin-

ing in the Early Days," *OP&D Reporter* (June 19, 1889), 27, 30.
98 *Derrick's Hand-Book*, I, 19, 20.
99 J. T. Henry, *Early History*, 364–65.
100 *Derrick's Hand-Book*, I, 991.
101 *Derrick's Hand-Book*, I, 27.
102 J. P. Leslie, "Coal Oil," *Report of the Commissioner of Agriculture for the Year 1862*, 37th Cong., 3rd Sess., H.R. Doc. 78 (Washington, D. C.: Government Printing Office, 1863), 44; J. T. Henry, *Early History*, 335–43; Giddens, *Birth of the Oil Industry*, 76–78.

103 J. T. Henry, *Early History*, 225, 331–34.
104 J. T. Henry, *Early History*, 227–28, 233.
105 J. T. Henry, *Early History*, 232–34; *Derrick's Hand-Book*, I, 24; Giddens, *Birth of the Oil Industry*, 80–82.
106 J. T. Henry, *Early History*, 109.
107 Herbert Asbury, *The Golden Flood, An Informal History of America's First Oil Field* (New York: Alfred A. Knopf, 1942), 269.
108 *Derrick's Hand-Book*, I, 711.

Chapter 6

1 For critiques of data, see *Derrick's Hand-Book*, I, 704–11, and S. S. Hayes, *Report . . . on Petroleum*, 5–7, 23–39.
2 See J. T. Henry, *Early History*, 113–39; also Hayes, *Report . . . on Petroleum*, 5, 6, 13, 14.
3 McLaurin, *Sketches in Crude Oil*, 121; Giddens, *Pa. Petroleum*, 315.
4 John F. Carll, Geological Survey of Pennsylvania, *Seventh Report on the Oil and Gas Fields of Western Pennsylvania, for 1887 and 1888* (Harrisburg: Board of Commissioners for the Geological Survey, 1890), 29.
5 *Derrick's Hand-Book*, I, 19.
6 Folger, *Pa. Industrial Statistics, 1892*, B.29.
7 McLaurin, *Sketches in Crude Oil*, 158, 180, 181.
8 Cone & Johns, *Petrolia*, 466, 473, 506.
9 J. T. Henry, *Early History*, 296–98, 304; Asbury, *The Golden Flood*, 280–83.
10 Cf. Giddens, *Birth of the Oil Industry*, 85.
11 *Derrick's Hand-Book*, I, 26, 707, 712.
12 *Venango Spectator*, Nov. 5, 1862; Giddens, *Pa. Petroleum*, 223–24; *Derrick's Hand-Book*, I, 26, 27.
13 *Derrick's Hand-Book*, I, 80, 91.
14 Giddens, *Birth of the Oil Industry* 87.
15 *Derrick's Hand-Book*, I, 707, 711.
16 *Derrick's Hand-Book*, I, 35, 707, 711, 712.
17 J. T. Henry, *Early History*, 225.

18 McLaurin, *Sketches in Crude Oil*, 127–30.
19 *Oil City Register*, Nov. 24, 1864; *Venango Spectator*, Apr. 27, 1864; Giddens, *Birth of the Oil Industry*, 121.
20 *Derrick's Hand-Book*, I, 41; Giddens, *Pa. Petroleum*, 224; Bone, *Petroleum & Petroleum Wells*, 33.
21 McLaurin, *Sketches in Crude Oil*, 149–52.
22 McLaurin, *Sketches in Crude Oil*, 149–52; J. T. Henry, *Early History*, 230–31.
23 McLaurin, *Sketches in Crude Oil*, 185; Asbury, *The Golden Flood*, 177–78.
24 *Oil City Register*, June 1, 1865; McLaurin, *Sketches in Crude Oil*, 159; Asbury, *The Golden Flood*, 186.
25 McLaurin, *Sketches in Crude Oil*, 159–60; Giddens, *Birth of the Oil Industry*, 131–32, citing *Oil City Register*, June 29, 1865.
26 J. T. Henry, *Early History*, 240–41; Asbury, *The Golden Flood*, 182.
27 Carll, Geological Survey of Pa., *7th Report on the Oil & Gas Fields of W. Pa., for 1887, 1888*, 29.
28 McLaurin, *Sketches in Crude Oil*, 159; *Titusville Morning Herald*, July, 29, 1865.
29 *Pithole Daily Record*, Sept. 29, 1865, cited in Giddens, *Birth of the Oil Industry*, 138–39.
30 Sept. 21, 1865.
31 *Nation*, Sept. 21, 1865.

32 Asbury, *The Golden Flood*, 181, 184, 193, 194.

33 McLaurin, *Sketches in Crude Oil*, 167, 168; Asbury, *The Golden Flood*, 193, 194.

34 Ellis Paxson Oberholtzer, *Jay Cooke: Financier of the Civil War* (Philadelphia: George W. Jacobs & Co., 1907), I, 615.

35 J. T. Henry, *Early History*, 232, 233.

36 *Crawford Journal*, Sept. 13, 1865; *Venango Spectator*, Feb. 22, 1865; Morris, *Derrick & Drill*, 258.

37 *Derrick's Hand-Book*, I, 42; S. Morton Peto, *The Resources and Prospects of America, Ascertained During a Visit to the United States in the Autumn of 1865* (London: Alexander Strahan, 1866), 204.

38 Theodore Clarke Smith, *The Life and Letters of James Abram Garfield* (New Haven: Yale University Press, 1925), II, 822–23.

39 Henrietta M. Larson, *Jay Cooke* (Cambridge: Harvard University Press, 1936), 153. Early in 1866 the Internal Revenue Commission estimated the amount of capital invested in the purchase and development of oil territory during the preceding two years to have exceeded $100 million. Cf. Hayes, *Report . . . on Petroleum*, 7.

40 Hayes, *Report . . . on Petroleum*, 31, 32; J. D. Henry, *History & Romance*, I, 286.

41 Asbury, *The Golden Flood*, 200.

42 *Titusville Morning Herald*, Jan. 22, 1866.

43 *Titusville Morning Herald*, June 20, 1868. In the 1870 Census Pithole City was too small to be included as an organized settlement.

44 *Titusville Morning Herald*, March 13, 1866.

45 *Derrick's Hand-Book*, I, 712.

46 *Titusville Morning Herald*, March 28 & 29, 1866; Giddens, *Birth of the Oil Industry*, 154–55.

47 *Derrick's Hand-Book*, I, 708.

48 *Derrick's Hand-Book*, I, 711.

49 *Derrick's Hand-Book*, I, 71.

50 Cone & Johns, *Petrolia*, 266–67.

51 *Titusville Morning Herald*, Jan. 10, 1866.

52 *Titusville Morning Herald*, July 30, 1866.

53 J. D. Henry, *History & Romance*, I, 301–5. c.f. George W. Brown, *Old Times in Oildom* (Oil City: Derrick Publishing Co., 1911), 16; Asbury, *The Golden Flood*, 218–19, and *passim*.

54 Asbury, *The Golden Flood*, 262–63.

55 Cone & Johns, *Petrolia*, 434–36; *Derrick's Hand-Book*, I, 57, 58, 68.

56 *Derrick's Hand-Book*, I, 57; Cone & Johns, *Petrolia*, 423–25, 428–36; *Titusville Morning Herald*, July 19, 1866.

57 *Titusville Morning Herald*, cited in *Derrick's Hand-Book*, I, 92.

58 Asbury, *The Golden Flood*, 263–64; Henry E. Wrigley, Second Geological Survey of Penn., 1874, *Special Report on the Petroleum of Pennsylvania, its Production, Transportation, Manufacture, and Statistics* (Harrisburg: Bd. of Commissioners for the Second Geological Survey, 1875), J.6; Cone & Johns, *Petrolia*, 464.

59 Asbury, *The Golden Flood*, 266–72; Giddens, *Birth of the Oil Industry*, 165–68.

60 McLaurin, *Sketches in Crude Oil*, 164–65; Asbury, *The Golden Flood*, 271–72.

61 Cone & Johns, *Petrolia*, 467–70.

62 *Derrick's Hand-Book*, I, 164.

63 *Derrick's Hand-Book*, I, 138–39.

64 McLaurin, *Sketches in Crude Oil*, 181–82; Asbury, *The Golden Flood*, 265–66.

65 *Derrick's Hand-Book*, I, 141, 149; A. R. Crum & A. S. Dungan (eds.), *Romance of American Petroleum & Gas* (New York: Romance of American Petroleum and Gas Co., 1911), I, 89.

66 Cone & Johns, *Petrolia*, 505–7; *Derrick's Hand-Book*, I, 97, 105, 107.

67 Morris, *Derrick & Drill*, 176–77.

68 J. R. G. Hazard, *New York Daily Tribune*, summer 1868, reprinted in *Titusville Morning Herald*, Aug. 8, 1868, and in Giddens, *Pa. Petroleum*, 336–44. Cone and Johns, informed petroleum journalists, reported the "oil belt" as "the all prevailing theory of 1868, the first . . . that has given general practical results . . . proved

by successful experiments, during the past season." Cone & Johns, *Petrolia*, 123–25.

69 McLaurin, *Sketches in Crude Oil*, 100, 101.

70 McLaurin, *Sketches in Crude Oil*, 100, 101; *Derrick's Hand-Book*, I, 145, 151, 152, 155.

71 C. E. Bishop, *New York Tribune*, 1871, in J. T. Henry, *Early History*, 482–93.

72 McLaurin, *Sketches in Crude Oil*, 103.

73 Asbury, *The Golden Flood*, 278; J. T.

Henry, *Early History*, 294–304, 492.

74 Carll, Geological Survey of Pa., *7th Report on the Oil & Gas Fields of W. Pa., for 1887*, 1888, 29.

75 J. T. Henry, *Early History*, 301–2; Asbury, *The Golden Flood*, 280.

76 J. T. Henry, *Early History*, 304.

77 Asbury, *The Golden Flood*, 284; Wrigley, *Special Report on the Petroleum of Pa.* (1874), J46.

78 J. T. Henry, *Early History*, 295.

79 Carll, Geological Survey of Pa., *7th Report on the Oil & Gas Fields of W. Pa., for 1887*, 1888, 29.

Chapter 7

1 J. T. Henry, *Early History*, 250.

2 Tait, *The Wildcatters*, 18, 19.

3 Wright, *The Oil Regions of Pennsylvania*, 65, 66.

4 Tait, *The Wildcatters*, 18, 19.

5 McLaurin, *Sketches in Crude Oil*, 392–93; *Derrick's Hand-Book*, I, 1055.

6 Eaton, *Petroleum*, 110–11.

7 Wright, *The Oil Regions of Pa.*, 101; *Scientific American* (April 22, 1865), 258.

8 Cone & Johns, *Petrolia*, 126, 149, 463, 627.

9 Wright, *The Oil Regions of Pa.*, 101; *Scientific American* (April 22, 1865), 258.

10 Cone & Johns, *Petrolia*, 85–86.

11 *Titusville Morning Herald*, Dec. 8, 1870.

12 Cf. Cone & Johns, *Petrolia*, 645.

13 Bone, *Petroleum & Petroleum Wells*, 41.

14 Wright, *The Oil Regions of Pa.*, 74–75.

15 Cone & Johns, *Petrolia*, 151.

16 Cone & Johns, *Petrolia*, 151–52, 645.

17 Cone & Johns, *Petrolia*, 645.

18 Cone & Johns, *Petrolia*, 463.

19 Louis C. Hunter, *Steamboats on the Western Rivers* (Cambridge: Harvard University Press, 1949), 149.

20 Hunter, *Steamboats on the Western Rivers*, 150.

21 Cf. Crum & Dungan (eds.), *Romance of American Petroleum & Gas*, I, 125.

22 Wrigley, *Special Report on the Petro-*

leum of Pa. (1874), J.53–54; J. H. Newton (ed.), *History of Venango County, Pa. . . .* (Columbus, Ohio: J. A. Caldwell, 1879), illustration facing 373.

23 McLaurin, *Sketches in Crude Oil*, 71.

24 Wrigley, *Special Report on the Petroleum of Pa.* (1874), J.54–56.

25 McLaurin, *Sketches in Crude Oil*, 391; Crum & Dungan (eds.), *Romance of American Petroleum & Gas*, I, 127.

26 Crum & Dungan (eds.), *Romance of American Petroleum & Gas*, I, 129; John W. Oliver, *History of American Technology* (New York: The Ronald Press Co., 1956), 334–35.

27 Eaton, *Petroleum*, 144, 155.

28 Eaton, *Petroleum*, 126–28.

29 Eaton, *Petroleum*, 138; Wright, *The Oil Regions of Pa.*, 70.

30 McLaurin, *Sketches in Crude Oil*, 392; Crum & Dungan (eds.), *Romance of American Petroleum & Gas*, I, 128–29; Peckham, *Census Report*, 87, 88; Boverton Redwood, *Petroleum: A Treatise on the Geographical Distribution and Geological Occurrence of Petroleum and Natural Gas, the Physical and Chemical Properties, Production, and Refining of Petroleum and Ozokerite, the Characters and Uses, Testing, Transport, and Storage of Petroleum Products, and the Legislative Enactments Relating Thereto; Together with a Description of the Shale-Oil and Allied*

Industries; and a Bibliography (London: Charles Griffin & Co., Ltd., 1922), II, 393, hereafter cited as B. Redwood, *Treatise on Petroleum; Titusville Morning Herald*, Dec. 8, 1870, quoted in Giddens, *Pa. Petroleum*, 353–54.

31 McLaurin, *Sketches in Crude Oil*, 139, 392; *Derrick's Hand-Book*, I, 50. A crude attempt at casing a wet-drilled well was made by Julius Hall at Tidioute in December 1861. *Derrick's Hand-Book*, I, 24.

32 *Derrick's Hand-Book*, I, 69.

33 McLaurin, *Sketches in Crude Oil*, 392. Cased wells allowed tubing to be taken up in a few hours and steamed to remove paraffin clogging. See Cone & Johns, *Petrolia*, 147.

34 Cone & Johns, *Petrolia*, 145–47, 624–27; *Titusville Morning Herald*, Dec. 8, 1870.

35 *Titusville Morning Herald*, Dec. 8, 1870; Crum & Dungan (eds.), *Romance of American Petroleum & Gas*, I, 128–29; Peckham, *Census Report*, chart opposite 83, 87–88, *and passim*.

36 Crum & Dungan (eds.), *Romance of American Petroleum & Gas*, I, 128–29; Cone & Johns, *Petrolia*, 147. This ease of steaming out paraffin in tubing became of critical importance in the great Bradford field in the second half of the 1870's, where the exceptionally high paraffin content of the crude caused considerable difficulties.

37 McLaurin, *Sketches in Crude Oil*, 392.

38 Report of George Boulton, superintendent of Story Farm, to the President and Board of Directors of the Columbia Oil Company, January 4, 1869, hereafter cited as *Superintendent's Report, Story Farm, Jan. 4, 1869*, reproduced in Cone & Johns, *Petrolia*, 624–28.

39 *Titusville Morning Herald*, Dec. 6, 1870, cited in Giddens, *Pa. Petroleum*, 353–54. The unpredictable reactions from torpedoing presented a greater menace in cased wells if the casing extended too near the producing formations, or if precautions had not been taken to provide a higher shoulder to which the casing could be lifted during torpedoing. C. E.

Beecher, "Production Techniques & Control" (MS, 1955, of chap. 12, American Petroleum Institute, *History of Petroleum Engineering*).

40 Crum & Dungan (eds.), *Romance of American Petroleum & Gas*, I, 127–28; McLaurin, *Sketches in Crude Oil*, 392.

41 See *Titusville Morning Herald*, Dec. 8, 1870, and J. T. Henry, *Early History*, 250; *Derrick's Hand-Book*, I, 137.

42 Peckham, *Census Report*, 88.

43 Eaton, *Petroleum*, 132–33.

44 McLaurin, *Sketches in Crude Oil*, 332.

45 *Derrick's Hand-Book*, I, 960–61; McLaurin, *Sketches in Crude Oil*, 190.

46 "Crocker's Oil Ejector," *American Gas-Light Journal*, VI (Jan. 16, 1864), 217.

47 McLaurin, *Sketches in Crude Oil*, 332–33.

48 McLaurin, *Sketches in Crude Oil*, 335–36.

49 U. S. Patent No. 59,936, Nov. 20, 1866.

50 Deposition of Elijah Brady, Sept. 28, 1868 (Marcus Sachet, Notary Public, New York City), *In the Matter of the Interference Declared Between the Patent Applications of William Reed for Letters Patent of the United States for Improvement in Torpedoes for Oil Wells and Letters Patent granted to Edward A. L. Roberts, November 20th, 1866*, before the Commissioner of Patents, reprinted in *Copy of Testimony Taken on the Part of Edward A. L. Roberts*, Bakewell and Christy, Attorneys for E. A. Roberts (Pittsburgh: Bakewell & Marthens, 1869), 5–7.

51 Letter of E. Mills to Col. E. A. L. Roberts, Titusville, May 20, 1865, quoted in *Articles of Association, By-Laws, and Prospectus of the Roberts Petroleum Torpedo Company* (New York: Sanford, Harroun & Co., 1865), hereafter cited as *Roberts Co. Prospectus* (1865).

52 *Roberts Co. Prospectus* (1865), 15, 16.

53 *Opinions of Board of Examiners in Chief and of Chief Justice D. K. Carter*, reprinted in *Decisions and Opin-*

ions of the Courts and Patent Office Under the Oil Well Torpedo Patents and Applications of Edward A. L. Roberts (Pittsburgh: Bakewell & Marthens, printers, 1873).

54 *Titusville Morning Herald,* Feb. 8, 1867.

55 "The Roberts Torpedo," *Scientific American,* XV (July 21, 1866), 54.

56 *Derrick's Hand-Book,* I, 67.

57 *Scientific American,* XV (July 21, 1866), 54.

58 J. T. Henry, *Early History,* 543; McLaurin, *Sketches in Crude Oil,* 334.

59 McLaurin, *Sketches in Crude Oil,* 334.

60 McLaurin, *Sketches in Crude Oil,* 338–42; *Derrick's Hand-Book,* I, 114.

61 Peckham, *Census Report,* 85.

62 Deposition of Lewis B. Silliman, Aug. 16, 1871, before R. Arthurs, *E. A. L. Roberts vs. Reed Torpedo Company, et al.* (U. S., Pennsylvania).

63 McLaurin, *Sketches in Crude Oil,* 336; J. T. Henry, *Early History,* 544.

64 Gale, *Wonder of the 19th Century,* 25, 31; Cone & Johns, *Petrolia,* 123.

65 Gale, *Wonder of the 19th Century,* 25; Eaton, *Petroleum,* 103–5, 112–14, 129–33.

66 Wright, *The Oil Regions of Pa.,* 74–78, 255–56.

67 Wright, *The Oil Regions of Pa.,* 99, 100. Cone & Johns, *Petrolia,* 146–47, reported in 1869 that few wells maintained their maximum output more than two or three months.

68 Cone & Johns, *Petrolia,* 463.

69 J. T. Henry, *Early History,* 219–21, 248–50. Henry drew upon data that the Petroleum Producers' Association had begun gathering after its organization in 1869. He was careful to point out the limitations of his findings, including the omission of dry holes from the calculations of costs.

70 Cf. J. T. Henry, *Early History,* 221.

71 Henry's estimates check closely with figures given by the *Titusville Morning Herald* on Dec. 8, 1870, that placed the typical cost range of all wells, shallow and deep, at between $4,100 and $6,225.

72 Hayes, *Report . . . on Petroleum,* 27, 28.

73 J. T. Henry, *Early History,* 221.

74 *Derrick's Handbook,* I, 711.

75 J. T. Henry, *Early History,* 221, 222.

76 *Derrick's Hand-Book,* I, 711.

77 *Derrick's Hand-Book,* I, 805.

78 Cf. J. T. Henry, *Early History,* 233–34; McLaurin, *Sketches in Crude Oil,* 136–38.

79 *Oil City Register,* May 11, 1865, reprinted in "New Era in Petroleum Mining-Boring Machinery," *American Gas-Light Journal,* VI (June 16, 1865), 369.

80 *Oil City Register,* May 11, 1865, quoted in *American Gas-Light Journal,* VI (June 16, 1865), 369.

81 *Superintendent's Report, Story Farm, Jan. 4, 1869,* in Cone & Johns, *Petrolia,* 624–28.

82 *Superintendent's Report, Story Farm, Jan. 4, 1869,* reproduced in Cone & Johns, *Petrolia,* 624–28.

83 *Superintendent's Report, Story Farm, Jan. 4, 1869,* reproduced in Cone & Johns, *Petrolia,* 624–28.

84 *Seventh Annual Report of the Columbia Oil Company,* Jan. 9, 1869, reproduced in Cone & Johns, *Petrolia,* 620–24.

85 McLaurin, *Sketches in Crude Oil,* 139, 392; *Derrick's Hand-Book,* I, 40, 65; Cone & Johns, *Petrolia,* 168, 169; Wright, *The Oil Regions of Pa.,* 261.

86 Wright, *The Oil Regions of Pa.,* 261.

87 *Conservation of Oil and Gas: A Legal History,* ed. Blakely M. Murphy (Chicago: American Bar Association, Section of Mineral Law, 1948), 431. Also see Appendix E.

Chapter 8

1 Cf. Giddens, *Birth of the Oil Industry,* 102–3; Crum & Dungan (eds.), *Romance of American Petroleum & Gas,* I, 91.

2 McLaurin, *Sketches in Crude Oil,* 260. During 1862 assessments of 2¢ a barrel were sufficient to meet the $100 average cost of each freshet.

This included the salary of the superintendent, the wages of two helpers, and the mill owners' charges for the use of their dams. During 1863 assessments rose to 4¢ a barrel as mill owners, taking advantage of higher crude prices, increased their charges and drove the cost of freshets to nearly $400. A. S. Dobbs, statement in *Oil Register*, quoted in *Derrick's Hand-Book*, I, 34.

3 *Warren Daily Mail*, Jan. 24, 1863.

4 McLaurin, *Sketches in Crude Oil*, 200.

5 McLaurin, *Sketches in Crude Oil*, 260–62.

6 *Venango Spectator*, May 21, 1862.

7 *Derrick's Hand-Book*, I, 27.

8 *Venango Spectator*, May 13, 1863. When several loaded boats at Noble farm on upper Oil Creek broke loose during a freshet in May 1864, they created a wreckage of moored boats all down the stream, with a loss of 20,000 to 30,000 barrels. Cone & Johns, *Petrolia*, 106, 107.

9 McLaurin, *Sketches in Crude Oil*, 262.

10 *Pittsburgh Gazette and Commercial Journal*, April 15, 1862.

11 Cone & Johns, *Petrolia*, 101. Glyde received a patent on his improvement but found it impossible to collect royalties from others who quickly adopted his methods. Cf. *Derrick's Hand-Book*, I, 645.

12 J. D. Henry, *History & Romance*, I, 259–60; *Derrick's Hand-Book*, I, 642–50.

13 Cone & Johns, *Petrolia*, 101; *Derrick's Hand-Book*, I, 645–46.

14 Cone & Johns, *Petrolia*, 108; McLaurin, *Sketches in Crude Oil*, 260–61.

15 *The Daily Commercial* (Pittsburgh), Feb. 22, 1872.

16 *U. S. Railroad & Mining Register*, Oct. 10, 1863; *American Railroad Journal*, XXXV (Nov. 8, 1862), 881.

17 *American Railroad Journal*, XXXVI (April 11, 1865), 338.

18 *U. S. Railroad & Mining Register*, Oct. 10, 1863.

19 *Derrick's Hand-Book*, I, 40, 76.

20 Maybee, *Railroad Competition*, 26.

21 Maybee, *Railroad Competition*, 6.

22 Giddens, *Birth of the Oil Industry*, 113; *American Railroad Journal*, July 11, 1863, Jan. 31, 1863.

23 Giddens, *Birth of the Oil Industry*, 113; *American Railroad Journal*, Feb. 25, 1865.

24 *Oil City Register*, March 15, 1866.

25 Maybee, *Railroad Competition*, 167.

26 *Derrick's Hand-Book*, I, 58, 63–64. Mention should also be made of the Oil Creek Lake & Titusville Mining and Transportation Company that started construction of a line to Union Mills in 1866 to compete with the Oil Creek road. This line, however, progressed very slowly and was not completed until 1871. Cf. Giddens, *Birth of the Oil Industry*, 150.

27 Maybee, *Railroad Competition*, 36–37, 51–52.

28 Maybee, *Railroad Competition*, 47–49.

29 Maybee, *Railroad Competition*, 125–30.

30 Taylor & Neu, *The American Railroad Network*, 68.

31 Maybee, *Railroad Competition*, 114–21.

32 Maybee, *Railroad Competition*, 40, 41, 167. One of numerous side effects if this completion was to stimulate the interest of producers in developing the territory around Tidioute and subsequently Tionesta. *Derrick's Hand-Book*, I, 72; Cone & Johns, *Petrolia*, 84–86.

33 Maybee, *Railroad Competition*, 167; *United States Railroad and Mining Register*, Oct. 28, 1865, citing *Oil City Register*, Oct. 19, 1865.

34 Cf. Maybee, *Railroad Competition*, 231–32.

35 Maybee, *Railroad Competition*, 60.

36 Maybee, *Railroad Competition*, 61.

37 Maybee, *Railroad Competition*, 52–53, 66.

38 Maybee, *Railroad Competition*, 67. It was also argued that the contract violated a Pennsylvania law of 1861 prohibiting such arrangements between lines that, because of gauge differences, could not be physically connected with each other.

39 Maybee, *Railroad Competition*, 72.

40 William Abbott, *Daily Tribune Republican* (Meadville, Pa.), (May 12, 1888), 164–65.

41 Cf. Asbury, *The Golden Flood*, 140; Henri Erni, *Coal Oil and Petroleum* (Philadelphia: Henry Carey Baird, 1865), 116.

42 *The Report of the Commissioner of Patents for the Year 1863*, Hse. Exec. Doc. No. 60, 38th Cong., 1st Sess., 23–26; *Scientific American* (Nov. 28, 1863), 344; *American Gas-Light Journal*, VI (March 16, 1863), 273–74.

43 "Inflammability of Refined Oil," *American Gas-Light Journal* (Nov. 1, 1862), 135; "Improved Mode of Lining Oil Barrels," *Scientific American*, XIII (Dec. 23, 1865), 399.

44 Asbury, *The Golden Flood*, 138.

45 Samuel Downer to W. H. L. Smith, Aug. 31 and Sept. 22, 1863, W. H. L. Smith Papers; Eaton, *Petroleum*, 280–81.

46 *Titusville Morning Herald*, Aug. 23, 1863, and Aug. 27, 1867.

47 Eaton, *Petroleum*, 159–60, 280; Joshua Merrill to W. H. L. Smith, July 3, 1863, W. H. L. Smith Papers.

48 Anonymous Account, Sept. 1860, reprinted in Morris, *Derrick & Drill*, 38; *American Gas-Light Journal* (April 15, 1862). The *Pittsburgh Commercial*, March 4, 1868, reported prices of $2–$2.50 for barrels. Though formerly priced $2.50–$3.50, barrel prices in Pittsburgh declined from $2.40 in September 1870 to $1.75 in January 1871. *Pittsburgh Evening Chronicle*, Sept. 24 and Nov. 1, 1870; Jan. 5, 1871, cited in Maybee, *Railroad Competition*, 277. But the *Titusville Morning Herald*, May 25, 1872, reported barrels for refined oil priced $2.25 in Titusville, $1.80 in Cleveland.

49 *Derrick's Hand-Book*, I, 50, 60.

50 *Derrick's Hand-Book*, I, 60; Wm. Lay, Oil City Chamber of Commerce, in S. S. Hayes, *Report . . . on Petroleum*, reports 50¢ for "cooperage, leakage, etc." in shipment Jan. 19, 1866. In August 1866 Oil Creek producers agreed to formalize a practice of long standing, giving the buyer an allowance of two gallons on every forty. *Derrick's Hand-Book*, I, 77.

51 *Scientific American*, March 4, 1865.

52 Alfred Wilson Smiley, *A Few Scraps, Oily and Otherwise* (Oil City: The Derrick Publishing Co., 1907), 145–46.

53 Maybee, *Railroad Competition*, 241. In March 1867 the *Titusville Morning Herald* noted that "a great part" of crude was being shipped to market in tanks, each with a capacity of 25 to 30 barrels, mounted two or three on a railroad flatcar. Most of these tanks were constructed of wood, "although many iron ones are coming along." *Titusville Morning Herald*, March 9, 1867.

54 *Derrick's Hand-Book*, I, 969.

55 *U. S. Railroad and Mining Journal*, Oct. 21, 1865, in Maybee, *Railroad Competition*, 240–41.

56 *Oil & Gas Journal*, Diamond Jubilee Issue (Aug. 27, 1934), 23.

57 *Philadelphia Commercial List and Price Current*, March 1, 1871; *Derrick's Hand-Book*, I, 969.

58 Maybee, *Railroad Competition*, 242, 251.

59 Carl Russell Fish, *The Common Man, A History of American Life*, VI (New York: The Macmillan Co., 1944), 104, 105, 331.

60 McLaurin, *Sketches in Crude Oil*, 265–66.

61 *Derrick's Hand-Book*, I, 27.

62 McLaurin, *Sketches in Crude Oil*, 266.

63 McLaurin, *Sketches in Crude Oil*, 266.

64 *Derrick's Hand-Book*, I, 52; McLaurin, *Sketches in Crude Oil*, 266, 267.

65 *Derrick's Hand-Book*, I, 52; "Oil Transportation," *Petroleum Age*, VII (1888), 34–36, cited in Giddens, *Birth of the Oil Industry*, 143.

66 Edwin C. Bell Memoirs, MS, 16–18, Drake Museum; *Titusville Morning Herald*, Dec. 11, 1865.

67 McLaurin, *Sketches in Crude Oil*, 267; Giddens, *Birth of the Oil Industry*, 147, 154, 155.

68 See Chapter 11.

69 Edwin C. Bell Memoirs, MS, 22, 23, Drake Museum; *Titusville Morning Herald*, Oct. 27, 1865.

70 Edwin C. Bell Memoirs, MS, 25, 26, Drake Museum.

71 *Titusville Morning Herald*, Aug. 3, 1865; *Oil City Register*, Aug. 24,

1865, cited in Giddens, *Birth of the Oil Industry*, 144.

72 Edwin C. Bell Memoirs, MS, 24, Drake Museum.

73 Edwin C. Bell Memoirs, MS, 28, Drake Museum.

74 J. T. Henry, *Early History*, 526–27; *Derrick's Hand-Book*, I, 83.

75 *Scientific American*, XV (Sept. 1, 1866), 144.

76 Smiley, *A Few Scraps, Oily and Otherwise*, 137–40, cited in Arthur Menzies Johnson, *The Development of American Petroleum Pipelines: A Study in Private Enterprise and Public Policy, 1862–1906* (Ithaca, N. Y.: Cornell University Press, 1956), 11, hereafter cited as Johnson, *Pipelines*.

77 Smiley, *A Few Scraps, Oily and Otherwise*, 133, cited in Johnson, *Pipelines*, 11.

78 *Titusville Morning Herald*, March 6, 1866, quoted in Giddens, *Pa. Petroleum*, 310–11.

79 Samuel Downer to W. H. L. Smith, June 20, 1862, W. H. L. Smith Papers.

80 Samuel Downer to W. H. L. Smith, March 30, May 29, and Sept. 27, 1863, W. H. L. Smith Papers.

81 *Thirty Second Annual Report of the Philadelphia Board of Trade* (Philadelphia: Collins, Printer, 1865), 90, 91.

82 Samuel Downer to W. H. L. Smith, June 2, 1862, March 30, May 29, May 30, 1863, W. H. L. Smith Papers.

83 Joshua Merrill to W. H. L. Smith, Dec. 7 and 23, 1861, Nov. 15, 1862, Aug. 30, 1863, W. H. L. Smith Papers.

84 *Oil & Gas Journal*, XV (Jan. 4, 1917), 31–33.

85 E. C. Bell, in *Titusville Morning Herald*, n.d., reprinted in *Oil & Gas Journal* (April 4, 1918).

86 Cone & Johns, *Petrolia*, 99; Asbury, *The Golden Flood*, 139.

87 Edwin C. Bell, in *Oil City Derrick*, n.d., reprinted in *Oil & Gas Journal* (Jan. 4, 1917), 32, 33; Bell, in *Titusville Morning Herald*, reprinted in *Oil & Gas Journal* (Apr. 4, 1918), 34.

88 *Derrick's Hand-Book*, I, 42.

89 *Scientific American* (Oct. 8, 1881), 224. There were doubts whether the tank would withstand pressure. Partly as a result, the iron plates forming its bottom and sides were heavier than those of 35,000-barrel tanks twenty years later.

90 Bell, in *Oil City Derrick*, n.d., *Oil & Gas Journal* (Jan. 4, 1917), 32, 33.

91 Bell, in *Oil City Derrick*, n.d., *Oil & Gas Journal* (Jan. 4, 1917), 32, 33; Cone & Johns, *Petrolia*, 534–40. Graff, Hasson & Company's 8000-barrel iron tank at the very center of the flood in Oil City was moved from its foundations but not lost.

92 *Titusville Morning Herald*, March 6, 1866; Joshua Merrill to W. H. L. Smith, Boston, Dec. 7, 1861, and Dec. 5, 1862, in W. H. L. Smith Papers.

93 Edwin C. Bell, in *Oil City Derrick*, n.d., reprinted in *Oil & Gas Journal* (Jan. 4, 1917), 32, 33; *American Gas-Light Journal*, VIII (July 16, 1866), 21; *Derrick's Hand-Book*, I, 72.

94 Cone & Johns, *Petrolia*, 111.

95 Asbury, *The Golden Flood*, 110, 111; J. T. Henry, *Early History*, 586.

96 Cone & Johns, *Petrolia*, 113.

97 Smiley, *A Few Scraps, Oily & Otherwise*, 137–40.

98 *Titusville Morning Herald*, Aug. 8, 1868, and Aug. 16, 1873.

99 The background and history of these early conventions are given in detail in chap. iv, Maybee, *Railroad Competition*, 79–100.

100 J. H. Rutter testimony in New York Assembly, *Proceedings of the Special Committee on Railroads, Appointed Under a Resolution of the Assembly to Investigate Alleged Abuses in the Management of Railroads Chartered by the State of New York* (8 vols.; Albany, 1879–1880), I, 260, hereafter cited as *Hepburn Hearings*.

101 See Julius Grodinsky, "Standard Oil-Erie Ring, A Point of View," *Mississippi Valley Historical Review*, XXXIII (March, 1947), 618.

102 U. S., Congress, Senate, *Report of the Select Committee on Transportation Routes to the Seaboard*, Report No. 307, Part 2, 43d Cong., 1st Sess., 1874, 36, 45, hereafter cited as *Windom Report*.

103 Maybee, *Railroad Competition*, 131–33.
104 Maybee, *Railroad Competition*, 134.
105 Charles F. Adams, Jr., "A Chapter of Erie," in *Chapters of Erie and Other Essays* (Boston: Fields, Osgood & Co., 1871).
106 Maybee, *Railroad Competition*, 163.
107 Maybee, *Railroad Competition*, 164.
108 Cf. George H. Burgess and Miles C. Kennedy, *Centennial History of the Pennsylvania Railroad Company* (Philadelphia: Pennsylvania Railroad Company, 1949), 199, hereafter cited as Burgess & Kennedy, *History of Pennsylvania Railroad*. The Pennsylvania was also successful in blocking Gould's attempts to extend the Erie's connections to Cincinnati and St. Louis. Cf. Maybee, *Railroad Competition*, 155–58; Burgess & Kennedy, *History of Pennsylvania Railroad*, 195–200.
109 Cf. Burgess & Kennedy, *History of Pennsylvania Railroad*, 318–19.
110 Maybee, *Railroad Competition*, 181–84.

Chapter 9

1 Gale, *Wonder of the 19th Century*, 40, 41.
2 Dampier, *History of Science*, 429.
3 See W. H. Perkin, "On the Aniline or Coal-Tar Colors" (Cantor Lectures delivered before the Society of Arts, London), reprinted in *The American Chemist*, I (Sept., 1870), 83–89, 164, 260. Erni, *Coal Oil and Petroleum*, 81.
4 Dampier, *History of Science*, 252–56, *passim*; L. F. Haber, *The Chemical Industry of the Nineteenth Century* (London: Oxford University Press, 1958), 62–71, 81, 82.
5 Dampier, *History of Science*, 254, 255; J. R. Partington, *A Short History of Chemistry* (London: Macmillan and Co., Ltd., 1951), 290–92, 294.
6 *American Gas-Light Journal* (Feb. 2, 1865), 233, (July 2, 1866), 2.
7 E. Lawson Lomax, "The Pyrogenesis of Hydrocarbons," *Journal of Industrial and Engineering Chemistry* (Sept., 1917), 870.
8 B. Silliman, Jr., *Report on Rock Oil*, 9; William T. Brannt, *Petroleum: Its History, Origin, Occurrence, Production, Physical and Chemical Constitution, Technology, Examination and Uses; Together with the Occurrence and Uses of Natural Gas* (Philadelphia: Henry Carey Baird & Co., 1895), 368, 370, hereafter cited as Brannt, *Petroleum*.
9 See Table 8–1. A. Gesner, *Treatise* (1861), appendix, 126–28. In 1860 Gesner attempted to introduce his own scale modifying the Baume, and Tagliabue briefly entered production of a new hydrometer carrying the Gesner scale.
10 C. F. Chandler, *Report on Petroleum as an Illuminator* (extract from *Annual Report of the Board of Health of the Health Dept. of the City of New York for 1870*), (New York: The New York Printing Co., 1871), 17.
11 E. L. Lomax, *Journal of Industrial and Engineering Chemistry* (Sept., 1917), 879.
12 U. S. Patent No. 36,488 issued to John Tagliabue Sept. 16, 1862.
13 C. F. Chandler, *Report on Petroleum as an Illuminator*, 36–38.
14 C. F. Chandler, *Report on Petroleum as an Illuminator*, 48, *passim*.
15 Leet, *Petroleum Distillation*, 17, 18; Gale, *Wonder of the 19th Century*, 39, 40.
16 Leet, *Petroleum Distillation*, 17, 18; Gale, *Wonder of the 19th Century*, 39, 40; Abraham Gesner, *A Practical Treatise on Coal, Petroleum and other Distilled Oils*, ed. G. W. Gesner (rev. ed.; New York: Balliere Bros., 1865), 166, 167, hereafter cited as A. Gesner, *Treatise* (1865).
17 Leet, *Petroleum Distillation*, 10, 17, 18.
18 U. S. Patent 32,706 issued to Joshua Merrill, July 2, 1861; A. Gesner, *Treatise* (1861), 40–45, 108–22; A. Gesner, *Treatise* (1865), 80–85, 158–67; Leet, *Petroleum Distillation*, 10, 17, 18, 22, 23, 31, 50, 51; Schieffelin Bros.

Circular, 1860, reprinted in Gale, *Wonder of the 19th Century,* appendix.

19 Leet, *Petroleum Distillation,* 22, 25, 50. Leet errs in ascribing horizontal stills as a reaction to the vertical type in the petroleum era.

20 I. I. Redwood, *Mineral Oils and Their By-Products,* 121.

21 Benjamin J. Crew, *A Practical Treatise on Petroleum* (Philadelphia: Henry Carey Baird & Co., 1887), 247.

22 Crew, *Practical Treatise on Petroleum,* 247.

23 "Visit to a Petroleum Refinery," *Scientific American,* XII (Feb. 18, 1865); *Scientific American,* XIII (Aug. 19, 1865), 168; Leet, *Petroleum Distillation,* 7, 8, 11, 18, 50, 112. To save time some refiners refilled the still after distillation without removing the residue but this was found to lead to excessive coke-encrustation, the distillation being retarded and the still injured by overheating. B Redwood, *Treatise on Petroleum,* II, 479.

24 B. Redwood, *Treatise on Petroleum,* II, 478, states that the use of steam lowers the boiling point of a high boiling fraction from 325° C to 275° C (600° F to 510° F).

25 Computed from I. I. Redwood, *Mineral Oils and Their By-Products,* 191–195. The data are projected from the British shale-oil industry in the final decade of the century, but still sizes and types are similar to those in common use in American coal-oil and petroleum refining in the first half of the 1860's. Actually the comparison probably favors direct-firing because of improvements in this method. In 1865, for example, it commonly required 1 ton of coal to distill a 25-barrel charge by direct-firing in American practice. See "Improvement in the Distillation of Petroleum Oil," *Scientific American,* XIII (Aug. 19, 1865), 168.

26 I. I. Redwood, *Mineral Oils and Their By-Products,* 116–18, 192–94; David T. Day (ed.), *A Handbook of the Petroleum Industry* (2 vols.; New York: John Wiley & Sons, 1922), II, 343; A. Gesner, *Treatise* (1865), 159.

27 I. I. Redwood, *Mineral Oils and Their By-Products,* 115, 116; E. H. Leslie,

Motor Fuels, Their Production and Technology (New York: The Chemical Catalog Company, Inc., 1923), 248.

28 Samuel Downer to W. H. L. Smith, New York, Aug. 15, 1862, W. H. L. Smith Papers.

29 Herbert W. C. Tweddle, U. S. Patent No. 34,324, issued Feb. 4, 1862 and Nos. 72,125 and 72,126, issued Dec. 10, 1867.

30 *Journal of the Franklin Institute,* XLI (March, 1861), 203; Wm. H. Patchell, "Steam Superheating," *Cassier's Magazine,* X (June, 1890), 115–28.

31 B. Redwood, *Treatise on Petroleum,* II, 478.

32 Leet, *Petroleum Distillation,* 8, 11. Leet states that the cooling time was 12 hours after each run, followed by entry of the stillmen to remove 1-foot coke deposits, for at least two years after petroleum refining became general.

33 *American Gas-Light Journal,* I (June 1, 1860), 255.

34 U. S. Patent No. 24, 324, issued Feb. 4, 1862, to H. W. C. Tweddle.

35 W. Miller, "Basic Changes in Refining Processes," *Petroleum Development and Technology in 1925* (New York: American Institute of Mining and Metallurgical Engineering, Inc., 1926), 390.

36 A. F. Dunstan *et al.* (eds.), *The Science of Petroleum,* II (London: Oxford Univ. Press, 1938), 1599; Crew, *Practical Treatise on Petroleum,* 252.

37 *Comptes Rendues des Séances de l'Académie des Sciences,* LXII (Jan.–June, 1866), 905, 947, and LXIII (July–Dec., 1866), 788, 834.

38 U. S. Patent No. 28,246, issued to Luther Atwood, May 15, 1860; Merrill's "Reminescences," *Derrick's Hand-Book,* I, 885.

39 Leet, *Petroleum Distillation,* 7, 8.

40 S. Downer to W. H. L. Smith, Boston, Jan. 7, 1862, W. H. L. Smith Papers.

41 Leet, *Petroleum Distillation,* 14–17.

42 Schieffelin Bros. Circular, reproduced in Gale, *Wonder of the 19th Century,* appendix.

43 S. Downer to W. H. L. Smith, New

York, Aug. 15, 1862, W. H. L. Smith Papers.

44 S. Downer to W. H. L. Smith, Boston, Jan. 7, 1862, W. H. L. Smith Papers.

45 S. Downer to W. H. L. Smith, New York, Aug. 15, 1862, W. H. L. Smith Papers.

46 S. Downer to W. H. L. Smith, New York, Aug. 15, 1862, W. H. L. Smith Papers

47 Cf. Leet, *Petroleum Distillation*, 13–16; Walter Miller and Harold G. Osborne, "History and Development of Some Important Phases of Petroleum Refining in the United States," in Dunstan *et al.* (eds.), *The Science of Petroleum*, II, 1466.

48 Philadelphia Coal Oil Circular, cited in *Scientific American*, IX (Nov. 28, 1863), 340–41.

49 Leet, *Petroleum Distillation*, 9; A. C. Ferris, *OP&D Reporter* (May 1, 1889), 10, and (June 19, 1889), 27–30.

50 Frederick Challenger, "The Sulphur Compounds of Bituminous Oils," in Dunstan *et al.* (eds.), *The Science of Petroleum*, II, 1042–46; S. F. Birch, "The Chemistry of the Refining of Light Distillates," in Dunstan *et al.* (eds.), *The Science of Petroleum* III, 1708–1758; Birch, "Sulphuric Acid Treatment," in Dunstan *et al.* (eds.), *The Science of Petroleum* III, 1769–1778.

51 A. Gesner, *Treatise* (1865), 81–85.

52 Leet, *Petroleum Distillation*, 99.

53 Cf. Leet, *Petroleum Distillation*, 87.

54 A. C. Ferris, *OP&D Reporter* (May 1,

1889),10; A. Gesner, *Treatise* (1861), 116.

55 Schieffelin Brothers' Circular, reprinted in Gale, *Wonder of the 19th Century*, 79, 80.

56 A. C. F., *OP&D Reporter* (June 19, 1889), 30.

57 A. C. F., *OP&D Reporter* (June 19, 1889), 30.

58 Leet, *Petroleum Distillation*, 17, 18, *passim;* A. C. F., *OP&D Reporter* (June 19, 1889), 30.

59 A. Gesner, *Treatise* (1861), 116, 120–22.

60 Downer to Smith, April 2, 1863, W. H. L. Smith Papers.

61 Merrill's "Reminiscences," *Derrick's Hand-Book*, I, 886.

62 Chandler, *Report on Petroleum as an Illuminator*, 17.

63 For information on Pittsburgh refineries see *Philadelphia Commercial List and Price Current* (Aug. 27, 1864), 138, and George H. Thurston, *Pittsburgh and Allegheny in the Centennial Year* (Pittsburgh: A. A. Anderson & Son, 1876), 205, 206.

64 Maybee, *Railroad Competition*, 198, 199.

65 Deposition of Joshua Merrill (Nov. 3, 1870), *Merrill* v. *Yeomans*, MS, Curtis *et al.* (U. S.), 1870, 26, 27.

66 P. H. Van der Weyde, "Petroleum and its Products," *Scientific American*, VIII (July 2, 1866), 2.

67 Wright, *The Oil Regions of Pa.*, 200, reported an 80 to 82 per cent refined oil yield for 100° fire test and 75 per cent for a 115° test required for export.

Chapter 10

1 U. S. Revenue Commission (David A. Wells, Chairman), *Distilled Spirits as a Source of National Revenue*, Special Report No. 5, 39th Cong., 1st Sess., 1865, House Ex. Doc. No. 62, 7–12; "Turpentine and Naphtha in Paints," *Scientific American*, XI (Sept. 10, 1864), 170.

2 "Extraction of Oils by Means of Solvents," *OP&D Reporter*, XVI (1879), 421; *Scientific American* (Oct. 11

1879), 225; "On the Value of Gasoline (Canadol) for the Extraction of Oil Seeds," *American Chemist*, II (Dec., 1871), 233.

3 For keroselene as a local and general anesthetic, see Ephraim Cutter, M.D., "On the New Anaesthetic Keroselene," *Pharmaceutical Journal*, III (Sept. 2, 1861), 216; "A New Anaesthetic Keroselene," *Boston Medical and Surgical Journal*, LIV (July 11,

1861), 494, 495 (Aug. 22, 1861), 62, 63 (Sept. 5, 1861), 101 and (Oct. 10, 1861), 202, 203. For accounts of the first commercial production of gasoline at the Downer plants in Boston and at Corry, Pa., see Joshua Merrill to W. H. L. Smith, Boston, July 24, 1863, W. H. L. Smith Papers. "Drake's Gas-Light Machine," *American Gas-Light Journal*, IV (Nov. 2, 1864), 164, reports the motivations to supplant benzole and naphtha with gasoline in gas machines.

4 Dr. William H. Wahl, "Gas Machines" (paper read March 1872 before the Franklin Institute) in *American Gas-Light Journal*, XVII (Oct. 16, 1872), 129, 130; Institutional advertisement, The Gilbert & Barker Man'f'g. Co., "Springfield Gas Machine," *American Gas-Light Journal*, XVI (May 16, 1872), 153.

5 "Petroleum and Its Products," *Scientic American*, Supplement (May 18, 1872), 340, 341.

6 "Naphtha in Gas Lighting," *Scientific American*, Supplement (May 18, 1872), 342.

7 "Rand's Pneumatic Gas Machine," *American Gas-Light Journal*, IX (May 16, 1868), 273.

8 U. S. Patents 28028–30, reissues 3872, 3873, cited in Henry H. Edgerton, "A Report on Petroleum Naphtha Gas, To the President and Board of Directors of the Fort Wayne, Indiana, Gas-Light Company," *American Gas-Light Journal*, XIV (April 17, 1871), 114.

9 Collins, *Consolidated Gas Company of New York*, 158.

10 "A Cheap Process of Manufacturing Gas From Naphtha," *National Oil Journal*, III (March, 1873), 6; *Scientific American* (May 18, 1872), 342; William Henry White, "On Naphtha —Its Economy and Value as a Gas Producer," *Proceedings of the American Gas-Light Association, 1874*, 149–51. Henry H. Edgerton, a gas engineer who installed 14 plants to make naphtha gas exclusively for city gas systems reported one gallon of ordinary commercial naphtha of average 70° B gravity made 80 cubic feet of 80 candle gas, composed of four

gases, and containing no free hydrogen, no sulphuretted hydrogen, carbonic acid, or carbonic oxide. There were four constituent permanent gases, one non-illuminant or diluent, one slightly illuminant, the other two —56 per cent of the mixture—highly illuminant, of 1.272 specific gravity. He equated the gas from one gallon of naphtha with 400 cubic feet of coal-gas of 15–16 candle power. Henry H. Edgerton, "On Petroleum Gas," *Proceedings of the American Gas-Light Association, 1874*, 182, 183, 225. W. Henry White, engineer of the Citizens Gas Company of Brooklyn, reported a somewhat lighter naphtha gas produced for enrichment, with an average specific gravity between .525 and .600, nearly .130 heavier than coal-gas of the same illuminating power. He obtained 90 cubic feet of 60 candle gas to the gallon of naphtha, which with a diluent of water gas he equated with 130–140 cubic feet of coal gas of 17 candle power. W. H. White, *Proceedings of the American Gas-Light Association, 1874*, 156, 223.

11 W. H. White, *Proceedings of the American Gas-Light Association, 1874*, 149–51, 155, 223.

12 "Important Improvement in Gas Manufacture," *American Gas-Light Journal*, XIII (July 16, 1870), 24, 25.

13 H. H. Edgerton, *American Gas-Light Journal*, XIV (April 17, 1871), 114.

14 W. H. White, *Proceedings of the American Gas-Light Association, 1874*, 166.

15 *Annual Report of the American Institute of the City of New York for the Years 1866–67* (Albany: Benthuysen & Sons, 1867), 501, 502.

16 J. R. Weist, M.D., "Local Anaesthesia," *Western Journal of Medicine*, II (Jan., 1868), 9–12; J. D. Jackson, "Rhigolene as a Local Anaesthetic," *Western Journal of Medicine*, III (April, 1868), 210–12; S. D. Hayes, "On the History and Manufacture of Petroleum Products," *American Gas-Light Journal*, XVI *American Chemist*, II (May, 1872), 157–58; C. F. Chandler, *Report on Petroleum as an Illuminator*, 14.

17 P. H. Van der Weyde, Letter to Editor, *Dental Cosmos*, cited in "Chimogene—A New Anaesthetic," *Scientific American*, XV (July 21, 1866), 50; *Scientific American*, XV (Oct. 20, 1866), 268, and (Nov. 2, 1866), 132, 133.

18 Oscar Edward Anderson, Jr., *Refrigeration in America: A History of a New Technology and Its Impact* (Princeton: Princeton University Press for the University of Cincinnati, 1953), 83–85; "History of Ice-Making," *Scientific American*, XXVI (Jan. 20, 1872), 47, 48.

19 Leet, *Petroleum Distillation*, 25, 26, 53, 54.

20 *American Gas-Light Journal*, XI (May 16, 1865), 311; Alban C. Stimers, Letter to Editor, *New York Times*, June 17, 1867, quoted in *Scientific American*, XVII (July 6, 1867), 3; Hector Orr, Letter to Editor, Sept. 6, 1867, *Journal of the Franklin Institute*, LIV (Oct., 1867), 230.

21 U. S. Patent No. 39,918, issued to G. B. Hill, Sept. 15, 1863.

22 R. A. Fisher, *Petroleum versus Coal: Report Upon Experiments Undertaken to Determine the Relative Cost of Generating Steam, by Petroleum and Anthracite Coal* (New York: John W. Amerman, Printer, 1864), 1–12.

23 "Petroleum For Steamship Boilers—The Report of the Navy Department," *Scientific American*, XVII (Dec. 21, 1867), 393; *Boston Commercial Bulletin*, quoted in "Coal Superseded as Fuel for Steam Engines—Successful Experiments at the Navy Yard—Petroleum as Fuel—Enormous Saving in Space—Utility, Safety, Economy—Immense Advantage to Steam Navigation," *American Gas-Light Journal*, VIII (June 3, 1867), 356.

24 *Boston Commercial Bulletin*, cited in *American Gas-Light Journal*, XIII (June 17, 1867), 372; *New York Times*, reprinted in *Venango Spectator*, June 28, 1867.

25 B. F. Isherwood, Chief of the Bureau of Steam Engineering, USN, communication dated Oct. 25, 1867, in *Report of the Secretary of the Navy for 1867*, House Executive Document,

No. 1, 40 Cong., 2 sess. (Washington: Government Printing Office, 1867), 173–75.

26 Wrigley, *Special Report on the Petroleum of Pennsylvania* (1874), J73–J75.

27 Samuel A. D. Sheppard, "On Vaseline and Petroleum Products," *The Pharmacist*, XIV (June, 1881), 211–16; Leet, *Petroleum Distillation*, 36; U. S. Patents issued to Robert Chesebrough: No. 48,367, dated June 27, 1865, No. 49,502 dated Aug. 22, 1865, No. 51,557 dated Dec. 19, 1865, No. 56,179 dated July 10, 1866, No. 77,959 dated May 19, 1868.

28 "Galena Oil Company," *Derrick's Hand-Book*, I, 998–1001.

29 Cf. Peckham, *Census Report*, 157; Robert W. Thurston, *A Treatise on Friction and Lost Work in Machinery and Millwork* (New York: John Wiley & Sons, 1885), 235, 236, 284, 285.

30 Depositions of Joshua Merrill (Nov. 2 and 4, 1870), *Merrill v. Yeomans*, MS, Curtis *et al.* (U. S.), 32, 42, 44; deposition of Joshua Merrill (June 24, 1874), *Merrill v. Yeomans*, MS, Curtis *et al.* (U. S.), 202, 204.

31 Depositions of Joshua Merrill (Nov. 2, 1870), *Merrill v. Yeomans*, MS, Curtis *et al.* (U. S.), 11, 12, 34; deposition of William Atwood (Oct. 27 and 28, 1870), *Merrill v. Yeomans*, MS, Curtis *et al.* (U. S.), 25, 29, 47.

32 Deposition of Joshua Merrill (Nov. 2, 1870), *Merrill v. Yeomans*, MS, Curtis *et al.* (U. S.), 11, 12, 34.

33 Deposition of Joshua Merrill (Nov. 2 and 4, 1870), *Merrill v. Yeomans*, MS, Curtis *et al.* (U. S.), 43–51, 13–15; U. S. Patent No. 90,264, issued to Joshua Merrill, May 18, 1869; S. D. Hayes, *American Gas-Light Journal*, XVI (May 16, 1872), 157–58.

34 U. S. Patent No. 90,264, issued to Joshua Merrill, May 18, 1869.

35 Deposition of Joshua Merrill (Nov. 2, 1870), *Merrill v. Yeomans*, MS, Curtis *et al.* (U. S.), 7–9.

36 Merrill's "Reminiscences," *Derrick's Hand-Book*, I, 887–88; S. D. Hayes, *American Gas-Light Journal*, XVI (May 16, 1872), 158; C. F. Chan-

dler, *Report on Petroleum as an Illuminator*, 16, 19–21.

37 Merrill's "Reminiscences," *Derrick's Hand-Book*, I, 888, 889.

38 Deposition of Joshua Merrill (Nov. 7, 1870), *Merrill* v. *Yeomans*, MS, Curtis *et al.* (U. S.), 76.

39 J. Lawrence Smith "Petroleum," United States Centennial Commission, *Reports and Awards* (Washington: Government Printing Office, 1880), IV, 151, 158.

40 A. Gesner, *Treatise* (1865), 156–57; Crew, *Practical Treatise on Petroleum*, 293; C. F. Chandler, *Report on Petroleum as an Illuminator*, 13; J. L. Smith, U. S. Centennial Commission, *Reports and Awards*, IV, 151, 152.

41 *Proceedings of the Polytechnic Association of the American Institute*, Sept. 15, 1864, cited in *Scientific American*, XI, 14 (Oct. 1, 1864), 213; Leet, *Petroleum Distillation*, 26, 31, 32.

42 *American Gas-Light Journal* (Sept. 1, 1862), 77; S. D. Hayes, *American Gas-Light Journal*, XVI (May 16, 1872), 157, 158; A. Norman Tate, *Petroleum and Its Products* (London: John W. Davies, 1863), 93.

43 *Scientific American* (April 22, 1865), Eaton, *Petroleum* 219; S. D. Hayes, *American Gas-Light Journal*, XVI (May 16, 1872), 157, 158.

44 "On the Preservation of Meats by Paraffin," *American Journal of Pharmacy*, XIV (July, 1866), 341; *Scientific American* (Sept. 3, 1870), 152; *National Oil Journal*, IV (Sept., 1874), 6; A. W. Miller, "Paraffin, Cosmoline and Vaseline," *American Journal of Pharmacy* (Jan., 1874), 2, 3.

45 Peckham, *Census Report*, 254; *Scientific American* (Sept. 3, 1870), 152; "Some of the Uses of Paraffin," *National Oil Journal*, IV (Sept., 1874), 6; A. W. Miller, *American Journal of Pharmacy* (Jan., 1874), 3.

46 A. W. Miller, *American Journal of Pharmacy* (Jan., 1871), 2, 3; *National Oil Journal*, IV (Sept., 1874), 6; "Petroleum Products," *American Gas-Light Journal*, VIII (Nov. 2, 1866), 132; *Scientific American* (Sept. 3, 1870), 152.

47 U. S. Patent No. 127,668, issued to Robert M. Chesebrough, June 4, 1872.

48 Robert A. Chesebrough, Letter to Editor, Spring Lake (N. J.) *Gazette*, Sept. 12, 1927; Chesebrough Manufacturing Company, "Chesebrough" (n.d., circa 1952), 4–6, 8.

49 E. F. Houghton & Co., advertisement, *American Journal of Pharmacy* (Oct., 1872), advertising section, B; Robert A. Chesebrough, "Petroleum—Its Origin and Relation to Medicine," *The Medical Union* (Feb., 1872), 45, 49, 50, 51; A. W. Miller, "Cosmoline and Paraffin Ointment," *The American Pharmaceutical Journal and Transactions* (Jan. 17, 1874), 551, 552; Joseph L. Lemberger, "On Cosmoline," *Proceedings of the American Pharmaceutical Association, 1875*, XXII, 384–390.

Chapter 11

1 U. S. Patents No. 43,706 and 33,955 issued to Joshua Merrill, July 2 and Dec. 17, 1861, respectively; Joshua Merrill to W. H. L. Smith, Boston, April 7, 1863, W. H. L. Smith Papers.

2 U. S. Patent No. 32,951, issued to Joshua Merrill, July 30, 1861; U. S. Patent No. 80,294 issued to Charles Lockhart and John Gracie, July 28, 1868. In 1869, Merrill used the identical still in his process to deodorize heavy lubricating oil, with the added option of using a coil to introduce superheated steam to permit closer control at lower working temperatures. See U. S. Patent No. 90,284 issued to Joshua Merrill, May 18, 1869.

3 U. S. Patent No. 32,701, issued to Joshua Merrill, July 2, 1861. Merrill also developed a way to re-use spent sulphuric acids from previous treatments of refined distillates. U. S. Patent No. 32,705, issued July 2, 1861.

4 U. S. Patents No. 40,632 and 80, 294

issued to Charles Lockhart and John Gracie Nov. 17, 1863, and July 28, 1868, respectively.

5 Leet, *Petroleum Distillation*, 39, 40; A. Gesner, *Treatise* (1865), 80, 152, 153, 166.

6 Peckham, *Census Report*, 162 and figure 41; Crew, *Practical Treatise on Petroleum*, 248, 257–59.

7 U. S. Patents No. 40,632 and 80,295 issued to Charles Lockhart and John Gracie Nov. 17, 1863, and July 28, 1868, respectively.

8 Edward Parrish, "On the Rectification of Petroleum," *Journal of the Franklin Institute*, LXI (Feb., 1871), 117–122.

9 *Oil and Gas Journal*, Supplement, XXX (May 24, 1934), 65, 66; R. J. Forbes and D. R. O'Beirne, *The Technical Development of the Royal Dutch Shell* (Leiden; E. J. Brill, 1957), 299, 300.

10 Day (ed.), *Handbook of the Petroleum Industry*, II, 343, 344; Cf. U. S. Patent No. 58,005 issued to P. H. Van der Weyde, Sept. 11, 1866, antedated Aug. 8, 1866; No. 72,125 issued to H. W. C. Tweddle, Dec. 10, 1867, No. 87,485 issued to Samuel Gibbons, March 2, 1869. On the plausible basis of his pioneer use of superheated steam to refine deodorized lubricating oils in small stills, Joshua Merrill is usually credited with introducing steam into distillation in large stills, but P. H. Van der Weyde and many other contemporaries were also working effectively in the same field.

11 Deposition of Joshua Merrill (Nov. 3, 1870), *Merrill* v. *Yeomans*, MS, Curtis *et al.* (U. S.), 1870, 26, 27.

12 S. D. Hayes, *American Gas-Light Journal*, XVI (May 2, 1872) 137; *Scientific American*, Supplement (May 18, 1872), 340; Crew, *Practical Treatise on Petroleum*, 247–49.

13 Testimony of Lewis Emery, *United States* v. *Standard Oil Company of New Jersey, et al.* (U. S. Circuit Court, Eastern Division of Eastern Judicial District of Missouri, 1906) *Transcript*, VI, 604, hereafter cited as *US* v. *SONJ* (1906); Peckham, *Census Report*, 162; Crew, *Practical Treatise on Petroleum*, 247–53.

14 Miller and Osborn, in Dunstan *et al.*

(eds.), *The Science of Petroleum*, II 1466–1477; W. A. Peters, "History of the Development of the Fractionating Tower," *Oil and Gas Journal*, XXIII (March 5, 1925), 76, 161, 162; Forbes, *Short History of Distillation*, 348, 349.

15 Boverton Redwood, *A Practical Treatise on Petroleum* (3rd ed.: London: Charles Griffin & Company, Limited, 1913), II, 51–58; 109–15.

16 C. F. Chandler, *Report on Petroleum as an Illuminator*, 48.

17 U. S. Patent No. 27,242, issued to D. S. Stombs and Julius Brace, April 10, 1860.

18 U. S. Patent No. 61,120, issued to Alexis Thirault, Jan. 8, 1867.

19 *Pithole Daily Record*, Dec. 15, 1865.

20 U. S. Patent No. 52,167 issued to Adolph Millochau, March 13, 1866.

21 Cf. U. S. Patent No. 58,512, issued to Peter Van der Weyde Oct. 2, 1866, antedated Sept. 21, 1866.

22 U. S. Patent No. 58,005 issued to P. H. Van der Weyde, September 11, 1860, antedated Aug. 8, 1866; U. S. No. 191,294 issued to Samuel Van Syckel, May 22, 1877.

23 U. S. Patent No. 62,096 issued to P. H. Van der Weyde, Jan. 20, 1867.

24 U. S. Patent No. 120,530 issued to Henry H. Rogers, Oct. 31, 1871.

25 "Van Syckel Continuous Process for Refining Petroleum," *National Oil Journal*, IV (June, 1874), 9.

26 Cf. U. S. Patent No. 191,294, issued to Samuel Van Syckel, May 22, 1877; "The Van Syckel Continuous Process for Refining Petroleum," *National Oil Journal*, IV (June, 1874), 9.

27 *Titusville Morning Herald*, April 15, 1871; McLaurin, *Sketches in Crude Oil*, 266–67.

28 Deposition of Joshua Merrill (Nov. 3, 1870), *Merrill* v. *Yeomans*, MS, Curtis *et al.* (U. S.), 1870, 26, 27; J. L. Smith, U. S. Centennial Commission, *Reports and Awards*, IV, 154, 155; *Scientific American*, Supplement (May 18, 1872), 340.

29 Cf. J. T. Henry, *Early History*, 315–21; *Titusville Morning Herald*, March 3, 1871; Parrish, *Journal of the Franklin Institute*, LXI (Feb., 1871), 117–21; Testimony of William K.

Harkness and Lewis Emery, *US v. SONJ* (1906), *Transcript,* I, 221, 222, 228.

30 A. Gesner, *Treatise* (1865), 161–67; J. L. Smith, U. S. Centennial Commission, *Reports and Awards,* IV, 154, 155.

31 *Titusville Morning Herald,* Sept. 1, 1871.

32 Parrish, *Journal of the Franklin Institute,* LXI (Feb., 1871), 117–22.

33 A. W. Miller, *American Journal of Pharmacy* (Jan., 1874), 2, 3.

34 Deposition of Joshua Merrill (Nov. 3, 1870), *Merrill* v. *Yeomans,* MS, Curtis *et al.* (U. S.) 1870, 26, 27, 29; J. L. Smith, U. S. Centennial Commission, *Reports and Awards,* IV, 149–52.

35 Deposition of Joshua Merrill (Nov. 3, 1870, and March 28, 1872) *Merrill* v. *Yeomans,* MS, Curtis *et al.* (U. S.), 1870, 26–29 and *passim;* 1872, 1–3; J. L. Smith, U. S. Centennial Commission, *Reports and Awards,* IV, 149–52.

36 J. L. Smith, U. S. Centennial Commission, *Reports and Awards,* IV, 147–52; S. D. Hayes, *American Gas-Light Journal,* XVI, (May 16, 1872), 157–58.

37 J. Lawrence Smith, U. S. Centennial Commission, *Reports and Awards,* IV, 147–52.

38 *Scientific American,* Supplement (May 18, 1872), 340; S. D. Hayes, *American Gas-Light Journal,* XVI (May 2, 1872), 137; J. T. Henry, *Early History,* 316; C. F. Chandler, *Report on Petroleum as an Illuminator,* 15, 17, 19.

39 Testimony of Lewis Emery, *US v. SONJ* (1906) *Transcript,* I, 228; *Na-tional Oil Journal,* IV (June, 1874), 9.

40 E. Parrish, *Journal of the Franklin Institute,* LXI (Feb., 1871), 117–21; J. L. Smith, U. S. Centennial Commission, *Reports & Awards,* IV, 154, 155.

42 *Titusville Morning Herald,* March 3, 1871.

43 *National Oil Journal,* IV (June, 1874), 9; *Derrick's Hand-Book,* I, 140.

44 Testimony of Lewis Emery, *US v. SONJ* (1906), *Transcript,* I, 228, VI, 604, 2738–39.

45 J. L. Smith, U. S. Centennial Commission, *Reports & Awards,* IV, 154, 155; *National Oil Journal,* IV (June, 1874), 9.

46 Testimony of William K. Harkness and Lewis Emery, *US v. SONJ* (1906), *Transcript,* I, 221, 222, 228.

47 *Titusville Morning Herald,* March 3, 1871.

48 Testimony of August H. Tack, *US v. SONJ* (1906), *Transcript,* VI, 212, 214, 215.

49 "Petroleum—Its Origin, Description, and History," *National Oil Journal,* II (Aug., 1874), 5. Average refining wages were reported as $17.50 a week in currency.

50 E. Parrish, *Journal of the Franklin Institute,* LXI (Feb., 1871), 117, 121.

51 Cf. J. L. Smith, U. S. Centennial Commission, *Reports & Awards,* IV, 154, 155; *Scientific American,* Supplement (May 18, 1872); E. Parrish, *Journal of the Franklin Institute,* LXI (Feb., 1871), 117, 121.

52 *National Oil Journal,* II (Aug., 1874), 5.

53 *Titusville Morning Herald,* April 15, 1871.

Chapter 12

1 Raymond Foss Bacon and William Allen Hamor, *The American Petroleum Industry,* II (New York: McGraw-Hill Book Co., Inc., 1916), 210.

2 Morris, *Derrick and Drill,* 48.

3 *Derrick's Hand-Book,* I, 884.

4 *Philadelphia Commercial List and Price Current* (Aug. 27, 1864), 138.

5 J. T. Henry, *Early History,* 101–2, 225–28, 331–34, 360–74, 494–502; *Derrick's Hand-Book,* I, 870. From long experience, Samuel Downer rated Funk the best and most reliable, while he thought Haldeman among

the "more shiftless class of Oil Creek merchants." Samuel Downer to W. H. L. Smith, Feb. 12, 1863, W. H. L. Smith Papers.

6 Boyle testimony, in Industrial Commission, *Preliminary Report on Trusts and Industrial Combinations* (Washington: Government Printing Office, 1900), I, 450, hereafter cited as *Industrial Commission Report*.

7 *Derrick's Hand-Book*, II, 28.

8 James Lee, *Industrial Commission Report*, I, 280.

9 *Cleveland Leader*, Jan. 5, 1865, cited in Maybee, *Railroad Competition*, 15.

10 Wright, *The Oil Regions of Pennsylvania*, 204–5.

11 Cf. J. T. Henry, *Early History*, 320; *Scientific American*, XV (Dec. 15, 1866), 490; Joshua Merrill to Samuel Downer, Dec. 23, 1861, and June 28, 1862, W. H. L. Smith Papers.

12 *Derrick's Hand-Book*, I, 807.

13 *Derrick's Hand-Book*, I, 780; J. T. Henry, *Early History*, 315–32.

14 *Derrick's Hand-Book*, I, 780.

15 McLaurin, *Sketches in Crude Oil*, 281–82.

16 Joseph W. Orr, "Razing of Oil City Oil Exchange Recalls Days of Speculative Frenzy," *Oil & Gas Journal*, XXIII (April 9, 1925), 92–96.

17 Orr, *Oil & Gas Journal*, XXIII (April 9, 1925), 94.

18 Maybee, *Railroad Competition*, 198, 200, 204.

19 Maybee, *Railroad Competition*, 211.

20 Maybee, *Railroad Competition*, 222.

21 Maybee, *Railroad Competition*, 237.

22 Maybee, *Railroad Competition*, 231–32.

23 *Buffalo Express* in *Titusville Morning Herald*, April 6, 1872.

24 Cf. Maybee, *Railroad Competition*, 235.

25 *New York Bulletin* in *Titusville Morning Herald*, May 9, 1871; Maybee, *Railroad Competition*, 314. "Crude equivalents" computed on a basis of 1.4 barrels of crude per barrel of refined.

26 Allan Nevins, *John D. Rockefeller: The Heroic Age of American Enterprise* (2 vols.; New York: Charles Scribner's Sons, 1940), I, 98.

27 Nevins, *John D. Rockefeller*, I, 191.

28 Nevins, *John D. Rockefeller*, I, 196.

29 Nevins, *John D. Rockefeller*, I, 292.

30 Ida M. Tarbell, *The History of the Standard Oil Company*, I (New York: McClure, Phillips & Co., 1904), 45–46.

31 Nevins, *John D. Rockefeller*, I, 257, 258.

32 These contracts, dated May 14 and 15, 1868, are reproduced by Chester Mc. Destler in "The Standard Oil, Child of the Erie Ring, 1868–1872," *Mississippi Valley Historical Review*, XXXIII (June, 1946), 103–4.

33 Destler, *Mississippi Valley Historical Review*, XXXIII (June, 1946), 107.

34 Destler, *Mississippi Valley Historical Review*, XXXIII (June, 1946), 108.

35 Destler, *Mississippi Valley Historical Review*, XXXIII (June, 1946), 108.

36 Destler, *Mississippi Valley Historical Review*, XXXIII (June, 1946), 111.

37 Destler, *Mississippi Valley Historical Review*, XXXIII (June, 1946), 113.

38 Maybee, *Railroad Competition*, 221.

39 *Report of the Commissioner of Corporations on the Petroleum Industry*, Part 1, *Position of the Standard Oil Company in the Petroleum Industry* (Washington: Government Printing Office, 1907), 48.

40 Nevins, *John D. Rockefeller*, I, 296.

41 Affidavit of James H. Devereux in the case of the *Standard Oil Company* v. *William C. Scofield et al.* in the Court of Common Pleas, Cuyahoga County, Ohio, reproduced in Tarbell, *History of Standard Oil*, I, 277–79.

42 Affidavit of Devereux, *Standard Oil Co.* v. *William C. Scofield et al.*, in Tarbell, *History of Standard Oil*, I, 278.

43 Affidavit of Devereux, *Standard Oil Co.* v. *William C. Scofield et al.*, in Tarbell, *History of Standard Oil*, I, 278.

44 Affidavit of Devereux, *Standard Oil Co.* v. *William C. Scofield et al.*, in Tarbell, *History of Standard Oil*, I, 279.

45 Tarbell, *History of Standard Oil*, I, 49.

46 *New York Bulletin*, reprinted in *Titusville Morning Herald*, Jan. 12, 1871.

47 Maybee, *Railroad Competition*, 276.

48 *Cleveland Leader* in *Titusville Morning Herald*, Feb. 28, 1872. With production of about 5.2 million barrels in 1871, this estimate would indicate a daily refining capacity for the industry of approximately 35,000 barrels.
49 Quoted in *Titusville Morning Herald*, July 28, 1871.
50 Feb. 26, 1872.
51 Quoted in *Titusville Morning Herald*, July 28, 1871.

52 Quoted in *Titusville Morning Herald*, Feb. 24, 1872.
53 *Derrick's Hand-Book*, I, 160.
54 Devereux, *Standard Oil Co. v. William C. Scofield et al.*, in Tarbell, *History of Standard Oil*, I, 277.
55 *New York Commercial Bulletin* in *Titusville Morning Herald*, May 9, 1871; *Cleveland Leader*, July 24, 1871.

Chapter 13

1 U. S. Dept. of Commerce, Bureau of the Census, with the cooperation of the Social Science Research Council, *Historical Statistics of the United States, 1789–1945: A Supplement to the Statistical Abstract of the United States* (Washington: Government Printing Office, 1949), 232, hereafter cited as *Historical Statistics of the U. S., 1789–1945.*
2 Samuel Downer to Joshua Merrill, New York, Feb. 24, 1862, W. H. L. Smith Papers.
3 *Shipping & Commercial List and New York Price Current,* Jan. 14, 1871.
4 *Scientific American,* II (Jan. 2, 1860), 3; "Kerosene Lamp Burners," *Scientific American,* VIII, (June 13, 1863), 373.
5 "The Dangers of Our Artificial Lights," *Scientific American,* XVI (May 4, 1867), 285.
6 *Scientific American,* X (May 24, 1864), 344; *Scientific American,* XIII (Aug. 19, 1865), 114; *Scientific American,* XV (Aug. 4, 1866), 86.
7 *Scientific American,* XVI (May 4, 1867), 225; Larry Freeman, *Light on Old Lamps* (rev. ed.; Waltham Glen, N. Y.: Century House, 1946), 26–30.
8 Freeman, *Light on Old Lamps,* 26–29.
9 C. F. Chandler, *Report on Petroleum as an Illuminator,* 33.
10 James C. Booth and Thomas H. Garrett, "Experiments on Illumination with Mineral Oils," *Journal of the Franklin Institute,* reprinted in *American Gas-Light Journal* (July 15, 1862), 26, 27.

11 Booth & Garrett in *American Gas-Light Journal* (July 15, 1862), 26, 27.
12 *Scientific American,* XVI (May 4, 1867), 285.
13 *American Gas-Light Journal* (Nov. 15, 1862), 154, (Dec. 1, 1862), 16.
14 C. F. Chandler, *Report on Petroleum as an Illuminator,* 56, 92, 93.
15 C. F. Chandler, *Report on Petroleum as an Illuminator,* 60, 85–86, 103, 104.
16 *Shipping and Commercial List and New York Price Current,* Jan. 6, 1864, Jan. 7, 1865, Jan. 23, 1867, Jan. 8, 1870; Ivan C. Michels, "Petroleum, Historically, Scientifically, and Commercially Reviewed," *Atlas of the Oil Region of Pennsylvania,* ed. F. W. Beers *et al.* (New York: F. W. Beers, A. D. Ellis & G. G. Soule, 1865), 3, 4, hereafter cited as *Atlas of the Oil Region.*
17 "Refined Quotations," *Derrick's Hand-Book,* I, 781.
18 "Refined Quotations," *Derrick's Hand-Book,* I, 781–82; Michels, *Atlas of the Oil Region,* 3, 4.
19 Joshua Merrill to W. H. L. Smith, Boston, Oct. 9, 1861, Wm. B. Merrill to Samuel Downer, Boston, Sept. 16, 1862, Samuel Downer to W. H. L. Smith, Boston, Oct. 31, 1862, W. H. L. Smith Papers.
20 Leet, *Petroleum Distillation,* 19–21.
21 Cf. Testimony of Robert Chesebrough, Jan. 30, 1868, *The National Filtering Company v. The Arctic Oil Company* (U. S.); Invoice, Capen, Sherman & Company, June 23, 1866,

Scheide Collection, Drake Museum. In some instances a large marketer might act as an agent for several small refiners. Rider and Clark, for example, with offices in New York and Pittsburgh, in 1865 represented a number of small refiners, including the Portland Kerosene Company and the New York Paraffin Candle Company. Cf. Advertisement, *Atlas of the Oil Region,* advertising section, 5–6.

22 Cf. Samuel Downer to W. H. L. Smith, 1861–1863, *passim,* W. H. L. Smith Papers.

23 *New York Shipping and Commercial List and Price Current,* Jan. 7, 1865.

24 Samuel Downer to W. H. L. Smith, Boston, Sept. 22, 1863, W. H. L. Smith Papers.

25 *Shipping and Commercial List and New York Price Current,* Jan. 20, 1866, Jan. 23, 1867.

26 *Chemical News,* X, 204, cited in Peckham, *Census Report,* 261.

27 *Thirty-First Annual Report of the Philadelphia Board of Trade, for the Year 1863,* 34; *Annual Report of the Cincinnati Chamber of Commerce and Merchants' Exchange for the Year Ending August 31, 1868,* 78.

28 *Annual Statement of the Cleveland Chamber of Commerce, for the Year 1866,* 51, . . . *for the Year 1870,* 25, . . . *for the Year 1871,* 97–99, 119.

29 Cf. *Historical Statistics of the U. S., 1789–1945,* 232.

30 Interview with Lewis Peterson, Sr., of Tarentum, Pa., *The Pittsburgh Dispatch,* Aug. 7, 1892.

31 *Shipping and Commercial List and New York Price Current,* Jan. 14, 1871; Tate, *Petroleum and Its Products,* 112.

32 Samuel B. MacDonnell, "Early Oil Export," *Derrick's Hand-Book,* I, 1023–25.

33 *Shipping and Commercial List and New York Price Current,* Jan. 14, 1871.

34 David Murray, "Petroleum, Its History and Properties" (paper read before the Albany Institute, Dec. 16, 1862).

35 Giddens, *Birth of the Oil Industry,* 96, 97.

36 *Scientific American,* VIII (June 13, 1863), 373.

37 *American Gas-Light Journal,* V (June 1, 1863), 361, (Sept. 1, 1863), 146.

38 Secretary of State, *Report on Commercial Relations of the United States with Foreign Countries for the Year ended September 30, 1863,* 457–58.

39 Secretary of Treasury, *Report of Domestic Export for the Year Ending Sept. 30, 1864,* 65, *Report for 1865,* 33.

40 "Petroleum," *The Daily Commercial* (Pittsburgh), March 5, 1868.

41 Secretary of State, *Report on Commercial Relations . . . for the Year ended Sept. 30, 1863,* 170–71.

42 S. S. Hayes, *Report . . . on Petroleum,* 17.

43 Giddens, *Birth of the Oil Industry,* 97, 98.

44 "Cap. 22: An Act to Amend the Laws Relating to the Customs," 15 May 1860, effective upon passage, *The Statutes of the United Kingdom and Ireland,* 23–24 Victoria, 1860 (London: Eyre and Spottiswoode, 1860), C, 80; "Cap. 22: An Act to Grant Certain Duties of Customs and Inland Revenue," 8 June 1863, effective 1 July 1863, *The Statutes of the United Kingdom of Great Britain and Ireland,* CIII, 73, 74.

45 See Chapter 8.

46 Prospectus of Petroleum Trading Company, Inc., cited in Tate, *Petroleum and Its Products,* 108, 109.

47 B. Martell, "On the Carriage of Petroleum in Bulk on Over-Sea Voyages," *Transactions of the Institution of Naval Architects,* XXVIII (1887), 6, 7.

48 *London Times* (Aug. 20, 1867), 8; I. I. Redwood, *Mineral Oils & Their By-Products,* 10, 38–45; Conacher, Geological Survey, Scotland, *The Oil-Shales of the Lothians,* 245, 246.

49 Maybee, *Railroad Competition,* 71, 72.

50 Cf. *Derrick's Hand-Book,* I, 1023–25; S. S. Hayes, *Report . . . on Petroleum,* 25, 26; Tate, *Petroleum and Its Products,* 100.

51 "Petroleum Annual Review," *Shipping and Commercial List and New York Price Current,* Jan. 20, 1866;

"Petroleum Brokers," *National Oil Journal*, III (May, 1873), 8.

52 *Derrick's Hand-Book*, I, 84; *American Artisan and Patent Record*, VI (May 11, 1868), 137.

53 S. S. Hayes, *Report . . . on Petroleum*, 3, 33; Giddens, *Birth of the Oil Industry*, 195, 196.

54 Dr. John Atterfield, Letter to Editor, *London Times* (June 3, 1866), 10; John Atterfield "On the Igniting Point of Petroleum," *Chemical News*, XIV (Nov. 30, 1866), 257–58; A. Southby, B. F. Paul, and Editor of *Oil Trade Review*, Letter to Editor, *London Times* (Oct. 2, 1866), 10.

55 A Chaplain of the Bishop of Gibraltar, Letter to Editor, *London Times* (Sept. 25, 1868), 4; "Petroleum—In Law, In Commerce, In Transit, and In Store," *The Engineer*, XXVIII (Nov. 12, 1869), 320.

56 Cf. C. F. Chandler, *Petroleum as An Illuminator*, 64–69; B. F. Paul, "Petroleum Oil," *The British Medical Journal*, II (Dec. 25, 1869), 67.

57 Boverton Redwood, Letter to Editor, *London Times* (Aug. 24, 1871), 7; Boverton Redwood, *A Practical Treatise on Petroleum* (4th ed., 1922), II, 979–85.

58 *The Engineer*, XXVIII (Nov. 12, 1869), 320.

59 *Titusville Morning Herald* (Sept. 11, 1871), 3.

60 S. S. Hayes, *Report . . . on Petroleum*, 17.

61 Secretary of State, *Report on Commercial Relations . . . for the Year Ending September 30, 1873*, 323.

62 Sheppard Bancroft Clough, *France, A History of National Economics 1789–1939* (New York: Charles Scribner's Sons, 1939), 337.

63 Folger, *Pa. Industrial Statistics, 1892*, 170B–172B; *Reports . . . on U. S. Commerce and Navigation for the Years 1864–1873*.

64 Charles Marvin, *The Region of the Eternal Fire* (London: W. H. Allen & Co., Ltd., 1891), 182, 196, 203–6, 253–55.

65 Marvin, *The Region of the Eternal Fire*, 258–63.

66 U. S. House of Representatives, Committee on Manufactures, *Report on Investigation of Trusts*, 50th Cong. 1 sess., House Report No. 3112 (Washington: Government Printing Office, 1888), 526, hereafter cited as *1888 Trust Investigation*.

67 Allan Nevins, *Study in Power* (2 vols.; New York: Scribner's, 1953), I, 224.

68 Henry Clay Pierce, *U. S. v. SONJ.* (1906), *Transcript*, III, 1074.

69 Leet, *Petroleum Distillation*, 20.

70 Hamlin Garland, *Boy Life on the Prairie* (New York: Harper & Brothers, 1899), 1, 4, 166, 167, 176, 178, *passim*.

71 Harriet Beecher Stowe and Catherine Beecher, *American Woman's Home or Principles of Domestic Science* (New York: J. B. Ford & Co., 1869).

72 Cf. Chandler, *Petroleum as an Illuminator*, 53, 56; *American Railroad Journal*, XLI (Feb. 22, 1868), 193.

Chapter 14

1 *Derrick's Hand-Book*, I, 807.

2 Cf. *Cleveland Leader* (Aug. 19, 1869, July 4, 1870). Cf. Lee Benson, *Merchants, Farmers and Railroads: Railroad Regulation and New York Politics, 1850–1887* (Cambridge: Harvard University Press, 1955), 37–39.

3 Cf. *Cleveland Leader* (Aug. 17, 1869, Oct. 25, Nov. 30, 1871).

4 *New York Commercial Bulletin* in *Titusville Morning Herald* (May 9, 1871).

5 Cf. Nevins, *Study in Power*, I, 102–5; Tarbell, *History of Standard Oil*, I, 53–54.

6 Quoted by Maybee, *Railroad Competition*, 298.

7 Nevins, *Study in Power*, I, 100.

8 Petroleum Producers' Union, *A History of the Rise and Fall of the South Improvement Company* (Lancaster, Pa.: Wylie & Griest, Printers, *ca.* 1872), 30, 31, hereafter cited as *History of South Improvement Company*.

9 Maybee, *Railroad Competition*, 405.
10 Nevins, *John D. Rockefeller*, I, 324–25.
11 Nevins, *John D. Rockefeller*, I, 229.
12 *Titusville Morning Herald* (Feb. 26, 1872).
13 *Titusville Morning Herald* (Feb. 22, 1872).
14 *Derrick's Hand-Book*, I, 807.
15 Maybee, *Railroad Competition*, 380.
16 Nevins, *John D. Rockefeller*, I, 332.
17 Nevins, *John D. Rockefeller*, I, 353.
18 *History of South Improvement Company*, 27.
19 Johnson, *Pipelines*, 20–23.
20 Nevins, *Study in Power*, I, 134–36.
21 Nevins, *Study in Power*, I, 136.
22 Nevins, *Study in Power*, I, 136, 415 n.; J. T. Henry, *Early History*, 315–21.
23 Cf. Tarbell, *History of Standard Oil*, I, 63–64.
24 *History of South Improvement Company*, 45.
25 Tarbell, *History of Standard Oil*, I, 64.
26 Tarbell, *History of Standard Oil*, I, 66–67.
27 Nevins, *Study in Power*, I, 138.
28 Nevins, *John D. Rockefeller*, I, 312.
29 Nevins, *John D. Rockefeller*, I, 370.
30 Quoted in Nevins, *John D. Rockefeller*, I, 321.
31 Nevins, *John D. Rockefeller*, I, 372.
32 *History of South Improvement Company*, 51.
33 Nevins, *Study in Power*, I, 145.
34 Nevins, *Study in Power*, I, 134.
35 Tarbell, *History of Standard Oil*, I, 99.
36 *Derrick's Hand-Book*, I, 177.
37 *Derrick's Hand-Book*, I, 179.
38 *Derrick's Hand-Book*, I, 185.
39 *Derrick's Hand-Book*, I, 907.
40 Nevins, *Study in Power*, I, 168.
41 *Derrick's Hand-Book*, I, 197–98.

42 *Derrick's Hand-Book*, I, 201.
43 *Derrick's Hand-Book*, I, 712, 807.
44 *Derrick's Hand-Book*, I, 780.
45 *Derrick's Hand-Book*, I, 780.
46 *History of South Improvement Company*, 27, 28.
47 William T. Scheide testimony, *Hepburn Hearings*, III, 2762.
48 Scheide, *Hepburn Hearings*, III, 2764–65.
49 Scheide, *Hepburn Hearings*, III, 2764–65.
50 Scheide, *Hepburn Hearings*, III, 2767.
51 H. M. Flagler, *1879 Ohio Legislative Hearings*, testimony quoted in Tarbell, *History of Standard Oil*, I, 333.
52 Flagler, *1879 Ohio Legislative Hearings*, testimony quoted in Tarbell, *History of Standard Oil*, I, 333.
53 Flagler, *1879 Ohio Legislative Hearings*, testimony quoted in Tarbell, *History of Standard Oil*, I, 333.
54 Cf. George R Blanchard testimony in *Hepburn Hearings*, III, 3398–3408, and Nevins, *John D. Rockefeller*, I, 445.
55 Blanchard, *Hepburn Hearings*, III, 3393–94.
56 Cf. Alexander Diven testimony in *Hepburn Hearings*, III, 2729; Scheide, *Hepburn Hearings*, III, 2777–85, 2801–5.
57 Cf. Blanchard, *Hepburn Hearings*, III, 3393–95.
58 Scheide, *Hepburn Hearings*, III, 2778–85, 2804–5.
59 Cf. Nevins, *Study in Power*, I, 147, 192, and footnote 8, p. 411.
60 Cf. Nevins, *Study in Power*, I, 175.
61 Nevins, *Study in Power*, I, 192.
62 Nevins, *Study in Power*, I, 176.
63 Nevins, *Study in Power*, I, 175.
64 Nevins, *Study in Power*, I, 99.

Chapter 15

1 Peckham, *Census Report*, 145. Also *Titusville Morning Herald* (Oct. 13, 1873).
2 *Titusville Morning Herald* (Sept. 26, 1873.) At Bradford one contemporary noted that the small risk of dry holes was a signally important consideration and warranted "hardware dealers in taking heavier risks than they could in a field where the percentage of dusters [was] high." *Derrick's Hand-Book*, I, 257.

3 See Wrigley, *Special Report on the Petroleum of Pa.* (1874), 149; Tarbell, *History of Standard Oil*, I, 111; *Oil & Gas Journal* (Aug. 23, 1934), 59; Peckham, *Census Report*, 144–47; *Derrick's Hand-Book*, I, 360–61.

4 See Peckham, *Census Report*, 143.

5 Arnold & Kemnitzer, *Petroleum in U. S.*, 40, 70. *Oil and Gas Journal* (Aug. 23, 1934), 61.

6 *Derrick's Hand-Book*, I, 257.

7 Tarbell, *History of Standard Oil*, I, 112, observed: "If their [producers] oil property had not paid for itself entirely in six months, and begun to yield a good percentage, they were inclined to think it a failure."

8 Wrigley, *Special Report on the Petroleum of Pa.* (1874), J8.

9 Carll, Geological Survey of Pa., *7th Report on the Oil & Gas Fields of W. Pa., for 1887, 1888*, 23.

10 Carll, Geological Survey of Pa., *7th Report on the Oil & Gas Fields of W. Pa., for 1887, 1888*, 24.

11 *Barnard v. Monongahela Gas Co.*, 365 Atl. 802 (1907).

12 *Scientific American* (Aug. 30, 1873), 128.

13 *Pennsylvania Industrial Statistics, 1892*, 85.

14 *Derrick's Hand-Book*, I, 208, 215.

15 *Mineral Resources, 1891*, 419.

16 Tarbell, *History of Standard Oil*, I, 115, 125, 341–43; *Derrick's Hand-Book*, I, 201.

17 See Chapter 14; *Derrick's Hand-Book*, I, 209, 215; *Titusville Morning Herald* (Apr. 3, 1875), reprinting article from *Pittsburgh Dispatch*.

18 Charles R. Fettke, Pennsylvania Geological Survey, *The Bradford Oil Field, Pennsylvania and New York*, Fourth Series, Bulletin M 21 (Harrisburgh, 1938), 5, hereafter cited as Fettke, *Bradford Field*.

19 McLaurin, *Sketches in Crude Oil*, 189.

20 Fettke, *Bradford Field*, 5.

21 Asbury, *The Golden Flood*, 286–87.

22 *Derrick's Hand-Book*, II, 277.

23 *Bradford New Era*, quoted in *Derrick's Hand-Book*, II, 228.

24 *Derrick's Hand-Book*, II, 228.

25 Quoted in *Derrick's Hand-Book*, II, 227; also *Bradford Era* in *Derrick's Hand-Book*, II, 229.

26 *Derrick's Hand-Book*, II, 239–40.

27 *Scientific American* (Nov. 1, 1879), 282.

28 *Scientific American* (Aug. 23, 1879), 180.

29 *Derrick's Hand-Book*, II, 325, 342.

30 Quoted in *Derrick's Hand-Book*, II, 261.

31 *Titusville Morning Herald* (Dec. 27, 1877).

32 *Derrick's Hand-Book*, I, 292, 299, 300; *Titusville Morning Herald* (Aug. 16, 1879).

33 Cf. B. B. Campbell, *Commonwealth of Pennsylvania v. Pennsylvania Railroad Company*, 1879, in *1888 Trust Investigation*, 130–32, 166. Hereafter this court case will be cited as *Pa. v. Pa. RR*.

34 Petroleum Producers' Union, *A History of the . . . Grand Council of the Petroleum Producers' Unions . . .* (n.p., 1880), 10–15, Scheide Collection, Drake Museum, hereafter cited as *Producers' Union History*. Also reproduced in *1888 Trust Investigation*, 690–716.

35 *Derrick's Hand-Book*, II, 313–14; *Titusville Morning Herald* (June 21, 1879).

36 *Derrick's Hand-Book*, II, 302.

37 *Titusville Morning Herald* (Aug. 16, 1879).

38 *Titusville Morning Herald* (Aug. 14, 1879).

39 Carll, Geological Survey of Pa., *7th Report on the Oil & Gas Fields of W. Pa., for 1887, 1888*, 29.

40 *Titusville Morning Herald* (Sept. 19, 1879).

41 *Titusville Morning Herald* (July 17, 1879).

42 *Titusville Morning Herald* (Aug. 6, 19, 28, 1879).

43 *Titusville Morning Herald* (Aug. 6, 14, 16, 28, 1879).

44 *OP&D Reporter* (July 30, 1879).

45 *Titusville Morning Herald* (July 29, 1879).

46 *Titusville Morning Herald* (July 25, 1879).

47 *Derrick's Hand-Book*, II, 167, 303.

48 Letter reprinted in *Titusville Morning Herald* (Aug. 5, 1879); also reported in *OP&D Reporter* (Aug. 5, 1879).

49 *Titusville Morning Herald* (Aug. 6, 1879).
50 *Derrick's Hand-Book*, I, 316.
51 *Derrick's Hand-Book*, I, 320.
52 *Titusville Morning Herald* (July 14, 24, 1879); *Derrick's Hand-Book*, II, 302–304, 306–309, 312.
53 *OP&D Reporter* (Oct. 12, 1881).
54 Carll, Geological Survey of Pa., *7th Report on the Oil & Gas Fields of W. Pa., for 1887*, 1888, 3.
55 *Mineral Resources, 1891*, 419.
56 *Derrick's Hand-Book*, I, 346.
57 *Derrick's Hand-Book*, I, 346.
58 *Derrick's Hand-Book*, I, 346.
59 *Derrick's Hand-Book*, I, 347.
60 *Derrick's Hand-Book*, I, 347.
61 *Derrick's Hand-Book*, I, 343, 354.

62 *Pa. Industrial Statistics, 1892*, 20B.
63 *Bradford Era*, in *Derrick's Hand-Book*, I, 384.
64 *Derrick's Hand-Book*, I, 376–77.
65 *Derrick's Hand-Book*, I, 378.
66 *Bradstreets*, XI (Jan. 10, 1885), 20; *OP&D Reporter* (Aug. 27, 1884), 5.
67 *Bradstreets*, V (May 13, 1882).
68 *Titusville Morning Herald* (July 26, 1883).
69 See Nevins, *John D. Rockefeller*, I, 503–4.
70 *Titusville Morning Herald* (Sept. 17, 1881).
71 The geologists were C. F. Ashburner and J. F. Carll in *OP&D Reporter* (Nov. 4, 1884), 11.

Chapter 16

1 *Pa. Industrial Statistics, 1892*, B87.
2 *Derrick's Hand-Book*, I, 209, 215.
3 *Derrick's Hand-Book*, I, 177, 204, 218.
4 *Titusville Morning Herald* (Oct. 25, Nov. 13, 18, 1873).
5 *Derrick's Hand-Book*, I, 203.
6 J. T. Henry, *Early History*, 285; *Titusville Morning Herald* (Oct. 14, 1873); *Pittsburgh Commercial* in *Titusville Morning Herald* (Sept. 20, 1877).
7 *Titusville Morning Herald* (Aug. 13, Oct. 14, 1873).
8 *Titusville Morning Herald* (Aug. 13, 1873).
9 Nevins, *John D. Rockefeller*, I, 443.
10 For 1872, see J. T. Henry, *Early History*, 285. For 1875, see *Titusville Morning Herald* (Apr. 3, 1875), reprinting article from *Pittsburgh Dispatch*.
11 *Titusville Morning Herald* (May 19, 1873).
12 *Derrick's Hand-Book*, I, 220, 223, 226.
13 Memorandum of Agreement between United Pipe Lines, Union Pipe Company, *et al.* (Sept. 4, 1874), *Hepburn Hearings*, III, 3431–37, hereafter cited as *1874 Pipe Agreement*.
14 *Derrick's Hand-Book*, I, 217.
15 *1874 Pipe Agreement; Titusville Morning Herald* (Nov. 2, 1874).

16 Cf. *1874 Pipe Agreement;* Wm. T. Scheide, *Hepburn Hearings*, III, 2793.
17 Melville J. Ulmer, *Trends and Cycles in Capital Formation by United States Railroads, 1870–1950*, Occasional Paper 43, Studies in Capital Formation and Financing (New York: National Bureau of Economic Research, Inc., 1954), 25–34.
18 Blanchard, *Hepburn Hearings*, III, 3171–72, IV, 3508–9.
19 *Baltimore Bulletin* in *Titusville Morning Herald* (March 2, 1871); *New York Tribune* (Feb. 6, 1875), in Johnson, *Pipelines*, 42.
20 Blanchard, *Hepburn Hearings*, III, 3398–3408. Text of the contract will be found on pages 3398–3402.
21 Cited in Thomas C. Cochran, *Railroad Leaders, 1845–1890* (Cambridge: Harvard Univ. Press, 1953), 468. Cochran notes, however, that despite such protests, railroad men often acquiesced in the demands of big shippers rather than surrender the freight to a rival line. Cf. pp. 155–56.
22 Benson, *Merchants, Farmers and Railroads*, 39.
23 Memorandum of Agreement between the Erie Railway Company, the Pennsylvania Railroad Company, *et al.*, October 1, 1874, in *Hepburn Hear-*

ings, III, 3415–21, hereafter cited as *1874 Railroad Agreement.*

24 Blanchard, *Hepburn Hearings,* III, 3425–26.

25 *1874 Pipe Agreement.*

26 *1874 Railroad Agreement.*

27 *Titusville Morning Herald* (Sept. 22, 1874). The advance notice, known as the Rutter Circular, can be found in *1888 Trust Investigation,* 363.

28 Quoted in Tarbell, *History of Standard Oil,* I, 142–43.

29 *Titusville Morning Herald* (Sept. 28, 1874).

30 *Titusville Morning Herald* (Sept. 29, 1874).

31 *Derrick's Hand-Book,* I, 234–35.

32 Johnson, *Pipelines,* 39.

33 *Derrick's Hand-Book,* I, 238.

34 Blanchard, *Hepburn Hearings,* III, 3446–47; Scheide, *Hepburn Hearings,* III, 2767–70.

35 *Derrick's Hand-Book,* I, 242.

36 *Pittsburgh Dispatch* in *Titusville Morning Herald* (June 15, 1876).

37 *Titusville Morning Herald* (Mar. 31, 1872).

38 Over its thirty-four year life, the firm averaged annual sales of $750,000. *OP&D Reporter,* XXXIV (Nov. 14, 1888), 8.

39 *OP&D Reporter,* XXXIV (Nov. 14, 1888), 8; Giddens, *Pa. Petroleum,* 351–53; Johnson, *Pipelines,* 34–35.

40 *Pittsburgh Commercial* (Aug. 11, 1874), in *Titusville Morning Herald* (Aug. 13, 1874).

41 *Titusville Morning Herald* (June 17, 1874).

42 *Baltimore American* (June 30, 1874,) in *Titusville Morning Herald* (July 2, 1874); *Pittsburgh Evening Leader* in *Titusville Morning Herald* (July 10, 1874).

43 *Pittsburgh Commercial* (Aug. 11, 1874), in *Titusville Morning Herald* (Aug. 13, 1874); *Pittsburgh Dispatch* in *Titusville Morning Herald* (Nov. 18, 1874).

44 *Pittsburgh Commercial* (Oct. 10, 1874), in *Titusville Morning Herald* (Oct. 12, 1874).

45 Johnson, *Pipelines,* 39–40; *Pittsburgh Chronicle* (Oct. 14, 1874), in *Titusville Morning Herald* (Oct. 16, 1874).

46 *Pittsburgh Commercial* (Oct. 25, 1874), in *Titusville Morning Herald* (Oct. 29, 1874).

47 *Titusville Morning Herald* (Nov. 30, 1874).

48 *Titusville Morning Herald* (Dec. 3, 5, 1874); Johnson, *Pipelines,* 40.

49 *Pittsburgh Dispatch* (Dec. 16, 1874), in *Titusville Morning Herald* (Dec. 18, 1874).

50 *Titusville Morning Herald* (Jan. 5, 1875).

51 *Titusville Morning Herald* (Jan. 25, 1875).

52 *Pittsburgh Dispatch* (Jan. 28, 1875), in *Titusville Morning Herald* (Jan. 30, 1875); Johnson, *Pipelines,* 42.

53 Johnson, *Pipelines,* 44–45; *Titusville Morning Herald* (Dec. 16, 1874).

54 *Pittsburgh Chronicle* in *Titusville Morning Herald* (Dec. 16, 1874).

55 *Titusville Morning Herald* (Mar. 5, 1875).

56 *Derrick's Hand-Book,* I, 242.

57 *New York Tribune* (May 8, 1875), in *Titusville Morning Herald* (May 13, 1875); Robert D. Benson, Address given at Tide Water Dinner, Jan. 17, 1913, in *A Brief History of the Tide Water Companies* (n.p., n.d.), 8, hereafter cited as *Tide Water History.*

58 *Pittsburgh Chronicle* (June 30, 1875), in *Titusville Morning Herald* (July 2, 1875); Benson address, *Tide Water History,* 8; Lewis Emery, *US v. SONJ* (1906).

Professor Nevins says that the railroad first tried to block the hauling by keeping long lines of cars standing on the crossing. While he offers no documentation of this point, such behavior was not inconsistent with the Pennsylvania's other reactions to Conduit. Nevins, *John D. Rockefeller,* I, 471.

59 *Pittsburgh Commercial* (July 29, 1875), in *Titusville Morning Herald* (July 30, 1875).

60 *Pittsburgh Chronicle* in *Titusville Morning Herald* (Mar. 4, 1876).

61 *Pittsburgh Chronicle* in *Titusville Morning Herald* (Mar. 4, 1876). Lewis Irwin, *US v. SONJ* (1906), *Transcript,* VI, 3018.

62 *Pittsburgh Chronicle* in *Titusville Morning Herald* (May 5, 1876).

63 *Titusville Morning Herald* (Feb. 26, 1877).

64 Johnson, *Pipelines,* 279, note 57; *Titusville Morning Herald* (Oct. 3, 1877).

65 *Titusville Morning Herald* (Aug. 18, 1873).

66 *Pittsburgh Commercial* in *Titusville Morning Herald* (July 22, 1876); *Pittsburgh Chronicle* in *Titusville Morning Herald* (Aug. 5, 1876).

67 *Pittsburgh Dispatch* in *Titusville Morning Herald* (June 15, 1876); *Titusville Morning Herald* (July 27, Aug. 26, 1876).

68 *Pittsburgh Commercial* in *Titusville Morning Herald* (July 22, 1876).

69 *Pittsburgh Chronicle* in *Titusville Morning Herald* (Aug. 5, 1876).

70 *Derrick's Hand-Book,* I, 262, 271 ff. Harley was later acquitted, but assessed court costs. *Derrick's Hand-Book,* I, 295.

71 Jabez A. Bostwick, *Hepburn Hearings,* III, 2689–92.

72 Blanchard, *Hepburn Hearings,* III, 3398–3408, 3413–14; Scheide, *Hepburn Hearings,* III, 2789–90.

73 Blanchard, *Hepburn Hearings,* III, 3424.

74 *1874 Pipe Agreement.*

75 Blanchard, *Hepburn Hearings,* III, 3403, 3463.

76 A. J. Cassatt, *Pa. v. Pa. RR.,* in *1888 Trust Investigation,* 196.

77 Blanchard, *Hepburn Hearings,* III, 3486.

78 Blanchard, *Hepburn Hearings,* III, 3486–87.

79 Blanchard, *Hepburn Hearings,* III, 3472–73.

80 On the New York Central, see *US* v. *SONJ* (1906), *Plea of Petitioner,* 217–19. On the Erie, see Blanchard, *Hepburn Hearings,* III, 3450. These mention the 10 per cent rebate. The contract with the Pennsylvania Railroad is mentioned, but its terms not specified by A. J. Cassatt, *Pa. v. Pa. RR.,* in *1888 Trust Investigation,* 196.

Apparently these leases were quite profitable to the railroads and the oil company alike. From $45,000 in 1874, the Erie's royalties climbed to more than $500,000 five years later.

(Blanchard, *Hepburn Hearings,* III, 3411). Standard Oil's profits on these transactions are not recorded, but in view of Rockefeller's reluctance to pay outsiders profits on operations that his firm could perform, it is likely that they were not less.

81 Cassatt, *Pa. v. Pa. RR.* in *1888 Trust Investigation,* 196. Cf. also James H. Rutter, *Hepburn Hearings,* III, 2548.

82 For details of these arrangements, see Nevins, *Study in Power,* I, 209–11.

83 Tarbell, *History of Standard Oil,* I, 148; Johnson, *Pipelines,* 42.

84 Nevins, *Study in Power,* I, 210.

85 Nevins, *Study in Power,* I, 212–13.

86 *Pittsburgh Chronicle* in *Titusville Morning Herald* (May 17, 1875).

87 Tarbell, *History of Standard Oil,* I, 149.

88 Nevins, *Study in Power,* I, 216.

89 Nevins, *Study in Power,* I, 228.

90 Nevins, *Study in Power,* I, 228.

91 Nevins, *Study in Power,* I, 228.

92 Nevins, *Study in Power,* I, 225.

93 Nevins, *Study in Power,* I, 225.

94 Nevins, *Study in Power,* I, 225, 226, 252.

95 Nevins, *Study in Power,* I, 226.

96 *New York Observer* in *Titusville Morning Herald* (May 9, 1877); *Titusville Morning Herald* (May 2, June 15, 1877).

97 *Titusville Morning Herald* (July 20, 1876); *Elmira Advertiser* (Feb. 12, 1877), in *Titusville Morning Herald* (Feb. 13, 1877).

98 *Titusville Morning Herald* (Jan. 5, Feb. 1, 1877). The *Bradford New Era* gave a more sophisticated rationale. Admitting the inferiority of Bradford crude, the paper nonetheless felt that "the geographical difference in our favor is amply sufficient to balance its inferiority." *Bradford New Era* in *Titusville Morning Herald* (May 10, 1877).

99 *Titusville Morning Herald* (Feb. 3, 1877); *New York Observer* in *Titusville Morning Herald* (May 9, 1877).

100 *Titusville Morning Herald* (June 15, 1877); *US* v. *SONJ* (1906), *Brief*

for Defendants on the Facts, I, 39, 48.

101 Joseph Potts, *1888 Trust Investigation,* 259–60.

102 John D. Archbold, *1888 Trust Investigation,* 322–23.

103 Cassatt, *Pa. v. Pa. RR.* in *1888 Trust Investigation,* 176, 198.

104 Cassatt, *Pa. v. Pa. RR.* in *1888 Trust Investigation,* 175, 181.

105 B. B. Campbell, *Pa. v. Pa. RR.* in *1888 Trust Investigation,* 134.

106 Cassatt, *Pa. v. Pa. RR.* in *1888 Trust Investigation,* 202.

107 *Railroad Gazette* (June 8, 1877), in *Titusville Morning Herald* (June 11, 1877).

108 Quoted in Burgess & Kennedy, *History of Pennsylvania Railroad,* 328.

109 Quoted in Burgess & Kennedy, *History of Pennsylvania Railroad,* 371.

110 *Philadelphia Times* in *Titusville Morning Herald* (Oct. 20, 1877); Nevins, *Study in Power,* I, 246; Burgess & Kennedy, *History of Pennsylvania Railroad,* 363.

111 Cassatt, *Pa. v. Pa. RR.* in *1888 Trust Investigation,* 180–81; Nevins, *Study in Power,* I, 245.

112 *Titusville Morning Herald* (Oct. 3, 1877).

113 *Titusville Morning Herald* (Nov. 27, 1877). The line was completed November 24. Since the line was designed to serve Standard's twenty-eight refineries in Pittsburgh, Rockefeller's later statement that Conduit "was just the thing we wanted for our refineries there" is difficult to understand. John D. Rockefeller, *US* v. *SONJ* (1906), *Transcript,* XVI, 3096. Obviously the purchase was made for purely competitive reasons, rather than for any service that might be derived.

114 Nevins, *Study in Power,* I, 248. According to Nevins, Conduit claimed a daily capacity of 1,500 barrels, but Standard "persuaded them to accept a quota of less than half this . . . and left them happy."

115 *Tide Water History,* 9.

116 Nevins, *Study in Power,* I, 250.

117 Benson, *Merchants, Farmers and Railroads,* 46–47.

118 Cassatt to R. W. Downing (comptroller, Pennsylvania Railroad) (May 15, 1878), in *1888 Trust Investigation,* 210.

119 Cassatt, *Pa. v. Pa. RR.* in *1888 Trust Investigation,* 191.

120 *Commonwealth of Pennsylvania* v. *United Pipeline Co.* in *Titusville Morning Herald* (Aug. 28, 1879).

121 Cassatt, *Pa. v. Pa. RR.* in *1888 Trust Investigation,* 190–94.

122 Nevins, *Study in Power,* I, 250.

123 Nevins, *Study in Power,* I, 248.

124 Nevins, *Study in Power,* I, 254.

125 Nevins, *Study in Power,* I, 250–51.

126 N. J. Camden to J. D. Rockefeller, Dec. 21, 1877. Quoted in Nevins, *Study in Power,* I, 252.

127 Nevins, *Study in Power,* I, 254.

128 Nevins, *Study in Power,* I, 255.

129 Nevins, *Study in Power,* I, 255.

Chapter 17

1 *Derrick's Hand-Book,* I, 288.

2 *Producers' Union History,* 9.

3 *Titusville Morning Herald* (Jan. 18 & 30, 1878).

4 *Titusville Morning Herald* (Feb. 1, 1878).

5 *Producers' Union History,* 9.

6 *Producers' Union History,* 24–28.

7 Quoted in Nevins, *John D. Rockefeller,* I, 570.

8 *Producers' Union History,* 28, 33–39.

9 Benson, *Merchants, Farmers, & Railroads,* 116.

10 Cassatt, *Pa. v. Pa. RR.* in *1888 Trust Investigation,* 205.

11 Blanchard, *Hepburn Hearings,* III, 3471.

12 Josiah Lombard, *Hepburn Hearings,* I, 710–11. The reader will recall that Standard Oil at this time leased the New York Central's oil yards. According to Rockefeller, the Central had 1,000 tank cars of its own, which Standard had promised to keep in steady use. (Nevins, *John D. Rockefeller,* I, 563.) It was, of course, all to the company's interest to observe this

contract to the letter during this period.

13 Cassatt, *Pa. v. Pa. RR.* in *1888 Trust Investigation*, 205.

14 *Hepburn Hearings, Report of the Committee*, 44.

15 Simon Sterne, *Hepburn Hearings*, IV, 3968–69.

16 *OP&D Reporter*, XIX (June 1, 1881), 764.

17 *Hepburn Hearings, Report of the Committee*, 68.

18 Blanchard, *Hepburn Hearings*, III, 3491.

19 Blanchard, *Hepburn Hearings*, III, 3492.

20 *Railroad Gazette*, X (Jan. 4, 1878), 7; *OP&D Reporter*, XVII (Mar. 10, 1880), 304.

21 For raw data, see *Railroad Gazette*, X (Jan. 4, 1878), 6, 10.

22 Cf. *New York Times* (Jan. to Mar., 1879) for extensive accounts of the producers' cases.

23 Reproduced in Benson, *Merchants, Farmers, & Railroads*, 136.

24 *OP&D Reporter*, XVI (Sept. 3, 1879), 224.

25 S. Sterne, *Hepburn Hearings*, IV, 3970–71.

26 *Baltimore American* in *Titusville Morning Herald* (Feb. 20, 1878). The opinion is reproduced in *Titusville Morning Herald* (Mar. 13, 1878).

27 *Titusville Morning Herald* (May 23, 1878). The *Herald* commented that Philadelphia's gain was Baltimore's loss, but it is not clear whether Benson planned to abandon the Baltimore terminus entirely, or intended to service both Baltimore and Philadelphia. The *Philadelphia Record* spoke of the "line to Baltimore with a branch at Columbia to Philadelphia." *Philadelphia Record* in *Titusville Morning Herald* (July 20, 1878). The *Oil City Derrick*, in June, also reported that the line would end at Baltimore. *Derrick's Hand-Book*, I, 300.

28 Benson, *Tide Water History*, 10.

29 David Jones (assistant controller, Reading Railroad), *Pa. v. Pa. RR.* in *Titusville Morning Herald* (Jan. 9, 1879).

30 *Derrick's Hand-Book*, I, 659–60.

31 *Titusville Morning Herald* (Mar. 21, 1878).

32 *Titusville Morning Herald* (May 3, 1878). Robinson's sympathy for railroad interests led the *Buffalo Express* to describe him as a "good railroad attorney accidentally elevated to office." *Buffalo Express* in *Titusville Morning Herald* (Apr. 26, 1878).

33 *New York Tribune* (Aug. 8, 1878), in *Titusville Morning Herald* (Aug. 10, 1878).

34 *Titusville Morning Herald* (July 17, 1878).

35 *New York Tribune* (Aug. 8, 1878), in *Titusville Morning Herald* (Aug. 10, 1878).

36 *Titusville Morning Herald* (Sept. 13, 1878).

37 *Titusville Morning Herald* (Aug. 7, Oct. 18), 1878.

38 Cassatt, *Pa. v. Pa. RR.* in *1888 Trust Investigation*, 190–91.

39 *Titusville Morning Herald* (Oct. 18, 1878).

40 Articles of association are reproduced in Tarbell, *History of Standard Oil* II, 295–97.

41 Marvin W. Schlegel, *Ruler of the Reading: The Life and Times of Franklin B. Gowen, 1836–1889* (Harrisburg: Archives Publishing Company of Pennsylvania, 1947), 179–82.

42 Johnson, *Pipelines*, 74.

43 *Titusville Morning Herald* (Apr. 9, 1879). It is not clear whether voting rights were conferred in proportion to the amount of subscriptions or whether each partner had one vote.

44 *Titusville Morning Herald* (Apr. 9, 1879).

45 A. W. Golden address, *Tide Water History*, 23–25.

46 *Titusville Morning Herald* (Apr. 9, 1879).

47 Golden, *Tide Water History*, 30–32.

48 Golden, *Tide Water History*, 31.

49 *Titusville Morning Herald* (Mar. 13, 1879).

50 *Titusville Morning Herald* (June 3, 1879).

51 Golden, *Tide Water History*, 26; *Titusville Morning Herald* (Apr. 9, 1879).

52 *Titusville Morning Herald* (May 22, June 6, 1879).

53 Benson, *Tide Water History,* 13–14. Lombard & Ayres were originally located in New York, on New York Central property bordering the railroad's 65th street freight depot. This plant was vacated early in 1879 after the Central condemned the land in order to expand its yards. The Bayonne works were not in operation until August, some two months after Tidewater began. *OP&D Reporter,* XVI (July 30, 1879), 102.

54 *Philadelphia Record* in *Titusville Morning Herald* (May 22, 1879).

55 Cf. *Titusville Morning Herald* (Mar. 6, 1879).

56 *Titusville Morning Herald* (June 6, 1879).

57 Blanchard, *Hepburn Hearings,* III, 3498.

58 Golden, *Tide Water History,* 28.

59 *Titusville Morning Herald* (Apr. 14, 1880).

60 *The Philadelphia Inquirer* in *Titusville Morning Herald* (Dec. 28, 1878); Tarbell, *History of Standard Oil,* II, 11.

61 William Vanderbilt, *Hepburn Hearings,* II, 1669–70.

62 *Titusville Morning Herald* (Aug. 16, 1879).

63 Schlegel, *Ruler of the Reading,* 183.

64 *Titusville Morning Herald* (June 21, 1879).

65 Nevins, *Study in Power,* I, 365–66.

66 Schlegel, *Ruler of the Reading,* 183. Gowen's estimate implies the improbable condition that this road lost one dollar on every barrel that it carried. There are no reliable figures of the cost to Standard's allies; if one uses cost figures already cited, they probably lost about 20¢ a barrel on shipments approaching one million barrels per month.

67 *Titusville Morning Herald* (Feb. 13, 1880); *OP&D Reporter,* XVII (Mar. 24, 1880), 343. A week later local gathering charges were restored to 20¢, bringing total transportation cost to New York to 80¢ per barrel.

68 *Producers' Union History,* 40–48.

69 *OP&D Reporter,* XVII (Feb. 25, 1880), 208.

70 *Titusville Morning Herald* (Feb. 19, 1880).

71 *Hepburn Hearings, Report of the Committee,* 45; *Producers' Union History,* 30–31.

72 H. M. Flagler, *1888 Trust Investigation,* 783.

73 Nevins, *Study in Power,* I, 349–50; *OP&D Reporter,* XVII (Jan. 14, 1880), 69. Professor Nevins reports that the offer guaranteed sales of 10,000 barrels per day; if so, this must have been made somewhat later, for Tidewater's capacity did not reach that amount until 1880.

74 Both letters quoted in Nevins, *Study in Power,* I, 349.

75 *New York Times* (Nov. 6, 1879), 2.

76 *New York Times* (Feb. 22, 1882), 1; *Derrick's Hand-Book,* I, 318.

77 *New York Times* (Feb. 10, 1880), 2.

78 Affidavits of Daniel Shurmer, John Teagle and John D. Rockefeller, *Standard Oil Company of Ohio* v. *William C. Scofield et al.* (Court of Common Pleas, Cuyahoga County, Ohio, 1880), cited as Petitioner's Exhibits 1003 and 1006, *US* v. *SONJ* (1906), *Transcript,* XXI, 170–78; 187–97. Despite Standard's action, Scofield, Shurmer & Teagle did not sell out to Rockefeller until 1901. During the 1880's the company achieved a measure of fame first by successfully fighting Standard's attempt to enforce the market-limitation contract in the courts and second by suing the Lake Shore railroad for discriminating in favor of Standard on product shipments.

79 Nevins, *Study in Power,* I, 358–59.

80 *OP&D Reporter,* XVII (June 23, 1880), 724.

81 *New York Times* (Sept. 30, 1880), 2.

82 *OP&D Reporter,* XVII (Apr. 28, 1880), 495.

83 *New York Times* (Aug. 28, 1880), 2.

84 *New York Times* (March 4, 1880), 1.

85 Nevins, *Study in Power,* I, 352–53.

86 *OP&D Reporter,* XVIII (Dec. 8, 1880), 716; *OP&D Reporter,* XIX (Apr. 27, 1881), 577; *OP&D Reporter,* XX (Dec. 21, 1881), 1137.

87 *OP&D Reporter,* XVII (Apr. 28, 1880), 495.

88 *OP&D Reporter,* XVII (Apr. 28, 1880), 495.

89 Schlegel, *Ruler of the Reading,* 186–

230. When the Reading was declared bankrupt in May of 1880, Gowen was named as one of the receivers. After a bitter proxie fight, his plan for re-organization was adopted and he was reelected president in January, 1882.

90 Nevins, *Study in Power*, I, 361–63. Gowen had been elected to Tidewater's board of directors in January 1880. *New York Times* (Jan. 31, 1880), 1.

91 *The New York Times* (Sept. 23, 1880), 2; *OP&D Reporter*, XVIII (Sept. 29, 1880), 387; *Titusville Morning Herald* (Oct. 1, 1880).

92 *OP&D Reporter*, XVIII (Oct. 6, 1880), 419.

93 Lewis Emery, *1888 Trust Investigation*, 236.

94 *OP&D Reporter*, XX (June 8, 1881), 791.

95 *OP&D Reporter*, XVIII (Dec. 8, 1880), 716. National Storage was an Empire division that had not been liquidated in 1877. In the interim it had apparently only provided tankage for oil. Potts and George B. Roberts, by this time president of the Pennsylvania, were large stockholders. Nevins, *Study in Power*, I, 367.

96 Roberts to Rockefeller, April 9, 1880, quoted in Nevins, *Study in Power*, I, 353.

97 *New York Times* (Jan. 7, 1881), 2.

98 *Industrial Commission Report*, I, 763.

99 *OP&D Reporter*, XIX (June 8, 1881), 791; Nevins, *Study in Power*, I, 368–70. The timing of the change in the Pennsylvania's share indicates that some tentative agreement on rates had been reached before the formal papers were signed in May.

100 *Industrial Commission Report*, I, 104; Maybee, *Railroad Competition*, 292.

101 *Derrick's Hand-Book*, I, 806.

102 *OP&D Reporter*, XX (Oct. 12, 1881), 629.

103 Nevins, *Study in Power*, I, 366. On Tidewater's tankage problems, cf. *Derrick's Hand-Book*, I, 330.

104 *OP&D Reporter*, XIX (May 11, 1881), 632.

105 *New York Times* (Jan. 24, 1882), 2.

106 *New York Times* (Aug. 24, 1881), 5.

107 *Titusville Morning Herald* (Oct. 18, 1880); John P. Herrick, *Empire Oil: The Story of Oil in New York State* (New York: Dodd, Mead & Co., 1949), 194.

108 *New York Times* (Jan. 24, 1882), 2; C. B. Matthews, *1888 Trust Investigation*, 425.

109 *OP&D Reporter*, XXI (Mar. 8, 1882), 436, (Apr. 26, 1882), 817.

110 Computed from raw data on pipeline shipments published monthly in *Petroleum Age*, November, 1881–February, 1882.

111 *New York Times* (Jan. 24, 1882), 2.

112 Benson, *Tide Water History*, 15.

113 Benson, *Tide Water History*, 14–15.

114 Nevins, *Study in Power*, I, 377–78. Nevins states that the transaction was "apparently not made in an atmosphere of intense hostility. The Tidewater evidently felt somewhat apologetic about the event, which was soon known, and seems to have indicated that the stock it sold . . . was non-voting stock. Archbold was at pains to correct this impression."

115 J. Lombard, *US* v. *SONJ* (1906), *Transcript*, I, 250.

116 Benson, *Tide Water History*, 15; Tarbell, *History of Standard Oil*, II, 16.

117 Opinion of the Court, *Tidewater Pipe Company Limited* v. *John Satterfield, et al.* (Court of Common Pleas, Crawford County, Pennsylvania, February Term, 1883), cited as Petitioner's Exhibit 23, *US* v. *SONJ* (1906), *Transcript*, VII, 89–100. Cf. also Benson, *Tide Water History*, 16–18.

118 *Titusville Morning Herald* (Jan. 22, 1883).

119 *Titusville Morning Herald* (Feb. 22, 1883).

120 Nevins, *Study in Power*, I, 375–76; R. D. Benson, *US* v. *SONJ* (1906), *Transcript*, I, 207–8.

121 Johnson, *Pipelines*, 118–21; Nevins, *John D. Rockefeller*, II, 105–6.

122 *OP&D Reporter*, XXIII (May 29, 1883), 1438.

123 *Titusville Morning Herald* (Jan. 22, 1883).

124 *US* v. *SONJ* (1906), *Plea of Peti-tioners,* Exhibit 13, 251–60.

125 *New York Times* (Oct. 17, 1883), 4.

126 *Petroleum Age,* III (April, 1884), 672. The capacity figures are esti-mated from testimony before the House Trust Investigation of 1888. Daniel O'Day placed the capacity of the New York lines at 25,000 bar-rels per day; Archbold testified that a six-inch line would run about 10,-000 barrels per day with the pump-ing power then available (*1888 Trust Investigation,* 285, 346). These fig-ures correspond closely with those of Tidewater's capacity. Although O'Day and Archbold were speaking of a period five years later, their esti-mates are at least roughly applicable to 1883 since the pumping power of the lines had not been notably in-creased in the interim. On the basis of these data, H. M. Flagler's esti-mate that the two lines had a com-bined daily capacity of 32,000 bar-rels per day in 1883 must be re-jected. *OP&D Reporter,* XXIII (Jan. 17, 1883), 129.

127 *Petroleum Age,* III (Apr. 1884), 682.

128 *OP&D Reporter,* XXIV (Oct. 31, 1880), 10.

129 *OP&D Reporter,* XXIV (Aug. 1, 1883), 12; *Petroleum Age,* III (Apr., 1884), 682.

130 *OP&D Reporter,* XXIV (Nov. 28, 1883), 13, (Dec. 11, 1883), 11.

131 *OP&D Reporter,* XXIV (Oct. 31, 1883), 10; Peckham, *Census Re-port,* 271.

132 Cf. *New York Times* (Jan. 24, 1882), 2.

133 O'Day, *1888 Trust Investigation,* 286.

134 Ralph W. Hidy and Muriel E. Hidy, *Pioneering in Big Business, 1882–1911: History of Standard Oil Com-pany (New Jersey)* (New York: Harper & Bros., 1955), 726, note 18, hereafter cited as Hidy & Hidy, *Pio-neering.*

135 *US* v. *SONJ* (1906), *Plea of Peti-tioners,* Exhibit 14, 262–68.

136 J. D. Archbold, *US* v. *SONJ* (1906), *Transcript,* XVII, 3442.

137 Archbold, *US* v. *SONJ* (1906), *Transcript,* XVII, 3445; Archbold, *1888 Trust Investigation,* 322.

138 O'Day, *1888 Trust Investigation,* 267; D. Kirk, *1888 Trust Investiga-tion,* 71.

139 C. B. Matthews, *1888 Trust Investi-gation,* 426.

140 Cf. I. E. Dean, *1888 Trust Investiga-tion,* 91; D. Kirk, *1888 Trust Investi-gation,* 71–72; L. Emery, *Industrial Commission Report,* I, 666.

141 Archbold, *1888 Trust Investigation,* 325; H. H. Rogers, *Industrial Com-mission Report,* I, 581, 588.

142 *Derrick's Hand-Book,* I, 347–48, 360.

143 Hidy & Hidy, *Pioneering,* 177–83.

144 J. D. Rockefeller, *US* v. *SONJ* (1906), *Transcript,* XVI, 3094.

145 Hidy & Hidy, *Pioneering,* 79; *Der-rick's Hand-Book,* I, 381.

146 Cf. Leslie Cookenboo, Jr., *Crude Oil Pipe Lines and Competition in the Oil Industry* (Cambridge, Mass.: Harvard University Press, 1955), 14–30.

147 O'Day, *1888 Trust Investigation,* 272.

148 *OP&D Reporter,* XXIV (Oct. 31, 1883), 10.

149 *US* v. *SONJ* (1906), *Brief for De-fendants on the Facts,* I, 176.

Chapter 18

1 See *Shipping & Commercial List & New York Price Current* (Jan. 9, 1882); *New York Times* (Jan. 24, 1882).

2 *OP&D Reporter* (Aug. 25 and Sept. 1, 1880).

3 Cf. Anne Bezanson, *Wholesale Prices in Philadelphia, 1852–1896* (Phila-delphia: Univ. of Pennsylvania Press, 1954), 224–25.

4 *Petroleum Age* (Aug., 1885), 1063–67.

5 Cf. *1888 Trust Investigation,* 232–35.

6 Hidy & Hidy, *Pioneering,* 100–101; *OP&D Reporter* (Sept. 1, 1880); *Shipping & Commercial List & New*

York Price Current (Jan. 9, 1882).

7 *1888 Trust Investigation*, 234.

8 *Petroleum Age* (Dec., 1881), 93–94. Evidence submitted by John D. Archbold to the Industrial Commission in 1899 further confirms the significance of by-products, specifically lubricating oils, in stimulating outside refining capacity. Cf. *Industrial Commission Report*, I, 541, 542.

9 Nevins, *Study in Power*, I, 382.

10 See Flagler, Archbold and Rockefeller testimonies, *1888 Trust Investigation*, 287–313, 314–50, 387–93.

11 This famous document, known as the Keith, Vilas, and Chester agreement, is reprinted in *US v. SONJ* (1906), *Transcript*, XIX, 618–20.

12 Cf. Nevins, *Study in Power*, I, 390.

13 *Shipping & Commercial List & New York Price Current* (April 5, 1882), 106.

14 Hidy & Hidy, *Pioneering*, 44.

15 Nevins, *Study in Power*, I, 389.

16 The original trust agreement is reprinted in *1888 Trust Investigation*, 307–13.

17 Cf. *1888 Trust Investigation*, 311.

18 See Hidy & Hidy, *Pioneering*, 3.

19 Cf. Hidy & Hidy, *Pioneering*, 60–61.

20 *Shipping & Commercial List & New York Price Current* (Jan. 9, 1882).

21 Nevins, *Study in Power*, II, 54.

22 *OP&D Reporter* (March 8, August 28, 1883).

23 See above, Chapter 17.

24 Quoted in Nevins, *Study in Power*, II, 56.

25 See Hidy & Hidy, *Pioneering*, 93–94; *Titusville Morning Herald* (Oct. 5 & 18, 1880, Jan. 21, 1881, & Feb. 12, 1883); *US v. SONJ* (1906), *Transcript*, XVII, 3352, 3354, 3358–59, 3395.

26 *US v. SONJ* (1906), *Brief of Facts and Argument for Petitioner*, I, 104–105. This was estimated by compiling the capacities of known acquisitions. See Table 18:3, "Estimated Capacity of Major Refining Areas."

27 Hidy & Hidy, *Pioneering*, 100–106.

28 See Table 18:3, "Major Standard Oil Company Refining Plants, 1882–1886;" *US v. SONJ* (1906), *Transcript*, XVII, 3359, 3395.

29 Dodd testimony, *Industrial Commission Report*, I, 798; Hidy & Hidy, *Pioneering*, 94.

30 Nevins, *John D. Rockefeller*, II, 77–79.

31 Cf. *US v. SONJ* (1906), *Brief of Facts for Petitioner*, I, 105; Hidy & Hidy, *Pioneering*, 105.

32 Merrill's "Reminiscences," *Derrick's Hand-Book*, I, 886–87.

33 Merrill's "Reminiscences," *Derrick's Hand-Book*, I, 886–87; *Merrill v. Yeomans*, 94, U. S. Supreme Court (Oct., 1876), 568–74; *Titusville Morning Herald* (June 12, 1878).

34 Hidy & Hidy, *Pioneering*, n. 32, 728.

35 *Industrial Commission Report*, I, 566–70.

36 U. S. Patents 191, 203, issued May 22, 1877.

37 *Titusville Morning Herald* (Oct. 5 & 18, 1880, and Feb. 12, 1883).

38 Hidy & Hidy, *Pioneering*, 423–25.

39 *OP&D Reporter* (Mar. 12, 1894).

40 See Peckham, *Census Report*, 162–63.

41 U. S. Centennial Commission *Reports and Awards*, IV, 200, 201, 209, and *passim*, 196–257; "Pease's Oils At The Recent Fair," *Scientific American*, XVII (Nov. 16, 1876), 305.

42 *OP&D Reporter* (Jan. 5, 1881).

43 *OP&D Reporter* (Jan. 12, 1881).

44 Cf. Report of Professor J. M. Ordway and Edward Atkinson, *Proceedings of the Semi-Annual Meeting of the New England Cotton Manufacturers' Association*, Boston, Massachusetts, Oct. 30, 1878.

45 Thurston, *A Treatise on Friction and Lost Work in Machinery and Millwork*, 194, 195, 198.

46 Cf. Marcel Mitzakis, *The Oil Encyclopedia* (London: Chapel & Hall, Ltd., 1922), 440, 514, 515; B. Redwood, *Treatise on Petroleum*, I, 212, 218, 279, 280.

47 Cf. *Titusville Morning Herald*, Oct. 18, 1880. It is impossible to make any estimate of the costs of refinery construction during the 1874–1884 period, although according to one contemporary, it was possible at this time to build a plant with a daily capacity of 500 barrels at a cost ranging between $30,000 and $50,000. This figure, he stated, was "very much

less than it used to be." *1888 Trust Investigation,* 149. In 1880 plans for the Solar works at Buffalo called for an expenditure of $25,000. In the same year the new and modern Atlas works with daily capacity between 1,600 and 2,000 barrels was built at Buffalo for about $200,000, while the Chester Oil plant of the Tidewater refinery group with 2,000 barrels per day crude capacity was built for less than $250,000. The latter plant was equipped for a wider by-products manufacture. *OP&D Reporter* (Jan.

26, 1880, Sept. 7 and Oct. 12, 1881). At the farthest end of the scale was Standard's Bayonne refinery which cost close to $1 million. Hidy and Hidy, *Pioneering,* 50, 101.

48 Cf. Testimony of Theodore B. Westgate and James W. Lee, *Industrial Commission Report,* I, 276, 371. Cf. Chapter 11.

49 Cf. *1888 Trust Investigation,* 553.

50 Hidy & Hidy, *Pioneering,* 101, 107.

51 Cf. *1888 Trust Investigation,* 245.

52 *Shipping & Commercial List & New York Price Current* (Jan. 9, 1884).

Chapter 19

1 *Special Report on the Petroleum of Pennsylvania,* 1874, 72.

2 *U. S. Consular Reports,* XI, No. 37 (Jan., 1884), 441, 442.

3 *U. S. Consular Reports,* XI, No. 37 (Jan., 1884), 444, 445, 469.

4 Great Britain, Parliament, House of Commons, Select Committee on Petroleum, *Report from the Select Committee on Petroleum; Together with the Proceedings of the Committee, Minutes of Evidence, Appendix, and Index* (London: Her Majesty's Stationery Office, 1896), 340, hereafter cited as House of Commons, *Report of Select Committee on Petroleum* (1896).

5 Peckham, *Census Report,* 238. According to one report the growth of kerosene consumption per thousand inhabitants in the United Kingdom rose from about 800 gallons (640 imperial gallons) in 1874 to over 1350 gallons (1084 imperial gallons) in 1884. House of Commons, *Report of Select Committee on Petroleum* (1896), 389.

6 *Engineering and Mining Journal,* LIV (Dec. 31, 1892), 630.

7 Peckham, *Census Report,* 276.

8 *U. S. Consular Reports,* XI, No. 37 (Jan., 1884), 442.

9 *OP&D Reporter* (Dec. 10, 1883), 10.

10 Hidy & Hidy, *Pioneering,* 126.

11 Hidy & Hidy, *Pioneering,* 126.

12 Hidy & Hidy, *Pioneering,* 128.

13 For a full account of this firm see Ar-

thur C. Wardle, *Benjamin Bowring and His Descendants: A Record of Mercantile Achievement* (London: Hodder & Sons, 1938).

14 Committee on Petroleum, *Annual Report of the Board of Managers of the New York Produce Exchange for the Year Ending June 1, 1876, 1877, 1878, 1879.*

15 House of Commons, *Report of Select Committee on Petroleum* (1897), 153.

16 *U. S. Consular Reports,* XI, No. 37 (Jan., 1884), 410.

17 Dept. of State, Foreign Commerce Bureau. *Report Upon the Commercial Relations of the United States with Foreign Countries for the Year 1876,* 708.

18 House of Commons, *Report of Select Committee on Petroleum* (1896), 112.

19 House of Commons, *Report of Select Committee on Petroleum* (1896), 120.

20 House of Commons, *Report of Select Committee on Petroleum* (1896), 123.

21 Oswald von Brackel and J. Leis, *Der dreiszigjährige Petroleumkrieg* (Berlin, 1903), 348, quoted in *Report of the Commissioner of Corporations on the Petroleum Industry,* Part 3, *Foreign Trade* (Washington, D. C.: Government Printing Office, 1909), 330.

22 "Rules Regulating the Petroleum Trade, Adopted July 10, 1873," *An-*

nual Report of the Board of Managers of the New York Produce Exchange for the Year Ending June 1, 1874.

23 See reports of *Annual Report of the Board of Managers of the New York Produce Exchange for the Year Ending June 1, 1874–83, passim.*

24 *Titusville Morning Herald* (Nov. 7, 1874).

25 *Titusville Morning Herald* (Mar. 4, 1875).

26 *Titusville Morning Herald* (Mar. 4, 1875).

27 *Titusville Morning Herald* (April 18, 1879).

28 Photostat copy of Report by Wilson King, Consul, Feb. 27, 1879, subject: Report of a meeting of delegates of the German Chambers of Commerce to discuss the petroleum question.

29 *Titusville Morning Herald* (April 18, 1879).

30 *OP&D Reporter* (Mar. 26, 1879).

31 Wilson King, Consul, Report, Feb. 27, 1879.

32 Wilson King, Consul, Report, Feb. 27, 1879.

33 *OP&D Reporter* (April 30, 1879).

34 *OP&D Reporter* (June 11, 1879), 573.

35 B. Redwood, *Treatise on Petroleum,* II, 557–58.

36 Marvin, *The Region of the Eternal Fire,* 282.

37 B. Redwood, *Treatise on Petroleum,* I, 7, 8.

38 Cf. Mineral Resources, 1896, 885.

39 This material on the Nobels has been drawn chiefly from Ragnar Sohlman & Henrik Schuck, *Nobel: Dynamite and Peace,* trans. Brian & Beatrix Lunn (New York: Cosmopolitan Book Corp., 1929), and Herta E. Pauli, *Alfred Nobel: Dynamite King —Architect of Peace* (New York: L. B. Fischer, 1942).

40 Pauli, *Alfred Nobel,* 161.

41 Marvin, *Region of the Eternal Fire,* 280–81. As had happened in Pennsylvania a few years earlier, this step was bitterly fought by the "teamsters." The Nobels had to set up watch towers every few hundred yards along the route of the pipeline to guard it against destruction.

42 Marvin, *Region of the Eternal Fire,* 280–81; B. Redwood, *Treatise on Petroleum,* II, 4–9.

43 Cf. B. Orchard Lisle, *Tanker Technique, 1700–1936* (London: World Tankship Publications, 1936), 19.

44 Lisle, *Tanker Technique,* 20; Marvin, *Region of the Eternal Fire,* 285.

45 Marvin, *Region of the Eternal Fire,* 286; Pauli, *Alfred Nobel,* 168.

46 Marvin, *Region of the Eternal Fire,* 287–88. It is interesting to note that the railroads, in return for a commission, acted as collection agents for the Nobels, releasing the oil from consignment only upon receipt of cash from petroleum dealers.

47 Pauli, *Alfred Nobel,* 167.

48 B. Redwood, *Treatise on Petroleum,* II, 492–96, 558–59. See Chapter 11, for discussion of early development of continuous distillation in the United States.

49 Marvin, *Region of the Eternal Fire,* 301, 302.

50 B. Redwood, *Treatise on Petroleum,* I, 15.

51 Peckham, *Census Report,* 151–53.

52 *Titusville Morning Herald* (May 19, 1874).

53 William Brough, quoted in Peckham, *Census Report,* 152.

54 Marvin, *The Region of the Eternal Fire,* 203, 209.

55 Marvin, *The Region of the Eternal Fire,* 339.

Chapter 20

1 Cf. National Industrial Conference Board, *The Economic Almanac, 1956: A Handbook of Useful Facts about Business, Labor and Government in the United States and Other Areas;* ed. F. W. Jones and Bess Kaplan (New York: Thomas Y. Crowell Co., 1956), 423, hereafter cited as *The Economic Almanac, 1956.*

2 U. S., Bureau of the Census, *Twelfth*

Census of the United States: 1900, X. *Manufactures*, Part 4, 351, hereafter cited as *Twelfth Census of Manufactures: 1900*.

3 *Historical Statistics of U. S., 1789–1945*, 186, 200.

4 *Titusville Morning Herald* (April 30, 1873).

5 *Titusville Morning Herald* (Nov. 16, 1874).

6 Peckham, *Census Report*, 236.

7 *OP&D Reporter* (Nov. 21, 1881), 901–2.

8 Thomas D. Clark, *Pills, Petticoats and Plows: The Southern Country Store* (Indianapolis: The Bobbs-Merrill Company, 1944), 51.

9 Cf. *Twelfth Census of Manufactures: 1900*, 705.

10 Arthur A. Bright, Jr., *The Electric-Lamp Industry: Technological Change and Economic Development from 1800 to 1947* (New York: The Macmillan Company, 1949), 21.

11 *OP&D Reporter* (April 2, 1885), 26.

12 *OP&D Reporter* (April 26, 1885), 26.

13 *The Petroleum Age*, II (Nov., 1883), 563.

14 Cf. *Historical Statistics of U. S., 1879–1945*, 29, 95.

15 See, U. S. Dept. of Commerce, Bureau of the Census, *Statistical Abstract of the United States, 1946* (Washington: Government Printing Office, 1946), 500.

16 See Kent Healy, "Development of Transportation," in Harold F. Williamson (ed.), *Growth of the American Economy* (2nd ed.; New York: Prentice-Hall, Inc., 1951), 380–81.

17 Gerald Carson, *The Old Country Store* (New York: Oxford University Press, 1954), 188.

18 *Petroleum Age*, I (April, 1882), 159, 160.

19 *Petroleum Age*, I (April, 1882), 159, 160.

20 Henry Clay Pierce, *US v. SONJ* (1906), *Transcript*, III, 1077.

21 David Morey, General Freight Agent of the southern lines of the Illinois Central Railway Company, *1888 Trust Investigation*, 378.

22 *Industrial Commission Report*, I, 717.

23 David Morey, *1888 Trust Investigation*, 381.

24 *1888 Trust Investigation*, 526–27, 651.

25 D. E. Byles, *US v. SONJ* (1906), *Transcript*, III, 1354–55.

26 D. E. Byles, *US v. SONJ* (1906), *Transcript*, III, 1354–55.

27 George Rice v. *L&N RR et al.* in *1888 Trust Investigation*, 670.

28 Harlow Dow, *1888 Trust Investigation*, 413, 420. Giddens also mentions that 400 barrels usually constituted a 30–60-day supply, Giddens, *Standard of Indiana*, 47.

29 Cf. *1888 Trust Investigations*, 553; Hidy & Hidy, *Pioneering*, 729 (footnote 66).

30 *US v. SONJ* (1906), *Transcript*, XVII, 3242–44.

31 Nevins, *Study in Power*, II, 38–39.

32 Nevins, *Study in Power*, I, 257–58.

33 Nevins, *Study in Power*, I, 258.

34 *1888 Trust Investigation*, 526.

35 Nevins, *Study in Power*, I, 181. Nevins quotes one Standard official who stated, "John D. Rockefeller loved a man with initiative and the spirit of cooperation. That's why they waltzed around the hall with Carley; their arms were about him all the time he came to Cleveland." Nevins, *Study in Power*, I, 181–82.

36 *1888 Trust Investigation*, 526.

37 *1888 Trust Investigation*, 527, 531.

38 *1888 Trust Investigation*, 527.

39 *1888 Trust Investigation*, 527.

40 *1888 Trust Investigation*, 530.

41 *1888 Trust Investigation*, 532.

42 *1888 Trust Investigation*, 534, 536.

43 *1888 Trust Investigation*, 731.

44 *1888 Trust Investigation*, 732.

45 *1888 Trust Investigation*, 735.

46 *1888 Trust Investigation*, 533.

47 *Industrial Commission Report*, I, 764.

48 *1888 Trust Investigation*, 536.

49 George Rice, *Ex-Po-Za of the Business Methods of the Standard Oil Company as per Sample Branch: The Chess Carley Company* (Marietta, Ohio: By the author, 1884), 5–7, Item #65, Scheide Collection, Titusville, Pennsylvania.

50 *1888 Trust Investigation*, 534–35.

51 Nevins, *Study in Power*, II, 41.

52 Nevins, *Study in Power*, II, 41.

53 *Petroleum Age*, II (June, 1883), 494–95.

54 Nevins, *Study in Power*, II, 41.
55 Ray Allen Billington, *Westward Expansion: A History of the American Frontier* (New York: The Macmillan Company, 1949), 703.
56 Cf. *Statistical Abstract of the U. S., 1946*, 7.
57 Henry Clay Pierce, *US v. SONJ* (1906), *Transcript*, III, 1074.
58 Pierce, *US v. SONJ* (1906), *Transcript*, III, 1074.
59 Pierce, *US v. SONJ* (1906), *Transcript*, III, 1074.
60 Pierce, *US v. SONJ* (1906), *Transcript*, III, 1073.
61 Pierce, *US v. SONJ* (1906), *Transcript*, III, 1073–74.
62 Pierce, *US v. SONJ* (1906), *Transcript*, III, 1067.
63 Pierce, *US v. SONJ* (1906), *Transcript*, III, 1078.

64 *1888 Trust Investigation*, 734.
65 *1888 Trust Investigation*, 734.
66 *1888 Trust Investigation*, 734.
67 Nevins, *Study in Power*, II, 42.
68 Nevins, *Study in Power*, II, 42.
69 Nevins, *Study in Power*, II, 43.
70 Nevins, *Study in Power*, I, 224.
71 Nevins, *Study in Power*, I, 257.
72 "Memorandum on Marketing," typescript dated February 24, 1926, from Standard of Indiana records.
73 *1888 Trust Investigation*, 732.
74 Nevins, *Study in Power*, II, 43.
75 Nevins, *Study in Power*, II, 39.
76 Nevins, *Study in Power*, I, 257.
77 Giddens, *Standard of Indiana*, 9.
78 Giddens, *Standard of Indiana*, 9.
79 Hidy & Hidy, *Pioneering*, 115.
80 Hidy & Hidy, *Pioneering*, 112.
81 Hidy & Hidy, *Pioneering*, 113.

Chapter 21

1 *Mineral Resources, 1886*, 450.
2 *Derrick's Hand-Book*, I, 380.
3 Computed from Arnold & Kemnitzer, *Petroleum in the U. S.*, 100–101; see also Appendix F.
4 *Derrick's Hand-Book*, I, 808–809.
5 *New York Tribune* in *Titusville Morning Herald* (Mar. 2, 1887).
6 Hidy & Hidy, *Pioneering*, 87.
7 For the full text of original version of the Billingsley bill, see Tarbell, *History of Standard Oil*, II, 357–60.
8 *Oil City Derrick*, in *Titusville Morning Herald* (Mar. 30, 1887); Emery testimony, *1888 Trust Investigation*, 236–39.
9 *Titusville Morning Herald* (Mar. 12, 1887).
10 *Titusville Morning Herald* (Mar. 1, 1887).
11 *Titusville Morning Herald* (Feb. 11, 1887).
12 Quoted in *Titusville Morning Herald* (Feb. 23, 1887).
13 "Heydrick's Argument," *Petroleum Age*, VI (Mar., 1887), 1575.
14 For the full text of the amended bill, see *Petroleum Age*, VI (April, 1887), 1591–92.
15 J. W. Lee, *Pure Oil Trust vs. Standard Oil Company* (Oil City, Pa.: Derrick Publishing Co., 1901), 76. This is allegedly a transcript of testimony, taken by Standard stenographers, before the U. S. Industrial Commission. Standard had this published separately, maintaining that the official record omitted or substantially altered portions of the testimony. Here, Lee estimates (local?) pipeage at 8¢ per barrel, excluding interest charges; the official report omits the qualification that interest charges are not included in the estimate.
The circumstances of the contract with the Pennsylvania railroad also suggest that the agreed-on rates may not have covered full unit cost. Oil was to be piped on the railroad's account only if the Pennsylvania failed to carry 26 per cent of combined crude and refined oil shipped from the Regions coastward. If the pipeline anticipated that such occasions would be temporary, it may have been willing to accept rates below cost in order to keep peace with the railroad. See Chapter 17.
16 Cf. A. Leo Weil, "The Billingsley Bill," *Petroleum Age*, VI (March,

1887), 1571, citing *Wabash & Co. Railroad* v. *Illinois,* 118 U. S., 577. Judicial interpretation on this point was changing. In the granger cases, the courts had held that an individual state could regulate railroad charges on traffic passing through its territory to or from points outside the state. The majority opinion on the Wabash case, however, maintained that the state could regulate only commerce that began and ended within its boundaries.

17 See U. S. Dept. of the Interior, Census Office, *Report on Manufacturing Industries in the United States at the Eleventh Census: 1890,* Part 3, 368, hereafter cited as *Eleventh Census of Manufacturing Industries: 1890.*

18 Weil, *Petroleum Age,* VI (March, 1887), 1573.

19 *Derrick's Hand-Book,* I, 447–48.

20 "The New York Compromise Meeting," *Petroleum Age,* VI (May, 1887), 1625.

21 *Petroleum Age,* VI (April, 1887), 1594.

22 *Philadelphia Times* in *Titusville Morning Herald* (May 2, 1887).

23 Nevins, *Study in Power,* II, 473–74.

24 Quoted in *Titusville Morning Herald* (May 2, 1887).

25 David Kirk, *1888 Trust Investigation,* 33.

26 Kirk, *1888 Trust Investigation,* 34.

27 New York State Senate, *Report of the Committee on General Laws on the Investigation Relative to Trusts* (Albany, 1888), 449.

28 Hidy & Hidy, *Pioneering,* 179–80.

29 Hidy & Hidy, *Pioneering,* 176.

30 Kirk testimony, *1888 Trust Investigation,* 34.

31 "Memorandum of Agreement, made this 1st day of November, 1887, between the Standard Oil Company of New York and the following named . . . producers of crude petroleum, Thomas W. Phillips and others, . . ." reprinted in *1888 Trust Investigation,* 69–70.

32 See testimony of Henry Webster and I. N. Bennett, in *1888 Trust Investigation,* 1–30. Bennett testified that he thought that 90 per cent of the "workers" connected with the drilling and operating of oil wells in the Pennsylvania Regions were members of the Well-Driller's Union.

33 *Titusville Morning Herald* (Nov. 1, 1887).

34 *Derrick's Hand-Book,* I, 468, 470; *Titusville Morning Herald* (Dec. 1, 1887).

35 *Bradstreet's,* XV (Dec. 3, 1887).

36 *Derrick's Hand-Book,* I, 474, 479–80, 484.

37 Patrick Boyle, *Industrial Commission Report,* I, 460–61; Kirk, *1888 Trust Investigation,* 36; *Derrick's Hand-Book,* I, 466.

38 Cf. *Bradstreet's,* XVI (Dec. 1, 1888).

39 *Derrick's Hand-Book,* I, 484, 487, 493.

40 *Derrick's Hand-Book,* I, 486.

41 Cf. *Industrial Commission Report,* I, 284.

42 Hidy & Hidy, *Pioneering,* 215.

43 *Derrick's Hand-Book,* I, 793.

44 *OP&D Reporter,* XXVIII (Aug. 12, 1885), 29.

45 W. W. Tarbell, *US* v. *SONJ* (1906), *Transcript,* I, 464; Archbold, *US* v. *SONJ* (1906), *Transcript,* XVII, 3399–3400.

46 J. W. Lee, *US* v. *SONJ* (1906), *Transcript,* VI, 3161; *Industrial Commission Report,* I, 270. Emery's claim that the subscribers "were poorer than snakes" may have been true of many producers. But contributions made four years later to back Pure Oil indicate there were significant exceptions. By January, 1895, T. W. Phillips alone claimed to have invested $60,-000 in the Association's activities. *Titusville Morning Herald* (Jan. 25, 1895).

47 Crude consumption estimated from monthly pipeline reports in *OP&D Reporter,* February–June, 1893. In contrast, National Transit alone was shipping between twenty and thirty times more than the Producers' & Refiners' pipeline. *Derrick's Hand-Book,* I, 556.

48 T. B. Westgate, *US* v. *SONJ* (1906), *Transcript,* VI, 2845.

49 *Rice* v. *Louisville & Nashville Railroad et al., Interstate Commerce Commission Reports* (*1888*), I, 722–23.

50 *Independent Refiners' Association of Titusville and Oil City* v. *Pennsylvania Railroad et al.* (hereafter cited as *Independent Refiners' Assoc.* v. *Pa. RR.*), *Interstate Commerce Commission Reports* (*1888–1890*), II, 319. Crude and refined oil carried the same tariff.

51 *Titusville Morning Herald* (May 16 and 17, 1889).

52 *Independent Refiners' Assoc.* v. *Pa. RR., Interstate Commerce Commission Reports* (*1892–95*), IV, 162–63; T. B. Westgate, *Industrial Commission Report,* I, 379–80.

53 *Titusville Morning Herald* (May 16 and 17, 1889).

54 Emery, *US* v. *SONJ* (1906), *Transcript,* VI, 2767–69.

55 Monthly pipeline reports in *OP&D Reporter,* June–Dec., 1891.

56 Emery, *Industrial Commission Report,* I, 650–51.

57 M. L. Lockwood, *Industrial Commission Report,* I, 398.

58 Archbold, *US* v. *SONJ* (1906) *Transcript,* XVII, 3352; Emery, *US* v. *SONJ* (1906), *Transcript,* VI, 2661, 2701; *OP&D Reporter,* XLII (Aug. 8, 1892), 9.

59 Emery, *Industrial Commission Report,* I, 656.

60 Cf. Emery, *Industrial Commission Report,* I, 653–54; *US* v. *SONJ* (1906), *Transcript,* VI, 2776; also *OP&D Reporter,* XLIX (March 16, 1896), 8A.

61 For Emery's claim of Standard influence on the Erie and Standard's denial, see *Industrial Commission Report,* I, 651–53, and *US* v. *SONJ* (1906), *Transcript,* VI, 2655.

62 *Titusville Morning Herald* (Jan. 25, 1895).

63 *Derrick's Hand-Book,* I, 562.

64 *Derrick's Hand-Book,* II, 153.

65 *OP&D Reporter,* XLVI (Dec. 3, 1894), 6; Lee, *Industrial Commission Report,* I, 267–68.

66 T. W. Phillips, *Industrial Commission Report,* I, 590; Hidy & Hidy, *Pioneering,* 273.

67 Westgate, *Industrial Commission Report,* I, 382.

68 Text of the agreement is found in *Industrial Commission Report,* I, 507–508.

69 Emery, *US* v. *SONJ* (1906), *Transcript,* VI, 2661; Lee, *US* v. *SONJ* (1906), *Transcript,* VI, 3166. The refineries were those of Borne, Scrymser, purchased in 1893, and the Mutual, National and International Oil Companies in the Regions, purchased two years later.

70 Archbold, *US* v. *SONJ* (1906), *Transcript,* XVII, 3326. Archbold said that his company had also made the purchase "as an investment," a purpose that was less successfully realized.

71 Lee, *US* v. *SONJ* (1906), *Transcript,* VI, 3164; Tarbell, *History of Standard Oil,* II, 173–74, 211–14; Nevins, *John D. Rockefeller,* II, 318–20.

72 Lee, *US* v. *SONJ* (1906), *Transcript,* VI, 3164–65.

73 Lee, *Industrial Commission Report,* I, 271.

74 Lee, *US* v. *SONJ* (1906), *Transcript,* VI, 3168–69.

75 The exceptions were John Fertig of National, S. Y. Ramage of Mutual and H. F. Burwell of Union. Cf. Tarbell, *US* v. *SONJ* (1906), *Transcript,* I, 468.

76 Lee, *US* v. *SONJ* (1906), *Transcript,* VI, 3169–70.

77 *Titusville Morning Herald* (Jan. 1, 1895). Text of Pure's trust agreement is found in *Industrial Commission Report,* I, 508, 513. The company's authorized capitalization was $1 million, divided into 200,000 shares.

78 Lee, *US* v. *SONJ* (1906), *Transcript,* VI, 3173–74.

79 *Derrick's Hand-Book,* I, 588–89.

80 Lee, *US* v. *SONJ* (1906), *Transcript,* VI, 3175.

81 Westgate, *Industrial Commission Report,* I, 381.

82 Archbold, *US* v. *SONJ* (1906), *Transcript,* VI, 3630, 3648.

83 Lee, *US* v. *SONJ* (1906), *Transcript,* VI, 3175–76; Emery, *Industrial Commission Report,* I, 617.

84 Westgate, *Industrial Commission Report,* I, 365.

85 *New York Times* (Oct. 29, 1895), 1. The Pennsylvania Railroad had also opposed the line at Belvidere. The courts, however, recognized the

pipeline's title to the land under the railroad and vacated the injunction. *OP&D Reporter*, L (July 13, 1896), 7; Emery, *US* v. *SONJ* (1906), *Transcript*, VI, 2658.

86 *OP&D Reporter*, L (Nov. 16, 1896), 6.

87 Archbold, *Industrial Commission Report*, I, 529; Bureau of Corporations, *Report of the Commissioner of Corporations on the Transportation of Petroleum* (Washington: Government Printing Office, 1906), 90.

88 Emery, *US* v. *SONJ* (1906), *Transcript*, VI, 2658.

89 Tarbell, *History of Standard Oil*, II, 187.

90 Lee, *US* v. *SONJ* (1906), *Transcript*, VI, 3179; Tarbell, *US* v. *SONJ* (1906), *Transcript*, III, 1347.

91 *OP&D Reporter*, LI (March 1, 1897), 25.

92 *Report of the Commissioner of Corporations on the Petroleum Industry*, Part 1, *Position of Standard Oil*, 131–32.

93 *Report of the Commissioner of Corporations on the Petroleum Industry*, Part 1, *Position of Standard Oil*, 133; Petitioner's Exhibit 371, *US* v. *SONJ* (1906), *Transcript*, VIII, 899.

94 Tarbell, *US* v. *SONJ* (1906), *Transcript*, VI, 1442.

95 *Report of the Commissioner of Corporations on the Petroleum Industry*, Part 3, *Foreign Trade*, 383, 516.

96 *Titusville Morning Herald* (Jan. 25, 1895). Other reports on Standard's 6 and 8 inch lines placed their capacities at 8,000 to 10,000 barrels daily; but this factor varied with pumping capacity.

97 *Report of the Commissioner of Corporations on the Petroleum Industry*, Part 1, *Position of Standard Oil*, 133.

98 For details, See *Report of the Commissioner of Corporations on the Petroleum Industry*, Part 1, *Position of Standard Oil*, 169–83.

99 Benson testimony, *US* v. *SONJ* (1906), *Transcript*, I, 207–8; Hidy & Hidy, *Pioneering*, 725.

100 Benson, *US* v. *SONJ* (1906), *Transcript*, I, 218–19; Hidy & Hidy, *Pioneering*, 129.

101 Golden, *Tide Water History*, 29.

102 Benson testimony, *US* v. *SONJ* (1906), *Transcript*, I, 217–18; *Derrick's Hand-Book*, II, 168. Percentages were computed from data in the following sources: monthly pipeline reports in *OP&D Reporter*, June, 1885–Dec., 1889; *Mineral Resources, 1889*, 298–99, 303–4. Prior to 1889 monthly figures of production and shipments refer to Pennsylvania and New York only. Since these averaged 97 per cent of Appalachian output during this period, the discrepancy is insignificant.

103 Petitioners' Exhibits 27 and 28, *US* v. *SONJ* (1906), *Transcript*, VII, 106–7; Benson, *US* v. *SONJ* (1906), *Transcript*, I, 216.

104 Quoted in Hidy & Hidy, *Pioneering*, 182.

105 D. P. Reighard, *US* v. *SONJ* (1906), *Transcript*, VI, 3134–36; monthly pipeline reports in *OP&D Reporter*, Dec., 1887–Dec., 1888.

106 William Larimer Mellon, *Judge Mellon's Sons* (privately printed, 1948), 162–63.

107 Monthly pipeline reports in *OP&D Reporter*, June–Aug., 1891.

108 Mellon, *Judge Mellon's Sons*, 165. Mellon was apparently shipping under some special rate before the change. There is no record of a general rate increase from the Regions to tidewater in either 1891 or 1892.

109 Mellon, *Judge Mellon's Sons*, 166, 170–72; *OP&D Reporter*, XLII (Nov. 14, 1892), 9.

110 Monthly pipeline reports in *OP&D Reporter*, January–June, 1893.

111 Mellon, *Judge Mellon's Sons*, 178–79. Presumably the products of these refiners were marketed in England, for the British marketing firm of T. C. Bowring shared ownership of the Bear Creek works with Mellon.

112 Hidy & Hidy, *Pioneering*, 238–41. For a more detailed discussion of the French markets, see Chapter 24.

113 Mellon, *Judge Mellon's Sons*, 181–82. Mellon does not mention the Russian threat in France as a factor influencing his decision, but it may be safely inferred that it played an important role.

114 Hidy & Hidy, *Pioneering*, 753, n. 12.
115 *OP&D Reporter*, XLIII (June 26, 1893), 9.
116 *Titusville Morning Herald* (March 4, 1895, and March 9, 1895).
117 Quoted in *Titusville Morning Herald* (March 7, 1895).
118 *Derrick's Hand-Book*, I, 430.
119 *Derrick's Hand-Book*, I, 515, 519.
120 Defendants' Exhibit 251, *US* v. *SONJ* (1906), *Transcript*, XIX, 614.
121 Emery testimony, *US* v. *SONJ* (1906), *Transcript*, VI, 2745. There were, however, avoidable lags. The first tank car shipment left Washington county in March, 1885, but National Transit began building tankage there only in the following September. *Derrick's Hand-Book*, I, 385, 400. The independent Ohio Transit Company laid the first lines into Macksburgh field in November, 1883, six months before National Transit started tankage construction there. Ohio, Report of the Geological Survey, VI, *Economic Geology*, by Edward Orton *et al.* (Columbus: The Westbote Co., 1888), 462, hereafter cited as Orton, *Economic Geology*.
122 Hidy & Hidy, *Pioneering*, 187; cf. also below Chapter 23.
123 Archbold, *1888 Trust Investigation*, 326; *Titusville Morning Herald* (May 17, 1889).
124 *Report of the Commissioner of Corporations on the Petroleum Industry*, Part 1, *Position of Standard Oil*, 220.
125 *Report of the Commissioner of Corporations on the Petroleum Industry*, Part 1, *Position of Standard Oil*, 239. The Bureau says that rail and pipeline rates from western Pennsylvania to the seaboard were equal. This is in error. Standard charged 45¢ to pipe crude to New York and 39¢ to Philadelphia; rail rates, increased to 55¢ in 1888, remained at that level until 1902.
126 *US* v. *SONJ* (1906), *Brief of the Law for Petitioners*, 179–80.
127 Archbold testimony, *US* v. *SONJ* (1906), *Transcript*, XVII, 3432.
128 Hidy & Hidy, *Pioneering*, 187; Defendants' Exhibit 267, *US* v. *SONJ* (1906), *Transcript*, XIX, 626. Total outside sales averaged 2,236,420 barrels annually; total outside sales by refineries, 2,209,602 barrels per year. The figures include sales of both Lima and Pennsylvania-grade crude, but no breakdown by type is given.
129 *Report of the Commissioner of Corporations on the Petroleum Industry*, Part 1, *Position of Standard Oil*, 156.

Chapter 22

1 Cf. Arnold & Kemnitzer, *Petroleum in U. S.*, 132, 147.
2 Orton, *Economic Geology*, 101.
3 Orton, *Economic Geology*, 126–30, 165–68.
4 Orton, *Economic Geology*, 166.
5 Orton, *Economic Geology*, 168.
6 Orton, *Economic Geology*, 126.
7 *Mineral Resources, 1886*, 459–60; *Mineral Resources, 1889 and 1890*, 289.
8 U. S., *Eighth Annual Report of the United States Geological Survey to the Secretary of the Interior 1886–87*, Part 2, *The Trenton Limestone as a Source of Petroleum and Inflammable Gas in Ohio and Indiana*, by Edward Orton (Washington: Government Printing Office, 1889), 626–27, hereafter cited as Orton, *Trenton Limestone, U. S. Geological Survey 1886–87*. The Geological Survey further called attention to the difficulty of removing the sulphur odor from kerosene which continued to plague refiners of Ohio oil.
9 Orton, *Trenton Limestone, U. S. Geological Survey. 1886–87*, 627.
10 Orton, *Economic Geology*, 174–75.
11 Edward Orton, *First Annual Report of the Geological Survey of Ohio* (Columbus: The Westbote Co., 1890), 171–73, 175.
12 *Engineering & Mining Journal*, XLIII (Feb. 12, 1887), 108; Orton, *Economic Geology*, 150; Orton, *Trenton*

Limestone, U. S. Geological Survey, 1886–87, 629.

13 See Appendix F; *Industrial Commission Report*, I, 661.

14 *Titusville Morning Herald* (April 23, 1887).

15 Orton, *First Annual Report of Ohio Geological Survey* (1890), 509. According to Lewis Emery, the average cost of his wells drilled in the lower Pennsylvania regions (probably the McDonald field) around 1890–1891 was $8,000. "The reason why wells are so expensive in the lower country," Emery testified, "is that you have to put in three lengths of casing. Some of that casing runs down into the earth 1,600 feet, 2,000 feet. I have loaded into a single hole 150 tons of iron." *Industrial Commission Report*, I, 661.

16 Orton, *First Annual Report of Ohio Geological Survey* (1890), 509.

17 *Petroleum Age* (Oct., 1887), 1756.

18 Orton, *First Annual Report of Ohio Geological Survey* (1890), 210.

19 Orton, *First Annual Report of Ohio Geological Survey* (1890), 211.

20 *Mineral Resources, 1886*, 459.

21 E. Orton, as quoted in Indiana, *Department of Geology and Natural Resources, Twenty-First Annual Report, 1896*, W. S. Blatchley, state geologist (Indianapolis: Wm. B. Burford, 1897), 11.

22 *Engineering & Mining Journal*, XLII (Feb. 12, 1887), 108; *Titusville Morning Herald* (Jan. 4, 1887); *OP&D Reporter*, XXXI (June 22, 1887), 10.

23 Whiteshot, *Oil-Well Driller*, 150.

24 Hidy & Hidy, *Pioneering*, 156–57; *OP&D Reporter*, XXIX (April 14, 1886), 8.

25 Hidy & Hidy, *Pioneering*, 158; *OP&D Reporter*, XXIX (April 28, 1886), 8.

26 Brewster to Rockefeller, June 30, 1886, quoted in Hidy & Hidy, *Pioneering*, 158.

27 *OP&D Reporter*, XXX (Sept. 22, 1886), 9 (Oct. 20, 1886), 9; *Petroleum Age*, VI (Oct., 1887), 1756.

28 Orton, *Economic Geology*, 178–79, 237–38.

29 *Petroleum Age*, VI (Oct., 1887), 1756; *OP&D Reporter*, XXXII (Nov. 30, 1887), 8 (Dec. 28, 1887), 9.

30 *Derrick's Hand-Book*, I, 808.

31 Hidy & Hidy, *Pioneering*, 158.

32 Hidy & Hidy, *Pioneering*, 158.

33 *Derrick's Hand-Book*, I, 435, 459.

34 *Petroleum Age* (July, 1887), 1681.

35 Toledo *Bee* reprinted in *Titusville Morning Herald* (July 12, 1887).

36 *Report of the Commissioner of Corporations on the Petroleum Industry*, Part 1, *Position of Standard Oil*, 222.

37 "Lima Oil," *OP&D Reporter*, XXXII (July 6, 1887), 20. The Trust's internal accounts roughly corroborate O'Day's estimate. After Buckeye's original $100,000 capital was exhausted in construction, National Transit loaned its subsidiary $1.5 million. Hidy & Hidy, *Pioneering*, 158–61.

38 *OP&D Reporter* (July 13, 1887), 9.

39 Hidy & Hidy, *Pioneering*, 162.

40 *Titusville Morning Herald* (July 4, 1887).

41 Orton, *Economic Geology*, 238; *Titusville Morning Herald* (July 28, 1887, and August 29, 1887).

42 *Petroleum Age* (Aug., 1887), 1702–3.

43 *The Iron Trade Review*, reprinted in the *OP&D Reporter* (Oct. 26, 1887), 8–9.

44 *OP&D Reporter* (Dec. 8, 1886), 8.

45 *OP&D Reporter*, XXXIII (Mar. 15, 1888), 9.

46 *Mineral Resources, 1895–96*, 676–77.

47 *1888 Trust Investigation*, 305–6.

48 *OP&D Reporter*, XXXIII (May 2, 1888), 8; *OP&D Reporter*, XXXIV (Aug. 8, 1888), 51.

49 *OP&D Reporter*, XXXIII (Jan. 25, 1888), 9; *OP&D Reporter*, XXXIV (Aug. 8, 1888), 51. It is not certain that the traction engine was used to distribute the pipe; other reports state that tong-gangs were distributing the sections. Cf. *New York Times* (April 16, 1888), 1.

50 *Chicago Tribune* (Aug. 5, 1888); *OP&D Reporter*, XXXIV (Sept. 30, 1888), 39.

51 *Chicago Tribune* (Aug. 14, 1888).

52 *Chicago Times* in *OP&D Reporter*, XXXIV (Sept. 30, 1888), 39.

53 *Engineering & Mining Journal*, XLV (Mar. 31, 1888), 202.

54 *Mineral Resources, 1895–96*, 676–77; *Chicago Times* in *OP&D Reporter*,

XXXIV (Sept. 30, 1888), 39; XXXIV (Oct. 10, 1888), 8; *Chicago Tribune* (Dec. 9, 1888), 25.

55 Cf. Hidy & Hidy, *Pioneering*, 166–67.
56 Hidy & Hidy, *Pioneering*, 163. For details of this development carried on under the direction of Herman Frasch see Chapter 23.
57 June 29, 1885, quoted in Hidy & Hidy, *Pioneering*, 178.
58 Hidy & Hidy, *Pioneering*, 176.
59 Cf. Hidy & Hidy, *Pioneering*, 172–75.
60 Cf. *Sixty Years of Progress: 1887–1947, A History of The Ohio Oil Company*, 60.
61 Orton, *Trenton Limestone, U. S. Geological Survey, 1886–87*, 630.
62 Quoted in *Titusville Morning Herald* (Feb. 5, 1889).
63 *Sixty Years of Progress: 1887–1947, A History of The Ohio Oil Company*, 39.
64 Orton, *First Annual Report of Ohio Geological Survey* (1890), 200.
65 Hidy & Hidy, *Pioneering*, 187.
66 Hidy & Hidy, *Pioneering*, 183–84.
67 Reprinted in *OP&D Reporter* (Mar. 13, 1889), 8–9.
68 Orton, *First Annual Report of Ohio Geological Survey* (1890), 201. One producer testifying before the Industrial Commission a decade later put it more forcibly, "Think of what the financial slaughter must have been to force over 55 per cent of the producers to transfer their property to the Standard . . . in little more than two years. In some places nearly whole townships were sacrificed—transferred to Standard. I know all about it, I was one of the producers who was obliged to sacrifice his property there." M. Lockwood testimony, *Industrial Commission Report*, I, 402–3.

69 *US v. SONJ* (1906), *Transcript*, XVII, 3424.
70 *Chicago Tribune* (April 20, 1889), May 12, 1889.
71 Cf. Giddens, *Standard of Indiana*, 10–21.
72 Giddens, *Standard of Indiana*, 21.
73 Giddens, *Standard of Indiana*, 25.
74 *Chicago Tribune* (April 19, 1889).
75 Cf. Hidy & Hidy, *Pioneering*, 187.
76 Rockefeller to John D. Archbold, Dec. 7, 1891, quoted by Nevins, *Study in Power*, II, 105.
77 Cf. *Report of the Commissioner of Corporations on the Petroleum Industry, Part 1, Position of Standard Oil*, 245.
78 Cf. *New York Times* (April 6, 1892), 2.
79 Orton, *First Annual Report of Ohio Geological Survey* (1890), 206.
80 *Report of the Commissioner of Corporations on the Petroleum Industry, Part 1, Position of Standard Oil* (Washington: Government Printing Office, 1907), 245, Part 2, *Prices and Profits*, 220–24.
81 *US v. SONJ* (1906), *Transcript* III, 1385–86.
82 *Report of the Commissioner of Corporations on the Petroleum Industry, Part 2, Prices & Profits*, 94–95.
83 Hidy & Hidy, *Pioneering*, 289.
84 Hidy & Hidy, *Pioneering*, 284.
85 *Titusville Morning Herald* (Feb. 5, 1889); *New York Times* (April 18, 1889).
86 Cf. Hidy & Hidy, *Pioneering*, 285. Apparently the Toledo refineries turned to Lima crude about the same time, although no date is available.
87 Hidy & Hidy, *Pioneering*, 287.

Chapter 23

1 Cf. *Shipping and Commercial List and New York Price Current* (Jan. 14, 1891), 15.
2 Hidy & Hidy, *Pioneering*, 160.
3 William Burton testimony, *US v. SONJ* (1906), *Transcript*, XVI, 2634–35.

4 U. S. Patent Nos. 378 and 246.
5 Hidy & Hidy, *Pioneering*, 163. William Burton, Frasch's assistant, stated that the redistillation of the oil with the oxide compounds and the reroasting of the sulphides back to their original form were the key steps in the

commercialization of Frasch's process. Cf. *US* v. *SONJ* (1906), *Transcript*, XVI, 2634–35.

6 Orton, *First Annual Report of Ohio Geological Survey* (1890), 202. Later Paragon was reported as employing the "Apex" process. Differences, if any, with the Pitt process are not known. *OP&D Reporter* (May 11, 1896).

7 *OP&D Reporter* (May 1, 1889 and June 11, 1890); Orton, *First Annual Report of Ohio Geological Survey* (1890), 204.

8 For some pre-Frasch evaluations of Lima crude by Eastern refineries, see, *Titusville Morning Herald* (Sept. 15, 1887); *Derrick's Hand-Book*, I, 463; *OP&D Reporter* (Feb. 1, 1888).

9 Giddens, *Standard of Indiana*, 33–35.

10 Hidy & Hidy, *Pioneering*, 87.

11 See D. Kirk and I. E. Dean, *1888 Trust Investigation*, 39–41, 76–78.

12 *Derrick's Hand-Book*, I, 774. Also see, *Oil and Gas Journal*, XXIII (Apr. 9, 1925), 92–96.

13 *Derrick's Hand-Book*, I, 508.

14 Hidy & Hidy, *Pioneering*, 159.

15 *Derrick's Hand-Book*, I, 510.

16 Reprinted in *Derrick's Hand-Book*, I, 775.

17 See *Petroleum Age* (Aug., 1885), 1063–67.

18 Lee testimony, *Industrial Commission Report*, I, 268–69, 278; Theodore Childs Boyden, "The Location of Petroleum Refining in the United States" (unpublished Ph.D. thesis, Dept. of Economics, Harvard University, 1955), 305. Lee estimated that capital costs for a new plant which could produce a full range of by-products at $500,000 in the late 1890's. But also see, Archbold testimony, *Industrial Commission Report*, I, 570–72. It would appear, however, that initial capital costs for building a refinery using Appalachian crude had not materially changed since 1888. About $100 per barrel of daily crude capacity, on the average, was a close approximation. See *Petroleum Age* (May, 1886), 1317; *OP&D Reporter* (July 24, 1888); B. B. Campbell testimony, *1888 Trust Investigation*, 144; Davis testimony, *Industrial Commis-*

sion Report, I, 355–56. This would have made the capacity of Lee's estimate 5,000 barrels daily, which on a simple average (total capacity divided by the number of plants) basis was close to actual conditions. The Whiting works were said to have cost $5 million. See *Chicago Tribune* (Sept. 22, 1889). With 36,000 barrels daily capacity, capital costs per barrel would have been $141. Capital costs for laying a 6-inch or 8-inch pipeline were estimated at between $6,000 and $8,000 per mile laid.

19 *Petroleum Age* (Oct., 1886), 1462; Hidy & Hidy, *Pioneering*, 101.

20 *OP&D Reporter* (Mar. 1, 1897).

21 Cf., Hidy & Hidy, *Pioneering*, 157.

22 Lee and Westgate testimony, *Industrial Commission Report*, I, 269–76, 368.

23 Rice testimony, *Industrial Commission Report*, I, 735.

24 *OP&D Reporter* (Dec. 3, 1890, July 20, 1896); *Report of the Commissioner of Corporations on the Petroleum Industry*, Part 2, *Prices and Profits*, 933–34. While nearly every refinery had a brand name for its product the general categories of lubricants were as follows: black oils, summer and winter grades (usually used in light machinery); cylinder oils (popularly used in heavy machinery such as turbines or heavy tooling machines); neutral oils and spindle oils; paraffin oils.

25 Hidy & Hidy, *Pioneering*, 289.

26 *Mineral Resources, 1905*, 883–84.

27 *OP&D Reporter* (Apr. 17, 1893).

28 *OP&D Reporter* (Apr. 17, 1893).

29 *OP&D Reporter* (Dec. 9, 1895).

30 *Report of the Commissioner of Corporations on the Petroleum Industry*, Part 2, *Prices and Profits*, 671.

31 See Table 23:1 and Hidy & Hidy, *Pioneering*, 289.

32 See Hidy & Hidy, *Pioneering*, 289.

33 Archbold testimony, *Industrial Commission Report*, I, 560.

34 *US* v. *SONJ* (1906), *Summary of Defendant's Brief*, 26–28; Hidy & Hidy, *Pioneering*, 188–89.

35 Hidy & Hidy, *Pioneering*, 284.

36 Hidy & Hidy, *Pioneering*, 286; *Derrick's Hand-Book*, II, 151–52.

Chapter 24

1 *Scientific American* (May 31, 1884), 342.
2 *Bradstreet's, A Journal of Trade, Finance, and Public Economy,* X (Oct. 25, 1884), 258.
3 J. D. Henry, *Baku: An Eventful History* (London: Archibald Constable & Co. Ltd., 1905), 114.
4 J. D. Henry, *Baku,* 117.
5 "History of the Russian Petroleum Industry," *Petroleum Industrial and Technical Review,* I (June 10, 1899), 271.
6 *Petroleum Industrial and Technical Review,* I (June 10, 1899), 271.
7 *Petroleum Review,* II (May 26, 1900), 314.
8 *OP&D Reporter* (July 22, 1885), 29.
9 *OP&D Reporter* (Dec. 23, 1885), 11.
10 Peckham, *Census Report,* 101.
11 James Dodd Henry, *Thirty-Five Years of Oil Transport: The Evolution of the Tank Steamer* (London: Bradbury, Agnew & Co., Ltd., 1907), 8, hereafter cited as J. D. Henry, *Thirty-Five Years;* Martell, *Transactions of the Institution of Naval Architects,* XXVIII (1887), 6.
12 Martell, *Transactions of the Institution of Naval Architects,* XXVIII (1887), 6. From the standpoint of subsequent tanker design the most serious limitations of these ships was the lack of provision for the expansion of oil or any means of carrying off accumulations of explosive gases.
13 J. D. Henry, *Thirty-Five Years,* 8. Aside from the comment by Henry that the *Stat* carried oil in "huge tanks," there are no details of her design.
14 J. D. Henry, *Thirty-Five Years,* 9; Martell, *Transactions of the Institution of Naval Architects,* XXVIII (1887), 6.
15 Martell, *Transactions of the Institution of Naval Architects,* XXVIII (1887), 3–5.
16 Martell, *Transactions of the Institution of Naval Architects,* XXVIII (1887), 10.
17 *OP&D Reporter* (Oct. 7, 1885), 5, 6.
18 Martell, *Transactions of the Institution of Naval Architects,* XXVIII (1887), 9; J. D. Henry, *Thirty-Five Years,* 25.
19 *OP&D Reporter* (July 11, 1880), 40.
20 *OP&D Reporter* (Dec. 2, 1885), 6.
21 *U. S. Consular Reports,* XJX, No. 63 (Apr., 1886), 168.
22 J. D. Henry, *Thirty-Five Years,* 11.
23 J. D. Henry, *Thirty-Five Years,* 13.
24 Cf. Martell, *Transactions of the Institution of Naval Architects,* XXVIII (1887), 10–11; Lisle, *Tanker Technique, 1700–1936,* 25.
25 Herbert Barringer, "The Evolution of the Oil Tank-Ship," *Journal of the Institute of Petroleum Technologists,* I (1914–15), 285–86.
26 J. D. Henry, *Thirty-Five Years,* 13. The *Gluckauf* remained in the petroleum trade until 1893 when, after being stranded on the coast of Long Island, it was sold for $350. *OP&D Reporter* (June 5, 1893), 9.
27 *Engineering & Mining Journal,* LIV (Dec. 31, 1892), 630.
28 *Bradstreet's,* XV (Aug. 27, 1887), 569; *Engineering & Mining Journal,* LIV (Dec. 31, 1892), 630.
29 *Engineering &.Mining Journal,* XLVI (Nov. 17, 1888), 416; *Engineering & Mining Journal,* LIV (Dec. 31, 1892), 630.
30 *Petroleum Review,* I (May 20, 1899), 250.
31 Henry C. Folger, "Petroleum: Its Production and Products," *One Hundred Years of American Commerce,* ed. Chauncey M. Depew, I (New York: D. O. Haynes & Co., 1895), 213.
32 Hidy & Hidy, *Pioneering,* 126.
33 *Report of the Commissioner of Corporations on the Petroleum Industry,* Part 3, *Foreign Trade,* 429.
34 Hidy & Hidy, *Pioneering,* 489.
35 *Petroleum Industrial and Technical Review,* III (1901–1902), 629.
36 Cf. *Report of the Commissioner of Corporations on the Petroleum Industry,* Part 3, *Foreign Trade,* 532.
37 *Petroleum Industrial and Technical Review,* III (1901–1902), 629.
38 Cf. Boverton Redwood, "The Russian Petroleum Industry," *Journal of the*

Society of Chemical Industry (Feb. 28, 1885), 77.

39 Hidy & Hidy, *Pioneering*, 138.
40 Hidy & Hidy, *Pioneering*, 140.
41 *OP&D Reporter* (Dec. 23, 1885), 6.
42 *Petroleum Review*, II (May 26, 1900), 314.
43 Hidy & Hidy, *Pioneering*, 140–41.
44 Hidy & Hidy, *Pioneering*, 142.
45 *Petroleum Review*, I (Mar., 1899), 78.
46 Hidy & Hidy, *Pioneering*, 144.
47 Hidy & Hidy, *Pioneering*, 147.
48 Hidy & Hidy, *Pioneering*, 148.
49 Hidy & Hidy, *Pioneering*, 148–49.
50 *Bradstreet's*, XIX (July 4, 1891), 429.
51 Hidy & Hidy, *Pioneering*, 150–51.
52 Hidy & Hidy, *Pioneering*, 149.
53 *Bradstreet's*, XIX (July 4, 1891), 429.
54 Hidy & Hidy, *Pioneering*, 151, 249.
55 Hidy & Hidy, *Pioneering*, 236.
56 D. Ghambashidze, "The Russian Petroleum Industry and Its Prospects," *Journal of the Institute of Petroleum Technologists*, IV (1917), 171.
57 *Berlin Vossische Zeitung* (June 12, 1891), quoted in *U. S. Consular Reports*, XXXVI, No. 131 (Aug., 1891), 649.
58 *Bradstreet's*, XXI (June 3, 1893), 345.
59 Hidy & Hidy, *Pioneering*, 237; *Bradstreet's*, XVI (Dec. 2, 1893), 764.
60 *US v. SONJ* (1906), *Brief of Petitioner*, I, 221.
61 W. L. Mellon, *Judge Mellon's Sons*, 177; Hidy & Hidy, *Pioneering*, 748.
62 *Report of the Commissioner of Corporations on the Petroleum Industry*, Part 3, *Foreign Trade*, 333–34.
63 Hidy & Hidy, *Pioneering*, 240.
64 Hidy & Hidy, *Pioneering*, 222, 241.
65 Hidy & Hidy, *Pioneering*, 241–42.
66 Hidy & Hidy, *Pioneering*, 250.
67 *Report of the Commissioner of Corporations on the Petroleum Industry*, Part 3, *Foreign Trade*, 334. According to Ida Tarbell, the general belief among "independent [oil] circles" was that when Poth learned the truth, he "literally died of grief." Cf. Tarbell, *History of Standard Oil*, II, 177.
68 *U. S. Consular Reports*, LV, No. 206 (Nov., 1897), 371.
69 Hidy & Hidy, *Pioneering*, 249.
70 House of Commons, *Report of the Select Committee on Petroleum* (1897), 91.
71 House of Commons, *Report of Select Committee on Petroleum* (1897), 100.
72 House of Commons, *Report of Select Committee on Petroleum* (1897), 93.
73 *Berlin Vossische Zeitung* (June 18, 1891), quoted in *U. S. Consular Reports*, XXXVI, No. 131 (Aug., 1891), 649.
74 House of Commons, *Report of Select Committee on Petroleum* (1897), 184.
75 House of Commons, *Report of Select Committee on Petroleum* (1897), 232.
76 Lewis Emery, Jr., *US v. SONJ* (1906), *Transcript*, VI, 2782–83.
77 Hidy & Hidy, *Pioneering*, 252–53.
78 *The Economist, Monthly Trade Supplement* (Feb. 21, 1890), 18.
79 *Report of the Industrial Commission*, I, 623.
80 Hidy & Hidy, *Pioneering*, 128, 258.
81 *Titusville Morning Herald* (Nov. 28, 1893); *OP&D Reporter* (Jan. 15, 1894).
82 Hidy & Hidy, *Pioneering*, 258.
83 *U. S. Consular Reports*, XIX, No. 63 (1886), 161–62.
84 *Bradstreet's*, XXI (Dec. 2, 1893), 771.
85 Cf. *Petroleum Industrial and Technical Review* (June 21, 1899), 309.
86 Hidy & Hidy, *Pioneering*, 152.
87 Cf. Kendall Beaton, *Enterprise in Oil, A History of Shell in the United States* (New York: Appleton-Century-Crofts, 1957), 38–39, hereafter cited as Beaton, *Enterprise in Oil*.
88 *OP&D Reporter* (Feb. 20, 1884), 35.
89 F. C. Gerretson, *History of the Royal Dutch*, II (Leiden: E. J. Brill, 1955), 146.
90 Cf. "Interview with Sir Marcus Samuel," *Petroleum Industrial and Technical Review*, I (1899–1900), 60.
91 Cf. J. D. Henry, *Thirty-Five Years*, 39–47. According to Samuel, Russian and American "packers" of Petroleum products also brought pressure on the canal authorities not to relax their regulations. Cf. *Petroleum Industrial and Technical Review*, I (1899–1900), 60.
92 Cf. F. C. Gerretson, *History of Royal Dutch*, I (Leiden: E. J. Brill, 1953), 216–17.
93 Gerretson, *History of Royal Dutch*, I, 225.

94 In 1899 the group had some 23 tankers in operation or under construction. Cf. *Petroleum Industrial and Technical Review,* I (1899–1900), 60.
95 Beaton, *Enterprise in Oil,* 46.
96 Gerretson, *History of Royal Dutch,* I, 59, 69.
97 Cf. Gerretson, *History of Royal Dutch,* I, 82–83; Beaton, *Enterprise in Oil,* 22–23.
98 Beaton, *Enterprise in Oil,* 24.
99 Gerretson, *History of Royal Dutch,* I, 127–33.
100 Gerretson, *History of Royal Dutch,* I, 196.
101 Beaton, *Enterprise in Oil,* 28.
102 Beaton, *Enterprise in Oil,* 30.
103 Gerretson, *History of Royal Dutch,* I, 251.
104 Cf. Gerretson, *History of Royal Dutch,* I, 278–79; Beaton, *Enterprise in Oil,* 33.
105 Gerretson, *History of Royal Dutch,* I, 280.
106 Beaton, *Enterprise in Oil,* 43.
107 Cf. *Petroleum Industrial and Technical Review,* I (1899–1900), 59.
108 Gerretson, *History of Royal Dutch,* I, 282–83.
109 Hidy & Hidy, *Pioneering,* 264–65.
110 Gerretson, *History of Royal Dutch,* I, 283–86. At the time they were preparing to reject the offer of purchase, the directors were aware that Standard was buying shares of Royal Dutch in the open market. To avoid the possibility of losing control in this manner, the directors decided to issue 1500 "preference shares." These shares, issued in 1898, not only carried the sole rights to vote for members of the board but none could be transferred without the consent of the remaining holders of this class of stock. Cf. Gerretson, *History of Royal Dutch,* II, 46.
111 Hidy & Hidy, *Pioneering,* 266. Royal Dutch's announced intention of expanding its operations in the Dutch East Indies was undoubtedly a factor in the attitude of the government toward Standard. Cf. Gerretson, *History of Royal Dutch,* II, 61–67.
112 Cf. Hidy & Hidy, *Pioneering,* 262, 267.
113 Gerretson, *History of Royal Dutch,* I, 251; II, 147.

Chapter 25

1 *Historical Statistics of the U. S., 1789–1945,* 25; *The Economic Almanac,* 1956, 423.
2 *Historical Statistics of the U. S., 1789–1945,* 95.
3 *Historical Statistics of the U. S., 1789–1945,* 29.
4 *Historical Statistics of the U. S., 1789–1945,* 179.
5 *Historical Statistics of the U. S., 1789–1945,* 179, 200.
6 Collins, *Consolidated Gas Company of New York,* 204.
7 Cf. *American Gas-Light Journal,* XXXI (Dec. 2, 1879), 246.
8 Harold C. Passer, *The Electrical Manufacturers, 1875–1900: A Study in Competition, Entrepreneurship, Technical Change, and Economic Growth* (Cambridge: Harvard Univ. Press, 1953), 196–97.
9 Passer, *The Electrical Manufacturers,* 197.
10 Abbott Payson Usher, *A History of Mechanical Inventions* (revised ed.; Cambridge: Harvard Univ. Press, 1954), 402–403.
11 Passer, *The Electrical Manufacturers,* 77, 114.
12 Passer, *The Electrical Manufacturers,* 132–37.
13 *U. S. Census of Manufactures, 1900,* 718–19.
14 Cf. Hidy & Hidy, *Pioneering,* 299.
15 *Historical Statistics of the U. S., 1789–1945,* 187; C. R. Daugherty, *The Development of Horsepower Equipment in the United States,* Water-Supply Paper 579, U. S. Geological Survey (Washington: Government Printing Office, 1928), 45.
16 Letter to the authors from Philip D. Croll, Sun Oil Company, July 24, 1958.
17 *US v. SONJ* (1906), *Brief for Petitioner on the Facts,* I, 125.

18 *US* v. *SONJ* (1906), *Transcript,* XI, 496–500.
19 *US* v. *SONJ* (1906), *Transcript,* XI, 491.
20 *Report of the Commissioner of Corporations on the Petroleum Industry,* Part 2, *Prices & Profits,* 672–73.
21 *Report of the Commissioner of Corporations on the Petroleum Industry,* Part 2, *Prices & Profits,* 671.
22 *Report of the Commissioner of Corporations on the Petroleum Industry,* Part 2, *Prices & Profits,* 932.
23 Hidy & Hidy, *Pioneering,* 110.
24 Nevins, *Study in Power,* II, 54.
25 Hidy & Hidy, *Pioneering,* 112.
26 Hidy & Hidy, *Pioneering,* 108.
27 Hidy & Hidy, *Pioneering,* 196–97; Nevins, *Study in Power,* II, 45.
28 Hidy & Hidy, *Pioneering,* 197.
29 *Pure Oil Trust vs. Standard Oil Co.,* 582.
30 *Pure Oil Trust vs. Standard Oil Co.,* 583.
31 Hidy & Hidy, *Pioneering,* 258.
32 *Report of the Commissioner of Corporations on the Petroleum Industry,* Part 2, *Prices & Profits,* 440–41.
33 Cf. Giddens, *Standard of Indiana,* 75; Nevins, *Study in Power,* II, 45.
34 Hidy & Hidy, *Pioneering,* 459.
35 James A. Moffett, *US* v. *SONJ* (1906), *Transcript,* III, 1142–43.
36 Nevins, *Study in Power,* II, 46–47.
37 *Report of the Commissioner of Corporations on the Petroleum Industry,* Part 1, *Position of Standard Oil,* 328–29.
38 Copy of original report, from archives of Sacony-Mobile Oil Company.
39 Ackert testimony, *US* v. *SONJ* (1906), *Transcript,* III, 1099.
40 Ackert testimony, *US* v. *SONJ* (1906), *Transcript,* III, 1097.

41 "Statement Number 672-R: Bulk Stations and Number of Tank Wagons in the United States, 1901–1907" (Archives of Sacony-Mobile Oil Company).
42 *Standard Oil Co. of New Jersey et al.* v. *United States of America,* Supreme Court of the U. S., Oct. term, 1910, *Oral Arguments on Behalf of Appellants,* 49.
43 *US* v. *SONJ* (1906), *Transcript,* III, 1051.
44 *US* v. *SONJ* (1906), *Brief of Facts and Argument for Petitioner,* II, 150.
45 Westgate testimony, *US* v. *SONJ* (1906), *Transcript,* II, 704–29.
46 *US* v. *SONJ* (1906), *Brief for Defendants on Facts,* III, 647.
47 James Lee testimony, *Pure Oil Trust vs. Standard Oil Co.,* 86.
48 E. M. Wilhoit testimony, *US* v. *SONJ* (1906), *Transcript,* III, 1037.
49 *Industrial Commission Report,* I, 343–44.
50 *US* v. *SONJ* (1906), *Brief of Facts and Arguments for Petitioner,* II, 358–428.
51 William Z. Ripley, *Railroad Rates and Regulation* (New York: Longmans, Green & Co., 1920), 202; Bureau of Corporations, *Report of the Commissioner of Corporations on the Transportation of Petroleum,* 151.
52 24 Stat. 379.
53 Hidy & Hidy, *Pioneering,* 198.
54 Bureau of Corporations, *Report of the Commissioner of Corporations on the Transportation of Petroleum,* 94–95.
55 Cf. Bureau of Corporations, *Report of the Commissioner of Corporations on the Transportation of Petroleum,* 316–17, 338–339.
56 Bureau of Corporations, *Report of the Commissioner of Corporations on the Transportation of Petroleum,* 188, 248.

Chapter 26

1 *Chicago Tribune* (Dec. 30, 1881) quoted in Hans B. Thorelli, *The Federal Antitrust Policy: Organization of an American Tradition* (Baltimore: The Johns Hopkins Press, 1955), 134.

2 Nevins, *John D. Rockefeller,* II, 53–54.
3 Nevins, *Study in Power,* II, 146–47, 149, 154. In 1884, a professor requested Standard to furnish data to prove that the Trust's economies of

scale were responsible for lower prices and improved product. Standard executives refused, maintaining that this could not be conclusively "proved," and that release of the data might lead to other inquiries, "some of which it might be undesirable to answer." G. Vilas to J. D. Rockefeller, Jan. 8, 1884, in Hidy & Hidy, *Pioneering*, 214.

4 Hidy & Hidy, *Pioneering*, 212–13.

5 Nevins, *Study in Power*, II, 151–54. The *Reporter's* comments on the House Trust Investigation of Standard Oil in 1888 are illustrative of the change. That many of the smaller refiners had failed "only shows that they lack the brains, capital and energy profitably to conduct" their business. Other firms, however, had been "started with the one object of compelling the Standard to buy them out at an enormous advance on the outlay . . . When they fail, they . . . resort to all manner of persecution," and the investigation showed that "the hands of the persecutors" were behind the committee. *OP&D Reporter*, XXXIII (May 2, 1888), 6.

6 J. N. Camden, "The Standard Oil Company," *North American Review*, CXXXVI (Feb., 1883), 181–90. Camden's mention of "politicians" probably referred to a tax suit brought in the previous year. The attorney general of Pennsylvania tried, unsuccessfully, to sue Ohio Standard for back taxes on its assets employed outside Pennsylvania. The suit was perhaps indicative of a rising trend of anti-Standard opinion. But since there was a good deal of local hostility in Pennsylvania, the case may not have been representative of feeling in areas less directly connected with the petroleum industry.

7 U. S. Senate, *Report of the Senate Select Committee on Interstate Commerce*, 49th Congress, 1st Session, Senate Report 46, Part 2 (Washington: Government Printing Office, 1886), 69, 122, 184, 530 *et passim*, hereafter cited as *Cullom Report*.

8 *Cullom Report*, Pt. 1, 2–3.

9 *New York Times* (Feb. 28, 1887), 4.

10 Quoted in Thorelli, *Federal Antitrust Policy*, 157.

11 Henry Seager & Charles Gulick, Jr., *Trust and Corporation Problems* (New York: Harper & Bros., 1929), 341–43.

12 See Thorelli, *Federal Antitrust Policy*, 136–43. Thorelli concludes that, while there was not "an irresistible tide of public opinion" building pressure on Congress, "public concern at the end of the 1880's was serious enough to make immediate federal actions against the trusts a clear desideratum. . . ."

13 *OP&D Reporter*, XXXIII (Feb. 29, 1888), 8, and (Mar. 7, 1888), 9–10.

14 Nevins, *John D. Rockefeller*, II, 121.

15 Thorelli, *Federal Antitrust Policy*, 40, 156–57; Nevins, *John D. Rockefeller*, II, 122.

16 Badcock to Rockefeller, undated, 1888, in Hidy & Hidy, *Pioneering*, 214. Italics as quoted.

17 Representative Smith, *1888 Trust Investigation*, 391.

18 Cf. H. M. Flagler, *1888 Trust Investigation*, 768–85.

19 Emery, *1888 Trust Investigation*, 249.

20 F. B. Gowen, *1888 Trust Investigation*, 54–55, 84.

21 J. D. Archbold testimony, *1888 Trust Investigation*, 332, 334.

22 *1888 Trust Investigation*, i–iii.

23 *1888 Trust Investigation*, Sugar Trust, 200–202.

24 Thorelli, *Federal Antitrust Policy*, 80–81. Having decided that the company surrendered its independence, the Court of Appeals deemed it needless to "advance into the wider discussion over monopolies and competition and restraint of trade. . . ." *People* v. *North River Sugar Refining Co.*, 121 N. Y. (1890), 582 ff. A similar decision against a California member of the Sugar Trust was handed down by the San Francisco Superior Court in 1889.

25 Nevins, *John D. Rockefeller*, II, 142. According to Miss Tarbell, Watson first conceived of prosecuting Ohio Standard after reading William W. Cook's book, *Trusts: The Recent Combinations in Trade*. The book called upon state attorney-generals to attack the trusts, and reprinted the Standard Oil Trust Agreement of

1882. Tarbell, *History of Standard Oil*, II, 142.

26 *New York Times* (July 28, 1890), 2; Thorelli, *Federal Antitrust Policy*, 80.

27 *New York Times* (July 28, 1890), 2.

28 Hanna to Watson, Nov. 21, 1890, in Nevins, *John D. Rockefeller*, II, 145–46.

29 *State ex rel.* v. *Standard Oil Company*, 49 Ohio St. 137 (1892), 185–86.

30 *OP&D Reporter*, XLI (March 28, 1892), 9.

31 Hidy & Hidy, *Pioneering*, 223.

32 Nevins, *John D. Rockefeller*, II, 150–51.

33 *Report of the Commissioner of Corporations on the Petroleum Industry*, Part 1, *Position of Standard Oil*, 76–81; Nevins, *John D. Rockefeller*, II, 343–44.

34 Thorelli, *Federal Antitrust Policy*, 260–63, 339.

35 Thorelli, *Federal Antitrust Policy*, 359–60.

36 Thorelli, *Federal Antitrust Policy*, 376–88. The decision in question was that of *U. S.* v. *E. C. Knight Co. et al.*, 156 U. S. (1895), 1.

37 Thorelli, *Federal Antitrust Policy*, 339.

38 Quoted in Johnson, *Pipelines*, 159.

39 On other errors, see Nevins, *John D. Rockefeller*, II, 335–39. Chester M. Destler, while admitting that Lloyd's account of this affair was wrong, maintains that Lloyd's version of several other points was nearer the truth than are Nevins' accounts. Chester M. Destler, "Wealth Against Commonwealth, 1894 and 1944," *American Historical Review*, L (Oct., 1944), 49–72.

40 Nevins, *John D. Rockefeller*, II, 337.

41 Nevins, *John D. Rockefeller*, II, 344.

42 *Report of the Commissioner of Corporations on the Petroleum Industry*. Part 1, *Position of Standard Oil*, 80.

43 Nevins, *John D. Rockefeller*, II, 344–51.

44 Thorelli, *Federal Antitrust Policy*, 262.

45 Hidy and Hidy, *Pioneering*, 306–9.

46 Thorelli, *Federal Antitrust Policy*, 294–301. This series, it should be noted, excluded consolidations whose markets were confined to relatively isolated cities and their environs.

47 Thorelli, *Federal Antitrust Policy*, 511.

48 Rockefeller, *Industrial Commission Report*, I, 797.

49 Cf. Lee, *US* v. *SONJ* (1906), *Transcript*, VI, 3185.

50 Cf. *Industrial Commission Report*, I, 382, 394, 670.

51 Archbold testimony, *Industrial Commission Report*, I, 517–26.

52 Emery, *Industrial Commission Report*, I, 670. Emery thought that the Sherman Act was "one of the best laws ever written," but complained that it could not, or would not, be enforced.

53 *Industrial Commission Report*, XIX, 609. The Commission never really faced up to the problems arising when economies of scale were great enough to preclude competition. Underlying much of its thought was the idea that competition would somehow prevail. Citing capital requirements upwards of $20 million for a complete steel plant, the *Report* admitted that "the necessity for so large an amount of money excludes anything like what is usually understood as free competition; but if adequate profits are indicated, it is probable that even such sums could be raised without difficulty."

54 *Industrial Commission Report*, XIX, 650–51.

55 Nevins, *John D. Rockefeller*, II, 500–502.

56 Tarbell, *History of Standard Oil*, II, 283, 292.

57 Hidy & Hidy, *Pioneering*, 663.

58 Gilbert Holland Montague, *The Rise and Progress of the Standard Oil Company* (New York: Harper & Bros., 1903), 33.

59 Montague, *Rise & Progress of Standard*, 105, 132–35. Montague wrote before the legality of Standard's Whiting–Grand Junction, Whiting–East St. Louis, and Rochester-Olean rates was contested.

60 Montague, *Rise & Progress of Standard*, 137–38.

61 Montague, *Rise & Progress of Standard*, 143.

Appendix E

1 12 Meeson and Welsby, 324.

2 George A. Blanchard and Edward P. Weeks, *The Law of Mines, Minerals and Mining Rights* (San Francisco: Summer Whitney and Co., 1877), 759–60, hereafter cited as Blanchard & Weeks, *Law of Mines.*

3 Blanchard & Weeks, *Law of Mines,* 761.

4 Blanchard & Weeks, *Law of Mines,* 360–61.

5 *Brown* v. *Vandergrift,* 80 Pennsylvania 142 (1875). Also, Northcutt Ely, "The Conservation of Oil," *Harvard Law Review,* LI (1938), 120–24.

6 See R. E. Hardwicke, *et al., Legal History of the Conservation of Oil and Gas, A Symposium* (Chicago: American Bar Association, Section of Mineral Law, 1938), 6; Blakley M. Murphy (ed.), *Conservation of Oil and Gas: A Legal History,* 427.

7 R. E. Hardwicke, "Rule of Capture and Its Implications as Applied to Oil and Gas," *Texas Law Review,* XIII (1935), 395; *Westmoreland Natural Gas Co.* v. *Dewitt,* 130 Pennsylvania 235 (1889); *Kelly* v. *Ohio Oil Co.,* 570 Ohio St. 317 (1897); *Barnard* v. *Monongahela Gas Co.,* 365 Atl. 802 (1907).

8 U. S. Department of the Interior, Census Office, *Report on Mineral Industries in the United States at the Eleventh Census:* 1890 (Washington: Government Printing Office, 1892), 434.

9 *Bradstreets* (Sept. 10, 1881), 165, attributed the consequences of excessive production to the character of oil leases, indicating that the problem had not escaped the concern of contemporaries.

10 J. H. Marshall & N. L. Meyers, "Legal Basis of Petroleum Production in the United States," *Yale Law Review,* LXI (1938), 43.

11 W. L. Summers, *The Law of Oil and Gas* (8 Vols.; Kansas City: Vernon Law Book Co., 1938–39), II, Sec. 334, 195.

12 Samuel H. Glassmire, *Law of Oil and Gas: Leases and Royalties* (St. Louis: Thomas Law Book Co., 1935), 17.

13 80 Pa. 142 (1875).

14 80 Pa. 147 (1875).

15 80 Pa. 147 (1875).

16 96 Pa. 307 (1880).

17 *Venture Oil Co.* v. *Fretts,* 152 Pa. 45, 25 Atl. 732 (1893); *Crawford* v. *Ritchie,* 42 W. Va. 252, aff'd. in 249 Fed. 675.

18 Summers, *The Law of Oil & Gas,* II, 211 ("unless" drilling clause in *Glasgow* v. *Chartier Gas Co.,* 152 Pa. 48, 25 Atl. 232 [1892]). Also see Marshall & Meyers, *Yale Law Review,* LXI (1938), 46–7.

19 Summers, *The Law of Oil & Gas,* II, sec. 284, 105 ff.

20 Summers, *The Law of Oil & Gas,* II, sec. 285, 107–108.

21 Summers, *The Law of Oil & Gas,* I, sec. 71, 195–97. Indiana, for example, with conservation legislation dating from 1891 did not establish a Department of Conservation until 1919.

22 *Ohio Oil Co.* v. *Indiana,* 177 U. S. 190, (1900).

23 *Ohio Oil Co.* v. *Indiana,* 210.

24 *Ohio Oil Co.* v. *Indiana,* 210.

25 Marshall & Meyers, *Yale Law Review,* XLI (1938), 49.

26 *People* v. *Assoc. Oil Co.,* 211 Cal. 93, 294 Pac. 717, 722.
    In one case, which may be considered somewhat exceptional because there was actually no statute involved, the court invoked the "correlative-rights" doctrine as a matter of common law. In this case, a gas company obtained an injunction restraining a rival gas company possessing other sources of supply from unnecessarily depleting the pressure of a common pool by burning gas at the well. *Louisville Gas Co.* v. *Kentucky Heating Co.,* 171 Ky. 71, 77 SW 368 (1903). On this case Marshall & Meyers remarked, ". . . the more flagrant infringements upon the property rights of collective owners in a gas pool can be restrained even with-

out legislation." Marshall & Meyers, "Legal Basis of Petroleum Production," *Yale Law Review*, XLI (1938), 50.

27 *Scientific American* (Oct. 10, 1863), 226.

28 *Scientific American* (April 22, 1865), 268.

29 Wrigley, *Special Report on the Petroleum of Pennsylvania* (1874), 77.

30 I. C. White, "The Geology of Natural Gas," *Science*, V, (June 26, 1885), 521–22; "The Mannington Oil Field and the History of its Development," *Bulletin of the Geological Society of America*, III (April 15, 1892), 193–200.

31 Marshall & Meyers, *Yale Law Review*, XLI (1938), 63.

32 *Wurts* v. *Hoagland*, 114 U. S. 606, 5 S.Ct. 1806 (1885).

Appendix F

1 *Historical Statistics of the U. S., 1789–1945*, 231–32.

2 *Derrick's Hand-Book*, I, 235.

Index

Grove Run: 128
Gushers. *See* Flowing wells

Haas & Brothers: 539
Haldeman, L.: 288, 448*n*
Hanford Oil Company: 688
Hanford, P. C.: 547, 688
Hanford (P. C.) & Company: 546–547
Hanna, Baslington & Company: 352, 354
Hanna, Marcus A. (Mark): 710
Hanna, Robert: 354
Harkness brothers (Stephen V. and William K.): 283, 302, 420, 470
Harley, Henry: 187–189, 365, 400, 402–403, 410–412, 421–422, 438
Harmonial wells: 114*n*, 130
Harper, John F.: 151
Harrisburg Car Works: 531*n*
Harrisburg refineries, deliveries to: 174
Hasson, William: 358
Hastings, Governor: 585
Haupt, Herman: 411, 438, 445
Havana, earns name of "Carine": 8
Havens, Rensselaer H.: 73–74
Hazard, J. R. G.: 132
Henry, J. T.: 136
Henry's Bend: 101, 185
Hepburn investigation: 433–437, 445, 702
Herodotus: 6
Heye, G.: 365*n*
Highways, growth of: 680
Hit, ancient exploitation of: 4
Hofman, J. P.: 204
Holden, Daniel L.: 240
Holmden (Thomas) farm: 122
Holmes, Charles: 111
Hope, Stanley: 217–218
Hopkins, James: 431, 438–440, 442, 453
Hopkins, Robert E.: 409
Horstmann & Company: 649
Hostetter, David: 405–412, 426
Houghton (E. F.) & Company: 251, 279
Housewives, acceptance of kerosene: 339–340
Hoyt, Governor: 433
Hubbard, O. P.: 65
Hudson River Railroad: 85, 182–183
Hughes River, discoveries at: 12*n*, 102
Humboldt refinery, deliveries to: 183, 228, 229, 240, 292
Hutchings, J. L.: 183–184
Hyde, Charles: 121
Hydrocarbons, action of heat on: 205
Hydrology. *See* Divining methods

Ice, manufacture of: 239–240
Illinois Central Railroad: 530, 700
Illuminating oils (*see also* Kerosene)
    Albert coal, yield from: 105
    camphene oil: 33–34
    coal oil, emergence of: 43–60
    consumption, world: 634–635
    dissatisfaction with: 505–509, 646–647
    electricity, eclipse by: 681–683
    Europe, industry in: 323
    exports: 328, 489, 491–496, 633–635, 660
    Far East markets: 635
    grade and quality standards: 508, 523–524
    kerosene established as: 371–372
    lard oil for: 34–37
    market prices, domestic: 36, 307, 484–485, 495, 521–523, 524, 528, 574–575, 623, 680
    market prices, foreign: 495, 574–575, 634
    naphtha: 51
    naphtha gas: 236
    paraffin oils: 38, 51
    petroleum for: 27–42, 64, 69–70, 246–248, 249
    production totals: 290, 489–491, 513, 615, 625–626, 633, 677–679
    profit margins: 621–623
    Russian output: 511–513, 517, 633, 660
    safety regulations: 312–316, 339–340, 523–524
    schist distillation for: 37
    shale distillation for: 37
    spermaceti, use of: 29
    taxes on: 310, 318
    whale oil, use of: 29
Illumination
    candles, use for: 29, 30–31, 34–37
    demand for: 33, 680–681
    lamps, use for: 29–42
    portable units for: 40–41
    revolution in: 25–60
"Immediate shipment": 383–390, 464
Imperial Oil Company: 591*n*, 689–690, 412, 419
Incas, use of bitumen: 9*n*
Income per capita: 521. *See* Economy, national
Independent Petroleum Refiners Bureau: 621
Indiana region, discoveries in: 601–613, 625
Indianapolis Gas Company: 685

Lubricants (*continued*)
  expansion of manufacture: 465–466
  exports: 489–491, 496, 635, 645
  market prices: 623, 680, 684
  from petroleum: 64, 71, 242–245
  production totals: 486, 489–491, 513,
      521, 614–615, 624, 633, 677–679
  profit margins: 621–623
  quality improvement: 480–482
  railroad uses: 244
  Russian output: 511–513, 633, 645
Lubricating Oil Import Company: 645*n*
Lucesco coal oil: 58, 108, 110
Lufkin, C. F.: 675
Luther, L. H.: 144

Macgowan, James G.: 644
Machinery, drilling: 137–143 (*see also*
    Equipment; Tools)
  coal as fuel for: 140–141
  costs of: 140, 157
  gas as fuel for: 140–141
  oil as fuel for: 140*n*
  steam for power: 137–143
  retrieving equipment: 144
  variable cutoff principle: 141–143
  wood as fuel for: 140
Macksburg: 589–590
Marginal Railroad Company: 477
Marietta refinery, deliveries to: 132*n*, 248
Manhattan Oil Company: 685
Manifold well: 556
Mannheim-Bremer Petroleum, A.G.: 653
Mansfield, C. B.: 40
Mantle, gas, development: 681
Manufactories. *See* Plants; Oil industry
Manufacturing costs. *See* Production costs
Mapleton Oil Company: 169
Marco Polo: 8
Marcus Graecus: 7–8
Marine shipments. *See* Tanker shipments;
    Waterways, inland; Overseas ship-
    ments
Market prices, domestic
  benzine: 485
  camphene: 47–48
  caustic soda: 284, 464
  coal: 241, 284
  coal gas: 48, 59
  coal oil: 47, 55, 58–59
  crude petroleum: 103–104, 108, 114–
      135, 158–163, 183, 240–242, 324,
      326, 344, 358–359, 372–373, 375,
      379–381, 384, 388, 390, 430–431,

Market prices (*continued*)
    464, 484–485, 548, 566–567, 569,
    575, 604, 606, 610–611
  gas oil: 680
  gasoline: 235, 238, 485
  illuminating oil: 36, 307, 484–485, 495,
      521–523, 524, 528, 574–575, 623,
      680
  kerosene: 109, 247, 322, 337, 728
  lard oil: 47
  lubricants: 623, 680, 684
  mineral sperm oil: 247
  naphtha: 238, 495, 524, 680
  naphtha gas: 237
  neutral oils: 246
  petroleum products: 549, 680
  quoting practices: 317–318
  sulphuric acid: 284, 464
  whale oil: 47–48
Market prices, foreign
  crude oil: 344*n*
  Germany: 495
  illuminating oils: 495, 574–575, 634
  kerosene: 325–326, 578
  petroleum products: 326, 332
  United Kingdom: 495
Markets, domestic (*see also* Shipping;
    Production; Retailers)
  areas of: 689
  by areas, Standard policy: 535
  bulk distribution, spread of: 690–693
  by-products: 684–687
  centers, location: 697
  channels for: 309–340, 520–551, 677–
      701
  coal oil: 55, 58–60
  competition in: 677–701
  cymogene: 239–240
  demand, growth of: 337–340, 372,
      521–528, 679–684
  development of: 309–340
  drugstore outlets: 318–319, 535
  expansion of: 338–340, 488–519, 520–
      551, 677–701
  fuel oils: 683
  gasoline: 235, 683*n*
  grade and quality standards: 318–320,
      508, 523–524
  grocery store outlets: 318–319, 535
  illumination, demand for: 680–681
  distributors, domination by Standard:
      535–547
  independents, threat to Standard: 686–
      690
  independents, Standard acquisition of:
      693–701

Safety measures: 339–340, 500–504, 523–524
Sahagun, discoveries by: 9
St. Petersburg, Pa.: 131, 379
St. Louis refineries, deliveries to: 172
Salamanca, N. Y.: 170, 176, 177
Sales. *See* Marketing
Sale prices. *See* Market prices
Salisbury, Lord: 667
Salt
 as food preservative: 14
 production, early: 14
Salt deposits
 drilling methods: 14–17
 exploitation of: 16
 medicine, connection with: 14–24
 natural gas, connection with: 14
 petroleum, connection with: 14
Samuel, Marcus: 664–668, 672–674
Sanders (August) & Company: 649, 654n
Saratoga producers' meeting: 402, 563
Satterfield, John: 454–455, 581
Saybolt's viscometer: 482n
Scandinavian countries, markets in: 336, 636, 649–650, 654
Scheele, Carl W.: 28
Scheide, William: 362
Schibaieff, Ltd.: 645n
Schieffelin brothers: 190, 220, 225, 318
Schists, distillation of: 37
Schütte & Sohn: 648–649
Scientific American: 55, 57, 107, 154, 187, 630–632
Schofield, Schurmer & Teagle: 449, 689, 693
Scott, Joseph: 11
Scott, Thomas A.: 173, 178, 196, 199, 297, 308, 346, 350–351, 364n, 402, 404, 406, 433n, 452n
Scrubgrass, Pa.: 134
Seaways, use of oil on: 10–11
Seep (Joseph) Purchasing Agency: 387, 558, 577, 597, 599, 619–620
Sefelder, A & H: 662
Selligue, discoveries by: 37, 38, 40, 44, 322
Seneca oil: 12–13, 16
Seneca Oil Company: 76–77, 88
Shaffer farm: 170, 184, 187, 189, 192
Shale, distillation of: 37
Shamburg, Pa.: 119, 126, 130, 157
Shannopin, Pa.: 556
Shaw, Thomas: 240
Sheldon, Anson: 67

Shell Transport & Trading Company: 674
Sherman Act: 712, 715n, 720, 722
Sherman, John W.: 93, 121, 706
Sherman well: 184, 191
Shipping methods (*see also* Marketing; Pipelines; Production)
 bulk shipments: 168–183, 690–693, 528–535, 653–657, 727
 bulk shipments by railroads: 528–535
 bulk shipments by tank cars: 528–535, 537
 centers, location: 697
 costs: 106, 166–167, 169, 170, 176, 344n. *See also* Railroads
 crude oil: 164–201
 difficulties of: 83–86, 104–114, 178–183
 facilities in Ohio-Indiana: 596–601
 facilities in Pennsylvania regions: 164–201, 438, 466, 621, 627
 gasoline: 235
 gathering lines: 183–189
 "immediate shipment" practices: 383–390, 464
 kerosene, costs: 428
 load sizes: 166
 losses during: 178
 overseas: 497–498, 503–504, 507–508, 580
 overseas rates: 330–331
 by pipeline: 383–390
  deliveries, totals: 581
  introduction of: 183–189, 548
  map of: 186
  rates: 185, 187–188, 400n, 403, 406, 407n, 445–449, 453, 458, 557–576, 577–578, 586–587
  totals: 625
 by pond freshets: 165–169
 railroad-pipeline pool: 400–405
 railroads
  competition for: 83–86, 170–183, 170–178, 194–201, 343–368, 396–429
  indifference to: 107
  rates: 291–292, 297–301, 303–306, 436, 440, 451–452, 477, 570–572, 586–587, 603, 611, 696–701
  rates, general: 527
 regulatory legislation: 558–562, 585
 safety measures: 500–504
 Standard's control of pipelines: 383–390
 Standard's entry into: 412–416